ADVANCES IN CATALYSIS

VOLUME 44

ADVANCES IN CATALYSIS

VOLUME 44

Edited by

WERNER O. HAAG†
Lawrenceville, New Jersey
†Deceased

BRUCE C. GATES
*University of California
Davis, California*

HELMUT KNÖZINGER
*University of Munich
Munich, Germany*

ACADEMIC PRESS
San Diego   London   Boston
New York   Sydney   Tokyo   Toronto

Academic Press
A Harcourt Science and Technology Company
525 B Street, Suite 1900, San Diego, California 92101-4495, USA
http://www.apnet.com

Academic Press
24-28 Oval Road, London NW1 7DX, UK
http://www.hbuk.co.uk/ap/

International Standard Book Number: 0-12-007844-9

Transferred to Digital Printing 2006

99 00 01 02 03 04 BB 9 8 7 6 5 4 3 2 1

Contents

Werner Otto Haag, 1926–1998

PAUL B. WEISZ

Charles Kemball, 1923–1998

FRANK S. STONE

John Turkevich, 1907–1998

MICHEL CHE AND MICHEL BOUDART

NMR Spectroscopy as a Probe of Surfaces of Supported Metal Catalysts

J. J. VAN DER KLINK

Applications of Photoluminescence Techniques to the Characterization of Solid Surfaces in Relation to Adsorption, Catalysis, and Photocatalysis

MASAKAZU ANPO AND MICHEL CHE

The Surface Science Approach toward Understanding Automotive Exhaust Conversion Catalysis at the Atomic Level

BERNARD E. NIEUWENHUYS

Experiments and Processes in the Transient Regime for Heterogeneous Catalysis

CARROLL O. BENNETT

Influence of Phosphorus on the Properties of Alumina-Based Hydrotreating Catalysts

RYUICHIRO IWAMOTO AND JEAN GRIMBLOT

Skeletal Isomerization of *n*-Butenes Catalyzed by Medium-Pore Zeolites and Aluminophosphates

PAUL MÉRIAUDEAU AND CLAUDE NACCACHE

Contributors

Numbers in parentheses indicate the pages on which the authors' contributions begin.

MASAKAZU ANPO, *Department of Applied Chemistry, College of Engineering, Osaka Prefecture University, Sakai, Osaka 599-8531, Japan* (119)

CARROLL O. BENNETT, *Laboratoire de Réactivité de Surface, Université Pierre et Marie Curie, 75005 Paris Cedex, France* (329)

MICHEL BOUDART, *Department of Chemical Engineering, Stanford University, Stanford, California 94305* (xxi)

MICHEL CHE, *Laboratoire de Réactivité de Surface, UMR 7609-CNRS, Université Pierre et Marie Curie, 75005 Paris Cedex, France* (xxi, 119)

JEAN GRIMBLOT, *Laboratoire de Catalyse Hétérogène et Homogène, URA CNRS 402, Université des Sciences et Technologies de Lille, 59655 Villeneuve d'Ascq Cedex, France* (417)

RYUICHIRO IWAMOTO, *Petroleum Refining Technology Center, Idemitsu Kosan Co. Ltd., Sodegaura, Chiba 299-0293, Japan* (417)

PAUL MÉRIAUDEAU, *Institut de Recherches sur la Catalyse—CNRS, 69626 Villeurbanne Cedex, France* (505)

CLAUDE NACCACHE, *Institut de Recherches sur la Catalyse—CNRS, 69626 Villeurbanne Cedex, France* (505)

BERNARD E. NIEUWENHUYS, *Leiden Institute of Chemistry, Gorlaeus Laboratories, Leiden University, Leiden 2300 RA, The Netherlands* (259)

FRANK S. STONE, *Department of Chemistry, University of Bath, Bath BA2 7AY, England* (xvii)

J. J. VAN DER KLINK, *Institut de Physique Expérimentale, Département de Physique, Ecole Polytechnique Fédérale de Lausanne, CH-101S Lausanne, Switzerland* (1)

PAUL B. WEISZ, *Foxdale Village, A-1, State College, Pennsylvania 16801* (xiii)

Preface

We note with sadness that this is the last volume of *Advances in Catalysis* edited by our colleague Werner Haag, who passed away last year. The catalysis community will miss him. His obituary appears in this volume.

This issue of the *Advances* reflects the expanding impact of experimental surface characterization on the understanding of catalysis. The catalysts emphasized here are representative of the complexity of today's technology; examples include catalysts for hydrocarbon re-forming, automobile exhaust conversion, and hydroprocessing to make clean-burning fossil fuels.

Nuclear magnetic resonance (NMR) spectroscopy continues to gain importance in catalyst characterization, and van der Klink contributes the first chapter on metal NMR to the *Advances;* the focus is on supported platinum catalysts, including those with adsorbed carbon monoxide or hydrogen. The method provides evidence of metal dispersions and distinguishes between hydrogen on the metal and hydrogen on the support.

Anpo and Che write about applications of photoluminescence techniques, which are powerful but only seldom used methods for the identification of surface sites and their local environments, particularly on oxide surfaces. The dynamics of energy and electron transfer processes are discussed in light of catalytic and photocatalytic phenomena.

Nieuwenhuys summarizes the understanding of supported metal catalysts for automobile exhaust abatement as it has developed from ultrahigh-vacuum surface science. The account is an impressive validation of the success of the surface science methods for elucidation of complex multicomponent catalysts.

The quantitative foundation of catalysis is kinetics, and lively recent developments in this field center around transient experiments, many quite elegant, as summarized in the chapter by Bennett.

Because of their widespread applications, hydroprocessing catalysts have been the subjects of chapters in two of the preceding three volumes of this series. Iwamoto and Grimblot add another chapter, focusing on the roles of phosphorus in these multicomponent catalysts.

Mériaudeau and Naccache conclude the volume with a concise description of skeletal isomerization of butenes catalyzed by medium-pore zeolites and molecular sieves. This isomerization is a relatively new industrial process, and it is remarkable how fast a good fundamental understanding of it has developed in a few years; the chapter is an account of catalysis by well-defined acidic groups in pores that exert a subtle control over catalyst performance, including selectivity. It is a story that was deeply appreciated by Werner Haag.

B. C. GATES
H. KNÖZINGER

Werner Otto Haag, 1926–1998

We have lost the presence of an exceptional scientist, but we have been enriched by the legacy of his contributions to basic catalytic science and to catalytic process technology. Werner Haag, born in Heilbronn, Germany, studied chemistry at the University of Tübingen and receiving his diploma (M.S. equivalent) in 1954. His doctoral work at Northwestern University with Professor Herman Pines provided him with an early introduction to heterogeneous catalysis (1958). He joined the research staff of the Mobil Oil Corporation in 1959, where he rose to the position of senior scientist in 1981 and laboratory advisor, the highest technical position. Upon his retirement in 1993, he remained active as a consultant. He became a research associate at the Fritz Haber Institute of the Max Planck Gesellschaft in Berlin, continuing his research activities as well as being a teacher to colleagues and students. He continued to be an active contributor to catalysis symposia and to interact with research colleagues internationally. Untiringly, he also found time to write two chapters for the *Handbook of Heterogeneous Catalysis* (G. Ertl, H. Knözinger, and J. Weitkamp, Eds., Wiley–VCH, Weinheim, 1997). He had been coeditor of *Advances in Catalysis* since 1994. His dedication went beyond that of the usual editorship. He often devoted time and effort to help authors expand and improve presentations by pointing to related issues and subjects.

Werner's contributions to catalysis were unique. They combined delving deeply into basic science with a keen awareness of technological relevance and utility. In addition, his ever readiness to share ideas, knowledge, and skills with others provided inspiration to research colleagues worldwide and patient guidance to new process development at Mobil. A proper recitation of his gifts to our science would fill a chapter of these *Advances*, but we must most certainly record the resolution he has provided to major and classical mysteries of heterogeneous catalysis. In addition, we must record his demonstration of new basic principles leading to the creation of new catalytic processes.

Two or three decades ago, no clear conceptual (mechanistic) or practical (experimental) bridge existed between homogeneous and heterogeneous catalysis. In some of his earliest work at Mobil, Werner demonstrated that hydroformylation, carbonylation, and other reactions catalyzed by transition metal complexes were accomplished equally by heterogeneous catalysis when these complexes were anchored to solid organic resins.

The seeds implanted by his early work with Herman Pines in acid catalysis grew into blossoming flowers: He developed an ever-growing and detailed map of the applicable carbonium and carbenium ion mechanisms in hydrocarbon conversion and their relationships to achievable product distributions. He provided the long-sought bridge between heterogeneous acid catalysis and the superacid chemistry of George Olah (recognized by the 1994 Nobel Prize in chemistry) when he, with his able associate Ralph Dessau, demonstrated the occurrence of direct protonation of paraffins to give pentacoordinated carbonium ions as a step initiating paraffin cracking.

The discovery that crystalline zeolites could be turned into active catalysts provided a plethora of challenges for research, basic and applied. Werner participated in both and brought findings in one activity to guide the other! The "active site," proposed by H. S. Taylor in 1925, had remained a prime mystery parameter in the science of heterogeneous catalysis. Could we identify or even count the active sites? Werner Haag was the center of generating and coordinating the research. By demonstrating the quantitative correlation of rate constants of hydrocarbon reactions with Al contents, Cs ion exchange capacity, as well as nuclear magnetic resonance quantification of tetrahedrally coordinated Al atoms in the silica structure of many ZSM-5 catalysts, he both identified and counted the active sites!

Another old "mystery" in heterogeneous catalysis, dating back to F. H. Constable (1925) and G.-M. Schwab (1929), was the "compensation effect" or "theta rule." The Arrhenius plots for similar reactants on the same catalyst or for the same reactant on similar catalysts would differ in slope across a common point of intersection. Approximately 70 years later, a First Workshop on the Compensation Effect was organized (DECHEMA, Berlin, 1997) to debate this enduring mystery. Werner Haag demonstrated that such an effect must necessarily result from the temperature dependence of reactant adsorption (and hence its site concentration) and that of the reaction rate of the adsorbed species, operating in opposite directions. A "second" workshop may never follow!

When published reports of the diffusivity of paraffins in ZSM-5 catalysts obtained from uptake rate measurements appeared grossly inconsistent with catalytic behavior, Werner participated in resolving the problem by determining diffusivities from catalytic behavior of catalysts of very different particle sizes. The analysis not only confirmed the many orders of magnitude higher true diffusivities but also allowed Werner to extend the technique to demonstrate that "shape selectivity" could occur due to lack of fit of a reactant (e.g., diffusion of a dimethyl paraffin) in the structure or lack of fit of a reaction complex (transition state) that must be created on the active site (e.g., the methyl paraffin/propyl cation complex).

This and Werner's extensive investigations of reactions of aromatics led to a comprehensive view of the three ways by which molecular shapes and sizes can lead to unusual product selectivities: These are, for the competing species participating, differences in diffusion rates, equilibrium sorption constants, and relative sizes of the transition state (active site complex).

Werner Haag was both a scientist and a technologist. This is somewhat reflected by more than 70 U.S. patents that bear his name. A better illustration may be his essential contributions to the currently practiced para-xylene production processes. They provide 30% or more of worldwide production of this raw material for polyester. Probably a majority of readers are wearing clothing made from these molecules.

We will miss the constant enthusiasm for exploration and innovation, generosity, modesty, inspiration, cooperation, and friendship that Werner brought to so many of us. However, his many solid and lasting contributions will live on and grow in catalytic science and technology.

PAUL B. WEISZ

Charles Kemball, 1923–1998

Charles Kemball, who died on September 4, 1998, will be long remembered for his wide-ranging research on deuterium exchange reactions and for the insight he provided into the mechanisms of hydrocarbon reactions on metal and oxide catalysts. He was a leading figure in academic life in the United Kingdom for almost 40 years, highly respected, instantly recognizable, and widely sought for his skills in administration and management. He was one of the most distinguished British physical chemists of his generation.

Born in Edinburgh on March 27, 1923, the son of a dental surgeon, he was educated at Edinburgh Academy until 1940, when he attended Trinity College, Cambridge. He graduated with First Class Honors in both parts of the Cambridge Natural Sciences Tripos in 1943 and in the same year began research in the Department of Colloid Science of the university with Professor E. K. Rideal. At that time during the war, Rideal was carrying out work for the Ministry of Aircraft Production on adhesives and bonding to surfaces. Clean metal surfaces had been a concern of Rideal and others in Cambridge in the 1930s and Kemball was given the job of designing an apparatus to study the adsorption of hydrocarbons on a uniquely clean metal surface—that of freshly distilled liquid mercury. He did this by measuring the change in the surface tension of sessile drops. Because the adsorption of hydrocarbons on mercury was reversible, it was amenable to thermodynamic treatment, and Kemball used the entropy change to diagnose the degrees of freedom lost on adsorption. This work, which led to his Ph.D. in 1946, inculcated a respect for thermodynamic argument which was to surface frequently in his papers in the future.

Kemball kept an active interest in physical adsorption for several years, but a defining change came in 1946 with the award of a fellowship from the Commonwealth Fund of New York for research and travel in the United States. Influenced by Eric Rideal's long-standing friendship with Hugh S. Taylor, who had known Rideal since World War I when they worked together in London on the catalysis of carbon monoxide oxidation, Charles chose to spend his fellowship year with Professor Taylor at Princeton University. Here he was introduced to experimental work on catalysis and, significantly, to mass spectrometric analysis. He worked on the hydrogenolysis and decomposition of ethane, an investigation which stemmed from the ethane–deuterium exchange studies pioneered by H. S. Taylor and K. Morikawa at Princeton shortly after the discovery of deuterium in 1933. Kemball was greatly inspired by this experience. Returning to Cambridge, where he had been appointed a fellow of Trinity College, he used a grant from the Rockefeller Foundation to build a mass spectrometer and began a study of the exchange reaction of methane with deuterium catalyzed by evaporated nickel films. A feature of his technique was the continuous analysis of the reaction mixture by bleeding small amounts via a capillary leak directly to the adjacent mass spectrometer. This method of working was to become the mainstay of his research on deuterium exchange reactions for the rest of his career, albeit it was complemented eventually by other methods of isotope analysis.

Kemball's work on methane exchange, published in 1951, paved the way for further studies, notably with J. R. Anderson, on the exchange reactions of hydrocarbons and their derivatives on metal films. Much of this work has become classical. He introduced procedures for calculating the yields of deuterated molecules for comparison with experimental results, a methodology which has proved extremely useful in defining the mechanisms of the surface reactions of hydrocarbons.

In 1951 Kemball was appointed junior bursar at Trinity College, and recognition of his ability to combine administrative and research endeavor came with his appointment in 1954, when only 31 years old, to the chair of physical and inorganic chemistry in the Queen's University of Belfast. The university, at that time the only one in Northern Ireland, was constructing an impressive new building for chemistry, and Kemball soon found himself endowed with ample space for research and with able staff and students ready to join him in lines of work linked to his own interests. He remained at Queen's for 12 years, which was a very successful period in which he established the laboratory as a leading center of catalysis research, one which became internationally known and extensively visited by both academics and industrial researchers. F. G. Gault, R. B. Anderson, J. W. Hightower, and S. Siegel were among those who came from abroad for research visits.

In Belfast Charles met Kathleen (Kay) Lynd and they were married in 1956. From 1957 to 1960 he was dean of the science faculty and from 1962 to 1965 vice president of the university. He was honored both at home and abroad. The Chemical Society awarded him its Corday–Morgan Medal and later a Tilden lectureship. In 1962 he was awarded the Ipatieff Prize of the American Chemical Society. In 1965 he was elected a fellow of the Royal Society.

His research in Belfast remained centered on deuterium exchange reactions, not only of hydrocarbons but also of derivatives such as ketones and mercaptans, but always with emphasis on the intermediates and how they were bonded to the surface. His mechanisms for reactions grew more sophisticated as the reactions studied became more complex. Together with J. J. Rooney, he enjoyed regular friendly sparring with R. L. Burwell, G. C. Bond, and others on the skillfully imagined (and still debated) roles played by π-bonded intermediates and by mechanisms of rollover and even, as he once suggested, rock and roll! Charles viewed the catalyst principally as an agent for providing adsorption sites able to give a particular type of bonding or a certain strength of bonding. He was content to leave others to debate what many viewed as equally great issues, such as the role of electronic, ionic, or crystallographic structure of the catalyst in activating or orientating chemisorbed molecules. Nevertheless, he investigated a wide range of metals as catalysts, the use of films being well suited to this, and established patterns of activity and selectivity differences between metals which remain very relevant for catalytic re-forming and the synthesis of fine chemicals.

In 1966 Kemball returned to Scotland to take up the chair of chemistry at the University of Edinburgh. His choice of catalysts became more versatile. With D. A. Whan he studied molybdenum obtained by carbonyl decomposition on alumina. Oxides and zeolites featured strongly among the solids whose catalytic properties he investigated, notably with assistance from H. F. Leach. Carbenium ion and carbanion mechanisms in deuterium exchange and isomerization became part of the repertoire, and there were often salutary reminders in his admirably clear papers of the interplay between thermodynamic and kinetic control in the catalytic reactions investigated.

His lucid and crisp style of writing had its counterpart in management. He was much in demand on this account and participated increasingly in chemistry on the national scene. He was president of the Royal Institute of Chemistry (1974–1976) and chairman of the publications board of the Chemical Society (1973–1981). He was a natural choice to act as chairman of the organizing committee for the 6th International Congress on Catalysis (ICC) held in London in 1976 and was subsequently president of the ICC Council, officiating at the congress in Tokyo in 1980. In Edinburgh he was dean of the science faculty from 1975 to 1978, a time of great financial trial for UK universities. His administrative skill and sense of fairness served the University of Edinburgh well.

A cardinal problem in the use of deuterium in catalytic reactions which remained to be addressed when Kemball moved to Edinburgh was that of locating the precise positions of the D atoms in the hydrocarbon or related products. By 1970 K. Hirota and his group in Japan had shown that the rotational spectra of C_3 and C_4 deuterated hydrocarbons obtained by microwave spectroscopy provided a viable solution. Kemball searched for an opportunity to followup this discovery and, together with C. S. John, teamed up with J. K. Tyler in Glasgow to apply this method of analysis. However, it involved transport of samples in all weathers from Edinburgh to Glasgow and the method was limited to fairly small hydrocarbons. A much better prospect soon emerged, namely, the exploitation of nuclear magnetic resonance (NMR) spectroscopy for which there were good facilities in Edinburgh and expertise on hand from I. H. Sadler. The successful application of NMR spectroscopy to determine deuterium positions, coupled with gas-chromatographic and mass-spectrometric analysis, was Kemball's most notable contribution in the 1980s. It enabled previously proposed mechanisms to be checked and, when necessary, amended, and it opened the way to an extensive scenario for the future in the field of isotope-assisted kinetic analysis.

Kemball retired from the Chair of Chemistry in 1983, and from 1988 to 1991 he enjoyed the distinction of being president of the Royal Society of Edinburgh. He continued his research activity for another 10 years or so after 1983 as a university fellow. Toward the end of his career he had renewed his long-time interest in hydrogenolysis in a joint Edinburgh–Yale study with G. L. Haller of alkane decomposition on supported alloys. Charles always appreciated his links with the catalysis scene in America, and it was fitting that one of the last conferences he attended was the inaugural H. S. Taylor Conference in Nottingham in 1996. Those of us who were his contemporaries remember with affection his many contributions at conferences, some of which twinkled with mischief but all of which were stimulating. Not least among these conferences were the triennial Rideal conferences which he initiated in Belfast in 1961 (and which still continue). However, perhaps the meetings which gave him the greatest personal pleasure were the informal annual conferences which he held for his local colleagues and privileged guests at Firbush on the shore of Loch Tay, among the Scottish mountains with whose Munro summits he was intricately familiar, and walking on their slopes was his favorite outdoor recreation.

FRANK S. STONE

John Turkevich, 1907–1998

John died peacefully in Lawrenceville, New Jersey, on March 25, 1998. With him, the catalysis community lost one of its strongest and most fascinating personalities.

Born in Minneapolis in 1907, he was the eldest son of the Primate of the Russian Orthodox Church in the United States. He was educated first at Dartmouth (B.S. and M.S. degrees) and then at Princeton where he obtained a second M.S. degree in 1932. He defended his Ph.D. thesis in 1934, on the adsorption of hydrocarbons on oxide catalysts; this work was performed under the supervision of Sir Hugh Taylor. During the academic year 1935–1936, he did postdoctoral research at Cambridge University with Sir J. E. Lennard Jones and at Leipzig University with K. Bonhoeffer on quantum chemistry. He was appointed to the Princeton faculty in 1936, became full professor in 1952, and was named to the Eugene Higgins Chair in 1955. He was a very talented teacher and for many years taught a most popular freshman chemistry course. He retired from full-time teaching in 1975 but continued his research at Princeton for 20 years. Many of us will remember his enthusiastic talk on cancer chemotherapy at the International Congress on Catalysis in Baltimore in 1996. He delivered his last lecture at Princeton on October 9, 1997, at the age of 90, in his seventh decade of teaching and lecturing.

John Turkevich was a fascinating individual who made major contributions to several fields of human endeavor. With his wife Ludmilla Buketoff Turkevich, John initiated Russian courses at Princeton and edited a monthly "Guide to Russian Scientific Literature" from 1947 to 1952. He wrote several authoritative publications and articles in popular magazines on Russian science, particularly chemistry. He also lectured on government and science at the famous Woodrow Wilson School at Princeton.

He contributed to public service in an exceptional way, serving both his country and science. He was an expert in Soviet science and frequently advised the U.S. government and congressional committees on atomic energy and foreign affairs. He served on the Manhattan Project for the development of atomic energy, providing the first measurement and analysis of the infrared spectrum of uranium hexafluoride, a key step for its purification and isotopic enrichment. During World War II, with Hugh Taylor he developed a method for the production of heavy water. With G. Joris, Turkevich built an early isotope ratio mass spectrometer and was the first to apply the new technique to the study of isotope distributions in chemical reaction mechanisms. In 1941, following the Japanese blockade of shipment of natural rubber to the United States, he became head of a research team that developed a catalyst for preparation of synthetic rubber by dehydrogenation of butane to give butadiene. He taught extensively at various military institutions in the United States and in England. Following the war, he founded, with R. W. Dodson, the Chemistry Department at Brookhaven National Laboratory. He was an active advocate for the peaceful uses of atomic energy and participated as a U.S. delegate in two United Nations conferences on this key subject. He founded the office of scientific attaché in the U.S. Embassy in Moscow. In his house was a picture of him in discourse with Nikita Khrushchev that attracted much attention from his visitors. As chairman of the first U.S. delegation of University Professors to the Soviet Union, he initiated the first scientific exchanges and later served as mentor for five Soviet scientists and as the organizer of three U.S.–Soviet scientific conferences.

Professor Turkevich had a distinguished research career, starting in 1936 with the discovery in Sir Hugh Taylor's laboratory of the re-forming of linear heptane on chromium oxide. At that time, no pure hydrocarbons higher than C_3 were available. Having heard that pure *n*-heptane was obtained from pine trees growing on the shores of Lake Tahoe in California, Turkevich selected this hydrocarbon as the starting material. They discovered that linear heptane could produce not only toluene but also substantial amounts of hydrogen. During the war, toluene was important not only because high-octane fuels were badly needed but also because it was required in munitions works for production of trinitrotoluene. Furthermore, this dehydrocyclization process became the first bridge between aliphatic and aromatic chemistry. This discovery was followed by the synthesis and characterization of many other catalysts for controlling the rate and course of various chemical reactions. In the late 1940s and the 1950s, Turkevich pioneered the first and currently used method for preparation of platinum and gold microparticles called colloids, possessing the most uniform size and catalytic properties then known. He developed the first method for obtaining high-resolution images of the surfaces of catalysts by introducing a technique known as electron microscopy. He introduced methods for the determination of the number and chemical identity of the active species on catalytic surfaces. He had an insatiable curiosity which made him grasp the importance of any new discovery in a nearby discipline, be it chemistry, materials science, or physics. He thus was in the forefront in the use of organometallic compounds such as ferrocene or new crystalline solids such as zeolites. However, perhaps one of the most characteristic aspects of his research was his tremendous ability to grasp the principles and potential applications in catalysis of any new spectroscopic method that was discovered. In doing so, he greatly helped catalysis to move from an art to science.

After retiring in 1975, he continued active research but moved to a new application of catalytic chemistry. He synthesized new platinum-containing molecules that are derivatives of the same family known as "*cis*-platinum," which was discovered to have antitumor activity. Turkevich's "candy-coated" *cis*-platinum derivatives were also useful clinical agents in cancer chemotherapy. Even at the end of his life, Turkevich was still doing experiments with colleagues in medicine and chemistry.

Professor Turkevich attracted many students and foreign scholars to Princeton University and expanded its reputation as a world center for catalysis research. As an Orthodox priest, Rev. Prof. Turkevich served the spiritual needs of the university for 25 years as the Orthodox chaplain and those of the Princeton community throughout his life. He was a spiritual father, counselor, and master of the sermon. He was loved by his colleagues, students, and parishioners, who also saw him as a role model for uniting the spiritual and scientific worlds.

MICHEL CHE
MICHEL BOUDART

NMR Spectroscopy as a Probe of Surfaces of Supported Metal Catalysts

J. J. VAN DER KLINK

Institut de Physique Expérimentale
Ecole Polytechnique Fédérale de Lausanne
CH-1015 Lausanne, Switzerland

The nuclear magnetic resonance (NMR) properties of bulk metals are usually dominated by a magnetic coupling between the nuclear spin and the spin of the conduction electrons. The coupling is proportional to the density of one-electron-states at the Fermi energy and results in NMR shifts that are large in comparison with the usual chemical shift range and also in a characteristic temperature dependence of the nuclear spin lattice relaxation rate. In systems that lack the translational invariance of the extended solid, these properties depend on the local electronic environment of the resonating nucleus (the local density of states). Examples of such systems are bulk random alloys and small metal particles, such as those in supported metal catalysts. In certain cases an adsorbate on a metal also acquires such metallic NMR properties; this often happens with carbon monoxide and probably also with hydrogen, the only adsorbates considered in this review. These metallic NMR characteristics are of interest in catalysis for several reasons. In investigations of the metal (typically ^{195}Pt), the local density of states on the surface is different from that in the bulk so that NMR spectroscopy can be used to determine the metal dispersion. The ^{1}H NMR shift of adsorbed hydrogen induced by the metal surface can be used to distinguish hydrogen on the metal from that on the support, e.g., in investigations of hydrogen spillover. The spin lattice relaxation rate of adsorbed ^{13}CO shows that the chemisorption bond has metallic character. On a theoretical level, the local density of electron states at the Fermi energy on metal surface sites plays a role in frontier orbital theories of chemisorption. This quantity can be measured by NMR of the metal for catalysts that have undergone various surface treatments, not necessarily under vacuum conditions, and the results can be qualitatively interpreted in a much simplified version of such a theoretical framework. This review provides a brief presentation of the relevant theory of metals

Abbreviations BCC, body centered cubic; DOS, density of states; ESR, electron spin resonance; EXAFS, extended X-ray absorption fine structure; FCC, face centered cubic; (a crystal structure), FID, free induction decay; FT, Fourier transform; FWHM, full width at half maximum; HCP, hexagonal close packed; HOMO, highest occupied molecular orbital; IR, Infrared or infrared spectroscopy; LDOS, local density of states; LUMO, lowest unoccupied molecular orbital; MAS, magic angle spinning; NMR, nuclear magnetic resonance; PVP, poly(vinyl pyrrolidone); RF, Radiofrequency; RT, room temperature; SEDOR, spin echo double resonance; Sf, sedor fraction; SMSI, strong metal-support interaction; TEM, transmission electron microscopy; TOSS, total suppression of sidebands.

1

and a thorough discussion of the metallic characteristics observed in the NMR spectroscopy of hydrogen, carbon monoxide, and platinum.

I. Introduction

A. Scope and Organization

Currently, nuclear magnetic resonance (NMR) spectroscopy is successfully applied in several fields of research in heterogeneous catalysis. The most widespread is probably the use of magic angle spinning (MAS) NMR to find the local compositions of aluminosilicates. Another important characteristic of zeolites, their acidity, can also be found from NMR. Two previous articles in *Advances in Catalysis* (*1a, 1b*) have described these applications in detail. A second category of investigation is concerned with the structure and dynamics of relatively simple adsorbates and reaction products on the surfaces of supported metals. Slichter (*2*) reviewed the results of such work in his laboratory, as well as the experimental techniques used (*3*). Duncan (*4*) has given special attention to the ^{13}CO NMR of adsorbed carbon monoxide, and Pruski's (*5*) review has an important section on NMR of chemisorbed hydrogen. Recently, controlled-atmosphere ^{13}C MAS NMR has been developed to investigate reaction pathways, starting from suitably labeled compounds (*6*). A third area of NMR research in heterogeneous catalysis is concerned with the supported metal particles themselves. The most successfully used nucleus has been ^{195}Pt, and early work has been reviewed in Ref. (*2*). This was mostly concerned with the NMR shift difference between nuclei in the surface of a metal particle and those in the interior and its possible use to determine the metal dispersion.

This chapter focuses on "metallic" NMR behavior, i.e., on properties that are governed by electrons and holes in extended states with small energy differences, such as typically found in metals. Such NMR properties obviously allow us to determine whether the supported "metal" particles are indeed metallic and not simply small molecules built from atoms that would form a metal in the bulk. In addition, from NMR of adsorbed molecules, some adsorbates become "a piece of the metal," (which tells us something about the nature of the chemisorption bond), as frequently happens with chemisorbed carbon monoxide and sometimes with hydrogen. This aspect of the NMR of these adsorbates is discussed later, but work related to their dynamics and reactions is only partially covered; other adsorbates are not treated at all.

The standard monograph on metal-NMR (*7*) is out of print, and the main results of NMR theory are given in Section I.H. The equations for NMR quantities in metals contain a generalized Pauli-type magnetic susceptibility, which depends on the local density of electron states at the Fermi level (LDOS). In this expression, the word density refers to the crowding of electronic states along the energy axis and not to the spatial variation of the LDOS, to which we refer with the word local. The Fermi level is the highest occupied electronic energy level at zero temperature. In a metal such as platinum, the NMR shift varies strongly with changing LDOS, and therefore the metal surface nuclei resonate at a frequency that is markedly different from that of the bulk. An immediate consequence is that ^{195}Pt NMR can be used to determine the dispersion of the supported metal catalyst.

With regard to theoretical catalysis, these NMR results can sometimes be related to frontier orbital theories for chemisorption on metals. Generally, such theories attempt to establish quantitative correlations between some attribute of a clean surface and its reactivity. In the simplest version, that attribute is just a single number: the local density of states at the Fermi energy on metal surface sites. This form of the theory is an extension of Fukui's ideas on the reactivity of aromatics (*8*), which focus on the spatial variation in the highest occupied and lowest unoccupied molecular orbitals (HOMO and LUMO, the frontier orbitals) rather than on the total electron density. Hoffmann (*9*) has written about their application to the bonding of adsorbate layers to metal surfaces, the metal frontier orbitals being those around the Fermi energy. Van Santen's (*10*) work has shown that symmetry considerations based on frontier orbital theory enable a good understanding of H_2 dissociation on transition metal clusters. That work also showed that calculated charges in the CO orbitals after chemisorption on Rh agree well with the familiar notions of donation and back-donation, first proposed by Blyholder (*11*) as a general scheme to describe the chemisorption of carbon monoxide on a transition metal. (Later, we follow many other authors in considering CO chemisorption as the paradigm of metal surface reactivity, but the question of the relevance of this paradigm to individual catalytic processes is not addressed). These points are described in detail in sections I.F and I.G.

This chapter consists of six sections of varying length, depth, and internal structure. The next subsection of the introduction describes the historical link between NMR and magnetic susceptibility experiments. The following three subsections consider relevant topics from the solid-state theory of metals. Two additional subsections introduce ideas about chemical bonding, and the last subsection enumerates the equations of metal-NMR, providing the basis for the interpretation of NMR results.

Section II, experimental considerations, is hardly more than a thesaurus of NMR terms used here. It is not a guide for those planning to enter this line of research nor a replacement for, or summary of, NMR textbooks, but it points out why in given circumstances and with a given objective a certain experimental method has been chosen.

Section III is concerned with hydrogen NMR. It begins by describing the differences between hydrogen on metals and hydrogen on oxidic supports and how these are used in the investigation of spillover. To show the complexity of the interpretation of the NMR of hydrogen/metal systems, I next discuss absorbed hydrogen in bulk palladium hydride. Currently, there is only a single clearly identified case of metallic NMR behavior for adsorbed hydrogen (hydrogen on copper). The systems for which ^{13}CO NMR data are available range from small metal carbonyl molecules to large supported metal particles covered with carbon monoxide. Therefore, their treatment is split into two sections. The first of these simply lists the experimental results metal by metal, with as little interpretation as possible. The second is a discussion of these data in terms of general (not metal-specific) phenomena.

The final section is concerned with the NMR of supported metal particles, predominantly ^{195}Pt NMR. The data and their interpretation are given in relation to a number of concepts in phenomenological (NMR spectrum and dispersion; NMR spectrum and chemisorption) or theoretical (electron deficiency; promoting effect) catalysis.

This review does not attempt an exhaustive survey of all NMR work on catalysts, and its reference list is not a suitable guide to the historical development of the general subject. Papers on adsorbate NMR have typically been selected because they address some specific aspect of metallic behavior (or a pointed lack of it). Due to the availability of additional information, the reasoning and the interpretation of results given here are sometimes different from those in the original papers.

B.   NMR Shifts

Nearly all NMR work reported measures couplings between the magnetic moment of the nucleus to which the spectrometer is tuned and the electron magnetism induced in the sample by the applied field. To a very good approximation, the spin and orbital magnetism of the electrons can be considered separately and their effects simply added. The chemical shift is the manifestation of orbital magnetism, and the Knight shift is that of spin magnetism. A simple example (12) is provided in Fig. 1; the temperature dependence of the ^{17}O resonance frequency in liquid oxygen is due to the unpaired spins in the two 2p-π^* orbitals of the molecule. The 1/T variation of the magnetic susceptibility of oxygen was first observed by Pierre Curie in 1895 (13) for gaseous samples at temperatures between 293 and 723 K (Fig. 1b). In metals without permanent magnetism, the Pauli spin susceptibility is (nearly) temperature independent and so is the Knight shift (14).

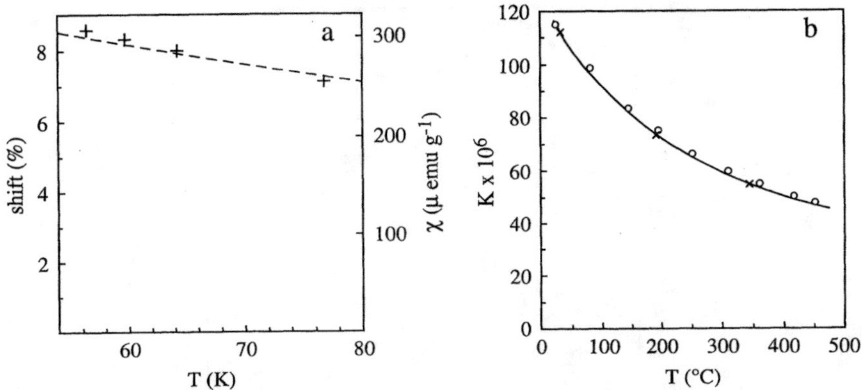

FIG. 1.   (a) The shift (left scale) of the ^{17}O NMR line (+) in liquid oxygen and as a function of the magnetic susceptibility (dashed line and scale to the right) as a function of temperature. The shifts are large (the chemical shift range of ^{17}O in different compounds is 0.1%) and proportional to the temperature-dependent electron-spin paramagnetism. [Reproduced with permission from Dundon (12). Copyright 1982 American Institute of Physics.] (b) The "specific magnetization" K of two samples of oxygen gas, as reported by Pierre Curie in 1895. The curve shows the $1/T$ behavior (with temperature in Kelvin, not degrees Celsius) for the susceptibility that now bears his name. (The symbol K, used by Curie, is no longer in use).

Palladium (Pd) is a nonmagnetic metal with a very strong spin susceptibility. Dilke, *et al.* (15) noticed 50 years ago that the magnetic susceptibility of finely divided Pd powder decreases when it is saturated with dimethyl sulfide gas, a powerful catalytic poison. We expect the ^{105}Pd Knight shift to also decrease under this treatment since the Knight shift generally very precisely tracks the change in susceptibility with temperature (16) (Fig. 2). (Contrary to the general rule, the Pauli susceptibility of palladium varies much with temperature. Unfortunately, the NMR parameters of ^{105}Pd are not suitable for experiments on small particles.) Since NMR is a local method, one might hope to observe whether the susceptibility (or the Knight shift) diminishes equally on all Pd atoms or perhaps more on the surface than in the interior, etc.

To demonstrate the local character of the Knight shift, Fig. 3 shows a ^{195}Pt NMR spectrum (absorption derivative) in a very slightly alloyed (7×10^{-4} atom fraction Mo) bulk platinum sample (17). The overwhelming majority of ^{195}Pt nuclei resonate at approximately 1.14 G/kHz, just as in pure platinum, but those that are nearest neighbors to a Mo impurity have a smaller Knight shift and resonate at approximately 1.127 G/kHz; second- and third-neighbor resonances are seen in between (the assignments follow from the intensity variations with Mo fraction). For higher concentrations

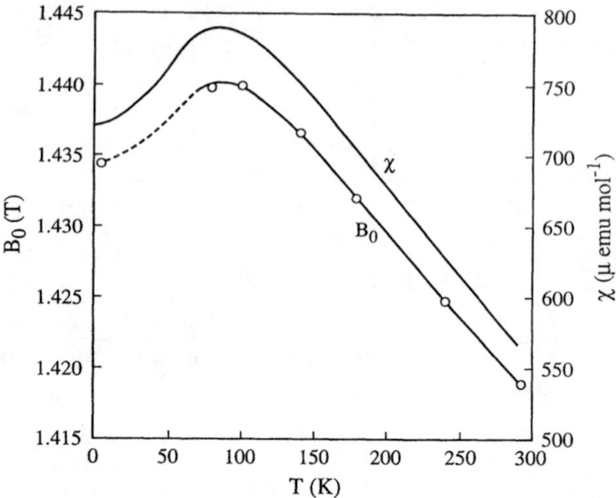

FIG. 2.   Temperature dependence of the magnetic susceptibility χ (right scale) and of the field for ^{105}Pd NMR at fixed Larmor frequency B_0 (left scale) for palladium. Similar to Fig. 1a, the NMR field shift and the susceptibility are proportional. Both are (mainly) caused by the Pauli-type paramagnetism of the d-like conduction electrons. (The temperature dependence is not predicted by the simple free-electron description of the susceptibility in metals.) [Reproduced with permission from Seitchik *et al.* (*16*). Copyright 1964 American Physical Society.]

of Mo, the magnetic susceptibility of such alloys is measurably smaller (*18*) than that of pure Pt; the NMR suggests that this diminution is localized around the impurity.

Usual theoretical expressions for the Knight shift in metals describe it as a property of the conduction electrons. It follows that the zero of the theoretical shift scale is not the resonance frequency of the bare nucleus but rather that of the nucleus plus all the core electrons (approximately the metal ion corresponding to the next lower noble gas configuration). Even when one is reasonably sure that the experimental shift scale corresponds to this ideal one, Knight shifts of either sign may be found, as shown in Fig. 4 for CuPt alloys of different composition (*19, 20*). These results are expected because s-like conduction electrons give positive shifts and d-like ones negative shifts. If both types are present, an accidental zero shift may be found. The metallic character of NMR remains visible through the very specific temperature dependence of the Korringa spin lattice relaxation mechanism (see Section I.H).

In determining whether a small particle is metallic, it is not practical to attach wires and measure a conductivity. However, conductivity, Pauli susceptibility, Knight shift, and Korringa relaxation all rely on the same

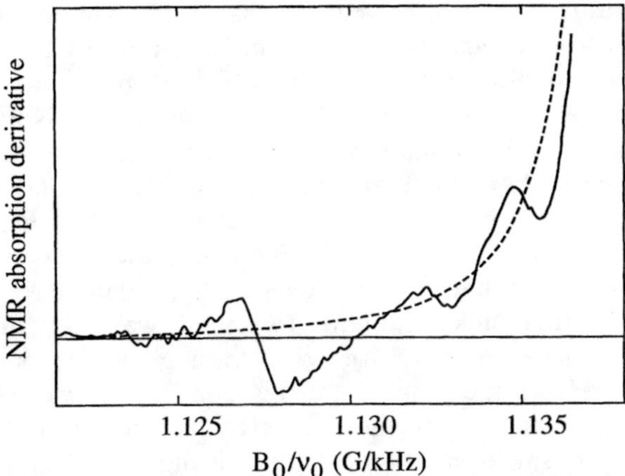

FIG. 3.   The derivative of the ^{195}Pt NMR absorption signal in a slightly alloyed platinum sample. (In this experiment the field is swept slowly while the sample is continuously irradiated at fixed frequency. Lock-in detection records the derivative.) Most of the ^{195}Pt nuclei resonate at a field/frequency ratio of approximately 1.14 G/kHz, off scale to the right (dashed curve). Those ^{195}Pt atoms that are first neighbors to an alloying impurity resonate at 1.127 G/kHz, and second and third neighbors give separate resonances as well. The results show that the susceptibility is site dependent. [Reproduced with permission from Weisman and Knight (*17*). Copyright 1968 American Physical Society.]

characteristic of a metal; in a one-electron model, the separation in energy between orbitals near the Fermi level is very small compared with the thermal energy kT. Therefore, the NMR properties of small particles can give information about their metallic character. For example, consider the Pt$_{13}$ cubooctahedral particle (*21*). From molecular structure calculations it is found that the highest molecular orbital is fully occupied and separated

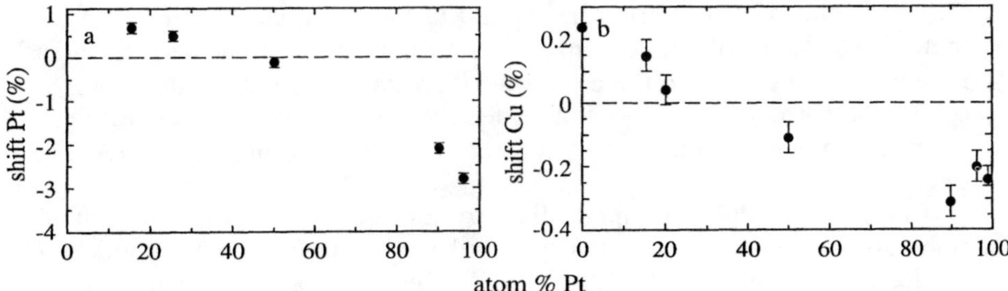

FIG. 4.   NMR shifts of ^{195}Pt (a) and ^{63}Cu (b) in copper–platinum alloys as a function of composition. Both shifts change sign (with respect to the usual shift standards), and zero shift does not mean that the samples are not metallic.

from the lowest empty orbital by more than 6000 K in energy. Therefore, thermal excitation to this level can be neglected at any temperature, and Pt_{13} should show neither Curie-nor Pauli-type magnetic susceptibility; the electronic spins are always paired off at any temperature. Now imagine that we add another 42 platinum atoms to form the next larger cubooctahedral molecule. The number of orbitals will increase correspondingly or, more precisely stated, in a given energy interval we will find more molecular orbital levels for Pt_{55} than for Pt_{13}. By increasing the size continuously, the number of orbitals (in that same energy interval) will increase continuously. Since we know that bulk platinum is a metal, we will achieve a size at which the separation in energy between the highest occupied and lowest unoccupied levels (in the ground state) is smaller than kT. This will result in Pauli-like paramagnetism and a NMR spectrum with a Knight shift; regarding magnetism, a particle of this size is metallic, even if the electron eigenfunctions do not have Bloch wave character, periodic in space.

C.   Band Structure of Solids

Much of theoretical solid-state physics is based on the model of a perfect crystal with a perfect translational periodicity of its lattice. The one-electron orbitals extend through the whole crystal, similar to the delocalization in aromatic molecules (22). To find an expression for these one-electron orbitals, it is advantageous to consider the electronic charge distribution in reciprocal space of wavevectors k, which is the Fourier transform of the distribution in real space of position vectors r. A one-electron orbital is identified completely by its wavevector k and an additional (band) index n. In a tight-binding (linear combination of atomic orbitals) situation, this index labels from which atomic orbitals (1s, 2p, 3d, etc.) or from which hybrids the delocalized orbital is formed. The energy of the orbital is denoted $E(n, k)$. The collection of orbitals with the same n is called a band. Different bands may overlap in energy; in fact, this is a requirement for metallic character of an elemental solid formed from atoms with an even number of electrons. On the energy scale, certain quasi-continuous ranges of E that correspond to possible one-electron energies (the bands) are separated by gaps (ranges of values that do not correspond to a possible one-electron energy).

If in a given solid at temperature zero all one-electron levels are filled right up to the beginning of a gap and empty above it, that solid is a semiconductor; otherwise it is a metal. In the latter case, the highest occupied level is called the Fermi level, and its energy is the Fermi energy E_f. The relation $E(n, k)$ for all n and k is called the electronic band structure of the solid; it is the central quantity in solid-state physics. For many metallic properties, we do not need the full $E(n, k)$ relation but only certain sums over (n, k) values. The Pauli susceptibility is large when there are many (n, k) combinations that have $E(n, k)$ close to the Fermi energy. More precisely, it is proportional to the density of one-electron states, (DOS) i.e., the number of such states per unit energy interval and per metal atom with energies approximately equal to the Fermi energy E_f.

At the time of the $Pd/(CH_3)_2S$ susceptibility experiment (*15*) referred to previously, the only conceptual framework available to describe the result was the rigid-band model: The adsorbate is supposed to dump some of its electrons into the lowest empty orbitals of the metal (or to pull some from the highest filled ones) without changing the energetic sequence of these orbitals. In other words, the adsorbate adds electrons but no orbitals (or the other way round). It is now known that this is usually a poor approximation (*23*), and the rigid-band model is no longer used.

The Fermi energy is an important quantity, and it is worthwhile to recall its origin (*24*). Quantum statistical mechanics asserts that half-integer spin particles are fermions and that in a system of weakly interacting fermions each orbital (a state of the Schrödinger equation for only one particle) is either singly occupied or empty. (I do not discuss the very important question of the extent to which electrons in a metal or in a molecule can be considered weakly interacting.) In general, the number of orbitals for the N-particle system is infinite, and at finite temperature at least some of them are sometimes occupied and sometimes empty. Consider now one of these levels and take the point of view that it "exchanges" particles with the rest; the number of particles in the level under consideration fluctuates, and in statistical–mechanical considerations some chemical potential must appear. Therefore, the notion of chemical potential for a weakly interacting N-fermion system can be arrived at without considering the transport of a particle from infinitely far to inside the system. This chemical potential is naturally expressed with respect to the lowest orbital energy. The Fermi energy E_f is the zero-temperature value of this chemical potential. In metals its value is much larger than kT at practical temperatures; therefore, all levels more than kT below E_f are occupied, and all those more than kT above it are empty.

D. WORK FUNCTION

With regard to the model of N electrons in a box, the Fermi energy as introduced previously does not say anything about how difficult it is to push an electron through the walls of the box. The minimum work required to bring an additional electron from infinity, push it through the walls of the box, and land it in an unoccupied level near the Fermi level is called the work function (W). To determine the difference between the Fermi energy and the work function, it is simplest to use the jellium model for a semi-infinite metallic solid (25). In this model, the nuclear charges are represented as a uniform ("background") charge distribution that decreases discontinuously to zero at the surface of the solid. All quantities vary only in a direction perpendicular to the surface; the model is effectively one-dimensional. In the independent-electron (or Hartree) approximation, any electron moves in the electrostatic potential $V_H(r)$ due to the uniform background charge and to the average charge distribution of all the other electrons. The Hartree approximation neglects the antisymmetry of the exact solution of Schrödinger's equation for an N-fermion wavefunction. To take this requirement into account, an "exchange and correlation" potential $V_{xc}(r)$ is added to the electrostatic one. (For our purpose, we need not be concerned about how this potential is obtained.) As shown in Fig. 5a, the effective one-electron potential (the sum of Hartree and exchange–correlation potentials) is such that the electrons "spill out" of the box; the electron charge density decreases continuously to zero (Fig. 5b).

It is a natural choice to set the value of the effective potential to zero at a very large distance from the box so that the value deep inside the solid

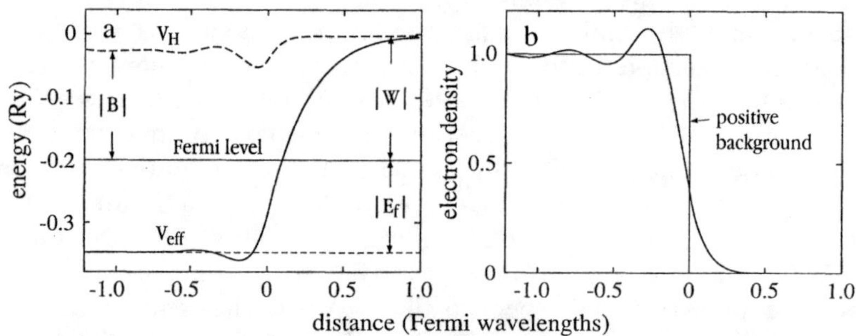

FIG. 5. Calculations for the semi-infinite jellium model of a metal surface. The ionic cores in the metal are represented as a continuous background of positive charge density (the value chosen here corresponds to potassium). Distances are in units of 0.866 nm, perpendicular to the surface. (a) The effective potential well V_{eff} in which the electrons move (continuous curve) has two contributions: the electrostatic energy, V_H (dashed curve), due to the spatial unbalance of positive and negative charges near the surface (right) and the exchange–correlation potential V_{xc} (the difference between the dashed and continuous curves). V_{xc} accounts for the many-body character of the N-electron ground state. The zero of the energy scale is the vacuum. All one-electron levels between the bottom of V_{eff} and the Fermi level are occupied; all states above the Fermi level are empty. Indicated are the definitions of the Fermi energy, E_f, the work function W, and the binding energy B. (b) Spill-out of the electron charge (curve) from the positive background (step function). [Reproduced with permission from Lang (25).]

is $-|V_{\text{eff}}|$. One-electron levels with different wavevectors $|k|$ (there is no band index because there are no atoms) have different kinetic energies; the lowest possible kinetic energy is zero and the highest E_f. With these conventions, the work function becomes

$$W = |V_H| + |V_{xc}| - |E_f|, \tag{1}$$

where the Hartree and exchange–correlation potentials are taken deep inside the solid. Two contributions to the work function are sometimes distinguished (26): One is a property of the infinite solid, the "binding energy" for a Fermi-level electron $B = |E_f| - |V_{xc}|$, and the other is the energy required to move the electron through an electrostatic double layer at the surface of the metal $|V_H|$ (in a real metal this may depend on the type of surface considered). The double layer represents the local imbalance between the positive background charge and the negative electronic charge over a small but finite range near the surface (Fig. 5b) and is often simplified to an electric dipole density P at the surface:

$$P(0) = |V_H|/(4\pi e). \tag{2}$$

The partition of the work function in Eq. (1) is useful to discuss the question of whether it can be changed by an electric field from "faraway" external charges. In the language of continuum electrostatics, the gradient of the electrostatic potential ϕ due to external charges disappears inside a conductor, with the continuity of ϕ at the surface being ensured by the appearance of a surface charge density. (In a real metal the spatial extension of this additional charge layer will be comparable to that of the original electrostatic double layer.) For a semi-infinite solid, the necessary charge can be pulled from its interior without perceptibly changing the occupation of levels; the Fermi energy and the exchange–correlation potential are unaffected. In other words, the internal binding energy B is a property of the infinite solid, on which no outside influence is possible. We must still consider the electrostatic contribution of the surface charge layer to the work function. To exclude work to be done in moving an electron through the vacuum, the work function is now defined as the work needed to go from "just inside" the surface to "just outside" it. (We do not need to consider here the precise definitions of these distances.) Since the electrostatic energy of an electron on one side of a sheet of charge is the same as that at the same distance on the other side, the additional charge sheet also has no effect on this contribution. This simply expresses the continuity of the electrostatic potential due to the outside sources at the surface of a conductor. Similar arguments can be used to show that the addition of electrons from the outside, which according to electrostatics will appear at the surface of the metal, have no effect on the work function, as long as their number is too small to change the Fermi energy.

This is clear for a macroscopic (even if not semi-infinite) crystal, but we may wonder about the limit in size to which it will remain valid: What is the work function of a nanocrystal? For a molecule, there is a difference between the amount of work required to pull an electron off (the ionization energy) and the work gained by pushing an extra electron on (the electron affinity). As a condition for the application of the work function concept we may require that the ionization energy of the system be reasonably nearly equal to its electron affinity (it is of course easier to propose the criterion than to verify it experimentally). In this chapter, I frequently use the existence of a Korringa-type of spin lattice relaxation, or that of a conspicuous Knight shift, as a criterion for metallic behavior, but this is not quite the same as the ionization/affinity criterion.

Two other jellium calculations are of interest. One (27) is a study of the local density of states in the spill-out region. It has been found that the farther out from the surface, the larger the contribution of electrons at the Fermi level to the total charge density (Fig. 6). The latter quantity decreases with distance; it decreases exponentially, with a characteristic length that is inversely proportional to the square root of the work function. The other

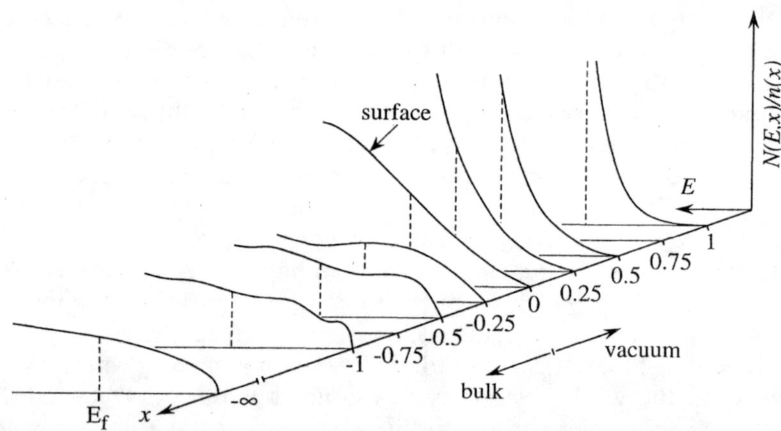

Fig. 6. $N(E, x)$ is the local density of states of energy E, at distance x from the surface of a semi-infinite jellium; $n(x)$ is the electron charge density at x (cf. Fig. 5, right). The quantity plotted along the vertical axis is the ratio $N(E, x)/n(x)$ for several values of x. Negative x are inside the solid and positive are outside. The dashed vertical lines show the LDOS of states near the Fermi level, normalized by the electron charge density; $n(x)$ falls exponentially outside the surface. A large fraction of the charge outside the surface is due to Fermi-level electrons. [Reproduced with permission from Werner et al. (27). Copyright 1975 Institute of Physics Publishing.]

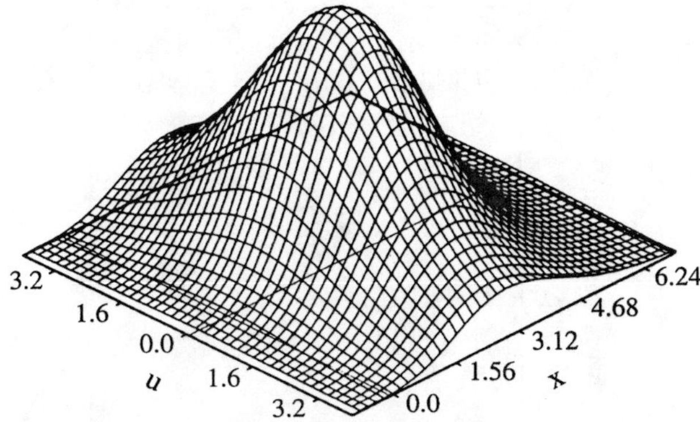

Fig. 7.   Screening electron charge density (vertical axis not shown) around a unit positive point charge just outside a jellium surface. The surface is at $x = 0$, and the point charge is at $u = 0$, $x = 4.5$ units (the unit of distance is 51 pm, and the positive background density in the jellium corresponds to aluminum). [Reproduced with permission from Smith *et al.* (*28*). Copyright 1973 American Physical Society.]

study (*28*) considers probably the simplest possible interaction between a jellium edge and an external charge distribution: a positive point charge (a proton) in the dipole layer region of a jellium. Figure 7 shows how the jellium electrons rearrange their spatial distribution to screen the proton. While all these jellium models are useful to introduce the phenomena, their neglect of atomic structure makes them less suitable for quantitative considerations.

The work functions for low-index surfaces of the 4d transition metals have been calculated by a full-potential linear-muffin-tin-orbital method using a slab geometry (a periodic arrangement of 7-layer metal slabs and 10-layer vacuum slabs) (*29*), and the results (Fig. 8) agree well with experimental results. This is a considerable improvement with respect to extended Hückel calculations for slab [*30* (see footnote 21)] or cluster [*10* (see Chapter 3)] geometries, which usually yield values closer to the atomic ionization energies. However, the shape of the DOS curves and the relative position of the Fermi level as found by the extended Hückel calculations are reasonably similar to those obtained by the more sophisticated methods. Therefore, it seems that the major error is in the determination of the dipole layer potential. (Further analysis of this topic would lead us beyond the scope of this chapter.)

E.   Local Density of States

Band theory has proved enormously successful for calculation of bulk properties of metals. It is less suitable for the determination of local proper-

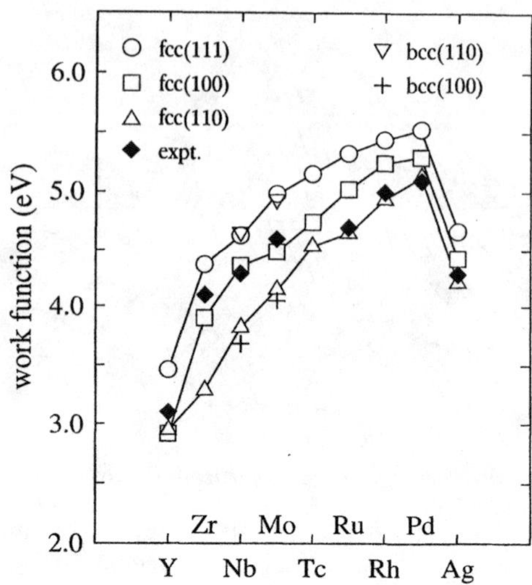

FIG. 8.   Calculated (open symbols) and experimental (solid diamonds) values of the work function for the 4d transition metals. These calculations are considerably more sophisticated than the older ones shown in Figs. 5–7, and they reproduce several properties of these metal surfaces very well. The calculated values are for different single-crystal faces, as indicated; the experimental values refer to polycrystalline specimens. [Reproduced with permission from Methfessel *et al.* (*29*). Copyright 1992 American Physical Society.]

ties, such as the charge density, bond order, and localized magnetic moments, because their theoretical expressions involve a summation over Avogadro's number of extended one-electron orbitals. For example, consider the ferromagnetic ordered alloy (intermetallic compound) Fe_3Al (*31*). It has a body-centered cubic (BCC) structure, just as ordinary iron does, except that every fourth Fe is replaced by Al. There are two types of Fe sites in Fe_3Al: one with four Fe and four Al nearest neighbors and a magnetic moment of 1.46 μ_B, and the other with all eight nearest neighbors Fe and a moment of 2.14 μ_B. The latter is very close to that of pure iron (2.20 μ_B); adopting an expression from theoretical catalysis, an *ensemble* of nine Fe atoms seems to be large enough to create bulk iron-like magnetic properties at its center. This similarity can be understood (*31*) by considering the LDOS, which is the density (on an energy scale) of states weighted by their local intensity (on a spatial scale). (It is of course equally valid to start by considering the charge density at a given site and break up this quantity along the energy axis.) Denoting the wavefunction of orbital (*n*, *k*) as $\psi_{nk}(r)$, the local density of states at point r at energy E is

$$LDOS(E, r) = N^{-1} \sum \delta(E - E_{nk})|\psi_{nk}(r)|^2, \tag{3}$$

where N is the number of atoms or molecules, and $\delta(E)$ the Dirac "function." The energy integral of the local density of states up to the Fermi energy will give the total electron density at site r:

$$\rho(r) = \int_{-\infty}^{E_f} \text{LDOS}(E, r)dE. \tag{4}$$

The spatial integral gives the usual DOS:

$$\text{DOS}(E) = \int_{\text{sample}} \text{LDOS}(E, r)dr = N^{-1} \sum \delta(E - E_{nk}). \tag{5}$$

Unless mentioned otherwise, LDOS and DOS in the following sections will mean $\text{LDOS}(E_f)$ and $\text{DOS}(E_f)$.

In a compound, the LDOS will distinguish which bands are more important on which atoms. At the surface of a solid, the LDOS may be different from that in the interior. Developing earlier work by Friedel (*32*), Heine (*31*) demonstrated an invariance property of the LDOS, that somewhat imprecisely can be rephrased as follows: The LDOS in the center of a small ensemble of similar atoms is like that in the extended solid, corrected for boundary effects. The larger the initial ensemble, the smaller the boundary corrections. For the energy integral of the LDOS, the convergence is more rapid than that for the LDOS itself. A magnetic compound has separate "spin-up" and "spin-down" LDOS (with a common E_f), and the difference of their energy integrals up to E_f gives the distribution of the magnetization in space. In the Fe_3Al example, the magnetization on the central atom in the nine-atom Fe cluster apparently needs little boundary correction compared with that in the infinite bulk iron.

The diminishing susceptibility in the Pd/dimethyl sulfide experiment showed that the total density of one-electron states with energies around E_f, the E_f DOS, diminished upon chemisorption. Furthermore, the hypothetical [105]Pd NMR experiment could indicate whether the charge density in such orbitals was everywhere the same, higher on surface sites than in the interior, etc.; it could give information about the E_f LDOS. Just as the Pauli susceptibility is sensitive only to the density of states in a small energy range (of the order kT) around the Fermi energy, the Knight shift probes the local density of states at those energies. (The NMR does not give information about changes in total charge density on a given site, which is proportional to the integral of the E LDOS over Evalues from minus infinity to E_f, [cf. Eq. (4)].

F. FRONTIER ORBITALS

Figure 9 shows the density of one-electron energy levels in the 4d transition metals, all drawn with respect to the potential at infinite distance from

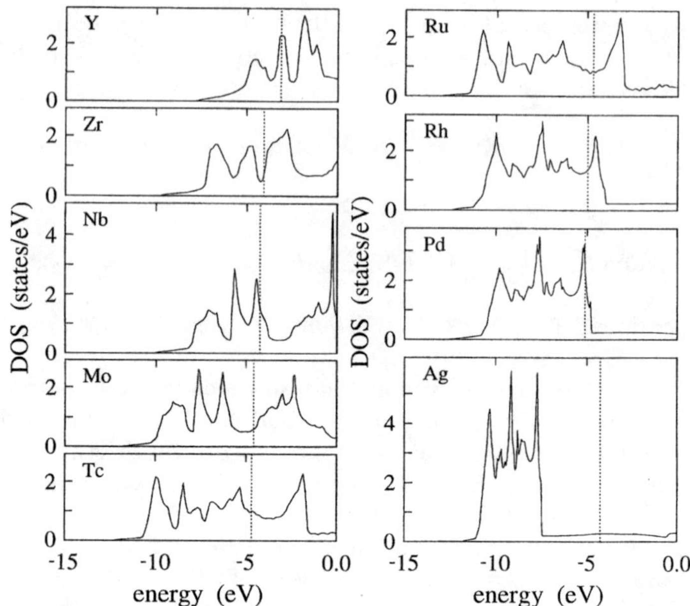

FIG. 9. Bulk densities of state (DOS) in the 4d transition metals. The energy scales have been adjusted to show the Fermi levels (dotted lines) in agreement with experimental work function values. (Experimental results included here are not the same as those in Fig. 8.) Note the hybrid character of this figure; it combines calculations for the bulk with experimental results for a (polycrystalline) surface. For comparison, the $2\pi^*$ levels of an isolated CO molecule are at -7.5 eV, and the 5σ level is at -13 eV (*33, 34*).

the metal surface. The figure combines the experimental work function data (*33*) with bulk DOS curves (*34*). This combination should be handled with care; there is plenty of evidence that the LDOS on a metal surface is different from that far in the interior. A particularly instructive example, shown in Fig. 10, is that of the W(001) surface, calculated for a seven-layer slab (*35*). The calculated E_f LDOS on the surface sites is particularly high, and the whole LDOS curves on the third and the central layers are very similar. The latter is clearly a manifestation of the Heine–Friedel invariance rule for the LDOS; the local environment of atoms in these layers is very similar. The high density of states calculated at E_f for the ideal surface is probably related to the fact that the real surface reconstructs. The reconstruction will of course also change the work function. Note that in Fig. 9, when going from Rh to Pd to Ag, the work functions vary much less than the densities of state at the Fermi energy.

The expression "frontier orbitals" was apparently used for the first time in Fukui *et al.*'s (*8*) famous paper on the reactivity of aromatic hydrocarbons,

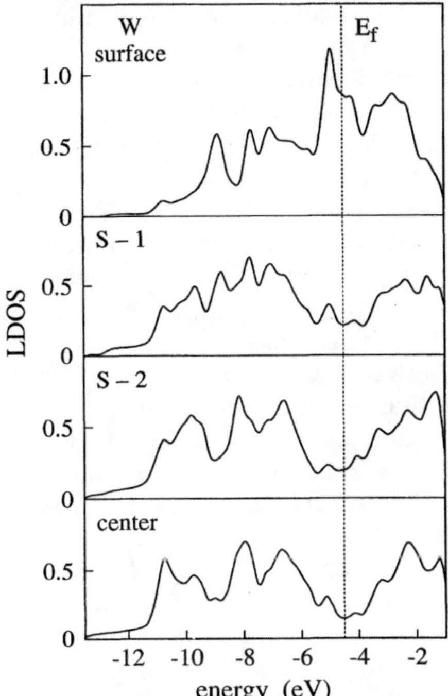

FIG. 10.   Local density of states on different atomic layers in a seven-layer tungsten slab. The (calculated) Fermi level is shown by the dotted line. Note that the two deepest layers have very similar LDOS curves but that the surface LDOS is markedly different. (The calculation is for the ideal BCC (100) surface; the real surface reconstructs.) Compare with the bulk-DOS of molybdenum in Fig. 9. [Reproduced with permission from Posternak *et al.* (*35*). Copyright 1980 American Physical Society.]

denoting the set of HOMO and LUMO. These authours pointed out that the site-specific reactivities in many unsubstituted aromatic hydrocarbon molecules cannot be understood from a consideration of the total electron density in the electronic π system since, by the Coulson–Rushbrooke theorem, this is the same on all sites. The new idea introduced was to consider not the total electron density but only that in the highest occupied π-orbital; the carbon sites with the largest charge density in this orbital will be attacked by electrophilic or oxidizing agents.

Fukui *et al.*'s (*8*) ideas were refined to extend their validity to other types of reactions, but here I need only mention an interesting remark in the original paper. The authors observed that triphenylene is more stable to oxidation than naphthacene and phenantrene more than anthracene; they concluded that, at a constant number of carbons and π electrons, those molecules with a higher absolute value of frontier orbital density will be the more reactive. Later, I will present an analogue of this idea, which states that among metal particles with similar surfaces, those with the highest average E_f LDOS will show the largest reactivity toward CO.

In the orbital-interaction scheme for chemisorption (*9, 10*), the incoming reagent (the CO molecule) will first meet, and therefore interact with, the electrons that spill out of the metal surface. This spilt-out electron cloud is very tenuous, compared with the density inside the metal, and, furthermore, it consists mostly of Fermi-level electrons (cf. Fig. 6) that can be rearranged into other, as yet empty, levels at the lowest possible energetic cost: They are the frontier orbitals of the metal. Once this "diffusive contact" for electrons between the metal and the molecule is established, there will be exchange of charge to establish a common chemical potential. (The Fermi energy of the isolated CO molecule is halfway between the 5σ and the $2\pi^*$ levels.) Supposing that in this process the CO molecule remains a recognizable entity, the Blyholder model (*11*) states that bonding/antibonding orbital pairs will be formed between the metal levels below E_f and the CO levels above it (a broadened version of the $2\pi^*$) as well as between metal levels above and CO levels below it (the 5σ). More precisely, the metal levels enter the model in the form of the local density of states on a surface site.

In further refinements of the theory, this LDOS is decomposed into contributions with different types of symmetry; it is also possible to incorporate the concept of ensembles, but the discussion of NMR data does not require these details. I also note the respective roles played by the surface E_f LDOS and the work function: a high LDOS ensures the availability of many electrons for the initial metal–adsorbate interaction, whereas a low work function makes the charge tail extend further in space. Once the metal–adsorbate bond has been formed, the LDOS at any energy on sites in or close to the surface can be different from that of the clean surface, but sufficiently far inside the particle the perturbation will be negligible (Heine–Friedel invariance).

The LDOS-based frontier orbital model is different from the simpler collective-electron model, in which all local information is averaged out. In the context of alloys, Ponec and Bond (23, p. 451), stated that "it must be clear to the reader that [the collective electron model of catalytic activity of alloys] has now been consigned to the trash can of science . . . [because of] the discovery that, more in tune with chemists' intuition, the atoms in an alloy retain their identity more or less completely." Such local properties disappear from the collective-electron model in which "it was supposed that the available electrons were equally shared by all the atoms present, and that . . . the density of states at the Fermi surface or some related quantity would determine catalytic activity." I discuss these objections in Section VI.D.

G. Bonding Classification

A question may be raised concerning the nature of the bond: Is it covalent, ionic, or metallic? In cases such as the proton in the dipole layer of a jellium (Fig. 7) we can probably always find a reasonably small volume such that the extra charge inside (proton plus electrons) is zero. More generally, most perturbing "point charges" in a metallic-like electron cloud will be very efficiently screened to zero effective charge, just as for an elementary cell in a metallic solid. The NMR of such point charges will also have metallic character because a large part of the screening will be done by Fermi-level electrons. In the simplest definition of covalent bonding, the basic unit is the two-center two-orbital bond. Simple molecules and certain covalent crystals (quartz and diamond) can be well described in these terms, but a bulk metal such as platinum, in which the number of nearest neighbors (12) is larger than the number of valence electrons (10), cannot. For our purposes, we observe that covalent crystals are semiconductors, whereas metals are not, and we distinguish them by their Knight shift and Korringa relation. The distinction will be used not only for the atoms of the metal particle but also for those of the adsorbate; at least in NMR terms, the adsorbate becomes a piece of the metal if the Fermi energy is inside a continuum of local density of states on the adsorbate molecule.

H. Knight Shift and Korringa Product

The tensor describing the magnetic interaction between the electron spins and the nuclear spins can be separated into isotropic and traceless symmetric parts. The isotropic part is the Fermi contact interaction, which has no analogue in classical magnetostatics (because, classically, two different magnetic dipoles cannot be in the same place). The traceless symmetric part represents a dipole–dipole coupling. In cubic symmetry its average value is zero, but its fluctuations in time contribute to the nuclear spin lattice relaxation. Since the dipolar interaction is traceless, it does not affect the center of gravity of the shift pattern.

The interaction between the orbital magnetic moment of the electrons and the nuclear moment can also be separated into an isotropic part (the chemical shift) and a traceless symmetric part (the shift anisotropy). The equation for the shift tensor contains two terms, usually called the dia- and paramagnetic contribution, but only the sum of the two corresponds to a physical quantity. Actually, the theory (36) is concerned with the shielding σ rather than the shift δ (the most important difference between the two is that they differ in sign). The standard way of partitioning σ into a diamagnetic part σ_{dia} and a paramagnetic part σ_{para} yields (in a notation appropriate for molecules)

$$\sigma_{\text{dia}} = (\mu_0/4\pi)(e^2/2m)\left\langle 0\left| \sum_k r_k^{-3}(r_k \cdot r_k \mathbf{1} - r_k r_k) \right|0\right\rangle \tag{6}$$

$$\sigma_{\text{para}} = -(\mu_{\text{o}}/4\pi)(e^2/2m^2) \sum_{\substack{n\neq 0 \\ k'k'}} \left(\frac{\langle 0|r_k^{-3}l_k|n\rangle\langle n|l_{k'}|0\rangle}{E_n - E_0} + \text{cc} \right), \tag{7}$$

where r_k is the position vector for electron k, l_k the angular momentum operator with respect to the observed nucleus, and cc is the complex conjugate. Equations (6) and (7) provide all the components σ_{zz}, σ_{xy}, etc. by use of appropriate components of r and l. The diamagnetic part corresponds to the Lamb shift for atoms, or to the Larmor diamagnetic susceptibility. The paramagnetic part corresponds to the van Vleck paramagnetic susceptibility (again, physically, there exists only a single orbital susceptibility, which is the sum of the two contributions).

A simple extension of the derivation of these equations to the case of metals yields the following for the isotropic part of the chemical shift δ, commonly called the "orbital Knight shift" K_{orb}:

$$K_{\text{orb}} = K_{\text{para}} + K_{\text{dia}} \approx K_{\text{para}},$$

since (in the standard partitioning) the diamagnetic contribution from the non-core electrons is usually small compared with the paramagnetic part. The non-core electrons are all electrons outside the next lower noble gas configuration, and it is assumed that the shielding due to this configuration is the same in all chemical environments of the atom under consideration. It is usual to write

$$K_{\text{orb}} = \chi_{\text{orb}}(\Omega H_{\text{hf,orb}}/\mu_{\text{B}}), \tag{8}$$

where χ_{orb} is the van Vleck contribution to the orbital susceptibility, Ω is the atomic volume, μ_{B} is the Bohr magneton, and $H_{\text{hf,orb}}$ is given by

$$H_{\text{hf,orb}} = (\mu_{\text{B}}/2\pi) \langle r^{-3}\rangle. \tag{9}$$

Here $\langle\,\rangle$ indicates some suitable average over electron orbitals, the precise definition of which we do not need to consider. From the analogy with Eq. (7), however, it is clear that all energy levels contribute (not only those within kT around E_{f}).

In metals, the orbital interaction also gives rise to nuclear spin lattice relaxation. Because of energy conservation, only electron energy levels close to the Fermi energy are of concern, and therefore the averaging replacing that in Eq. (9) is over a different set of orbitals. However, it is customary to use the same $H_{\mathrm{hf,orb}}$ in the equations for K_{orb} and for $T_{1,\mathrm{orb}}$.

Returning to the interaction between electron spin and nuclear spin, we consider here only the isotropic part. The effect of the Fermi contact interaction on the NMR parameters (Knight shift, relaxation rate, and indirect coupling) in paramagnetic systems can be described in quite a general way (valid in metals, semiconductors, or molecules) through the complex generalized electron spin susceptibility $\chi(r, r'; \omega)$, $= \chi'(r, r'; \omega) - i\chi''(r, r'; \omega)$ that describes the local response of the spin magnetization in r to a time-varying magnetic field $H(r')\cos(\omega t)$ applied in r'. In NMR we are usually concerned with the $\omega \to 0$ limit of $\chi(r, r'; \omega)$, and the ω argument will not be written in the following. The uniform susceptibility of the sample $\bar{\chi}'$, the Knight shift K of a nucleus at site R, and its spin lattice relaxation rate T_1^{-1} at temperature T are given by

$$\bar{\chi} = V^{-1} \iint \chi'(r, r')\mathrm{d}r\, \mathrm{d}r' \tag{10}$$

$$K = (2/3) \int \chi'(R, r')\, \mathrm{d}r' \tag{11}$$

$$S(T_1 T)^{-1} = 2\mu_{\mathrm{o}}(2\mu_{\mathrm{B}}/3)^2 \chi''(R, R)(\pi\hbar\omega)^{-1}. \tag{12}$$

To obtain expressions in the centimeter-gram-second (cgs) system, replace the factors 2/3 in the right-hand sides of Eqs. (12) and (13) by $8\pi/3$. In Eq. (12), S is the Korringa constant $(\gamma_e/\gamma_n)^2(\hbar/4\pi k_{\mathrm{B}})$. Its values are $^{195}S = 5.77$ μs K for ^{195}Pt, $^1S = 263.3$ ns K for ^{1}H, and $^{13}S = 4.165$ μs K for ^{13}C.

Furthermore, the indirect coupling between two nuclear moments I_1 and I_2 and has an isotropic part with a Hamiltonian

$$H = J_{12}I_1 \cdot I_2 \tag{13}$$

where the coupling energy J_{12} can be expressed as

$$J_{12} = \mu_{\mathrm{o}}(2\hbar/3)^2 \gamma_1\gamma_2\chi'(R_1, R_2), \tag{14}$$

where γ is the gyromagnetic ratio. Since both of its arguments are exactly on nuclear sites, $\chi'(R_1, R_2)$ contains mainly contributions from s-like electrons. For nearest neighbors in platinum metal this coupling amounts to $J_{12}/h = 4$ kHz.

To use these equations in the analysis of experimental results, some simplifying assumptions are usually made. For cubic transition metals, it is assumed that the wavefunctions at the Fermi level can be decomposed into s-like and d-like parts and that their exchange interactions are mostly s–s and d–d like, so that s–d interactions can be neglected. The equations become (*37*)

$$\chi = \mu_0\mu_B^2\Omega^{-1}[D_s(E_f)/(1 - \alpha_s) + D_d(E_f)/(1 - \alpha_d)] + \chi_{orb}$$
$$= \chi_s + \chi_d + \chi_{orb} \tag{15}$$

$$K = \chi_s(\Omega H_{hf,s}/\mu_B) + \chi_d(\Omega H_{hf,d}/\mu_B) + \chi_{orb}(\Omega H_{hf,orb}/\mu_B)$$
$$= K_s + K_d + K_{orb} \tag{16}$$

$$S(T_1T)^{-1} = k(\alpha_s)K_s^2 + k(\alpha_d)K_d^2R_d + (\mu_0\mu_BD_d(E_F))^2H_{hf,orb}^2R_{orb}, \tag{17}$$

where $D_s(E_f)$ and $D_d(E_f)$ are the s-like and d-like densities of state (twice the number of energy levels per unit energy interval and per atom) at the Fermi energy; α_s and α_d Stoner factors that can be written as

$$\alpha_s = I_sD_s(E_f), \quad \alpha_d = I_dD_d(E_f), \tag{18}$$

where I_s and I_d are exchange integrals; the three H_{hf}'s are the hyperfine fields; $k(\alpha)$ is a function of α that accounts for the difference in static and dynamic enhancement effects and R_d and R_{orb} are factors that take into account the relative amount of t_{2g}- and e_g- type wavefunctions.

Equations (10)–(17) are given in the MKSA system, although it is still usual practice to quote experimental values in the Gaussian cgs system (in which these equations have a slightly different form). Furthermore, it is common to attribute to magnetic fields H the units of gauss (instead of oersted).

For bulk platinum and palladium, these equations describe the experimental results very well when I_s, I_d, and $H_{hf,s}$ are considered as fittable parameters, all other quantities being either known or estimated from calculations or from other experiments (*37*). In the application to the [195]Pt NMR of small platinum particles, it has been assumed that only $D_s(E_f)$ and $D_d(E_f)$ vary from site to site in the particle, all other quantities remaining as in the bulk. This assumption neglects the lower symmetry of surface sites (the anisotropy in the shift) and variations in the orbital susceptibility. In the cases of Pt and Pd, the absolute values of the orbital contributions are not large and the neglect of their site variation is probably not important, but for metals more nearly in the center of the transition series, the approximation is poor. On the other hand, the hyperfine fields are atomic-like quantities, and the exchange integrals are integral properties and therefore reasonably site independent. This method of data analysis has been called "the LDOS description" (*38*).

The LDOS description for the metal NMR assumes that variations in Knight shift K and in spin lattice relaxation time T_1 among nuclear spin sites are mainly determined by variations in the intensity of electronic wavefunctions at the Fermi energy. (The intensity is averaged over the atomic cell containing the nucleus.) It is furthermore assumed that s- and d-like local densities of state (and their associated hyperfine fields) can be distinguished. As will be shown, the ^{13}CO NMR data for adsorbed CO can in many cases be analyzed in a very similar way, but now in terms of the 5σ- and $2\pi^*$-like hyperfine fields and LDOS. The extended electron wavefunctions are projected locally on molecular (rather than atomic) wavefunctions because after chemisorption the CO still appears to be more like a molecule than like a collection of atoms. For the same reason, the orbital hyperfine field should be defined on the basis of molecular orbitals. The values found for the LDOS on CO are small, and therefore the exchange integrals I_σ and I_π can be set to zero. It is perhaps not surprising that the ^{1}H NMR cannot be reliably and systematically described within this framework; "The hydrogen atom can happily occupy a variety of sites on a metal surface, it can diffuse, migrate, dissolve, spill over, and generally behave in a most undisciplined manner" (*23*, p. 458).

II. Experimental Considerations

A. Spectrum Acquisition

It may be that the radio frequency (RF) field B_1 available during the pulse in the NMR probe is not strong enough to excite the whole spectrum uniformly. In this case, the spectrum must be constructed point by point, shifting the RF (or the static field B_0) between points (*39*). The amplitude of the spectrum in a given point is obtained as the integral of the second half of a spin echo signal, created by a $(\pi/2) - (\pi)$ pulse sequence. In the opposite case, if the excitation is uniform, the spectrum can simply be obtained by Fourier transformation (FT) of the second half of a spin echo signal or of the free induction decay (FID) after a single pulse, at a fixed value of the RF. It is experimentally impossible to create a strong enough RF field for metal NMR spectra of transition metal catalysts, and the point-by-point method is always used. In the case of ^{13}CO NMR, much of the early work was also done by that method, but technical improvements have made it obsolete.

Shifts are reported in this chapter using the IUPAC convention for δ, which coincides with the usual definition of the Knight shift; in both cases signals that (in constant field) are at a higher frequency than the reference are assigned positive shifts, but they are plotted on the left-hand side of the spectrum. The conventional zeroes of the ^{1}H and ^{13}C scales are the signals from tetramethylsilane (TMS). Secondary standards have often been used, and in this review no attempts are made to account for occasional differences of a few ppm between different reports. The zero of the [195]Pt scale is H_2PtI_6, but again in practice secondary standards are used. Linewidths will be given as full width at half maximum (FWHM) values. The abbreviation "RT" stands for room temperature.

B. SPIN ECHO DECAY

At zero temperature, when there is no lattice motion, the nuclear spin relaxation rates are zero. However, even then the lifetime of the FID after a single RF pulse is not infinite (in other words, the lineshape in the frequency domain is not infinitely sharp). A typical source of this rigid lattice linewidth is the static, time-independent dipole–dipole coupling between the nuclei. A more trivial cause, but one that is frequently encountered when studying catalysts, is a distribution of shifts. In many cases of interest, the shift broadening is more important than that due to (dipolar or scalar) couplings between the nuclei, although we actually want to retrieve the latter because of the geometric information (distances between pairs of nuclei) that they contain. This can be done by using a spin echo sequence instead of a single $\pi/2$ pulse because the second π pulse "refocuses" the effect of shifts and of heteronuclear couplings; at the moment of the echo maximum, the signal behaves as if these interactions did not exist. Therefore, the decay of the amplitude of the echo as a function of the waiting period τ between the two pulses is mainly determined by the homonuclear couplings. The distinction between homo- and heteronuclei is based on whether they are present or not in the RF Hamiltonian (homonuclei all "feel" the RF pulse; heteronuclei do not). Within the group of homonuclei, it is often necessary to distinguish "like" and "unlike" nuclei, on the basis of whether they have the same shift Hamiltonian ("resonate at the same frequency") or not. (Sometimes the expressions like and unlike are used for homo- and heteronuclei.)

The spin echo decay under homonuclear couplings only (as opposed to decay by relaxation) is often called the "slow beat," even if no periodic modulation is observed. A real beat usually occurs when a scalar coupling between unlike nuclei is the main decay mechanism, e.g., for [195]Pt in dilute bulk alloys. Dipolar coupling usually leads to a monotonic decay, which initially is Gaussian (at the origin all its odd time derivatives are zero) and becomes more exponential at longer times.

At finite temperature, the couplings between nuclei may become time dependent due to atomic motion (diffusion) or nuclear relaxation. Such time dependencies obscure the desired information (the dipolar couplings) in the echo decay. Certain modifications of the simple two-pulse Hahn echo sequence can minimize these unwanted effects. Carr–Purcell sequences (a single $\pi/2$ pulse followed by many equidistant π pulses, where after every π pulse an echo is formed), or rather variants thereof (*40*, p. 76; *41*), are often used for this purpose. The goal of these Carr–Purcell-based multipulse sequences is the opposite of that of WAHUHA-based sequences (*40*). The former are used to recover small dipole–dipole couplings when they are masked by chemical shifts. The latter are designed to recover small shifts in the presence of large dipole–dipole couplings, which is rarely needed in studies of catalysts.

C.   Spin Echo Double Resonance

This pulse technique is used when it is desired to measure the heteronuclear couplings selectively. As an example, we consider the ^{195}Pt spin echo amplitude, which is insensitive to the heteronuclear dipolar coupling with a ^{13}CO that might be on the surface of the platinum particle. This insensitivity can be destroyed selectively (i.e., while maintaining the refocusing of the shifts) by an additional pulse on the heteronucleus (the ^{13}C in this example). Two variants of the experiments have been used. In the first, one simply determines as a function of ^{195}Pt frequency whether or not a coupling exists (keeping all timing parameters of the pulse sequence constant). In the second, the actual value of the coupling at constant frequency is measured by varying time parameters of the pulse sequence. To avoid artifacts related to ^{195}Pt relaxation, the latter experiment should be done at a constant value of 2τ (the time between first ^{195}Pt pulse and its echo signal) and with variation of the position t of the ^{13}C pulse in the τ window. Relaxation of the ^{13}C appears as an extra mechanism of destruction of the refocusing of the heteronuclear coupling in addition to that due to the ^{13}C pulse. If an experimenter finds that (over a certain 2τ period) there is no effect of the ^{13}C pulse on the ^{195}Pt echo amplitude, then either the coupling is not present (no ^{13}CO molecule is close to the ^{195}Pt being measured) or its refocusing has been destroyed by ^{13}C spin flips. There is another possible origin for such insensitivity which is of a purely instrumental character: insufficient B_1 amplitude on the ^{13}C frequency to invert all the ^{13}C magnetization so that for certain ^{13}C–^{195}Pt pairs the coupling is not sufficiently perturbed. This possibility can be checked by deliberately diminishing the B_1 value further (in practice, the test is rarely performed; at maximum available B_1 the experiment usually suffers from poor sensitivity, and weakening it may lead to unrealistic requirements in the signal averaging).

The Slichter group (see Section IV.F) uses the following terminology: the spin echo double resonance (SEDOR) fraction (Sf) for a ^{13}C pulse applied at time t, with the ^{195}Pt π pulse applied at τ ($\tau \geq t$), is defined as

$$Sf = 1 - (^{195}\text{Pt echo with } ^{13}\text{C pulse})/(^{195}\text{Pt echo without } ^{13}\text{C pulse}).$$

In some of their papers, they use "relative intensity" instead, which is $(1 - Sf)$. If there is no effect of the ^{13}C pulse on the ^{195}Pt echo amplitude, then the Sf is zero and the relative intensity is 1.

D. MAS AND TOTAL SUPPRESSION OF SIDEBANDS

In MAS, the sample is rotated rapidly by mechanical means around an axis that makes a prescribed angle with the static field. If the static spectrum (at rotation speed zero) has contributions from second-order tensor interactions such as the chemical shift anisotropy, then MAS breaks these contributions into a series of sidebands, separated from the average resonance frequency by multiples of the rotation frequency (*1*). The envelope of these sidebands traces out the original second-order tensor. At high rotation speeds, all sidebands are outside the static spectrum and therefore have zero amplitude so that only the average resonance survives. It is not always technically possible to reach such speeds; in such cases the intensity of the sidebands can be diminished by applying certain pulse sequences [total suppression of sidebands (TOSS)] (see section IV.D).

If there is a distribution of isotropic resonance positions, then the MAS spectrum and the static spectrum may look identical. To have a distribution of shifts that is effectively static, the exchange time for a nucleus between regions of the particle surface with different shifts must be large compared the inverse of the shift difference.

For molecules adsorbed on colloidal metals, the tumbling motion of the metal particles can be rapid enough to average out the chemical shift anisotropy so that the lines are narrow enough for a simple "liquids" NMR experiment. Using this method, the ^{13}CO resonance has been observed in colloidal solutions of Pd and Pt particles (see Sections IV.C and IV.D).

E. EXCHANGE NMR

Exchange NMR measures the simultaneous probability that the resonance frequency of a nucleus lies initially in the interval $d\omega_1$ around ω_1 and some fixed time (the exchange interval) later in $d\omega_2$ around ω_2 (*42*). Such frequency jumps occur through a change in molecular surroundings during the exchange interval. It is a form of 2D NMR spectroscopy; the spectrum $S(\omega_1, \omega_2)$ depends on two frequency variables. The raw data are acquired as free induction decays in a three-pulse experiment as a function of two time variables: the usual FID decay time and the time between the first two pulses, called the preparation time τ_{prep}. The shape of the exchange spectrum depends parametrically on the value of the exchange interval (the time between second and third pulses): $S(\omega_1, \omega_2; \tau_{\text{exch}})$. Just as a T_1 experiment consists of a series of spectra obtained for different values of the relaxation interval between pulses, the 2D experiment is repeated for a number of exchange times.

In most applications of exchange NMR there are a small number of possible environments, which give a small number of distinct peaks in the usual 1D spectrum (see Section III.F); however, it can also be applied to cases for which the number of sites is so large that the 1D spectrum is an unresolved superposition (see Section IV.C). Of course, in the latter case it is much harder to associate a given frequency with a specific molecular environment, and this often limits the usefulness of the method.

F.   Saturation Transfer Spectroscopy

Originally, the goal of this technique was to determine the rates of exchange of a nucleus between two environments in solution, when the lifetimes are comparable to the spin lattice relaxation times and long with respect to the inverse of the frequency difference (*43*). However, it has also been used for indirect detection of a broad resonance in one of the two environments by monitoring a narrow resonance in the other environment. The exchanging unit was the ^{13}CO molecule, which occurred either as a solute in the solvent or as an adsorbate on a colloidal metal particle. The broad line (the adsorbed state) could not be detected directly with the liquid high-resolution equipment used (see Section IV.D).

A low-level RF pulse of long duration at a trial frequency is applied, which saturates any NMR signal (even an unobservable one) that might be present at that frequency. The low level ensures that only a small band around the trial frequency can be affected by this pulse. After a suitable waiting time, short compared with the spin lattice relaxation time but comparable to the lifetime in the undetected environment, the signal in the other environment is inspected by a $\pi/2$ pulse. If indeed there has been exchange between the saturated (undetected) and the unsaturated (now inspected) signal, then the latter will be weaker with the long low-level pulse than without it. If the low-level pulse has no effect, then either the time parameters of the experiment are inadequate or the amplitude of the broad resonance is zero at this trial frequency. The latter is varied and the experiment repeated to trace out the broad resonance.

G.   Self-Diffusion

Self-diffusion in liquids (and similar systems) can be measured by the pulsed-field gradient technique. For a hypothetical sample without molecu-

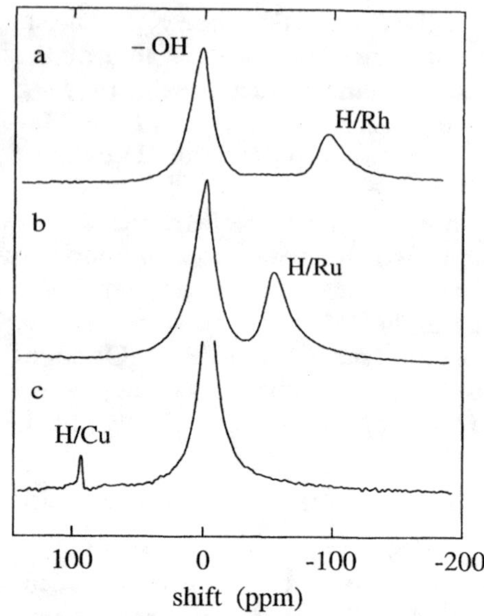

Fig. 11. ^{1}H NMR spectra of several supported metal catalysts under 30 Torr of H_2 gas. The peak around 0 ppm is due to hydroxyl groups on the surface of the carrier oxide; the other peak is hydrogen on the metal surface. The shift of the latter can have either sign (but H/Cu is the only known case of positive shift for chemisorbed ^{1}H). [Reproduced with permission from Pruski (5). Copyright 1996 Wiley.]

lar motion, placed in an inhomogeneous magnetic field, the decay of the FID signal due to the inhomogeneity can be refocused by a second pulse, forming an echo. If there is molecular motion, the instantaneous resonance frequency of a nucleus will vary in time, and the refocusing effect is incomplete. For quantitative measurements, a linear gradient is superposed on an otherwise homogeneous field during a brief period between the $\pi/2$ and π pulses and again between the π pulse and the echo. From the echo attenuation as a function of the strength of the field gradient, the diffusion coefficient can be found (an application is mentioned in Section III.B).

III. ^{1}H NMR

A.   LINE INTENSITIES: SPILLOVER

In oxide-supported catalysts, the ^{1}H NMR signal from the hydrogen on the metal is distinct from that due to hydroxyl groups on the oxide surface, although often the wings of the two signals have a strong overlap (Fig. 11)

(5). The intensities of these lines (more exactly, their integrals, calibrated in terms of number of nuclei) can be used to investigate spillover. Schematically, the experiment starts with an evacuated catalyst in the spectrometer, and the changes in intensity of ^{1}H/support and of ^{1}H/metal are recorded with increasing hydrogen dosing. Actually, these changes are not large compared with the intensity in the spectrum of the evacuated catalyst; the overlap complicates the separation of the spectrum into two signals, and the relaxation times of the support protons vary with the dosing. Consequently, it is not easy to obtain quantitative reliability.

In the case of Pt/SiO$_2$, it was initially believed that 87% of the added hydrogen ended up on the support (44); however, after improvement of the NMR procedures, no spillover was found (45). For Ru/SiO$_2$, at first a very good agreement was found between the signal from ^{1}H "strongly bound" to the metal and the volumetrically measured "irreversible" hydrogen (46), but later work showed a discrepancy attributed to ^{1}H "strongly bound" to the silica (47). The distinction between strongly bound or irreversible hydrogen and weakly bound or reversible hydrogen in ^{1}H NMR is similar to that in volumetric work; the reversible hydrogen (be it on the metal or on the support) has an NMR signal that can be "easily" made to disappear in a few minutes (evacuation at room temperature); the irreversible hydrogen requires several hours (evacuation at elevated temperatures). As has been pointed out by several authors, the classification is not strict; most irreversible hydrogen will also have disappeared after overnight evacuation at room temperature.

It is clear from recent work that for a quantitative discussion of spillover, only those spectra should be used in which all signals are in full spin lattice equilibrium. This restriction implies low repetition rates and lengthy experiments. Since small changes in the spectrum must be analyzed, the setup should enable *in situ* dosing and evacuation to allow reproducible determination of difference spectra (with respect to the state after high-temperature evacuation), as e.g., in Fig. 12 (48). The calibration of spectral intensity in terms of number of spins can be done either explicitly or implicitly. In the explicit method, a reference sample with a known amount of ^{1}H (e.g., a small quantity of water) is used. Although this is simple in principle, in practice good accuracy can be obtained only when attention is paid to spectrometer linearity (if the reference sample has the same size as the catalyst, it will usually give a considerably stronger signal) and coupling of the sample to the RF coil in the probe (if the reference has a different size, it may couple differently). In the implicit method, it is simply assumed that at low enough pressures all added hydrogen contributes only to the ^{1}H/metal signal. For further improvements in accuracy, the support protons are often exchanged for deuterons before the initial

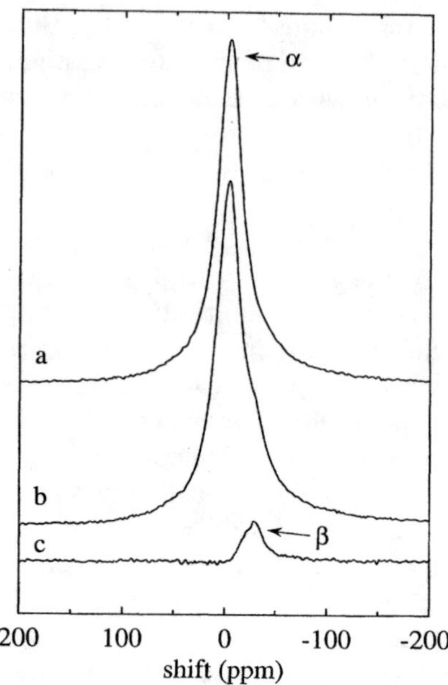

FIG. 12. ^{1}H NMR spectra of EuroPt-1 (platinum on silica). (a) Reduced and evacuated (hydroxyl peak α only). (b) In equilibrium with 0.08 Torr of H_2 gas. (c) Difference spectrum b − a; the α peak is not visible (no spillover); only the H/Pt or β peak remains. These spectra are quantitative (in complete spin lattice equilibrium) so that the quantities of hydrogen can be deduced from the intensity (i.e., the integral over the line) of the signal. [Reproduced with permission from Chesters *et al.* (*48*). Copyright 1996 Royal Society of Chemistry.]

high-temperature evacuation. If this is done, however, the ^{1}H dosing experiment must be carried out on a time scale of approximately 1 h since otherwise the slow exchange between the support and on-metal hydrogens will be significant.

The pressure needed to create high coverages is such that the ^{1}H in the gas phase in the pores of the sample starts to contribute to the NMR signal. (Only ortho-H_2 can be detected by NMR, which at RT is about three-fourths of the total amount of H_2.) If the exchange with the adsorbed hydrogen is slow, this gives a separate sharp line at typically 3.5 ppm, which does not interfere with the determination of the amount of ^{1}H/metal, as shown in Fig. 13 (*49*). However, if the exchange is rapid, only a single NMR line will occur, with an intensity I_{obs} corresponding to the sum of adsorbed and gaseous hydrogen, and a shift δ_{obs} that is a weighted average of the intrinsic shifts of the two "sites":

$$I_{obs} = I_{ads} + I_{gas} \tag{19}$$

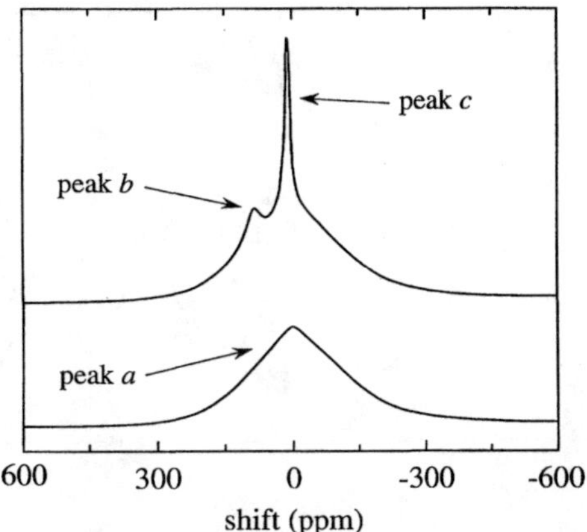

FIG. 13.   Quantitative ^{1}H NMR spectra of a Cu/MgO catalyst. The lower trace is for the reduced and evacuated sample, showing only ^{1}H on the MgO surface (peak a). The upper trace represents a sample under 475 Torr of H_2 gas and shows two peaks in addition to peak a; b is hydrogen on copper, and c is hydrogen gas. The gaseous and chemisorbed hydrogen lines are distinct because the exchange between the two is slow. In Fig. 11c the pressure is not high enough to allow observation of the gas signal. [Reproduced with permission from Chesters *et al.* (*49*). Copyright 1997 American Chemical Society.]

$$I_{obs}\delta_{obs} = I_{ads}\delta_{ads} + I_{gas}\delta_{gas}. \tag{20}$$

In the simplest case, the intrinsic shifts δ_{ads} and δ_{gas} are coverage and pressure independent, and when the pressure varies there is a direct relationship between the changes in δ_{obs} and I_{obs}. On the other hand, if this relationship is not found experimentally, then the shift δ_{ads} most likely depends on the coverage.

When *in situ* dosing onto two different samples of Pt/silica (*45, 48*) and onto Cu/MgO was used (*49*), no evidence for spillover was found from ^{1}H NMR. Only one detailed study based on fully relaxed spectra led to observation of a non zero spillover (*47*). In a Ru/SiO$_2$ catalyst, the silanol protons were exchanged for deuterons, the sample was evacuated at 623 K, and a reference ^{1}H NMR spectrum was taken at room temperature. The sample was then exposed to 20 Torr of H_2, an NMR spectrum was taken, and the difference with respect to the reference was calculated (line in Fig. 14). This represents the sum of reversible and irreversible hydrogen on the metal (resonating at -65 ppm) and spilled over on the support (at about 3 ppm). Then the sample was pumped out at room temperature for 10 min, and again a difference spectrum with the reference state was obtained (dashed line in Fig. 14); this represents irreversible hydrogen both on the support and on the metal. Similar *in situ* NMR techniques were used

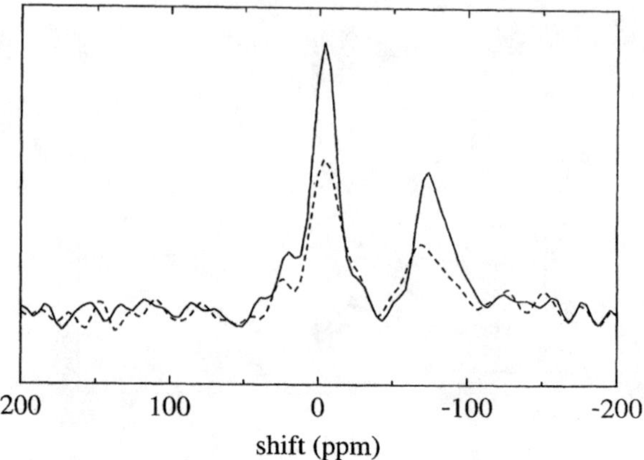

FIG. 14. ^1H NMR spectra of Ru/SiO$_2$. The initial number of silanol protons has been reduced by exchange with deuterium. Both traces are difference spectra with respect to the state after initial evacuation. The continuous line represents a sample under 20 Torr of H$_2$ gas, and the dashed line represents a sample after pumping away the "reversible" hydrogen. There is both reversible and irreversible spillover to the support (signal at 3 ppm), and there is reversible and irreversible chemisorption on the metal (signal at −65 ppm). [Reproduced with permission from Uner *et al.* (*47*).]

to monitor the time evolution of the "total" ^1H (the sum of ^1H/Ru and ^1H/silica) and of the total ^1H/Ru after the start of the exposure as well as the time evolution toward "strongly bound" ^1H and strongly bound ^1H/Ru after starting a room-temperature evacuation. The NMR intensities were compared with volumetric data on the basis of an explicit calibration with a water sample. The results (Fig. 15) show that the amount of hydrogen on the metal saturates within a few minutes after exposure, but the total uptake of hydrogen (determined according to both NMR and volumetric measurements) evolved much more slowly, not even reaching a steady state after 24 h (data not shown). There appears to be a large and rapid initial spillover, followed by a slower process involving a smaller quantity of gas. Similarly, on evacuation, the ^1H/Ru signal is stable after a few minutes, but the spilled-over hydrogen continues to decrease with time. Given the good agreement between volumetric and total NMR quantities, obtained without adjustable proportionality factors, it is clear that, at least for this system, the volumetric data cannot be interpreted by themselves; in contrast, the ^1H/Ru NMR data stabilize rapidly after a change of conditions and are unequivocal.

For a Rh/TiO$_2$ (anatase) catalyst (dispersion 0.12, loading 2.5 wt%), an important increase of the ^1H/TiO$_2$ signal with increasing hydrogen pressure was found (*50*). Under 100 Torr of gas, the intensity of the support signal

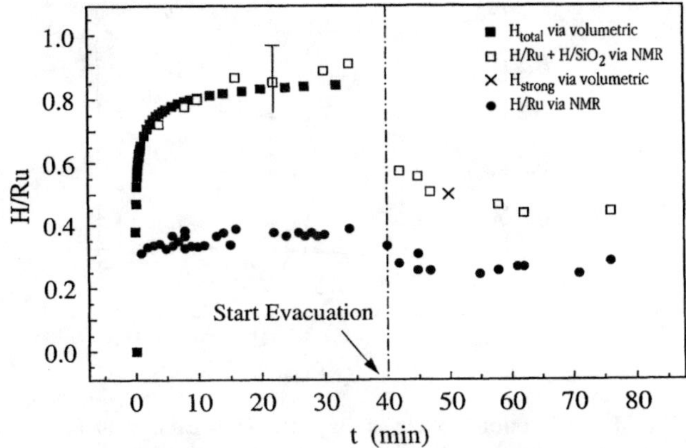

FIG. 15.   Time evolution of hydrogen chemisorption and spillover on the catalyst of Fig. 14. At time $t = 0$ the sample was exposed to 20 Torr of H$_2$ gas. The solid circles show that the quantity of ^{1}H/metal (intensity of the peak at -65 ppm in Fig. 14) equilibrates rapidly, whereas the total quantity of hydrogen (the sum of both intensities in Fig. 14; open squares) continues to evolve in time. This pattern is confirmed by a separate volumetric experiment (solid squares). At time $t = 40$ min, the pumping away of the reversible hydrogen starts. The ^{1}H/metal signal again equilibrates rapidly (solid circles), whereas the total quantity of ^{1}H diminishes more slowly (open squares). A single volumetric experiment (cross) confirms the value found by the open squares. [Reproduced with permission from Uner et al. (47).]

was found to be twice that obtained after evacuation and the quantity of spilled-over hydrogen larger than that on the metal surface (Fig. 16). This surprising result has not been analyzed in detail. For a catalyst of similar dispersion and loading on strontium titanate (51), the spillover was found to be less important but still measurable (Fig. 17).

Two other kinds of NMR experiments are also related to the spillover question, but neither has found an easy interpretation. In the first, the sample is exposed to a few Torr of deuterium gas after the initial high-temperature evacuation. The metal surface is rapidly covered with ^{2}H, which is not detected. The only ^{1}H observed is in the surface hydroxyl groups. Over a period of time, however, the signal typical of ^{1}H/metal appears, showing that some protons that were initially on the support moved to the metal. It has been reported that the exchange occurs in tens of minutes on platinum (52), (Fig. 18) but requires many hours on ruthenium (53). No explicit attempts have been made to duplicate the platinum result, but exchange of silanol hydrogens with deuterons has been used in two studies of the ^{1}H/Pt NMR (45, 54) to diminish the background signal. Apparently, the ^{1}H/Pt signals were not affected by exchange with the silanol deuterons so that the exchange process was perhaps slower than initially

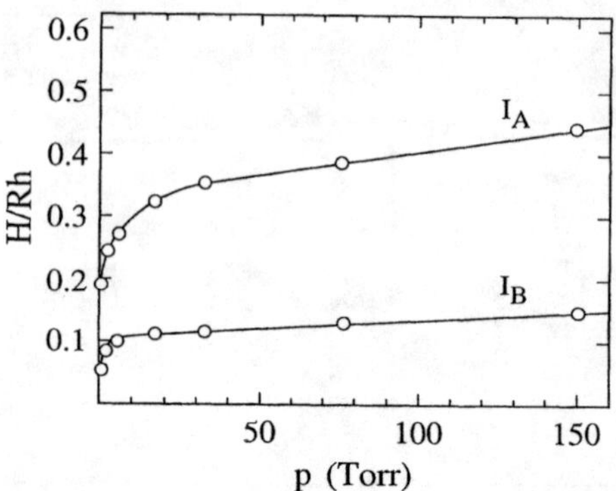

FIG. 16. ^{1}H NMR adsorption isotherms for a Rh/TiO$_2$ catalyst of fairly low dispersion. I_A is the intensity of the line (at shift around 0 ppm) corresponding to hydrogen on the support; I_B is the intensity of the line (shift -130 ppm) for ^{1}H/metal. Both intensities are calibrated in terms of the H/Rh ratio. It is remarkable that I_A increases more rapidly with hydrogen pressure than I_B; more hydrogen spills over than stays on the metal. [Reproduced with permission from Sanz and Rojo (50). Copyright 1985 American Chemical Society.]

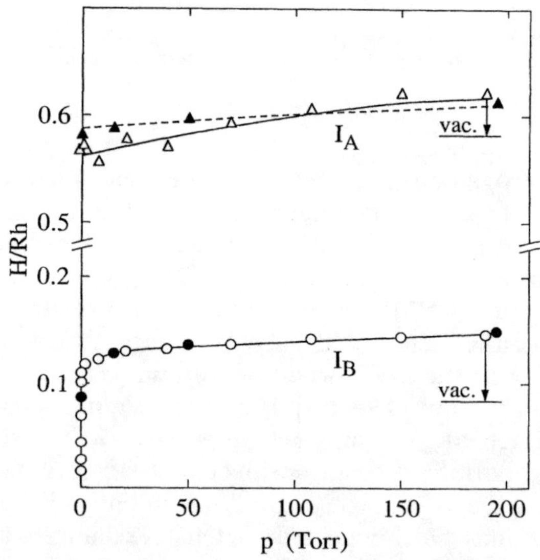

FIG. 17.   A plot similar to that of Fig. 16 but for Rh/SrTiO$_3$. The open symbols are the adsorption after the initial evacuation. After the highest pressure point was taken, the reversible hydrogen was pumped away, and the intensities I_A and I_B decreased to the level indicated by "vac." Next, the adsorption experiment was repeated (solid symbols). The increase of I_A with pressure (the spillover) is less marked than indicated in Fig. 16. [Reproduced with permission from Rojo et al. (51). Copyright 1994 American Chemical Society.]

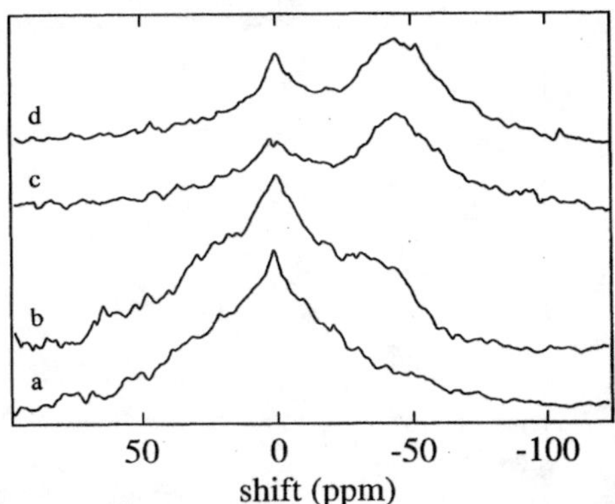

FIG. 18. ^{1}H NMR spectra of a Pt/SiO$_2$ catalyst in a reverse-spillover experiment. Initially (trace a), the sample was evacuated and only the silanol ^{1}H NMR signal deserved at about 0 ppm. Next, deuterium gas was admitted to the sample. Traces b, c, and d were taken after 15 min, 15 h, and 200 h, respectively. The presence of the deuterium gas leads to the appearance of ^{1}H on the metal (signal at approximately -50 ppm). It is also known from other experiments that, if the reversible deuterium gas is pumped off, the process becomes unmeasurably slow. [Reproduced with permission from Sheng and Gay (52).]

reported (52). The experiment with Ru has been confirmed (46), and it was also shown that no detectable proton–deuteron exchange on the metal surface occurs if the initial coverage is only by strongly bound deuterium under very low pressure. This result was explained by unfavorable thermodynamics for spillover of strongly bound hydrogen, the supposed mechanism being surface diffusion and atomic jumps from the metal to the silica and back. Another possible mechanism is desorption from the silica and adsorption onto the metal. At very low gas pressures (46), this process would be slow. The appearance of protons on the metal by this mechanism would not be due to diffusion, and its relation with spillover would be indirect.

The other type of experiment, which also was reported in early work (52) and confirmed later (45), is the study of silanol ^{1}H spin lattice relaxation time as a function of dosing (Fig. 19). Even in pure silicas, the observed relaxation is not a simple exponential and depends on the reduction temperature (55): Schematically, geminal OH pairs have a short T_1, independent of the quantity of such pairs present, whereas single silanols have a longer T_1, which depends on their concentration because of changing proton–proton distances. In a catalyst, spillover from the metal onto the silica could decrease the T_1 of the single silanols (by increasing their concentration) or decrease the apparent T_1 by increasing the fraction of rapidly relaxing geminal silanols. However, a change of silanol T_1 with coverage is also

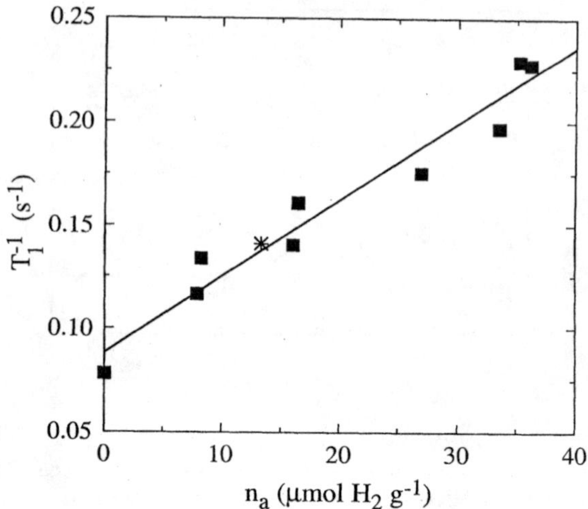

FIG. 19. Spin lattice relaxation rate T_1^{-1} of ^{1}H in the support of a Pt/SiO$_2$, catalyst as a function of hydrogen uptake n_a. The mechanism by which the coverage influences the T_1 is not well understood. (The value marked by an asterisk was obtained by reducing the pressure from an initial value of 500 Torr, corresponding to an uptake off scale to the right.) [Reproduced with permission from Chesters *et al.* (45). Copyright 1995 Royal Society of Chemistry.]

observed in systems for which the spectrum indicates no measurable spillover, e.g., in Pt/SiO$_2$. In fact, the ^{1}H/silica spin lattice relaxation for a given Pt/SiO$_2$ catalyst in an *in situ* dosing system depends on the hydrogen pressure but not on whether that pressure was reached from below (by adding H$_2$) or from above (by desorption by evacuation) (56). No entirely satisfactory description of this relaxation mechanism is available.

It is frequently suggested that no (hydrogen) mass spills over from the metal but only (proton) nuclear magnetization, by a process known as "spin diffusion." For a simple explanation of spin diffusion, imagine a line of dipolar-coupled spins in thermal equilibrium. In a zero applied field, every spin spends half its time in the "up" and half its time in the "down" state. In a finite field, the fraction of time spent in the up state increases by an amount determined by the field and by the temperature, but it is the same for all spins. Now suppose that we heat only the spin at one extreme end of the line; it will spend a smaller fraction of its time in the up state than does its neighbor. The neighbor will sense this through the dipole–dipole coupling, and its temperature will rise. This in turn will be sensed by the next neighbor and so on. A similar mechanism can perturb measurements of spin lattice relaxation time; imagine that we have inverted the magnetization of the whole line of spins at time zero and want to investigate its recovery to equilibrium. If for some reason the spin at one end has a faster intrinsic spin lattice relaxation than the others, there will be transport of nonequilibrium magnetization to the rapidly relaxing site so that the rest of the chain acquires a (pseudo-) spin lattice relaxation that is faster than its intrinsic spin lattice relaxation process. Now, the ^{1}H atoms on the metal have fast relaxation channels (to the conduction electron spins) that are not available to the protons on the oxide surface. The idea has not been put to a serious experimental test, however; in fact, some authors believe that it cannot quantitatively explain the observed relaxation rates of the hydroxyl protons.

B.   Line Shifts: Bulk Palladium Hydride

As shown in Fig. 11, ^{1}H chemisorbed on metals can have positive or negative shifts, with fairly large absolute values. This observation is not specific for metallic systems, however; many molecular hydrides (compounds with a hydrogen directly bonded to an atom of a metallic element) have shifts in solution between -7 and -20 ppm, and values as low as -50 and as high as 20 ppm occur in some of these molecules. The negative values result from the paramagnetic current density induced in the d-like orbitals of the metal fragment by the applied field; this current creates a shielding field at the site of the hydrogen (57, 58). When going from a mononuclear metal hydride such as $[HCr(CO)_5]^-$ to the binuclear $[HCr_2(CO)_{10}]^-$, the paramagnetic part of the absolute shielding approximately doubles, whereas the diamagnetic part stays constant (58). The result is an upfield shift from -6.9 ppm for the mononuclear species to -19.5 ppm for the binuclear species. Simply extrapolating to six neighboring metal fragments, as in an octahedral interstitial of a face-centered cubic (FCC) metal, gives $\delta = -70$ ppm for such a hypothetical hexanuclear species. This illustrates how difficult it is to make a reasonable a priori estimate for the zero of the Knight shift scale of a metallic hydride (or of a chemisorbed hydrogen). As a specific example, I discuss the ^{1}H NMR of bulk palladium hydrides PdH_x; $0 < x < 1$.

The FCC lattice of Pd accommodates hydrogen in the octahedral interstitial sites. The main part of the phase diagram (Fig. 20) looks like that of a fluid with a critical point at a temperature of approximately 565 K, a pressure of 20 bar, and a density $x = H/Pd$ of 0.25. [At temperatures below 50 K, ordered hydrides exist, which are not considered here (59).] Above the critical temperature, single-phase samples can be obtained for all x by loading Pd under suitable pressures of hydrogen. At RT, the "gas" (low x) α phase is stable for $0 \leq x \leq \alpha_{max} = 0.015$, and the "liquid" (high x) β phase is stable for $0.61 = \beta_{min} \leq x \leq 1$. The β phase is stable to low temperatures for $x > 0.65$. The only structural difference between the α

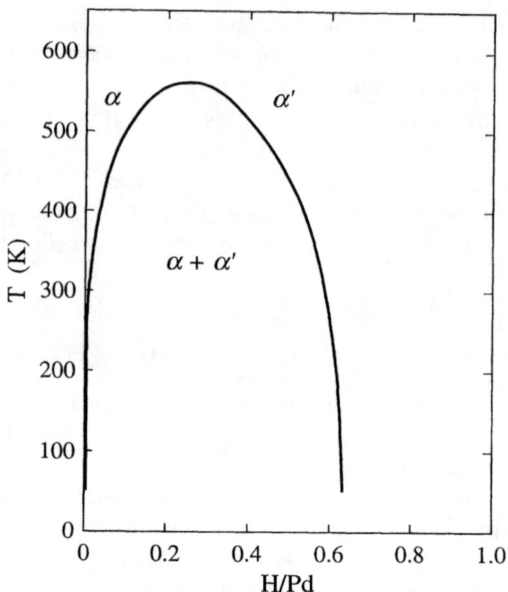

FIG. 20. Part of the phase diagram of bulk palladium hydride. At low temperatures, additional ordered phases exist which are not shown. The α- and α' phase diagram is analogous to that of a fluid; the α phase is low density (small H/Pd) "gas," and the α' phase is high-density "liquid." In the NMR literature the α' is usually, but not quite correctly, called β.

and β phases is the lattice parameter, which increases proportionally to the H/Pd ratio. (Here we use the word hydride to also designate disordered solid solutions of hydrogen in metals. In a more precise classification, one distinguishes between hydrides, which are regular arrangements of hydrogen and metal atoms, and random solid solutions. Different kinds of solid solution should be denoted α, α', etc., whereas β, γ, etc. are reserved for ordered hydride structures. In the NMR literature it is common to find "β phase" instead of the more correct "α' phase.")

For $x < 0.4$, the magnetic susceptibility is temperature dependent like that of pure palladium, but it has lower values. For $x > 0.6$, this temperature dependence disappears and the system is diamagnetic (60), like that of silver, a neighbor of Pd in the periodic table. This change is related to a change in position of the Fermi level with respect to the d-band. For low x the Fermi level crosses the d-band, as in pure Pd (cf. Fig. 9); at $x = 1$, E_f is above the d-band (Fig. 21a) (61). When the H/Pd ratio increases, the total DOS at E_f (and therefore the susceptibility) decreases, but it is calculated that the LDOS at E_f on the hydrogen site should increase (62) (Fig. 21b).

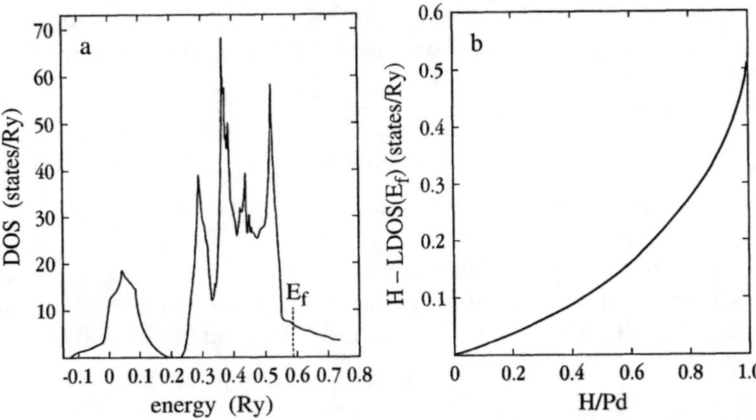

FIG. 21. (a)   The density of states in stoichiometric bulk PdH. The zero of the energy scale is arbitrary. 1 Ry = 13.6 eV. Compare the shape to those given for Pd and for Ag in Fig. 9, and note the low-lying band around 0.05 Ry. This is a strong bonding band between d-states on the palladium and s-states on the hydrogen. [Reproduced with permission from Gupta and Freeman (61). Copyright 1978 American Physical Society.] (b) The local density of states at the Fermi energy on the hydrogen site as a function of hydride composition. The calculation was done disregarding the α/α' phase separation. The LDOS was normalized per occupied hydrogen site. Note that when the bulk E_f DOS is high (at $x = 0$; Fig. 9), the corresponding LDOS on hydrogen is low; and when the bulk value is low (a) the local value is high. [Reproduced with permission from Papaconstantopoulos et al. (62). Copyright 1978 American Physical Society.]

At RT and higher temperatures, the motion of the hydrogen through the palladium lattice is rapid enough to average the proton–proton dipolar couplings and to give fairly narrow ^{1}H NMR lines. The linewidth in the β phase is 1 or 2 ppm, and the α signal is 10 times broader. The self-diffusion coefficient of ^{1}H for H/Pd = 0.70 has been measured at temperatures between 295 and 415 K by NMR, by use of the pulsed field gradient method (63). The room-temperature value is about 1.5×10^{-7} cm^2 s^{-1}; the values found in the α phase (by other methods) are approximately twice this amount.

There is only a single pressure/NMR shift isotherm for the α phase in the literature; it was taken in a Pd black at 350 K (64). The measured shifts range from $\delta(p) = -20$ ppm for $p = 80$ Torr to -60 ppm at 2 Torr; extrapolation to zero pressure yields $\delta(0) = -65 \pm 5$ ppm. Although in general a NMR shift may be a function of pressure, temperature, and composition, it is usually expected that composition is the determining variable (although in the particular case of palladium and its α hydrides it must be kept in mind that the bulk susceptibility varies with temperature).

The pressure/NMR shift isotherm can be translated into a composition/ shift isotherm with the aid of the pressure/composition isotherm. [Here we use data for a Pd foil (65), which is not what the original authors did.] The 350-K isotherm is a simple straight line:

$$\delta(x) = \delta(0) + x\delta_1, \tag{21}$$

where $\delta_1 = 2860$ ppm for $0 \leq x \leq 0.03$. There can be little doubt that in this case the shift variation reflects a continuous change in the electronic structure with composition, but in general this is not the only possibility. Another common interpretation of composition-dependent shifts leads to rewriting Eq. (21) in the same form as Eq. (20) (remember that $0 < x < 1$):

$$\delta(x) = (1 - x)\delta(0) + x[\delta(0) + \delta_1], \tag{22}$$

where $\delta(x)$ is interpreted as an average, resulting from rapid exchange of the nuclei between two sites, each with a composition-independent shift— one with $\delta(0)$ and the other with $[\delta(0) + \delta_1]$. The factors $(1 - x)$ and x are the relative occupations of these two sites. Of course, both effects expressed by Eqs. (21) and (22) may exist simultaneously.

In the β-hydrides with low susceptibility, explicit temperature effects are probably unimportant. Indeed, all four pressure/NMR shift isotherms taken on the Pd black between 273 and 350 K give the same composition/NMR shift relationship (64): Equation (21) with $\delta(0) = -34$ ppm and $\delta_1 = 85$ ppm ($0.60 \leq x \leq 0.75$). The NMR isotherms have also been studied in a 0.2-μm powder at temperatures between 273 and 383 K (66). Only the two lowest temperature isotherms yield a common shift/composition relationship, with $\delta(0) = -76$ ppm and $\delta_1 = 152$ ppm. The results for the Pd black and the powder agree for compositions at the low end of the experimental range (approximately H/Pd = 0.61), but they differ at the high end (approximately H/Pd = 0.73).

The temperature independence of the Korringa product T_1T is a better indicator of the metallic character of an NMR signal than the value of its shift. In most metal hydrides, the ^{1}H spin lattice relaxation at room temperature contains important contributions from the diffusive motion of the proton in the hydride lattice. To investigate the Korringa product one must work at relatively low temperatures; for the palladium hydrides only values of $x > 0.65$ are accessible (Fig. 20). For $x = 1$, $T_1T = 46$ sK (67), but no values have been given for the shift. Supposing that the total shift $\delta(1)$ can be derived from Eq. (21) with the parameters given previously for the palladium black, we obtain $\delta(1) = 51$ ppm. From the T_1T value and the s-like part of Eq. (17), with $\alpha_s = 0$, we find $K_s(1) = 74$ ppm. [The calculated value (61) is $K_s(1) = 81$ ppm, in very good agreement.] From these values for $\delta(1)$ and $K_s(1)$, we estimate the zero of the Knight shift scale as $51 - 74 = -23$ ppm. This negative chemical shift should have the same origin as that in the molecular hydrides mentioned previously. Relaxation data are also available (*68, 69*) for x in the 0.7–0.8 range (Fig. 22). Using the same reasoning, we find $\delta(0.75) = 30$ ppm, $K_s(0.75) = 61$ ppm, and the zero of the Knight shift scale at -31 ppm. The drop of 20% in K_s when x decreases from 1 to approximately 0.75 can be understood from a decreasing local density of states at E_f on the proton site (Fig. 21b). (The two are not simply linearly proportional, however, probably because there is no well-defined, composition-independent hyperfine field.)

For the low-x compositions of PdH_x, no Korringa-product data are available. Therefore, we may simply assume that the observed negative shifts are the same kind of (x-dependent) chemical shifts as just proposed for the high-x case since calculations (*62*) predict that the E_f-LDOS on the hydrogen site decreases to zero when x goes to zero (Fig. 21b). Another possibility would be to invoke a "core polarization" mechanism. There is a low-lying band (below the d-band) in the palladium hydrides (Fig. 21a) which contains electrons on both the hydrogen and the palladium sites. Those on the hydrogens could play the role of "core" electrons and be polarized by (hypothetical) electrons at the Fermi energy that would have d-like character on the hydrogen site. However, from calculations (*61, 62*) it seems unlikely that such electrons exist. Another possible origin for a negative shift can be found from the calculated spin density in pure platinum

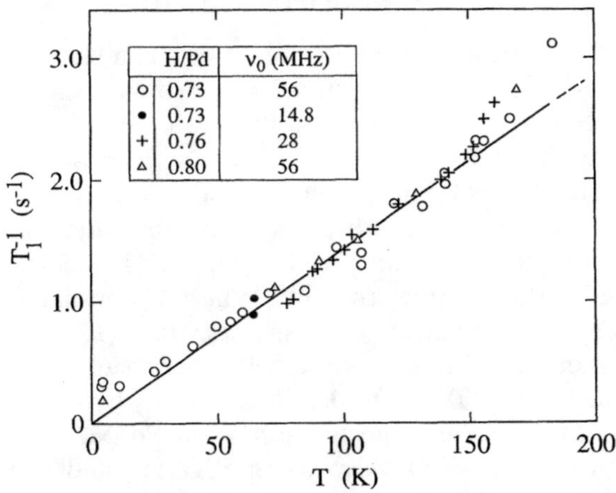

FIG. 22. Spin lattice relaxation rate T_1^{-1} of H in bulk PdH_x with x in the 0.7 to 0.8 range as a function of temperature and for several Larmor frequencies ν_0. The straight line indicates a temperature-independent Korringa product T_1T, characteristic of metallic behavior; there is a nonzero LDOS on the ^{1}H at the Fermi energy, in qualitative agreement with Fig. 21b. [Reproduced with permission from Schoep *et al.* (*68*). Copyright 1974 Elsevier Science.]

under an applied magnetic field (*70*). Its value is negative in the octahedral interstitials, and perhaps similar tails of negative spin density exist at the site of ^{1}H in PdH$_x$ with x being small (compositions close to pure palladium). The effect is a local expression of the exchange enhancement of the suscepti- bility (which is the reason for the high bulk susceptibility of palladium) and does not appear in the band-type calculations (*61, 62*) performed for the hydrides. Contrary to the negative hyperfine fields created in the "ionic cores" of transition metals, which are largely independent of the environ- ment in which we put the "ion" (*70*), this negative spin density at the hydrogen site is completely created by the environment and therefore not a property of the hydrogen ion.

For completeness, I mention the ^{105}Pd Knight shift measured in microme- ter-sized powders with $0 < x < 0.025$ (*64*). The shift changes linearly with the bulk susceptibility as $\Delta K/\Delta\chi = -41$ mol emu^{-1}. In pure Pd, this value is -61 mol emu^{-1}, indicating that in the hydride both d-like and s-like susceptibilities change with x. Fitting the observed slope using parameters derived for bulk Pd gives a decrease in χ_s of 2.8 μemu mol^{-1} and in χ_d of 67.8 μemu mol^{-1} when x increases from 0 to 0.025. Because of the very strong enhancement effects in the magnetic susceptibility of palladium, this change in susceptibility corresponds to a change in the E_f DOS of less than 2% of the bulk value.

C.   Hydrogen on Palladium

The pressure/composition relationship of small palladium particles under hydrogen has been investigated extensively. From *in situ* X-ray diffraction (*71*) it was concluded that in nanocrystalline palladium (a compact obtained by compressing a powder of particles with diameters of 10–20 nm or less) at room temperature (RT), α_{max} is larger and β_{min} is smaller than those in coarse-grained Pd. In a volumetric and neutron scattering study of such material it was argued that the excess hydrogen is incorporated in grain boundaries and surfaces rather than in a phase of unusually high α_{max} (*72*). An early extended X-ray adsorption fine structure (EXAFS) study of Pd/ SiO$_2$ catalysts concluded that catalysts with dispersions higher than 0.5 do not form hydrides at RT and under 760 Torr of hydrogen (*73*); in later work carried out with the same technique on Pd/NaY zeolite, however, the opposite was found (*74*). Also using EXAFS, and working with Pd/ Al$_2$O$_3$ containing 2- or 3-nm particles, it was concluded (*75*) that in this system at RT, $\beta_{min} = 0.44$ (instead of 0.61 for bulk β-hydride). It is noted that from the experimental value for the self-diffusion coefficient in the bulk hydride at RT (*63*), one expects a hydrogen to diffuse over a distance of 3 nm in less than 1 μs. Theoretical work on single-crystal faces suggests that at zero temperature the Pd surface is loaded first with hydrogen, and that subsurface hydride is formed only afterward (*76*). A Monte Carlo simulation of the interaction of a 500-atom palladium cluster with hydrogen at 300 K gave the same result and in addition showed no miscibility gap (*77*).

All ^{1}H/Pd (and ^{2}H/Pd) NMR experiments reported to date have been performed on samples in sealed NMR tubes. The results of four different experiments [with Pd/SiO$_2$ (*53*), Pd/Al$_2$O$_3$ (*66*), Pd/NaY zeolite (*78*), and ^{2}H/Pd/Al$_2$O$_3$ (*79*)] agree that under hydrogen pressures in the 100- to 500-Torr range, the shifts of the ^{1}H and ^{2}H NMR lines are 28 ± 2 ppm. This is the value one expects for a bulk β-hydride with concentration $x \approx 0.75$ [if the palladium black data (*64*) are used] or $x \approx 0.68$ [with the palladium powder data (*66*)]. The dispersions given for these samples range up to 0.6, so the NMR data do not confirm the conclusion in Ref. (*73*) that for dispersions >0.5 no hydrides are formed. In a direct comparison, it has been shown that at 295 K the pressure/NMR isotherms in a catalyst with low dispersion (0.13) and in a bulk powder are reasonably similar (*66*) (Fig. 23). Two different interpretations can be given for the two lowest pressure points for the catalyst. If it is assumed that these NMR signals are due to

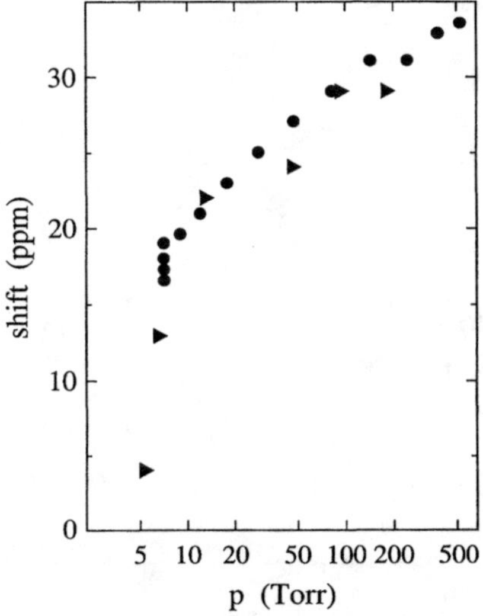

FIG. 23. ^{1}H NMR shift for a coarse-grained (200-nm) samples (circles) and for a low-dispersion Pd/alumina catalyst (estimated metal particle diameter 9 nm) (triangles) under various pressures of H$_2$ gas. At pressures exceeding 10 Torr, the hydriding behavior of the two samples seems to be the same; the catalyst particles form the β-hydride. The two smallest shift points might indicate that for the catalyst the lowest possible value of H/Pd in the β phase is smaller than that for the bulk. [Reproduced with permission from Barabino and Dybowski (*66*). Copyright 1992 Elsevier Science.]

hydrogens in a β phase, then the shift can be interpreted as evidence of a lowering of $\beta_{\min}$, in line with the results of Ref. (75). The other interpretation is that the signal is due to hydrogen on the surfaces of the particles, as discussed later for Pd/NaY zeolite catalysts.

The Pd/SiO$_2$ (53) and Pd/NaY zeolite (78) catalysts have also been investigated at low pressures and RT. Both reports describe a signal with a larger linewidth than that of the β-hydride, which shifts with hydrogen concentration and metal dispersion. The shift ranges of this signal (-80 ppm at H/Pd = 0.12 to -5 ppm at H/Pd = 0.51 for the Pd/SiO$_2$ sample) in the two reports agree well. This signal could be observed simultaneously with the β-hydride signal for some concentrations of hydrogen in the Pd/ NaY zeolite. This, however, was not the case for the Pd/SiO$_2$ sample. This high-field signal is ascribed to hydrogen on the particle surface, in rapid exchange with the gas phase and possibly with an α-hydride (the relative amount of the latter should be small for highly dispersed samples at room temperature). If the total hydrogen uptake in the different Pd/NaY zeolite samples is normalized by the dispersion, all shift data at 300 K fall approximately on one shift/coverage isotherm (78) (Fig. 24); the shift increases

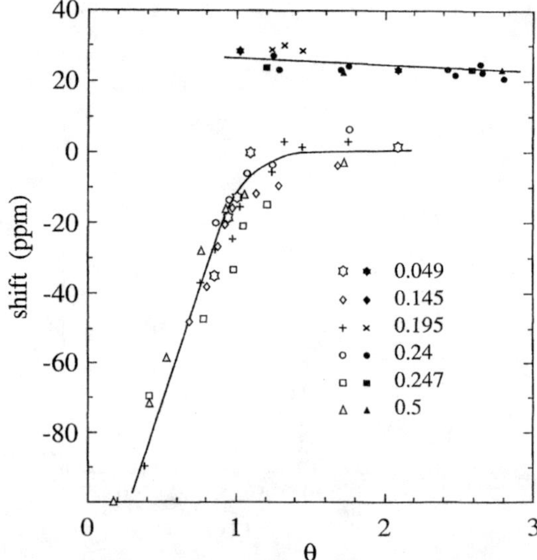

FIG. 24. ^{1}H NMR shift as a function of coverage θ for six Pd/NaY zeolite catalysts of different dispersions. Two distinct lines are observed, one corresponding to the β-hydride (solid symbols) and the other probably to hydrogen on the surface of the metal particles (open symbols). At low concentrations, the hydrogen appears to stay at the metal surface; once a monolayer coverage has been attained, the hydride starts to form. [Reproduced with permission from Polisset and Fraissard (78). Copyright 1993 Elsevier Science.]

from -110 ± 20 ppm at coverage $\theta = 0$ to -10 ± 10 ppm at $\theta = 1$. When H/Pd increases further, the shift remains in this range, but at the same time the intensity diminishes with respect to the β-hydride signal. The conclusion is that for coverages <1, there is only hydrogen on the particle surface [as predicted by calculations (76)], with a coverage-dependent, dispersion-independent shift. Increasing the overall H/Pd ratio further converts a growing number of particles completely to the β-hydride with $x \approx$ 0.67, and the surface sites of this hydride cannot be distinguished from bulk sites, either because of an intrinsic similarity or because of rapid exchange between them. The ^{2}H/Pd NMR results support these observations (79) (Fig. 25,22). If these data points are normalized by the dispersion, then Fig. 25b and Fig. 24 become very similar. The plateau region shown in Fig. 25b could be explained as an unresolved superposition of the two signals in Fig. 24; alternatively, it could be explained as a β-hydride signal for low β_{min} (cf. the discussion of Fig. 23).

The interpretation of Fig. 24 may turn out to be more complicated than stated previously, because of an experimental indication of an explicit effect of temperature on the shift of the "surface" signal. For one Pd/NaY zeolite sample with a low dispersion (78), the adsorption and the NMR measurement were done at a higher temperature (358 K instead of 300 K for the others), and it was concluded that at constant coverage the increase in

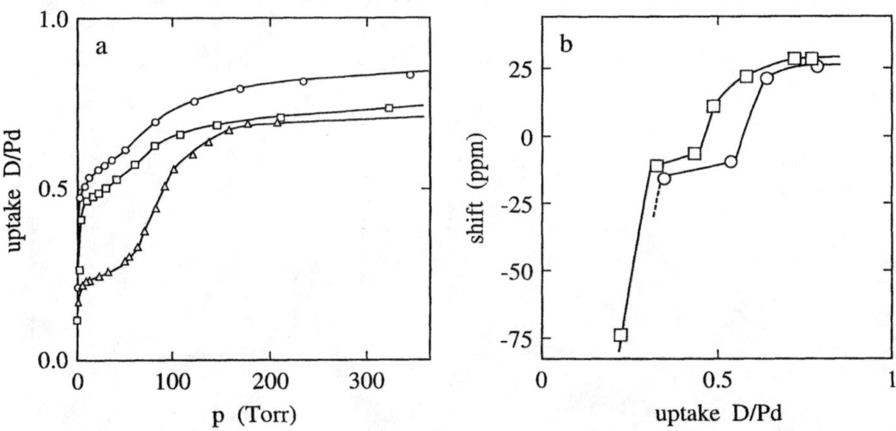

FIG. 25.   (a) ^{2}H uptake as a function of pressure and (b) ^{2}H NMR shift as a function of uptake for Pd/Al$_2$O$_3$ catalysts. From the knee in the curves in a, the dispersions are estimated as 0.48 (squares) and 0.57 (circles). (The third sample in a has a dispersion of 0.22.) Note the similarity of Fig. 25b and Fig. 24, especially when the uptake D/Pd is converted into coverage θ. ($\theta = 1$ at D/Pd = 0.48 for the squares and at 0.57 for the circles.) [Reproduced with permission from Chang *et al.* (79). Copyright 1994 Royal Society of Chemistry.]

temperature shifts the NMR line to higher field by more than 30 ppm or, at constant NMR shift, it increases the required coverage by 0.25. This is a huge effect, and it is difficult to explain quantitatively and certainly merits further experimental study. These questions, as well as the possible interpretation of the low-pressure points in Fig. 23 and the plateau in Fig. 25, in terms of a lowering of β_{min}, should be investigated by pressure/composition and pressure/shift isotherms obtained *in situ*.

D.   HYDROGEN ON PLATINUM

There is a fair amount of experimental information in the literature for ^{1}H NMR of hydrogen on platinum catalysts, and in general it is in good agreement. I consider mainly data for the standard Pt/SiO$_2$ catalyst EuroPt-1 reported by two groups, but results obtained with alumina supports, other silica-based catalysts, or Pt/NaY zeolite are not qualitatively different.

Before examining the NMR, it is useful to consider some adsorption and desorption studies of this standard catalyst. Hydrogen chemisorption data have been modeled by a Langmuir isotherm for pressures between 1 and 80 Torr (Fig. 26a) (*80*) but also by a Temkin isotherm at pressures between 0.1 and 100 Torr (Fig. 26b) (*48*). The Langmuir isotherms in Fig. 26a give H/Pt ratios for monolayer adsorption of 1.17 (circles) and 1.06 (triangles). The break point in the Temkin isotherm of Fig. 26b corresponds to H/Pt = 0.82 (*48*). (The part of the isotherm below 0.1 Torr has not often

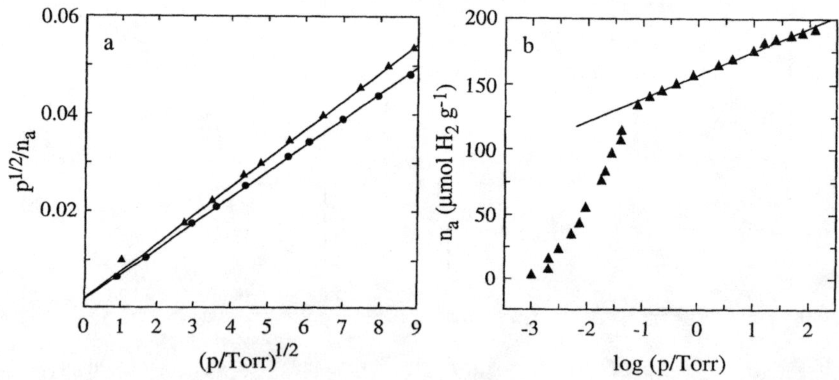

FIG. 26.   Volumetric hydrogen chemisorption isotherms for EuroPt-1. The three sets of data are independent. (a) The two lines are for slightly different pretreatments; the circles provide the more representative data. The lines represent Langmuir isotherms; (b) the line represents a Temkin isotherm. In the common pressure range, both fits are equally plausible. [(a) Reproduced with permission from Bond and Lou (*80*); (b) reproduced with permission from Chesters *et al.* (*48*). Copyright 1996 Royal Society of Chemistry.]

been measured with any precision.) In the Langmuir model, a finite number of equivalent adsorption sites is supposed to exist, and full coverage is reached asymptotically with increasing pressure. In the Temkin model, the number of available sites is unlimited, but they are not identical; the enthalpy of adsorption is supposed to become progressively less negative with increasing coverage.

Volumetric thermal desorption experiments have been performed at temperatures between 77 and 900 K (81). The existence of two species is indicated by a break in the desorption curve at around 250 K; at this point, 18% of the hydrogen (fixed at 300 K under 100 Torr) has desorbed. From an analysis of their complete data, the authors proposed that an incoming hydrogen molecule is first dissociated on a pair of on-top sites. To settle finally in the C_{3v} or C_{4v} sites on the platinum surface, the atoms must cross a barrier of 5 kcal mol^{-1} (217 meV). Adopting a very similar hypothesis, the Nottingham group assumes that at pressures below 0.1 Torr (Fig. 26b) sites are occupied where the hydrogen is strongly held, whereas at higher pressures on-top sites are filled, where the hydrogen is more weakly bound (56). From ^{195}Pt NMR spectroscopy (Section VI.B), the dispersion of the EuroPt-1 catalyst was found to be 0.60 ± 0.05. Assuming a 1:1 H/surface Pt stoichiometry, and with a Pt loading of 6.3% by weight, the monolayer coverage should occur at 97 ± 8 μmol of H_2 molecules per gram of catalyst. This agrees well with another estimate (82).

The NMR probe used by the Nottingham group (49) has an unusually large sample volume (7 cm^3) so that *in situ* volumetric and NMR spectroscopic measurements can be made simultaneously. Spectra were taken as a function of coverage, and the differences from the spectrum after high-temperature evacuation were calculated. The NMR intensities were calibrated by the "implicit method," described in Section III.A. The difference spectra show a peak at a constant shift of −31 ppm (with respect to the silanol peak) at all coverages below 120 μmol g^{-1}; when the coverage increases, the line shifts to lower field (Fig. 27). This result is in satisfactory agreement with earlier results by the Paris group with samples in sealed tubes (83), which show a line at −29 ppm (with respect to silanol) for coverages up to 85 μmol g^{-1}, and then shifting to lower field. Both groups found evidence that this line actually consists of two signals (48, 54) that shift together, but their intensities could not be evaluated separately.

For these very similar data, widely different interpretations have been given. On the basis of additional experiments (83), the Paris group concluded that the coverage independence of the shift is a kind of "deep-bed" effect; when hydrogen is dosed into the NMR tubes, only the platinum particles in the top layer of the sample become covered with hydrogen—those in the bottom layer remain bare. Therefore, the effective coverage

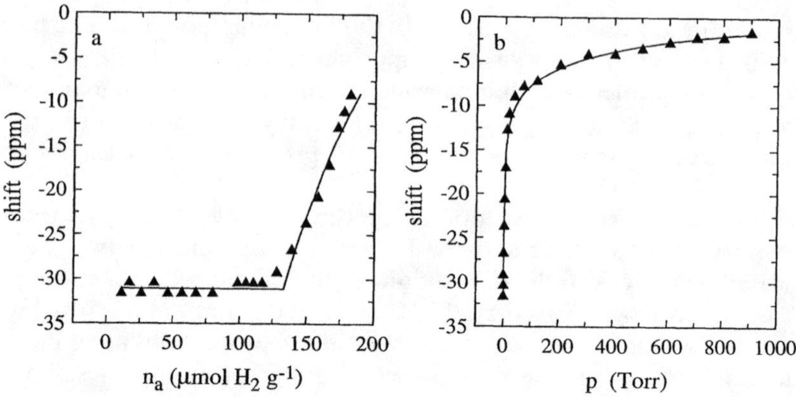

FIG. 27.   NMR shift of the ^{1}H/metal signal on EuroPt-1 (peak β in Fig. 12) for different coverages (a) or under different pressures (b). Compare the knee in Fig. 27a with that in Fig. 26b. The lines are a fit to a three-site rapid-exchange model. At low coverage, only one site on the metal is occupied (horizontal part in Fig. 27a). At intermediate values there are two metal sites with different shifts (curved part in Fig. 27a), and at high pressure there is a third contribution from hydrogen in the gas phase (right-hand part of Fig. 27b). [Reproduced with permission from Chesters *et al.* (*48*). Copyright 1996 Royal Society of Chemistry.]

on those particles that contribute to the ^{1}H NMR signal remains constant while the average coverage increases, and the observed shift is characteristic of the effective rather than the average coverage. Similar effects have been described by another group in some of their ^{13}CO NMR experiments, referred to in Section IV.C. The Nottingham group reported having sought evidence of this phenomenon in their setup but did not find any (*56*).

It seems therefore that even when deep-bed artifacts can be avoided, the NMR shift of ^{1}H on EuroPt-1 at RT is coverage independent for coverages up to approximately a monolayer (97 μmol g^{-1}). [At temperatures exceeding 373 K, the shift varies continuously with coverage (*48*).] The narrowness of the NMR line shows that the ^{1}H atoms are in rapid motion over the surface so that the observed NMR shift corresponds to an "average bonding site." Initially, increasing the coverage simply results in increasing numbers of hydrogen in very similar sites; the average bonding site does not change with coverage, and the NMR shift is constant. It is believed (*48*) that these are bridge-bonding sites. At approximately a monolayer of coverage, no further bridge sites are available, and another type of site (believed to be on-top sites) starts to fill up with a different characteristic shift. Because of exchange, only a single line is observed, with an average shift that is coverage dependent, as per Eq. (20).

The initial independence of coverage of the ^{1}H NMR shift is remarkable since the "average surface platinum" ^{195}Pt NMR parameters change continuously in this coverage range (see Section VI.E) (*84*). The latter change is usually explained by the long-range perturbation of the metal's E_f LDOS by adsorbates; however, the ^{1}H NMR suggests a slightly different interpretation (*85*). Suppose that the perturbation is much larger in the immediate neighborhood of a hydrogen than at some distance away; an individual ^{1}H atom senses mainly the perturbation that it creates—and that it drags with it during the motion over the metal surface. The platinum atoms, on the other hand, are either strongly perturbed (when they have a hydrogen in the immediate neighborhood) or only mildly so (when the hydrogen is farther away). The average situation on the platinum is mainly determined by the fraction of time in which it has a hydrogen close by, and this fraction increases with coverage. According to this scenario, whereby the short-range effects are more important than the long-range ones, the ^{1}H shift could be coverage independent, whereas the ^{195}Pt NMR changes with coverage.

From the earliest NMR experiments carried out with hydrogen on platinum catalysts by the Paris group (*86*), it is known that the absolute value of the shift at low coverage is large for large particles and small for small particles. The clean-surface ^{195}Pt NMR, on the other hand, has a shift that does not measurably vary with particle size, but the linewidth attributed to the surface signal is larger for small particles (see Section VI.B). It is believed that the average E_f LDOS on a clean surface is more or less size independent but that the distribution of surface atoms over different geometrical types of sites (faces, edges, and corners) becomes more diverse when particle size decreases. Similarly, the (bridge) sites sampled by ^{1}H on the surface of a large particle can on average be different from those on a small particle. Data from the Nottingham group give the ^{1}H shift in bridge sites on EuroPt-1 as -31 ± 1.6 ppm and for a low-dispersion Pt/SiO$_2$ as -48 ± 2 ppm. The shift in the on-top sites for the two catalysts was found as 45.3 ± 3.8 ppm and 37 ± 10 ppm, respectively (*48*). In the EuroPt-1 catalyst, the total coverage under 760 Torr is of the order of two monolayers at room temperature.

At high coverage, the hydrogen atoms exchange rapidly (on the NMR shift scale, of the order of microseconds) between the bridge sites, the on-top sites, and (as molecules) the gas phase. The lines drawn in Fig. 27 represent a fit to a three-site rapid exchange model (*48*). By going to low temperatures, it has been possible to freeze out the exchange with the gas phase, which then gives a separate peak. However, the bridge and on-top signals have not been seen separately. It would of course be a great experimental confirmation of the rapid-exchange model if this could be done. On the other hand, the resolved structure that can be found on the ^{1}H/metal peak (especially at low coverage) indicates the existence of two populations that do not exchange rapidly. In the deep-bed model (*83*), the result is ascribed to a bimodal particle size distribution; the Nottingham group believes that it may be interfacial hydrogens, such as Si-O-H-Pt (*48*).

E. Hydrogen on Rhodium

Much NMR work in these systems has been directed toward the strong metal support interaction (SMSI) question, but here I review only the more general results. The shifts of ^{1}H on Rh are the largest (in absolute value) observed on any transition metal—approximately -130 ppm at low coverage on catalysts with low metal dispersions (*50, 51, 53*). In a study of the shifts of ^{2}H on Rh/Al$_2$O$_3$ it was reported that highly dispersed catalysts (^{2}H/Rh $\approx$ 1 from chemisorption) show signals as far upfield as -200 ppm (*87*). This might be another example (in addition to platinum) of different shift values on "small" and on "large" particles. Similar large ^{1}H shifts for Rh/NaY zeolite have also been reported by the Iowa State group (*88*). From the intensity of the ^{1}H NMR it was concluded that at saturation coverage H/Rh $= 1 \pm 0.2$. The preparation method is believed to yield very small particles. The ^{1}H NMR shift has a strong temperature dependence, moving from -210 ppm at 300 K to -275 ppm at 500 K. Furthermore, the linewidth is temperature dependent and considerably larger than that observed for the catalysts with low dispersions. It is thought that these particles are too small to show metallic behavior, and that the shift and linewidth are due to Curie paramagnetism (*88*).

For a Rh/TiO$_2$ catalyst (reduced at temperatures between 573 and 673 K) an *in situ* pressure/NMR intensity isotherm, using explicit calibration, yielded (at $p = 10$ Torr) a dispersion of 0.12, compared with 0.13 ± 0.1 from volumetric and titration methods (*50*). The agreement is somewhat surprising since the NMR dispersion is based on the ^{1}H/metal signal only, whereas volumetric methods cannot distinguish the different types of hydrogen (Fig. 16). Evacuation at 473 K eliminates the ^{1}H/Rh signal. The shift moves downfield with increasing coverage to -25 ppm at 1.5 monolayer. An attempt was made to fit the changes of shift and intensity to Eqs. (19) and (20). It was concluded that at coverages up to one monolayer, two sites with coverage-independent shifts of -130 and 210 ppm could represent the curve, but that at higher H/Rh ratios a third site would be needed.

It is possible that the third site is gaseous hydrogen, not explicitly considered in the analysis. As shown in Fig. 28a, the shift vs intensity curve for the ^{1}H/Rh signal of this sample is flat up to approximately one monolayer (*50*), and could possibly be represented by the Nottingham model for ^{1}H/Pt (described in Section III.D). This is not the case, however, for a very similar Rh/SrTiO$_3$ catalyst (*51*) (Fig. 28b).

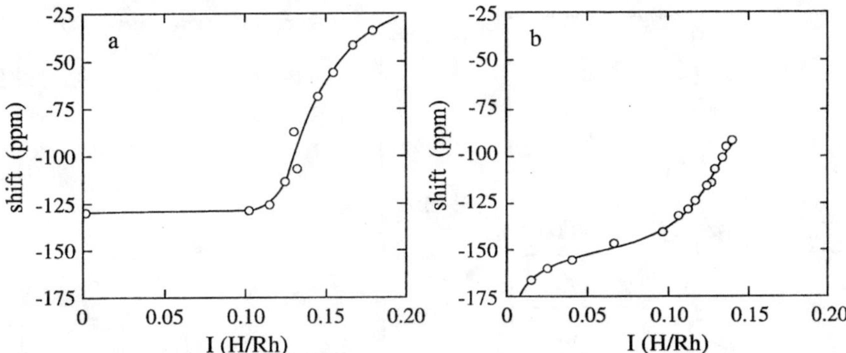

FIG. 28. ^{1}H/metal NMR shift as a function of coverage for two Rh catalysts. The coverage is measured from the calibrated intensity I of the NMR lines. (a) Rh/TiO$_2$, same sample as Fig. 16. Note the similarity with Fig. 27a. [Reproduced with permission from Sanz and Rojo (*50*). Copyright 1985 American Chemical Society.] (b) Rh/SrTiO$_3$, same sample as Fig. 17. In this case there is no clear horizontal plateau. [Reproduced with permission from Rojo *et al.* (*51*). Copyright 1994 American Chemical Society.]

F.   HYDROGEN ON RUTHENIUM

^{1}H NMR adsorption isotherms for Ru/SiO$_2$ catalysts have been obtained using explicit calibration (*89*). Although the pressure over the sample could be adjusted *in situ,* no volumetric data were taken simultaneously, probably because of the important spillover effects in this catalytic system (see Section III.A). The NMR study was performed at pressures between 10 and 760 Torr and at temperatures between 323 and 473 K (only the 323-K results are reviewed here). The dispersion of the catalyst was determined from the irreversible ^{1}H NMR signal as 0.29. The metal loading was 8 wt% so that a monolayer coverage on 1 g of catalyst corresponds to 2.8 cm^3 of H$_2$ under standard conditions. It is typical for an NMR sample to contain 0.5 g of material in a 1-cm^3 sample volume, and the pores in the powder make up about half the volume. If such a sample of this catalyst is under 760 Torr of hydrogen, the gas phase corresponds to one-third of a monolayer, and it can make a detectable contribution to the NMR signal.

At pressures >100 Torr, two distinct lines, both ascribed to ^{1}H/metal, were observed in addition to the silanol signal (Fig. 29a). One of the peaks was observed even after evacuation, at −65 ppm; the other appears in the 100-Torr range at −55 ppm, and both shift to low field with increasing pressure (*89*). No separate line for gaseous hydrogen was observed, but the summed intensity of the two ^{1}H/metal lines increases with pressure (Fig. 29b). It was initially believed that only one of the two lines (the low-field line) could represent hydrogen in rapid exchange with the gas phase;

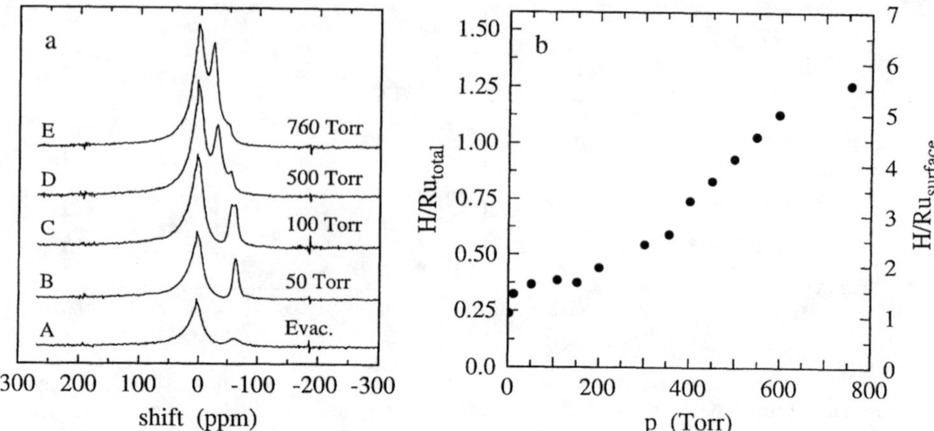

FIG. 29. ^{1}H NMR of a Ru/SiO$_2$ catalyst with a volumetrically determined dispersion of 0.29. (a) Spectra under different pressures of H$_2$ gas. The trace marked "Evac" was taken after the reversible hydrogen had been pumped off. The intensity of the ^{1}H/metal NMR line at -60 ppm (the 0 ppm line is from the silanol groups) in this trace has been used to establish the right-hand scale in Fig. 29b. For pressures above 100 Torr, two separate ^{1}H/metal lines are observed. Initially, they were believed to represent two distinct nonexchanging sites on the same metal particle, but recently it has been proposed that they are on different particles: one kind in large pores of the silica and another in small pores. (b) The total intensity of the two ^{1}H/metal lines as a function of pressure. It is unlikely that (at high pressure) there can really be five adsorbed hydrogen atoms per surface site; rather, this curve indicates that adsorbed and gas phase ^{1}H cannot be distinguished. As a consequence of this rapid exchange, both ^{1}H/metal lines in Fig. 29a shift toward the gas position with increasing pressure. [Reproduced with permission from Bhatia *et al.* (*89*).]

actually, this was an important clue in the data analysis (*89*). However, recent work, done with an NMR technique related to saturation transfer spectroscopy (see Section II.F), has shown that under 0.5 Torr of hydrogen gas there is a rapid exchange between the gas and the adsorbate (*90*). It seems, therefore, that each of the two ^{1}H/metal peaks in Fig. 29 represents hydrogen in rapid exchange with gas, but there is nevertheless no fast exchange between the two adsorption sites.

This seemingly inconsistent behavior is probably due to a property of the silica support (*91*); it has a bimodal pore distribution. For simplicity, assume that there are exactly two pore sizes, large and small, and that the exchange of gas between the pores is very slow. The gas inside any pore is in rapid exchange with the adsorbate on the metal particles in that pore, and the intensity I and shift δ vary according to Eqs. (19) and (20). In a large pore, I_{gas} is more important than in a small pore, and $\delta_{\text{gas}} = 3.5$ ppm (see Section III.G); therefore, the δ_{obs} in large pores is shifted to low field relative to that in small pores (*91*). At very low pressures, the mechanism

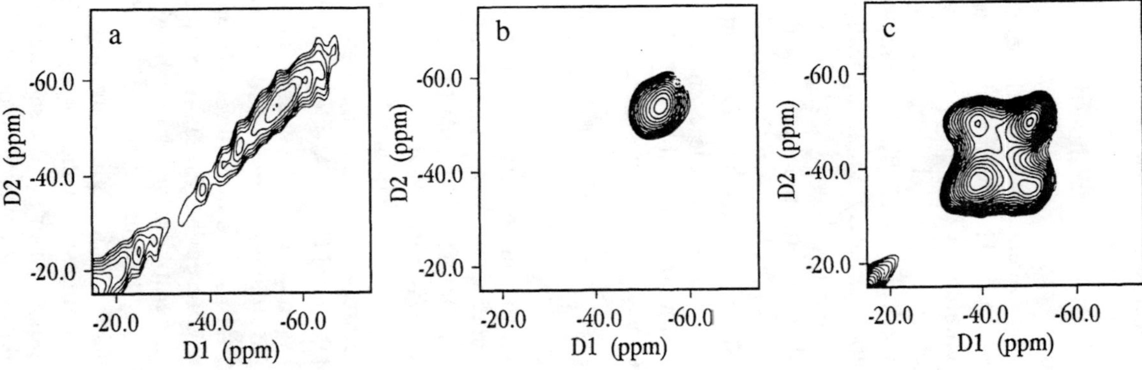

FIG. 30. Two-dimensional ^{1}H exchange NMR spectra at 400 K for the same sample as shown in Fig. 29 (a) after removal of reversible hydrogen, (b) under 80 Torr, and (c) under 300 Torr. The lines are contours of equal amplitude after a two-dimensional Fourier transformation of two time variables into two frequency variables, the latter are drawn in shift units. In the absence of any exchange, the peaks (contours of maximum amplitude) in an exchange spectrum are only along the main diagonal (as in a). If exchange occurs, then "cross-peaks" (off-diagonal contours of local amplitude maxima) appear (as in c). Here, exchange between the two ^{1}H/metal signals (the two peaks along the main diagonal) is detected on a time scale of several hundred microseconds. [Reproduced with permission from Engelke *et al.* (*92*). Copyright 1994 American Physical Society.]

gives line broadening before a visible splitting occurs. The spin-lattice relaxation times of both signals are found to be nearly independent of temperature between 300 and 473 K. The results suggest that the relaxation is dominated by the exchange processes rather than by the Korringa mechanism.

The 2D-NMR exchange spectra (*92*) in Fig. 30a show the inhomogeneous and homogeneous linewidths after evacuation of the sample. The width measured along the diagonal is inhomogeneous and includes variations in shift and heteronuclear dipolar couplings. The width along the antidiagonal is homogenous, mainly due to homonuclear couplings, and its value corresponds to a nearly rigid arrangement of the ^{1}H atoms at this coverage. When the pressure is increased to 80 Torr (Fig. 30b), the homogeneous and inhomogeneous linewidths are approximately equal, indicating motional averaging. At 300 Torr (Fig. 30c), the two types of ^{1}H/Ru appear in the spectrum along the diagonal, and the off-diagonal cross-peaks show that there is exchange between these sites [i.e., between the two types of pores (*91*)]. From the results of additional experiments, it is found that the populations of the two sites are about equal and that the exchange time is 700 μs. The two relaxation times are similar, compatible with the result that exchange occurs on a time scale shorter than the observed T_1 (15 $\pm$ 3 ms).

G.   Hydrogen on Copper

Hydrogen on copper powder was among the first adsorbates on a metal observed by NMR (*93*). The resonance was found to be shifted to low field, at 94 ppm (with respect to a rubber sample). A detailed *in situ* volumetric and flowing-gas ^{1}H NMR study (*49*) was performed on a Cu/MgO catalyst with a dispersion of 0.06 (as measured by oxygen uptake from N_2O), with the metal loading being 10 wt%. Assuming 1:1 stoichiometry, this corresponds to a monolayer capacity of 94 μmol H_2 g^{-1}. Modeling the volumetric adsorption curves with Langmuir isotherms gave values of approximately 30 μmol g^{-1}, or a 1:3 H/surface Cu stoichiometry. In addition to the peak indicating hydroxyls on the support, two lines were observed: one from the gas phase at 3.5 ppm and another representing ^{1}H/Cu at 85 ppm (Fig. 13). These results show that at room temperature the exchange between surface and gas hydrogen is slow, presumably because of the activated nature of the adsorption. At temperatures below 273 K, equilibrium was not attained over a period of 1 h. No increase in intensity of the ^{1}H/MgO NMR signal (i.e., no spillover) was observed upon exposure of the sample to hydrogen, and the position of the ^{1}H/Cu peak was only slightly dependent on surface coverage. The spin lattice relaxation of the ^{1}H/Cu peak was measured at temperatures between 200 and 400 K (Fig. 31a), and a value of $T_1T = 6.54$ s K was found (*49*).

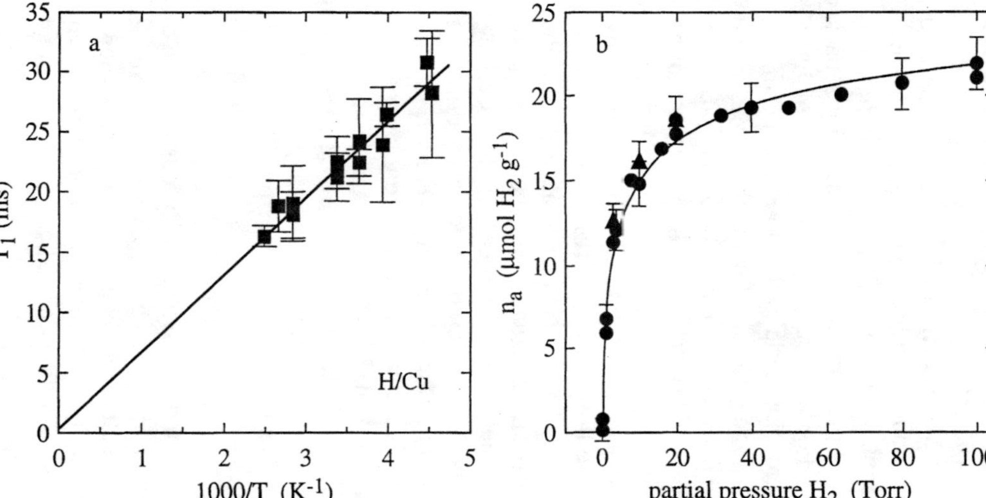

Fig. 31. (a) Spin lattice relaxation time of the ^{1}H/metal line in a Cu/MgO catalyst (peak b in Fig. 13) as a function of reciprocal temperature. The constant Korringa product (cf. Fig. 22) indicates a metallic character of the ^{1}H site. (b) Volumetric (line) and ^{1}H NMR intensity (points) adsorption isotherms for Cu/MgO. The volumetric and NMR scales have been aligned by fitting a proportionality constant in a static NMR experiment. The data points were taken under flowing gas conditions in a ^{1}H$_2$/^{2}H$_2$ stream. Circles, increasing partial pressure of ^{1}H$_2$; triangles, decreasing partial pressure. [Reproduced with permission from Chesters *et al.* (49). Copyright 1997 American Chemical Society.]

This result seems to indicate the only Korringa-type spin lattice relaxation for adsorbed hydrogen reported so far. If we assume that it is due to contact interaction with the spins of s-like electrons (by using reasoning similar to that applied for PdH in Section III.B), its value gives $K_s = 200$ ppm and places the zero of the Knight shift scale at -115 ppm. Whereas such values cannot be ruled out, it is more likely that this one-component analysis of shift and relaxation is insufficient.

The features of the ^{1}H NMR spectra of hydrogen in static equilibrium with the catalyst were reproduced under flowing-gas conditions. The total flow rate of separately regulated nitrogen, hydrogen, and deuterium flows was kept constant, and the pressure was maintained by a back-pressure regulator at the gas exit. For a given partial pressure of hydrogen (the deuterium flow being zero), the static and dynamic experiments give the same ^{1}H/Cu NMR intensity and position (49). Furthermore, the flow experiments showed full reversibility (Fig. 31b). By using implicit calibration (adjusting a single proportionality constant), the authors found that the ^{1}H NMR intensity/pressure isotherm correlated well with the volumetric Langmuir isotherm. The hydrogen/deuterium exchange reaction was investigated after a switch of the flow from 100 Torr of hydrogen to 100 Torr of deuterium partial pressure. The ^{1}H NMR signal of the gas disappeared almost instantaneously; the ^{1}H/Cu signal diminished over a period of 1 h, and the hydroxyl/MgO signal remained unaffected. Assuming that the decrease in ^{1}H/Cu intensity is due to associative desorption of HD and of H_2 (with the same rate constants) in a Langmuir–Hinselwood mechanism, the desorption rate constant is equivalent to one monolayer (as determined from the adsorption isotherm) in 5 h at 260 K and in a few minutes at 320 K. This experiment shows that, for the type of Cu/MgO catalyst used here at about RT, the MgO support does not participate in the exchange reaction, contrary to what has been reported for pure MgO at 78 K (94).

In an NMR investigation of a methanol synthesis catalyst based on Cu/Zn/Al oxides (95), a ^{1}H resonance line was found at 85 ppm (with respect to H_2O) and attributed to dissociatively adsorbed hydrogen on copper metal. Its intensity was found to be highly variable, even for similar sample preparations and treatments. The hypothesis was therefore proposed that the actual availability of copper metal in these systems is very sensitive to preparation and handling details, e.g., oxygen contamination.

H. HYDROGEN ON BIMETALLICS

Several combinations of the metals discussed previously have been studied by ^{1}H NMR. As shown, it is usual for the hydrogen to move rapidly over the metallic surface, but exchange with the gas phase is not always fast on the NMR time scale. Therefore, a situation can occur in which we observe a sum, rather than the average, of NMR spectra of ^{1}H on individual particles. In bimetallic systems, the surface composition may vary from particle to particle, leading to line broadening or even to the appearance of several distinct peaks.

Ruthenium and copper are inmiscible in the bulk, and this is believed to be a factor in the catalytic behavior of supported $Ru_{1-x}Cu_x$ catalysts. For $x > 0.45$, two signals attributed to ^{1}H/metal are observed (*96*), neither of which shows an important shift variation when x increases up to 0.8, with the pressure being maintained at 60 Torr (most of the signal then comes from reversible hydrogen). One of the signals occurs close to the ^{1}H/Cu position and disappears when the sample is evacuated, just as in monometallic Cu/SiO_2 (*96*). The other signal remains visible after evacuation but becomes so broad that its position is hard to define. This broadening should reflect a particle-to-particle heterogeneity because after evacuation of the sample the exchange of hydrogen is extremely slow, and the "strongly held" hydrogen on monometallic Ru/SiO_2 does not have a particularly large linewidth (*89*). However, under 60 Torr of H_2, at least the reversible hydrogen is in rapid exchange with the gas phase, thereby averaging over many particles and narrowing the line. The hydrogen that causes the ^{1}H/Cu signal is reversible, but not in rapid exchange with the gas, just as in Cu/MgO (*49*). (In ^{1}H spectra of catalysts on SiO_2 carriers, the gas line should be less conspicuous than on MgO, and it has not been searched for explicitly.)

Under 60 Torr of hydrogen, the resonance position of the reversible hydrogen shifted 10 ppm downfield when the first 20 atom% of ruthenium was replaced by copper, and it shifted an additional 3 ppm when increasing from 20 to 80% (Fig. 32a); overall, the linewidth increased from 16 to 40 ppm. The initial 10 ppm shift has been interpreted according to a two-site fast-exchange model. As long as the copper atom fraction x is smaller than the dispersion, all Cu atoms are supposed to sit on the surface of the bimetallic particles. The shift δ_{Cu} of ^{1}H in such a site is considered a parameter to be determined from the experimental results. The other site is ^{1}H on Ru, with a shift $\delta_{Ru} = 62$ ppm, equal to that for $x = 0$. To find the value of δ_{Cu}, it is assumed that all catalysts with $0 < x < 0.2$ have the same dispersion (about 0.29) and that under 60 Torr of hydrogen the number of hydrogen atoms adsorbed per surface copper atom is the same as that per surface ruthenium atom. The shift δ_{Cu} determined in this way is -49 ppm (different from that on pure copper particles).

Once δ_{Cu} has been found, the same fast-exchange model can be used to derive the average surface composition $Ru_{1-y}Cu_y$ from the measured variation of the shift with overall composition $Ru_{1-x}Cu_x$. It is then found

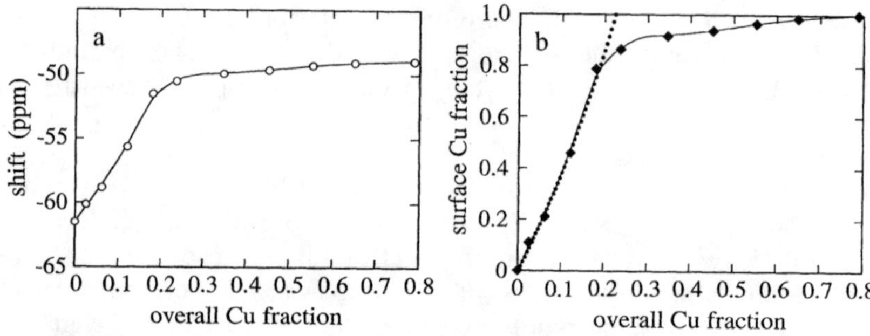

FIG. 32. (a) ^{1}H NMR shift for hydrogen on Cu/Ru bimetallic particles as a function of overall sample composition. For Cu fractions exceeding 0.5, the samples also contain (nearly) pure Cu particles, as indicated by a second ^{1}H/metal signal with a shift near 95 ppm (not plotted). The bimetallics are believed to have all their Cu atoms at the surface (at most a monolayer). The NMR shift data are fitted to a two-site rapid-exchange model as a basis for estimating the surface composition of the bimetallic particles as a function of the overall composition of the sample (plotted in b). In the simplest model of surface enrichment, initially all Cu is used to cover the surface of pure Ru particles. Once a monolayer coverage is reached, all additional Cu atoms form pure copper particles (this model is represented by the dotted line in b). [Reproduced with permission from Wu *et al.* (*96*).]

that at $x = 0.2$, $y = 0.81$, and at $x = 0.8$, $y = 1$ (Fig. 32b). These data cannot be very precise because of the difficulty in determining changes of a few ppm in the position of lines that are 16–40 ppm wide and because of the hypotheses of constancy of dispersion and number of adsorption sites; if at $y = 0.81$ and $x = 0.2$ all the copper were on the surface, the dispersion would have to be 0.25. In this example, the fraction of Ru atoms that is on the surface must be $(0.25 - 0.20)/0.8 = 0.06$. Such values have been used to discuss the variation of specific activity of these catalysts in ethane hydrogenolysis (*97*).

The ^{1}H NMR of hydrogen on $Pt_{1-x}Rh_x$ catalysts shows a single ^{1}H/metal line for all values of x, and the shift varies monotonically (but not linearly) with x (*98*) (Fig. 33). In a simple model, one assumes that the shift of a given ^{1}H is completely determined by the nature of the closest metal neighbor on the surface (Pt or Rh), independent of composition, and that the observed shift is an average over all surface sites due to rapid exchange. According to this model, the shape of the shift vs composition curve indicates that the surface is enriched in Rh. As discussed in Section IV.B, the ^{13}CO shift/composition curve for these systems is similar to that for ^{1}H; however, a ^{195}Pt/^{13}CO double-resonance experiment for a sample with $x = 0.5$ did not support the idea of Rh enrichment at the surface (see Section IV.F). The simple model is probably not valid; the ^{1}H shift does not depend only on

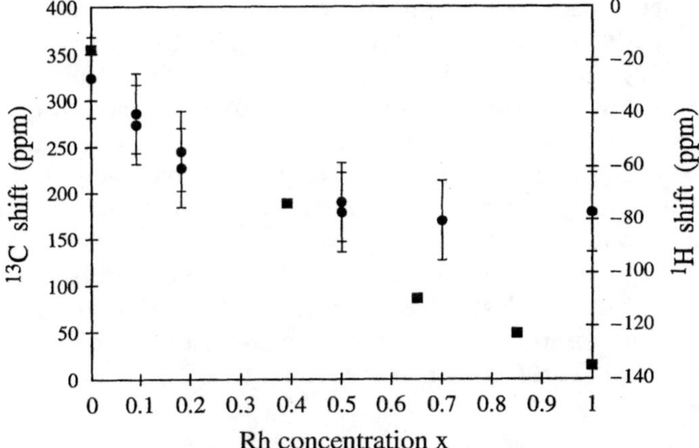

FIG. 33. Shift data for two adsorbates on Pt/Rh catalysts as a function of overall Rh concentration. Squares and right-hand scale, ^{1}H; circles and left-hand scale, ^{13}CO. In both cases application of a simple two-site rapid-exchange model (as in Fig. 32) would lead to the conclusion that the surface is enriched in Rh. At least for the ^{13}CO-covered samples, additional NMR experiments do not confirm this conclusion (98, 105).

the identity of the nearest neighbor but also on the composition of a larger region around the binding site (see Section VI.G).

IV. ^{13}CO NMR: Survey of Results

A.   SAMPLE PREPARATION

Among the bulk transition metals of group VIIIB (Fe to Pt), only Ni reacts very rapidly with CO (in fact, the reaction is used in the preparation of ultrapure Ni metal). At the other extreme, no free, pure carbonyl derivatives of zerovalent Pd or Pt have been reported (but see Section IV.B), although it is thought that they are stable (99). Cycles of adsorption at RT and desorption at 673 K of CO on very small Pd particles lead to slight disproportionation, $2CO \rightarrow CO_2 + C$ (100). On Ru and Rh particles some disproportionation occurs even upon adsorption at RT. In some cases the $CO_2(g)$ is detected by NMR.

According to X-ray photoelectron spectroscopy data, the adsorption of CO on supported Ru of rather low-dispersion (0.27 by hydrogen chemisorption) at 310 K leads to the formation of some oxidized Ru, which can be re-reduced by hydrogen treatment (101). Similarly, an EXAFS study of the effect of RT chemisorption of CO on very highly dispersed Rh (1.70

from hydrogen adsorption) supported on γ-alumina showed the formation of isolated $Rh(CO)_2$ species on the support. The CO/Rh ratio was found to be 1.9 (*102*); very small Rh particles are completely fragmented in CO.

Saturation coverage samples for NMR are usually made by exposing the sample for a period of hours to a sufficient amount of CO gas and pumping the excess off afterwards. Preparation of partial-coverage samples is more tricky (see the last paragraph of Section IV.C). Usually, samples are sealed under a controlled atmosphere in glass or quartz ampoules of 1 or 2 cm^3 volume. The quantity of CO in a typical NMR sample would exert a pressure of a few bar if the catalyst were not present. Therefore, desorption upon increasing temperature is much less in such closed sample tubes than in ultrahigh vacuum surface-science experiments.

B. ^{13}CO AND C^{17}O NMR: GENERAL

The shielding tensors for ^{17}O and ^{13}C in the CO molecule are anisotropic with zero asymmetry, reflecting the axial symmetry of the molecule. The isotropic shift of ^{13}C with respect to TMS is 185 ppm, and the anisotropy in the solid is 406 ppm. Upon linear coordination in transition metal carbonyl molecules, these values typically change by tens of ppm (*103*). In the few bridge-coordinated compounds for which data are available the isotropic shifts are around 230 ppm and the anisotropies 175 ppm. (No data for solid-state platinum carbonyls have been reported, but there is no reason to believe that they are an exception.) In the CO molecule, the isotropic shift of ^{17}O is 350 ppm with respect to H_2O, and the anisotropy is 675 ppm. Again, these values change by a few tens of ppm in solid molecular metal carbonyls with linear coordination (*104*).

The Slichter group found that for all their transition metal (Ru, Rh, Pd, Os, Ir, and Pt) catalysts, the ^{13}CO lineshapes, obtained with the point-by-point method at 77 K, are wide and featureless. They can be parameterized as Gaussians with FWHM values of between 300 ppm (Os) and 600 ppm (Pd) and positions of the maxima between 180 ppm (Rh) and 540 ppm (Pd) (*2*). It is thought that the particles in these samples are large enough to show metallic NMR for the adsorbate. Wang *et al.* (*105*) studied the shift as a function of the overall composition of bimetallic catalysts $Pt_{1-x}Rh_x$, finding that between $x = 0.5$ and $x = 1$ the shift does not vary (see Fig. 33). This result might be considered a demonstration that in this composition range the outer layers of the particles are pure Rh (cf. Section III.H; see Section IV.F).

A coexistence of metallic and molecular metal carbonyls has been found in ^{13}CO/Rh catalysts (*106*) and for Pt in L zeolites (*107*). When the CO adsorbs on large particles, the resulting carbonyls have metallic character,

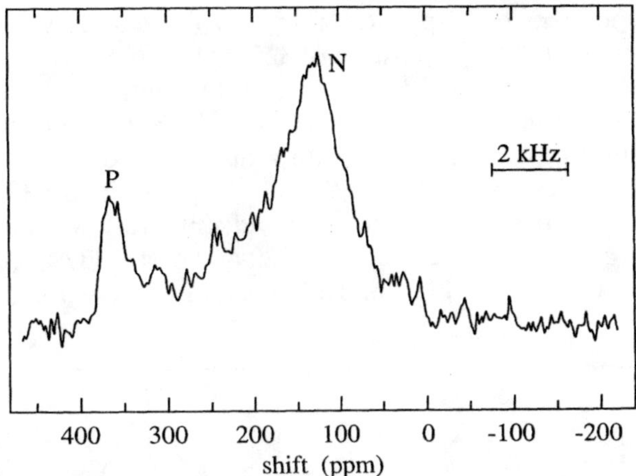

FIG. 34. ^{13}C NMR spectrum of a Cu/Zn/Al oxides based catalyst after exposure to 17 Torr of ^{13}CO at 500 K. The line marked P, at about 350 ppm, has been attributed to ^{13}CO on metallic copper; the line N, in the 110–130 ppm range, is believed to be related to some form of adsorbed ^{13}CO$_2$. These data have been presented as preliminary, but they still seem to be the only ones possibly related to ^{13}CO on copper. [Reproduced with permission from Dennison *et al.* (*95*). Copyright 1989 Royal Society of Chemistry.]

but on very small particles they are molecular. This molecular behavior may be local (with the interior of the particle staying metallic), but it is also possible that the adsorbate induces nonmetallic behavior throughout an initially metallic particle. In metal cluster molecules, for example, the ^{195}Pt NMR is interpreted by assuming that the platinum atoms on the outer layer of the platinum core of the molecule are nonmetallic, whereas the inner Pt atoms show a more metallic NMR (*108*). The surface Pt atoms are nonmetallic because of the large number (>1 per metal atom) of ligands coordinated to them. In line with this idea, the ^{31}P NMR of the phosphine ligands has molecular character (*109*).

Preliminary results for ^{13}CO on a Cu/Zn/Al methanol synthesis catalyst show a fairly narrow (a few tens of ppm) resonance at 350 ppm (Fig. 34) (*95*). It has been ascribed to Knight-shifted CO on copper, but its relative sharpness might indicate that it is actually the low-field divergence of a chemical-shift powder pattern (the corresponding high-field discontinuity being overlapped by the rest of the spectrum). In principle, a study of the spin lattice relaxation time could resolve the issue.

C. ^{13}CO ON PLATINUM

At 77 K, point-by-point ^{13}CO/Pt spectra have been obtained by several members of the Slichter group (*110–114*), but only a few have been published. Independent of dispersion, they can reasonably well be fitted to Gaussians, with an average position of 325 ($\pm$25) ppm and average FWHM of 360 ($\pm$40) ppm. An unpublished analysis by Rudaz (*111*, p. 92) suggests that on larger particles the line is narrower. An extreme example of such behavior is provided by a sample with a dispersion of 0.04, for which the line position is 310 ppm and the FWHM 260 ppm. Figure 35a shows how the line narrows (from 390 to 50 ppm) and shifts to lower field (from 330 to 400 ppm) at temperatures between 320 and 490 K (*112*). (These values vary only weakly in the high end of the temperature range.) Later FT

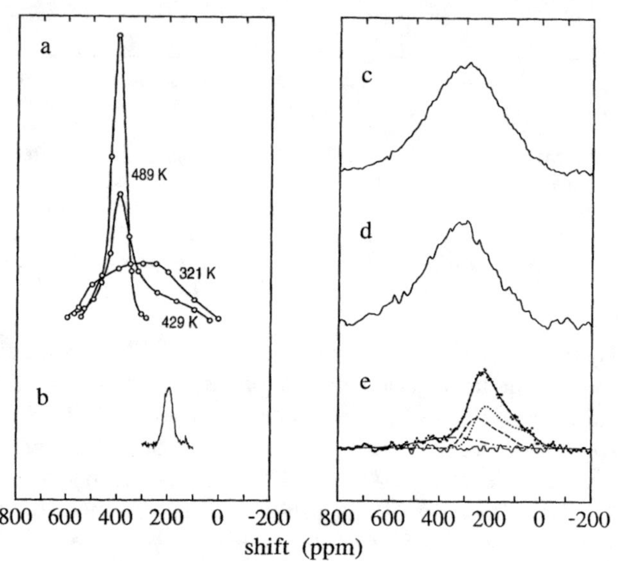

FIG. 35.   Spectra of ^{13}CO on platinum. Data in a were obtained with the point-by-point method at different temperatures as indicated; the others are from Fourier transforms at room temperature. The vertical scales are all different. (a) High-dispersion (0.76) Pt/Al$_2$O$_3$. The narrowing of the line as seen here (three spectra with a common vertical scale) has also been observed for lower dispersion catalysts at practically the same temperatures. (b) Platinum particles of sub-nanometer size in colloidal suspension. There seems to be no Knight shift (the line is at 192 ppm), but the linewidth is appreciable (comparable to the 489 K trace in a), although rotational diffusion of the particles should be rapid on this time scale. [Reproduced with permission from Bradley *et al.* (*116*).] (c) High-dispersion (0.80) Pt/Al$_2$O$_3$. Note the similarity with the 321-K trace in a. (d) Same sample as in c, but with magic angle spinning. The spin rate corresponds to 112 ppm on the horizontal scale, but the spectrum remains the same as in c. [Reproduced with permission from Zilm *et al.* (*117*). Copyright 1990 American Chemical Society.] (e) Very highly dispersed particles in L zeolite (CO/Pt = 0.87). The spectrum has been deconvoluted into three contributions: powder spectra for bridge and on-top sites in molecular-type carbonyls (dotted and dashed lines) and a Gaussian for metallic carbonyls (–·–). [Reproduced with permission from Sharma *et al.* (*107*).]

spectra characterizing a highly dispersed sample at 77 and 337 K have given similar results (*115*).

Bradley *et al.* (*116*) obtained room-temperature NMR signals of ^{13}CO absorbed by suspensions of subnanometer Pt particles in methylcyclohexane solutions of isobutylaluminoxane. The position is 195 ppm and the FWHM 50 ppm (Fig. 35b). The absence of a Knight shift is attributed to the smallness of the particles and the linewidth to a distribution of binding sites. The linewidth (but not the position) is the same as the high-temperature linewidth for supported catalysts with much larger particles (Fig. 35a).

For two highly-dispersed samples, Zilm *et al.* (*117*) found that at RT the spectrum has position 340 ppm and FWHM 300 ppm. The results do not change under MAS (Figs. 35c and 35d), even when the sample temperature is lowered to 150 K. The lack of narrowing by MAS shows that anisotropic mechanisms of broadening are small compared with the static distribution of isotropic shifts. This observation has not yet been convincingly explained, although several hypotheses have been put forward (see Section V.B). These authors also verified that the linewidth is linear in the applied field B_0.

In L zeolites with a CO/Pt ratio of 0.87, a 14% fraction of the ^{13}CO signal was ascribed to a metallic carbonyl at 360 ppm with FWHM 240 ppm (Fig. 35e) and a spin lattice relaxation time at RT of 0.8 ± 0.6 s ($T_1 T$ between 60 and 400 s K) (*107*). The remaining signal was attributed to platinum carbonyls, species not yet found in the free form (without encaging by the zeolite).

Spin lattice relaxation for ^{13}CO/Pt in the high-temperature region (290–500 K) is single exponential with $T_1 T = 46 \pm 3$ s K (*118*). The behavior at lower temperatures is complicated (*119, 120*) and not considered here. Zilm *et al.* (*117*) reported some variation in T_1 across the line, with the low-field part relaxing faster. The relatively small variation of the T_1 across the line shows that, if sites with clearly different intrinsic T_1's do exist, then the exchange between them cannot be slow compared with the difference in T_1. It is also known from Fig. 35a that at RT the linewidth is not motionally narrowed. If the exchange time τ_{exch} between sites with distinct T_1's is the same as that between sites with distinct shifts, the observed values of the linewidth and of the T_1 give the following for the condition to obtain T_1 averaging (by hopping from site to site) without line narrowing: $10 \ \mu s \ll \tau_{\text{exch}} \ll 100$ ms.

A simplified version of exchange NMR has been applied to study diffusion of ^{13}CO on platinum clusters between 300 and 400 K (*115*). The delay τ_{prep} (see Section II.E), which corresponds to the frequency variable ω_1, was kept fixed, and in the ω_2 spectrum only the amplitude A at a suitably chosen frequency $\omega_2 = \omega_0$ was monitored as a function of the parametric variable

τ_{exch}. The relation between this A and the 2D exchange spectrum S is given by

$$A(\tau_{\text{prep}}, \omega_0; \tau_{\text{exch}}) = \int S(\omega_1, \omega_2 = \omega_0; \tau_{\text{exch}}) \sin(\omega_1 \tau_{\text{prep}}) \, d\omega_1. \qquad (23)$$

The data were modeled as a diffusion in ω_2 space during the exchange interval τ_{exch}, and the resulting diffusion coefficient was converted to a coefficient for spatial diffusion on the surface of the platinum particles by a simple geometric argument. On the time scale of the experiment, of the order of the spin lattice relaxation time T_1, about half of the CO molecules (identified with on-top CO) are immobile. For a low-dispersion sample at saturation coverage, it was found that at RT for the remaining nuclei $D = 1.8 \times 10^{-11}$ cm^2 s^{-1}; for a high-dispersion sample, this value is 2.4×10^{-14} cm^2 s^{-1} (Fig. 36). The activation energies were approximately 7 kcal mol^{-1} (304 meV) for the low-dispersion sample and 10 kcal mol^{-1} (434 meV) for the high-dispersion sample.

The spin echo decay of ^{13}CO has been measured at 77 K by using RF fields of typically 50 G so that the homonuclear condition was fulfilled (121). The decay rate observed for samples under saturation coverage was compared with that calculated for several regular overlayers of CO on extended low-index platinum surfaces. It was shown that the actual packing of the molecules is denser than that of some of the model overlayers. Partially covered samples, prepared by admitting a calculated quantity of ^{13}CO at room temperature to the NMR tube containing the catalyst, show a dipolar spin echo decay rate very similar to that characterizing samples at saturation coverage. This method of preparation obviously does not lead

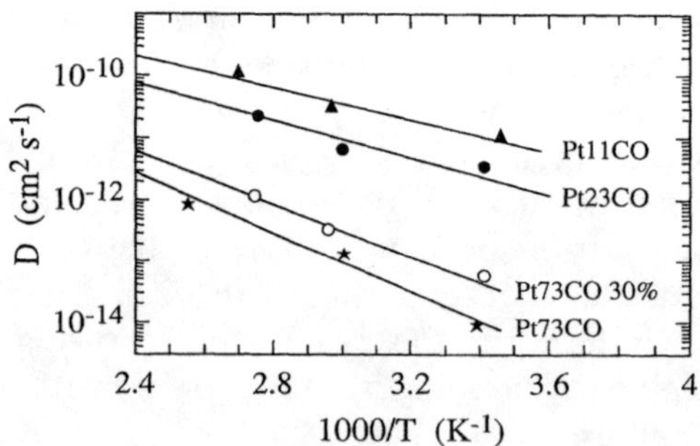

Fig. 36. Diffusion coefficient for CO on Pt/Al$_2$O$_3$, derived from a variant of ^{13}CO exchange NMR. The numbers in the sample designations give dispersions in percentages. The "30%" indicates partial coverage with CO. These data indicate that at room temperature the diffusion on large particles is three orders of magnitude faster than on small particles. [Reproduced with permission from Becerra et al. (115). Copyright 1993 American Chemical Society.]

to uniform distribution of ^{13}CO over all the available metal surface, but annealing at 550 K for 2 days distributes the molecules more evenly. Similar phenomena are known from infrared studies of chemisorbed CO (*122*).

D. ^{13}CO ON PALLADIUM

In the initial work (*123*), spectra were acquired by the point-by-point method. In later studies, only the FT method was used. On low-dispersion catalysts and at RT, the ^{13}CO spectrum is not affected by MAS, and it is hardly affected by a MAS–TOSS combination (*117*) (no low-temperature results are available). The linewidth is about 85 ppm (field independent) and the position +750 ppm. The results agree with the result of Shore *et al.* (*123*) that the static lineshape for ^{13}CO on low-dispersion Pd narrows at temperatures between 200 and 300 K and is temperature-independent in the range between 300 and 400 K.

The results are more complicated for highly dispersed catalysts. Both groups found that around RT the lines are broad, with FWHM of the order of 600 ppm. Zilm *et al.* (*124*) found no evidence for line narrowing up to 400 K, whereas the data of Becerra *et al.* (*125*) showed clear narrowing between 300 and 500 K. Just as in the low-dispersion sample (*117*), MAS by itself does not affect the spectral shape (Fig. 37a), but adding a TOSS pulse sequence breaks the spectrum into three features (Fig. 37b) (*124*); a sharp high-field line at 192 ppm is ascribed to on-top ^{13}CO molecules, and a broader doublet at 450 and 700 ppm is ascribed to bridge-bonded ones. For the static spectrum, the ^{13}CO spin lattice relaxation rate is estimated to vary an order of magnitude across the spectrum, increasing toward lower field (*124*).

The Slichter group found that both on high- (*125*) and on low-dispersion Pd (*123*), the maximum of the ^{13}CO line shifts to lower field when the temperature is increased from 200 to 300 K. On the basis of ^{13}CO T_2 measurements, they proposed that for saturation coverage on low-dispersion particles at low temperatures (77 K), about half of the ^{13}CO is in on-top sites, with a NMR line centered at 400 ppm, and that a rearrangement to bridge sites (line centered at 650 ppm) occurs upon increasing the temperature above 250 K (*125*). Their T_1 experiments suggest that the ^{13}CO resonance for on-top CO occurs at 500 ppm and that for bridge-bonded CO at 850 ppm (*126*). The Korringa products T_1T measured at these two positions in the line are different, but they are constant from 77 K upwards (no lower temperature results are available), indicating metallic character.

Two types of ^{13}CO-covered Pd colloids have been investigated. One has 1.8-nm particles but is otherwise similar to the colloid mentioned in Section IV.C (*127*). The other was a colloid in methanol, protected by polyvinyl

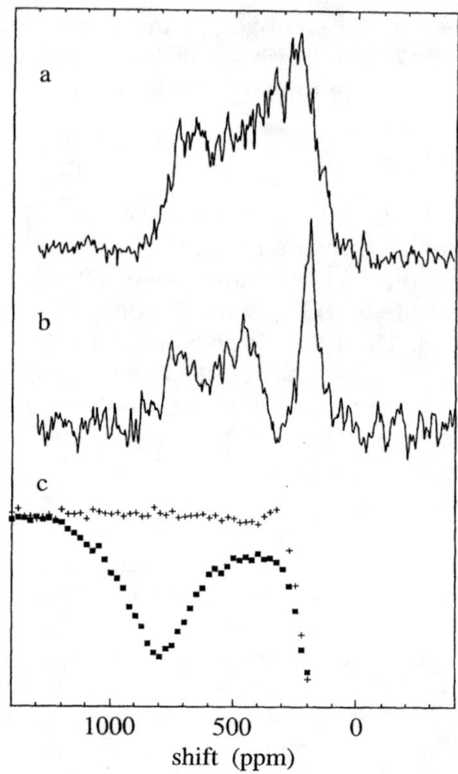

FIG. 37.   Spectra of ^{13}CO on palladium. The three vertical scales are different. (a) Medium-dispersion Pd/SiO$_2$ (H/Pd = 0.47). The center of gravity (first moment) of the line is at 410 ppm. The spectrum is temperature independent up to 400 K. In a low-dispersion sample (0.16), only a narrow peak at 750 ppm was observed (not shown). (b) The same sample as in a but with magic angle spinning, at a rate corresponding to 120 ppm, and with a sideband-suppressing pulse sequence. The sharp feature is at 192 ppm (cf. Fig. 35b). [Reproduced with permission from Zilm *et al.* (*124*). Copyright 1990 American Chemical Society.] (c) A colloidal suspension with particles in the 3- to 15-nm range. The ^{13}CO on the Pd was detected at 800 ppm by spin-saturation transfer (squares) to the free ^{13}CO in solution. The baseline (+) was established by irradiating at a supposedly "empty" frequency. Near 200 ppm there is a perturbation due to ^{13}C in the protecting agent of the colloid. [Reproduced with permission from Bradley *et al.* (*128*). Copyright 1991 American Chemical Society.]

pyrrolidone (PVP), with 7-nm average particles (3–15 nm observed) (*128*). On the first sample, a resonance was observed at 195 ppm, with a FWHM of 20 ppm. The absence of a Knight shift is remarkable. In the second colloid, a wide and Knight-shifted signal was detected by the spin saturation transfer method (Fig 37c); the values found (position 800 ppm, FWHM 300 ppm) may be influenced by details of the molecular exchange process that underlies the method, but the results are compatible with the previously mentioned values for a high-dispersion supported catalyst.

E. ^{13}CO ON RHODIUM AND RUTHENIUM

The vast majority of ^{13}CO/Rh studies have been carried out with samples with CO/Rh > 1, which contain mostly molecular metal carbonyls. The spectrum for $Rh_6(CO)_{16}$ (Fig. 38a) consists of two axially symmetric powder patterns—one for on-top and the other for bridging carbonyls. The spectra in Figs. 38b–38d contain these same components plus a signal attributed to di- or multicarbonyls (an additional line at 124 ppm is due to spurious $^{13}CO_2$). These higher carbonyls are most likely distinct from the remaining "agglomerates" (CO-covered particles that contain more than a few atoms of Rh). The CO/Rh ratio for the agglomerates only can be found from the ^{13}CO NMR spectra (*129*). The spectrum in Fig. 38b has the value 1.01 for this ratio, and that in Fig. 38d is 0.52. It is surprising that the spectral fit of the latter does not require a metallic component. The spectrum for CO/Rh = 0.87 (Fig. 38c) has been analyzed in terms of four components, one of which, at 277 ppm, is attributed to metallic carbonyls (*106*). The spectrum in Fig. 38e has a FWHM of only 90 ppm, and its center is close to 270 ppm. These values (de Koster and Duncan, cited in Ref. *4*), taken at RT, are different from those in Fig. 38f, taken at 77 K (*2*). Figure 38f is for a sample with a dispersion determined by hydrogen chemisorption of 0.58 but no measured CO uptake (one expects CO/Rh < H/Rh for steric reasons).

In the case of supported ruthenium, there are not many ^{13}CO spectra for low-dispersion materials. One is for a sample with a dispersion determined by hydrogen chemisorption of 0.27 and another for a sample with a dispersion of 0.23. One of the spectra was taken at RT (*130*), and the other at 77 K (*2*); apart from the $^{13}CO_2$ peak, the two are similar (Fig. 39).

F. $^{195}Pt^{13}CO$ AND $^{13}C^{17}O$/Pd DOUBLE RESONANCE

This kind of experiment has been performed only by the Slichter group (Fig. 40). Their first study was of ^{13}CO/^{195}Pt, with ^{195}Pt being the observed nucleus (*131*). The B_1 of ^{13}C was about 11 G, and the FWHM ^{13}C linewidths were of the order of 30 G. In these early experiments, $t = \tau$ (see Section II.C), and both were varied (instead of the preferred method, whereby τ is constant and only t varies). For long times τ $(= t)$ the value of Sf was found to reach a plateau on the order 0.4 or 0.5. It was argued that this is due to insufficient B_1 on the ^{13}C frequency rather than to platinum atoms without ^{13}CO neighbors or to ^{13}C spin flips. The Sf was measured as a function of ^{195}Pt frequency, and the experimental values were multiplied

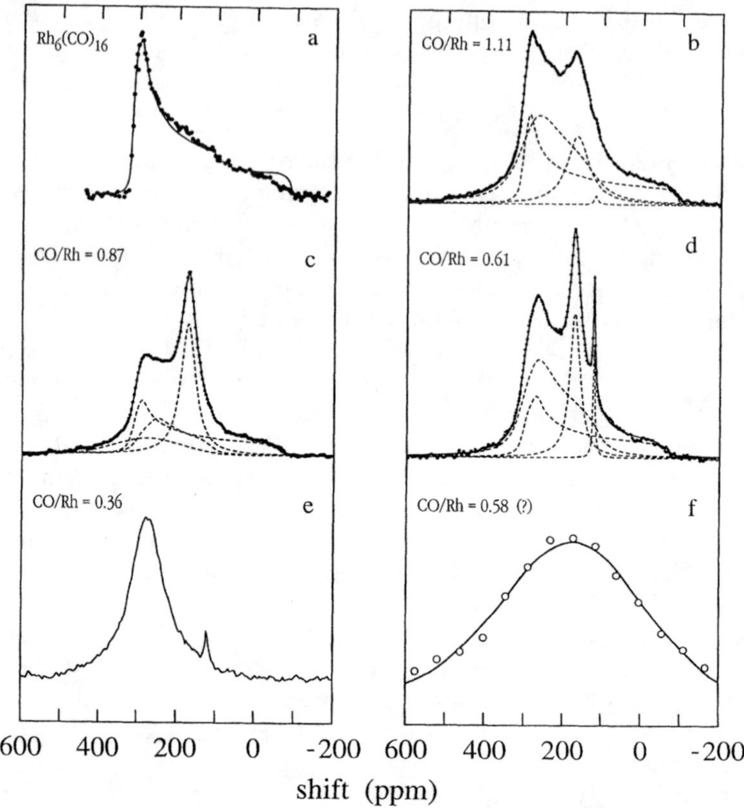

FIG. 38.   Spectra of ^{13}CO on rhodium. The vertical scales are all different. All spectra were obtained by the Fourier transform method at room temperature, except for f, at 77 K, obtained with the point-by-point method. (a) A molecular compound. The spectrum is a superposition of bridge and on-top powder spectra. Such powder spectra have also been fitted in b–d. [Reproduced with permission from Duncan (4). Copyright 1990 Elsevier Science.] (b) Rh/SiO_2 of very high dispersion (CO/Rh = 1.11). The spectral component fitted at 175 ppm is ascribed to (rotating) dicarbonyls; the small peak at 124 ppm is due to CO_2 gas. These two components are also fitted in d; in c, no CO_2 gas is seen. [Reproduced with permission from Duncan and Root (129). Copyright 1988 American Chemical Society.] (c) Highly dispersed Rh/SiO_2 (CO/Rh = 0.87). A metallic component has been fitted to the spectrum at 277 ppm; its relative area is 0.13. [Reproduced with permission from Compton and Root (106).] (d) Rh/SiO_2 with CO/Rh = 0.61. Except for the CO_2 peak, the overall spectrum looks similar to that of c, but no metallic component has been fitted. [Reproduced with permission from Duncan and Root (129). Copyright 1988 American Chemical Society.] (e) Rh/SiO_2 with CO/Rh = 0.36. Compare the shift to that of the metallic component in d. [Reproduced with permission from Duncan (4). Copyright 1990 Elsevier Science.] (f) Rh/Al_2O_3 with dispersion H/Rh = 0.58. The CO/Rh ratio has not been determined. Note that this line is broader and occurs at higher field than that in e.

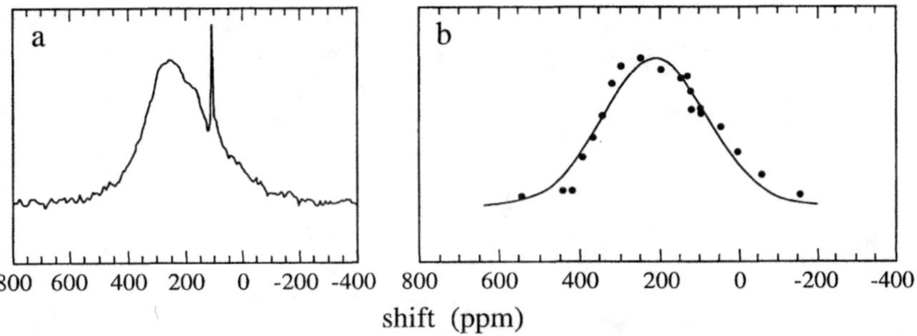

FIG. 39.   Spectra of ^{13}CO on Ru. Note the similarity of the two spectra. (a) Ru/SiO_2, dispersion H/Ru = 0.27, room temperature, Fourier transform spectrum. (The sharp peak is $^{13}CO_2$.) [Reproduced with permission from Duncan *et al.* (*130*). Copyright 1983 Elsevier Science.] (b) Ru/Al_2O_3, dispersion H/Ru = 0.23, 77 K, point-by-point method.

by a calculated, frequency-independent factor to correct for the insufficient B_1. After this correction, the Sf was approximately 1 in the ^{195}Pt spectral region below 1.096 G/kHz, and it decreased to zero when the field/frequency ratio was increased to 1.11 G/kHz. Therefore, no ^{195}Pt nucleus that resonates above 1.11 G/kHz has a ^{13}CO neighbor; these nuclei are inside the particles. If the $^{13}C–B_1$ correction factor has been determined correctly, then all nuclei resonating below 1.096 G/kHz have a ^{13}CO neighbor, which may be either in a bridge or in an on-top position. In the intermediate range, resonances are due to either type of ^{195}Pt environment. (Note that this statement implies that ^{195}Pt nuclei in different environments may have the same Knight shift.)

From the relative area of the ^{195}Pt spectrum that is sensitive to the double resonance, the fraction of Pt atoms that are bonded to a CO can be found. This fraction was 40% for a sample with a dispersion of 0.26 measured by hydrogen adsorption [0.22 from transmission electron microscopy (TEM)] (Fig. 40a), and it was 81% for a sample with dispersion of 0.76. No clean-surface ^{195}Pt NMR spectra have been published for these samples, but it was stated that the change in ^{195}Pt lineshape upon CO chemisorption is similar to that caused by hydrogen chemisorption (see Section VI.C).

The same experiment has been performed by Wang *et al.* (*105*) for ^{13}CO on a bimetallic catalyst with overall composition $Pt_{0.5}Rh_{0.5}$. The dispersion as determined from hydrogen chemisorption was 0.40, and from TEM micrographs it was 0.67. The experimental value of the SEDOR fraction Sf was corrected to give a value of 1 at the low-field maximum of the ^{195}Pt NMR spectrum. After this correction, it was found that 49 ± 7% of the platinum atoms were bonded to CO. The inverse double-resonance experiment (varying the observing frequency on ^{13}C, keeping the frequency of

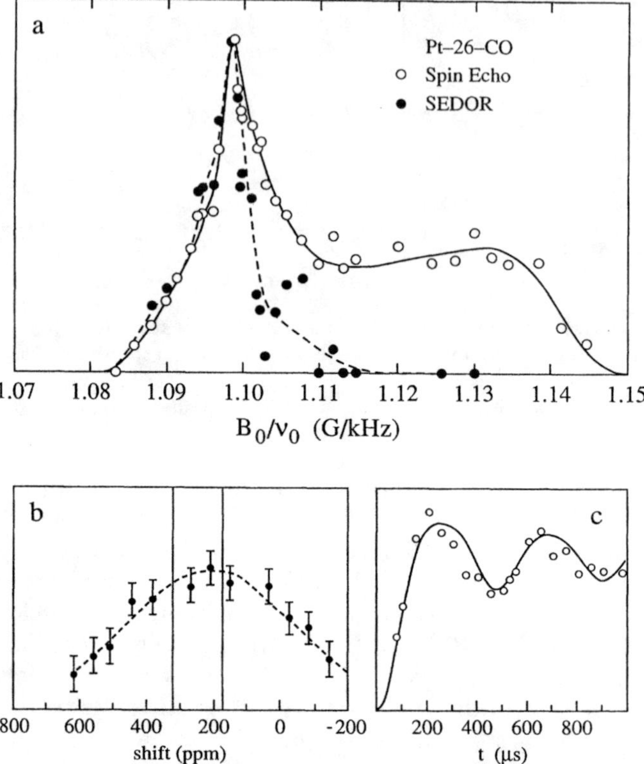

FIG. 40.　Three spin-echo double-resonance experiments (except the open circles in a, which show a single-resonance point-by-point spectrum for comparison). The spin-echo amplitude of the "observed" nucleus is recorded twice: with and without an additional pulse on the "decoupled" nucleus. Plotted is the difference between the two echo amplitudes (multiplied by an overall scaling factor). (a) Pt/Al_2O_3, covered with ^{13}CO. The observed nucleus is ^{195}Pt; the decoupled nucleus is ^{13}CO. The solid circles indicate that only part of the ^{195}Pt NMR spectrum is sensitive to the decoupling; these are ^{195}Pt close to a ^{13}C, i.e., in the surface of the particle. [Reproduced with permission from Makowka *et al.* (*131*). Copyright 1982 American Physical Society.] (b) $Pt_{0.5}Rh_{0.5}/Al_2O_3$, covered with ^{13}CO. The observed nucleus is ^{13}C; the decoupled nucleus is ^{195}Pt in the surface region of the ^{195}Pt NMR spectrum. This plot represents the spectrum of only those ^{13}CO groups that are near a ^{195}Pt atom in the metal surface. The center of this spectrum is closer to the center of the single-resonance ^{13}CO spectrum on pure Rh (right line) than on pure Pt (left line). [Reproduced with permission from Wang *et al.* (*105*). Copyright 1988 Royal Society of Chemistry. (c) $^{13}C^{17}O$ on Pd/Al_2O_3. Here the frequencies of both pulses are fixed, but the parameter t of the pulse timing is varied. The observed nucleus is ^{17}O, and the decoupled nucleus is ^{13}C. From the frequency of the oscillation the nuclear dipole–dipole coupling between ^{17}O and ^{13}C is determined, and from this the distance between the two nuclei is determined.

the ^{195}Pt pulse fixed at 1.10 G/kHz) (Fig. 40b) showed that the spectrum of ^{13}CO attached to platinum is essentially the same as the "single-resonance" spectrum of all the ^{13}CO in the sample (105). The shift of the single-resonance ^{13}CO line at 77 K is different on particles of pure platinum (325 ppm, left vertical line in Fig. 40b) and particles of pure rhodium (180 ppm, right vertical line) of similar dispersion. The double-resonance experiment (data points in Fig. 40b) detects only ^{13}CO close to a ^{195}Pt atom that resonates at 1.10 G/kHz. Nevertheless, the shift of this signal is closer to that of the single resonance on Rh (or on $Pt_{0.5}Rh_{0.5}$; Fig. 33) than that on Pt. For further discussion of these bimetallic catalysts, see Section VI.G.

Shore *et al.* (123) used ^{13}C^{17}O double resonance (^{17}O being observed) to determine the dipolar coupling between the two nuclei (Fig. 40c). Since any oxygen is much nearer to its bonding carbon than to the other molecules, the C–O bond length can be derived from this coupling; a value of 120 ± 3 ppm was found, which was slightly longer than the usual 112–116 ppm for on-top carbonyls and typical of bridging carbonyls. The catalyst was 5% Pd on η-Al$_2$O$_3$, with a dispersion found from hydrogen chemisorption of 0.19. The strengths of the RF fields and the position in the lines of the irradiation frequencies were not given, but it is likely that the fields were a fraction of the linewidths and that the irradiations were at the maxima of the lines as found by the point-by-point method: 540 ppm for ^{13}C and 470 ppm for ^{17}O (113). The experiment was performed at 4 K with the fixed-τ, variable-t method.

G. ^{13}CO NMR Spectroelectrochemistry

Slezak and Wieckowski (132) initiated NMR investigations of molecules (mainly ^{13}CO) adsorbed on an electrode formed of unsupported aggregates of platinum particles, which is part of an electrochemical cell incorporated in the NMR probe. This arrangement allows them to investigate *in situ* the effect of changing electrode potentials on the NMR spectra of the adsorbates (Fig. 41) (133, 134). Additional work has been done on *ex situ* samples (powdered electrode, adsorbates, and electrolyte in a sealed NMR ampule) or with open electrodes on the electrochemical cell (the cell was not connected to the potentiostat while in the NMR magnet, which reduces the noise introduced into the NMR experiment by the electrochemical setup). It has been shown that the adsorbate resulting from the potentio-static catalytic decomposition of methanol on such a platinum black electrode has the same NMR characteristics as ^{13}CO directly adsorbed on the electrode from gas dissolved for the purpose in the electrolyte (135). Moving the electrode potential beyond the CO oxidation threshold (0.4 V) creates

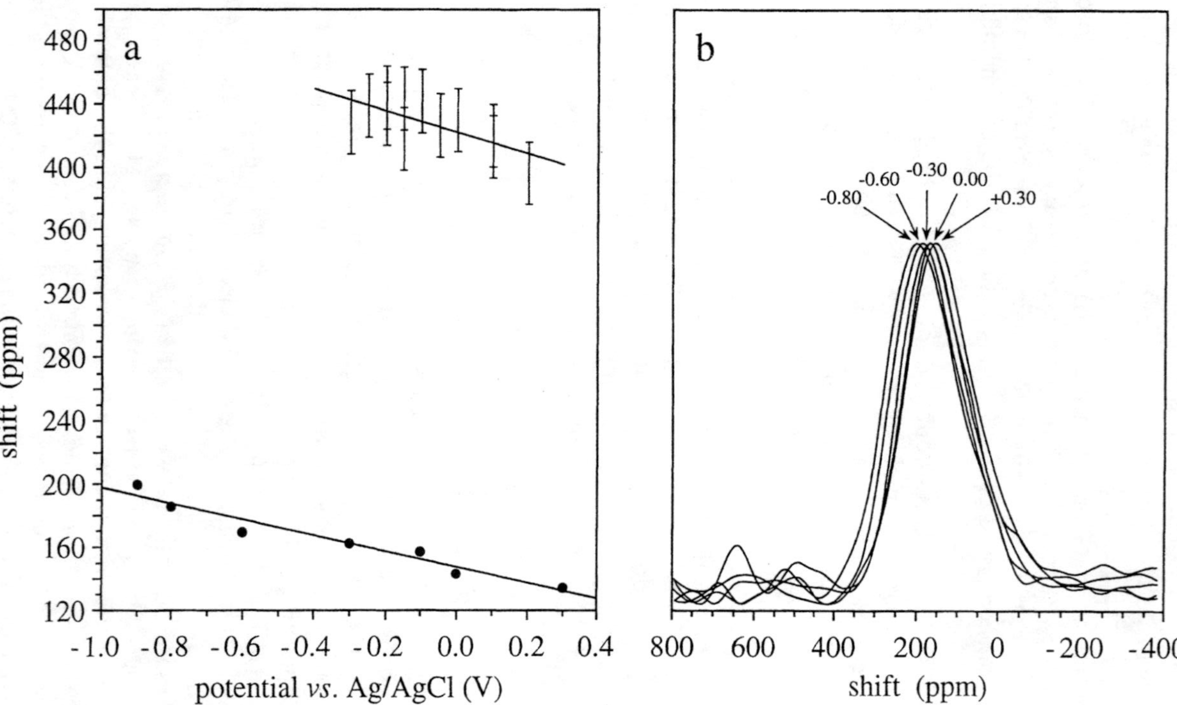

FIG. 41. ^{13}C NMR spectroelectrochemistry. The NMR probe is equipped with an electrochemical cell, and the ^{13}C shift of adsorbates on the working electrode (in this case platinum) is determined as function of the potential. The solid circles represent the shift of the ^{13}CN line (shown in b). The vertical bars represent ^{13}CO, obtained by electrochemical oxidation of methanol. The CO is stable only over a limited potential range (see Fig. 42), but the data suggest that in both cases decreasing the potential shifts the signal to a lower field. [Reproduced with permission from Wu *et al.* (*133*). Copyright 1997 Royal Society of Chemistry.]

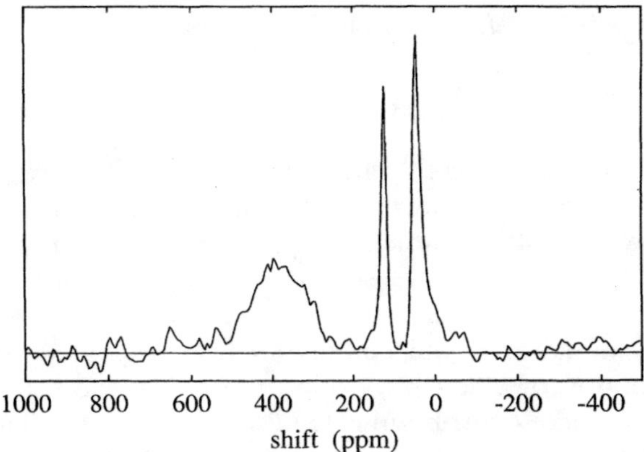

FIG. 42. ^{13}C NMR spectroelectrochemistry: Pt black electrode in 0.5 M H$_2$SO$_4$ with ^{13}CH$_3$OH at a potential of 200 mV with respect to Ag/AgCl. ^{13}CH$_3$OH at 49 ppm is used as an internal shift reference; the ^{13}CO/Pt signal is at 391 ppm, and some CO has been oxidized to ^{13}CO$_2$ (gas) at 133 ppm.

bubbles of ^{13}CO$_2$, with a narrow NMR line at 133 ppm (Fig. 42) (*134*). Over the potential range in which the ^{13}CO is stable, a small downfield shift of the NMR frequency seems to occur with decreasing electrode potential. The effect is clearer for adsorbed ^{13}CN, which is stable over a much larger potential range. The Korringa product T_1T was measured between 10 and 250 K in *ex situ* samples obtained with different decomposition times and potentials (*133*). All the results are single-exponential relaxation curves, with $T_1T = 80 \pm 10$ s K (the margins give the sample-to-sample variation rather than the uncertainty in the individual results). It is remarkable that even the ^{13}CO adsorbed from solution on this platinum black electrode has a single-exponential relaxation with $T_1T = 88 \pm 11$ s K over this temperature range, in sharp contrast to results for ^{13}CO adsorbed on supported platinum catalysts, for which the relaxation is nonexponential below room temperature, and $T_1T = 46$ s K above room temperature.

Yahnke *et al.* (*136*) obtained spectra for ^{13}CO adsorbed from solution onto a commercial platinum on carbon fuel cell electrode. The surface coverage was modified by galvanostatic oxidation in the NMR electrochemical cell, but the spectra were measured under open-cell conditions. At saturation coverage, the maximum of the line is at approximately 450 ppm; below half that coverage, the maximum shifts to 330 ppm. The value at saturation agrees well with those in Fig. 41.

V. ^{13}CO NMR: Discussion

A.   MOLECULAR AND METALLIC CARBONYLS

Calculations representing the chemisorption of CO on extended surfaces of Ni, Pd, or Pt indicate that both the surface metal atoms and the CO molecules have metallic character (*137*); the one-electron states on the chemisorbed CO molecule are spread over a continuum in energy and the Fermi energy, E_f, that separates the occupied states from the unoccupied states, lies in the continuum (*30*) (Figs. 43a and 43b). The spatial distribution of electrons around the CO site can be decomposed into π-like and σ-like contributions (*138*) (Figs. 43c–43f). The experimental ratio, (CO molecules)/(metal surface atoms), on extended surfaces is not larger than 0.5 (*139*). Interaction of CO with highly dispersed supported Rh or Ru yields molecular carbonyls with CO/(metal atom) ratios higher than 1 (*88, 102*). Gentle reduction of PdNaY zeolite, obtained by ion exchange of NaY zeolite, leads to the formation of very small Pd particles. After CO chemisorption, the IR spectra suggest the presence of (molecular) carbonyl clusters, possibly $Pd_{13}(CO)_n$ (*140*). No NMR results are available for these systems. The ^{13}CO spectrum of carbon monoxide chemisorbed on platinum in L zeolite has both metallic and molecular components (Fig. 35e). CO absorbed by platinum (Fig. 35b) or palladium (*127*) particles in cyclohexane solutions of isobutylaluminoxane also gives rise to molecular carbonyls, as judged from the values of the shift. The platinum result can be explained by the small ("subnanometer") size of the particles, but the palladium result (average size 1.8 nm) is surprising. A MAS–TOSS experiment characterizing silica-supported ^{13}CO/Pd of similar size has also resolved a sharp spectral feature at the same shift (Fig. 37b). It would be interesting if the nonmetallic character of these carbonyls could be confirmed by studies of the Korringa product T_1T.

When CO/(metal atom) ratios are available, the data suggest that, as a rule of thumb, values >1 indicate molecular carbonyls, values <0.5 mainly metallic carbonyls and in between are probably mixtures. From this rule we do not expect to find metallic carbonyls in the agglomerates (particles containing more than a few Rh atoms) in the sample of Fig. 38b, but we do for those in the sample of Fig. 38d (agglomerates CO/Rh $\approx$ 0.56). The spectrum, however, has been analyzed (*129*) without considering a possible metallic component, although it appears very similar to the CO/Rh = 0.87 spectrum in Fig. 38c (except for the ^{13}CO$_2$ peak). A T_1 analysis was difficult to perform (*129*), but it is speculated that if it could be done a metallic component would be found.

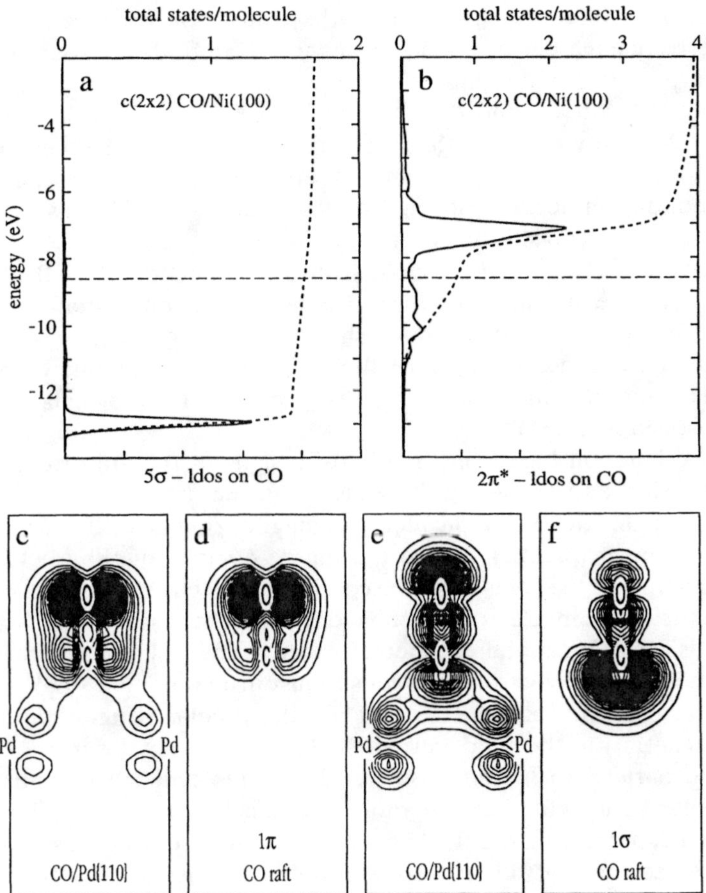

FIG. 43. Calculations for adsorbed CO. (a and b) Local densities of state on CO in a regular overlayer on a Ni(100) surface, full curves and bottom scales; energy integral of that LDOS, dashed curves and top scales. The Fermi level is indicated by the horizontal dashed line; there is a finite LDOS of both types at E_f, which will lead to metallic NMR behavior of the ^{13}CO. [Reproduced with permission from Hoffmann (*9*). Copyright 1988 American Institute of Physics.] (c–f) Charge densities for the 1π and 5σ states for a free CO layer (d and f) and the same layer on a Pd(110) surface (c and e). The symmetry of the wavefunctions remains clearly molecular-like after chemisorption. [Reproduced with permission from Hu *et al.* (*138*). Copyright 1995 Elsevier Science.]

B. Shift Mechanisms

Except for the fact that Knight shifts are often much larger than chemical shifts, there is no experimental distinction between them. [This is not quite true; in a very few favorable cases the distinction has been made by a double-resonance electron spin resonance (ESR)/NMR experiment.] For the interpretation of the shift values, however, it is convenient to distinguish between the contact and dipolar Knight shifts and the orbital chemical shift. It is usually assumed that the orbital contribution to the NMR shift of chemisorbed CO is approximately equal to the chemical shift in the corresponding molecular carbonyl, as described by Eq. (7). Any experimentally observed extra shift must then be due to the K_σ and K_π contributions, similar to Eq. (16); this is discussed further in Section V.D. If we accept that the average orbital shift is the same in molecular and metallic carbonyls, then their orbital shift anisotropies should also be equal. How is it possible that (in the absence of motion) the ^{13}CO line for metallic carbonyls is narrower than the chemical-shift powder pattern for molecular carbonyls (cf. Figs. 38a and 38e)?

An explanation based on the opposite signs of the anisotropies of the chemical shift and of the dipolar Knight shift has been proposed (*4*). This dipolar Knight shift is often called a demagnetizing field. It is a symmetric traceless tensor quantity due to the dipole–dipole coupling between the nuclear spin and the spins of all the conduction electrons, with the electrons being outside the nucleus (in a cubic environment, e.g., ^{195}Pt in bulk platinum, it is zero). We model the metal surface as an ideal geometrical surface, with a continuous electron spin density just inside it and a single localized ^{13}CO nuclear spin just outside it. Then the dipolar Knight shift is more paramagnetic for the field perpendicular to the metal surface than for the field parallel to it. In an isolated linear molecule, the chemical shift representing the field along the molecular axis is always more diamagnetic than that representing the field perpendicular to it. For purposes of discussion, we assume that this pattern still holds for the chemisorbed CO and that the molecule is absorbed with its axis perpendicular to the metal surface. In this case, the field direction for the most diamagnetic chemical shift coincides with that for the most paramagnetic Knight shift, thus diminishing the overall shift anisotropy. However, perhaps the difference in the bonding situation simply invalidates the comparison between orbital shifts of molecular and metallic carbonyls.

C. TEMPERATURE-DEPENDENT LINESHAPES

When the temperature increases, the ^{13}CO/Pt (Fig. 35a) and ^{13}CO/Pd lines become narrower and shift to lower field. There are indications that the ^{13}CO/Rh spectra show the same characteristics. The metallic resonance inferred from the spectral fit in Fig. 38c occurs at 277 ppm and has a FWHM of 86 ppm and a spin lattice relaxation time of 0.18 s (*106*). This is very similar to the single resonance observed in Fig. 38e [de Koster and Duncan cited in Ref. (*4*)]: position, 270 ppm; FWHM, 90 ppm; and $T_1 = 0.2$ s. Comparison with the 77-K spectrum in Fig. 38f (position, 180 ppm; FWHM, 400 ppm) suggests that on Rh, just as on Pt and on Pd, the ^{13}CO line narrows and shifts to lower field when the temperature increases. The two ^{13}CO/Ru spectra for Ru of low dispersion (Fig. 39) are very similar at 77 K and at RT. This near agreement may be due simply to the fact that on Ru all densities of states are lower than on the other metals and that therefore all effects are much smaller, or else the temperature may have to be raised.

For a symmetric lineshape, the maximum of the line and its first moment coincide. Increasing the motion, but without changing the sites among which motion occurs, cannot change the first moment, but it can shift the maximum of an asymmetric line as a side effect of line narrowing. If, on the other hand, a change in temperature causes a change of sites, then the first moment will change as well. In practice, it is not easy to decide whether a first moment varies with temperature without making some assumption about the shape of the line. However, in at least one sample (*125*), the temperature-dependent shift occurs with little change in lineshape (Fig. 44). This observation strongly favors the proposal of Becera *et al.* (*125*) that generally the shift is due to a change in the type of bonding of the ^{13}CO molecules to the metal.

The sharpening of the line with increasing temperature shown in Fig. 35a is ascribed to the onset of a molecular motion, which may be unrelated to the overall change in bonding. Increasing the temperature from 489 to

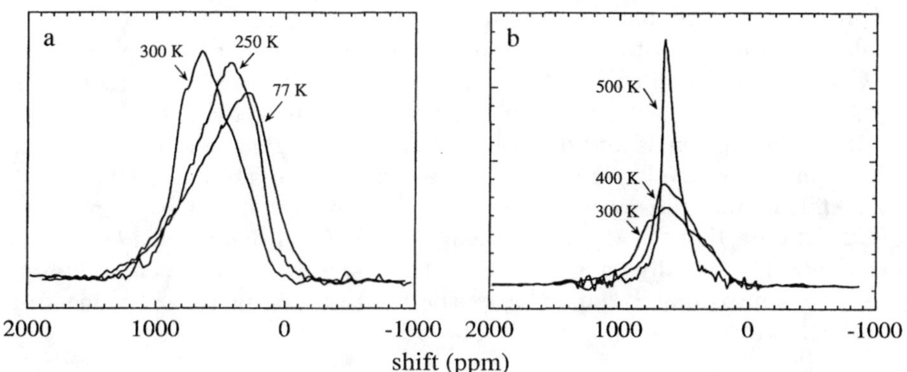

FIG. 44.   Spectra of ^{13}CO on Pd/alumina (dispersion = 0.52) as a function of temperature. In this sample the shift of the line (a) occurs in a different temperature range than that of the narrowing (b) (cf. Fig. 35a). [Reproduced with permision from Becerra *et al.* (*125*). Copyright 1993 American Chemical Society.]

550 K does not narrow the line further. It is thought that at these temperatures the exchange of CO between different metal particles is still slow (*112*) but that each molecule diffuses rapidly over the surface of the particle to which it is attached. Different particles (especially those differing in size) have different collections of possible adsorption sites and therefore give different average Knight shifts for the ^{13}CO line. Even the (nonmetallic) ^{13}CO resonance in the subnanometer Pt methylcyclohexane colloid has a 50-ppm linewidth (Fig. 35b), and the saturation-transfer linewidth on the large Pd particles in the PVP colloid (Fig. 35c) is comparable to that on static supported samples (Fig. 44). Since the tumbling motion of the metal particles in colloids should be fast enough to average out shift anisotropies, there is again a distribution of shifts, specifically of chemical shifts in the case of Pt.

For temperatures above 300 K, the time scale for motion of ^{13}CO on Pt can be found from several experiments (see Section V.E). From the lack of narrowing of the lineshape and the single-exponential relaxation, the hopping time is estimated to be between 10 μs and 100 ms (see Section IV.C). At 150 K, however, the motion should be frozen out so that any shift anisotropy that might exist should show up in the low-temperature MAS experiment (*117*). The rotor frequency was about 110 ppm of the Larmor frequency, and the (static) FWHM ^{13}CO linewidth was slightly more than 300 ppm. Nevertheless, the MAS did not change the spectrum, showing that there is a wide distribution of isotropic shifts—but without detectable shift anisotropies. This situation is reminiscent of the discussion of Figs. 38a and 38e given in Section V.B.

Several different experimental results [MAS–TOSS lineshape (*124*), T_2 relaxation (*125*), and T_1 relaxation (*126*)] indicating that the ^{13}CO molecules occur in at least two groups of environments have been reported for ^{13}CO/Pd. The data have been interpreted by assigning the group resonating at lower field (450–700, 650, or 850 ppm, according to the three methods) to bridging carbonyls and that resonating at the higher field (192, 400, or 500 ppm, respectively) to on-top carbonyls. The two-site interpretation of the shift of the maximum of the ^{13}CO line on Pd with a low dispersion (*125*) implies that the C–O bond length of 0.12 nm determined by Shore *et al.* (*123*) is actually an average over bond lengths in bridges and on-top sites in approximately equal proportions instead of a typical value for bridged carbonyls, as originally proposed.

D. LOCAL DENSITY OF STATES

The local density of states near E_f on chemisorbed CO can be projected on one kind of σ-like and two kinds of π-like orbitals (Figs. 43a and 43b).

Its value is relatively low so that we neglect enhancement effects in the equations for the Knight shift and the relaxation rate. For ^{13}CO the equivalents of Eqs. (16) and (17) are as follows:

$$^{13}K = \mu_0\mu_B(D_\sigma(E_f)\,^{13}H_{hf,\sigma} + D_\pi(E_f)\,^{13}H_{hf,\pi}) + {}^{13}K_{orb} \tag{24}$$

$$^{13}S(^{13}T_1 T)^{-1} = {}^{13}K_\sigma^2 + {}^{13}K_\pi^2/2 + {}^{13}X_{orb}. \tag{25}$$

The dipolar contributions and the anisotropy of the orbital parts have been neglected; the notation X_{orb} simply reflects our ignorance of the details of the expression for the orbital relaxation. If the applied field B_0 is parallel to the molecular axis, the relaxation is associated with jumps of an electron from a σ or π orbital just below E_f to a π or σ orbital just above E_f. If the field is perpendicular, we must consider jumps between a σ orbital and one type of π orbital as well as those between the two different types of π orbitals. (All these jumps change only the orbital symmetry, without electron spin flip.) Currently, no values are known for these σ–π and π–π transition matrix elements. It is clear, however, that a higher E_f-LDOS will lead to a higher value for $^{13}X_{orb}$.

As mentioned in Section V.B, it is usually assumed that K_{orb} can be estimated from data representing the corresponding molecular carbonyl. The hyperfine fields $H_{hf,\sigma}$ and $H_{hf,\pi}$ can be estimated theoretically from calculations for the molecule or experimentally from ESR hyperfine couplings in radicals that have been identified as CO$^+$ (an unpaired electron in the 5σ) (141) or CO$^-$ (an unpaired electron in the $2\pi^*$) (142). From the calculated charge density at the nucleus for one direction of spin in the isolated CO molecule, $H_{hf,\sigma}$ is 398 kG at the carbon and 5.24 kG at the oxygen (143). The π-hyperfine field is zero by virtue of symmetry on both sites. The experimental values for the ^{13}C site are $H_{hf,\sigma} = 363$ kG and $H_{hf,\pi} = 17$ kG. No signs have been experimentally determined; in the following we suppose that the experimental π-hyperfine field is different from zero due to core polarization (the effect in a magnetic field of electron–electron interactions between valence and core electrons), which (usually) has a minus sign.

To show some orders of magnitude, suppose that $K_\sigma = 200$ ppm. This value corresponds to a susceptibility of 3.1 μemu mol^{-1} (i.e., per mol of CO) or a σ LDOS of 1.3 states per CO molecule and per Rydberg. From Eq. (25) we find $T_1 T = 104$ s K. If $K_\pi = -200$ ppm, then the susceptibility is 65.6 μemu mol^{-1}, the π LDOS is 27.6 Ry^{-1}, and $T_1 T = 208$ s K. With the reasonable hypothesis that the π-like shift should be much smaller in absolute value than 200 ppm, we estimate the total density of states at E_f as between 1 and 10 states Ry^{-1}. The extended Hückel calculations (30) for high-coverage CO on Ni(100) give a total density of states at the Fermi energy of approximately 1.4 Ry^{-1}, with a fraction of σ-like states ≈ 0.2 [these are very approximate values obtained by estimating the slopes in a figure of Ref. (*30*)], which is approximately the same order of magnitude. In this estimate, the π-like density of states is about four times larger than the σ-like density of states. Since $H_{hf\pi}$ is 20 times smaller than $H_{hf,\sigma}$, the π contribution to the Knight shifts and to the Korringa product is predicted to be small. However, the orbital relaxation, denoted $^{13}X_{orb}$ in Eq. (25), may be considerably enhanced by the high π LDOS at E_f.

To get a sense for the values of the susceptibilities involved, a comparison can be made with the change of the magnetic susceptibility of Pt surface atoms upon chemisorption of hydrogen, as derived from ^{195}Pt NMR (corresponding data for CO are not available, but it is believed that they should be similar). From data for K_s and K_d given in Ref. (*38*) (see Section VI.C), we calculate a diminution of 18 μemu mol^{-1} (i.e., per mole of surface Pt atoms, not of all the atoms in the sample). Results based on a more complicated data analysis (*84*) suggest a change of about twice this value. Therefore, we expect that CO chemisorption on Pt will not be detectable by susceptibility measurements.

If it is assumed that the 200 ppm shift is entirely due to the spins of σ-like electrons, then the calculations (*143*) predict a Knight shift of 2.6 ppm on the ^{17}O. Unpublished point-by-point low-temperature data for C^{17}O on Pd (*113*) show a peak with position 480 ppm and an FWHM of 300 ppm. The position for bridging carbonyls in solutions of $Co_4(CO)_{12}$ is 501 ppm; therefore, the Knight shift of C^{17}O on Pd is small compared with that of ^{13}CO.

E. DIFFUSION

The phenomenon of surface heterodiffusion on metals is very complex and cannot be properly introduced here (*144*). Simply stated, there is no such thing as "the" diffusion coefficient of the systems of interest here, molecules at high coverages on very heterogeneous surfaces. On extended platinum surfaces several types of CO diffusion coefficient have been measured by different techniques (*145–152*), and the results for D (at RT, and for coverages typically between 0.1 and 0.5 monolayer) fall in the range $3 \times 10^{-7} \geq D \geq 3 \times 10^{-10}$ cm^2 s^{-1}, and the activation energies E_a are between 6 and 13 kcal mol^{-1}. In contrast, a field-emission experiment (*145*) gives a D value of approximately 3×10^{-14} cm^2 s^{-1} with $E_a = 14.5$ kcal mol^{-1}.

The NMR results for ^{13}CO/Pt in Fig. 36 give values at the low end of the range for D, and the two values for E_a have the expected magnitude. In general, the diffusion must, at a well-chosen time scale, be slower on small particles than on single-crystal faces (because of the difficulty of hopping over edges), but it would be difficult to propose a mathematical form for the diffusion equation. The numerical results for D derived from the experiment will depend on the assumed properties of the diffusion equation; the values of E_a are less subject to this restriction. An uneasy feature of the NMR result is the assumption (*115*) that some nuclei do not change their frequency during the waiting period τ_{exch}. Since the experimental τ_{exch} is of the order of the spin lattice relaxation time T_1, this means that these nuclei do not participate in the "averaging" of the T_1, which is believed to be the reason for the observed single-exponential relaxation curves (*118*).

F.   Spectroelectrochemistry

The spectroelectrochemical results show interesting parallels and differences with respect to those on heterogeneous catalysts. The foremost result probably is that the detected COs are metallic, not molecular, carbonyls. This result is of interest in the description of the well-known effect of the electrode potential ϕ on the infrared vibration frequency ν of an adsorbed molecule; for CO on polycrystalline platinum electrodes in aqueous electrolyte, the shift $(d\nu/d\phi)$ is 30 cm^{-1} V^{-1}. This has been attributed both to the effect of ϕ on the COs bonding and to the interaction of the molecule with the local electrostatic field (the vibrational Stark effect). It has been argued that these are different ways of describing the same phenomena (*153*). However, if the COs are really "a piece of the metal," then the local electrical field should be zero and the bonding interpretation should be favored. Increasing the electrode potential amounts to pulling out electrons near the Fermi energy of the neutral conductor. Pulling out electrons from orbitals that (near the CO) have 5σ character decreases the stretching frequency; pulling them from $2\pi^*$ orbitals increases it. The sign of $(d\nu/d\phi)$ implies that the change in intramolecular C–O bond strength is dominated by the depletion of $2\pi^*$ states near the Fermi energy; this, however, is a "total charge" effect, in principle independent of the E_f LDOS. However, according to a phenomenological theory (*154*), the intrinsic infrared linewidth (the inverse of the vibrational lifetime) representing adsorbed CO is proportional to the $2\pi^*$ partial density of states at the Fermi energy. Within experimental precision, the variation $(d\tau/d\phi)$ of lifetime with electrode potential is undetectable in an aqueous electrolyte, but it is slightly negative in an acetonitrile electrolyte, in which a larger range of potentials are accessible (*155*). The effect of the electrolyte is not expected to be significant (*155*). According to this theory, the sign of $(d\tau/d\phi)$ shows that an increase in electrode potential increases $D_\pi(E_f)$. Therefore, the $2\pi^*$ contribution to the Knight shift of ^{13}CO should be more negative with increasing potential, which correlates with the tendency of the ^{13}CO and ^{13}CN results in aqueous electrolytes (Fig. 41). This effect is difficult to detect because it involves the $2\pi^*$ hyperfine field, which is much smaller than the 5σ one.

The observed linewidths are smaller than those for CO chemisorbed on highly dispersed supported catalysts (*133*). This comparison probably shows that the heterogeneity in adsorption sites is less on the electrode materials than on the catalysts, as can be expected when the grain size of the platinum black in the former is sufficiently large. As mentioned in Section IV.C, there are indications that lines characterizing ^{13}CO on oxide-supported platinum with a low dispersion are narrower than those characterizing ^{13}CO on oxide-supported platinum with a high dispersion material. Similarly, the "surface peak" in ^{195}Pt NMR becomes wider with increasing dispersion (Section VI.B).

VI. ^{195}Pt NMR

A. ^{195}Pt NMR Data Analysis

It follows from the LDOS description (Section I.H) that at a single resonance frequency (fixed value of K) one might find signals from nuclei with many different combinations of s- and d-like LDOS. Each such combination would give rise to the same K but a different T_1. Therefore, generally, the spin lattice relaxation curves measured at a certain resonance frequency should be nonexponential. To within the experimental precision of the earlier work, such nonexponentiality has not been observed, and the data have been discussed in terms of a "most likely site" that dominates the NMR behavior at a given resonance frequency. More precise T_1 measurements sometimes show nonexponential decays for the signals from platinum surfaces. The Korringa mechanism for spin lattice relaxation by conduction electrons in a metal requires that such decay curves obey time–temperature scaling because the spin lattice relaxation rate T_1^{-1} for each individual nucleus is proportional to temperature T so that $T_1 T = C$. At a given resonance position and temperature, we measure a series of recovered signal amplitudes A_i as a function of the relaxation interval τ_i. When the A_i are normalized by the fully relaxed amplitude, the Bloch equation for a single site is

$$1 - A_i \exp(-\tau_i/T_1), \tag{26}$$

and using the Korringa relation,

$$1 - A_i = \exp(-\tau_i T/C). \tag{27}$$

If the relaxation curve is a sum of N different exponentials (corresponding to different sites), there are N different constants C, but when plotted as normalized A_i vs $(\tau_i T)$, curves taken at different temperatures collapse into one (*84*), as shown in Fig. 45. It is usually impossible to determine a value of N from the data. Most relaxation curves can be described by a sum of

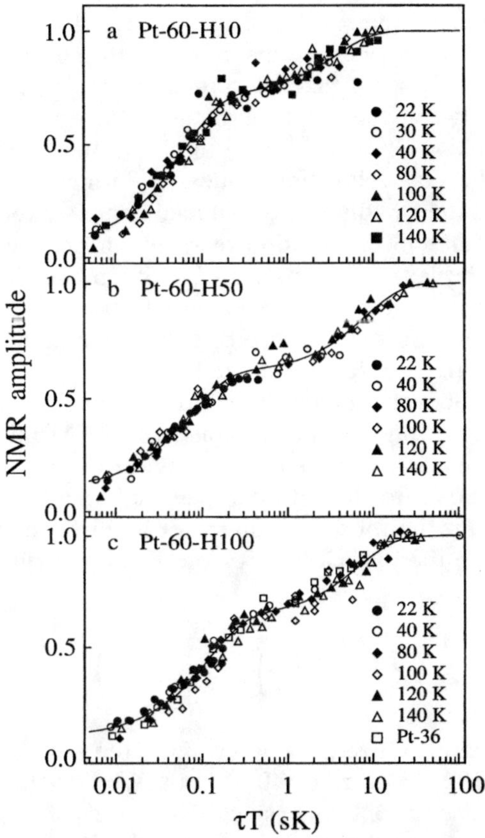

FIG. 45. Time/temperature scaling of ^{195}Pt spin lattice relaxation curves in a saturation-recovery experiment. At each temperature the raw amplitude data are first normalized by the amplitude of the fully relaxed signal so that all values fall between zero (saturation at short times) and 1 (full recovery at long times). The time points τ are multiplied by the temperature at which the relaxation curve was obtained. If the relaxation is governed by the Korringa process, the scaled points fall on a temperature-independent curve, even if the relaxation is not simply exponential, as in the cases shown here. The sample is Pt/TiO_2 of dispersion 0.60 (determined by electron microscopy) at several hydrogen coverages (calculated from the dispersion): 0.1, 0.5, and 1.0 monolayers. The squares in c show data at 110 K for another Pt/TiO_2 catalyst of dispersion 0.36.

two exponentials, having temperature-independent amplitude ratios. While it is clear that nuclei in at least two different environments resonate at the frequency under consideration, it is impossible to demonstrate that there are not more than two environments. Therefore, one should be very cautious in interpreting the amplitude ratio of the two exponential decays as a ratio of "site occupation."

To derive an estimate for the widths and for the average values of the LDOS distribution on the surface sites, we consider the relaxation curve measured at the frequency of the maximum in the spectrum as representative of the surface sites; we take the range of their spin lattice relaxation times as lying between the "fast" and the "slow" values of the double-exponential fit. We use Eqs. (15)–(17) to obtain the LDOS corresponding to the fast and the slow relaxation values. The range between the LDOSs of the slowly and the rapidly relaxing nuclei is now taken as a measure of the width of the LDOS distribution over all surface sites. To obtain the average values, the fast and the slow LDOSs are weighted by the corresponding fractions of the double-exponential fits.

The relation between particle size distribution and NMR spectrum has been modeled in the NMR layer model (*156*). Here I describe a simple version; a more elaborate one has been proposed by Makowka *et al.* (*157*). We start by constructing size histograms from TEM micrographs. To do this, it is customary to consider the images on the micrograph as circles with diameter d equal to that of effectively spherical metal particles that cause the image. Furthermore, this diameter is converted to the total number N_T of atoms in the particle by (for the FCC structure, four atoms per unit cell)

$$N_T = \frac{2\pi}{3}\left(\frac{d}{a}\right)^3 = \frac{\pi\sqrt{2}}{6}\left(\frac{d}{2r}\right)^3, \tag{28}$$

where a is the (bulk) lattice parameter and r the hard-sphere radius. For platinum, $a = 0.392$ nm and $2r = 0.277$ nm. While both hypotheses seem reasonable for particles containing several hundreds of atoms, a justification for their use in very small particles is lacking (*158*). It is general practice to ignore these complications and to accept some uncertainty in the true diameter of the smallest particles. In the extreme, Eq. (28) states that a single atom will yield an image with $d = 2.2r$. In a hard-sphere model for both zeolite and platinum, the biggest FCC particle that fits into the supercage (cage diameter 1.3 nm) contains 31 atoms (*159*); its image, according to Eq. (28), has a diameter of 0.96 nm. Under our experimental conditions (*158*), comparison of electron micrographs for untreated and platinum-loaded zeolites (Fig. 46) shows that the smallest features that can be reason-

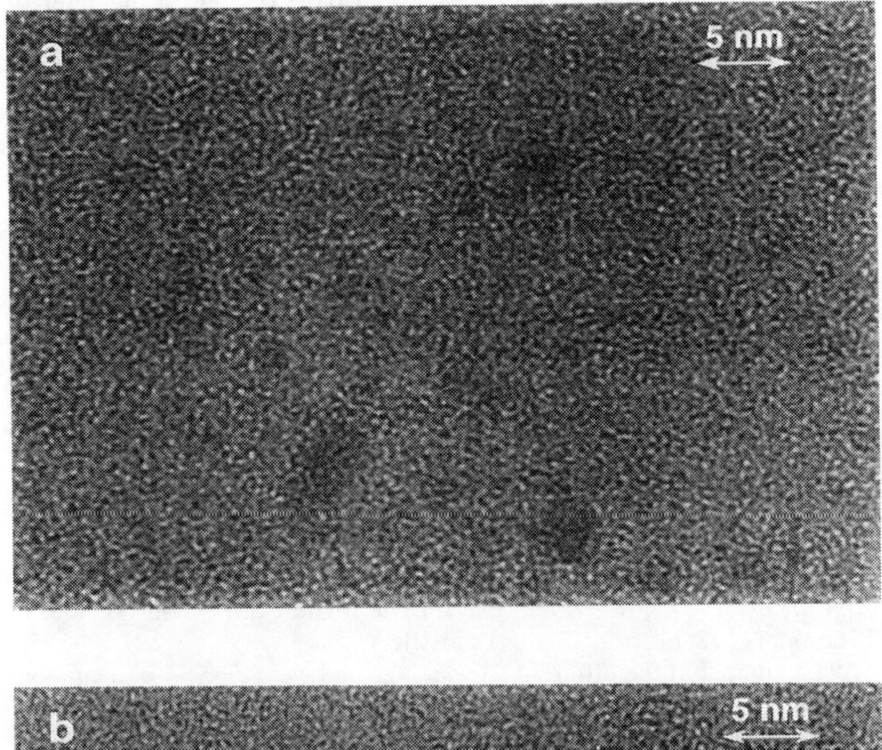

FIG. 46. Electron micrographs of Pt-loaded and unloaded zeolite. (a) Sample containing 8.1% by weight of Pt with mean particle diameter of 1 nm. (b) As-received zeolite; the largest features are smaller than 0.5 nm.

ably attributed to the presence of platinum particles have a diameter of approximately 0.5 nm. For oxide-supported particles contrast is often a problem, and the lower limit is closer to 1 nm.

For the NMR layer model, the atoms in a small-particle sample are divided into groups belonging to different atomic layers: the surface layer, the subsurface layer, and so on. To find the fraction of atoms in each group, the particle size histograms obtained by electron microscopy are interpreted in terms of FCC cubooctahedra. The smallest such particle contains 13 atoms; the next larger one contains 55. To convert a histogram into layer statistics, we use Eq. (28) to find an average N_T for each class of the histogram; next we determine the corresponding (noninteger) number of cubooctahedral layers l from the equation (*156*)

$$N_T = \frac{10}{3} l^3 - 5 l^2 + \frac{11}{3} l - 1 \tag{29}$$

and the number of surface atoms from

$$N_S = 10 l^2 - 20 l + 12. \tag{30}$$

The number of atoms in subsurface layers is obtained by replacing l in Eq. (30) by $(l - 1)$, $(l - 2)$ and so on. The highest dispersion possible in this model is 12/13 for a 13-atom particle.

We assume that the different sites in a given layer (Fig. 47) of a cubooc-

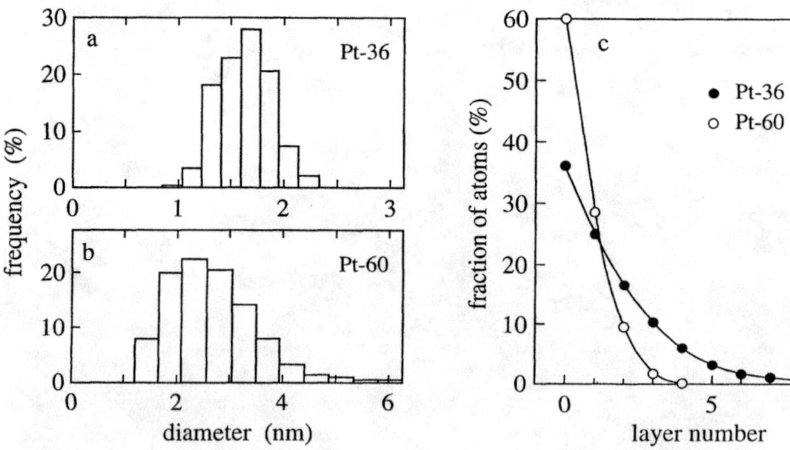

FIG. 47.   Particle size distribution and layer statistics. (a and b) Size histograms for two samples of Pt/TiO$_2$. (c) Distribution of Pt atoms over the different atomic layers. Layer number 0 is the surface, and the fraction of atoms in layer 0 is the dispersion of the sample. Layer number 1 is the subsurface layer, and so on.

tahedral particle are sufficiently similar that the resonance frequencies of all nuclei in the same layer are relatively close to each other on the scale of the total spectrum width. The superposition of NMR signals from a given layer is called a peak; its (inhomogeneous) width is supposed to be of the order of 1 MHz. A similar assumption is the basis for the correlation between low-field NMR intensity and particle size, already found in the early work of Rhodes *et al.* (*160*). For convenience, these peaks are taken to be Gaussians, completely characterized by the positions of their maxima in the spectrum, by their widths, and by their integrals. The integral must be proportional to the relative number of atoms in the corresponding layer (given in Fig. 47). For the position of the maximum as a function of layer number, we impose a behavior similar to that found for the resonance frequency of near neighbors to an impurity in very dilute alloys (*18*) (see Eq. 31). Finally, the width of the Gaussian is considered to be a freely fittable parameter, but it must be sample independent for not too different samples (as in the case of Fig. 48).

The maximum of the peak corresponding to the nth layer is assumed to

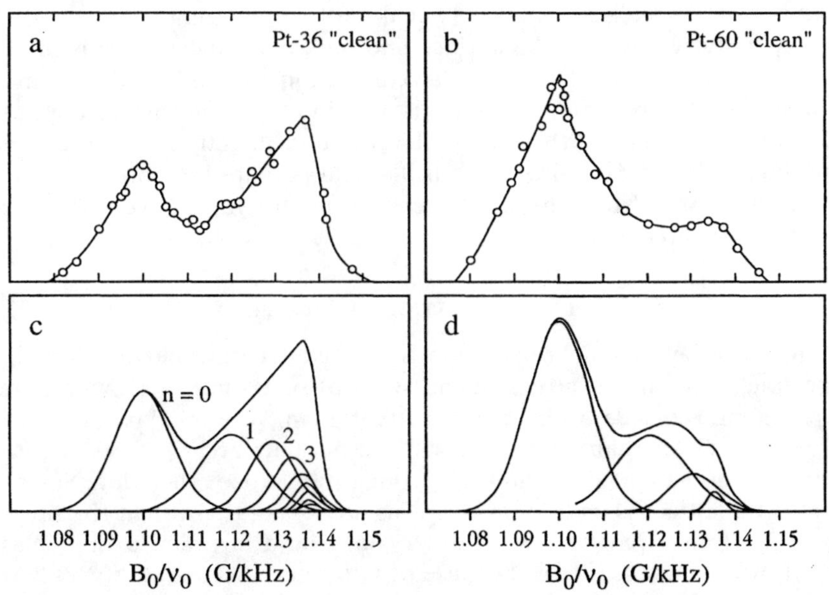

FiG. 48. ^{195}Pt NMR spectrum and layer statistics. (a and b) Point-by-point ^{195}Pt NMR spectra (under clean-surface conditions) for the samples in Fig. 47. (c and d) Fits of the spectra in a and b by a superposition of Gaussians, each representing the NMR line of nuclei in a given layer. The area of each Gaussian is given by the corresponding fraction in Fig. 47c.

occur at a Knight shift K_n (K_0 is the Knight shift of the surface, and K_∞ is that of the infinite solid) obeying the relation

$$K_n - K_\infty = (K_0 - K_\infty) \exp(-n/m), \tag{31}$$

where the dimensionless constant m represents the "healing length" for the Knight shift, expressed in units of a layer thickness (0.23 nm). According to this assumption and to the data in Fig. 47c, the NMR spectrum of Fig. 48b should consist mainly of a superposition of three Gaussian peaks, with relative areas 0.60, 0.29, and 0.09. The spectrum of Fig. 48a contains these same Gaussians (having the same positions in the spectrum and the same widths), but now with relative areas 0.36, 0.25, and 0.17, and several more Gaussian peaks. Fits according to this principle are shown in Figs. 48c and 48d. They correspond to $K_0 = 0$ and $m = 1.35$ (the corresponding healing length is 1.35×0.23 nm) in Eq. (31). The experimental spectrum of Fig. 48a is well reproduced. The agreement with the experimental results of Fig. 48b is only qualitative, no doubt because of the small number of parameters in the fit. However, for our main purpose here (to derive a simple description of the effect of the particle size distribution on the NMR spectra), the agreement is sufficient.

In our fit, the subsurface ($n = 1$) peak of the clean sample falls approximately halfway between the surface and bulk resonances. This is in very good agreement with a five-layer slab calculation (*70*) and shows that more than half of the spectrum contains information from the surface region. It should be noted that NMR layer model considers directly the layer-to-layer variation of the NMR shift, whereas it would perhaps be more reasonable physically to start from a hypothesis concerning the variation of the density of states (Section VI.C).

B. Determination of Dispersion

The NMR layer model can obviously be used to determine the dispersion (the fraction of metal atoms in the surface) of a sample. However, two experimental considerations are important. First, very small particles (e.g., those with fewer than 25 atoms) will not be metallic in the NMR sense; they have no Knight shift and no Korringa spin lattice relaxation. Since all other spin lattice relaxation mechanisms are less efficient than Korringa's, the T_1's of such particles will be very long, and in practice their NMR signals will be unobservable because of saturation. Second, the [195]Pt NMR spectra are obtained by measuring the amplitude of spin echoes, created by a pair of RF pulses. The echo amplitude depends on the pulse spacing through an (effective) transverse relaxation time T_2, which varies across the spectrum. Usually, spectra are plotted without correcting for this effect, but in quantitative work this must be done. Since an explicit determination of T_2 across the spectrum is time-consuming, one often applies the rule of thumb that states that the measured amplitudes must be scaled linearly with frequency, with the high-field end being multiplied by a factor of 1.20 for typical values of the pulse spacing.

Results of the NMR method have been compared (*161*) with those from hydrogen chemisorption, oxygen titration, and electron microscopy for four oxide-supported platinum catalysts prepared by different methods in different laboratories (Table I, Fig. 49). The one labeled Al 1.5.1 was prepared at the ICP-EPFL, EuroPt-1 was surface treated at the Shell Research and Technology Center Amsterdam (SRTCA) and provided in a sealed ampoule, Pt-46-clean is based on published data (*160*) by Slichter's group, and PR2 is a highly dispersed catalyst prepared at the Technical University of Eindhoven. These catalysts, of very different origin and methods of preparation, give excellent correlation between the values found by the two physical methods of dispersion determination. The agreement between the hydrogen and oxygen chemisorption results is not impressive, and several explanations for such discrepancies have been proposed (*81*). The NMR surface peak in very small particles (Fig. 49d) is broader than in larger ones (Fig. 49c); this difference probably reflects the frequency distribution of different surface sites (corners, edges, and faces).

When zeolite Y is loaded with platinum by generally accepted methods of preparation, the ^{195}Pt NMR signal usually indicates the presence of particles larger than the supercage (Fig. 50). Electron microscopy of 40-nm-thick slices prepared by ultramicrotomy has shown that these particles are inside the zeolite matrix, and the histograms are given in Fig. 51. From Figs. 51a and 51f, and with the Gaussian peaks fitted in Fig. 48, we "predict" the spectra in Fig. 52. The samples contained between 0.5 and 1 Pt atom per supercage so that the fraction of cages damaged by the growth of the

TABLE I

Dispersions of Four Catalysts, Measured by Four Different Methods: H_2 Chemisorption, H_2/O_2 Titration, TEM, and NMR (Spectra shown in Fig. 49)[a]

				Dispersion			
Sample	Preparation	Support	Loading	H_2	H_2/O_2	TEM	NMR
Pt-46-clean	Impregnation	η-Alumina	10	46	—	61	52–62
EuroPt-1	Ion exchange	Silica	6	91	56	60	55–65
Al1.5.1	Citrate reduction	β-Alumina	5	34	15	30	30–40
PR2	Impregnation	Alumina	3	131	65	>80	85–100

[a] From Bucher *et al.* (*161*). Dispersions in atom%; loadings in weight%.

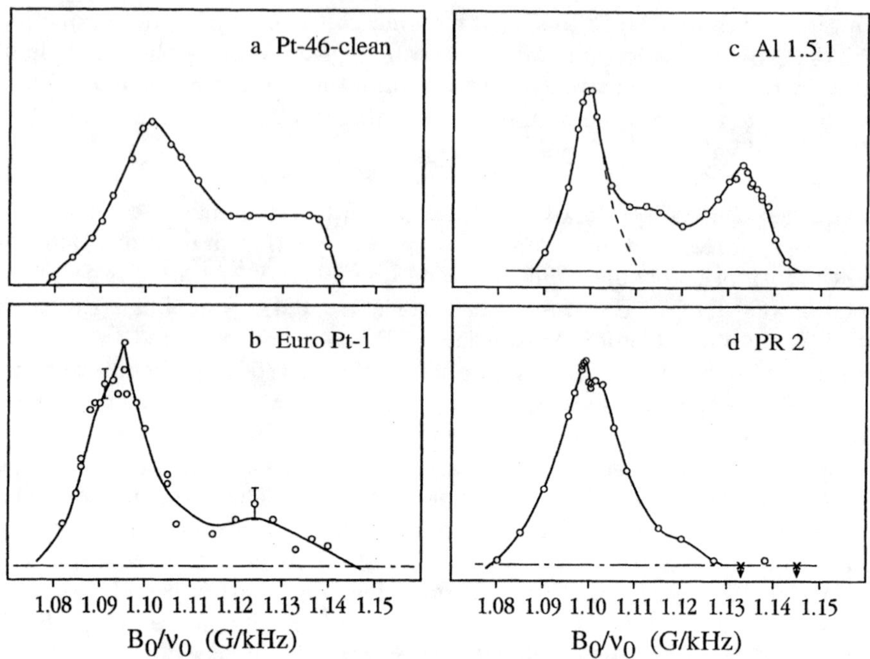

FIG. 49.   Point-by-point ^{195}Pt NMR spectra of four different catalysts, used in the determination of dispersion in Table I.

large particles was at most a few percent, and the overall zeolite structure was retained. As noted previously, the NMR cannot exclude the simultaneous presence of much smaller, nonmetallic clusters in undamaged supercages; however, if a large fraction of Pt would have gone undetected, the remaining signal should have been measurably weaker than expected on the basis of the known Pt content. Six samples were investigated, with some variation in the parameters of their preparation. The ordering of the average particle sizes was the same according to hydrogen chemisorption, electron microscopy, and NMR, although the three methods correspond to different types of averaging. There was no clear correlation between the parameters of the preparation and the particle size (158).

Yu and Halperin (162) observed a ^{195}Pt surface resonance from platinum particles initially prepared on silica gel; the carrier was removed afterwards with a solution of sodium hydroxide, thus forming a "self-supported" powder sample. Its average particle diameter determined by TEM was 4 nm; assuming a log-normal distribution, this corresponds to a dispersion of 0.31. The NMR samples were extensively washed with water, dried, and left exposed to the atmosphere. A signal detected at 1.089 G/kHz was attributed

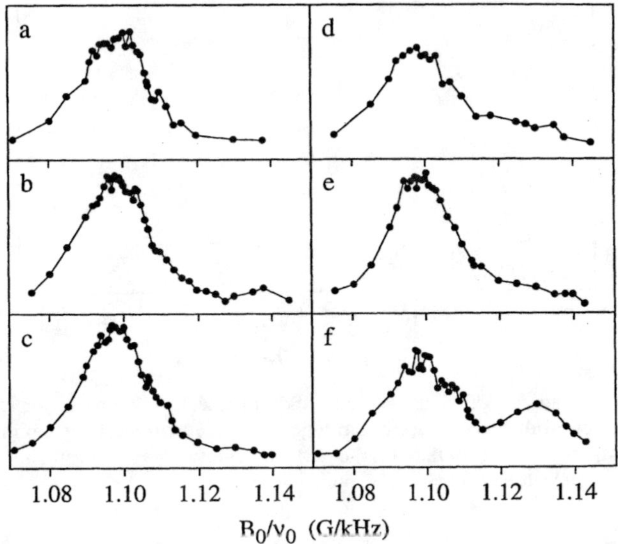

FIG. 50.   Point-by-point ^{195}Pt NMR spectra of six Pt/NaY zeolite samples measured under clean-surface conditions. Most of these samples have a detectable intensity in the high-field region of the spectrum, indicating the presence of particles larger than the supercage (cf. Fig. 51).

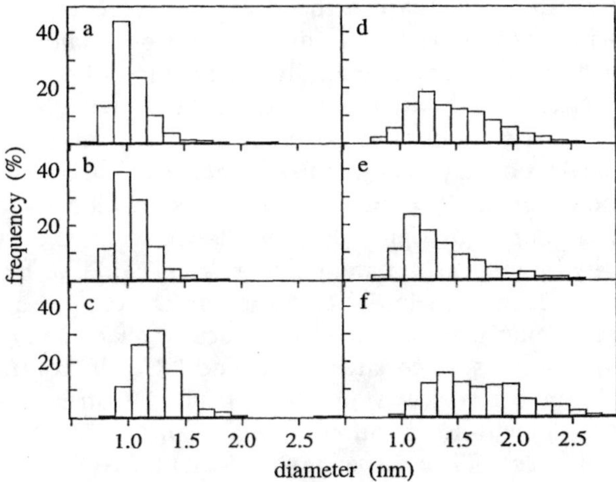

FIG. 51.   Particle size histograms for the samples in Fig. 50 from electron micrographs of ultramicrotome slices. These particles are all inside the zeolite matrix, but most do not fit in a supercage. Some rare particles larger than 3 nm (see Fig. 46) cannot be counted with any statistical precision.

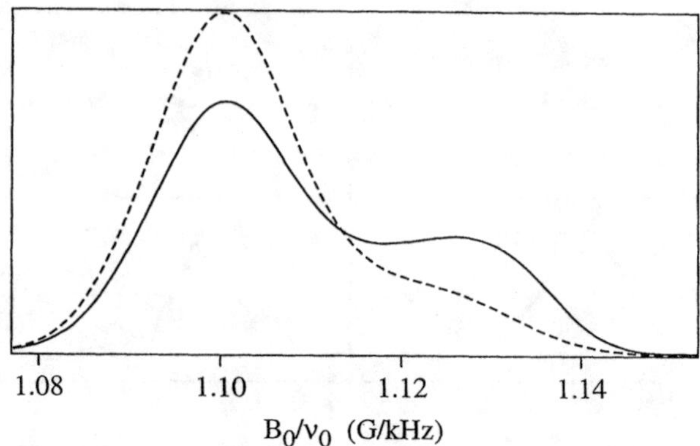

FIG. 52. The Gaussians in Figs. 48c and 48d and the histograms in Fig. 51 predict the spectra drawn here: full curve, spectrum in Fig. 50f; dashed curve, spectrum in Fig. 50a. The agreement is sufficient to accept that it is the particles counted in Fig. 51 (and not the occasional large ones) that yield the spectra in Fig. 50.

to surface Pt atoms; its relative area indicated a dispersion of 0.47. Rhodes *et al.* (*160*) identified the signal at this position in their samples as due to $H_2Pt(OH)_6$ or a similar compound, formed by interaction with the atmosphere. It disappears after oxidation/reduction treatment, which moves the peak position to approximately 1.10 G/kHz. For the self-supported sample, no surface NMR signal was found after hydrogen treatment.

An NMR method to measure the dispersion of copper catalysts was devised by King *et al.* (*163*). The traditional hydrogen chemisorption does not work (Section III.G), and generally the reaction of copper atoms with nitrous oxide N_2O is used, with monitoring of the evolving N_2. The assumed stoichiometry is one N atom for one surface Cu atom. The ^{63}Cu NMR spectra of silica-supported copper particles consisted of a single peak, with a width of the order of 100 ppm (6 kHz) and its position corresponding to that of bulk copper. This signal did not change when the catalysts were treated with hydrogen, carbon monoxide, nitrous oxide, or hydrogen chloride. It was therefore concluded that the surface copper atoms did not contribute, presumably because of strong nuclear quadrupolar interactions in the low-symmetry surface sites (^{63}Cu and ^{65}Cu, in contrast to ^{103}Rh, ^{109}Ag, or ^{195}Pt, have a nuclear spin $I > 1/2$). The HCl treatment, however, resulted in the appearance of an additional peak in the ^{63}Cu spectrum at the position of ^{63}CuCl. The dispersions calculated from the ratio (^{63}CuCl signal)/(^{63}CuCl + ^{63}Cu^0 signal) are in good agreement with the values determined by the nitrous oxide method.

Similar experiments have been attempted without success for alumina-supported silver catalysts (*164*). The dispersions were measured by oxygen chemisorption, by subsequent hydrogen titration, and by X-ray line broadening. Samples with nominally clean surfaces showed only the [109]Ag signal when they had wide, possibly bimodal, size distributions, and TEM characterization revealed particles larger than 50 nm. The [109]Ag NMR of one such sample was measured again after chemisorption of oxygen, chlorine gas, or hydrogen chloride; none of these had a measurable effect (except after oxygen chemisorption on highly loaded, Fe-doped Ag/SiO_2, whereby a line broadening occurred). Spin-1/2 nuclei in ionic compounds lack the efficient nuclear spin lattice relaxation mechanism of quadrupolar nuclei such as [63]Cu; this may be the reason why, for example, the [109]Ag NMR of solid silver oxides has not been observed. A similar absence of a signal from small silver particles has been reported in Ref. (*165*). The lack of effect of oxygen absorption on the [109]Ag spectrum has been confirmed for an (undoped) Ag/Al_2O_3 sample with a mean Ag particle size of 20 nm, as measured by X-ray line broadening (*166*).

C.   Qualitative Effects of Chemisorption

It was shown in early work (*160*) that the shape of the [195]Pt NMR spectrum of platinum catalysts varies strongly with surface conditions. In Fig. 53 two sets of spectra are shown: one from the Illinois group (Fig. 53a is the same as Fig. 49a) and one from the Lausanne group (Fig. 53c is the same as Fig. 48b). The two catalysts were prepared by different methods, and the spectra were taken in different laboratories. However, the particle sizes of the samples, as measured by TEM, are very similar, as are the spectra, with or without chemisorbed hydrogen. Note that the sample Pt-46 has a dispersion of 0.46 measured by hydrogen chemisorption but 0.61 according to electron microscopy and 0.52–0.62 measured by the clean-surface NMR (Table I). It has been stated (*157*) that the changes in the [195]Pt spectrum resulting from chemisorption of CO are similar to those resulting from chemisorption of hydrogen; furthermore, the [195]Pt T_1 in the surface region of the spectrum changes in a similar way for both adsorbates. The Korringa product $T_1 T$ is temperature independent (Fig. 54), showing that even after the adsorption of hydrogen (*38*) carbon monoxide (*3*) the platinum surface has metallic character. It was found that $T_1 T$ increases after adsorption, indicating a decrease in the local density of states at the Fermi energy on the surface sites.

In the interpretation of these spectra it is assumed that the geometry (i.e., the layer statistics in the NMR layer model) of the particles does not change as a result of chemisorption but that the surface LDOS and the healing length are affected. In principle, this suggestion can be tested experimentally by comparing the lineshape in a double-resonance experiment on

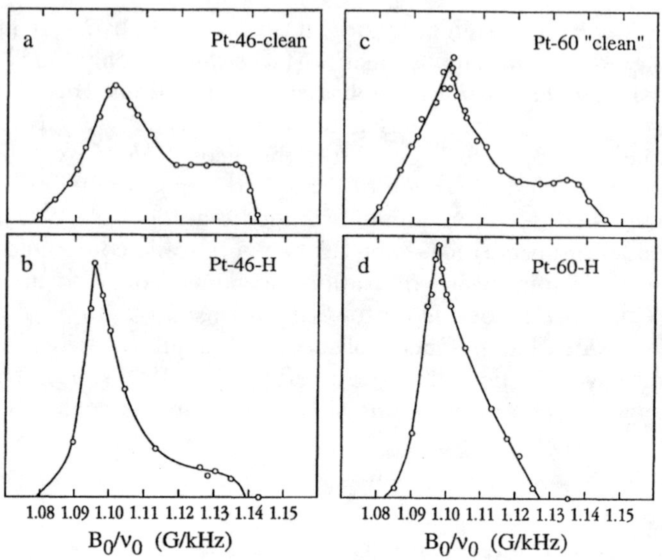

FIG. 53.   Point-by-point ^{195}Pt NMR spectra of Pt/Al$_2$O$_3$ (a and b; from the Illinois group) and Pt/TiO$_2$ (c and d; from the Lausanne group) under clean-surface (a and c) and hydrogen-covered (b and d) conditions. The two catalysts were prepared by different methods and on different carriers, but their dispersions measured by electron microscopy are very similar—and so are their spectra. For the effect of hydrogen on the shape of the spectra, see Figs. 55, 56a–56d. [a and b: Reproduced with permission from Rhodes *et al.* (*160*). Copyright 1982 American Institute of Physics.]

a covered sample with the lineshape of the clean surface (but unfortunately this has not been done).

The double-resonance data (Fig. 40a) show that the signal from ^{195}Pt in the surface of the particles is symmetric around a center position of approximately 1.096 G/kHz. This is an inhomogeneous linewidth; there are many different types of surface ^{195}Pt, each resonating at a slightly different frequency, and the observed signal is an unresolved superposition of all these elementary resonances. In the NMR layer model (Fig. 48), this idea is expanded further; it is supposed that the difference in resonance frequency, and therefore also in LDOS, between all elementary surface resonances is less than the difference between typical surface and subsurface resonances. The important distribution of surface LDOS values has its equivalent in the MAS–NMR of adsorbed ^{13}CO; the MAS does not narrow the signal (Figs. 35c and 35d) because its width is due to a distribution of isotropic shifts rather than to shift anisotropies.

It is clear that NMR can give surface information only in some average sense and that it is unlikely that the more detailed data required by, e.g.,

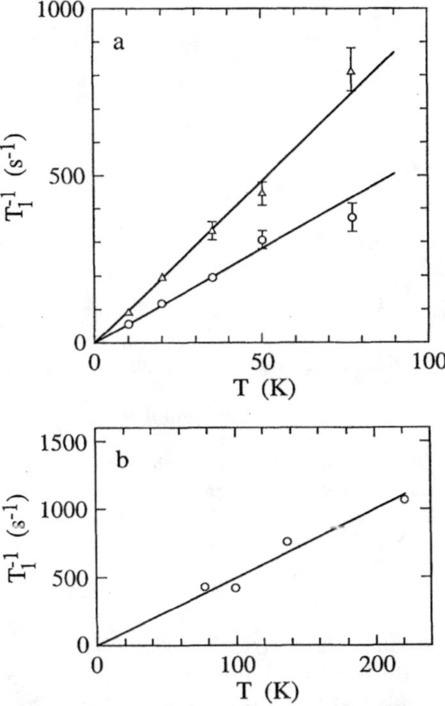

FIG. 54.   Korringa relationship for the ^{195}Pt spin lattice relaxation in the surface peak of the spectrum. The straight lines show that the platinum surface keeps its metallic character, but the smaller slope for the adsorbate-covered surfaces means that the E_f LDOS has diminished. (a) Pt/TiO$_2$: clean surface, triangles; hydrogen covered, circles. (b) Pt/Al$_2$O$_3$: CO covered. [b. Reproduced with permission from Ansermet *et al.* (*3*). Copyright 1990 Elsevier Science.]

ensemble theories can be extracted from it. On the other hand, if we accept the resonance positions of the fitted Gaussians in Fig. 48 as being characteristic of atoms in the respective layers, we can, by combining shift and T_1 data, find the variation in LDOS when moving from layer to layer (*156*), as shown in Fig. 55. The increase in healing length upon hydrogen chemisorption can be interpreted as indicating that the hydrogen forms a bond not only with surface Pt atoms but also with subsurface Pt atoms and beyond.

As shown earlier (Fig. 49), the surface resonance of highly dispersed catalysts is broader than that of catalysts of intermediate dispersions, and the fit parameters clearly cannot be transposed between the two groups; in very small particles an atom in a given surface site senses the presence

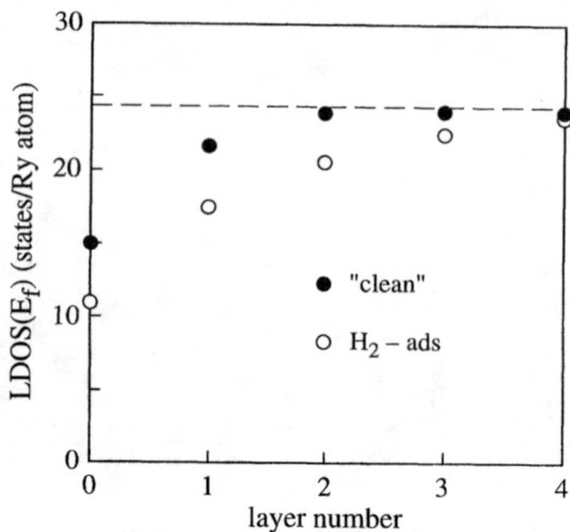

FIG. 55.   The E_f LDOS variation with layer number (cf. Fig. 48) derived from NMR data for ^{195}Pt/TiO$_2$ with clean and hydrogen-covered surfaces. For the clean sample, the subsubsurface layer ($n = 2$) is already bulk-like, but the perturbation by the hydrogen reaches deeper. In terms of the NMR layer model, the healing length increases upon hydrogen adsorption.

of the atom in the surface diametrically opposite. The exponential healing model is probably only useful for particles large enough to have bulk-like interiors. A more formal objection to the parameterization scheme is that it starts from a hypothesis about the layer dependence of the shift rather than of the LDOS. A given layer could be characterized by certain distributions of the s- and d-like LDOS. This would imply that a given resonance position does not correspond to a unique value of T_1, as indeed has been found in later, more precise work, whereby the changes in the surface LDOS with hydrogen coverage were investigated (84). The evolution of the spectra is shown in Figs. 56a–56d; the nonexponential relaxation curves at the position of the spectral maximum are given in Fig. 45.

The effect of chemisorbed oxygen on the ^{195}Pt spectrum is less drastic (167) (Figs. 56e–56h), but the behavior of the spin lattice relaxation is more complicated. For Pt/oxide samples at intermediate oxygen coverages, a fraction of the ^{195}Pt surface signal no longer has a temperature-independent Korringa product T_1T. Therefore, on some surface sites the density of Fermi-energy electrons is very low. At saturation dosing of oxygen, measured from chemisorption isotherms to correspond to a ratio (O atoms)/ (surface Pt atoms) = 0.75, a Korringa-type of behavior is recovered.

In principle, when T_1T is temperature dependent, one expects to observe temperature-dependent features in the spectrum as well, but these have not been detected. The reason for this is not clear, but the same discrepancy

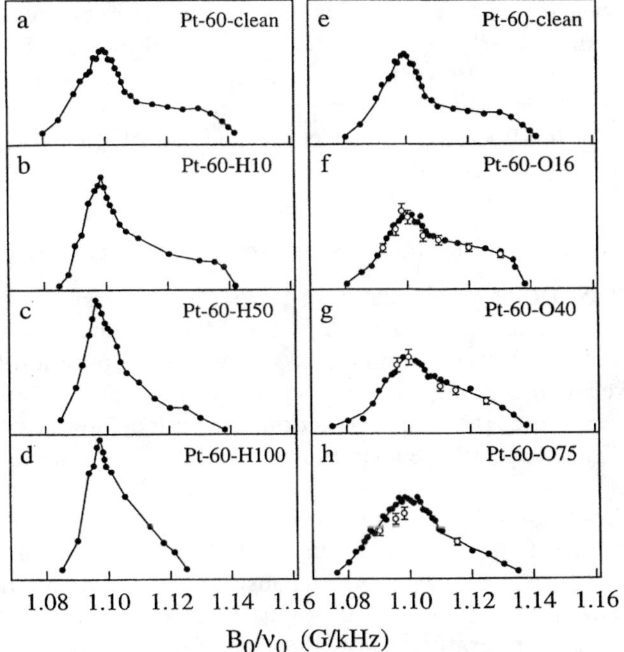

FIG. 56.   Effects of chemisorption on the ^{195}Pt NMR spectrum of Pt/TiO$_2$. Spectra taken at 80 K. Panels b–d are the same samples as those in Fig. 45; panels f–h are for samples under oxygen. The open symbols are scaled from experiments at 20 K. The changes under hydrogen can be described as a gradually increasing healing length (Fig. 55), which effectively sweeps all Gaussians in Fig. 48 to low field. The oxygen case is more complicated; in f and g, the spin lattice relaxation at the surface peak does not obey time/temperature scaling (Fig. 45) so that some platinum must be in a nonmetallic environment.

has been found experimentally in the ^{195}Pt NMR of other systems (*108, 168*); it seems that spin lattice relaxation is a more sensitive monitor of nonmetallic behavior than the spectral shape.

A remarkable result is the metallic to nonmetallic to metallic series of transitions observed locally for some surface sites as a function of oxygen coverage. [A well-known bulk metal to nonmetal transition as a function of oxide composition occurs in VO$_x$; see the references given in Ref. (*169*).] At the lowest coverage studied (0.16 O atoms per surface Pt atom), approximately three nonmetallic Pt sites are found for each oxygen atom; at a coverage of 0.40 this number is approximately halved, and at 0.75 coverage it is zero. Under UHV conditions the saturation coverage on the Pt(111) surface is one O atom per 4 Pt atoms (*170*). For a coverage of 3 O atoms per 4 Pt atoms, which is the saturation value under our experimental conditions, oxygen–oxygen direct bonding becomes important and restores the metallic character of the underlying platinum surface. At the lower coverages, the oxygen–oxygen interaction is mainly indirect, the platinum–oxygen interaction is localized, and the NMR properties of atoms in the deeper layers are essentially unaffected. However, the LDOSs of the metallic surface sites are modified by the presence of oxygen.

D.   ELECTRON DEFICIENCY AND FRONTIER ORBITALS

The [195]Pt NMR of small platinum particles on classic oxide supports shows that the clean-surface LDOS is largely independent of the support (silica, alumina, and titania) and of the method of preparation (impregnation, ion exchange, and deposition of colloids). At a given resonance position, one always finds the same relaxation rate, independent of particle size or support. The shape of the spectrum is related to the sample dispersion. The same is true for particles protected in films of PVP. [However, samples prepared under conditions giving strong SMSIs behave differently (*171*).]

The [195]Pt NMR behavior of platinum particles in zeolites is more complicated (*168*). In this review I do not consider the low-temperature data, which indicate a transition toward nonmetallic character. Experiments at 80 K and higher temperatures show that the spin lattice relaxation at fixed resonance position is still independent of particle size for a given zeolite, but at constant dispersion it is different in different zeolites. As remarked in Section VI.B, most of the particles in these samples do not fit inside an undamaged supercage. There is a systematic variation of the spin lattice relaxation in the surface region of the [195]Pt NMR spectrum as a function of the framework acidity of the encaging zeolite. In a very approximate, qualitative manner we can say that the relaxation curves (the normalized amplitude of the nuclear signal as a function of the recovery time after the initial saturation pulses), which are not simple exponentials and therefore cannot be easily compared, give an indication that the relaxation is slower when the zeolite acidity increases (*172*) (Fig. 57). To make the statement more quantitative, the relaxation curves have been fitted to double exponentials, each of the relaxation rates has been converted into values for an s-like and a d-like LDOS, and average values have been formed by weighting with the relative amplitudes in the curves, as described in Section VI.A. It is clear that other fitting procedures could have been adopted because there is no reason to think that there are really two groups of different surface sites; the operation is simply one of convenience. Other procedures would give different values, but it is probable that the ordering of the samples according to LDOS would be preserved.

It is well-known that the IR stretch frequency of CO adsorbed on

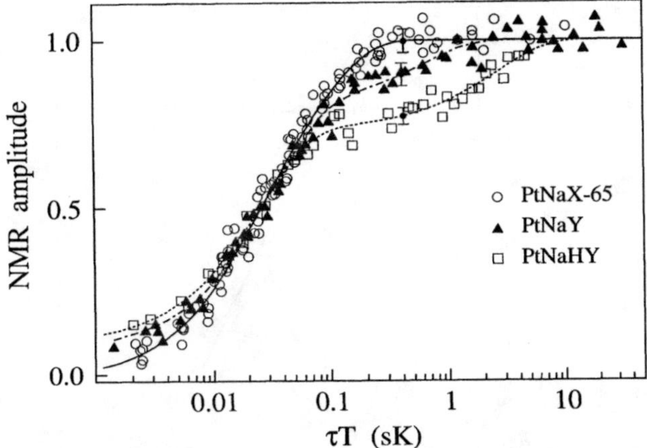

FIG. 57.   Time/temperature-scaled ^{195}Pt spin lattice relaxation data at the surface peaks of nominally clean platinum particles in six different zeolites. Only the data for Pt/NaX zeolite (one dispersion: 0.65) have actually been obtained at different temperatures (between 80 and 250 K); all other data are at 80 K only. The triangles represent data for three different dispersions (0.56, 0.64, and 0.77) in Pt/NaY zeolite; the squares represent two dispersions (0.72 and 0.79) in Pt/NaHY zeolite. The small black dots with error bars represent a double-exponential fit and its errors. Apparently, the relaxation varies with zeolite type but much less with dispersion. In general, the circles represent a faster relaxation than the squares; therefore, in Pt/NaX zeolite the E_f LDOS on the platinum surface is higher than that in Pt/NaHY zeolite. An increase in zeolite acidity decreases the metal surface E_f LDOS. The double-exponential fits (curves) are used to quantify this statement (Fig. 58).

Pt/zeolite varies with zeolite acidity. Precautions must be taken to eliminate the CO–CO interaction from the experimental IR spectra. In the work under discussion (*172, 173*), this was done by measuring spectra as a function of coverage θ and extrapolating to $\theta = 0$. After saturation of a sample with CO at room temperature (exposure to 15 Torr), a reference IR spectrum was taken. Next, the sample was evacuated at a temperature above ambient, and a new spectrum was taken. The new coverage was estimated from the relative value of the integral of the IR band, compared with the saturation value. The procedure was repeated several times, at increasingly higher evacuation temperatures. The precision is thought to be sufficient to obtain a reasonable extrapolation of the measured frequencies to zero coverage, which should give the stretch frequency of a single CO molecule on a clean Pt surface.

Figure 58 shows that when the ^{195}Pt spin lattice relaxation in the surface region of the spectrum of a clean-surface sample is faster, the (extrapolated) stretch frequency of a CO molecule adsorbed on that same sample is lower (*173*). The faster relaxation means a higher density of states at the Fermi

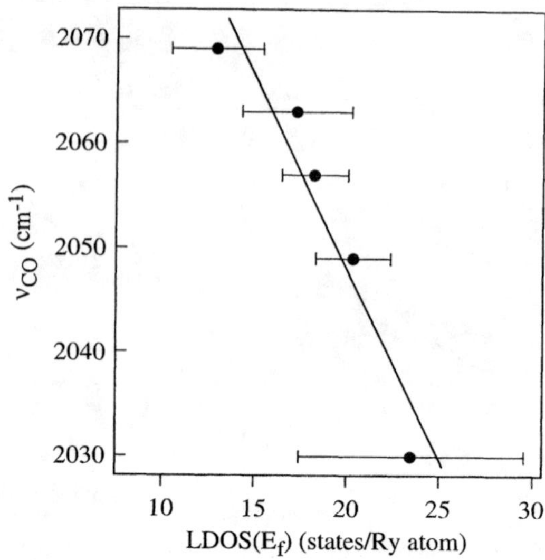

FIG. 58.   Local density of states on the clean metal surface at the Fermi level and (extrapo-lated) IR stretch frequency of CO adsorbed on the same surfaces. For samples of Pt/zeolite with different zeolite acidities, the clean metal-surface E_f LDOS was found from curves as in Fig. 57. Next, CO was adsorbed on these same samples, and the stretch frequency was measured as a function of coverage. The value plotted here is an extrapolation to zero coverage. This plot indicates that the C–O intramolecular bond after chemisorption will be weaker if the initial clean-surface E_f LDOS is higher.

energy; the lower stretch frequency means a weakening of the intramolecu-lar C–O bond. The correlation states that if a single CO molecule adsorbs onto a clean platinum particle (in a zeolite), its intramolecular bond will be more weakened when the surface LDOS on the metal particle is higher. The result can be immediately related to the frontier orbital interpretation of the Blyholder model (Section I.F): A higher LDOS means that more metal electrons are available to dump into the LUMO of the CO molecule and also that more metal holes can pull electrons out of the HMO. Since the LUMO is antibonding and the HOMO is bonding, the result is an increased weakening of the C–O bond in the molecule. (The elongation of the C–O bond after chemisorption has been determined by $^{13}C^{17}O$ double resonance; Fig. 40c.)

This process is often described not in terms of Fermi-level quantities but in terms of total charge; if the metal particle is electron deficient, it has fewer electrons available for the Blyholder mechanism. Initially, the expression referred to the experimental fact that catalysts of platinum or palladium in zeolites, when compared to the same metals on oxides, often seemed to behave like their neighbors in the periodic system, iridium and rhodium.

In a rigid-band picture, the principal difference is that these have one electron less available to fill the band. For a view beyond the rigid-band model, compare Rh and Pd in Fig. 9. The calculated density of states at E_f decreases as well when a step is made to the left (*34*)—from 32.2 per Ry and per atom in Pd to 18.7 in Rh and from 29.9 in Pt to 12.7 in Ir. A recent calculation for three-layer slabs of Pd and Rh shows differences for the surface LDOS at the Fermi energy similar to those in the bulk DOS (*174*).

The idea of sizable charge transfers between metal particles of the dimensions considered here and their support has been severely criticized (*23*, Chapter 5). On the other hand, the collective electron model of catalytic activity was "sent to the trash can of science" because it does not account, e.g., for the fact that metal atoms in an alloy retain distinct properties (*23*, p. 458). This model does not distinguish between DOS (density of states on the energy scale, a global property of a particle) and LDOS (a local property). Similarly, the charge-deficiency models failed to recognize that not only charge but also the energy cost at which it can be made available are important. Again, the E_f LDOS is a quantity that can express this requirement.

The NMR data can give only some average values for the E_f LDOS on surface sites, and in the averaging all information required by ensemble theories is lost. It is also true that the data cannot give values for the partial LDOS of different symmetries (σ, π, and δ). This is not due to a failure of the frontier orbital model but simply to the fact that the NMR experiment yields just two numbers, a shift and a relaxation rate.

The ^{195}Pt NMR experiments with zeolites of different acidities (*168*, *172*, *173*) can be said to show that the metal surfaces have different electron deficiencies at the Fermi energy, but they provide no information concerning the total charge of the metal particles. The mechanism that governs these variations has not been identified, but in the NMR context it is tempting to suggest that the metal–support interaction has similarities with promoting and poisoning. Indeed, whereas it is certain that most of the metal particles do not fit in undamaged supercages, the low-temperature NMR experiments (which are not discussed further in this review) have been explained by supposing a tight fit of the metal particles in the matrix (*168*), rendering plausible the suggestion that the surface condition even of a large metal particle varies with certain properties of the matrix.

E.   THE SURFACE LDOS, PROMOTION AND POISONING

In the ^{195}Pt NMR spectra of platinum catalysts with different surface conditions, nuclei in surface sites resonate at low field and those deep in the interior of the particles at high field. Whereas, according to the

Heine–Friedel invariance idea (Section I.E), all sites deep inside a sufficiently large particle must have very similar DOS curves, the surface sites will be different according to their positions: in the center of a facet, on an edge, at a corner, etc. The NMR experiment measures simultaneously the signals from a large collection of particles; the results can be interpreted in terms of the probability of finding an atomic site with a given local density of states at the Fermi energy. As was shown, much modeling is needed to transform the NMR data (essentially the spin lattice relaxation curve, including the equilibrium amplitude, at a given spectral position) into these LDOS values. On the other hand, if we assume that the d-like electrons are the most important, it is easy to show from Eqs. (15)–(17) that faster relaxation at a constant spectral position means a higher DOS (and the model is needed only to find a numerical value for the change). Therefore, it is thought that, although the LDOS values must be used with some caution, the trends should be correct.

A simple test of this suggestion is the comparison of a five-layer slab calculation for the Knight shift in platinum (*70*) with the spectral fits of the layer model (Fig. 48). In both cases the surface resonance is shifted about 4% to low field with respect to the bulk signal, and the subsurface signal is found at approximately the halfway point. Another test is qualitatively to compare experimental results for hydrogen chemisorption on platinum (Fig. 55) with a calculation for hydrogen on palladium (*175*); in both cases an important diminution of the surface LDOS on the metal is found.

It has been argued on the basis of calculations for a two-layer Rh(001) film that the catalytic effect of small amounts of promoters such as alkali metals cannot be understood from charge-density perturbations alone since these are screened over a short distance. The perturbation of the Fermi-level LDOS, extends much farther, and its value just outside the metal increases with respect to the clean-surface case (*176*). This increase was connected to the promoting effect by use of the frontier orbital method. A qualitative experimental test of these ideas was performed with Pt/TiO_2 catalysts impregnated with lithium or potassium salts and re-reduced afterwards (*177*). It is not known how much of the alkali ends up on the metal particles, but the change in ^{195}Pt NMR parameters in the surface region of the spectrum corresponds to an average increase of the surface LDOS of 26%; the bulk-like parameters did not change. The NMR spectra are not visibly modified by the impregnation, but the spin lattice relaxation becomes faster on the surface and is unchanged in the bulk (Fig. 59a). These relaxation data could be described by simple exponentials, and Fig. 59b shows that the ^{195}Pt in the impregnated surface still has metallic character. The impregnation increases the relaxation rate (at fixed position in

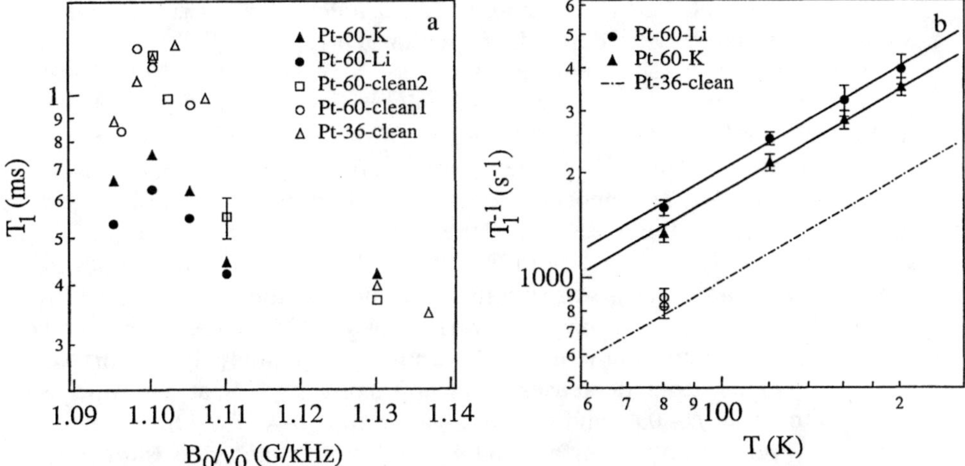

Fig. 59. Effect of alkali impregnation on ^{195}Pt spin lattice relaxation for Pt/TiO$_2$. (a) Spin lattice relaxation time across the NMR spectrum for several clean-surface (open symbols) and alkali-impregnated (solid symbols) samples at 80 K. The changes are important near 1.10 G/kHz (the surface signal) and undetectable at 1.13 G/kHz. (b) Korringa relationship for the spin lattice relaxation at the surface peak of the spectrum. The alkali impregnation does not change the metallic character but increases the E_f LDOS on the metal surface. (An extrapolation of earlier clean-surface data, obtained at lower temperature, is shown by the dashed line.)

the spectrum); this is direct evidence that the E_f LDOS in the surface has increased.

Information about the alkali adsorbate has been obtained by Ebinger *et al.* (*178*). They produced atomic beams of ^{8}Li or ^{6}Li with a very high nuclear polarization and deposited these atoms on (among others) Ru(0001) substrates. The electronic structure of such alkali–atom/transition-metal surface complexes has long been debated. The atomic-beam NMR experiments showed that at temperatures between 200 and 1250 K the nuclear spin relaxation has a contribution with a Korringa product T_1T that is independent of temperature. (Another contribution is related to diffusion of the lithium atoms on the surface and is not of interest here.) This result shows that, at least in this system, the alkali is in a metallic environment; the local density of electron states is continuous in energy at E_f. It is surprising that up to a coverage of $\theta = 0.15$ the value of the Korringa product does not change. Apparently, any given Li atom does not have its LDOS perturbed by the presence of the other atoms. (The work function, on the other hand, decreases steeply, by more than 2 eV.) This situation seems to be similar to that of ^{1}H NMR, whereby the coverage dependence of the shift is explained by assuming the existence of sites with intrinsic shifts that are coverage independent (Section III.D).

In a speculative extension of the connection between surface LDOS and promotion effects, the difference in chemisorption behavior of hydrogen and oxygen has been related to [195]Pt surface NMR as a function of coverage (*179*). When the amount of hydrogen increases, the surface LDOS, averaged over both occupied and unoccupied sites (the hydrogens probably move around rapidly, and the NMR represents an average) decreases monotonically. Values derived from data in Fig. 45 are shown by the solid circles in Fig. 60. (This data analysis was slightly different from the one used for Fig. 55.) In the case of partial oxygen coverage, two kinds of sites seem to be created: metallic and nonmetallic. The number of nonmetallic Pt surface sites per oxygen atom decreases from approximately 2.7 at a coverage $\theta = 0.16$ to 1.5 at $\theta = 0.40$ and to zero at saturation, $\theta = 0.75$. On the other hand, it is known that the surface LDOS on the remaining sites, shown by the open circles in Fig. 60, is enhanced over the clean-surface value (*179*). In most platinum catalysts, the oxygen chemisorption isotherm becomes flat at pressures below 1 Torr, whereas the hydrogen isotherm joins the asymptote only at pressures above 100 Torr. It is speculated that the continuous decrease of the surface LDOS with increasing hydrogen coverage makes the remaining surface increasingly less reactive, whereas the enhanced LDOS on the oxygen-free sites leads to an autopromotion effect for oxygen chemisorption.

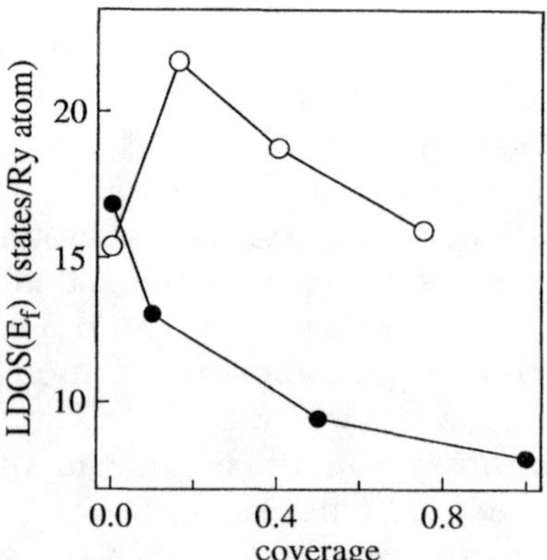

FIG. 60. Variation of the metal surface E_f LDOS of Pt/TiO$_2$ with hydrogen (solid circles) and oxygen (open circles), derived from Figs. 56 and 45 together with additional relaxation data for the oxygen-covered surface. The determination of the hydrogen points is straightforward, but the reasoning for the oxygen case is more complicated.

F. ELECTROCATALYSTS

Recently, the ^{195}Pt NMR of commercial fuel cell electrode material has been observed (*180, 181*) (Fig. 61). This material consists of platinum supported on carbon black and pressed into graphitized-carbon cloth. (Similar material has been used to study ^{13}CO NMR; see Section IV.G.) Because of the conducting nature of the carrier, one might expect to see differences with respect to ^{195}Pt NMR of particles supported on oxides. Furthermore, if an electrolyte is present in the NMR sample, the electric double layer at the metal/electrolyte interface might influence the ^{195}Pt surface signal.

The as-received material shows the NMR peak at 1.089 G/kHz that is characteristic of platinum particles exposed to the atmosphere (Fig. 61a); it has been attributed to $H_2Pt(OH)_6$ or PtO_2. For oxide-supported catalysts, it is removed by thermal treatment under oxygen and hydrogen. The electrocatalysts are cleaned by electrochemical methods, whereby the sample is used as the working electrode in a three-electrode cell; as a reference electrode, 1 M NaCl | AgCl | Ag was used. In the first experiments, the as-received material was subjected to extensive potential cycling, but later it

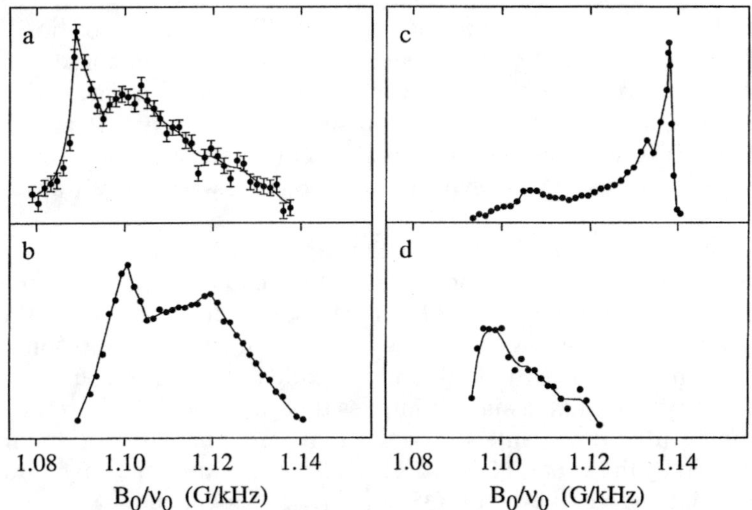

FIG. 61. ^{195}Pt NMR spectra of fuel cell electrode material (Pt/graphite) with platinum particles of different size. The pair a and b have the same vertical scale, as do the pair c and d. The gross features are the same as those for oxide-supported catalysts (Fig. 48), but here the healing length is smaller (for the same particle size, there is less intensity at the high-field end). (a) 2.5-nm particles, as-received material; (b) as in a, but after electrochemical cleaning; (c) 8.8-nm particles, electrochemically cleaned; (d) 2.0-nm particles, electrochemically cleaned. [Reproduced with permission from Tong *et al.* (*180*). Copyright 1997 American Chemical Society.]

was found that results are more reproducible when the potential is held fixed within the double-layer region (250 mV) until the reduction current falls below μA values (and can no longer be measured—this takes several hours) (Fig. 61b). After this cleaning procedure, the fuel cell electrode material, together with some electrolyte (0.5 M H_2SO_4), is transferred into an NMR ampule.

Three samples with different particle sizes were investigated (Figs. 61b–61d). The platinum loadings and average particle diameters were provided by the manufacturer of the material. From the hydrogen adsorption/desorption profiles in cyclic voltammetry experiments, the total quantity of adsorbed hydrogen was determined; together with the value of the platinum loading, this yields a ratio for (atoms of H)/(total atoms of Pt), the dispersion. An independent value of dispersion can be obtained from the average diameters, assuming the particles to be cubooctahedra. Reasonable agreement between these values and the intensities in the ^{195}Pt NMR spectra was obtained by considering the spectral region below 1.11 G/kHz in Fig. 61 as due to surface platinums. (No correction for T_2 effects was applied.)

A remarkable difference in the ^{195}Pt spectra of oxide- and carbon-supported platinum is especially clear for the 2.5-nm sample; the fuel cell material shows much less intensity at the bulk resonance position (1.138 G/kHz). A similar difference is shown by the spectrum for the 2.0-nm sample. In terms of the NMR layer model, this comparison means that the healing length is larger in the carbon-supported material. It is not clear whether this result is related to the conducting nature of the carrier or to the presence of the electrolyte; comparisons between "wet" and "dry" samples are needed.

Only very preliminary values for the surface LDOS are available. They are based on an analysis of the decay of the echo amplitude in terms of a modulation by the J-coupling of Eq. (14). Marginally better fits to the data are obtained by assuming that the value of J for the surface resonance in Pt/carbon-black is larger than that in Pt/oxide (*180*). If it is confirmed, this result is very remarkable since it implies that the s-like LDOS is higher on the surface than in the interior of the particle. (The value of J is mainly determined by the s-like DOS at E_f.) In all other platinum catalysts investigated to date, the s-like E_f LDOS is nearly site independent.

G. Bimetallics

The NMR lines in bulk alloys are considerably broader than those in pure metals; this fact simply reflects the distribution of atomic environments and, thus, of E_f LDOS, in an alloy. The lines in highly dispersed bimetallic catalysts have additional broadenings according to the NMR layer model.

Furthermore, if segregation occurs, the average composition of each layer may be different. Therefore, the interpretation of ^{195}Pt NMR in the two alloys studied, $Pt_{1-x}Rh_x$ and $Pt_{1-x}Pd_x$, is only qualitative.

For a $Pt_{0.8}Rh_{0.2}$/alumina with a low dispersion (0.14), the position of the high-field peak in the ^{195}Pt NMR spectrum (*105*) was compared with the value in the bulk alloy (Fig. 62a). The NMR spectrum does not give any indication of a bimetallic character of the particles, and for this particular sample no other checks were made. The ^{195}Pt NMR spectrum in Fig. 62a shows that there are no particles with interior composition $Pt_{0.8}Rh_{0.2}$. Therefore, if they are really bimetallic, these particles must have surfaces enriched in rhodium (the other possibility is that there are separate platinum and rhodium particles).

An analysis of a $Pt_{0.5}Rh_{0.5}$ with dispersion 0.40 by energy-dispersive X-ray spectroscopy showed that the individual particles had approximately the overall composition. The $^{13}CO/^{195}Pt$ double-resonance spectrum of CO close to Pt in the $Pt_{0.5}Rh_{0.5}$ surface is shown in Fig. 40b. The mere existence of this double-resonance signal shows that there is platinum in the surfaces of these particles. Its position, however, is different from that of CO on a pure platinum surface, showing that these particles are alloys. From the analysis of the $^{195}Pt/^{13}CO$ double-resonance spectrum of platinum in Fig. 62b, it is found that a fraction 0.49 ± 0.07 of the Pt atoms are attached to CO, whereas the dispersion is estimated to be between 0.40 and 0.67 (Sec-

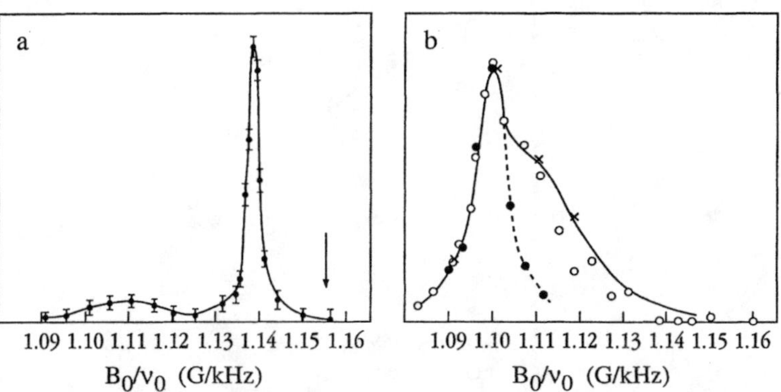

FIG. 62. ^{195}Pt NMR spectra of $Pt_{1-x}Rh_x/Al_2O_3$ samples. (a) $x = 0.2$; dispersion 0.14; the arrow indicates the position of the ^{195}Pt resonance in the corresponding bulk alloy; the actual peak is close to that for pure platinum; (b) $x = 0.5$; dispersion 0.40; the open circles and crosses are the usual point-by-point ^{195}Pt spectrum (the crosses take corrections for relaxation effects into account); the solid circles are results of a spin-echo double-resonance experiment, as in Fig. 40a; in fact, the two spectra ($x = 0$ in Fig. 40a and $x = 0.5$ in Fig. 62b) are very similar. [Reproduced with permission from Wang *et al.* (*105*). Copyright 1998 Royal Society of Chemistry.]

tion IV.F). This result means that there is at most a moderate difference between the surface and the overall composition of the particles. All these results taken together lead to the conclusion that, although the single-resonance ^{13}CO line position is close to that for CO on pure rhodium (Fig. 40b), the surface is nevertheless bimetallic. Therefore, the simple "local" interpretation of the shift vs composition data in Fig. 33 (Section III.H) cannot be correct; the shifts must depend on the composition of a larger surface region around the atom to which the ^{1}H or ^{13}CO is bonded.

The $Pt_{1-x}Pd_x$ samples (*182*) were prepared as colloids, protected by a PVP polymer film. Layer statistics according to the NMR layer model (Eqs. 28–30) for samples with $x = 0, 0.2$, and 0.8 are shown in Fig. 63. The metal/polymer films were loaded into glass tubes and closed with simple stoppers. The NMR spectrum and spin lattice relaxation times of the pure platinum polymer-protected particles are practically the same as those in clean-surface oxide-supported catalysts of similar dispersion. This comparison implies that the interaction of the polymer with the surface platinums is weak and/or restricted to a small number of sites. The spectrum predicted by using the layer distribution from Fig. 63 and the Gaussians from Fig. 48 shows qualitative agreement with the observed spectrum for $x = 0$ (Fig. 64a).

No metal NMR data are available for supported palladium catalysts.

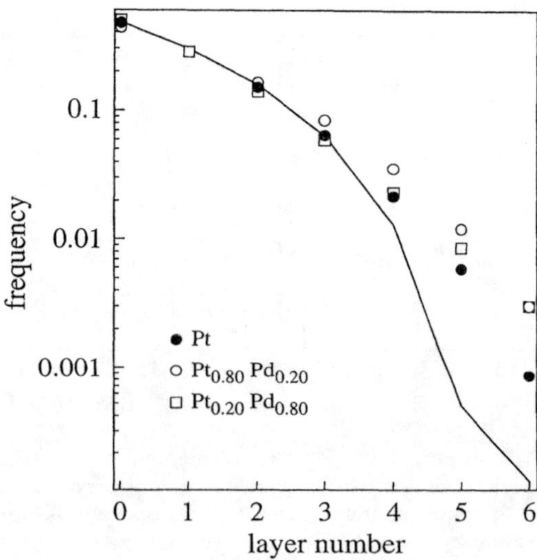

FIG. 63. Layer statistics (as in Fig. 47c) for three samples of $Pt_{1-x}Pd_x$ in PVP film. The line was calculated for a monodispersed sample with particle diameter 2.35 nm. For purposes of the NMR layer model, the three samples have identical size distributions.

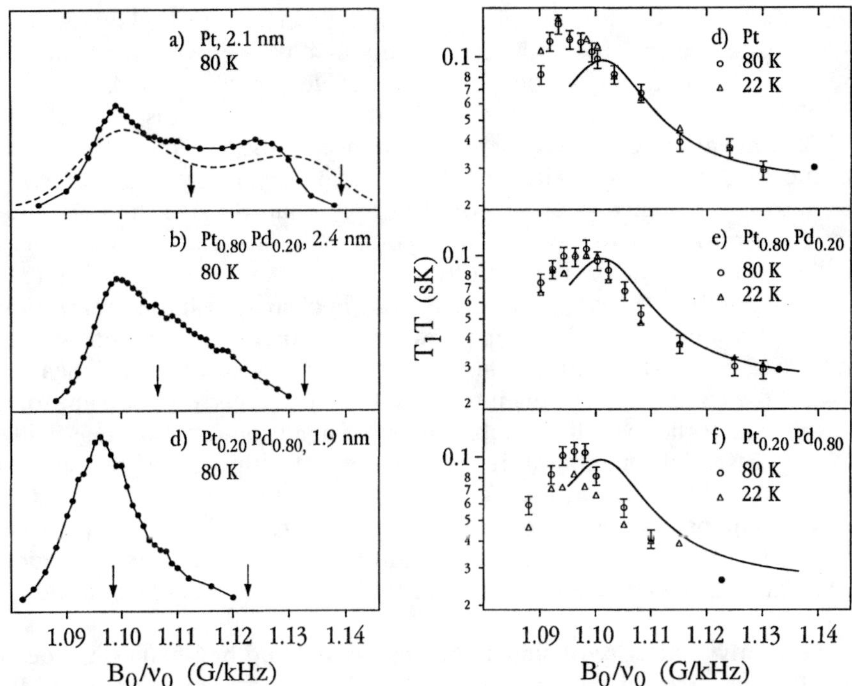

FIG. 64.   [195]Pt NMR spectra and spin lattice relaxation times across the spectra for the three $Pt_{1-x}Pd_x$ samples in Fig. 63. (a–c) Spectra: The right arrow is the resonance position in the bulk alloy. The left arrow is the center of the spectra (divides the integral in halves). For $x = 0$ the dashed line is the prediction using the solid circles in Fig. 63 and the Gaussians from Fig. 48. The protecting PVP film does not have a significant influence. The surface peak does not seem to shift with composition (cf. Fig. 62b for a rhodium alloy). (d–f) Korringa product T_1T across the spectrum at two temperatures (open symbols). The [195]Pt atoms are in a metallic environment. The full curve sketches the values found for Pt/TiO_2 catalysts. The open symbols extrapolate well to the solid circle in each panel, which is the corresponding bulk value.

From magnetic susceptibility studies it has been found that the susceptibility of surface atoms is less than that of the bulk (*183*). A similar conclusion was reached for Pd cluster molecules (*184*), which means that in Pd particles the E_f LDOS is lower on surface sites than on bulk sites, just as for Pt. Therefore, we assume that the [195]Pt NMR spectra of $Pt_{0.2}Pd_{0.8}$ and $Pt_{0.8}Pd_{0.2}$ particles can also be interpreted with an exponential healing model (Eq. 31).

In [195]Pt NMR spectra of catalysts, the nuclei in a bulk Pt-like environment (those that have approximately two layers of platinum atoms around them) resonate in the range 1.12–1.14 G/kHz. The fraction of such nuclei in Figs. 64a–64c decreases with increasing Pd content. Regarding [195]Pt NMR, the size distributions of these samples are completely characterized by the data in Fig. 63. Since the site statistics of layers 0–4 are virtually identical and the absolute values for the deeper layers are very small, the effective size distributions are the same (and very close to that for a monodisperse sample of 2.35- nm particles). Therefore, the differences in the NMR spectra cannot be due to differences in size distribution and they must be an effect of the alloying. The arrows on the right in Figs. 64a–64c give the (average) resonance position in the corresponding bulk materials. The high-field edges of the spectra follow these positions very well, as expected from Eq. (31) for layers with $n \geq 2m$. Furthermore, the nuclear spin lattice relaxation rates at the high-field edges tend toward the corresponding bulk values (Figs. 64d–64f). This pattern shows that on the scale of 1 or 2 healing lengths, the composition of the interior of the particles is to a good approximation that of the overall formula. The Korringa products $T_1 T$ (Figs. 64d–64f) are essentially temperature independent at all points of the spectra for $x = 0$ and $x = 0.2$. In the surface region of the spectrum for $x = 0.8$, the T_1 is comparatively shorter at low temperature, indicating an increase of the effective DOS. The order of magnitude of $T_1 T$ in the surface region of all three spectra is compatible with the existence of atoms in a metallic environment.

The catalytic activity of similar polymer-protected bimetallics has been found to vary strongly with composition (*182*). Because of the lack of NMR data for the palladium sites, we cannot quantitatively relate this variation to the surface LDOS, but a qualitative discussion can be given. The [195]Pt NMR has shown that the interior of the alloy particles is bulk-like. In the bulk alloys the E_f LDOS on both Pt and Pd sites varies strongly with composition around $x = 0.8$ (*184*). It is supposed, but not proven, that on the surfaces of the alloy particles the E_f LDOS changes strongly with composition as well and that this explains the variation in catalytic activity.

H.   NMR of Metals: Miscellaneous

Small-particle [103]Rh spectra have been published for just two samples (*186*): one supported on titania and the other a colloid in a PVP film (Fig. 65). The particle size distributions were not particularly narrow, and it was reported that the TEM micrographs showed the presence of some large particles. Even if such particles were too few in number to appear in a histogram of particle diameters, they may have made a large contribution to the bulk-like NMR signal since the NMR is a volume-sensitive technique. The PVP sample seemed to have fewer of these large particles than the titania-supported sample, and this difference was thought to be the reason for the Rh/TiO$_2$ spectrum's larger peak at the bulk metal position (7.4425

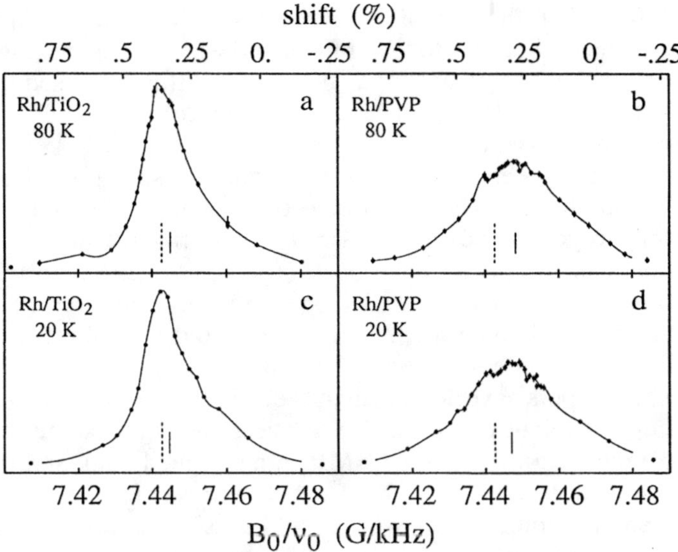

Fig. 65. Point-by-point [103]Rh NMR spectra for a sample of Rh/TiO$_2$ and another of Rh in PVP film at two temperatures. The size distributions peak in the 2- or 3-nm region, but they are not particularly narrow. The Rh/TiO$_2$ has more large particles than does the Rh/ PVP. The long, dashed lines indicate the bulk metal [103]Rh resonance position (the shift scale on top assumes that this is 0.36%); the positions of the short lines divide the integrals into halves. The remarkable result is that the spectrum broadens in both directions and not only to low field, as for platinum. The broadening is more important for the sample with fewer large particles.

G/kHz). Both samples, however, showed considerable NMR intensity on both sides of the bulk position, contrary to what was found for [195]Pt. Regarding Eqs. (15)–(17), the large difference between the NMR of the two metals is the influence of the orbital contributions. Whereas for [195]Pt they are relatively unimportant so that in the NMR layer model the orbital shift is taken as site independent, this is probably not the case for [103]Rh. Indeed, it is well-known that the orbital shifts are large for metals in the center of the transition series, whereas the (pure) Knight shift is more important toward the end.

Theoretically, on surface sites of a rhodium slab the Fermi-level DOS is higher than in the bulk and the bandwidth is smaller (*187*). Therefore, the nuclei in these sites should have a more negative value for K_d and a more positive value for K_{orb} as compared to the bulk values. In small particles (that have lower symmetry than slabs), there is a wide variety of surface sites, which can be expected to have, statistically speaking, independently varying values for the orbital and for the d-like spin contributions to their

Knight shifts. When the orbital contributions are more important, the resulting resonance will occur to lower field, and when the spin value is larger the resonance shifts to higher field. In principle, this increased sensitivity to the orbital shift could be interesting in chemisorption studies. No such work has been published to date.

Preliminary results characterizing ^{119}Sn (another spin-1/2 nucleus) have shown the potential of its NMR for investigation of the modifying function of tin in dehydrogenation catalysts based on supported platinum, palladium, and possibly nickel. Spectra have been obtained for Sn/Ni and Sn/Pd on silica, (with tin in excess in each) (*188*), and they show resonances that have not been clearly identified, in addition to those attributable to β-tin. No additional reports have been published.

A few short papers have dealt with the zero-field NMR of small ferromagnetic nickel and cobalt particles. The physical basis of such measurements is more complicated than that of NMR in an applied field, and this is not discussed here. Suffice it to say that for ^{61}Ni a considerable size-dependent line broadening is found (*189*), which could possibly be used to investigate the effect of chemisorption on the surface magnetism of such particles. The ^{59}Co zero-field NMR can distinguish between superparamagnetic, FCC, and hexagonal close packed (HCP) cobalt (*190*).

ACKNOWLEDGMENTS

During the preparation of the manuscript many colleagues were kind enough to explain their work to me. I particularly thank Jean-Philippe Ansermet, Ken Packer, Marek Pruski, and Jesús Sanz for their detailed remarks on adsorbate NMR. Over the years, our work on metal NMR has often benefitted from discussions with Charles Slichter, who initiated this line of research. I am most grateful to Walter Knight, who got us to think about local densities of state by pointing out the relevance of Fig. 3 to small-particle NMR. Michael Graetzel's suggestions for the preparation of our NMR samples have been of great importance. Our NMR group in Lausanne is partially funded by the Swiss National Science Foundation, recently under Grant 20-53637.98. I thank André Châtelain for the favorable conditions for our work in the Institut de Physique Expérimentale. I thank the following colleagues for kindly allowing me to use material from their publications: Jean-Philippe Ansermet, Geoffrey Bond, John Bradley, Helmut Bross, Michael Duncan, Cecil Dybowski, Jacques Fraissard, Ian Gay, Michèle Gupta, Roald Hoffmann, Vincent Jaccarino, Walter Knight, W. Kohn, Norton Lang, Ken Packer, D. A. Papaconstantopoulos, M. C. Payne, Michel Posternak, Nico Poulis, Marek Pruski, Thatcher Root, Jesús Sanz, Matthias Scheffler, Charles Slichter, Andrzej Wieckowski, Chuin-tih Yeh, and Kurt Zilm.

REFERENCES

1a. Thomas, J. M., and Klinowski, J., *Adv. Catal.* **33,** 199 (1985).
1b. Haw, J. F., and Xu, T., *Adv. Catal.* **42,** 115 (1998).
2. Slichter, C. P., *Annu. Rev. Phys. Chem.* **37,** 25 (1986).
3. Ansermet, J. P., Slichter, C. P., and Sinfelt, J. H., *Prog. NMR Spectrosc.* **22,** 401 (1990).

4. Duncan, T. M., *Colloids Surf.* **45,** 11 (1990).
5. Pruski, M., in "Encyclopedia of Magnetic Resonance" (D. M. Grant and R. K. Harris, Eds.), p. 4636. Wiley, Chichester, UK, 1996.
6. Derouane, E. G., Abdul Hamid, S. B., Pasau-Claerbout, A., Seivert, M., and Ivanova, I. I., *Stud. Surface Sci. Catalysis* **92,** 123 (1995).
7. Winter, J., "Magnetic Resonance in Metals." Clarendon, Oxford, 1971.
8. Fukui, K., Yonezawa, T., and Shingu, H., *J. Chem. Phys.* **20,** 722 (1952).
9. Hoffmann, R., *Rev. Mod. Phys.* **60,** 601 (1988).
10. van Santen, R., "Theoretical Heterogeneous Catalysis." World Scientific, Singapore, 1991.
11. Blyholder, G., *J. Phys. Chem.* **68,** 2772 (1964).
12. Dundon, J. M., *J. Chem. Phys.* **76,** 2171 (1982).
13. Curie, P., *Ann. Chim. Phys.* **5,** 289 (1895).
14. Townes, C. H., Herring, C., and Knight, W. D., *Phys. Rev.* **77,** 852 (1950).
15. Dilke, M. H., Eley, D. D., and Maxted, E. B., *Nature* **161,** 804 (1948).
16. Seitchik, J. A., Gossard, A. C., and Jaccarino, V., *Phys. Rev.* **136A,** 1119 (1964).
17. Weisman, I. D., and Knight, W. D., *Phys. Rev.* **169,** 373 (1968).
18. Inoue, N., and Sugawara, T., *J. Phys. Soc. Jpn.* **45,** 450 (1978).
19. Itoh, J., Asayama, K., and Kobayashi, S., *Proc. Coll. Ampere* **13,** 162 (1964).
20. Narath, A., and Weaver, H. T., *J. Physique (Paris)* **32,** C1–992 (1971).
21. Messmer, R. P., Knudson, S. K., Johnson, K. H., Diamond, J. B., and Yang, C. Y., *Phys. Rev. B* **13,** 1396 (1976).
22. Atkins, P. W., "Physical Chemistry," 4th ed. Oxford Univ. Press, Oxford, 1990.
23. Ponec, V., and Bond, G. C., "Catalysis by Metals and Alloys." Elsevier, Amsterdam, 1995.
24. Kittel, C., and Kroemer, H., "Thermal Physics." Freeman, New York, 1980.
25. Lang, N. D., *Solid State Phys.* **28,** 225 (1973).
26. Bardeen, J., *Phys. Rev.* **49,** 653 (1936).
27. Werner, C., Schulte, F. K., and Bross, H., *J. Phys. C* **8,** 3817 (1975).
28. Smith, J. R., Ying, S. C., and Kohn, W., *Phys. Rev. Lett.* **30,** 610 (1973).
29. Methfessel, M., Hennig, D., and Scheffler, M., *Phys. Rev. B* **46,** 4816 (1992).
30. Sung, S.-S., and Hoffmann, R., *J. Am. Chem. Soc.* **107,** 578 (1985).
31. Heine, V., *Solid State Phys.* **35,** 1 (1980).
32. Friedel, J., *Adv. Phys.* **3,** 446 (1954).
33. Michaelson, H. B., *J. Appl. Phys.* **48,** 4729 (1977).
34. Papaconstantopoulos, D. A., "Handbook of the Band Structure of Elemental Solids." Plenum, New York, 1986.
35. Posternak, M., Krakauer, K., Freeman, A. J., and Koelling, D. D., *Phys. Rev. B* **21,** 5601 (1980).
36. Mason, J., (Ed.), "Multinuclear NMR." Plenum, New York, 1987.
37. van der Klink, J. J., *J. Phys. Condens. Matter* **8,** 1845 (1996).
38. Bucher, J. P., and van der Klink, J. J., *Phys. Rev. B* **38,** 11038 (1988).
39. Clark, W. G., *Rev. Sci. Instr.* **35,** 316 (1964).
40. Haeberlen, U., "High Resolution NMR in Solids: Selective Averaging." Academic Press, New York, 1976.
41. Lizak, M. J., Gullion, T., and Conradi, M. S., *J. Magnetic Res.* **91,** 254 (1991).
42. Jeener, J., Meier, B. H., Bachmann, P., and Ernst, R. R., *J. Chem. Phys.* **71,** 4546 (1979).
43. Forsén, S., and Hoffman, R. A., *J. Chem. Phys.* **39,** 2892 (1963).
44. Chesters, M. A., Dolan, A., Lennon, D., Williamson, D. J., and Packer, K. J., *J. Chem. Soc. Faraday Trans.* **86,** 3491 (1990).

45. Chesters, M. A., Packer, K. J., Lennon, D., and Viner, H. E., *J. Chem. Soc. Faraday Trans.* **91,** 2191 (1995).
46. Wu, X., Gerstein, B. C., and King, T. S., *J. Catal.* **118,** 238 (1989).
47. Uner, D. O., Pruski, M., and King, T. S., *J. Catal.* **156,** 60 (1995).
48. Chesters, M. A., Packer, K. J., Viner, H. E., Wright, M. A. P., and Lennon, D., *J. Chem. Soc. Faraday Trans.* **92,** 4709 (1996); *ibid.* **93,** 2023 (1997).
49. Chesters, M. A., Packer, K. J., Viner, H. E., and Wright, M. A. P., *J. Phys. Chem. B* **101,** 9995 (1997).
50. Sanz, J., and Rojo, J. M., *J. Phys. Chem.* **89,** 4974 (1985).
51. Rojo, J. M., Belzunegui, J. P., Sanz, J., and Guil, J. M., *J. Phys. Chem.* **98,** 13631 (1994).
52. Sheng, T.-C., and Gay, I. D., *J. Catal.* **71,** 119 (1981).
53. Sheng, T.-C., and Gay, I. D., *J. Catal.* **77,** 53 (1982).
54. Rouabah, D., Benslama, R., and Fraissard, J., *Chem. Phys. Lett.* **179,** 218 (1991).
55. Hwang, S.-J., Uner, D. O., King, T. S., Pruski, M., and Gerstein, B. C., *J. Phys. Chem.* **99,** 3697 (1995).
56. Chesters, M. A., Packer, K. J., Viner, H. E., Wright, M. A. P., and Lennon, D., *J. Chem. Soc. Faraday Trans.* **91,** 2203 (1995).
57. Buckingham, A. D., and Stephens, P. J., *J. Chem. Soc.,* 2747 (1964).
58. Ruiz-Morales, Y., Schreckenbach, G., and Ziegler, T., *Organometallics* **15,** 3920 (1996).
59. Fukai, Y., "The Metal–Hydrogen System." Springer, Berlin, 1993.
60. Frieske, H., and Wicke, E., *Ber. Bunsenges. Phys. Chem.* **77,** 48 (1973).
61. Gupta, M., and Freeman, A. J., *Phys. Rev. B* **17,** 3029 (1978).
62. Papaconstantopoulos, D. A., Klein, B. M., Faulkner, J. S., and Boyer, L. L., *Phys. Rev. B* **18,** 2784 (1978).
63. Seymour, E. F. W., Cotts, R. M., and Williams, W. D., *Phys. Rev. Lett.* **35,** 165 (1975).
64. Brill, P., and Voitländer, J., *Ber. Bunsenges. Phys. Chem.* **77,** 1097 (1973).
65. Wicke, E., and Nernst, G. H., *Ber. Bunsenges. Phys. Chem.* **68,** 224 (1964).
66. Barabino, D. J., and Dybowski, C., *Solid State Nucl. Magnetic Resonance* **1,** 5 (1992).
67. Wiley, C. L., and Fradin, F. Y., *Phys. Rev. B* **17,** 3462 (1978).
68. Schoep, G. K., Poulis, N. J., and Arons, R. R., *Physica* **75,** 297 (1974).
69. Cornell, D. A., and Seymour, E. F. W., *J. Less-Common Metals* **39,** 43 (1975).
70. Weinert, M., and Freeman, A. J., *Phys. Rev. B* **28,** 6262 (1983).
71. Eastman, J. A., Thompson, L. J., and Kestel, B. J., *Phys. Rev. B* **48,** 84 (1993).
72. Stuhr, U., Wipf, H., Udovic, T. J., Weissmueller, J., and Gleiter, H., *J. Phys. Condens. Matter* **7,** 219 (1995).
73. Nandi, R. K., Georgopoulos, P., Cohen, J. B., Butt, J. B., Burwell, R. L., and Bilderback, D. H., *J. Catal.* **77,** 421 (1982).
74. Moraweck, B., Clugnet, G., and Renouprez, A., *J. Chim. Phys.* **83,** 265 (1986).
75. McCaulley, J. A., *J. Phys. Chem.* **97,** 10372 (1993).
76. Paul, J.-F., and Sautet, P., *Phys. Rev. B* **53,** 8015 (1996).
77. Wolf, R. J., Lee, M. W., and Ray, J. R., *Phys. Rev. Lett.* **73,** 557 (1994).
78. Polisset, M., and Fraissard, J., *Colloids Surf. A* **72,** 197 (1993).
79. Chang, T. C., Cheng, C. P., and Yeh, C. T., *J. Chem. Soc. Faraday Trans.* **90,** 1157 (1994).
80. Bond, G. C., and Lou, H., *J. Catal.* **147,** 346 (1994).
81. Candy, J. P., Fouilloux, P., and Renouprez, A. J., *J. Chem. Soc. Faraday I* **76,** 616 (1980).
82. Frennet, A., and Wells, P. B., *Appl. Catal.* **18,** 243 (1985).
83. Rouabah, D., and Fraissard, J., *Solid State Nucl. Magnetic Resonance* **3,** 153 (1994).
84. Tong, Y. Y., and van der Klink, J. J., *J. Phys. Chem.* **98,** 11011 (1994).
85. Packer, K. J., personal communication.
86. de Menorval, L. C., and Fraissard, J., *Chem. Phys. Lett.* **77,** 309 (1981).

87. Chang, T. C., Sheng, C. P., and Yeh, C. T., *J. Catal.* **138,** 457 (1992).
88. Rao, L. F., Pruski, M., and King, T. S., *J. Phys. Chem. B* **101,** 5717 (1997).
89. Bhatia, S., Engelke, F., Pruski, M., Gerstein, B. C., and King, T. S., *J. Catal.* **147,** 455 (1994).
90. Engelke, F., Vincent, R., King, T. S., and Pruski, M., *J. Chem. Phys.* **101,** 7262 (1994).
91. Pruski, M., personal communication.
92. Engelke, F., Bhatia, S., King, T. S., and Pruski, M., *Phys. Rev. B* **49,** 2730 (1994).
93. Ito, T., and Kadowaki, T., *Jpn. J. Appl. Phys.* **14,** 1673 (1975).
94. Boudart, M., Delbouille, A., Derouane, E. G., Indovina, V., and Walters, A. B., *J. Am. Chem. Soc.* **94,** 6622 (1972).
95. Dennison, P. R., Packer, K. J., and Spencer, M. S., *J. Chem. Soc. Faraday Trans. I* **85,** 3537 (1989).
96. Wu, X., Gerstein, B. C., and King, T. S., *J. Catal.* **121,** 271 (1990).
97. Smale, M. W., and King, T. S., *J. Catal.* **119,** 441 (1989).
98. Savargaonkar, N., Khanra, B. C., Pruski, M., and King, T. S., *J. Catal.* **162,** 277 (1996).
99. Calderazzo, F., personal communication.
100. Ichikawa, S., Poppa, H., and Boudart, M., *J. Catal.* **91,** 1 (1985).
101. Robbins, J. L., *J. Catal.* **115,** 120 (1989).
102. van't Blik, H. F. J., van Zon, J. B. A. D., Huizinga, T., Vis, J. C., Koningsberger, D. C., and Prins, R., *J. Phys. Chem.* **87,** 2264 (1983).
103. Gleeson, J. W., and Vaughan, R. W., *J. Chem. Phys.* **78,** 5384 (1983).
104. Oldfield, E., Keniry, M. A., Shinoda, S., Schramm, S., Brown, T. L., and Gutowsky, H. S., *J. Chem. Soc. Chem. Commun.,* 791 (1985).
105. Wang, Z., Ansermet, J.-Ph., Slichter, C. P., and Sinfelt, J. H., *J. Chem. Soc. Faraday Trans. I* **84,** 3785 (1988).
106. Compton, D. B., and Root, T. W., *J. Catal.* **136,** 199 (1992).
107. Sharma, S. B., Laska, T. E., Balaraman, P., Root, T. W., and Dumesic, J. A., *J. Catal.* **150,** 225 (1994).
108. van der Putten, D., Brom, H. B., Witteveen, J., de Jongh, L. J., and Schmid, H., *Z. Phys. D* **26,** S21 (1993).
109. Kolbert, A. C., de Groot, H. J. M., van der Putten, D., Brom, H. B., de Jongh, L. J., Schmid, G., Krautscheid, H., and Fenske, D., *Z. Phys. D* **26,** S24 (1993).
110. Makowka, C. D., PhD thesis, University of Illinois, Urbana (1982).
111. Rudaz, S. L., PhD thesis, University of Illinois, Urbana (1984).
112. Ansermet, J.-Ph., PhD thesis, University of Illinois, Urbana (1985).
113. Shore, S., PhD thesis, University of Illinois, Urbana (1986).
114. Wang, Z., PhD thesis, University of Illinois, Urbana (1987).
115. Becerra, L. R., Klug, C. A., Slichter, C. P., and Sinfelt, J. H., *J. Phys. Chem.* **97,** 12014 (1993).
116. Bradley, J. S., Millar, J. M., Hill, E. W., and Behal, S., *J. Catal.* **129,** 530 (1991).
117. Zilm, K. W., Bonneviot, L., Hamilton, D. M., Webb, G. G., and Haller, G. L., *J. Phys. Chem.* **94,** 1463 (1990).
118. Rudaz, S. L., Ansermet, J.-Ph., Wang, P. K., Slichter, C. P., and Sinfelt, J. H., *Phys. Rev. Lett.* **54,** 71 (1985).
119. Durand, D. J., PhD thesis, University of Illinois, Urbana (1989).
120. Dean, E. J., PhD thesis, University of Illinois, Urbana (1997).
121. Ansermet, J.-Ph., Slichter, C. P., and Sinfelt, J. H., *J. Chem. Phys.* **88,** 5963 (1988).
122. Eischens, R. P., *Acc. Chem. Res.* **5,** 74 (1972).
123. Shore, S. E., Ansermet, J.-Ph., Slichter, C. P., and Sinfelt, J. H., *Phys. Rev. Lett.* **58,** 953 (1987).

124. Zilm, K. W., Bonneviot, L., Haller, G. L., Han, O. H., and Kermarec, M., *J. Phys. Chem.* **94,** 8495 (1990).

125. Becerra, L. R., Slichter, C. P., and Sinfelt, J. H., *J. Phys. Chem.* **97,** 10 (1993).

126. Becerra, L. R., Slichter, C. P., and Sinfelt, J. H., *Phys. Rev. B* **52,** 11457 (1995).

127. Bradley, J. S., Millar, J. M., Hill, E. W., and Melchior, M., *J. Chem. Soc. Chem. Commun.,* 705 (1990).

128. Bradley, J. S., Millar, J. M., and Hill, E. W., *J. Am. Chem. Soc.* **113,** 4016 (1991).

129. Duncan, T. M., and Root, T. W., *J. Phys. Chem.* **92,** 4426 (1988).

130. Duncan, T. M., Winslow, P., and Bell, A. T., *Chem. Phys. Lett.* **102,** 163 (1983).

131. Makowka, C. D., Slichter, C. P., and Sinfelt, J. H., *Phys. Rev. Lett.* **49,** 379 (1982).

132. Slezak, P. J., and Wieckowski, A., *J. Magnetic Resonance A* **102,** 166 (1993).

133. Wu, J., Day, J. B., Franaszczuk, K., Montez, B., Oldfield, E., Wieckowski, A., Vuissoz, P. A., and Ansermet, J.-Ph., *J. Chem. Soc. Faraday Trans.* **93,** 1017 (1997).

134. Vuissoz, P. A., Ansermet, J.-Ph., and Wieckowski, A., *Electrochim. Acta,* **44,** 1397 (1998).

135. Day, J. B., Vuissoz, P. A., Oldfield, E., Wieckowski, A., and Ansermet, J.-Ph., *J. Am. Chem. Soc.* **118,** 13046 (1996).

136. Yahnke, M. S., Rush, B. M., Reimer, J. A., and Cairns, E. J., *J. Am. Chem. Soc.* **118,** 12250 (1996).

137. Wong, Y. T., and Hoffmann, R., *J. Phys. Chem.* **95,** 859 (1991).

138. Hu, P., King, D. A., Lee, M.-H., and Payne, M. C., *Chem. Phys. Lett.* **246,** 73 (1995).

139. Engel, T., and Ertl, G., *Adv. Catalysis* **28,** 1 (1979).

140. Sheu, L. L., Knözinger, H., and Sachtler, W. M. H., *J. Am. Chem. Soc.* **111,** 8125 (1989).

141. Vedrine, J. C., and Naccache, C., *Chem. Phys. Lett.* **18,** 190 (1973).

142. Lunsford, J. H., and Jayne, J. P., *J. Chem. Phys.* **44,** 1492 (1966).

143. Vuissoz, P. A., and Reuse, F., personal communication.

144. Gomer, R., *Rep. Progr. Phys.* **53,** 917 (1990). [See Section 4.3.2]

145. Lewis, R., and Gomer, R., *Nuovo Cimento Suppl.* **5,** 506 (1967).

146. Poelsema, B., Verheij, L. K., and Comsa, G., *Phys. Rev. Lett.* **49,** 1731 (1982).

147. Seebauer, E. G., Kong, A. C. F., and Schmidt, L. D., *J. Chem. Phys.* **88,** 6597 (1988).

148. Kwasniewski, V. J., and Schmidt, L. D., *Surf. Sci.* **274,** 329 (1992).

149. von Oertzen, A., Rotermund, H. H., and Nettesheim, S., *Surf. Sci.* **311,** 322 (1994).

150. Froitzheim, H., and Schulze, M., *Surf. Sci.* **320,** 85 (1994).

151. Croci, M., Félix, C., Vandoni, G., Harbich, W., and Monot, R., *Surf. Sci. Lett.* **290,** L667 (1993).

152. Wang, L., Ge, Q., and Billing, G. D., *Surf. Sci. Lett.* **304,** L413 (1994).

153. Lambert, D. K., *Electrochim. Acta* **41,** 623 (1996).

154. Persson, B. N. J., and Persson, M., *Solid State Commun.* **36,** 175 (1980).

155. Schmidt, M. E., and Guyot-Sionnest, P., *J. Chem. Phys.* **104,** 2438 (1996).

156. Bucher, J. P., Buttet, J., van der Klink, J. J., and Graetzel, M., *Surf. Sci.* **214,** 347 (1989).

157. Makowka, C. D., Slichter, C. P., and Sinfelt, J. H., *Phys. Rev. B* **31,** 5663 (1985).

158. Tong, Y. Y., van der Klink, J. J., Clugnet, G., Renouprez, A. J., Laub, D., and Buffat, P. A., *Surf. Sci.* **292,** 276 (1993).

159. Kleine, A., and Ryder, P. L., *Catal. Today* **3,** 103 (1988).

160. Rhodes, H. E., Wang, P.-K., Stokes, H. T., Slichter, C. P., and Sinfelt, J. H., *Phys. Rev. B* **26,** 3559 (1982).

161. Bucher, J. P., Buttet, J., van der Klink, J. J., Graetzel, M., Newton, E., and Truong, T. B., *J. Mol. Catal.* **43,** 213 (1987).

162. Yu, I., and Halperin, W. P., *Phys. Rev. B* **47,** 15830 (1993).

163. King, T. S., Goretzke, W. J., and Gerstein, B. C., *J. Catal.* **107,** 583 (1987).

164. Plischke, J. K., Benesi, A. J., and Vannice, M. A., *J. Phys. Chem.* **96,** 3799 (1992).

165. Mastikhin, V. M., Mudrakovsky, I. L., Goncharova, S. N., Balzhinimaev, B. S., Noskova, S. P., and Zaikovsky, V. I., *React. Kinet. Catal. Lett.* **48,** 425 (1992).
166. Bercier, J. J., Jirousek, M., Graetzel, M., and van der Klink, J. J., *J. Phys. Condens. Matter* **5,** L571 (1993).
167. Tong, Y. Y., and van der Klink, J. J., *J. Phys. Condens. Matter* **7,** 2447 (1995).
168. Tong, Y. Y., Laub, D., Schultz-Ekloff, G., Renouprez, A. J., and van der Klink, J. J., *Phys. Rev. B* **52,** 8407 (1995).
169. Edwards, P. P., and Rao, C. N. R., (Eds.), "The Metallic and Nonmetallic States of Matter." Taylor & Francis, London, 1985.
170. Luntz, A. C., Williams, M. D., and Bethune, D. S., *J. Chem. Phys.* **89,** 4381 (1988).
171. Bucher, J. P., van der Klink, J. J., and Graetzel, M., *J. Phys. Chem.* **94,** 1209 (1990).
172. Tong, Y. Y., Billy, J. J., Renouprez, A. J., and van der Klink, J. J., *J. Am. Chem. Soc.* **119,** 3929 (1997).
173. Tong, Y. Y., Mériaudeau, P., Renouprez, A. J., and van der Klink, J. J., *J. Phys. Chem. B* **101,** 10155 (1997).
174. Mannstadt, W., and Freeman, A. J., *Phys. Rev. B* **55,** 13298 (1997).
175. Louie, S. G., *Phys. Rev. Lett.* **42,** 476 (1979).
176. Feibelman, P., and Hamann, D. R., *Surf. Sci.* **149,** 48 (1985).
177. Tong, Y. Y., Martin, G. A., and van der Klink, J. J., *J. Phys. Condens. Matter* **6,** L533 (1994).
178. Ebinger, H. D., Jänsch, H. J., Polenz, C., Polivka, B., Preiss, W., Saier, V., Veith, R., and Fick, D., *Phys. Rev. Lett.* **76,** 656 (1996).
179. Tong, Y. Y., Renouprez, A. J., Martin, G. A., and van der Klink, J. J., *Stud. Surface Sci. Catal.* **101,** 901 (1996).
180. Tong, Y. Y., Belrose, C., Wieckowski, A., and Oldfield, E., *J. Am. Chem. Soc.* **119,** 11709 (1997).
181. Rice, C., Tong, Y. Y., Oldfield, E., and Wieckowski, A., *Electrochim. Acta,* **43,** 2825 (1998).
182. Tong, Y. Y., Yonezawa, T., Toshima, N., and van der Klink, J. J., *J. Phys. Chem.* **100,** 730 (1996).
183. Ladas, S., Dalla Betta, R. A., and Boudart, M., *J. Catal.* **53,** 356 (1978).
184. van Leeuwen, D. A., van Ruitenbeek, J. M., Schmid, G., and de Jongh, L. J., *Phys. Lett. A* **170,** 325 (1992).
185. Toshima, N., Yonezawa, T., and Kushihashi, K., *J. Chem. Soc. Faraday Trans.* **89,** 2537 (1993).
186. Vuissoz, P. A., Yonezawa, T., Yang, D., Kiwi, J., and van der Klink, J. J., *Chem. Phys. Lett.* **264,** 366 (1997).
187. Eichler, A., Hafner, J., Furthmüller, J., and Kresse, G., *Surf. Sci.* **346,** 300 (1996).
188. Chang, T. H., Cheng, C. P., and Yeh, C. T., *Mat. Chem. Phys.* **22,** 503 (1989).
189. Kohara, T., Yamaguchi, M., and Asayama, K., *J. Phys. Soc. Jpn.* **54,** 1537 (1985).
190. Zhang, Z., Suib, S. L., Zhang, Y. D., Hines, W. A., and Budnick, J. I., *J. Am. Chem. Soc.* **110,** 5569 (1988).

Applications of Photoluminescence Techniques to the Characterization of Solid Surfaces in Relation to Adsorption, Catalysis, and Photocatalysis

MASAKAZU ANPO

Department of Applied Chemistry
College of Engineering, Osaka Prefecture University
Sakai, Osaka 599-8531, Japan

and

MICHEL CHE

Laboratoire de Réactivité de Surface
UMR 7609- CNRS
Université Pierre et Marie Curie
75252 Paris Cedex 05, France and
Institut Universitaire de France

This review deals with the applications of photoluminescence techniques to the study of solid surfaces in relation to their properties in adsorption, catalysis, and photocatalysis. After a short introduction, the review presents the basic principles of photoluminescence spectroscopy in relation to the definitions of fluorescence and phosphorescence. Next, we discuss the practical aspects of static and dynamic photoluminescence with emphasis on the spectral parameters used to identify the photoluminescent sites. In Section IV, which is the core of the review, we discuss the identification of the surface sites and the following: coordination chemistry of ions at the surface of alkaline–earth and zirconium oxides, energy and electron transfer processes, photoluminescence and local structure of grafted vanadium oxide, and photoluminescence of various oxide-

Abbreviations: S_0, ground singlet state; S_1, excited singlet state; T_1, excited triplet state; $R_{if} = [\varphi_i|R|\varphi_f]$, transition moment integral; UV, ultraviolet light; E_{sg}, surface band gap; Φ_e, quantum efficiency (or yield) of photoluminescence; Φ^*, concentrations of the excited state; τ, lifetime; ΔD, dead-layer thickness, i.e., the change in depletion width; φ_i, φ_f, initial and final wave functions, respectively; I_e, emission intensity; I_a, absorption intensity; Σ_0, extinction coefficient; α, absorptivity for the exciting light (*227*); β, absorptivity for the emitted light (*227*); k_q, absolute quenching rate constant; λ_{max}, wavelength at the maximum intensity of the spectrum; transition metal ion; LC, low coordination; EA, electron affinity; CVD, chemical vapor deposition; PVC, porous Vycor glass; EXAFS, extended X-ray absorption fine structure; XPS, X-ray photoelectron spectroscopy; TPD, temperature-programmed desorption; DRS, diffuse reflectance spectroscopy.

119

based catalysts. In Section V, we present a summary of the close relatiohship between the dynamics of the photoluminescence and the reactivities of the catalysts. Sections VI and VII deal with the applications of photoluminescence to adsorption and catalysis and discuss the case of both bulk and supported oxides. Section VIII is devoted to photocatalysis, again on bulk and supported catalysts. Sections IX and X briefly review the application of photoluminescence to chemical detection and its relation with other techniques. The last section concludes the work covered in the review and gives the most likely future directions of research work involving photoluminescence spectroscopy.

I. Introduction

An important goal of investigations of the fundamental nature of catalysis is to provide a detailed analysis and characterization of the active surface sites and their role in the dynamics of adsorption and catalysis at the molecular level, whether induced by thermal energy (catalysis) and/or by photonic energy (photocatalysis).[1]

One approach toward this goal has been to use oxide solid solutions, as proposed by Vrieland and Selwood (*1*) and Cimino *et al.* (*2, 3*). The principle, as applied to oxide catalysts, is to select a diamagnetic oxide of high crystal symmetry as the matrix and to dissolve in it a second oxide which contains metal ions of catalytic interest (*3–6*). In many cases, the solute oxide is a transition metal oxide. Solid solutions allow the investigation of the catalytic activity of a selected transition metal ion (TMI) as the catalyst changes from insulating and paramagnetic behavior in very dilute solutions to semiconducting and magnetic behavior for increasing TMI concentrations, and they also allow the observation of how adsorbed molecules are bonded to the TMI relative to its location (face, edge, or corner) and coordination at the surface.

Another approach involves the preparation and use of well-defined supported oxide catalysts. Because the method of preparation, such as simple impregnation, leads to active sites which may be isolated from one another, particularly at low concentrations of the supported oxide, the sites become relatively easy to identify and characterize at the molecular level (*7*). Im-

---

[1] Photocatalysis can be defined as the field combining photochemistry and catalysis and indicates that light and a catalyst are necessary to induce chemical reactions. For example, semiconducting catalysts absorb photons with an energy higher than their band gap, producing electrons and holes in the conduction and valence band, respectively. These electrons and holes can produce surface active sites or species, respectively, which induce catalytic reactions in their transient lifetimes, 10^{-9}–10^{-6} s. On the other hand, highly dispersed catalysts absorb photons corresponding to their absorption bands, producing excited states with high reactivity. Therefore, these excited states induce catalytic reactions in their transient lifetimes of 10^{-7}–10^{-5} s.

proved methods such as anchoring, grafting, competitive ion exchange, and deposition precipitation have been developed, and the effects of changing the preparation method can now be better understood. When concentrations typically <0.1 wt% are involved, unusual and unique catalytic or photocatalytic properties due to active phase–support interactions, dispersion effects, and/or quantum size effects can be observed. However, most conventional characterization techniques are not sensitive enough for such low concentrations.

Therefore, photoluminescence spectroscopy, involving lifetime measurements and photoluminescence intensity and line shape determination, has been applied to obtain significant information about the structures of active surface sites, especially for oxide loadings <0.1 wt%. When combined with laser or time-resolved spectroscopy, photoluminescence measurements provide insight not only into the dynamics of catalysis and photocatalysis,[2] including molecular migration and electron transfer processes at surfaces, but also into the molecular-level discrimination and identification of active sites on catalysts or photocatalysts regardless of their morphology or porosity.

Because of the significant contribution of fluorescence and phosphorescence studies to gas-phase and solution photochemistry, photoluminescence investigations of oxide catalysts have also been carried out. Valuable information characterizing the surface structures of oxides and their excited states has been obtained by workers including Vol'kenshtein *et al.*(8), Tench and Pott (9), Pott and Stock (10) Kazansky, *et al.* (11, 12), Coluccia (13), Anpo *et al.* (14–17), Texter *et al.* (18), Iwasawa (19, 20), Iwamoto *et al.* (21), Moser and Grätzel (22), Becker and Bard (23), Weller *et al.* (24), Tanguay and Suib (25), and others (26–32).

Originally, photoluminescence spectroscopy was applied to characterize the local coordination of metal ions as well as to probe structural perturbations that occur due to alkaline earth and rare earth metal ions in oxides such as silica and alumina. Emphasis has turned to elucidating the mechanisms of catalytic and photocatalytic reactivity, i.e., the characterization, at the molecular level, of the active surface sites as well as the significant role of these sites in catalysis and photocatalysis.

The sensitivity and selectivity of catalytic and photocatalytic reactions to small electronic and structural perturbations in the chemical surroundings of the active sites are crucial to understanding the mechanisms involved.

---

[2] Dynamics of photoluminescence refers to the behavior of photoluminescence in the presence of reactant molecules as the pressures of reactant molecules and/or diffusion rates are changed. It is possible to obtain detailed information about the excited state of the emitting sites as a function of time during their transient lifetimes.

Although there are numerous techniques which can be used to detect chemical perturbations, they often lack sensitivity. Photoluminescence spectroscopy can be considered ideal because of its high sensitivity and nondestructive nature (*14, 33, 34*).

This review covers adsorption, catalysis, and photocatalysis that can be investigated and understood by photoluminescence spectroscopy. Most of the results discussed in this review have been obtained by photoluminescence techniques, but other, complementary techniques, are also discussed to emphasize the originality and potential value of photoluminescence spectroscopy, particularly with regard to anion coordination chemistry, excited states, and reaction dynamics. The latter field is of utmost importance in chemistry (*35*). Additional applications of photoluminescence spectroscopy to the study of solid surfaces are reviewed in the books "Photochemistry on Solid Surfaces"(*36*) and "Surface Photochemistry"(*37*).

The determination of the photoluminescence parameters (excitation frequency, emission frequency, Stokes shift, fine structure parameter, and lifetime) can lead to information which, at the simplest level, indicates the presence of an electronically excited state of a species, but which can be sufficiently detailed so as to lead to a clear identification and characterization of the photoluminescent sites(*36–44*). Moreover, measurements of the variations in the intensity and positions of the bands as a function of time (time-resolved photoluminescence) provide valuable kinetic data representing the reactions occurring at the surface. Although most of the photoluminescence measurements have been carried out at low temperatures for specific reasons (see Section III.C.2), there is much evidence that some of the excited states involved are present even at higher temperatures and that they play an important role in catalytic and photocatalytic reactions. Therefore, it is clear that the information obtained by photoluminescence techniques is useful and important for the design of new catalysts and photocatalysts.

II. Basic Principles of Photoluminescence

Photoluminescence can be defined as the radiation emitted from a molecule or a solid which, after it has absorbed energy from an external source and been transferred into an electronic excited state, returns to its ground electronic state. Although it can be said that photoluminescence consists of both fluorescence and phosphorescence, the distinction between the two is generally phenomenological, but, as shown later, a more theoretical distinction can be proposed. This review is presented in the context of photoluminescence as an analytical tool, the basic principles of which can be found in physical chemistry (*45–48*) and photochemistry books (*49–53*), and, whenever necessary, reference has been made to solid surfaces, essentially oxides, typically represented by alkaline earth oxides.

A. ABSORPTION SPECTRUM, FRANCK–CONDON PRINCIPLE, AND VIBRATIONAL STRUCTURE

The principles of photoluminescence applied to solid oxide surfaces can be most easily understood by assuming some simplifications. For example, we can start by considering the Morse potential energy curves (Fig. 1) related to an ion pair such as $M^{2+}O^{2-}$, taken as a harmonic oscillator to represent an oxide, typically an alkaline earth oxide. The absorption of light close to the fundamental absorption edge of this oxide leads to the excitation of an electron in the oxide ion followed by a charge-transfer process to create an *exciton* (an electron–hole pair), which is essentially

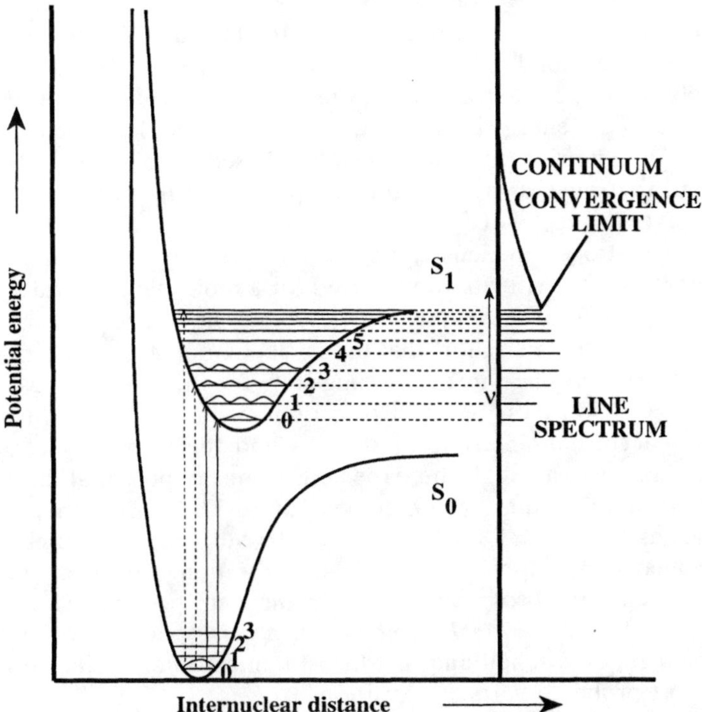

FIG. 1.   Potential energy curves for the ground (S_0) and excited states (S_1) which account for the absorption spectrum with a vibrational structure shown on the right-hand vertical side [reproduced with permission from Barrow (*47*)].

free to migrate through the lattice:

$$M^{2+}(ns^0)\ O^{2-}(2p^6) \uparrow \downarrow \quad (S_0) \rightarrow M^{2+}(ns^0)\ O^{2-}(2p^53s^1) \uparrow \downarrow \tag{1}$$

$$M^{2+}(ns^0)\ O^{2-}(2p^53s^1) \uparrow \downarrow \quad \rightarrow M^{+}(ns^1) \uparrow O^{-}(2p^53s^0) \downarrow \quad (S_1) \tag{2}$$

The horizontal lines in the potential wells shown in Fig. 1 represent the possible vibrational states of the oscillator, either in the ground electronic singlet state (S_0) in which all the electrons are paired or in the electronic excited singlet state (S_1) in which the charge-transfer shown in Eq. (2) occurs and the unpaired electrons remain with opposite spins. The vibrational states (v) are labeled by increasing energy, starting with the vibrational ground state $v = 0$, of the electronic ground state, followed by the vibrational excited states v_1, v_2, etc. Labeling is performed for the electronic excited state, starting again with the vibrational ground state $v' = 0$, followed by the vibrational excited states v_1', v_2', etc. It is readily understood that the potential energy curve of the excited state (S_1) is located at higher energy than that of the ground state (S_0), with its potential well located at larger internuclear distance because, electrostatically, the charge transfer leads to a M^+O^- bond, i.e., a weakening of the $M^{2+}O^{2-}$ bond.

The observed spectral transitions, which lead to the absorption spectrum, are related to the energy diagram such as that of Fig. 1 on the basis of the Franck–Condon principle. This principle is based on the fact that electrons move and rearrange at a much higher speed than that of the vibrational movement of the nuclei of a molecule. For example, the time for an electron to circle a hydrogen nucleus can be calculated from Bohr's model to be about 10^{-16} s. A typical vibration period for a molecule is only about 10^{-13} s. The comparison of these times suggests that an electronic configuration changes within a time so short that the nuclei do not change their positions during absorption. The same reasoning state for the molecule also applies to the $M^{2+}O^{2-}$ ion pair. Spectral transitions with the highest probability then occur at constant internuclear distance and, therefore, should be drawn vertically and not, as might be expected, from the potential minimum of the lower curve to that of the upper curve. The transition probability decreases as the transition departs from the vertical and is highest when starting near the midpoint of the lowest vibrational level of the ground electronic state. In absorption, an electronic transition may show a series of lines, called the *vibrational* or *fine structure,* corresponding to the different vibrational states reached and in which the most intense transition refers to the most probable vertical transition.

B.   THE FATE OF ELECTRONIC EXCITATION ENERGY

In absorbing light, the system acquires the associated excitation energy and moves to an upper electronic excited state. There are many ways by which the system can return to the lower energy state. The excess energy can be lost to the vibration, rotation, and translation of the surrounding molecules or ions. This loss of energy is most efficient for gases and liquids and less so for solids, in which vibrations essentially are predominant since rotations and translations are hindered. This thermal degradation transforms the excitation energy into thermal motion of the environment, i.e., heat. Another, more interesting, possibility is that the excess energy becomes involved in a chemical reaction, and the result is photochemistry (*49–53*).

Another possibility is radiative decay (as opposed to nonradiative or radiationless decay), occurs when the molecule or ion pair loses its excitation energy as a photon. Two different radiative decay mechanisms are possible: fluorescence and phosphorescence. The essential distinction between these two emissions was not clearly understood until the work of Lewis *et al.* (*54*) and Terenin (*55*), who proposed that a triplet state was responsible for the long-lived emission. The radiation emitted in a transition between states of the same spin multiplicity (i.e., singlet–singlet or triplet–triplet transitions) is called fluorescence, and the radiation emitted in a transition between states of different spin multiplicity (i.e., triplet–singlet transitions) is called phosphorescence. The fluorescence lifetimes are usually very short (e.g., about 10^{-9} s for organic molecules), whereas phosphorescence lifetimes are about 10^{-3} s to minutes or longer because transitions between states of different spin multiplicity are forbidden and thus have very low probabilities.

1. *Excitation, Emission (Fluorescence), and Stokes Shift*

The sequence of steps involved in fluorescence is described in Fig. 2. The initial absorption takes the molecule or ion pair from the ground electronic singlet state (S_0) to the electronically excited singlet state (S_1), and the resulting absorption (*excitation*) spectrum should resemble that shown in Fig. 3a.

In the excited state, the ion pair is subjected to the influence of its lattice environment, and as it gives up energy (by radiationless decay or a vibrational relaxation process) it steps down the ladder of vibrational levels. The lattice environment, however, may be unable to accept the larger energy difference needed to lower the ion pair to the ground electronic state; therefore, the ion pair may survive long enough to undergo spontaneous emission, releasing the remaining excess energy as radiation. The downward electronic transition is vertical in accordance with the Franck–Condon principle, and the fluorescence spectrum shown in Fig. 3b has a vibrational structure characteristic of the lower electronic state.

This mechanism accounts for the observation that fluorescence occurs

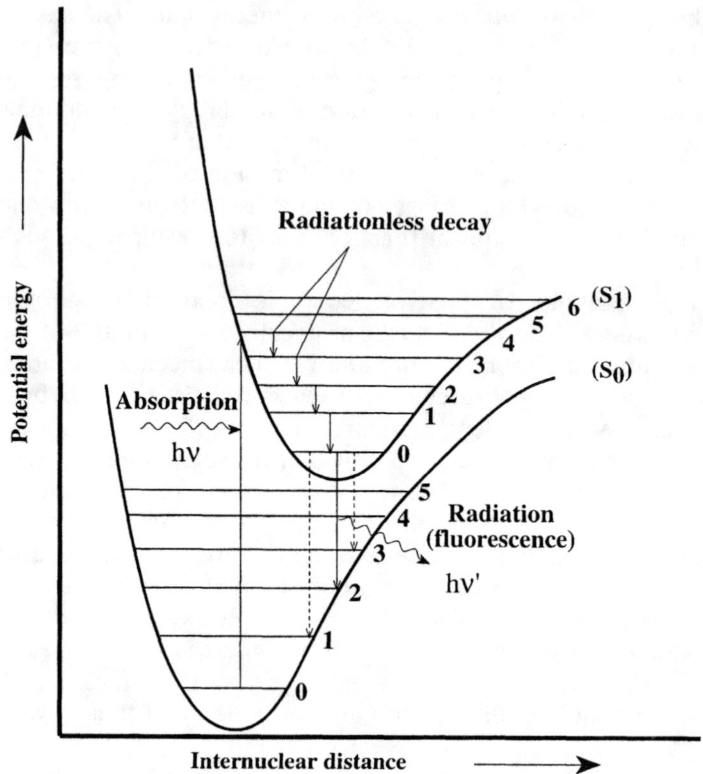

FIG. 2. The sequence of steps leading to fluorescence. After the initial absorption, the upper vibrational states in the excited state (S_1) undergo radiationless decay (vibrational relaxation) by giving up energy to the lattice environment and/or surrounding gas-phase molecules, when the ion pair is located at the solid–gas interface. A radiative transition then occurs from the vibrational ground state (v_0) of the electronically excited state (S_1) to the ground state (S_0) [reproduced with permission from Atkins (46)].

at a frequency lower than that of the incident light. The difference between the two frequencies is referred to as the *Stokes shift*. The emission occurs after some vibrational energy has been dissipated into the surroundings. The mechanism also suggests that the intensity of the fluorescence should depend on the ability of the lattice environment (or the surrounding gas-phase molecules if the ion pair is located at the solid–gas interface) to accept the electronic and vibrational quanta. It has indeed been observed that molecules with widely spaced vibrational levels, such as molecular oxygen, may be able to accept the large quantum of electronic energy and extinguish or quench the fluorescence. Also, the environment is less likely

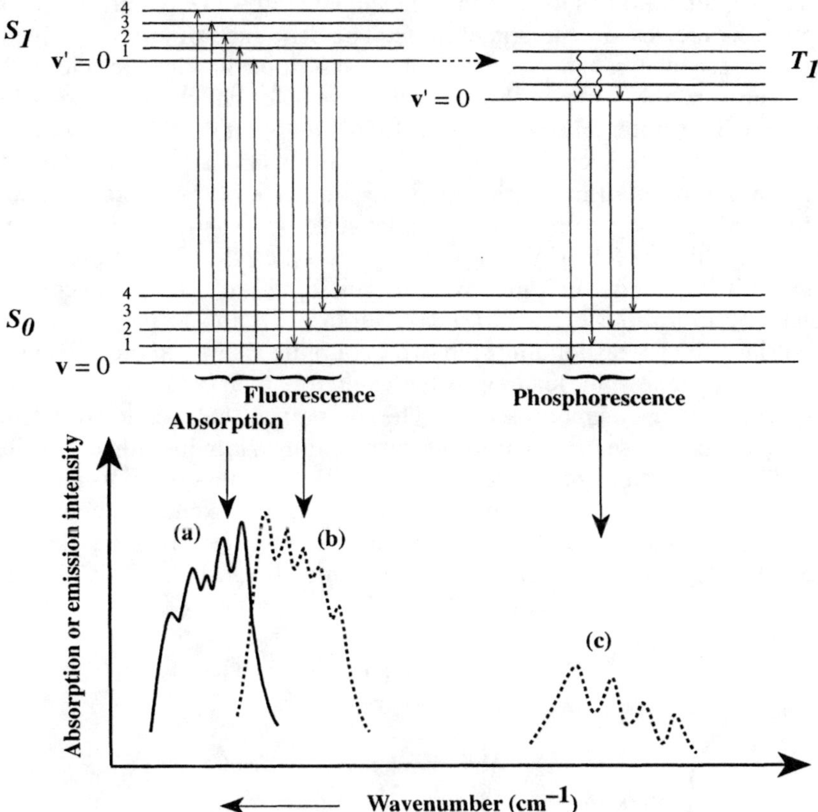

Fig. 3.   Schematic diagram showing the origin of absorption (a), fluorescence (b), and phosphorescence (c) spectra. The absorption spectrum (a) shows a vibrational structure characteristic of the upper excited state (S_1). The fluorescence spectrum (b) shows a structure characteristic of the lower ground state (S_0); it is also displaced to lower frequencies (the 0–0 transitions show a coincidence) and resembles a mirror image of the absorption [reproduced with permission from Daniels and Alberty (45)].

to accept the excitation energy as the temperature is lowered since lattice vibrations are less favored at low temperatures (49–53).

2. Excited Triplet State, Intersystem Crossing, Phosphorescence, and Selection Rules

When a spin flip occurs for an electron in the S_1 state, an excited triplet state (T_1) is created:

$$M^+(ns^1) \uparrow \ O^-(2p^53s^0) \downarrow \ (S_1) \rightarrow M^+(ns^1) \uparrow \ O^-(2p^53s^0) \uparrow \ (T_1) \qquad (3)$$

The triplet state lies lower than the singlet state since electron–electron repulsions are less in the triplet state. The sequence of steps involved in phosphorescence can be understood as described in Figs. 3c and 4. The first steps are the same as those for fluorescence, but the presence of the excited triplet state plays a decisive role. At the point at which the potential energy curves intersect (Fig. 4), the two states share a common geometry. If there is a mechanism for changing the spin state of the S_1 state, then the ion pair M^+O^- may cross into the triplet state by the so-called intersystem crossing. This singlet–triplet transition may take place in the presence of spin orbit coupling. Such intersystem crossing is expected to occur more efficiently in a molecule with a heavy atom or in an ion pair in an oxide solid since the corresponding spin orbit coupling is large and electrons can acquire the same spin, leading to the excited triplet state (50–52).

Now consider phosphorescence. The ion pair in the excited triplet state continues to release energy into the surroundings, moving down the vibrational ladder, until it is trapped in the lowest vibrational state. Its surroundings cannot accept the large quantum of electronic excitation energy, and the ion pair cannot radiate its energy because the return to the electronic ground state involves a forbidden triplet–singlet transition. This transition

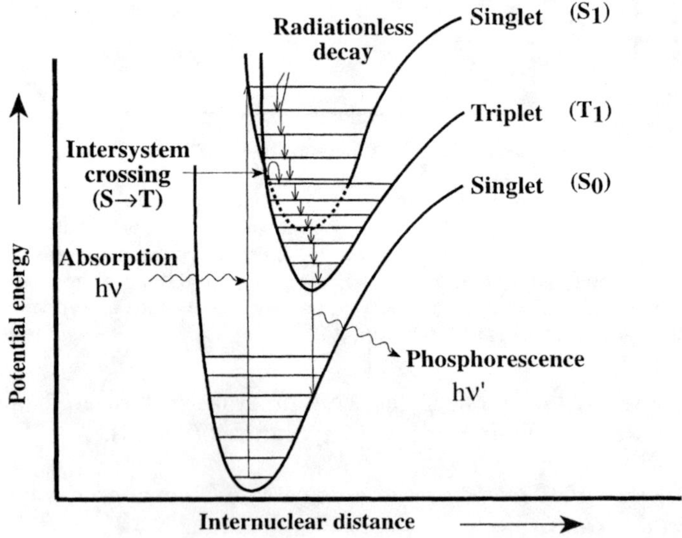

Fig. 4. The sequence of steps leading to phosphorescence. The important step is the intersystem crossing, the passage from the excited singlet (S_1) to triplet state (T_1) induced by spin orbit coupling. The triplet state acts as a slowly radiating reservoir since return to the ground state is spin forbidden [reproduced with permission from Atkins (46)].

is not totally forbidden, however, because the spin-orbit coupling that was responsible for the intersystem crossing also breaks the selection rule. The ion pairs are able to emit phosphorescence weakly, and the emission may continue long after the original excited state is formed.

This mechanism is in agreement with the experimental observation that the excitation energy appears trapped in a reservoir that slowly leaks. Generally, the phosphorescence is most intense when catalysts have highly dispersed metal ions and solid materials have highly localized emitting sites because the energy transfer becomes less efficient and there is enough time for intersystem crossing to occur, as the singlet excited state steps slowly past the intersection point. The mechanism also suggests that the phosphorescence efficiency should depend on the presence of heavy atoms (with strong spin-orbit coupling) in molecules and on the bonding strength of the ion pair with its surroundings. Finally, it also predicts that, due to the unpaired spins in the triplet, magnetic properties should be involved either as a $S = 1$ system, if the two unpaired electrons are present, or as a $S = 1/2$ system, if only one unpaired electron remains after reaction (e.g., with gas-phase molecules) (49–62). A good illustration of the latter case is the activation of CH_4 by surface O^- ions, produced by processes shown in Eqs. (1)–(3), following the reaction

$$O^- + CH_4 \rightarrow OH^- + \dot{C}H_3 \tag{4}$$

The transitions involved in the photochemical processes discussed so far are electronic in nature and thus associated with electric dipole changes during either absorption or emission. From a theoretical standpoint, the dipole moment change which occurs during the transition of a molecule moving from a state of energy E_i (wavefunction: φ_i) to a state of higher energy E_f (wavefunction: φ_f) is represented by the transition-moment integral, $R_{if} = \langle \varphi_i | R | \varphi_f \rangle$. The probability of the transition (absorption or emission) is proportional to the magnitude of R_{if}^2. If $R_{if} = 0$, the corresponding transition is forbidden, whereas if R_{if} differs from zero it is allowed (49–53).

From a more practical point of view, electronic transitions follow two types of selection rules because of the orbital and spin nature of the electronic wavefunction. The first, called the Laporte rule, requires that $\Delta l = \pm 1$ for the orbitals involved in the transition. It predicts, for instance, that electronic transitions for transition metal ions in T_d symmetry (involving orbitals with d–p character) should be more intense than "Laporte-forbidden" d–d transitions in O_h symmetry involving orbitals of the same character thus leading to $\Delta l = 0$. By contrast, charge-transfer transitions are essentially Laporte-allowed since they concern orbitals involving different atoms with different characters. In the case of centrosymmetric complexes, this rule implies a change of parity: u $\rightarrow$ u and g $\rightarrow$ g transitions (as, for instance, $t_{2g} \rightarrow e_g$ transitions in octahedral complexes) are forbidden, whereas transitions such as g → u are allowed. The second wavefunction referred to as the spin rule, states that the spin multiplicity $2S + 1$ of the levels involved must be the same, i.e., $\Delta S = 0$. Thus, singlet–singlet transitions such as those involved in fluorescence are spin-allowed, whereas singlet–triplet transitions such as those involved in phosphorescence are spin–forbidden.These selection rules may be partially or even totally relaxed by various mechanisms, such as vibronic coupling, spin orbit coupling, and exchange interaction in the case of polymetallic systems (*49–53*).

3. *Vibrational Deactivation and Internal Conversion*

As stated previously, the loss of the excitation energy of the system can occur by a nonradiative process. Figure 5 shows a representative arrangement of potential energy curves which lead to such deactivation. The excess energy is first lost by vibrational deactivation, internal conversion; note that this process refers to electronic excited states with the same spin multiplicity within the highest potential energy curve of the excited singlet state S_1. At the crossover point (C) with the excited triplet state, the geometry and potential energies of the two electronic states are equal so that the system can transfer to the lower potential energy curve T_1 by intersystem crossing (this process

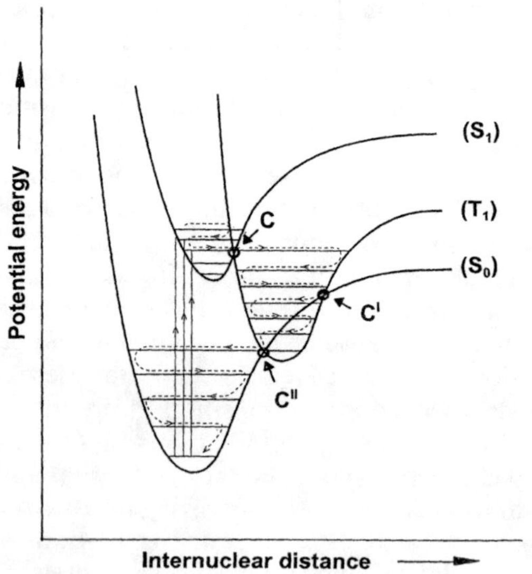

FIG. 5. Energy dissipation (dashed arrow) by vibrational deactivation and internal conversion [reproduced with permission from Barrow (*47*)].

involves states of different spin multiplicities). Other crossover points (C′ and C″) allow the systems to reach their original ground states. This case is commonly encountered with polyatomic molecules for which potential curves often exist with suitable relative positions so that combinations of internal conversion and intersystem crossing return the molecule to its ground state before another emission process has had a chance to occur. The previous points are schematically represented in Fig. 6, which summarizes the processes described so far between various energy levels (*49–53*).

III. Practical Aspects of Photoluminescence

Photoluminescence analysis generally involves various molecular spectroscopic techniques used to observe the sensitivity and selectivity of the

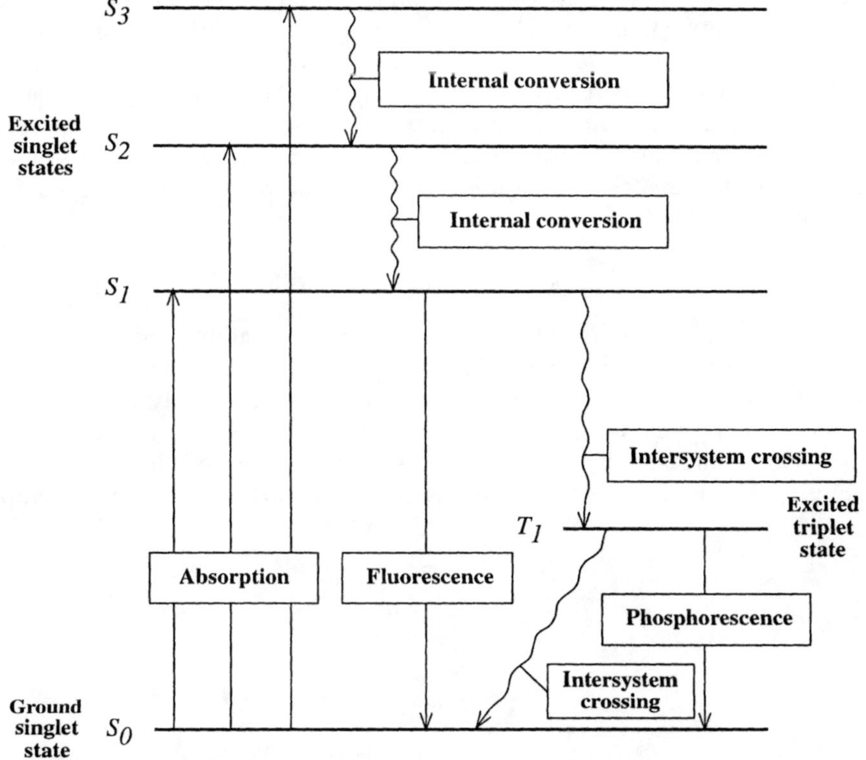

FIG. 6.   Schematic representation of changes in energy levels which may occur upon absorption of radiation. Nonradiative processes are represented by wave arrows and radiative processes by straight line arrows [reproduced with permission from Daniels and Alberty (*45*)].

emitting species of molecules and/or materials. The high sensitivity of photoluminescence analysis has allowed the detection and determination of extremely low concentrations of emitting materials. Lower limits of detection are in the parts per million (ppm) to parts per billion (ppb) range. The selectivity of photoluminescence analysis has become possible using variations in excitation and/or analytical wavelengths, allowing the simultaneous determination of emitting components in many mixtures. Potentially, phosphorescence can be observed with a higher degree of sensitivity and selectivity than fluorescence due to its longer lifetime, which offers the possibility of investigations with differing spectroscopic parameters.

For a better understanding and more efficient application of photoluminescence spectroscopy, it is essential to gain insight into the nature of the excited states of the target species and of the various photochemical processes (*33, 34, 36–38, 51, 52*). A molecule in an excited state can be considered a new entity, only remotely related to the parent molecule in the ground state. In the excited state, the molecule will have a completely different electron distribution from that of the molecule in the ground state as well as a different geometry. Also, it usually undergoes chemical reactions that are quite different from those of the molecule in the ground state. These photochemical and photophysical processes can be used to explain the basic principles of the photocatalytic reactivity of catalysts(*40–44*).

A. Instrumentation

A variety of commercially available autocompensating spectrophotofluorometers (photoluminescence instrumentation) have been employed to measure and analyze the photoluminescence spectra of catalysts. The three principal components of all spectrophotofluorometers are the excitation light source, the chamber to set the sample, and the emitted photon detector (Fig. 7). The light source is usually either a mercury or a xenon arc lamp.

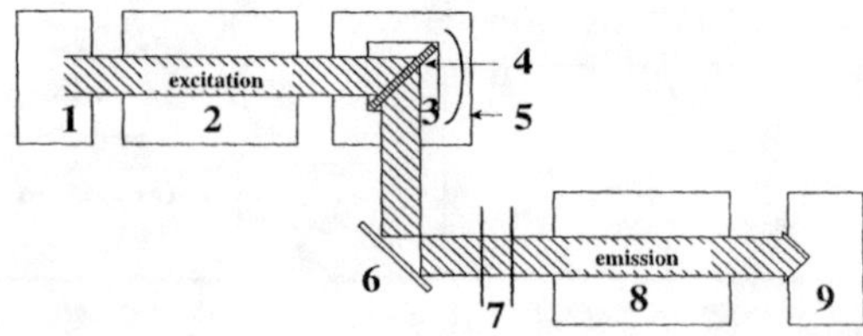

FIG. 7.   Placement of optical components in a conventional spectrophotofluorometer. 1, lamp; 2, excitation monochromator; 3, cell; 4, sample; 5, light trap; 6, mirror; 7, filters; 8, emission monochromator; 9, detector.

The light passes through a monochromator (equipped with a grating or a prism) to select the excitation radiation to be focused on the sample cell. Interference filters can also be used to provide greater selectivity and resolution for analysis of the photoluminescence spectra. The emission leaving the sample cell usually passes directly into the emission monochromator and then into a high gain photomultiplier tube. Although the photoluminescence output can be displayed in a variety of ways, the spectrum is recorded with the correct compensation to cover small variations in the excitation light power (*33, 34, 36–38, 49–53*).

B.   SAMPLE PREPARATION

Although the design of the cell holder may determine the amount of scattered and stray light from the samples, glass filters can be used to minimize scattered light. Some kinds of quartz may fluoresce under ultraviolet (UV) excitation causing errors in the photoluminescence measurements. Therefore, secondary filters may be used, and the sample cell should be made from high-quality nonfluorescent quartz. It is not possible to eliminate the Raman emission from the sample, which is sometimes used to interrelate the fluorescence spectra, but the fluorescence spectrum of the blank which comes from the sample cell can be used to detect the background signals due to scattering.

The presence of oxygen is critical because the interaction of oxygen with the sample can cause serious errors, and its ability to quench not only the fluorescence but also phosphorescence makes deaeration necessary. Also, the photoluminescence intensity usually decreases when the temperature of the sample is increased; this is attributed to the higher probability for other nonradiative deactivations in the excited state of the molecule. Therefore, to minimize such temperature effects, the photoluminescence spectra are often measured at liquid-nitrogen temperature and even liquid-helium temperature (*33, 34, 36–38, 49–53*).

C.   SPECTRAL PARAMETERS TO IDENTIFY PHOTOLUMINESCENT SITES

Measuring the fluorescence intensities of fluorescent reference compounds for the purpose of identification and standardization is essential in monitoring the emission yields of the sample and calibrating the number of photons emitted from the excitation source. The ideal is a series of reference compounds that cover the visible to near-UV regions of the spectrum. The reference compound should also have its absorption in the same region as that of the sample compound, and its emission spectrum should be in the same general region as the emission of the compound under investigation.

The most important parameters of the photoluminescence spectrum which are necessary to identify the emitting species are its spectral shape, from which the emission intensity I_e is obtained as a function of wavelength λ; its quantum efficiency Φ_e, by which the intensity is measured relative to the absorption photointensity I_a; and its lifetime τ, determined from the decay curves of variation of the intensity measured as a function of time (*33, 34, 36–38, 49–53*).

1. Wavelength and Spectral Shape

In a photoluminescence spectrum, the emission intensity, I_e, is plotted as a function of the frequency μ (or wavelength λ) of the emitted light I, for which the excitation frequency and the intensity of the exciting light are fixed at constant values. The photoluminescence spectrum may be well-defined but it is also often partially resolved or even unresolved. This can be explained by the molecular electronic transition which does not always correspond to a well-defined quantum of energy because different nuclear geometries are associated with the initial or final electronic states, leading to partially resolved or unresolved spectra. Such a situation is most likely found for molecules in solution or for solids. In certain cases, however, some vibrational fine structure is observed in the band corresponding to the electronic transition, subsequently leading to the prominent vibrational progression of the emission band of the vibrations associated with this emission, the nuclear equilibrium positions are most dramatically changed by the radiative electronic transition.

As shown in Fig. 8, the vibrational bands are clearly resolved in the photoluminescence spectrum of vanadium oxide supported on Vycor glass (*33, 34, 36–38*). The vibrational separation of about 1040 cm^{-1} is in good agreement with the energy of the $V=O$ stretching vibration of the ground state of the vanadyl group of the oxide as measured by IR or Raman spectroscopies.

On the other hand, in the excitation spectrum, the emission intensity I_e, at the monitored emission band, is plotted as a function of the wavelength λ of the excitation light, which varies as the extinction coefficient ε_a of the absorbing molecules. Therefore, the excitation spectrum exhibits the same spectral appearance as that of the absorption spectrum. The advantage of measuring the excitation spectrum in addition to the emission spectrum is the greater sensitivity even for low concentrations of photoluminescent material compared to standard absorption measurements.

2. Quantum Efficiency

The quantum efficiency or yield of the photoluminescence, Φ_e, is defined as the ratio of the number of photons emitted from the excited molecules

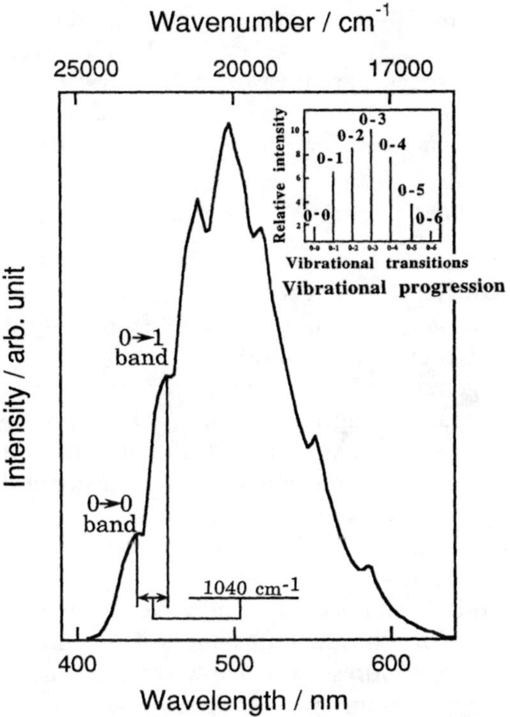

FIG. 8. Vibrational fine structure of the photoluminescence of vanadium oxide supported on porous Vycor glass at 77 K; inset shows its vibrational progression [reproduced with permission from Kubokawa and Anpo (*63*)].

to the number of photons absorbed by the ground state molecules and can be expressed as

$$\Phi_e = k_e/(k_e + k_{IC}) = \Phi^* k_e \tau, \tag{5}$$

where k_e and k_{IC} are the radiative rate constant and the sum of the unimolecular rate constants of the nonradiative deactivation processes from the radiative excited state, respectively, and Φ^* and τ are the concentrations of entities in the excited state and the measured experimental lifetime of the excited state, respectively (*33, 34, 36–38, 49–53*). All excited states emit a certain number of photons, but it has been found that quantum yields of $<10^{-5}$ are difficult to measure accurately. The experimental photoluminescence quantum yield, Φ_e, depends on k_{IC}, which in turn is sensitive to measurement conditions and molecular electronic properties. Thus, to observe a photoluminescence spectrum with a high efficiency, it is usually necessary to minimize k_{IC}. This can be accomplished by cooling the sample to low temperatures, e.g., 77 K or even 4 K. However, even when the photoluminescence spectra are measured at 77 or 4 K, the quantum yields are generally less than 1.00 since radiationless processes occur even at such low temperatures (*33, 34, 36–38, 49–53*).

3. Lifetimes and the Stern–Volmer Expression

The inherent radiative lifetime, τ_0, of an excited state is the mean time it would take to deactivate the excited state when no radiationless process has occurred. The measured lifetime of an excited state is generally not equal to the true radiative lifetime because of competing radiationless deactivations. In general, lifetimes are longer for the upper states reached by a more weakly absorbing transition, and they are also longer as the energy separation between the ground and the excited states becomes larger. Any decrease in the efficiency of the collisional deactivation, e.g., by freezing in a glass or cooling to liquid-nitrogen temperature, causes the suppression of the radiationless deactivation, leading to significant radiative decay (*33, 34, 36–38, 49–53*).

For liquid or solid in solutions, the radiative lifetime is shortened by the presence of a variety of collisional substances. These are believed to quench the radiative process (fluorescence or phosphorescence). The lifetime in the absence of the quenching species (the so-called quencher) represented by τ_0 has been expressed as

$$\tau_0 = 1/(k_e + k_{IC}). \tag{6}$$

In the presence of the quenching molecules, the lifetime of the excited state τ has been given as

$$\tau = 1/(k_e + k_{IC} + k_q[Q]), \tag{7}$$

where k_q is the specific rate constant for the quenching process and $[Q]$ is the concentration of quencher molecules. When τ_0 cannot be measured directly, the relative efficiencies of the photoluminescence spectrum in the presence or the absence of the quencher molecules [i.e., Φ_e and $(\phi_e)_0$], respectively, can be used in place of τ and τ_0 (*49–53*).

Under conditions of steady and constant illumination, with no irreversible photochemical processes or chemical reactions, the concentrations of the excited state in the presence and the absence of the quenchers can be expressed as $(\Phi) = I/(k_e + k_{IC} + k_q[Q])$ and $(\Phi)_0 = I_0/(k_e + k_{IC})$, respectively. When the illumination intensity and concentration of the molecules are maintained constant in the absence or the presence of the quenching species, the photoluminescence signal yield (intensity) is proportional to the concentration of the excited molecules. Thus, the following equation is derived:

$$(\Phi)_0/(\Phi) = 1 + k_q[Q]/(k_e + k_{IC}) = 1 + \tau_0 k_q[Q]$$

This is known as the Stern–Volmer expression. A plot of $(\Phi)_0/(\Phi)$ versus $[Q]$ gives a straight line with an intercept equal to 1.0 and a slope equal to $k_q \tau_0$ (*51, 52, 56–58*).

D.   ENERGY TRANSFER AND MIGRATION

If a molecule A is in the vicinity of an excited molecule D^* and the energy of the exicted state of A lies below that of D^*, then the energy transfer from the donor molecule D^* to the acceptor molecule A occurs, (*51, 52*). The energy transfer process is schematically represented by the equations

$$D_0 + h\nu \rightarrow D^* \tag{8a}$$

$$D^* + A_0 \rightarrow A^* + D_0 \tag{8b}$$

where the asterisks indicate the excited states. The energy transfer from D^* to A_0 must occur within the lifetime of D^*, τ, and only within this time does molecule D remains in the excited state. The energy migration describes the energy transfer between the same kind of molecules, causing the excitation energy to delocalize.There are two possibilities by which an energy transfer or migration occurs; (i) by the radiative transfer involving the emission of a photon by the donor molecule and the subsequent reabsorption by the acceptor molecule and (ii) by the radiationless transfer caused by collisional interactions between donor and acceptor molecules:

$$D^* + A_0 \rightarrow A^* + D_0 \tag{9}$$

E.   RELEVANCE OF PHOTOLUMINESCENCE TO SURFACE PHENOMENA

Photoluminescence spectroscopy was not originally a surface technique; however, as is the case for several other spectroscopies, it has been widely applied to investigate surfaces as well as surface chemical phenomena with a high degree of sensitivity (*59–62*). This technique has been successfully applied

(a) When the photoluminescent sites are on bulk oxides having sufficiently large surface to volume ratios.

(b) When the photoluminescent sites are located on the surface of a support on which they have been deposited by conventional techniques, such as impregnation, ion exchange, anchoring, grafting, deposition, and precipitation. This is typically the case in oxide-supported catalysts (*33, 34, 36–38*).

(c) When the photoluminescent sites of the two preceding types of materials are modified by treatment such as calcination, evacuation, and reduction, which all affect the surface.

(d) When probe molecules change the environment of the photoluminescent sites.

IV. Photoluminescence and the Nature of Surface Sites

When a solution of a transition metal complex which interacts with a solid surface to form a catalytic system is further subjected to thermal treatments, a solid–fluid (liquid or gas) interface is produced. The presence of interfaces defines a new type of transition metal chemistry, which can be characterized by using TMIs as probes of their own interaction with the oxide surface. This is possible because their partly filled d orbitals make them extremely sensitive to any change in their environment, which can be monitored by their specific optical and magnetic properties. This approach has led to the establishment of interfacial coordination chemistry. At solid–fluid interfaces, mixed complexes are formed which not only involve conventional mobile ligands (water, ammonia, ethylenediamine, etc.) but also involve the rigid surface ligand. Because of oxide ions available to bind TMIs, the silica support, via its surface SiO^- oxide ions, has been found to be a weak σ donor–π donor ligand which can be introduced in the spectrochemical series of ligands, as follows: $SiO^- < H_2O < NH_3$ (*7*). Other oxides have been investigated, and a spectrochemical series of oxide supports has been proposed which can be included within the spectrochemical series of ligands (*64*):

$$Cl^- < AlO^- < ZO^- < SiO^- < H_2O < NH_3 < en,$$

where AlO^- and ZO^- represent surface oxide ions of γ-alumina and NaY zeolite, respectively.

The surface oxide ions play a key role in their binding to oxide supports. However, comprehending their coordination chemistry and that of surface nontransitional cations and anions, with empty d orbitals, is difficult. Photoluminescence techniques have been found to be effective for elucidating this chemistry, particularly for alkaline earth oxides, which are largely ionic while having a simple cubic structure and can be regarded as good model systems. They also provide the advantage of allowing the study of oxide ions in the absence of the complicating effects of variable-valent TMIs. A wide range of material sources with added impurities have been investigated so that it has been concluded that the effects described are not likely to arise from adventitious impurities.

A.   COORDINATION CHEMISTRY AND STABILIZATION OF SURFACE ANIONS AND CATIONS IN THE CASE OF UNSUPPORTED ALKALINE EARTH AND ZIRCONIUM OXIDES

1.   *Surface Oxide Ions*

There is strong evidence that O^{2-} ions in positions of low coordination (LC) at oxide surfaces have unusual electron donor properties. At this point it is useful to define what is meant by ions of low coordination.

a.   *Representation of the Coordination of Oxide Ions.*   Discussions of oxygen species at oxide surfaces often exclude the O^{2-} ion (although there are many such ions present on oxide surfaces) because surface oxide ions are difficult to study since they are not easily distinguishable from oxide ions of the bulk. However, from a representation of the MgO(100) surface (Fig. 9), it is evident that surface oxide ions have very different environments.

A typical coordination of the ion in the bulk lattice can be defined by the number of nearest neighbor ions of the opposite sign; for the octahedrally coordinated alkaline earth oxides this number is 6. Ions of lower coordination are designated by the subscript LC to denote a low coordination. For

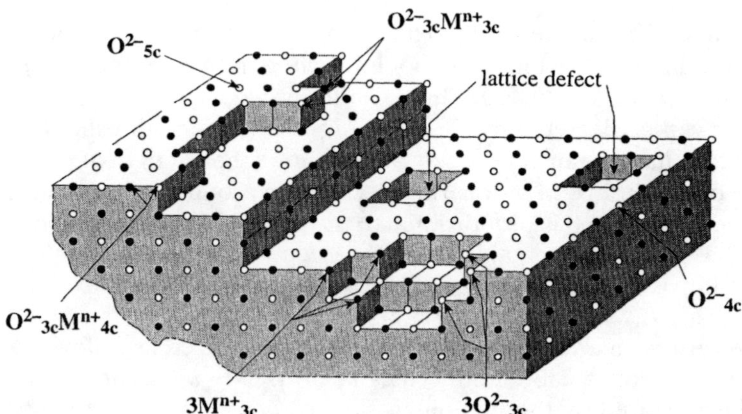

FIG. 9.   Representation of a surface plane (100) of MgO showing surface imperfections such as steps, kinks, and corners which provide sites for ions of low coordination [adapted from Coluccia (65)].

instance, the surface oxide ions on an MgO(100) plane can be described as O_{LC}^{2-}, with a coordination of 5 or O_{5C}^{2-}, whereas generally LC can take the values of 5, 4, or 3 for the different coordinations, as shown in Fig. 9. Also, within a given coordination number such as O_{3C}^{2-}, there is a wide range of oxide ions which differ in their environments as demonstrated in Fig. 9 which shows two types of O_{3C}^{2-} ions (66).

Whereas these oxide ions have the same number of nearest neighbor cations, they differ in the number of the next nearest neighbors as well as those farther removed, leading to different Madelung constants, showing that there can be more than two types of O_{3C}^{2-} ions. Superoxide O_2^- ions produced on MgO by the adsorption of probe molecules such as dioxygen have been shown to be very sensitive to the difference in the Madelung constants of surface ions relative to that of bulk ions, as shown by their differing EPR parameters (67). In some cases, coordination numbers less than three can be found, particularly for supported ions deposited by ion exchange or grafting. For example, O_{2C}^{2-} ions can be found in dimers such as $M-O_{2C}-M$, and O_{1C}^{2-} ions can be found with "yl" oxygen such as in terminal molybdenyl $Mo=O$ or vanadyl $V=O$ groups (63, 68, 69).

b. *Electron Donor Properties.*   The electron donor properties of surface O_{LC}^{2-} ions have been investigated both chemically and spectroscopically by means of charge-transfer reactions and EPR. Organic molecules were characterized by electron affinities (EA) enabling the following reactions to take place:

$$S + A \rightarrow S^+ + A^-, \tag{10}$$

where S is the surface of an oxide support and A an electron acceptor molecule involving a nitro compound such as nitrobenzene (NB) (68) or tetracyanoethylene (70, 71). The A^- radicals produced upon such a charge transfer can be detected by EPR techniques (61). The threshold of EA at which the charge-transfer process occurs shows the strength of the reducing power of the surface sites donating electrons to produce A^-, and the number of radicals produced is proportional to the number of such reducing surface sites. NB and other organic electron-acceptor (A) molecules have been used to investigate the electron-donor properties of ionic oxides such as MgO (72, 76).

The results characterizing MgO can be summarized as follows: Adsorption of electron-acceptor molecules such as NB *in vacuo* onto a MgO powder prepared by thermal outgassing *in vacuo* exhibited an EPR signal that could be assigned to NB^- or the corresponding negative radical ions A^-. The concentration of the radicals was monitored as a function of the outgassing temperature of MgO *in vacuo*. Two maxima were found in the concentration of the radicals, at about 490 K and at about 970 K. It was shown that the source of electrons leading to NB^- or A^- involved OH^- groups for the low-temperature maximum and O_{LC}^{2-} ions for the high-temperature maximum, respectively. It appears that oxide ions O_{LC}^{2-} have a potential for ionization that is weakened as their coordination number decreases. To be released from an O_{3C}^{2-} surface site, the electron has to counterbalance to the first order the three neighboring Mg^{2+} and this number increases to four or five when the electron is released from the O_{4C}^{2-} or O_{5C}^{2-} ions, respectively, leading to an increasing ionization potential and, hence, to a lower electron donor strength as the coordination number increases. These arguments can also be applied to OH^- groups.

Measurements of the spin concentration indicate that relatively high concentrations of A^- (equivalent to about 0.5% of the surface oxide ions) can be formed. Other sources of electrons such as TMI impurities may be ruled out because of their low concentrations. The formation of NB^- radical ions was also found to be much greater on MgO smoke that had been etched by water vapor than on normal MgO smoke (65, 72). Etching of the regular cubic particles composing MgO smoke led to a considerable increase in the lowest coordination surface oxide ions, i.e., O_{3C}^{2-}.

It was also found that the strength of the electron-donor property varies in the same way as the basic strength (i.e., Lewis basicity) of the surface (66). Studies carried out with Al_2O_3 and TiO_2 catalysts using electron-acceptor molecules having electron affinities ranging from 1.77 to 2.84 eV indicate that there are distinct variations in the donating strength among the surface electron-donor sites(66).

The coordination of surface oxide ions has also been explored by using the adsorption of halogens on MgO (73). Oxygen was found to evolve as surface halide ions were formed according to the following reaction:

$$2(X_2)_g + 2(O^{2-})_s \rightarrow 4(X^-)_s + (O_2)_g \tag{11}$$

From the amount of oxygen that evolved, it appears that 20, 10, and 1% of the surface oxide ions were replaced by the chloride, bromide, and iodide ions, respectively. By considering the heat of formation of the solid oxide MgO and halides (MgO, 143 kcal mol^{-1}; $MgCl_2$, 153 kcal mol^{-1}; $MgBr_2$, 125 kcal mol^{-1}; MgI_2, 87 kcal mol^{-1}) (74), it was concluded that the liberation of gaseous oxygen from the bromine indicates the destabilization by at least 18 kcal mol^{-1} (i.e., 143–125 kcal) of some Mg^{2+}–O^{2-} ion pairs. This result can be interpreted to mean that iodine can only react with those O_{LC}^{2-} ions in very low (possibly threefold) coordination, whereas bromine reacts with O_{LC}^{2-} in four- and threefold coordination and chlorine reacts with a wide range of surface oxide ions. The modified and/or changed

Coulombic and polarization effects present at the oxide surface compared to bulk can be considered to be the cause of these reactions.

c.   *Spectroscopic Studies.*   While chemical evidence showing that oxide ions in a state of low coordination may act as electron donors has accumulated, spectroscopic studies have shown that highly dispersed alkaline earth oxides have optical absorption bands that are not representative of pure single crystals. This result is surprising since the energy required for electronic excitation of bulk MgO corresponds to the frequency in the vacuum UV regions, i.e., 8 or 9 eV.

The absorption of light close to the fundamental absorption band edge of a bulk oxide leads to the excitation of an electron in the oxide ion followed by a charge-transfer process to create an exciton such as has been described in Eqs. (1) and (2) (see Section II.A). These processes lead to electronic absorption bands of lowest energy which can be observed in pure undamaged single crystals, occurring at 16 nm (7.68 eV) for MgO and 182 nm (6.8 eV) for CaO (76). On surfaces, the ions experience a reduced Madelung potential because of their reduced coordination, and this leads to the absorption of light at energies lower than the band edge.

The various states of coordination are associated with the surface excitons at different energies. If the oxide ion were completely isolated from the lattice, the optical absorption process would correspond to the simple electronic excitation of the oxide ion Eq. (1). However, the experimental results shown in Table I (66) indicate that the energy required for this process decreases as the atomic number of the cation increases, which is consistent with the observation that a charge transfer is involved [(Eq. 2)]. Such a process may involve more than one cation because of the delocalization of the electron, but it may be increasingly regarded as localized as the coordination of the ions is reduced.

TABLE I

Peak Positions (in eV) of the Absorption, Excitation, and Photoluminescence Spectra of the Alkaline Earth Oxide Powder Catalysts[a]

Oxide	Peak position		
	Absorption	Excitation	Photoluminescence
MgO	5.70, 4.58	>5.40, 4.52	3.18
CaO	5.52, 4.40	>5.40, 4.40	3.06
SrO	4.64, 3.96	4.43, 3.94	2.64
BaO	3.60, 3.22	3.70	2.67

[a] From Che and Tench (66).

Intrinsic surface exciton absorption can be seen in high-surface-area MgO, provided that the surface is thoroughly outgassed to remove surface carbonate and hydroxide. Nelson and Hale (*76*) were the first to report that high-surface-area MgO, CaO, and SrO could lead, after quenching by oxygen and the removal of the strong fluorescence, to the reflectance spectra with bands in the range 5.7–3.9 eV, which they ascribed to the surface states. This fluorescence was confirmed for MgO by Tench and Pott (*9*), who were the first to report its photoluminescence spectrum with the presence of a doublet at about 400 (3.099 eV) and 440 nm (2.817 eV). The spectrum was found to be quite sensitive to the addition of molecules such as O_2 and air, leading to the conclusion that the observed photoluminescence was a surface phenomenon.

Two separate experimental approaches, diffuse reflectance and photoluminescence spectroscopy, were then taken; both led to similar results. The latter technique is the more sensitive, and well-resolved spectra can often be observed, but only when a radiative decay of the excited state occurs. The diffuse reflectance spectra are broader in scope but the absorption bands appear as shoulders. The reflectance spectra of alkaline earth oxides were examined by Zecchina *et al.* (*77, 78*), Garrone *et al.* (*79*), and Zecchina and Scarano (*80*), but an overpressure of a quenching gas (usually oxygen) had to be used to suppress the fluorescence and to allow observation of the reflectance absorption bands (Fig. 10). In addition to usual bands in the UV region due to bulk excitations (bulk excitons), new absorption bands which correspond to excitations localized on the surface ions are present.

Although many different configurations for surface sites may be expected in high-surface-area oxides, the reflectance spectra show only a few bands for each oxide (Table I) (*66*). Each band may be considered as representing a particular local surface coordination of the oxide ions, and these bands are changed by the adsorption of gases; those of lowest frequency, which correspond to the ions in the lowest coordination, are most affected. The adsorbing gases convert LC surface ions to ions of higher coordination, and the higher Madelung potential arising from this increase in coordination displaces the surface absorption toward a higher frequency. For example, in SrO (*81*) the absorption bands associated with the bulk ions can be detected, and the bands do not change upon the adsorption of gases, in contrast to the absorption bands associated with the surface ions. These assignments have been confirmed by the sintering effect on the optical properties of some alkaline earth oxides. The results show that the larger the dimension and perfection of the microcrystals, the lower the intensity of the peaks corresponding to oxide ions in states of lower coordination (*66, 77–83*).

A parallel approach involves investigating the photoluminescence spectra

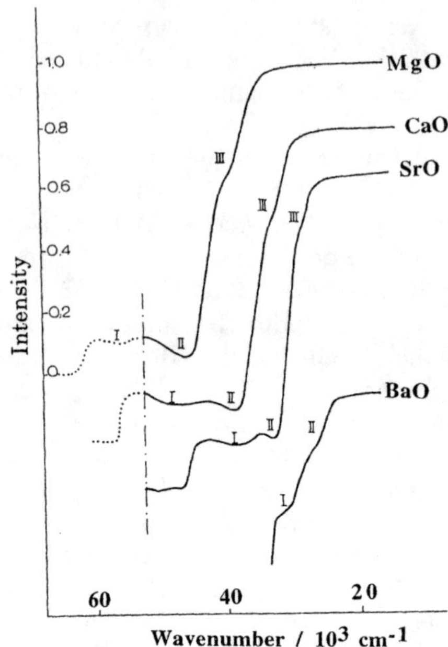

FIG. 10.   Diffuse reflectance spectra of MgO, CaO, SrO, and BaO showing exciton absorption (I, II, and III) due to the surface ions. Solids were outgassed at 1073 K, and the spectra were measured under conditions of quenched fluorescence. The reflectance scale is displaced vertically to avoid overlap of spectra. The dotted portions to the left of the vertical dashed line at $n > 52{,}000$ cm⁻¹ (the vacuum UV) are extrapolations (redrawn figure showing the O_{LC}^{2-} attached to each band) [reproduced with permission from Garrone *et al.* (*79*)].

of alkaline earth oxides (*81–83*). Although bulk excitations leading to free excitons undergo a radiationless thermal decay into phonons, the surface excitations result in a radiative decay of photons (photoluminescence). In the absence of a quenching gas such as oxygen, the exciton, previously formed by the absorption of light at frequency ν, may undergo radiative decay, and light is emitted at a lower frequency ν, due to the Stokes shift (see Section II.B.1):

$$(M^{+}\text{–}O^{-}) \rightarrow (M^{2+}\text{–}O^{2-}) + h\nu' \tag{12}$$

The difference in energy $(h\nu - h\nu')$ is lost to the vibration modes of the crystal. The emitting site may not be the original site of the absorption of light since the exciton may hop to other sites of low coordination, but the lower the coordination the less likely this will occur.

The advantage of photoluminescence techniques, i.e., the measurement of the emission and the excitation spectra (see Section II.B.1), is that it can provide information about both the absorbing and the emitting sites. The excitation spectra are linked to the absorbing sites and correspond to the absorption bands obtained from the reflectance data, which can be obtained without complications that occur with added gases such as oxygen; the emission spectra provide information about the emitting sites.

The formation of oxide surfaces occurs during the decomposition of the precursor hydroxide surface *in vacuo* (*81*). For MgO (*82*), two emissions with peaks at about 400 nm (3.1 eV) can be observed when the oxide is excited at 230 nm (5.4 eV) and 274 nm (4.5 eV), and the intensities of these emissions reach a maximum at temperatures between 1073 and 1273 K, being associated with the O_{LC}^{2-} ions. These photoluminescence spectra are immediately quenched by oxygen, and the emitting sites are destroyed by CO_2. The absorption bands measured in the reflectance spectra and the corresponding bands in the excitation spectra show a good coincidence for all the alkaline earth oxides (Table I).

Excitation data could not be obtained for the highest energies because of the low light intensities. The position of maximum intensity is given for the emission peak, but the shape changes with the excitation energy, clearly showing that there is more than one emitting component involved (*33, 34, 36–38*).

From comparisons of the absorption and excitation spectra for the oxides, as shown in Table I (*66*) it appears that the energy decreases with an increase in the cation size from Mg to Ba in the alkaline earth metal cation series. This pattern has been satisfactorily explained by using the approach of Levine and Mark (*84*), whereby ions located on an ideal surface are considered to be equivalent to the bulk ions, except for their reduced Madelung constants. A more detailed analysis has been carried out by Garrone *et al.* (*60, 79*), who reinterpreted earlier reflectance spectra and suggested that there is evidence of three absorption bands corresponding to ions in five, four, and three coordination—all three for MgO, CaO, and SrO.

If ions exist on the surface in various degrees of coordination, then the relative population of ions in each state of coordination should be strongly dependent on the surface topography of the particle and should change if the crystal imperfections change. For MgO, a comparison has been carried out with MgO smoke, which is formed in nearly perfect small cubes (500–1000 Å). This MgO smoke has a much smaller than normal concentration of O_{LC}^{2-} ions, as confirmed by the photoluminescence spectrum (*77–83, 85*) which shows a much smaller than usual excitation peak at 274 nm. Electron micrographs show that etching in water vapor leads to the erosion of the corners and edges of the MgO crystallites together with the creation of surface pits. The photoluminescence spectra showed a sharp increase in the usual excitation peak at 274 nm, which is attributed to the O^{2-}_{LC} ions in the lowest coordination (i.e., threefold coordination), thus confirming the general predictions of the model (*81–83*).

The absorbing sites have been discussed in terms of the cation–anion couples in low coordination on the surface, but no hypothesis has been made regarding the nature of the emitting sites. The emitting sites could be (i) extrinsic impurities on the surface, e.g., TMIS or organic materials; (ii) point defects such as trapped electron or hole centers; or (iii) sites identical or similar to the absorbing sites. These possibilities have been considered by Coluccia (*13*).

The first two possibilities have been shown to be unlikely. Deliberate dopings and treatments known to increase the population of surface defects lead to the quenching of the photoluminescence (*86*). Moreover, surface defects are annealed at high temperatures, whereas sample treatment at high temperatures is required to observe the photoluminescence (*87*). Although it has been suggested that trapped electron centers (F_s^+) could play a role (*88, 89*), it has been concluded by Coluccia (*13*) that the photoluminescent sites are of the same nature as the absorbing sites and the emission process is as described by Eq. (12).

The most important features of both the reflectance and the photoluminescence spectra have been explained by the preceding model since it is based on ideal surface structures essentially determined by (001) planes. Thus, several likely possibilities, such as the presence of surface defects, impurities, and remaining adsorbates, the relaxation of the planes exposed at the surface, the impurity-induced reconstruction of the surfaces, and changes in the force constants, have been excluded (*80*). A more detailed model is needed in which the ion pair of the metal cation and oxygen anion can be taken into account on the basis of such experimental evidence as the hydrogen adsorption on MgO obtained by Coluccia and Tench (*65*) and Ito *et al.* (*90*).

2. *Photoluminescence Spectra of Oxides with Coordinatively*
    Unsaturated Surface Sites

   a. *Photoluminescence Properties of Microcrystalline MgO.*   For octahedrally coordinated alkaline earth oxides such as MgO, the normal coordination number of ions in the bulk is 6. However, on microcrystalline MgO, in which (001) faces are largely predominant, the coordination number of the surface ions is no longer as high as 6. Generally, as shown in Fig. 9, they are expressed as Mg^{2+}_{LC} or O^{2-}_{LC} with such lower coordination numbers as 5, 4, or even 3 (*13, 66*). Furthermore, many researchers have reported direct spectroscopic evidence to indicate that such surface ions in low coordination play a significant role in the catalytic and photocatalytic activities of MgO.

Tench *et al.* (*9, 73, 91*) found that high-surface-area powdered MgO microcrystals degassed at high temperatures exhibit photoluminescence when excited by UV light at approximately 260–400 nm with an energy (about 4 or 3 eV) much lower than the band gap of MgO single crystals (about 9 or 8 eV)—i.e., pure bulk single crystals show no absorption in such near-UV regions. These authors also found that such photoluminescence is sensitively quenched by the addition of O_2 or air, leading to the conclusion that the photoluminescence of such high-surface-area MgO powders is associated with a surface phenomenon. They showed that the values of the surface band gap, E_{sg}, for different surface planes of MgO, which were calculated from the expression of Levine and Mark by considering the surface Madelung constants, are in good agreement with the excitation energies of the observed photoluminescence spectra of well-degassed MgO powders. Zecchina and coworkers (*77, 78, 92, 93*) found that abnormal absorption bands appear in the UV reflectance spectra of well-degassed powdered MgO (or CaO, SrO, or BaO) at much lower frequencies than those of bulk MgO single crystals. From more extensive studies, together with theoretical (*94*) and EPR measurements (*66, 95*), such abnormal absorption bands and the observed short lifetime (10^{-3}–10^{-6} s) of the photoluminescence spectra have been described by the charge-transfer processes [Eq. (13)] involving surface ions in low coordination at the surface of well-degassed MgO microcrystals:

$$(Mg_{LC}^{2+}-O_{LC}^{2-}) \xrightarrow{h\nu} (Mg_{LC}^{+}-O_{LC}^{-})^* \tag{13}$$

Anpo *et al.* (*96–98*) found that microcrystalline MgO evacuated at high temperatures for a sufficient time (defined as well degassed) exhibits two different types of photoluminescence; an emission with a short lifetime (about 10^{-5} s) and another with a long lifetime (about 1–10^4 s). The former emission is associated with the charge-transfer processes involving surface ions in low coordination and the latter with the presence of surface F_s^+ centers. Consequently, it has been confirmed that both the charge-transfer processes (charge-transfer mechanism) and the radiative process involving the surface F_s^+ center (the defect mechanism) account for the photoluminescence of MgO degassed at temperatures higher than 673 K. It has been shown that the yields of the photoluminescence with a short and a long lifetime change from sample to sample and depend on the pretreatment temperature. Anpo *et al.* (*96–98*) also found that the *trans–cis* isomerization of 2-butenes is photocatalyzed on well-degassed MgO microcrystals and

148   M. ANPO AND M. CHE

that a parallel exists between the intensity of the short lifetime photoluminescence and the rate of the photocatalyzed isomerization. This result
clearly indicates that coordinatively unsaturated surface ions play a significant role in the photocatalytic activity of well-degassed MgO catalysts.

On the other hand, Ito *et al.* (*99*) found that the oxidative dimerization
of methane to yield ethylene and ethane can be achieved with a high yield
and good selectivity on Li-doped MgO catalysts. Since this pioneering work,
many oxidic systems have been studied. Anpo *et al.* (*100*) found that surface
sites of low coordination produced by the incorporation of Li^+ into MgO
play a vital role in the methane oxidative coupling reaction. Thus, although
it was known that MgO acts as an acid-base catalyst, both the catalytic and
photocatalytic activities of the MgO catalysts seem to be associated with the
existence of surface ions in low coordination located on MgO microcrystals.

As shown in Fig. 11 (spectrum 1), the well-evacuated MgO catalyst was
found to exhibit a photoluminescence spectrum at approximately 340–450

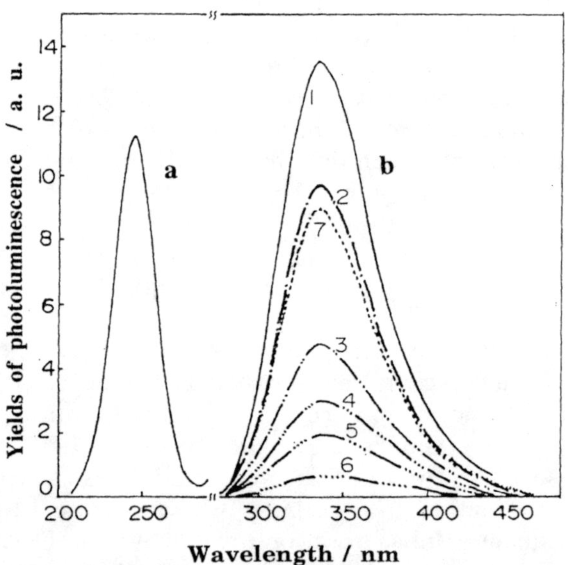

FIG. 11. Photoluminescence spectrum of the powdered MgO degassed at 1273 K for 2 h
(1) and the effect of the addition of O_2 on the photoluminescence spectrum at 295 K (2–7).
The excitation spectrum (a) and photoluminescence spectra (b) are shown. Excitation beams
at 240 ± 10 nm, emission monitored at 340 nm, spectra ± 2 recorded at 298 K, and O_2 added
at 298 K. Pressure of O_2 (Pa) determined (1) under vacuum, (2) 1.064, (3) 4.389, (4) 8.246,
(5) 13.97, (6) 52.54, and (7) after evacuation of sample 6. [reproduced with permission from
Anpo *et al.* (*98*)].

nm when it was excited with UV light at 240–280 nm. Figure 12 shows that increasing the degassing temperature of MgO increases the intensity of the photoluminescence, which passes, through a maximum at approximately 1173 K and then decreases due to sintering of the catalyst. The peak shifts toward shorter wavelength regions, leveling off at around 335 nm at degassing temperatures higher than 1073 K. During the degassing treatment, the major desorbed gas was found to be H_2O with minor amounts of CO_2. Moreover, H_2O desorption was observed at temperatures higher than 473 K, mainly at approximately 673–873 K.

It has also been shown that the photoluminescence spectrum of MgO observed at approximately 330–420 nm appears upon surface dehydroxylation (98). In other words, the intrinsic surface sites which appear upon removal of the surface OH^- groups [Eq. (14)] are linked to the photoluminescence observed with the well-degassed powdered MgO catalysts (96–98):

$$2OH^- \rightarrow O^{2-}_{LC} + H_2O \tag{14}$$

Figure 11 also shows that the addition of O_2 at 298 K onto MgO degassed at 1273 K leads to an efficient quenching of the photoluminescence without

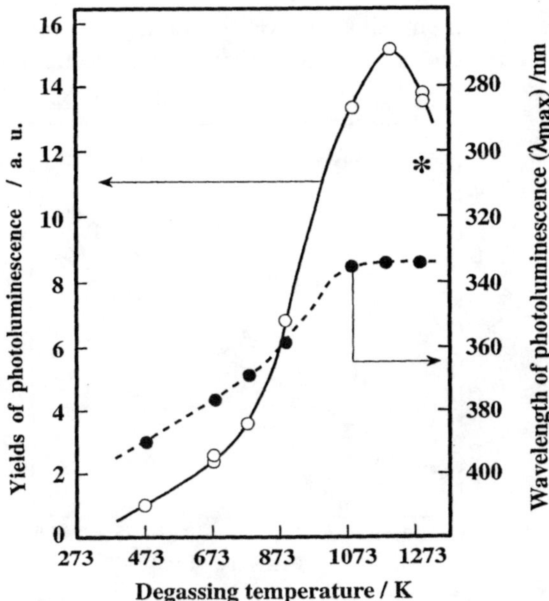

FIG. 12.   Effect of the degassing temperature on the photoluminescence spectrum of powdered MgO. *MgO degassed for 3 h rather than 2 h [reproduced with permission from Anpo *et al.* (98)].

any change in the spectrum shape (spectra 2–7). The evacuation of O_2 at the same temperature for about 20 min, after complete quenching, leads to the recovery of most of the emission. The percentage of recovery was found to depend on the time of MgO exposure to UV excitation light in the presence of O_2. The quenching of the photoluminescence by the addition of O_2 involves two mechanisms: (i) the collisional quenching (due to a weak interaction), whereby O_2 molecules interact with the active emitting sites in their excited state, and (ii) quenching due to the formation of an adsorbed complex between the adsorbed molecules and the excited active emitting sites (even in their ground state). These mechanisms result in the formation of a nonradiative deactivation pathway and/or the destruction of the radiative pathway through the formation of adsorbed complexes such as O_2^- on the emitting sites. For microcrystalline MgO, the former mechanism is predominant since the reversiblity suggests that the added molecules interact weakly with the emitting sites.

From the Stern–Volmer plots shown in Fig. 13, the values of k_q were determined to be 57, 71, and 85 $Torr^{-1}$ when the MgO catalyst was degassed at 1273, 1173, and 873 K, respectively. Although the exact values of the photoluminescence lifetime of MgO and of the absolute quenching rate constant are unknown, the assumption that the quenching rate constant of O_2 toward the excited emitting sites on the catalyst is approximately equal to the value toward the similar charge-transfer excited triplet state of the

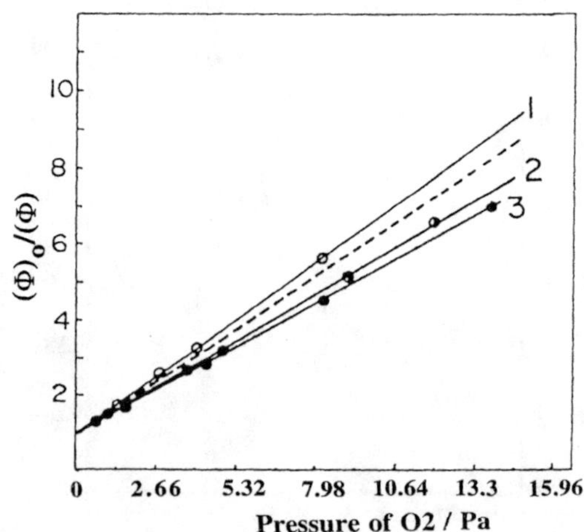

FIG. 13. Plots of the $(\Phi)_0/(\Phi)$ value at various pressures of O_2 in the photoluminescence spectrum of MgO degassed at various temperatures. Spectra were recorded at 298 K, excitation at 240 nm. Under degassing temperature (K): 1, 873; 2, 1173; 3, 1273 [reproduced with permission from Anpo *et al.* (*98*)].

supported vanadium oxide catalysts leads to estimates of the lifetimes of MgO degassed at 1273, 1173, and 873 K of 2.2×10^{-4}, 2.4×10^{-4}, and 3.6×10^{-4} s, respectively (*96–98*).

b. *Photoluminescence Properties of ZrO_2.*   Figure 14 shows the photoluminescence spectrum 1, its corresponding excitation spectrum, and UV-reflectance spectrum of the active ZrO_2 catalyst before (D) and after (E) evacuation at 973 K. After evacuation of the catalyst at high temperatures, a new absorption band appears at approximately 310 nm which is completely different from the fundamental absorption band of bulk ZrO_2 (4.96 eV = 250 nm). Figure 14 shows the photoluminescence spectrum 1, its corresponding excitation spectrum, and the UV-reflectance spectrum of the active ZrO_2 catalysts before (E) and after evacuation at 973 K (f). The photoluminescence spectrum is observed at approximately 400–600 nm with

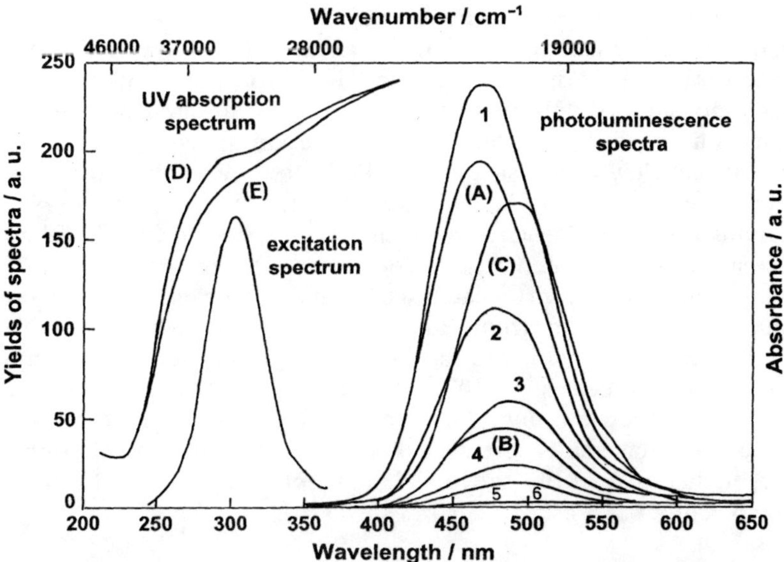

FIG. 14.   Photoluminescence spectrum of the powder ZrO_2 at 77 K (1) and its excitation spectrum and spectra after the addition of H_2 to the ZrO_2 at 298 K (2–6), and the diffuse reflectance absorption spectrum of the ZrO_2 degassed at 400 K (UV). Spectrum A was obtained by evacuation of the sample after complete quenching by added H_2 (spectrum 6). Spectrum B is the differential spectrum obtained by subtracting spectrum A from the original spectrum (spectrum 1). Spectrum C is the photoluminescence spectrum not affected or quenched by the added H_2 but quenched by the added CO. Pressure of added H_2 (Pa) determined (1) under vacuum, (2) 1.33, (3) 6.65, (4) 13.3, (5) 133, and (6) 2660 [reproduced with permission from Anpo *et al.* (*101*)].

an intensity detectable only when the catalyst was degassed at temperatures >574 K. The photoluminescence was efficiently and completely quenched by the addition of O_2 or air, indicating that the photoluminescence is a surface rather than a bulk phenomenon (*101–104*). These results show that the appearance of a new absorption band at around 260–350 nm and the photoluminescence spectrum at around 400–600 nm can be attributed to the charge-transfer processes [Eq. (15)] involving coordinatively unsaturated surface sites in a manner similar to those for powdered MgO and SrO (*77–80, 82, 96–98*):

$$(Zr^{4+}-O^{2-})_{LC} \xrightarrow{h\nu} (Zr^{3+}-O^{-})_{LC} \tag{15}$$

The addition of H_2 also leads to an efficient and irreversible quenching of the photoluminescence, as shown in Fig. 14 (2–4), and its extent depends on the pressure of H_2 and on the type of emitting sites on the surface. On the other hand, the nonactive ZrO_2 catalysts, i.e., those which do not lead to the formation of hydrocarbons from $CO-H_2$ mixtures, did not exhibit a photoluminescence spectrum due to surface sites of low coordination or IR absorption bands due to the dissociative adsorption of hydrogen. These results indicate not only that coordinatively unsaturated surface ions are formed on active ZrO_2 catalysts by evacuation at temperatures >600 K but also that they play a significant role in the reversible and irreversible dissociative adsorption of H_2. It is likely that the hydrogen species obtained upon adsorption onto such surface sites correspond to the photoluminescence spectrum associated with the formation of hydrocarbons from $CO-H_2$ mixtures on the active ZrO_2 catalysts. The quantum yield of the irreversible photoluminescence spectrum (Fig. 14, B) indicates that the number of surface sites corresponding to the irreversible adsorption of H_2 is approximately seven molecules per 10^5 $(Zr^{4+}-O^{2-})_{surf}$ pair sites, and the quantum yield of the reversible photoluminescence (Fig. 14, A) indicates that the number of surface sites corresponding to the reversible adsorption of H_2 is larger than that of the irreversible adsorption by at least 10 times, i.e., 100 molecules per 10^5 $(Zr^{4+}-O^{2})_{surf}$ sites (*101*).

    i. *Adsorption of H_2.* Upon adsorption of H_2 onto ZrO_2 evacuated at 750 K, the IR absorption bands were obtained at approximately 3782, 3668, and 1562 cm^{-1} due to the formation of the Zr–O–H, Zr–O(H)–Zr, and Zr–H species, respectively. These findings show that there are three different types of hydrogen adsorption; (1) a homolytic dissociative adsorption leading to the species with bands observed at about 1560 cm^{-1}, (2) a heterolytic dissociative adsorption leading to the species with bands at about 1562 and 3668 cm^{-1}, and (3) adsorption leading to the formation of OH groups with bands at about 3668 and 3782 cm^{-1}. As shown in Table II, it is of interest to correlate the type of emitting site (A–D) with the corresponding adsorption type 1, 2, or 3 (*101, 102*). Thus, photoluminescence and IR investigations show that surface sites of low coordination (or coordinatively unsaturated surface sites) formed on active ZrO_2 catalysts by evacuation at temperatures >600 K not only play a major role in the activation of H_2 but also are closely associated with the formation of branched hydrocarbons in catalytic $CO-H_2$ reactions.

ii.   *Adsorption of CO.*   UV-reflectance spectra can be obtained upon the addition of CO onto the ZrO_2 catalyst evacuated at 1073 K. A band at approximately 33,000 cm^{-1} appears immediately and can be assigned to the formation of the $(CO)_2^-$ species. Absorption bands at approximately 26,000 and 20,000 cm^{-1} gradually increase in intensity after the addition of CO. IR absorption bands can also be observed upon the addition of CO. Being in good agreement with the UV-reflectance results, the intensities of the new IR absorption bands gradually increase with the time of contact of CO with the catalyst, suggesting that the $(CO)_n^-$ species and also cyclic polymer species of CO [i.e., $(CO)_5^{2-}$ and $(CO)_6^{2-}$] are formed on surface sites of low coordination (*104*). Thus, it was found that the adsorption of CO on the active ZrO_2 catalyst immediately led to the formation of the $(CO)_2^-$ species, which gradually transformed into $(CO)_5^{2-}$ and then into $(CO)_6^{2-}$ upon reaction with CO, whereas UV irradiation of the catalyst in the presence of CO led to the selective formation of the $(CO)_6^{2-}$ species.

iii.   *Hydrogenation of CO.*   The addition of H_2 onto the ZrO_2 catalyst that has a preadsorbed CO species led to a decrease in the intensities of

TABLE II

Characterization of the Photoluminescent Sites on the Active Powder ZrO_2 Catalysts[a]

	Site			
	A	B	C	D
Excitation (nm)	310	310	310	310
Emission (nm)	490	490	460	495
Lifetime (s)	1.2×10^{-3}	1.2×10^{-3}	1.1×10^{-2}	1.1×10^{-5}
Adsorption of H_2	Heterolytic Reversible	Homolytic Irreversible	Homolytic Irreversible	None None
Infrared bands (cm^{-1}) after H_2 adsorption	3668; 1562	3668	3782	

[a] The characterization of the emitting sites was carried out by means of photoluminescence, H_2 adsorption measurements at various temperatures, as well as the effect of the addition of H_2 on the spectra [Anpo *et al.* (*101*); Moon *et al.* (*102*)].

the IR, UV, and EPR spectra assigned to the adsorbed anionic species of CO, i.e., $(CO)_2^-$, $(CO)_5^{2-}$, and $(CO)_6^{2-}$, respectively. Simultaneously, the appearance of a new species formed by reaction with H_2 was observed by IR and EPR. In this system, CH_4 and CH_3OH were produced as the major products (*102, 104*). A good correlation was found between the intensity of the EPR signal assigned to the adsorption species formed from CO, i.e., $(CO)_2^-$, $(CO)_5^{2-}$, and $(CO)_6^{2-}$ [the $(CO)_6^{2-}$ is the major species], and the yield of CH_3OH produced by the reaction of H_2 with the preadsorbed CO species. It is emphasized that the formation of CH_3OH was not observed to result from the addition of H_2 onto the ZrO_2 catalysts incorporating preadsorbed cyclic polymer species of CO such as $(CO)_5^{2-}$ and $(CO)_6^{2-}$ (*101, 104*).

Thus, the adsorption of CO on active ZrO_2 catalysts led to the formation of various types of adsorption species of CO having different reactivities toward H_2, and these species were found to play a significant role in the hydrogenation of CO. Moreover, it is likely that CO is adsorbed on the active surface sites of low coordination and that an electron transfer from the other surface sites to this CO species leads to the formation of the dimeric adsorbed species $(CO)_2^-$. These dimeric species react, step by step, with CO molecules from the gas phase to from a relatively stable cyclic polymer species of $(CO)_5^{2-}$ and then $(CO)_6^{2-}$. Such adsorbed CO species easily react with hydrogen and are also activated through the dissociative adsorption of hydrogen on surface sites of low coordination or the coordinatively unsaturated surface sites on the catalyst.

B. ENERGY AND ELECTRON TRANSFER PROCESSES ON CATALYSTS

1. *Energy Transfer*

a. *Energy Transfer among Sites with Different Coordination Numbers.* Well-degassed alkaline earth metal oxide catalysts, such as MgO and SrO, exhibit photoluminescence spectra when they are excited at wavelengths corresponding to the abnormal absorption bands due to the formation of surface sites in low coordination, as described previously. Coluccia *et al.* (*13, 65, 81, 85*) found that, as shown in Fig. 15, the photoluminescence spectra observed at 300 and 77 K for SrO and which have different coordination numbers (i.e., five-, four-, and three-coordinated sites) are remarkably different from each other. The peaks observed at 300 K can always be seen at approximately 470 nm with small differences in their intensities, and the photoluminescence peaks observed at 77 K can clearly be seen at approximately 400, 440, and 470 nm depending on the excitation wavelengths. The photoluminescence spectra at approximately 400 and 440 nm are associated with surface ions existing in five- and four-coordination,

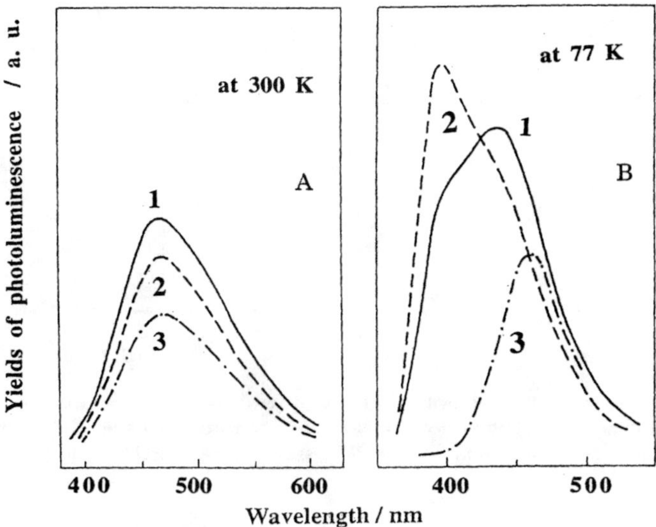

Fig. 15. Photoluminescence spectra of SrO at 330 K (A) and at 77 K (B). Excitation wavelengths (nm): 1, 285; 2, 315; 3, 330 [reproduced with permission from Coluccia (*13*)].

respectively. The spectrum observed at 470 nm is associated with surface ions having three-coordination. Figure 15 shows that the photoluminescence spectra at approximately 400 and 440 nm can be seen when the oxide is excited at either approximately 285 or 315 nm, whereas the photoluminescence spectrum at approximately 470 nm assigned to the three-coordinated surface sites can be observed only when the oxide is directly excited by its absorption band at approximately 330 nm.

These results can be interpreted in terms of the energy transfer which occurs at 300 K from the five- and four- to the three-coordinated surface sites. Being in agreement with the differences in mobility of the surface excitons, radiative decay pathways are more efficient for three-coordinated surface ions than for four- or five-coordinated ions. The surface ions existing in high coordination and having a larger number of lattice bonds to oxides interact more strongly with the phonon transitions of the lattice, leading to a high probability for nonradiative decay on such sites. The mobility of three-coordinated surface ions, which are normally located at the corner sites, is more restricted and constrained, and these sites act as sinks for the photon energy absorbed by the other surface centers. However, as shown in Fig. 16, such an energy transfer to sites of lower coordination requires a relatively high activation energy, resulting in the large suppression of the energy transfer at 77 K (*13*).

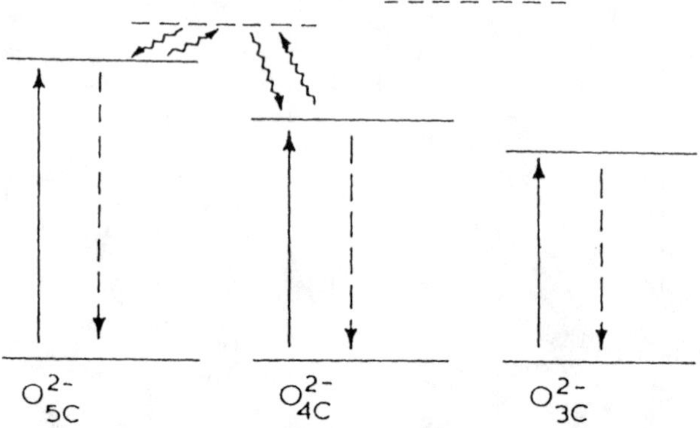

FIG. 16. Energy diagram showing energy transfer on SrO at 77 K from five-coordinated to four-coordinated sites and vice versa. No energy transfer occurs at 77 K from and to the three-coordinated sites [reproduced with permission from Coluccia (*13*)].

b. *Resonance Energy Transfer in Zeolites.* Copper(I) ions anchored within zeolite cavities and in SiO_2 matrices exhibit photoluminescence at approximately 400–500 nm upon excitation at approximately 300 nm (*105, 106*). Strome and Klier (*107, 108*) found that coexchanged ions (i.e., Ni^{2+}, Co^{2+}, and Mn^{2+}) lead to the quenching of the photoluminescence in intensity and time of decay or lifetime. The time of decay became shorter when the concentration of nickel(II) ions was increased and it exhibited a multiexponential decay character, although analysis of the photoluminescence decay curve of the Cu^+ species in the absence of Ni^{2+} showed a single exponential character.

The presence of coexchanged cations results in the following phenomena: (i) an increase in the decay rate and (ii) a decrease in the relative quantum efficiency of the Cu^+ photoluminescence. These findings suggest that the presence of second cations such as nickel(II) provides additional nonradiative deactivation pathways for the excited state of the Cu^+ species. The magnitude of such effects depends strongly on the type of coexsistent cations. It was found that with the $Cu_{0.94}/Ni_{1.1}$ Y zeolite, the photoluminescence due to the Cu^+ species overlaps with the absorption of the ion-exchanged Ni^{2+}, suggesting that the effect of the second cations is closely associated with the spectral overlap between the photoluminescence of the Cu^+ species and the absorption of the coexchanged second Ni^{2+} ions (*107, 108*).

Strome and Klier (*107, 108*) applied the Förster–Dexter theory of resonance energy transfer to explain these experimental observations, i.e., the energy transfer from the excited state of the Cu^+ species to the coexistent

Ni^{2+}, Co^{2+}, or Mn^{2+} ions. Using a value of the characteristic decay time of the excited state of the Cu^+ species (i.e., $\tau = 120 \times 10^{-6}$ s), the values obtained for the resonance energy transfer radii R_0 for Ni^{2+}, Co^{2+}, and Mn^{2+} were as follows:

$$R_0(Cu^+ \rightarrow Ni^{2+}) = 1.45 \text{ nm}$$

$$R_0(Cu^+ \rightarrow Co^{2+}) = 2.40 \text{ nm}$$

$$R_0(Cu^+ \rightarrow Mn^{2+}) = 1.13 \text{ nm.}$$

These findings suggest that a long-range excitation energy transfer proceeds with a high efficiency even across distances >1 nm.

2. Dynamics of Electron Transfer Processes on Semiconducting Catalysts

a. *Electron Transfer from Adsorbed Molecules to Semiconductors*   The charge separation of photoformed electrons and holes as well as their recombination are important chemical processes in semiconducting materials, such as TiO_2 and ZnO, when these are used as photocatalysts (*22–24, 33, 34, 109*). The photochemical properties of such semiconductors are critically dependent on the lifetimes of the photoinjected electrons and holes. Consequently, the rate or efficiency of the charge separation of the photogenerated electrons and holes and of their recombination play a crucial role in the photocatalytic reactivities of semiconductor photocatalysts. One of the most important and interesting subjects concerning such chemical processes is the electron transfer phenomenon from the photoexcited state of dye molecules adsorbed on a semiconductor into its conduction band (*22, 109*). Figure 17 shows the decay *in vacuo* of the photoluminescence of $Ru(bpy)_3^{2+}$ complexes adsorbed on several powdered metal oxide semiconductors that have different electrochemical properties. The decay of the photoluminescence depends strongly on the kind of oxide and on the photoluminescence of the $Ru(bpy)_3^{2+}$ species adsorbed on semiconducting oxides such as TiO_2 and SnO_2 decaying faster than that on insulating materials such as SiO_2 (*110, 111*).

Thus, the decay rates of the photoluminescence of $Ru(bpy)_3^{2+}$ complexes adsorbed on oxides depend on the electronic energy position of the conduction band edges of the oxides. A schematic energy-level diagram of the semiconductors and the $Ru(bpy)_3^{2+}$ complex is shown in Fig. 18. The decay rate of the photoluminescence of $Ru(bpy)_3^{2+}$ complexes adsorbed on the oxides increases on semiconductors which have a more positive electronic energy position of the conduction band edge. These findings of the remark-

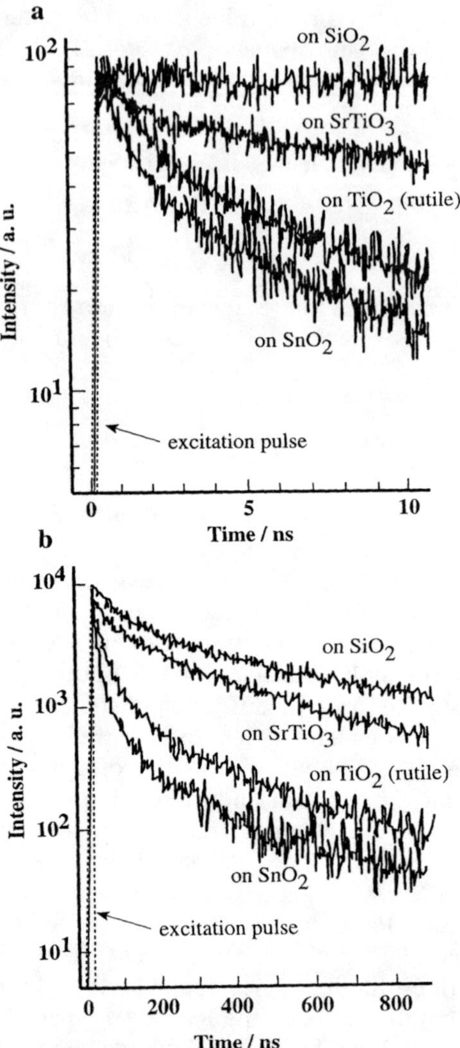

FIG. 17. Initial decay of the photoluminescence of $Ru(bpy)_3^{2+}$ complexes adsorbed on various powder metal oxides at 295 K. (a) Picosecond laser excitation at 540 nm and the emission monitored at 540 nm. (b) Nanosecond laser excitation at 450 nm and the emission monitored at 580 nm [reproduced with permission from Hashimoto *et al.* (*111*)].

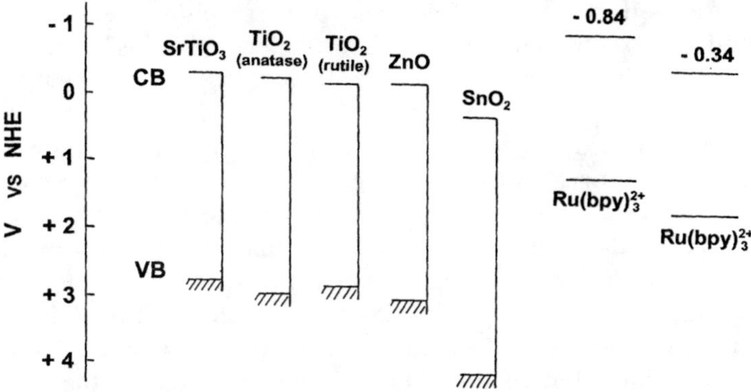

FIG. 18.   Energy level diagram of various metal oxides at pH = 0 and $Ru(bpy)_3^{2+}$ complexes in H_2O. NHE, normal hydrogen electrode [reproduced with permission from Hashimoto *et al.* (*111*)].

able quenching of the photoluminescence of $Ru(bpy)_3^{2+}$ by adsorption onto semiconductors have reasonably been attributed to the electron transfer from the excited state of $Ru(bpy)_3^{2+}$ to the conduction band of the semiconductor. This hypothesis of electron transfer has been supported by the observation of the photocatalytic production of H_2 on the TiO_2 system involving the adsorbed $Ru(bpy)_3^{2+}$ species *in vacuo* and by the photocurrent in the TiO_2 or SnO_2 electrode systems involving adsorbed $Ru(bpy)_3^{2+}$ complexes (*110, 111*).

Electron transfer from the conduction band of semiconducting catalysts across the interface into the accepting species in solution has also been investigated extensively with nanosecond laser-flash photolysis and time-resolved fluorescence spectroscopy (*109*).

b.   *Charge Separation in Semiconductors.*   One of the limiting factors controlling the efficiency of photocatalysis is the rate of recombination between the photogenerated electrons and the holes in the semiconductors. The platinization of TiO_2 or CdS particles with small amounts of Pt led to an enhancement of the efficiency of the photocatalytic reactions (see Section VII.A.2) (*109, 112*). Moreover, CdS colloids prepared in acetonitrile exhibited a red photoluminescence at wavelengths longer than 550 nm. This red emission, which occurs as a result of sulfur vacancies, is readily quenched by electron acceptors such as viologens and methylene blue (*113*). The addition of TiO_2 and AgI colloids also quenches the CdS emission, which can be attributed to the injection of an electron from the photoexcited CdS colloid into the conduction band of the AgI or TiO_2 colloid. The primary photochemical processes associated with such charge separation or electron transfer in CdS–TiO_2 binary semiconductor systems have been investigated by means of transient absorption spectroscopic measurements (*114*). The trapped holes (h^+) on the CdS side in the CdS–TiO_2 binary systems initiate chemical reactions generating S^- radical species. Such S^- species exhibit a broad absorption band in the visible regions having a maximum at approximately 480 nm. An enhancement of the S^- formation was observed when additional colloidal TiO_2 was introduced into the system, indicating that the hole trapping efficiency on CdS particles in the CdS–TiO_2 systems was increased. Injection of electrons from the conduction band of the CdS particles into that of the TiO_2 particles resulted in the retardation of the back reaction between the photogenerated electrons and holes.

C. PHOTOLUMINESCENCE AND LOCAL STRUCTURE OF ANCHORED VANADIUM OXIDE

1. *Vibrational Fine Structure Assigned to the Surface $V=O$ Bond*

Chemical vapor deposition (CVD) with various reactive transition metal compounds has been extensively applied in the preparation of anchored catalysts (*7, 20*). The photochemical activation of various transition metal compounds adsorbed on support surfaces is also useful for preparing highly dispersed, well-defined anchored catalysts. For the utilization of $VOCl_3$ as a reactive compound, since HCl was generated during UV irradiation of adsorbed $VOCl_3$ and the $V=O$ band was retained, the following reaction [Eq. (16)] was proposed for the anchoring reaction of $VOCl_3$. The facile control of the content of the anchored vanadium ions on the support by controlling the UV irradiation time and/or concentration of the isolated surface OH groups on the support is a great advantage of this photo-CVD method (*56*):

$$\text{Si–OH} + \text{VOCl}_{3(\text{ads})} \xrightarrow{h\nu} \text{Si–O–V} \overset{\displaystyle \diagup \text{Cl}}{\underset{\displaystyle \diagdown \text{Cl}}{=}} \text{O} + \text{HCl} \qquad (16)$$

Figure 19 shows the typical photoluminescence spectrum of the anchored vanadium oxide catalyst prepared by photo-CVD methods (a), its corresponding excitation spectrum (b), and the UV absorption spectrum of the catalyst (c) (*56, 115, 116*). These absorption and photoluminescence spectra (phosphorescence) are attributed to the following charge-transfer processes on the surface vanadyl group ($V=O$) of the tetrahedrally coordinated VO_4 species involving an electron transfer from O^{2-} to V^{5+} and a reverse radia-

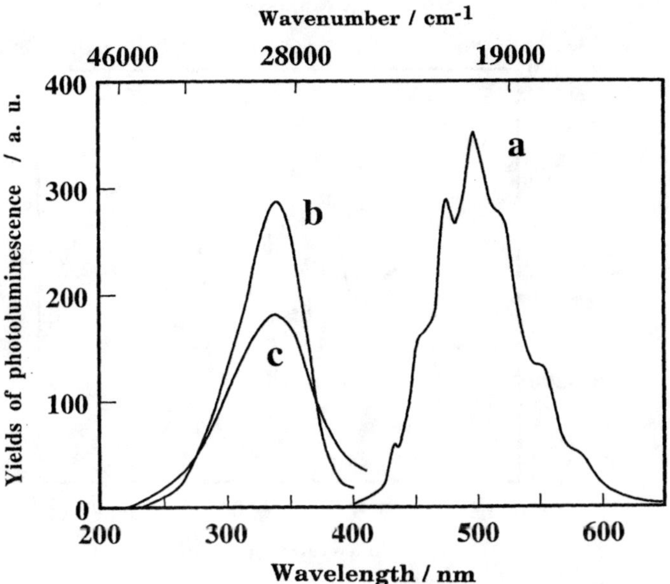

Fig. 19.   Photoluminescence spectrum (a), its excitation spectrum (b), and UV absorption spectrum (c) of the vanadium oxide catalyst anchored to Vycor glass by the photo-CVD method. Photoluminescence spectrum recorded at 77 K; excitation wavelength, 280 nm; emission monitored at 500 nm; UV absorption spectrum measured at 295 K; vanadium content, 0.1 wt% [reproduced with permission from Anpo *et al.* (*69*)].

tive decay from the charge-transfer excited triplet state, respectively (*117–120*):

$$\begin{array}{ccc} O^{2-} & & O^{-} \\ \| & & | \\ V^{5+} & \xrightarrow{h\nu} & V^{4+} \\ / | \backslash & & / | \backslash \\ O \quad O \quad O & & O \quad O \quad O \\ | \quad | \quad | & & | \quad | \quad | \end{array} \qquad (17)$$

Figure 20 shows the second-derivative spectrum of the photoluminescence shown in Fig. 19. The well-resolved fine structure indicates that the energy gap between the $(0 \to 0)$ and $(0 \to 1)$ vibrational transition bands is 1040 cm⁻¹, in good agreement with the vibrational energy of the surface $V=O$ bond obtained by IR or Raman measurements. These findings indicate that the photon energy absorbed by the vanadium oxide/SiO_2 catalyst is mainly localized on the $V=O$ surface bonds. The Raman spectrum of the anchored vanadium oxide/SiO_2 catalyst observed at 78 K exhibits a peak at 1050 cm⁻¹, in good agreement with the results obtained by Went

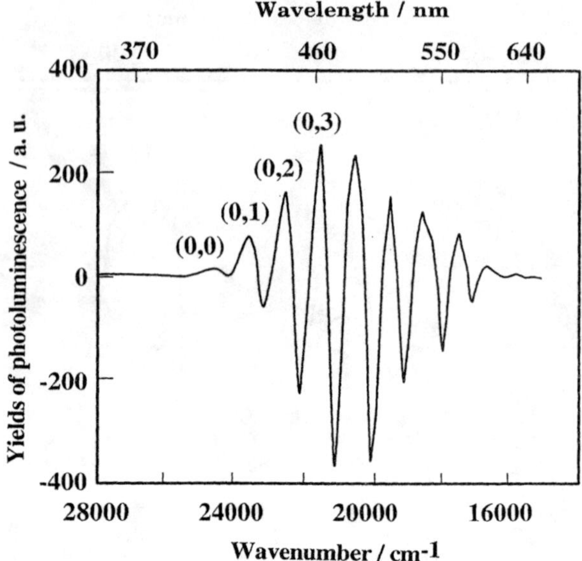

FIG. 20.   Second-derivative photoluminescence spectrum of the vanadium oxide catalyst anchored to Vycor glass shown in Fig. 19 [reproduced with permission from Anpo *et al.* (*69*)].

et al. (*121*) for vanadium oxide/SiO_2 having a loading of 1.4 wt% of V. They assigned the peak at 1050 cm^{-1} to the monomeric vanadyl species bonded to the SiO_2 support shown in Eq. (17). Cristiani *et al.* (*122*) and Iwamoto *et al.* (*21*) also showed that the Raman and IR peaks in the region of 1035–1049 cm^{-1} can be attibuted to a similar structure, i.e., a monoxo ($V{=}O$) species of the supported vanadium oxide. Thus, a good agreement between the Raman value of 1050 cm^{-1} and the energy separation of 1040 cm^{-1} determined from the $(0 \rightarrow 0)$ and $(0 \rightarrow 1)$ vibrational transition bands of the second-derivative photoluminescence spectrum indicates not only that Eq. (17) is in good agreement with the models proposed by Went *et al.* (*121*), Cristiani *et al.* (*122*), and Iwamoto *et al.* (*21*) but also that the photon energy absorbed by the catalyst is mainly localized on the vanadyl group, i.e., the $V{=}O$ bonds of the catalyst.

2. *Franck–Condon Analysis of the Photoluminescence Spectrum*

According to the Franck–Condon principle (*123*), to evaluate the Franck–Condon overlap integral the vibrational wavefunctions for the ground and excited electronic states are expressed in terms of the Hermite polynomials, which include the displacement of the origin of the normal coordinates of the ground and excited electronic states. The ratio of an integral corresponding to $(v' = 0) \rightarrow (v'' = 1)$ to that corresponding to $(v' = 0) \rightarrow (v'' = 0)$ can be expressed as follows (124):

$$R(1, 0)/R(0, 0) = -2\Delta/(B^2 + 1), \tag{18}$$

where Δ is a parameter proportional to the difference in the internuclear equilibrium separation between the ground and excited electronic states and B is the square root of the value of the energy difference between the excited state and the ground electronic state. The ratio of $R(n, 0)$ to $R(0, 0)$ can be determined from the following relationship for the Hermite polynomials:

$$R(n + 1, 0)/R(n, 0) = 2\Delta/((B^2 + 1)(n + 1)^{1/2})$$
$$+ n^{1/2}/(n + 1)(B^2 - 1)/(B^2 + 1)R(n - 1, 0)/R(n, 0). \tag{19}$$

The Franck–Condon overlap integrals are related to the relative intensities for the vibronic emission peaks represented in Eq. (19):

$$\begin{array}{ll} \text{Intensity of transition for } (0 \rightarrow n): & E(n, 0)^4 R(n, 0)^2 \\ \text{Intensity of transition for } (0 \rightarrow 0): & E(0, 0) \; R(0, 0), \end{array} \tag{20}$$

where $E(n, 0)$ is the energy of the vibronic peak corresponding to the transition from $(v' = 0)$ to $(v'' = n)$.

If the internuclear equilibrium distance of the excited electronic state (r'_e) shifts by the value Δ from the internuclear equilibrium distance of the ground state (r''_e), the Franck–Condon principle allows transitions to many excited vibrational levels. The shapes of the harmonic potentials also have an effect on the magnitude of the Franck–Condon integral. In this case, the theoretical intensities have been calculated as a function of Δ and B. The parameters B and Δ were varied until the theoretical intensities showed the closest match to the experimental intensities. In Fig. 21, the best fit for the progression obtained from the photoluminescence spectrum for the anchored vanadium oxide/SiO_2 catalyst and theoretical Franck–Condon analysis is represented (125).

Since the average vibrational energy in the progressions is >1000 cm^{-1}, it can be assumed that most of the vibrational states populate the $v'' = 0$ vibrational level in the excited electronic state while the emission occurs from the lowest $v'' = 0$ vibrational level to the vibrational levels of the ground state for which the Franck–Condon integral is not zero. The results obtained by the Franck–Condon analysis show that the excited state potential is displaced from the ground state potential by 0.013 nm, in agreement with the assignment of the electronic transition from O^{2-} to V^{5+} ions of the surface vanadyl groups (V=O) (115–120), and that, as shown later, the order of the photoreactivity of the anchored vanadium oxide/SiO_2

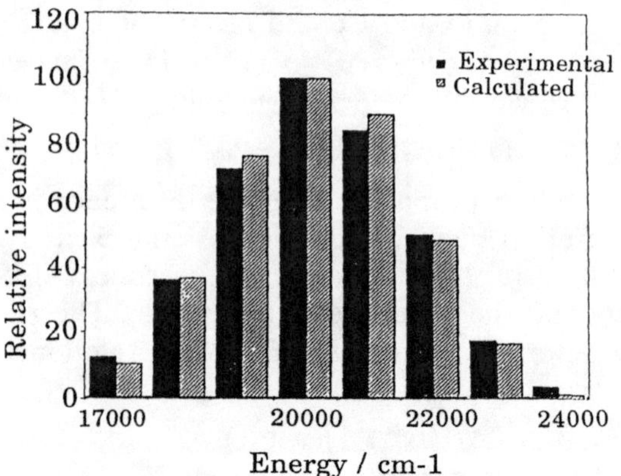

FIG. 21.   Progression of the vibrational fine structure of the photoluminescence spectrum of vanadium oxide catalyst anchored to SiO_2 at 77 K. Theoretical results from a Franck–Condon analysis (striped bars) and experimental results (solid bars) [reproduced with permission from Patterson *et al.* (*125*)].

catalysts toward CO molecules is in good accordance with that of the values of Δ (a, it was also found that the anchored vanadium oxide/SiO_2 catalyst having the highest photoreactivity exhibits the largest value of Δ).

It is interesting to compare the Franck–Condon results with recent *ab initio* Hartree–Fock molecular orbital results for the supported vanadium oxide/SiO_2 catalysts. Kobayashi *et al.* (*126*) calculated that the V=O double bond of the oxide is elongated in its lowest charge-transfer excited state by 0.3 Å compared with that in the ground state. Their potential energy curves for the ground state versus the lowest excited triplet state have indicated a sizable change in the V=O vibrational energy, in agreement with the results of the Franck–Condon analysis. Thus, it was found that the Franck–Condon analysis of the progression of the vibrational fine structure for the photoluminescence spectrum of anchored vanadium oxide/SiO_2 catalysts was in good qualitative agreement with the MO predictions.

3. *Relationship between the Results of the Franck–Condon Analysis and Photoreduction of the Catalyst with CO*

UV irradiation of anchored vanadium oxide/SiO_2 catalysts at 280 K in the presence of CO leads to the photoformation of CO_2 (i.e., the photoreduction of the catalyst by CO) accompanied by small amounts of a photoinduced adsorption (photouptake) of CO (*56*). After photoreduction of the anchored catalysts and further evacuation of excess CO, O_2 was introduced onto the catalysts at pressures less than 530 Pa and at 280 K for a few minutes, followed by evacuation. After this O_2 treatment, the photoreaction with CO proceeded with the same efficiency as on the original oxide. The original surface state of the oxide was completely regenerated by the contact with O_2 at 280 K (i.e., the photoreduced oxide surfaces of the anchored vanadium oxide/SiO_2 catalyst were easily reoxidized by O_2), in good agreement with the results obtained with anchored molybdenum oxide/ SiO_2 catalysts (*127–129*).

Figure 22 shows the relationship between the yields of the photoformed CO_2 and photouptake of CO as well as the phosphorescence yields of the catalysts. These results indicate that the charge-transfer excited triplet state of the surface vanadyl group plays a significant role in the photoformation of CO_2. The intensity and lifetime of the phosphorescence of the anchored vanadium oxide/SiO_2 catalyst decrease in the presence of CO; the extent depends on the amount of CO added. The Stern–Volmer plots for phosphorescence quenching of the intensity and lifetime as a function of the amounts of added CO molecules were determined to be almost linear, showing that the emitting sites are responsible for the photoreaction with CO. The Stern–Volmer plots for the catalyst at 77 K, together with the lifetime of the charge-transfer excited triplet state of the surface vanadyl group (τ), showed that the absolute quenching rate constant of the excited state toward CO molecules, k_q, was 4.1×10^{10} g catalysts/mol at 77 K. This value is

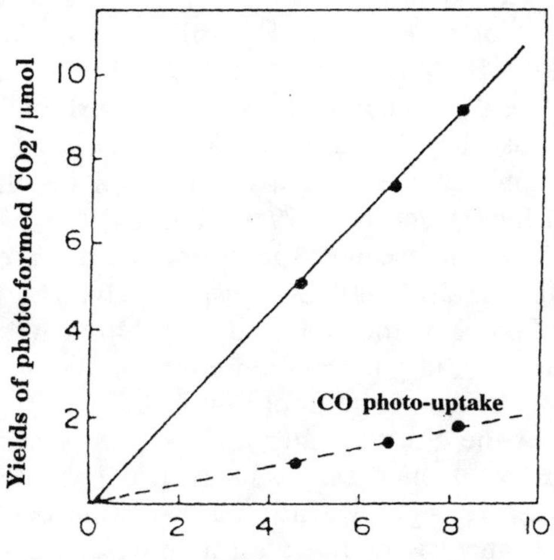

FIG. 22.   Relationship between yields of photoformed CO_2 on vanadium oxide catalysts anchored to SiO_2 and the relative yields of the photoluminescence spectra of catalysts having different vanadium contents at 280 K [reproduced with permission from Patterson *et al.* (*125*)].

larger than those determined for various alkene molecules such as C_2H_4 (3.5×10^{10}) and C_4H_8 (2.4×10^{10}) on the catalyst (*56*).

The photoinduced reduction of the oxide catalysts with CO was observed only for the highly dispersed supported oxides having a metal–oxygen surface double-bond character (Me=O) such as for supported vanadium, molybdenum, tungsten, and chromium oxides (*116, 117*). The photoreducibility of the highly dispersed supported oxides with CO decreases in the order V > Mo > W > Cr, i.e., with decreases in the lifetime of the charge-transfer excited triplet state of the supported oxides at 295 K. The observed parallel relationship between the yield of the photoformed CO_2 and the yield of the phosphorescence, as well as the quenching of the phosphorescence in its yield and lifetime by the addition of CO, clearly indicates the direct participation of the charge-transfer excited triplet state of the oxide in the photoinduced reduction of the oxide catalysts or the photoformation of CO_2 from CO.

The Franck–Condon analysis of the vibronic fine structure of the phosphorescence of the anchored vanadium oxide/SiO_2 catalysts at 77 K suggests that the equilibrium bond distance of V–O of the vanadyl group in the charge-transfer excited triplet state is elongated by 0.013 nm compared with that in the ground state. Since the photoformation of CO_2 from CO is accompanied by the removal of oxygen from the oxide (i.e., the photoreduction of the catalyst), such an elongation of the equilibrium distance of the V–O bond in the excited state is closely associated with the fact that the photoformation of CO_2 proceeds easily on the anchored vanadium oxide/SiO_2 catalysts. In other words, the O^- hole center in the charge-transfer excited state of the complex $(V^{4+}-O^-)^*$ exhibits a high reactivity similar to that of adsorbed O^- anion radicals (*66*).

The powdered ZnO catalyst (Kadox 25) exhibits a photoluminescence with a vibrational fine structure at 77 K (Fig. 23) (*129*). Four peaks (Fig. 23a–d) in the photoluminescence spectrum of ZnO are separated from each other by 420, 620, and 560 cm^{-1} respectively. The Raman spectrum of the catalyst at 300 K in the range of 200–1400 cm^{-1} was observed at approximately 420 cm^{-1}, in good agreement with the energy separation in the vibronic fine structure of the photoluminescence at 77 K. Although it is not clear whether the emitting sites on ZnO are located on the surface and/or near the surface, from these results peak *a* in the photoluminescence can be assigned as a $(0 \rightarrow 0)$ vibrational transition of the ZnO catalyst. A Franck–Condon analysis of the photoluminescence spectrum of ZnO indicates no change in the equilibrium distance Zn–O in the excited state, i.e., there is no displacement of the excited state potential relative to the ground state potential.

UV irradiation of powdered ZnO at 298 K in the presence of CO, even

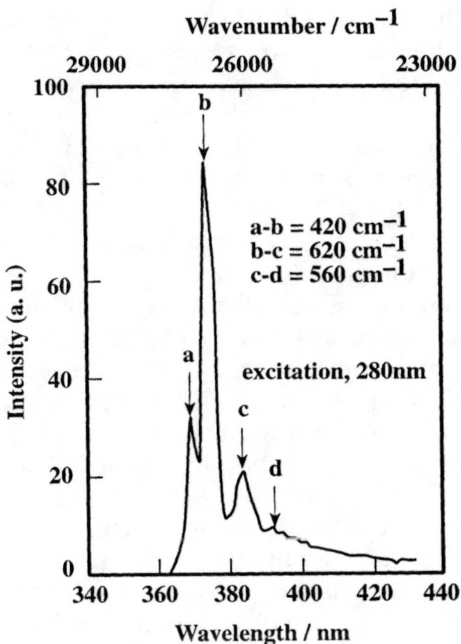

Fig. 23.   Photoluminescence spectrum of the powder ZnO catalyst at 77 K. Excitation wavelength, 280 nm [reproduced with permission from Anpo and Kubokawa (*129*)].

after prolonged UV irradiation, showed no evolution of CO_2 (*129*). Only the photouptake of CO could be detected (*125*). In good agreement with the predictions of the Franck–Condon analysis, these experimental results clearly indicate that the removal of oxygen from the ZnO catalyst scarcely proceeds under UV irradiation at 298 K and that the catalyst has a strong Zn–O bond even in the excited state of $(Zn^+–O^-)^*$. The situation for ZnO is quite different from that for the anchored vanadium oxide/SiO_2 catalyst in which the V–O bond in the charge-transfer excited $(V^{4+}–O^-)$ complex is sufficiently weakened to react with CO to form CO_2. From these results, it has been concluded that not only the electronic nature of the excited state but also the elongation of the Me–O bond in the excited state (i.e., the displacement for the excited state potential relative to the ground state potential) are important in affecting the photoreduction of the catalyst with CO at normal temperatures.

a.   *Vanadium Oxide/SiO$_2$ Catalysts.*   The phosphoresence and its corresponding excitation spectra (i.e., absorption bands) of the vanadium oxide/ SiO_2 catalyst prepared by an impregnation method exhibit a good correspondence with those obtained with the anchored vanadium oxide/SiO_2 (Fig. 19) and vanadium oxide/porous Vycor glass (PVG) catalysts prepared by an anchoring method (see Section IV.C.1). The absorption and phosphorescence spectra are attributed to the charge-transfer processes involving an electron transfer from O^{2-} to V^{5+} and a reverse radiative decay [Eq. (17)], respectively, on the surface vanadyl group of an isolated tetrahedral VO_4 unit species.

The wavelength of the spectra scarcely changed when observed at 77 and 298 K, whereas the yield (or intensity) of the phosphorescence at 77 K (Fig. 19, spectrum a) was remarkably high compared with the yield at 298 K. This result can be attributed to the fact that the lifetime of the charge-transfer excited triplet state is markedly affected by temperature-dependent radiationless processes. Such radiationless deactivation from the excited triplet state becomes less efficient as the temperature decreases, leading to an elongation of the phosphorescence lifetime from $\tau_{298} = 140$ ms at 298 K to $\tau_{77} = 5.8$ ms at 77 K (125).

As shown in Fig. 20, a vibrational fine structure of the phosphorescence due to the V=O double bond of the vanadyl group is clearly observed at 77 K. However, the fine structure is not observed at 298 K because of the significant contribution of an efficient radiationless deactivation arising from various types of vibrational interaction on the surfaces. From an analysis of the vibrational fine structure, the energy gap between the $(0 \rightarrow 0)$ and $(0 \rightarrow 1)$ vibrational transitions is determined to be about 1035 cm^{-1}, in good agreement with the vibrational energy of the surface V=O bond obtained by IR and Raman measurements (121, 122).

b. *Vanadium Oxide/Al$_2$O$_3$ Catalysts.*   Figure 24 shows the phosphorescence spectrum and its corresponding excitation spectrum of the vanadium oxide/Al_2O_3 catalyst at 4.2 K (130). The phosphorescence intensity of the hydrated vanadium oxide/Al_2O_3 catalyst was much lower than that of the comparable vanadium oxide/SiO_2 and vanadium oxide/Vycor glass catalysts. As shown in Fig. 24, the wavelengths at the maximum intensities for the absorption (excitation) and phosphorescence spectra of the vanadium oxide/Al_2O_3 catalyst can be seen at approximately 330 and 550 nm, respectively. The peak positions and the shapes of the emission changed upon variation of the excitation wavelengths, indicating that there are at least two different types of emitting species on the surface. In fact, the decay curve was nonexponential, indicating more than one emitting species. Thus, the vanadium oxide/SiO_2 (V/Vycor) and vanadium oxide/Al_2O_3 catalysts have different phosphorescence properties. The excitation edge for vanadium oxide/Al_2O_3 is located at longer wavelength regions than for vanadium oxide/SiO_2. There is only one emitting site on the vanadium oxide/SiO_2

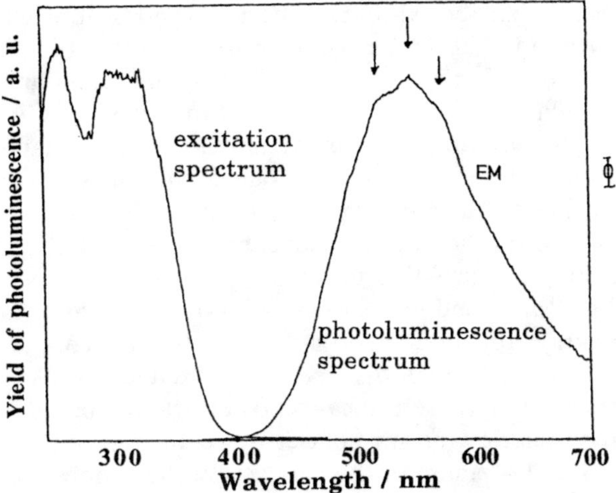

FIG. 24.   Photoluminescence spectrum and its excitation spectrum of the vanadium oxide catalyst supported on Al_2O_3 (vanadium oxide/Al_2O_3) at 4.2 K. Excitation wavelength, 300 nm; emission monitored at 520 nm [reproduced with permission from Hazenkamp (*130*)].

catalyst, whereas there are at least two different types of emitting sites on the vanadium oxide/Al_2O_3 catalyst, even at the lower loadings of V (i.e., there is a support effect).

These results for the dehydrated vanadium oxide/SiO_2 and vanadium oxide/Al_2O_3 catalysts obtained by photoluminescence measurements show a good correspondence with those obtained by Raman spectroscopy, especially at the lower loadings (*121, 122*). Consequently, these findings suggest the presence of one oxo-vanadium species on SiO_2 and more than one oxo-vanadium species on Al_2O_3. Raman investigations of the vanadium oxide/Al_2O_3 catalyst with lower loadings suggest that several monomeric tetrahedral vanadium species of similar structure as well as polymeric vanadate chains with tetrahedral coordination can be formed on SiO_2.

The peak position of the excitation spectrum of the vanadium oxides at longer wavelengths as well as the structureless photoluminescence spectrum can be attributed to polymeric vanadate chains with tetrahedral coordination since, upon connection of VO_4 tetrahedral units to form three-dimensional polyvanadates, the absorption energies (i.e., excitation energies) are more or less delocalized over the vanadate chains. Such delocalization is usually accompanied by a redshift in the absorption and photoluminescence spectra compared with the monomeric species (*40, 130*). Also, the deactivation paths of the photon energies along these chains result in a shortening of the decay time of the emission. Moreover, ESCA investigations of these supports indicate that the Al–O to V bond has a less ionic character than the Si–O to V bond (*130, 131*) since the positive shift in the binding energy with respect to the reference compounds is less for vanadium oxide/Al_2O_3 than for vanadium oxide/SiO_2. Theory confirms these ESCA results since the ionicity of the bonding between Al and O is greater than that between Si and O because of the smallest electronegativity of Al. Therefore, it can be concluded that the excited triplet state of the vanadium oxide species is poorly localized on the Al_2O_3 surface, having a short V–O bond due to the lower ionicity of the Al–O to V bond. A shorter decay time can also be observed for the vanadium oxide species on Al_2O_3 since localization of the photon energy absorbed on the short V–O bond is a major requirement for the observation of a long lifetime of the excited state.

It is emphasized that these delocalization effects attributed to the connection of the monomeric emitting species by chains can be used to evaluate the extent of the dispersion of the catalyst on the support surfaces.

4. *Thermal Activation and Photoactivation of the Surface V=O Bonds*

Thermal reduction of the surface V=O or Mo=O species proceeds in a manner similar to the photochemical reduction of these oxides with H_2 at lower temperatures. Such a similarity suggests that the limiting step in thermal reduction also involves an electron transfer from the lattice oxygen anion to the corresponding metal cation, expressed as follows:

$$(V^{5+}=O^{2-}) \xrightarrow{h\nu} (V^{4+}-O^-)$$

It is known that the activation energy for the thermal reduction of various transition metal oxides lies in the range of 50–100 kJ/mol, whereas the energy of the light quanta involved in photochemical reduction processes varies from 300 to 400 kJ/mol (*132*).

The formation of an electronically excited state of the V=O bonding plays a vital role in the photochemical reduction of surface V=O species in the presence of H_2, as described previously. Moreover, the vertical transition in the photon absorption process occurs from the zero vibrational level of the ground electronic state to the higher vibrational levels in the electronically excited state (see Section II.B). In other words, the photochemical reduction of the oxide is initiated by the electron transfer from the lattice oxygen to the corresponding metal cation, leading to a new V–O distance. Thus, it could be said that the electronic excitation process of the oxide is directly associated with the photochemical reduction of the catalyst.

On the other hand, in thermal reduction, the vibrational stepwise excitation processes play a vital role in reaching higher vibrational levels. Therefore, it can be said that the thermal reduction of the oxide is initiated by the movement of the nuclei (i.e., an elongation of the V–O bond), whereas the electron movement immediately follows to adjust to the new V–O nuclear configuration (*132*).

These hypotheses raise the question of whether it is possible to establish the same electronic excited state that is formed by photochemical excitation by thermal activation as well. As shown in Fig. 25, in the case of tetrahedrally coordinated vanadium oxide/SiO_2, the potential curves of the ground singlet state S_0 and the triplet state T_1 do not intersect. The electronic state corresponding to any elongation of the V=O bond is always separated by a considerable energy interval. As a result, thermal reduction through vibrational stepwise activation is considerably more difficult in the tetrahedrally coordinated oxide catalysts than in octahedrally coordinated oxide catalysts. Furthermore, in such tetrahedrally coordinated oxides, the photoluminescence (i.e., phosphorescence) can easily be observed, demonstrating the absence of any intersection of the potential curves of the ground state and the excited triplet state for the tetrahedrally coordinated oxide species (*132*).

On the other hand, as shown in Fig. 25, for octahedrally coordinated oxides the triplet state is separated from the ground singlet state by only a small energy interval, even with the mixing of the potential curves of the ground singlet state and triplet state, enabling the transition from the ground state to the excited triplet state by means of thermal excitation. Thus, for the octahedrally coordinated oxide, the V=O bond can be reduced by

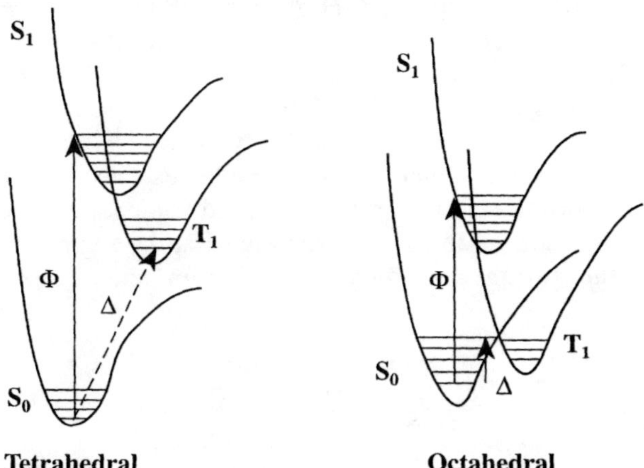

FIG. 25.   Quantum-chemical calculated potential energy curves for the ground singlet (S_0), electronically excited singlet (S_1), and excited triplet (T_1) states of the tetrahedral and octahedral complexes of vanadium oxide [reproduced with permission from Kazanskii (*132*)].

vibrational thermal activation resulting in the dissociation of the $V=O$ bond. In the octahedrally coordinated oxides, the reactive triplet state can be approached through either a photochemical process or thermal activation pathway, showing a similar pattern in reduction mechanisms (*132*).

Figure 25 also shows the threshold energies for both the thermal and the photochemical reduction processes. In general, a high selectivity can be achieved in photochemical reactions due to the highly selective nature of light absorption, which allows the injection of energy into particular bonds or molecules.

From these arguments, it is concluded that in the photochemical reduction reactions in which the charge-transfer excited triplet states of the metal oxides play a vital role the tetrahedrally coordinated metal oxides are reduced much more easily than in octahedrally coordinated oxides. Furthermore, the phosphorescence of the catalysts can be observed easily with tetrahedrally coordinated metal oxides since radiationless deactivation is much less efficient in the oxides than in the octahedrally coordinated oxides due to the larger energy gap between the excited triplet state and the ground state (*126, 133*).

D. PHOTOLUMINESCENCE OF VARIOUS OXIDE-BASED CATALYSTS

1. *Silica-Supported Oxides and Binary Oxide Catalysts*

a. *Molybdenum Oxide/SiO$_2$.* A typical photoluminescence and its excitation spectrum for the molybdenum oxide/SiO$_2$ system, prepared by reacting MoCl$_5$ with the surface OH groups of SiO$_2$ under air and water-free conditions at 77 K, as well as its diffuse reflectance spectrum are shown in Fig. 26 (*126*). These optical transitions are attributed to charge-transfer transitions involving a hexavalent molybdenum species. The photoluminescence of the highly dispersed supported molybdenum oxide catalyst (molybdenum oxide/SiO$_2$) is assigned to the phosphorescence, i.e., the radiative decay process from the charge-transfer excited triplet state, induced by the excitation of the charge-transfer absorption band at approximately 310–340 nm of the tetrahedral dioxomolybdenum species [Eq. 21] (*126, 134–149*):

$$
\begin{array}{ccc}
O^{2-} \quad O^{2-} & & O^{2-} \quad O^{-} \\
\diagdown \quad \diagup & & \diagdown \quad \diagup \\
Mo^{6+} & \xrightarrow{\; h\nu \;} & Mo^{5+} \\
\diagup \quad \diagdown & & \diagup \quad \diagdown \\
O^{2-} \quad O^{2-} & & O^{2-} \quad O^{2-} \\
| \qquad | & & | \qquad |
\end{array}
\tag{21}
$$

In the reflectance spectrum, the onset of the main absorption band can be observed at about 317 nm (31,500 cm^{-1}), in good agreement with the

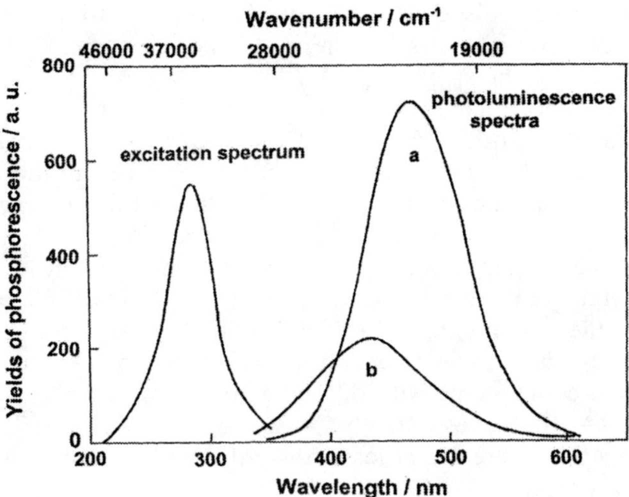

FIG. 26.   Photoluminescence spectra of the molybdenum oxide catalyst anchored to SiO_2 (molybdenum oxide/SiO_2) at 77 K (a) and at 298 K (b) and their excitation spectrum. Excitation wavelength at 280 nm; emission monitored at 520 nm [reproduced with permission from Anpo *et al.* (*127*)].

onset of the excitation spectrum of the phosphorescence. No photoluminescence could be observed for the oxo-molybdenum species which exhibits a weak absorption band at approximately 444 nm (22,500 cm^{-1}). The phosphorescence and its corresponding excitation spectra show peak maxima at approximately 486 nm (20576 cm^{-1}) and 286 nm (35,000 cm^{-1}), respectively. The phosphorescence spectrum does not change upon varying the excitation wavelength, which indicates that there is only one kind of photoluminescent species in the catalyst.

The positions of the photoluminescence (i.e., phosphorescence) and its corresponding excitation bands are nearly identical to those of the MoO_4^{2-} complex in $CaMoO_4$. The energetic positions of the peak maxima of the excitation and photoluminescence spectra of five- and six-coordinated oxo-molybdenum complexes are known to exist at lower energy levels, i.e., at approximately 400–333 nm (25,000–30,000 cm^{-1}) for the excitation and at 714–588 nm (14,000–17,000 cm^{-1}) for the photoluminescence, respectively (*130*). Hence, the photoluminescence spectrum shown in Fig. 26 can be attributed to the presence of an isolated four-coordinate molybdenum oxide species.

Depending on the samples, a weak absorption band can sometimes be observed at approximately 444 nm having relatively high molybdenum loading and/or poor dispersion. This band is ascribed to the absorption of the oxo-molybdenum species, i.e., the crystalline or amorphous MoO_3 which may be present in the samples (134–140).

The photoluminescence spectra of anchored molybdenum oxide/SiO_2 and molybdenum oxide/PVG catalyst recorded at 77 K decay as a single exponential, as evidenced by the straight line obtained in the logarithmic plots. The lifetime (i.e., τ at 77 K) of the charge-transfer excited triplet state was determined to be 1.9×10^{-3} s (33, 63, 126).

On the other hand, the decay curves of the photoluminescence spectra of the molybdenum oxide/SiO_2 and molybdenum oxide/PVG catalyst prepared by the impregnation method with an aqueous solution of $(NH_4)_6Mo_7O_{24}$ show some deviation from a single exponential decay, indicated by two components with different lifetimes (i.e., $\tau_1 = 1.6 \times 10^{-3}$ s and $\tau_2 = 0.8 \times 10^{-3}$ s, respectively), in the logarithmic plots. These results clearly show that there are at least two different kinds of emitting dioxo-molybdenum species (33, 126).

i. *Temperature Dependence.* The temperature dependence of the radiative decay time was investigated, and it was found that at temperatures below 30 K, the decay time is strongly temperature dependent, whereas at temperatures above 30 K a plateau is reached. The value of the plateau (at 50 K) of the radiative decay time was determined to be 1.6 ms. This decay time is much longer than that of the isolated tetrahedral molybdate complex in $CaMoO_4$, for which a decay time of about 0.25 ms has been reported (33). Unusually long decay times have also been observed for the photoluminescence of supported vanadium and chromium oxide catalysts (117).

It has been shown that a lengthening of the decay time can be linked to a special kind of distortion of the isolated transition metal complex (130, 150). If one or two metal–oxygen bonds within the complex become shorter and more covalent, whereas the remaining metal–oxygen bonds become longer and more ionic, the radiative decay time of the photoluminescence usually becomes longer (150–152). This is a result of the localization of the excited state in the shorter metal–oxygen bond. It is known that the same features are observed if the more inonically bound oxygen ligands are replaced by fluorines.

On the basis of these considerations, the photoluminescence is attributed to an isolated tetrahedral dioxo-molybdenum species:

$$\begin{array}{ccc} O & & O \\ \diagdown\!\!\!\!\diagdown & & \diagup\!\!\!\!\diagup \\ & Mo & \\ \diagup & & \diagdown \\ O & & O \\ | & & | \end{array}$$

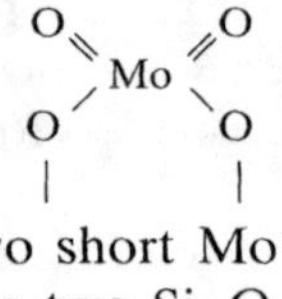

The surface complex has two short $Mo{=}O$ bonds (dioxo species) extending out of the surface. The two Si–O groups bonded to Mo may be ionic in nature due to the strong polarizing effect of the Si^{4+} ion. Iwasawa (20) reported that molybdenum oxide anchored to SiO_2 and formed by a chemical reaction consists of tetrahedrally coordinated molybdenum ions only. Extended X-ray absorption fine structure (EXAFS) results with anchored molybdenum oxide catalysts on SiO_2 (anchored molybdenum oxide/SiO_2) indicate that the monomeric tetrahedral molybdenum oxide species are the most important surface species. Raman spectroscopy also confirmed the presence of the monomeric tetrahedral molybdenum oxide species (20).

ii. *Effect of Mo Loading.*   Ono *et al.* (138) found that the concentration of the tetrahedral molybdenum ions in the molybdenum oxide/SiO_2 catalyst essentially corresponds to the intensity of the phosphorescence of the catalysts. As shown in Fig. 27, increasing the molybdenum content leads to an increase in the intensity of the phosphorescence, which passes through a maximum at about 1 atom% of Mo. As shown in Fig. 27, the concentrations of both the tetrahedrally coordinated molybdenum oxide species obtained by photoluminescence intensity and of the X-ray amorphous molybdenum oxide pass through maxima at similar molybdenum contents (about 2 atom% Mo). As can also be seen in Fig. 27, the photosphoresence intensity decreases remarkably in the region >2 atom% Mo, suggesting that a change occurs in the coordination of the molybdenum ions from tetrahedral to octahedral. An additional increase in the molybdenum content results in the crystallization to MoO_3, although the X-ray amorphous phase is still present at 3 or 5 atom% Mo.

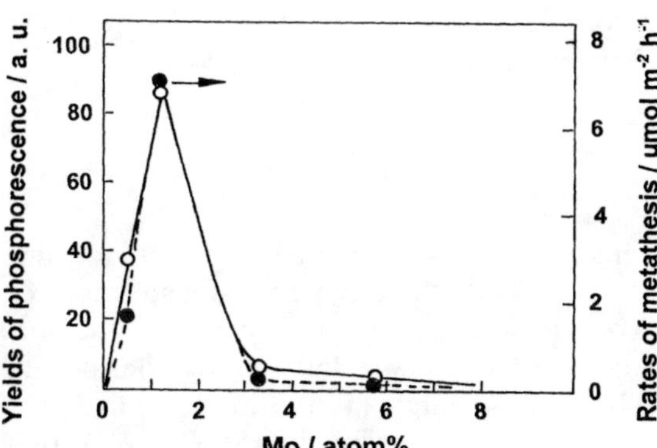

FIG. 27.   Yields of the photoluminescence (phosphorescence) spectra of molybdenum oxide catalysts supported on SiO_2 (molybdenum oxide/SiO_2) (O) and the rate of the metathesis reaction of C_3H_6 (●) on the catalysts at 473 K (pressure of C_3H_6, 4 kPa) [reproduced with permission from Ono *et al.* (138)].

It has been found that supported molybdenum oxide catalyzes an efficient catalytic metathesis reaction. The metathesis of C_3H_6 at 473 K on the molybdenum oxide/SiO_2 catalyst led to the products ethene, 1-butene, and *cis*- and *trans*-2-butene, the ratio of the three butenes being near the equilibrium ratio at the reaction temperature of 473 K. As shown in Fig. 27, with an increase in the Mo content the rate of the metathesis reaction increases, passing through a maximum at approximately 2 atom% Mo. This feature closely resembles the change in the concentration of the tetrahedrally coordinated dioxo-molybdenum oxide species (*138, 153*).

   b.   *Tungsten Oxide/SiO₂*.   Silica-supported tungsten oxide prepared by impregnation exhibits a broad photoluminescence spectrum at approximately 510–555 nm (19,600–18,000 cm^{-1}) with a maximum which depends on the excitation wavelength, suggesting at least two different types of sites (*154–158*). The latter can be distinguished with an excitation maximum at approximately 250 nm (40,000 cm^{-1}) for type I and 290 nm (34,5000 cm^{-1}) for type II.

The high-energy position of the excitation band of emitting species I and the corresponding large Stokes shift (i.e., the magnitude of about 20,000 cm^{-1}) are attributed to the typical value of the Stokes shift for isolated WO_4^{2-} complexes. Considering that (i) for $SrWO_4$ having the isolated WO_4^{2-} tetrahedral species, the excitation maximum is observed at approximately 245 nm (40,800 cm^{-1}) and the value of the Stokes shift is approximately 19,100 cm^{-1}, and that (ii) the lifetime of the photoluminescence (i.e., τ_I = 450 ms at 50 K) is much longer than the lifetime of 60 ms for tungstates, species I is associated with a strongly distorted monomeric dioxo-tungsten complex having a structure similar to that of the molybdenum oxide/SiO_2 catalyst (*40, 130, 154–158*):

$$\begin{array}{ccc} O & & O \\ \diagdown\!\!\!\diagdown & & \diagup\!\!\!\diagup \\ & W & \\ \diagup & & \diagdown \\ O & & O \\ | & & | \end{array}$$

Such dioxo-tungsten oxide species have two short-terminal $W\!=\!O$ bonds. Among various possibilities, the other emitting species (i.e., species II) can be suggested to be a distorted monomeric complex, the photoluminescence lifetime of which is very long (τ_{II} = 450 ms). By considering the low energy of its excitation band, this complex is likely to have five or six coordinations so that a square pyramidally coordinated $(Si–O)_4$ WO complex having one short-terminal $W\!=\!O$ bond can be proposed (*130, 153–160*).

   c.   *Niobium Oxide/SiO₂*.   Niobium oxide catalysts supported on SiO_2 by the impregnation method exhibit photoluminescence at approximately

400–650 nm (25,000–15,300 cm^{-1}) with a maximum at approximately 18,800 cm^{-1} when excited at approximately 235–290 nm at 4.2 K. These photoluminescence and excitation spectra as well as the absorption spectrum observed at approximately 240–330 nm are attributed to the typical oxoniobium complexes involving the pentavalent niobium species (*40, 151, 152, 160–164*). These absorption and photoluminescence spectra are associated with the charge-transfer processes on the oxo-niobium species. It has been reported (*160–164*) that with the niobium oxide/SiO$_2$ catalyst there are at least two different kinds of emitting species on the surface. The maximum peak position of the photoluminescence due to emitting species I is located at approximately 505 nm (19,800 cm^{-1}) and that of emitting species II at approximately 538 nm (18,600 cm^{-1}), with the corresponding excitation bands at approximately 244 nm (41,000 cm^{-1}) and 278 nm (36,000 cm^{-1}), respectively.

Blasse *et al.* (*160–164*) proposed that the measured lifetime of the photoluminescence (i.e., $\tau_1 = 1700$ ms at 50 K) is abnormally long for the photoluminescence of niobates. Normally, the lifetime of a niobate is not longer than 130 ms at 50 K. These findings indicate that the long lifetime may be associated with the presence of isolated dioxo-niobium complexes since such long lifetimes have also been observed for the photoluminescence of solid-state compounds such as K$_2$NbOF$_5 \cdot$ H$_2$O with a distorted and isolated NbOF$_5^{2-}$ species as the emission center. In view of the low energy position of the excitation maximum of species II, this species is attributed to the five- or six-coordinated niobate species (*151, 165*).

d.   *Chromium Oxide/SiO$_2$*.   The photoluminescence spectra of chromium oxide/SiO$_2$ and chromium oxide/PVG catalysts were observed at approximately 500 nm, and the spectra were attributed to the charge-transfer transition processes involving an electron transfer from O^{2-} to Cr^{6+}, which constitutes the dioxo-chromium species. Moreover, Blasse *et al.* (*40, 160–164*), observed the vibrational fine structure in the photoluminescence spectrum of chromium oxide/SiO$_2$ at 4.2 K. Such a vibrational structure in the photoluminescence spectrum and the unusually long lifetime (800 ms at 50 K) suggest that the excited state of chromium oxide/SiO$_2$ is localized in the short Cr–O bond in a manner similar to that for vanadium oxide/SiO$_2$.

The vibrational energy separation between the $(0 \rightarrow 0)$ and $(0 \rightarrow 1)$ vibrational transitions in the photoluminescence spectrum was found to be 960 cm^{-1}, indicating that the Cr–O bond in chromium oxide/SiO$_2$ is shorter than that in the tetrahedral CrO$_4^{2-}$ cluster ion (i.e., 850 cm^{-1}). From these considerations, a distorted monomeric four-coordinated dioxo-chromium species has been proposed as the emitting species (*117, 166, 167*):

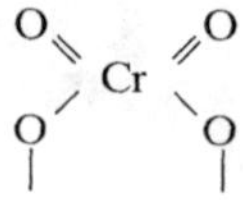

$$\begin{array}{ccc} O_\diagdown & & _\diagup O \\ & Cr & \\ O^\diagup & & ^\diagdown O \\ | & & | \end{array}$$

The species has two short $Cr{=}O$ bonds, in good agreement with the Raman results.

    e. Rhenium Oxide/SiO₂. The photoluminescence spectrum and the corresponding excitation spectrum of rhenium oxide/SiO_2 were observed at 4.2 K, although the intensity was very weak even at this low temperature (*130*). The excitation peak maxima were observed at 280 and 320 nm and the emission peak at about 556 nm. Raman investigations of the dehydrated rhenium oxide/SiO_2 having a low loading showed the presence of only one kind of ReO_4 species. However, compounds containing the ReO_4^- complex usually do not exhibit any photoluminescence spectrum. The only related photoluminescent compound is Bi_3ReO_8, in which the rhenium(VII) ion plays a vital role in the photoluminescence, which can be attributed to the $Bi^{3+}-Re^{7+}$ charge-transfer transition. Therefore, observing a photoluminescence spectrum for rhenium oxide/SiO_2 with a low Re loading suggests that the electronic state of the rhenium oxide species supported on SiO_2 may be different from that of the rhenium compounds containing the ReO_4^- complex. It is not clear why the tetrahedrally coordinated ReO_4^- complex exhibits only very weak photoluminescence compared with those of TiO_4^-, VO_4^-, NbO_4^-, and WO_4^-, which have similar energies for absorption as well as high photoluminescence yields (*130*).

    f. Titanium Oxide/SiO₂. Figure 28 shows the absorption spectra of anchored titanium oxides which were prepared by the facile reaction of gas-phase titanium(IV) chloride with the surface hydroxyl groups of the support at about 453–473 K, followed by treatment with gaseous H_2O to replace the chlorine atoms (*57, 168*). The absorption spectrum of the catalyst having one Ti–O layer can be observed at a shorter wavelength than that of bulk TiO_2 crystallites, its absorption maximum appearing at approximately 4.1 eV (about 300 nm). This absorption spectrum is attributed to the charge-transfer process involving an electron transfer from the molecular orbital t_{2u}, which is localized at the oxygen, to the vacant t_{2g} orbital of the titanium ion.

When the number of Ti–O layers is increased, the absorption spectrum of the anchored titanium oxide shifts to a longer wavelength region, approaching the absorption band of bulk TiO_2 (anatase) at four or five Ti–O layers. The anchored titanium oxide is X-ray amorphous up to three Ti–O layers, whereas at higher loadings the weak diffraction lines due to anatase

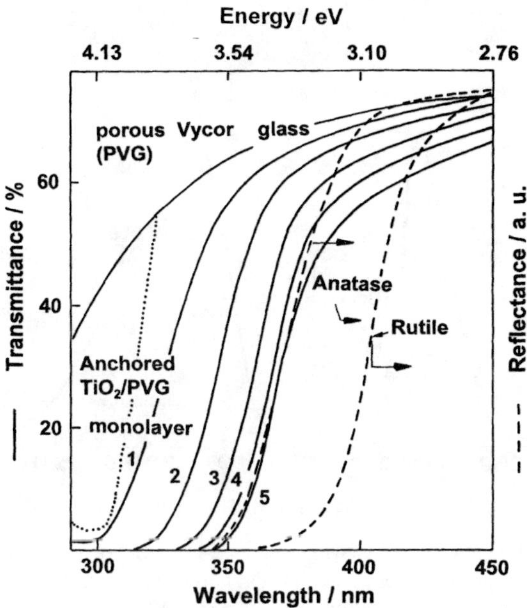

Fig. 28.   UV-visible absorption spectra of porous Vycor glass (PVG) and the titanium oxide catalyst anchored to PVG having one to five Ti–O layers. Dotted line is the absorption spectrum of the anchored titanium oxide having one Ti–O layer obtained by subtracting the PVG spectrum from the spectrum of the anchored titanium oxide/PVG catalyst [reproduced with permission from Anpo *et al.* (*168*)].

can be observed. With an increase in the number of Ti–O layers, the interactions of the individual titanium–oxygen complexes become stronger. Consequently, the energy levels of these complexes can no longer be defined strictly and a band structure emerges, which explains the shift of the absorption spectrum to longer wavelengths with increasing numbers of Ti–O layers.

Figure 29 shows the photoluminescence spectrum of the anchored titanium oxide catalyst having one Ti–O layer and the corresponding excitation spectrum (i.e., absorption spectrum) monitored at 460 nm of the emission at 77 K as well as the photoluminescence spectrum of bulk TiO_2 powder. Although there is no marked difference in the peak positions of the emission spectra of the bulk TiO_2 and anchored titanium oxides, the intensity is much higher for the latter. The photoluminescence excitation spectrum of the anchored titanium oxide having one Ti–O layer is essentially the same as its absorption spectrum (Fig. 28).

The vibrational fine structure of the photoluminescence spectrum of the

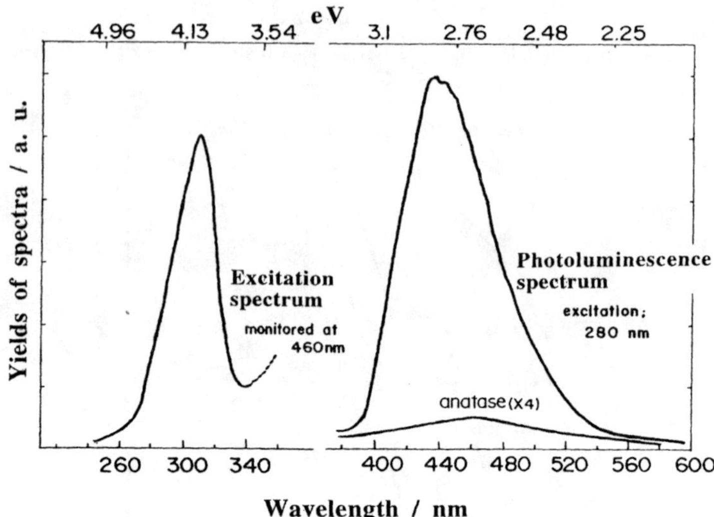

FIG. 29. Photoluminescence spectrum and its excitation spectrum at 77 K of titanium oxide anchored onto porous Vycor glass. (Excitation wavelength and its slit width were 280 nm and 10 nm, respectively. Emission slit width was 10 nm.) The anchored catalyst was degassed at 423 K and anatase TiO_2 powder was degassed at 573 K.

anchored titanium oxide having one Ti–O layer was observed at 77 K (*168*). Such a fine structure can be observed only when the slit width of the emission side is restricted to small values. The energy separation between the $(0 \rightarrow 0)$ and $(0 \rightarrow 1)$ transitions of the vibrational transition bands of the catalyst which corresponds to the energy obtained by the IR absorption spectrum of the Ti–O bond is observed at approximately 720–820 cm^{-1}. This value is in good agreement with the value obtained by the IR absorption band at approximately 780 cm^{-1} for the titanium oxide anchored to SiO_2. The absorption of light corresponds to the charge transfer from the O^{2-} to than Ti^{4+} ion and results in the formation of the pair state of the hole center (O^-) and trapped electron (Ti^{3+}); the photoluminescence is attributed to the radiative reverse process, i.e., the recombination of the hole and electron. Therefore, the appearance of such a vibrational fine structure indicates that the photon energy absorbed by the oxide is mainly localized on the Ti–O bonds of the surface species. The $(O \rightarrow 3)$ vibrational transition is the strongest. The Franck–Condon principle also suggests that the internuclear distance of Ti–O elongates in the excited state when compared with that in the ground state.

g. *Magnesium Oxide/SiO₂.* As shown in Fig. 30, highly dispersed magnesium oxide supported on SiO_2 exhibits a photoluminescence spectrum centered at approximately 430 nm upon excitation at approximately 240 nm (*169*). The emission is effectively quenched by the addition of CO or O_2. Figure 30 shows that the shape and the wavelength of the magnesium oxide/SiO_2 photoluminescence spectrum depends strongly on the magnesium loading as well as the pretreatment temperature. The magnesium oxide/SiO_2 having 1 wt% Mg evacuated at 1073 K exhibits a vibrational fine structure (Fig. 30, spectrum d) at 77 K.

According to Tanaka *et al.* (*169*), the photoluminescence spectrum of magnesium oxide/SiO_2 is associated with the excitation of specific bonds of the MgO diatomic moiety. Since neither SiO_2 nor bulk MgO exhibit, any photoluminescence spectrum at such long wavelengths, the origin of the photoluminescence of the magnesium oxide/SiO_2 can be related to specific Mg–O bonds that interact with the SiO_2 surfaces. The energy gap between the $(0 \rightarrow 0)$ and $(0 \rightarrow 6)$ vibrational transitions is about 900–1000 cm^{-1}, and the Franck–Condon analysis of the fine structure having the highest peak at approximately 520 nm suggests that the Mg–O bond is elongated in the excited state of the catalyst.

h.   *Zirconium Oxide/SiO_2 and Zirconium–Silicon Binary Oxides.*   As shown in Fig. 14, the bulk ZrO_2 oxide catalyst having surface ions in low

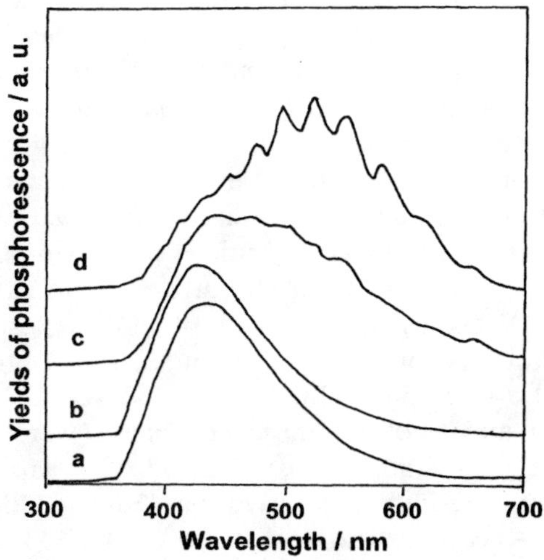

FIG. 30.   Photoluminescence spectrum of the magnesium oxide/SiO_2 catalyst at 77 K. MS20 (Mg/Si = 20/100) evacuated at 773 K (a), MS1 (Mg/Si = 1 : 90) evacuated at 773 K (b), MS20 evacuated at 1073 K (c), and MS1 evacuated at 1073 K (d) [reproduced with permission from Tanaka *et al.* (*169*)].

coordination (or coordinatively unsaturated surface sites) exhibits a photoluminescence at about 500 nm upon excitation at about 285 nm. Both the zirconium oxide/SiO_2 prepared by an impregantion method and the zirconium–silicon binary oxide prepared by a sol-gel method exhibit characteristic photoluminescence in the wavelength regions 350–550 nm when their absorption bands are excited at about 300 nm (*101–104*). The peak position of the emissions depend strongly on the Zr loading. These results, together with the efficient quenching of the photoluminescence by the addition of O_2, indicate that the photoluminescence of zirconium oxide/SiO_2 and zirconium–silicon binary oxides can be attributed to the following charge-transfer processes associated with surface sites in low coordination (coodinatively unsaturated surface sites) of the zirconium oxide species highly dispersed SiO_2 matrices or on SiO_2 surfaces (*170, 171*):

$$(Zr^{4+}\text{–}O^{2-})_{LC} \xrightarrow{h\nu} (Zr^{3+}\text{–}O^{-})^{*}_{LC} \tag{22}$$

2. *Supported Cation Catalysts*

a. *Cu(I)ZSM-5 Zeolite.* Only with Cu(I)ZSM-5 catalysts in which the EPR signal assigned to Cu^{2+} becomes weak as a result of evacuation at temperatures higher than 673 K can we observe a photoluminescence spectrum upon excitation at about 300 nm. Figure 31 shows a typical photoluminescence spectrum of Cu(I)ZSM-5 prepared by evacuation of Cu(II)ZSM-5 at 1173 K (Fig. 31a) and its corresponding excitation spectrum (Fig. 31b), i.e., its absorption spectrum. The excitation band at about 280–300 nm and the photoluminescence band at about 400–500 nm are attributed to the electronic excitation of an isolated Cu^+ monomer, i.e., to the ($3d^{10} \rightarrow 3d^9 4s^1$) electronic transition and its reverse radiative deactivation ($3d^9 4s^1 \rightarrow 3d^{10}$), respectively (*172–189*), and this electronic excitation involves the characteristics of small charge transfer from the lattice O^{2-} to Cu^+ ion (*183*).

With the Cu(I)Y zeolite and Cu(I)SiO_2 catalysts, especially with high copper loadings, an absorption band at about 300–320 nm and a photoluminescence band at about 450–560 nm were observed (*173–180*). The additional absorption and photoluminescence bands were attributed to the presence of a (Cu^+–Cu^+) dimer, i.e., to the ($3d\sigma^* \rightarrow 3d\sigma$) electronic excitation and its reverse radiative deactivation ($3d\sigma \rightarrow 3d\sigma^*$), respectively. XANES and EXAFS investigations of the Cu(I)ZSM-5 showed that the Cu^+ species anchored within the zeolite cavities exist as isolated Cu^+ monomers with two bonds with the lattice O^{2-} ions of the zeolite. On the other hand, with the Cu(I)Y zeolite and Cu(I)SiO_2 catalysts, the XANES and EXAFS spectra showed the presence of a Cu–O–Cu bond, suggesting the

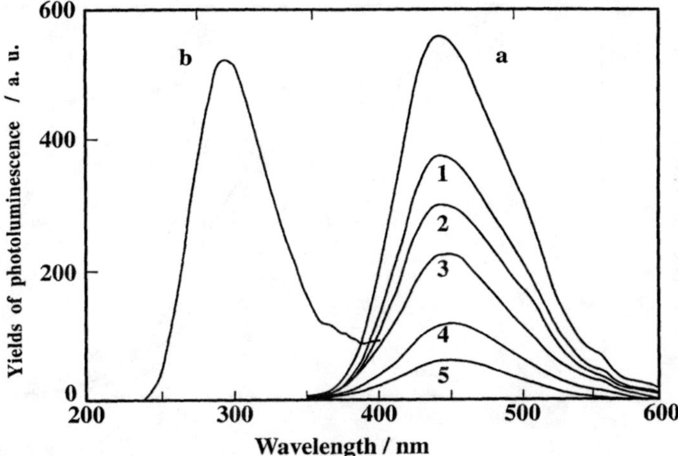

FIG. 31.   Photoluminescence spectrum of the Cu(I)ZSM-5 catalyst at 77 K (a) and the effect of addition of NO on the photoluminescence spectrum (1–5). (Cu(I)ZSM-5 prepared by evacuation of Cu(II)ZSM-5 at 1173 K. The addition of NO was carried out at 285 K. NO pressure (Pa): 1, 133; 2, 39.9; 3, 66.5; 4, 133; 5, 2660. The excitation spectrum (b) was monitored at 450 nm emission [reproduced with permission from Anpo *et al.* (*179*)].

formation of a $(Cu^+–Cu^+)$ dimer species (*180*). Thus, the results obtained by XANES, EXAFS, and photoluminescence investigations showed a good coincidence in the assignment of the copper species on the support surfaces (*172, 180*).

i.   *Effects of Various Kinds of Supports.*   As shown in Fig. 32, the yields of the photoluminescence spectra due to the Cu^+ monomer and the $(Cu^+–Cu^+)$ dimer species depend on the type of zeolite used as the support. The emission band at about 430–460 nm is the major component of the Cu(I)ZSM-5 (Fig. 32, spectrum a) and that of Cu(I) mordenite (not shown). The Cu(I)Y zeolite exhibits two different intense emission bands at about 450 and 525 nm. These results indicate that in the zeolites ZSM-5 and mordenite most of the copper cations are present as isolated Cu^+ monomers, but in the Y zeolite the copper species are aggregated to form the $(Cu^+–Cu^+)$ dimers. Thus, it has been shown that photoluminescence observations agree with the results obtained by X-ray absorption measurements (*178, 180*).

The values of the void space/volume of the supercages in zeolites are 0.28 (mordenite), 0.29 (ZSM-5), and 0.49 (Y zeolite), respectively. The zeolite channel has connections in two dimensions in ZSM-5 and in mordenite and in three dimensions in Y zeolite. Considering the volume of the supercages and the types of channel connections, we infer that the copper

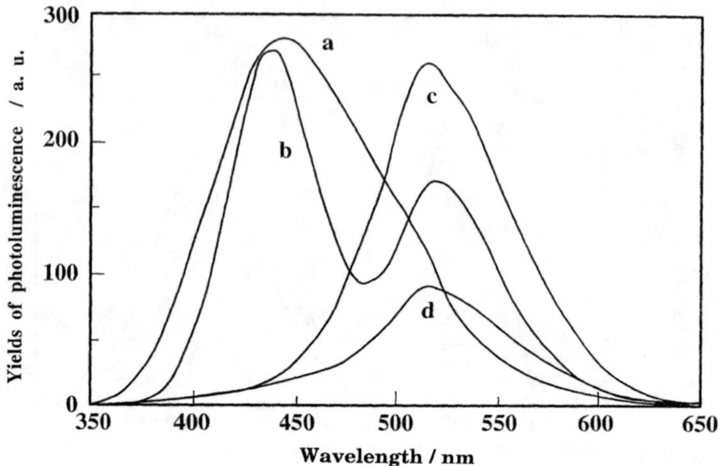

FIG. 32. Typical photoluminescence spectrum of Cu(I)ZSM-5 (a), Cu(I)Y zeolite (b), Cu(I)SiO$_2$ (c), and Cu(I)Vycor glass (d) at 77 K, prepared by the evacuation of Cu(II)ZSM-5 at 1173 K, Cu(II)Y zeolite at 973 K, Cu(II)SiO$_2$ at 973 K, and Cu(II)Vycor glass at 973 K, respectively [reproduced with permission from Anpo *et al.* (*175*)].

cations can diffuse relatively easily in the Y zeolite to form Cu$^+$–Cu$^+$ dimer species, whereas in the narrow channels of ZSM-5 and mordenite the low mobility of the copper cations suppresses the formation of the dimer species. In addition to this difference in of the copper cation, the density of the ion-exchangable sites within the zeolites must be considered. In the case of Y zeolite with a high Al/Si ratio, the high density of ion-exchangable sites causes the ion-exchanged copper cations to exist close to each other and, as a result, they aggregate easily during heat treatment (*180*).

b. *AgZSM-5 Zeolites.* Figure 33 shows the photoluminescence spectra (Fig. 33A) and the corresponding excitation spectrum (Fig. 33B) of the AgZSM-5 catalyst (*186*). The excitation band at about 220 nm shows a good coincidence with the absorption band attributed to the electronic excitation transition of $(4d^{10}) \rightarrow (4d^95s^1)$ of Ag$^+$ ions on the catalyst. Moreover, the photoluminescence band at approximately 320 nm can be attributed to the reverse radiative deactivation process of $(4d^95s^1) \rightarrow (4d^{10})$. Figure 33 also shows the effect of the addition of NO on the photoluminescence of the AgZSM-5 catalyst. The addition of NO leads to an efficient quenching of the photoluminescence. Figure 33 (A) shows that after complete quenching, the evacuation of the system leads to the recovery of the photoluminescence to its original intensity. These results clearly suggest

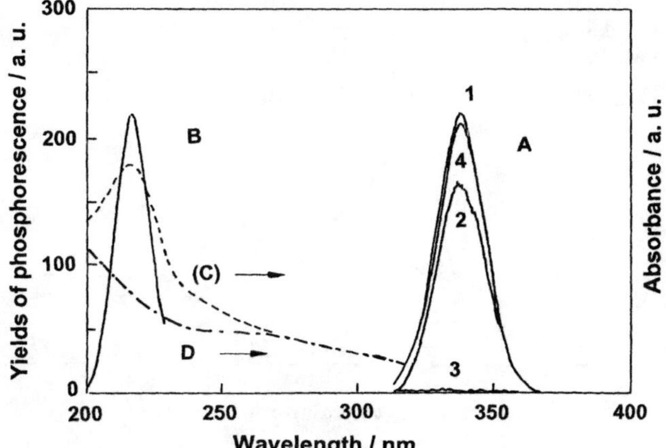

FIG. 33.   Photoluminescence spectrum of the AgZSM-5 catalyst (A) and its excitation spectrum (B), the effect of the addition of NO on the photoluminescence (2–4), and UV absorption spectrum of AgZSM-5 (C) and HZSM-5 (D), respectively. The addition of NO was carried out at 298 K. NO pressure (Pa): 1, 0.0; 2, 26.6; 3, 532; 4, after degassing of NO of sample 3 at 298 K. Excitation spectrum (B) monitored at 342 nm emission [reproduced with permission from Matsuoka *et al.* (*186*)].

that the interaction of NO with Ag^+ ions is weak, but its interaction occurs in the ground as well as the excited state of the Ag^+ ions. The photoluminescence of the AgZSM-5 was more easily quenched by the addition of NO than that of Cu(I)ZSM-5, suggesting that NO interacts more efficiently with the excited electronic state of Ag^+ ions than with that of Cu^+ ions (*186–189*).

c.   *Eu(III)Y Zeolites.*   As shown in Fig. 34, the photoluminescence spectra of the Eu^{3+} ion-exchanged Y zeolite is observed near 400–500 nm after sample degassing at various temperatures. After evacuation at room temperature, the Eu(III)Y zeolite exhibits an emission band near 570–680 nm with peaks at 590, 615, and 655 nm upon excitation at about 390 nm. This emission band is attributed to the transition between the ^5D and ^7F levels of the Eu^{3+} ion ($^5D_0$–$^7F_1$, $^5D_0$–$^7F_2$, and $^5D_0$–$^7F_3$ transitions, respectively (*130*). When the Eu(III) Y zeolite was degassed at temperatures 573 K and excited with a 345-nm beam, the photoluminescence band decreased in intensity; however, a new emission band appeared near 455 nm, which could be attributed to the presence of the Eu^{2+} ion, i.e., the 4f–5d electronic transition. The relative intensity of the emission increased as the evacuation temperature increased suggesting that the reduction of Eu^{3+} to Eu^{2+} in the zeolite cavities occurs as follows (*190*):

$$[2Eu^{3+}, nH_2O]/Y - (n - 1)H_2O \rightarrow [2Eu^{2+}, 2H^+]/Y + 1/2O_2. \quad (23)$$

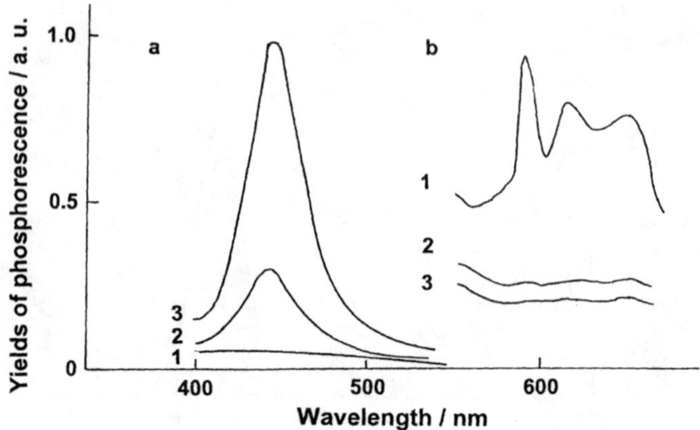

FIG. 34.   Photoluminescence spectra of the europium ion-exchanged Y zeolite after degassing at (1) 25, (2) 300, and (3) 500°C. Excitation at 345 nm (a) and 390 nm (b) [reproduced with permission from Arakawa *et al.* (*190*)].

d.   *Ce(III)/Porous Vycor Glass.*   The photoluminescence spectrum and its corresponding excitation spectrum (i.e., absorption) of Ce^{3+} supported on PVG (porous vycor glass) were observed at about 350–400 and 250–320 nm, respectively (*130, 191*). The two absorption bands near 250 and 310 nm are attributed to transitions from the ground level to two different crystal field components of the 5d levels. The photoluminescence spectrum consists of a broad band centered at 370 nm with a shoulder at about 335 nm, which are attributed to transitions from the lowest 5d level to the $^2F_{7/2}$ and $^2F_{5/2}$ ground levels of the $4f^1$ configuration of Ce^{3+}, respectively. The energy gap between the position of the shoulder and the peak of the emission was found to be about 2200 cm^{-1}.

3.   *Other Catalysts*

a.   *Supported CdS.*   The photochemistry of small particles of CdS, usually in the colloidal form, has been the focus of much attention. Much of this work stems from the search for photosystems that can be excited in the visible part of the spectrum, thereby initiating electron-transfer reactions. Recent work on the production of small CdS particles indicates that the colloid size has a remarkable effect on the semiconductor properties compared to those of a single crystal or large particles of CdS; this is often referred to as a quantum size effect (*192–195*).[3]

---

[3] When the particle sizes of semiconducting materials become smaller than about 100 Å, their band gap energies become larger. The band structure collapses, and discrete energy levels appear. Therefore, such small (and size quantized) semiconducting particles show a remarkable blue shift and clear structure in their absorption spectra.

Figure 35 shows the photoluminescence spectrum of CdS supported on PVG with a relatively high loading (*196*). Peaks are observed near 520, 560, and 680 nm. The 680-nm peak is associated with the sulfur vacancy since the presence of excess sulfide ions quenches the photoluminescence; however, the presence of excess cadmium has no effect on the emission. The 520- and 560-nm photoluminescence are associated with the major bulk emission (*197–199*). The 520-nm emission is attributed to the band-to-band transition, and the 560-nm emission is attributed to a typical radiative electron-hole recombination at the particle surface. As shown in Fig. 35 (b), the addition of H_2O to the catalyst has a significant effect on the spectrum. The 560-nm photoluminescence is completely quenched, as expected if the radiative recombination of electrons and holes occurs at the surfaces where H_2O molecules easily interact with these electrons and holes, thereby reducing the energy and intensity of the photoluminescence. On the other hand, the 520-nm emission from the bulk emitting sites is not affected by the addition of H_2O. The photoluminescence

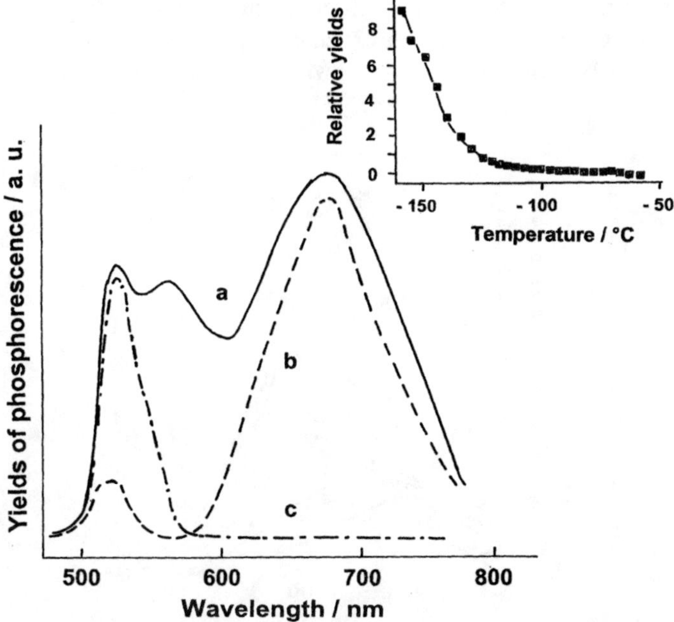

Fig. 35. Photoluminescence spectrum of CdS supported on porous Vycor glass in the absence (a) and presence of water (b) or MV^{2+} (c). Inset describes the relative intensity of the emission as a function of temperature, while the emission intensity remains virtually constant from $-60°C$ to room temperature [reproduced with permission from Kuczynski and Thomas (*196*)].

attributed to the presence of a sulfur vacancy also remains unaffected by the addition of H_2O.

b. *Titanium–Silicon Binary Oxides.* Some binary oxide catalysts show great potential for various promising photocatalytic systems since binary oxides often exhibit higher catalytic activities and selectivity than those predicted from the properties of their individual components. In this section, the characteristic photoluminescence spectra of titanium–silicon binary oxides (i.e., TiO_2–SiO_2) (*200*) and titanium–aluminum binary oxides (i.e., TiO_2–Al_2O_3) (*201*) prepared by a coprecipitation method are reviewed.

With the titanium–silicon binary oxides, a unique photoluminescence spectrum can be observed at about 500 nm (for Ti : Si = 1 : 1 binary oxide), and the λ_{max} position of the corresponding spectra shifts steadily toward shorter wavelength regions with decreasing titanium content, as shown in Fig. 36. Such behavior shows good agreement with the blueshift observed in the reflectance spectra of these oxides (*200*). As can also be seen in Fig. 36, the addition of O_2 leads to an efficient quenching of the photolumines-cence. A steady increase was observed in the extent of the emission parts quenched by O_2 at the same pressure when the silicon content of the oxides

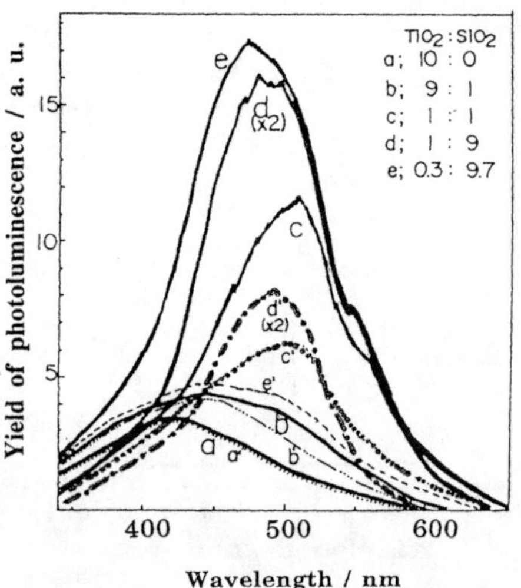

FIG. 36.   Photoluminescence spectra of titanium–silicon binary oxide catalysts with different Ti/Si ratios at 77 K and the effect of the addition of O_2. Excitation wavelength, 300 nm; slit width for emission, 7.0 nm; pressure of added O_2, 13.3 Pa (a'–e'); photoluminescence spectra in the absence of O_2 (a–e) [reproduced with permission from Anpo *et al.* (*200*)].

increased, suggesting not only that the emitting sites are present near the surface but also that the lifetime of the excited state of the emitting species becomes longer when the binary catalysts have lower Ti contents.

X-ray patterns of titanium–silicon binary oxides showed only the diffraction lines due to anatase. With decreasing titanium content, the diffraction lines of the catalyst decreased in intensity, broadened, and finally disappeared. These results indicate that with titanium–silicon binary oxides, the crystallinity of TiO_2 decreases dramatically and disappears at lower titanium contents (200). As shown in Table III there is a steady increase in the intensity of $Ti(2p_{3/2})$ X-ray Photoelectron Spectroscopy (XPS) signals with increasing silicon content. The higher ratio of the $Ti(2p_{3/2})$ to $Si(2p_{1/2})$ XPS signal intensity than what is expected from the initial composition, especially in the lower titanium content regions, suggests the enrichment of the titanium ions near the catalyst surface. With a mechanical mixture of TiO_2 and SiO_2, similar features were observed (200); however, their extents were much smaller than in the case of the the binary oxides.

All these results indicate that, with titanium–silicon binary oxides having low titanium contents the Ti^{4+} ions are enriched near the surface regions, separated from each other, and present as tetrahedral species in the SiO_2 carrier matrices. In such species, the radiationless transfer of photon energy absorbed by TiO_2 is suppressed because of the low coordination of the ions. As a result, the formation of the (electron–hole) ion pairs, i.e., the excited state of $(Ti^{3+}- O^-)^*$ complexes, is facilitated (200, 201).

TABLE III

Binding Energies of $Ti(2p_{3/2})$, $Si(2p_{1/2})$, and $Si(2s)$ and the Ratio of the Signal Intensity of $Ti(2p_{3/2})$ versus $Si(2p_{1/2})$ in Titanium–Silicon Binary Oxide Catalysts[a]

Bulk composition of catalyst	Surface characteristics of catalysts			
		Binding energy (eV)[b]		
$TiO_2:SiO_2$ (mass ratio)	$Ti(2p_{3/2})/Si(2p_{1/2})$	$Ti(2p_{3/2})$	$Si(2p_{1/2})$	$Si(2s)$
100:0	100% TiO_2	458.7_6		
90:10	5.85 (-35%)[c]	458.6_8	101.8_4	153.0_8
50:50	1.03 $(+3.0\%)$	459.3_2	102.4_8	153.4_0
10:90	0.15 $(+32\%)$	460.2_8	102.9_6	154.0_4
3:97	0.046 $(+48\%)$	460.5_4	103.0_2	154.0_6
0:100	100% SiO_2		102.9_2	154.0_8

[a] From Anpo *et al.* (200). [b] The binding energies of the $Ti(2p_{3/2})$, $Si(2p_{1/2})$, and $Si(2s)$ lines were referenced to the $Au(4f_{7/2})$ line (83.8 eV). The Au reference was introduced onto the sample surface by vacuum deposition. [c] Parentheses indicate the magnitude of the difference from the bulk composition. +, the enrichment of Ti ions in the surface region.

c. *Vanadium Silicalite.* Microporous crystals provide fascinating and also useful surfaces for both catalytic and photocatalytic reactions. Metallosilicates are one of the most attractive and potentially promising classes of microporous catalysts.

The vanadium silicalite (VS-2) degassed at 473 K exhibits a characteristic photoluminescence with a vibrational fine structure (due to the $V{=}O$ bonds) at about 450–550 nm upon excitation of the absorption band at about 280 nm (Fig. 37) (*202, 203*). These photoluminescence spectra and their corresponding excitation spectra (absorption) coincide well with those observed for the highly dispersed vanadium oxide species anchored to SiO_2 or Vycor glass and having a tetrahedral VO_4 unit structure; the spectra are attributed to the charge-transfer processes on the VO_4 structure. As shown in Fig. 37 (E), the second-derivative photoluminescence spectrum of the VS-2 catalyst indicates that the energy gap between the $(0 \rightarrow 0)$ and $(0 \rightarrow 1)$ vibrational transitions is about 970 cm^{-1}, in good agreement with the IR absorption band assigned to the vibrational mode of the $V{=}O$ bond combined with the $Si{-}O^-$ moiety.

Figure 38 shows that the XANES (left) of the catalyst exhibits a spectral pattern similar to that of $VO(O\text{-}iso\text{-}C_3H_7)$ complexes having C_{3v} symmetry (*202, 203*). The EXAFS spectrum (Fig. 38, right) of the catalyst exhibits only one peak at (0.12 nm), attributed to the presence of the neighboring O atoms. This observation supports the photoluminescence results indicating that V^{5+} ions in a highly dispersed state in VS-2 are present in a slightly distorted tetrahedral coordination.

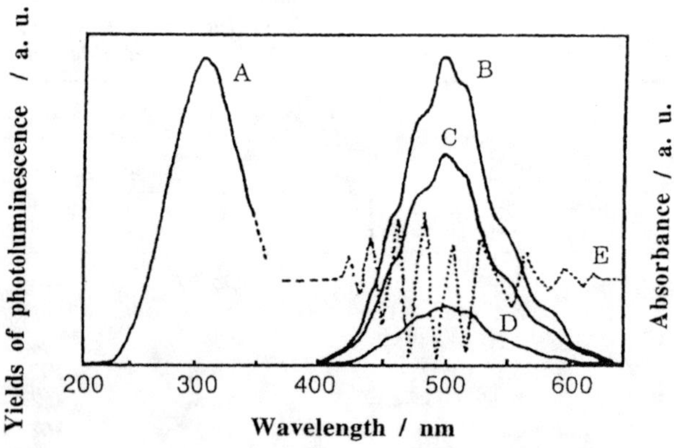

FIG. 37.   Photoluminescence spectrum (B) and its excitation spectrum (A) of the evacuated vanadium silicalite catalyst (VS-2) at 77 K and the effect of the addition of NO on the photoluminescence spectrum (C and D). Spectrum E is the second-derivative photoluminescence spectrum of spectrum B [completed from Anpo *et al.* (*202*)].

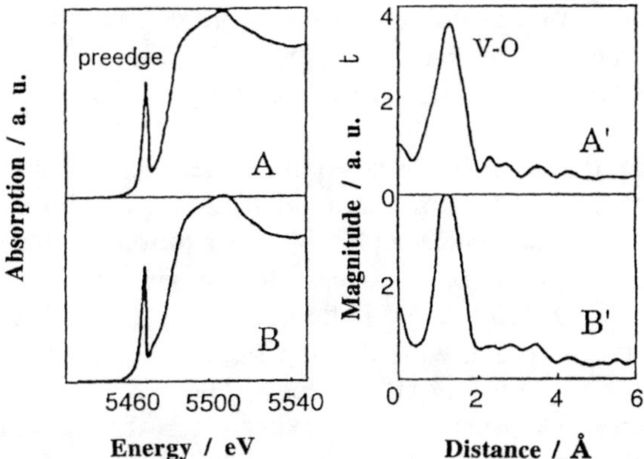

FIG. 38. XANES (A and B) and EXAFS spectra (A' and B') of the evacuated vanadium silicalite catalyst (VS-2) (A) and of VO(O-i-Pr)$_3$ (B) [reproduced with permission from Anpo *et al.* (*202*)].

d. *Titanium Silicalite.* Titanium silicalites referred to as 24-TS-1 (TS-1 incorporating 24 wt% Ti) and 12-TS-1 (TS-1 incorporating 12 wt% Ti) show a weak photoluminescence spectrum between 400 and 500 nm different from the wavelength of the photoluminescence band observed, for 85-TS-2 (TS-2 incorporating 85 wt% Ti) (Fig. 39) (*203–206*). The TS samples with higher Si/Ti ratios exhibit a strong photoluminescence band in the order of the yield of the photoluminescence: 85-TS-2 > 24-TS-1

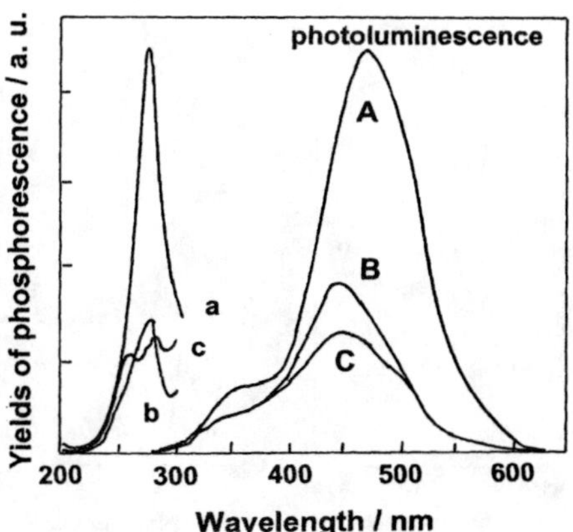

FIG. 39. Photoluminescence spectra (A–C) and the corresponding excitation spectra (a–c) of titanium silicalite catalysts (TS zeolites) at 77 K. (a, A) TS-2 (Si/Ti = 85), (b, B) TS-1 (Si/Ti = 24), (c, C) TS-1 (Si/Ti = 12) [reproduced with permission from Zhang *et al.* (*203*)].

> 12-TS-1. These photoluminescence spectra can be effectively quenched by the addition of O_2, N_2O, or H_2O, indicating that the photoluminescent-active species are located at surface positions accessible to small reactant molecules.

A strong preedge peak is observed in all the XANES data characterizing the TS catalysts indicating titanium oxide species with Four- or five-coordination (Fig. 40). A small EXAFS peak attributed to the neighboring Ti atoms (i.e., Ti–O–Ti) is observed as well as a peak due to the neighboring O atoms (i.e., Ti–O) (203–206). The intensity of the peak attributed to the Ti–O–Ti bonds increases with decreasing Si/Ti ratios, indicating partial aggregation of the titanium oxide species. These results indicate that most of the titanium oxide species in TS exist in tetrahedral coordination and exhibit photoluminescence; however, in the TS with a low Si/Ti ratio, aggregated titanium oxide species coexist with the tetrahedral species.

e. *Silica, Porous Vycor Glass, and Alumina.*   Silica, alumina, and Vycor glass are often used as supports for highly dispersed supported catalysts such as transition metal oxides, metals, and metal cations. The characteristics of the photoluminescence spectra of these supported catalysts have already been described. In this section, the photoluminescence properties of supports such as silica, Vycor glass, and alumina are discussed.

Normally, these supports exhibit neither absorption nor photoluminescence spectra in the wavelength regions 230–600 nm. However, the supports sometimes exhibit abnormal absorption and photoluminescence bands.

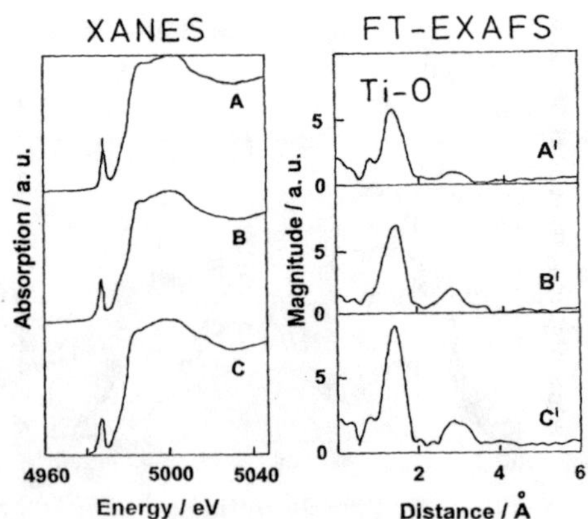

FIG. 40.   XANES (A–C) and EXAFS (A'–C') spectra of the titanium silicalite catalysts (TS zeolites). (A, A') TS-2 (Si/Ti = 85), (B, B') TS-1 (Si/Ti = 24), (C, C') TS-1 (Si/Ti = 12) [completed from Yamashita *et al.* (203) and unpublished results].

Both silica and Vycor glass degassed at temperatures >673 K exhibit such abnormal absorption bands at about 230–270 nm and photoluminescence spectra at about 380–580 nm (*14–16, 207*). The intensities of the absorption and photoluminescence spectra depend on the evacuation temperature.

The addition of O_2, N_2O, or C_3H_6 leads to an efficient quenching of the photoluminescence. Evacuation of the system after complete quenching of the photoluminescence results in a partial recovery of the photoluminescence (*14–16*). Heating samples that exhibited abnormal absorption and photoluminescence spectra at 673 K in the presence of O_2 leads to a complete disappearance of these abnormal absorption and photoluminescence bands. From these findings, such spectra observed with the well-degassed high-surface silica, Vycor glass, and alumina can be attributed to the charge-transfer processes on the surface sites (or coordinatively unsaturated surface sites), similar to those discussed in Section IV.A.2. Thus, these results show that it is vital to monitor the photoluminescence spectra of the catalysts supported on silica, Vycor glass, and alumina and to minimize the effects from the support used.

V. Dynamics of Photoluminescence and the Reactivities of Catalysts

A. Quenching and Absolute Rate Constants

1. Supported Vanadium Oxide

As shown in Fig. 19, vanadium oxide supported on Vycor glass exhibits a photoluminescence spectrum at about 400–600 nm upon excitation of the absorption band at about 320 nm (*33, 34, 63, 69, 115, 116*). The absorption and photoluminescence spectra are represented by Eq. (12). The addition of O_2, CO, N_2O, C_2H_4, C_3H_6, or C_4H_8 to the catalyst led to the quenching of the photoluminescence with differing efficiencies but without any changes in the shape of the spectrum.

Figure 41 shows the Stern–Volmer plots,

$$\Phi_0/\Phi = 1 + \tau k_q[Q] \qquad \text{(Stern–Volmer equation)} \qquad (24)$$

which can be obtained for the quenching of the photoluminescence yield with various quencher molecules by applying the steady-state treatment as in the following (*56, 69, 115–117*):

$$(V^{5+}\!-\!O^{2-}) \text{ photoluminescence } (k_p)$$
$$(V^{5+}\!-\!O^{2-}) \xrightarrow{h\nu} (V^{4+}\!-\!O^-)^* \rightleftharpoons (V^{5+}\!-\!O^{2-}) \text{ radiationless deactivation } (k_d)$$
$$(V^{5+}\!-\!O^{2-}) \text{ deactivation by quenching } (k_q)$$
$$(25)$$

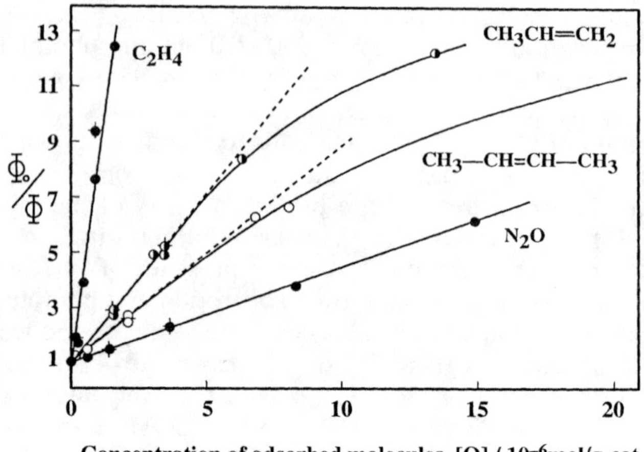

Concentration of adsorbed molecules, [Q] / 10⁻⁶mol/g-cat

FIG. 41.   Stern–Volmer plots of Φ_0/Φ values for the quenching by various molecules of the phosphorescence of the vanadium oxide catalyst supported on SiO_2 (vanadium oxide/ SiO_2; 0.03 V wt%) prepared by an impregnation method; data taken at 298 K [reproduced with permission from Anpo *et al.* (*116*)].

where Φ_0 and Φ are the yields of the photoluminescence of the catalyst in the absence and presence of the quencher molecules, respectively, and τ k_q, and $[Q]$ are respectively the lifetime of the charge-transfer excited state of the tetrahedrally coordinated vanadium oxide species in vacuum, the quenching rate constant (i.e., reaction rate constant for the charge-transfer excited state of the catalyst), and the concentration of the quencher molecules on the surfaces. As shown in Fig. 41, Φ_0/Φ is a linear function of the concentration of the quencher molecules, especially at low concentrations of the quenchers. These results indicate that the quenching of the photoluminescence occurs by the interaction of the added molecules with the charge-transfer excited state of the vanadyl bond, which leads to nonradiative deactivation pathways (*56, 69, 115–117*).

The results obtained from such dynamic photoluminescence studies shown in Fig. 41, together with the results obtained for O_2 and CO, permit calculation of the absolute rate constants of quenching for the various molecules, as follows: 9.34×10^{12} for O_2, 3.52×10^{10} for C_2H_4, 2.24×10^{10} for *trans*-2-butene, and 1.51×10^{9} for N_2O, all in units of (g/mol s), respectively (*33, 34, 56, 69, 115–117*). Consequently, the reactivities of these molecules toward the charge-transfer excited state of the vanadyl species decrease in the order $O_2 > CO > C_2H_4 > C_3H_6 > $ *trans*-2-$C_4H_8 > N_2O$ (*120, 208–210*).

2.   *Anchored Molybdenum Oxide*

Anchored molybdenum oxide exhibits a photoluminescence spectrum at about 370–550 nm when excited by a photon with a wavelength of about 280 nm (Fig. 26) (*37, 63, 69, 126, 127*). The absorption and photoluminescence spectra of the oxide are represented by Eq. (21). The addition of O_2, C_3H_6, or N_2O on the oxide leads to an efficient quenching of the photoluminescence, reducing the spectral intensity without any changes in the shape. The effect of the addition of C_3H_6 on the photoluminescence spectrum of the molybdenum oxide at 295 K is shown in Fig. 42 (*2–9*). The photoluminescence decayed as a single exponential as evidenced by the straight-line logarithmic plots. The addition of O_2, C_3H_6, or N_2O onto the oxide also led to the efficient quenching of the photoluminescence. Thus, the addition of O_2, C_3H_6, or N_2O onto the anchored molybdenum oxide decreases the emission not only in intensity but also in lifetime with almost the same effectiveness. Therefore, the photophysical and photochemical processes on the oxide in the presence of these quencher molecules can be described as follows (*33, 34, 126, 127, 133–140, 208–210*):

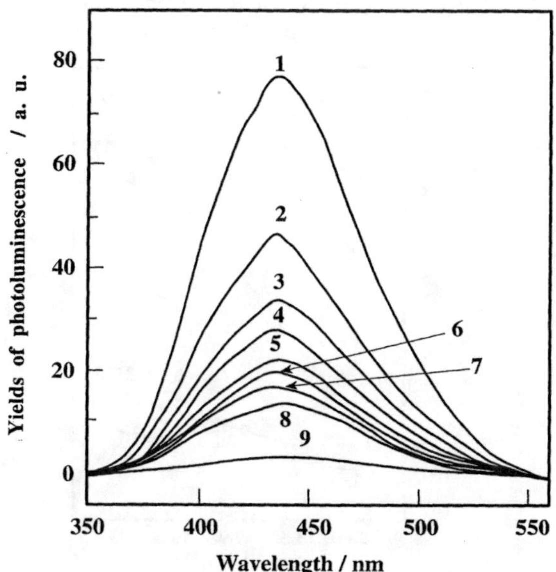

FIG. 42.   Phosphorescence spectrum of the molybdenum oxide catalyst anchored to SiO_2 (molybdenum oxide/SiO_2) (1) and its quenching by added C_3H_6 at 298 K (2–9). Excitation at 280 nm; pressure of C_3H_6 (Pa): 2, 3.72; 3, 7.45; 4, 11.17; 5, 14.89; 6, 18.62; 7, 29.93; 8, 37.11; 9, 74.21 [completed from Anpo *et al.* (*210*)].

$$(Mo^{6+}-O^{2-}) \xrightarrow{h\nu} (Mo^{5+}-O^{-})* \begin{cases} \text{photoluminescence } (k_p) \\ \rightarrow \text{radiationless deactivation } (k_d) \\ \text{deactivation by quenching } (k_q) \end{cases} \quad (26)$$

The Stern–Volmer equation (Eq. 24) has also been observed to pertain to both the yields (i.e., intensity) and the lifetimes of the photoluminescence spectra of the anchored molybdenum oxide.

The Stern–Volmer plots for the quenching of the photoluminescence yields at 295 K resulting from the addition of O_2, C_3H_6, or N_2O are linear functions of the pressure of the quencher molecules (Fig. 43). From the slopes of the Stern–Volmer plots for the quenching of the yield of the photoluminescence shown in Fig. 43, together with the lifetime of the excited triplet state of the tetrahedral dioxo-molybdenyl species (i.e., 48 μs at 298 K), and the absolute quenching rate constants of O_2, C_3H_6, and N_2O for the charge-transfer excited state of the anchored molybdenum oxide are determined to be 1.44×10^4, 3.23×10^3, and 1.49×10^3 in s^{-1} Pa^{-1} at 295 K, respectively (126, 210).

These absolute quenching rate constants are larger than the values for the excited triplet state of the tetrahedral oxo-vanadyl species of the supported vanadium oxide but decrease in the same order: 2.56×10^2, 4.35×10^2,

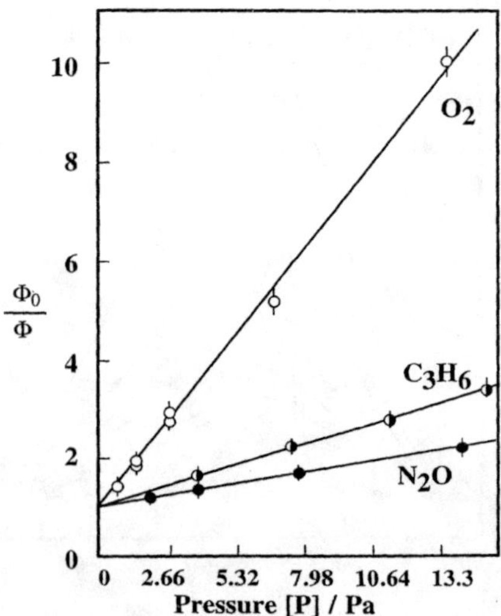

FIG. 43. Stern–Volmer plots of Φ_0/Φ values for the quenching of the phosphorescence intensity of the molybdenum oxide catalyst anchored to SiO_2 (molybdenum oxide/SiO_2) at various pressures of added quencher molecules at 298 K [reproduced with permission from Anpo *et al.* (210)].

and $1.21 \times 10^2\,\mathrm{s}^{-1}\,\mathrm{Pa}^{-1}$ for O_2, C_3H_6, and N_2O, respectively. The differences in the rate constants between the molybdenum and vanadium oxides reflect the different reactivities of the oxides in their charge-transfer excited triplet states. (*210*).

B. DYNAMICS OF THE EXCITED STATES AND THEIR ROLE IN PHOTOCATALYTIC REACTIONS

1. *Reactivity of Anchored Vanadium Oxides*

a. *Photoreduction of the Catalyst with CO.*   The photoluminescence spectrum of the anchored vanadium oxide catalyst prepared by the photo-CVD method, the intensity of which increased linearly with the concentration of the anchored vanadium oxides, was found to decay as a single exponential. The lifetime was determined to be 8.2 ms at 77 K and 450 μs at 298 K, respectively. These values changed little with an increase in the concentration of the anchored vanadium oxides up to 2 wt% V, whereas the lifetimes were found to be much longer than those of the supported vanadium oxides prepared by conventional impregnation methods, whereby the lifetimes become shorter with increasing vanadium oxide content. In agreement with these results, the quantum yield of the photoluminescence of the anchored vanadium oxide at 77 K was 0.35 and that of the impregnated vanadium oxide was 0.1 (*56, 115, 116*).

The Franck–Condon analysis of the vibrational fine structure of the photoluminescence spectrum of the anchored vanadium oxide observed at 77 K indicates that the equilibrium V–O bond distance of the vanadyl group is elongated in the charge-transfer excited state by 0.013 nm compared with the ground state value (*125*). UV irradiation of the anchored vanadium oxides at 280 K in the presence of CO led to the photoformation of CO_2. Since the photoformation of CO_2 from CO is accompanied by the removal of oxygen from the oxide (i.e., the photoreduction of the oxide), such an elongation of the equilibrium nuclear distance of the V–O bond in the excited state is closely associated with the facile photoformation of CO_2 on the anchored vanadium oxides. In other words, the O^- hole trapped centers in the electron–hole pair state of the $(V^{4+}\text{–}O^-)^*$ complex exhibit a high reactivity similar to O^- anion radicals (*66*).

b. *Photocatalytic Isomerization of C_4H_8.*   UV irradiation of anchored vanadium oxide catalysts in the presence of *trans*-2-C_4H_8 led to *trans* → *cis* (geometrical) and 2 → 1 (double bond shift) isomerization reactions even at temperatures lower than 273 K (*67, 116*). The yields of these reactions increased with the UV irradiation time. The yields of both the photocatalytic isomerization reactions and the photoluminescence of the catalysts increased with increasing vanadium loading. The rate of photocatalytic isomerization shows a good linear relationship with the yield of the photoluminescence (i.e., relative yields of the photoluminescence), indicating that the charge-transfer excited state of surface vanadyl groups plays an important role in the photocatalytic isomerization reactions (Fig. 44).

The lifetimes, which were measured in parallel with the photoluminescence quenching, became shorter when the pressure of the $trans$-2-C_4H_8 molecules increased. This result indicates that the quenching of the photoluminescence occurs through the interaction of $trans$-2-C_4H_8 with the charge-transfer excited state of the ($V^{5+}=O^{2-}$) vanadyl species (*17, 33, 34, 56, 63, 69, 115, 116*). These investigations of the dynamics of the catalyst photoluminescence spectra in the presence of reactant molecules show clearly that the interaction of the charge-transfer excited state of the ($V^{4+}-O^-$)* (electron–hole) paired state of the catalyst with $trans$-2-C_4H_8 results in the opening of its $C=C$ bond, leading to isomerization.

The photocatalytic isomerization reactions were markedly inhibited by the addition of O_2, and the photooxidation of $trans$-2-C_4H_8 occurred instead of isomerization. Analysis of the slope of the Stern–Volmer plots as a function of pressure of added O_2 (and $trans$-2-C_4H_8) showed the reactivity

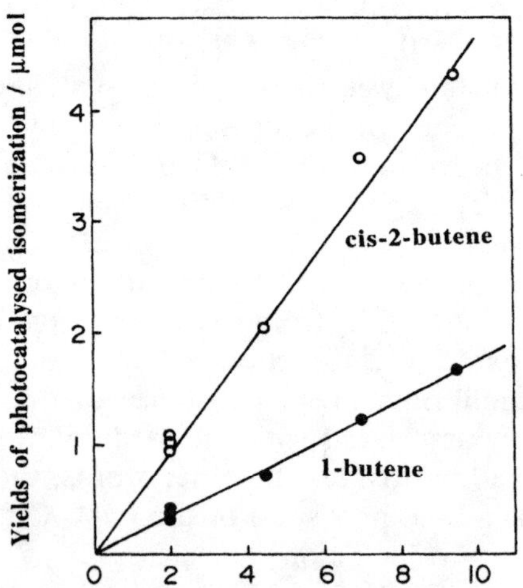

FIG. 44.   Relationship between the yields of the phosphorescence spectrum of the vanadium oxide catalyst anchored to Vycor glass and the yields of the photocatalytic isomerization of *trans*-2-butene on this catalyst. Phosphorescence yields measured at 280 K; photocatalytic reaction temperature, 298 K. ○, yields of *cis*-2-butene; ● yields of 1-butene. Initial pressure of *trans*-2-butene about 399 Pa [reproduced with permission from Anpo *et al.* (*69*)].

of O_2 with the charge-transfer excited state to be much higher than that of *trans*-2-C_4H_8 (*120*). Considering the fact that the O_2^- anion radicals adsorbed on the V^{5+} ions were formed upon UV irradiation of the catalyst in the presence of oxygen, we inter that the butene molecules interact with the O^- hole trapped center of the excited state of the $(V^{4+}-O^-)^*$ paired state to form cationic species from butene while no longer reacting with the V^{4+} electron trapped center. This result demonstrates that the O_2 molecules react rapidly with the V^{4+} electron trapped centers to form the $V^{5+}-O_2^-$ species (*120*). In other words, under UV irradiation of the catalyst in the presence of a mixture of O_2 and butene, both O_2^- anion radicals and the cationic species from butene are formed while simultaneously the V^{4+} ions and O^- ions return to the original ground state of the $(V^{5+}=O^{2-})$ vanadyl species. Subsequently, the reaction of these O_2^- anion radicals with the cationic species results in the photocatalytic oxidation of the alkene.

2.   *Reactivity of Anchored Molybdenum Oxide*

UV irradiation of the molybdenum oxide anchored on SiO_2 at 295 K in the presence of C_3H_6 initiates the formation of C_2H_4 and 2-C_4H_8 with the formation of minor amounts of CH_3CHO and $HCHO$ (*33, 34, 37, 127, 134–140, 211*). The yields of the photoinduced C_2H_4 and 2-C_4H_8 are almost equal, indicating that the metathesis of C_3H_6 takes place under UV irradiation in the presence of C_3H_6 (Table IV). In contrast, no reaction occurred in the dark at 295 K. From Table IV, it can be seen that the rate of the photoinduced metathesis reaction increases with increasing molybdenum concentration. The dependence of the lifetime and of the relative intensity of the photoluminescence spectra of these catalysts on the molybdenum content at 295 K are shown in Table IV. The lifetimes are nearly constant, i.e., 75–78 μs for molybdenum contents varying from 0.01 to 0.37 wt% Mo (*134–140*).

Table IV also includes values of the ratio $[M]/[P]$, where $[M]$ refers to the initial rate of the photoinduced metathesis and $[P]$ represents the yields of the photoluminescence spectra of the catalysts at 295 K. These values are important identifying the reactivity of the active site since the reaction proceeds only when a propene molecule interacts with the charge-transfer excited state of the catalyst that acts as the precursor, and this interaction corresponds to the quenching of photoluminescence of the catalyst with C_3H_6 (*140*).

The reactivity of the excited state of the $(Mo^{6+}=O^{2-})$ molybdenyl species measured by the $[M]/[P]$ ratio gradually decreases as the molybdenum concentration of the catalysts increases. According to the mechanism of the photoinduced metathesis (*140*), the rate of the reaction depends both

TABLE IV

Effect of Mo Content on the Excited State of Mo/SiO$_2$ Catalysts and the Rate of Photoinduced Metathesis of C$_3$H$_6$ at 298 K[a]

Mo content (wt%)	P[b]	$\tau/\mu s$[c]	Yield of photoproduced product				M/P
			C$_2$H$_4$[d]/nmol g^{-1} min^{-1}	2-C$_4$H$_8$/nmol g^{-1} min^{-1}	CH$_3$CHO[e]/nmol g^{-1} h^{-1}	M^f/μmol g^{-1} min^{-1}	
Anchored Mo/SiO$_2$ catalysts							
0.01	1.00	78 ± 5	36.9	41.5	n.d.[g]	36.9	37
0.02	1.90	77 ± 5	55.6	63.9	n.d.	55.6	29
0.043	2.60	75 ± 5	75.8	87.9	n.d.	75.8	29
0.07	3.02	75 ± 5	85.2	98.8	10.2	85.2	28
0.37	8.60	75 ± 8	124.8	156	15.8	124.8	15
Impregnated Mo/SiO$_2$ catalysts							
0.013	0.95	74 ± 5	21.6	27.2	n.d.	21.6	23
0.053	1.62	72 ± 5	22.5	28.3	n.d.	22.5	14
0.10	2.57	68 ± 8	22.5	31.7	n.d.	25.5	9.9

Note: Initial pressure of C$_3$H$_6$ was 665 Pa.

[a] From Anpo *et al.* (*140*). [b] Relative yield of phosphorescence at 298 K. Excitation wavelength, 280 ± 5 nm. The yields of phosphorescence are given relative to that of the anchored sample containing 0.01 wt% Mo. [c] Phosphorescence at 298 K; lifetime measured by monitoring 450-nm emission. [d] Yields of C$_2$H$_4$ are smaller than those of 2-C$_4$H$_8$ owing to the higher reactivity of C$_2$H$_4$ toward the excited Mo complex. [e] UV irradiation, 60 min. CH$_3$CHO was desorbed at temperatures up to 473 K. [f] Initial rate of photoinduced metathesis of C$_3$H$_6$ at 298 K. [g] Not detected.

on the rate of the reaction of the charge-transfer excited state $(Mo^{5+}-O^-)^*$ with C_3H_6 and on the reactivity of the photoformed carbene which acts as a chain carrier in the metathesis reaction. Assuming that the rate of the reaction of the excited state $(Mo^{5+}-O^-)^*$ with C_3H_6 is proportional to the concentration of the charge-transfer excited state (i.e., to the yield of the photoluminescence), the reactivity of the carbene species can be determined by dividing the rate of the metathesis $[M]$ by the yield of the photoluminescence $[P]$ (i.e., $[M]/[P]$). As shown in Table IV, the $[M]/[P]$ ratio gradually increases as the molybdenum concentration of the catalysts decreases. This result is associated with the higher dispersion of the $(Mo^{6+}=O^{2-})$ molybdenyl species, which leads to a less efficient deactivation of the carbene species through mutual interaction with neighboring species (140).

3. Reactivity of Anchored Titanium Oxide

As shown in Fig. 29, highly dispersed titanium oxide anchored onto Vycor glass exhibits a photoluminescence spectrum that has a peak near 485 nm when excited by UV light at about 300 nm; this is attributed to the radiative deactivation of the charge-transfer excited state of the titanium oxide species $(168, 212)$:

$$(Ti^{3+}-O^-)^* \xrightarrow{h\nu'} (Ti^{4+}-O^{2-}) \tag{27}$$

The addition of O_2 onto the anchored titanium oxide catalyst led to an efficient quenching of the photoluminescence at 77 K. The addition of N_2O also led to the quenching of the photoluminescence with an efficiency lower than that of O_2. Such an efficient quenching of the photoluminescence by the addition of O_2 or N_2O is expected when the emitting sites are dispersed on the support surfaces due to the efficient interaction of the emitting sites with the quencher molecules $(168, 212)$.

UV irradiation of the catalyst in the presence of N_2O at 295 K led to the evolution of N_2, its yield increasing with the irradiation time. Such an evolution indicates that the photocatalytic decomposition of N_2O proceeds on the catalyst under UV irradiation at 295 K. On the other hand, UV irradiation of the catalyst in the presence of N_2O at 77 K led to the appearance of an EPR signal attributed to N_2O^- anion radicals. These are thermally stable at 77 K, decomposing when the temperature increases from 77 to 298 K with the evolution of N_2 and the formation of the O^- species. The EPR signal arising from N_2O^- disappeared immediately upon the addition of O_2 at pressures of about 133–330 Pa at 77 K, while simultaneously the signal due to the O_2^- anion radicals appeared with no change observed in the overall spin concentration. These results indicate that O_2^- anion radicals are easily formed by an electron-transfer reaction between

N_2O^- and O_2 on the anchored titanium oxide catalyst (*168, 212, 213*). N_2O dissociates into N_2 and O^- on the activated oxide surfaces through an electron transfer from the oxide to N_2O; however, the N_2O^- species has not been detected by EPR on oxide surfaces.

UV irradiation of the anchored titanium oxide catalyst in the presence of an alkyne such as $CH_3C\equiv CH$ and water leads to the photocatalytic formation of hydrogenation products accompanied by the fission of the carbon–carbon bond of the alkyne (hereafter referred to as photocatalytic hydrogenolysis). The yields of the photocatalytic hydrogenolysis reaction of $CH_3C\equiv CH$ with H_2O change in a manner similar to the change in the intensity of the photoluminescence spectra of the catalyst (*57, 168, 214*). Since the photoluminescence originates from the charge-transfer excited state of the $(Ti^{3+}-O^-)^*$ paired state, the observations indicate that this excited complex plays a vital role in the photocatalytic hydrogenolysis reaction on the anchored titanium oxide catalyst.

The initial rate of the photocatalytic hydrogenolysis reaction on the anchored titanium oxide was determined to be about 88 nmol (h · g of catalyst), whereas with the bulk rutile powder catalyst the value was about 13 nmol (h · g of catalyst). The concentration of the effective Ti^{3+} or O^- species in the anchored catalyst was only 3.2×10^{-5} mol/(g of catalyst), whereas for bulk TiO_2 it was 1.25×10^{-2} mol/(g of catalyst) (*168*). As a result, the photocatalytic reactivity of the anchored titanium oxide can be considered to be higher than that of bulk TiO_2 catalyst by about two or three orders, of magnitude.

4. Reactivity of Anchored Copper(I) Cations in Zeolites

As shown in Fig. 45, UV irradiation of the copper(I) ion species anchored ZSM-5 zeolite in the presence of NO at 275 K leads to the formation of N_2 and O_2 with a good linear relationship between the UV irradiation time and the conversion of NO as well as a good stoichiometry during the long UV irradiation period (*172–180*).

As shown in Fig. 31, the Cu(I)ZSM-5 catalyst exhibits a photoluminescence spectrum at about 400–600 nm upon excitation at about 280–300 nm, attributed to the radiative deactivation pathway from the excited state of the isolated Cu^+ monomeric species to its ground state. With Cu(I)ZSM-5 catalysts having high copper loadings, another absorption band near 300–320 nm and a weak photoluminescence band near 500–600 nm were observed. These additional absorption and photoluminescence bands are attributed to the presence of the (Cu^+-Cu^+) dimeric species, i.e., to the $(3d\sigma^* \rightarrow 3d\sigma)$ electronic excitation and its reverse radiative deactivation $(3d\sigma \rightarrow 3d\sigma^*)$, respectively (*29, 181*).

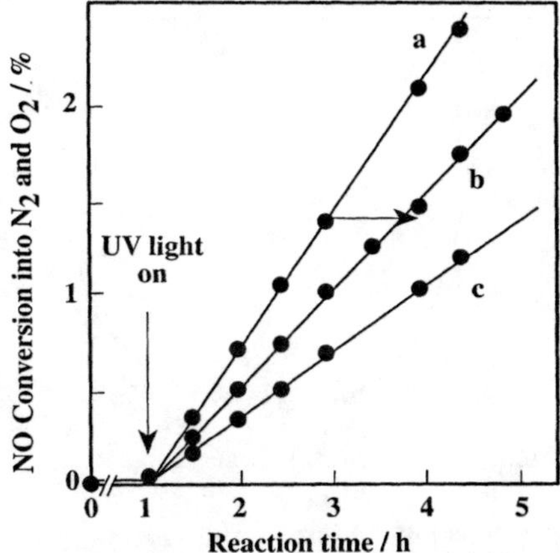

FIG. 45.   Reaction time profiles of the photocatalytic decomposition of NO into N_2 and O_2 at 275 K on Cu(I)ZSM-5 (a), Cu(I)Y zeolite (b), and Cu(I)SiO$_2$ catalysts (c). No reaction could be observed at 275 K without UV irradiation [reproduced with permission from Anpo *et al.* (*179*)].

As shown in Fig. 31 (1–4), the addition of NO to the Cu(I)ZSM-5 catalyst leads to an efficient quenching of the photoluminescence attributed to the Cu^+ species. The lifetime of the photoluminescence was shortened by the addition of NO, becoming even shorter with an increase in the NO pressure—the lifetime changed from 85 μs in vacuum to 50 μs in the presence of NO at a pressure of 133 Pa. Evacuation of the system after quenching of the photoluminescence led to a complete recovery of the photoluminescence intensity and lifetime. These findings clearly suggest that the interaction of NO with the catalyst is weak and that the added NO interact readily with the copper(I) species, in both their ground and excited states (*172–177*).

As shown in Fig. 46, the photocatalytic reactivity of the copper(I) ion species anchored within ZSM-5 increases with the evacuation temperature, passing through a maximum at 1173 K. Figure 46 also shows that the yields of the photoluminescence, i.e., the photoluminescence yield attributed to the copper(I) species, change in a similar manner. Such a good parallel between the yields of the photoluminescence and the yields of the photocatalytic decomposition of NO clearly indicates that the excited state of the isolated copper(I) species plays a decisive role in the decomposition of NO into N_2 and O_2 under UV irradiation of the catalyst at 275 K (*172–180*).

The direct observation of the $(Cu-NO)^+$ species and its behavior under UV irradiation by means of EPR and IR spectroscopies shows that the local electron transfer from the excited state of the copper(I) ion to the

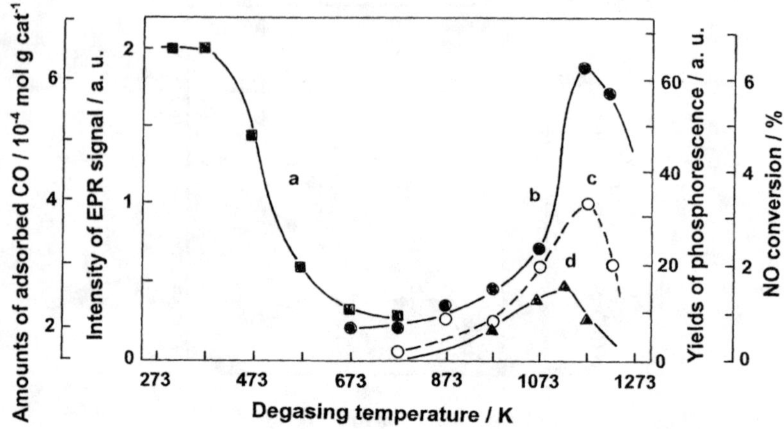

FIG. 46.   Effects of the evacuation temperature of Cu(II)ZSM-5 on the relative intensity of the EPR signal due to Cu^{2+} (a), yields of the photoluminescence spectrum due to Cu^+ (b), relative conversions (yields) of the photocatalytic decomposition of NO at 275 K (c), and the amount of CO selectively adsorbed on Cu^+ ions of the catalyst at 285 K (d) [reproduced with permission from Anpo *et al.* (*179*)].

anti-π-bonding orbital of NO and the simultaneous electron transfer from the π-bonding orbital of another NO to the vacant orbital of the copper(I) ion($3d^9\ 4s^0$) leads to the direct decomposition of two NO molecules on the Cu^+ site, selectively producing N_2 and O_2, even at 275 K (*172–177*). These findings, together with similar results obtained with AgZSM-5 photocatalysts, clearly show that the direct decomposition of NO into N_2 and O_2 on the Cu(I)ZSM-5 catalyst under UV irradiation is a new and unique type of photocatalytic reaction achieved within the microcavities of the zeolite (*186–188*).

5.   *Reactivity of Vanadium Silicalite*

As shown in Fig. 37, the VS-2 catalyst degassed at 473 K exhibits a characteristic photoluminescence spectrum at about 450–550 nm with a vibrational fine structure attributed to the $V{=}O$ bonds when the absorption band is excited at about 280 nm (*202*). As shown in Fig. 47, the addition of NO to the VS-2 catalyst leads to an efficient quenching of the photoluminescence in intensity and in lifetime, with the extent of both depending on the pressure of NO. These quenching results clearly indicate that the added NO molecules can approach the vanadium oxide sites located in the zeolite framework and that NO interacts readily with the vanadium oxide species in their excited state.

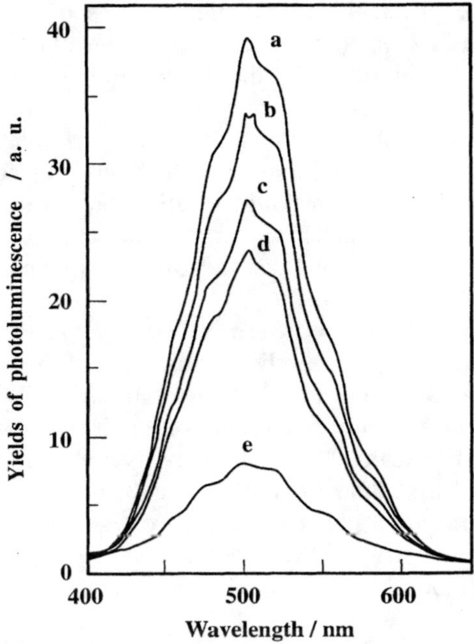

FIG. 47.   Photoluminescence spectrum of VS-2 catalyst at 77 K (a) and effect of the addition of NO on the photoluminescence spectrum (b–e). NO pressure (Pa): a, 0; b, 5.32; c, 7.98; d, 9.31; e, 53.2 [completed from Anpo *et al.* (*202*) and unpbulished data].

UV irradiation of the VS-2 catalyst in the presence of NO ($\lambda > 280$ nm) at 295 K leads to the formation of N_2, O_2, and N_2O. The yields of the photoformed N_2 increased linearly with the UV irradiation time, and the reaction with ceased immediately when UV irradiation was discontinued. After prolonged UV irradiation, the number of photoformed N_2 molecules per total vanadium ion present in the catalyst exceeded 1.0, indicating that the decomposition of NO proceeds photocatalytically on the catalyst, even at 298 K (*202*).

VI.   Application to Adsorption

A.   ACID-BASE PROPERTIES OF SURFACES

1.   *Adsorption to Probe Surface Acidic Properties*

Identification of the acid sites (especially the Brønsted acid sites) on supports such as Vycor glass, zeolites, silica, and alumina is important in understanding the physical and chemical nature of the catalysts as well as the supports. Lin *et al.* (*215*) investigated the photoluminescence and excitation spectra of quinoline, 2,3-dichloroquinoxaline, and 9,10-diazaphenanthrene embedded in calcined Vycor glass in which the molecules bind to physisorbed water (sample was exposed to the atmosphere after calcination) with a π-system or through lone pair electrons. By comparison of the observed spectra with those in liquid solutions at various temperatures, it was found that the spectral characteristics and photoluminescence (i.e., phosphorescence) lifetimes of these adsorbed molecules are identical to those in 0.5-*N* H_2SO_4 solution.

These results indicate that a proton transfer occurred from the Brønsted acid surface site to the heterocyclic N atom of the adsorbed molecule. It was suggested that, as one of the defining characteristics of the surface, an increase in the delocalization of oxygen lone pair electrons occurs through π bonding with the vacant d orbitals of both the bonded Si atom of the Si–OH group and an adjacent nonbonded Si atom. An increase in the delocalization of the lone pair electrons of oxygen causes a larger shift in the bonding electrons toward the oxygen of the OH moiety of the surface Si–OH groups (*215*).

a. *Acidic Properties of Various Oxide Supports.* Although the acidic properties of widely used supports, such as alumina, magnesia, silica, and zeolites, have been investigated extensively, the existence and chemical nature of intrinsic Brønsted acid sites, especially weak Brønsted acid sites, are not understood well because of the difficulty in making accurate measurements by conventional techniques such as temperature-programmed desorption (TPD), IR, NMR, and X-ray photoelection spectroscopy (XPS) or by model catalytic reactions. Here, the merits of photoluminescence investigations in comparison with such conventional methods are discussed in detail. Figure 48 shows the phosphorescence spectra of benzophenone dissolved in a CCl_4 solution and adsorbed on SiO_2 at 4.2 K as well as their corresponding excitation spectra (i.e., absorption spectra) (*216*). The phosphorescence spectra of benzophenone adsorbed on various oxides such as HY zeolite and aluminosilicate are shown in Fig. 49 and Table V. Upon adsorption of benzophenone on MgO, γ-Al_2O_3, and SiO_2, shifts in the phosphorescence spectra toward shorter wavelength regions can be observed when compared with the solution spectra. Such blue shifts are also observed for the corresponding excitation spectra.

In the cases of the aluminosilicate and HY zeolite, the phosphorescence spectra represent a superposition of two different spectra. To separate them, each sample was excited in the region of least excitation spectra overlap. Thus, for the phosphorescence with a λ_{max} of approximately 490

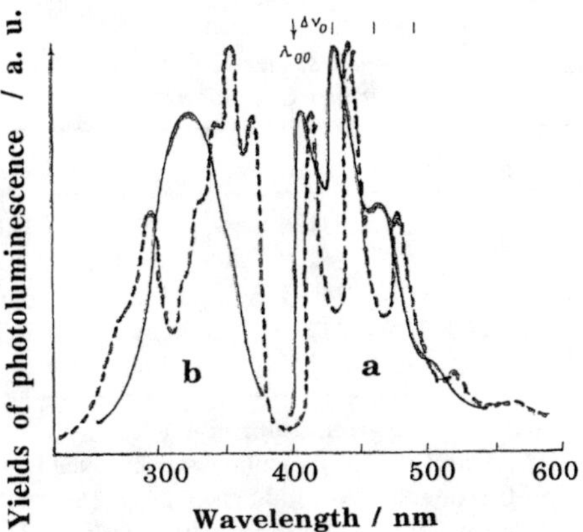

FIG. 48.   Phosphorescence spectrum (a) and its excitation spectra (b) of benzophenone at 4.2 K. Dashed line, solution in CCl_4; continuous line, adsorbed on SiO_2 [reproduced with permission from Fenin *et al.* (*216*)].

nm, excitation was carried out at about 390 nm, whereas for the emission at about 440 nm the excitation was carried out at about 310 nm (Fig. 49).

The phosphorescence spectra of benzophenone dissolved in CCl_4 and adsorbed on MgO, γ-Al_2O_3, and SiO_2 (all of which have a similar phosphorescence character) include four separable maxima (Table V) (*216*). The presence of these maxima, which are attributed to the electronic transitions from the excited state to lower vibrational levels of the ground electronic

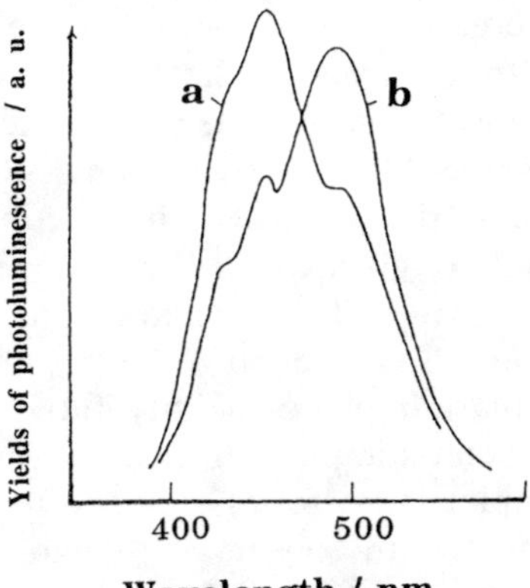

FIG. 49.   Phosphorescence spectrum of benzophenone adsorbed on aluminosilicate (composition; 75% SiO_2, 25% Al_2O_3) upon excitation at 310 nm (a) and at 390 nm (b) [reproduced with permission from Fenin *et al.* (*216*)].

TABLE V

Peak Positions ($\lambda_{\max}$) of the Phosphorescence Spectra of Benzopheneone (BP) under Various Conditions[a]

Sample	λ_{00} nm, λ_{00} cm^{-1}	$\Delta\nu_0$	$\Delta\nu_1$	$\Delta\nu_2$	λ_{excit}
In CCl$_4$ solution	413.5, 24184	1620	3235	4800	355
On SiO$_2$ (1223 K)[b]	412, 24272	1620	3250	4780	350
On SiO$_2$ (823–873 K)	405, 24691	1480	2915	4320	320
On MgO (823–873 K)	415, 24096	1560	3087	4500	350
On γ-Al$_2$O$_3$ (823–873 K)	412, 24272	1540	3085	4500	350

[a] Corresponding to transitions to the ground vibrational level (λ_{00}) and the lowest vibrational excited levels ($\Delta\nu_0$, $\Delta\nu_1$, $\Delta\nu_2$) and in the phosphorescence excitation spectra (λ_{excit}). From Fenin *et al.* (*216*). [b] The pretreatment involved heating in O$_2$ (66.7 kPa) for 1.5–2 h and then in vacuum at 1.33×10^{-3} Pa for 0.5 h at the temperature indicated.

state of benzophenone (*217*), makes it possible to determine the positions of these vibrational levels. A significant decrease in all frequencies of the valence vibration and the overtones ($\Delta\nu_0$, $\Delta\nu_1$, $\Delta\nu_2$) of the carbonyl group in the adsorbed benzophenone can be observed when the spectrum is compared with the solution spectrum. Furthermore, as shown in Table V and Fig. 48, there is a good coincidence of the phosphorescence spectra of the solution and of the gaseous benzophenone (*216–218*), suggesting that there is no interaction between the benzophenone molecules.

These results indicate a significant change in the lower part of the potential energy curve for the C=O bond upon the adsorption of benzophenone, including a broadening of the bottom of the potential curve which is a consequence of the decrease in the distance between the vibrational levels, i.e., decreases in the frequencies of the valence vibration of the C=O group and the overtones. This decrease correlates with the increase in proton acidity in the series MgO, γ-Al$_2$O$_3$, and SiO$_2$. The changes in the potential curve can be attributed to the interaction of the chromophore groups involving the C=O group with the adsorption centers, which are mainly surface hydroxyl groups. Table V shows that the shift increases in the series MgO, γ-Al$_2$O$_3$, and SiO$_2$ reaching 10 nm (800 cm^{-1}). This therefore correlates with the increase in proton acidity in the same series (*216*). Also the presence of a characteristic phosphorescence spectrum for benzophenone shows that a complete transfer of a proton to a benzophenone molecule does not occur and that the observed changes are associated with the formation of a hydrogen bond (*216, 217*).

IR techniques have been used to show that the frequency of the valence vibration of the C=O bond decreases when ketone molecules are adsorbed on catalyst surfaces. The results obtained from phosphorescence spectra are in good agreement with the IR measurements. However, the phosphorescence spectra contain overtones in addition to the fundamental vibrations so that, together with its significantly greater sensitivity, the photoluminescence method has some advantages over the IR spectroscopic method (*33, 34, 216, 217*).

A large shift in the high-frequency phosphorescence maximum (λ_{00}, Table V) resulting from the transition from a low-lying electronic state to the ground vibrational level can be observed. Table V shows that the shift increases in the series MgO, γ-Al$_2$O$_3$, and SiO$_2$ with a maximum shift magnitude of 10 nm (800 cm^{-1}), which correlates with the increases in proton acidity for this series (*216*).

The excitation spectra of benzophenone adsorbed on MgO and Al$_2$O$_3$ differ only slightly from the excitation spectrum in CCl$_4$ solution. However, the excitation spectrum of benzophenone adsorbed on SiO$_2$ shows a shift toward shorter wavelengths and does not have a clearly delineated vibrational structure even at 4.2 K. The excitation spectra confirm that the reaction of adsorbed benzophenone with a SiO$_2$ surface is stronger than with MgO and Al$_2$O$_3$ surfaces. The observed short-wavelength shift shows that the energy of the interaction between the benzophenone ground state and the SiO$_2$ surface is greater than the interaction energy of the excited state.

Upon adsorption of benzophenone on oxides with strongly acidic properties, the phosphorescence spectrum exhibits a structureless band with a λ_{max} at about 490 nm in addition to the normal phosphorescence of benzophenone. The λ_{max} of the excitation spectrum of this band was observed at approximately 380 nm, and its intensity increased in the order of the aluminosilicate, H-mordenite, and HY zeolite. In the spectrum of HY zeolite containing benzophenone, only one structureless phosphorescence band could be observed. A similar phosphorescence band could be observed for benzophenone dissolved in CHCl$_3$, which also involves dry HCl. We can therefore assign phosphorescence at about 490 nm to the protonated form of benzophenone. These findings correspond with studies of the photoluminescence of benzophenone dissolved in various concentrated acidic solutions (*217*). Consequently, since the presence of a phosphorescence spectrum at about 490 nm with benzophenone adsorbed on the aluminosilicate, H-mordenite, or HY zeolite is associated with the presence of the protonated form of benzophenone, the data indicate the existence of proton-donor centers on these oxides with acid strengths < pK_a for benzophenone (about −5.6) (*216*). On HY zeolite, almost all the adsorbed benzophenone changes into protonated benzophenone. On aluminosilicate surfaces, the relative intensities of the phosphorescence spectra attributed to the protonated and unprotonated forms are approximately the same.

These results show that the phosphorescence spectra of adsorbed benzophenone can be used to determine the fine differences between the abilities of hydroxyl groups on different solid surfaces to form benzophenone complexes that are different in nature from the benzophenone molecules. The short-wavelength shift of the phosphorescence maximum λ_{00} and the decrease in the vibrational frequency of the C=O group ($\Delta\nu_0$) as well as its overtones ($\Delta\nu_1$, $\Delta\nu_2$) in the surface complexes enable us to order the adsorbents in a series that reflects the changes in their acidic properties. The presence of proton-donor centers having proton-donor strengths less then that of benzophenone is manifested in a phosphorescence with a maximum at about 490 nm, enabling the differentiation of centers with varying strengths on the same support (38). The high sensitivity of the photoluminescence method associated with the use of probe molecules permits the determination of the concentration of the centers to as little as $\pm\sim10^{16}\ g^{-1}$.

Shen *et al.* (*219*) reported on the photoluminescence and IR spectra of pyridine adsorbed on γ-Al$_2$O$_3$ to measure the surface acidities of alumina. They observed that neither the photoluminescence nor the IR spectra of adsorbed pyridine showed any evidence of Brønsted acidity on γ-Al$_2$O$_3$ pretreated at 673 K, but they showed four different weak OH bands when the alumina was treated at 873 K.

i. *Acidic Properties of Alkali Cation-Exchanged Zeolites.* Photoluminescence measurements of the adsorbed probe molecules provide useful information about the nature of the adsorption sites, as mentioned previously. When benzophenone molecules are adsorbed on oxides which do not have strong acid sites, they adsorb on the surface OH groups only through hydrogen bonding, and their phosphorescence peaks can be observed at about 430 nm with excitation peaks at about 330 nm.

Figure 50 shows the effect of the exchanged alkali metal cations on the phosphorescence of benzophenone adsorbed on zeolites (*217*). The data for ZSM-5 show that the phosphorescence spectra shift to shorter wavelengths as the radii of the exchanged cations increase. This pattern can be explained by the fact that the smaller the radius of the exchange cations, the more it withdraws electrons from the surface OH groups that are present nearby. In other words, the cation on the surface remarkably changes the surface acidity, with the acidity depending on the exchange cation. As a result, the protonated benzophenone species may easily be formed on cation-exchanged zeolites such as HZSM-5 and NaZSM-5.

Moreover, the larger the radius of the exchange cation, the less predominant the protonated benzophenone species become and the more predominant the hydrogen-bonded benzophenone species become on the surface

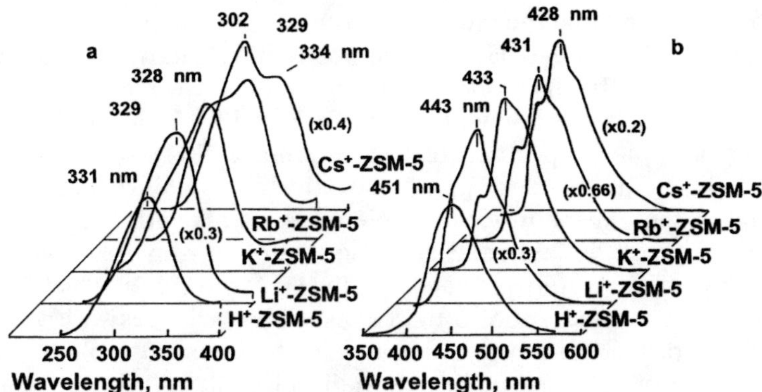

FIG. 50.   Effect of the exchanged cations on the excitation spectrum (a) and phosphorescence spectrum at 77 K (b) of benzophenone adsorbed on ion-exchanged ZSM-5. Phosphorescence spectra excited at 300 nm; excitation spectra monitored at 450 nm; 90% exchange for Na⁺–ZSM-5 and 40% exchange for Cs⁺-ZSM-5 [reproduced with permission from Okamoto *et al.* (*217*)].

(*217*). The protonated benzophenone species exhibit a much longer lifetime than the hydrogen-bonded species. The lifetimes of the protonated benzophenone species are as follows: in H_3PO_4 solution, 0.65 s; on HZSM-5, 0.41 s; and on NaZSM-5, 0.28 s; the lifetime of the hydrogen-bonded benzophenone species in ether-isopentane alcohol is 0.006 s; that on NaZSM-5 is 0.021 s; and that on CsZSM-5 is 0.017 s.

b. *Acidic Properties of Titanium–Aluminum Binary Oxides.*   Anpo, *et al.* (*220, 221*) investigated characteristics of the phosphorescence spectra of benzophenone adsorbed on Al_2O_3, TiO_2, and titanium–aluminum binary oxides to elucidate the surface acidic properties of these oxides. The phosphorescence spectra indicate at least two different emitting species. From the lifetime measurements of the deconvoluted emissions, a comparison of the emissions with those in various acidic solutions, and IR investigations of the adsorbed benzophenone species, it was concluded that the observed phosphorescence spectra indicate the phosphorescence from the excited triplet state of benzophenone adsorbed on surface OH groups by hydrogen bonding (A) and the phosphorescence from the excited triplet state of the protonated benzophenone formed on Brønsted acid sites (B).

From measurements of the phosphorescence spectra of benzophenone adsorbed in titanium–aluminum binary oxides with different compositions, it was found that the contribution of the two emissions (i.e., emission moieties A and B to the observed total photoluminescence changes remarkably as the Ti : Al ratio changes (*220, 221*).

It was also found that the rates of the *trans*-2-C_4H_8 →*cis*-2-C_4H_8 (geometrical) and 2-C_4H_8 → 1-C_4H_8 (double-bond shift) catalytic isomerization reactions on these binary oxides change dramatically when the Ti : Al ratio changes. Although it is known that the Brønsted acid sites and the Lewis acid sites both play important roles in the *trans*-2-C_4H_8 → *cis*-2-C_4H_8 and 2-C_4H_8 → 1-2-C_4H_8 isomerization, respectively, parallel relationships can be found in the phosphorescence moieties A and B with the geometrical isomerization and double-bond migration reaction isomerization, respectively. Thus, these results indicate that the surface acidity, which includes not only the Brønsted but also the Lewis acidity, changes with variations in the Ti : Al ratio. As a result, monitoring the features of the photoluminescence spectra of the adsorbed probe molecules allows us to evaluate the nature of the surface acidity, such as the type and number of acid sites as well as the strength of these sites, without any modification or destruction of the catalysts (*220–222*). (Measurement of the number of adsorbed molecules and quantum yields of the photoluminescence enables the determination of the number of acidic sites. Changing the probe molecules with different pK_a provides information on the type and strength of the acidic sites.

B. Adsorption on Semiconducting Catalysts

1. *Adsorption of O_2 and H_2O on TiO_2*

The TiO_2 catalyst exhibits photoluminescence at about 450–550 nm when excited with light having energies greater than the band gap of the catalyst, as shown in Fig. 51 (*223, 224*). With an increased adsorption of O_2 onto the TiO_2, the photoluminescence decreases markedly in intensity. However, the addition of excessive amounts of O_2 does not quench the photoluminescence completely, and about 15% of the photoluminescence remains unquenched with the sample under 1 bar of O_2. After sufficient quenching, the intensity recovered only partially upon evacuation of the sample at 298 K. EPR measurements indicated that the addition of O_2 at 298 K onto TiO_2, which had been evacuated at 473 K, led to the formation of O_2^- anion radicals adsorbed on Ti^{4+} sites. Therefore, the irreversible quenching of the photoluminescence is attributed to the formation of O_2^- since the superoxide is thermally stable and remains on TiO_2 even after exhaustive evacuation at 298 K. The addition of N_2O also leads, to a quenching of the photoluminescence but with a much lower efficiency than with O_2. Moreover, it was found that small amounts of N_2O decompose into N_2 and O^- through electron transfer from the TiO_2 to N_2O molecules.

Increasing the amounts of added 1-C_4H_8 onto the TiO_2 catalyst caused the photoluminescence to increase in intensity (*223, 224*). Similarly, the

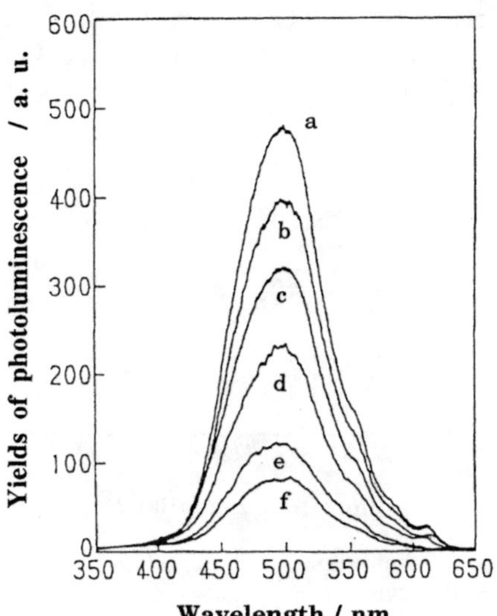

FIG. 51.   Photoluminescence spectra of powder TiO_2 at 77 K in the absence (a) and presence (b–f) of added O_2. Excitation wavelength, 300 nm; temperature, 77 K. Amounts of added O_2 (in 10^{-6} mol/g of catalyst): a, 0; b, 1.1; c, 4.7; d, 11.8; e, 38.2; f, 72–150 [reproduced with permission from Anpo *et al.* (*223*)].

addition of various unsaturated hydrocarbons, such as C_3H_6, C_2H_5 $C≡C–H$, $CH_3C≡CH$, C_2H_4, and $CH≡CH$, as well as H_2O and H_2 enhanced the photoluminescence. As shown in Fig. 52, the extent of photoluminescence enhancement depends strongly on the ionization potential of the added compounds (i.e., the lower the ionization potential of the added compound, the larger the photoluminescence intensity). The addition of N_2 was responsible for only a negligible change in the photoluminescence intensity.

These results suggest that the formation of negatively charged adducts through electron donation by the TiO_2 surface causes the quenching of the photoluminescence, whereas the formation of positive adducts through hole trapping results in an enhancement of the photoluminescence. These changes in the observed photoluminescence intensity have been explained in the dead-layer model proposed by Meyer *et al.* (*225–227*) which describes the photoluminescence enhancement of *n*-CdS and *n*-CdSe by the adsorption of amines or unsaturated hydrocarbons, resulting in the reduction in work function of a semiconductor which corresponds to a thinner depletion width of a semiconductor by donating eletron density for these adsorbents.

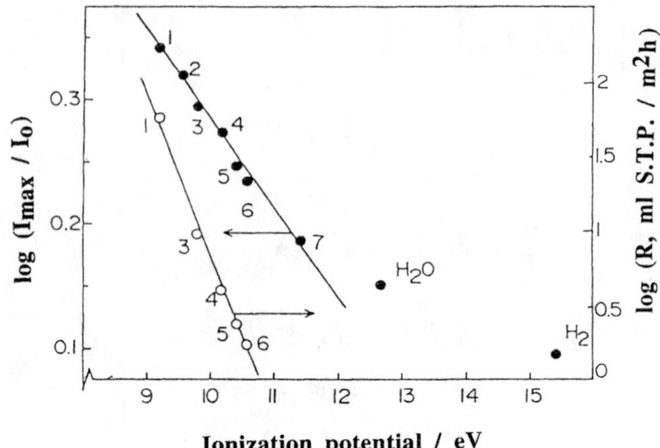

FIG. 52. Effect of additive ionization potentials on the photoluminescence intensity ($\bullet$) and the rate (R) of the photocatalytic hydrogenation of the added unsaturated hydrocarbons with H_2O on TiO_2 ($\bigcirc$). I_0 and I_{max} are maximum photoluminescènce intensities, respectively, under vacuum (or in N_2) and in the presence of added compounds; 1, 1,3-butadiene; 2, 1-butylene; 3, propylene; 4, 1-butyne; 5, 1-propyne; 6, ethylene; 7, acetylene. Photoluminescence spectra recorded at 77 K; photocatalytic reactions carried out at 298 K [reproduced with permission from Anpo et al. (223)].

2. Adsorption of O_2 and H_2O on ZnO

As shown in Fig. 53, the photoluminescence of the powder ZnO semiconducting catalyst consists of an UV emission (λ_{max} at 380 nm) and a visible emission (λ_{max} at about 500 nm) (33, 34, 228–230). The former is very intense and sharp, whereas the latter is broad and weak, in good agreement with results of earlier studies. It was found that both emissions can be observed when the ZnO is excited with light having energies greater than the band gap of ZnO (3.3 eV = 370 nm). The decrease in intensity of the excitation spectrum in the wavelength range of <360 nm was attributed to the decrease in the percentage of the oxide activated by UV irradiation due to the increase in the absorption coefficient in this region. Since band-gap irradiation is necessary for the appearance of the photoluminescence spectra, the photoformed electrons and holes play a direct and significant role in these emissions (33, 34, 228–230).

As shown in Fig. 23, a vibrational fine structure of the UV photoluminescence of the ZnO catalysts can be observed at 77 K. No fine structure can be observed for the visible photoluminescence. The values of the energy separation sequence of the vibrational bands in the system (i.e., 420, 620, and 560 cm^{-1}) are in good agreement with the vibrational energies of the

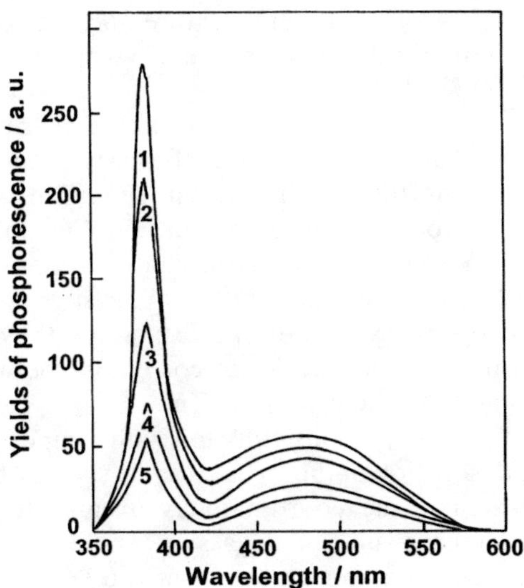

FIG. 53.   Effects of O_2 on the photoluminescence of powder ZnO at 298 K. Pressure of added O_2 (Pa): 1, 0.0; 2, 66.5; 3, 466; 4, 1596; 5, 2660 or excess amounts [reproduced with permission from Anpo *et al.* (*224*)].

Zn–O group since the Zn–O bond stretching of the catalyst can be observed in the range of 530, 500, and 440 cm^{-1} by IR measurements (*231*). An investigation of the effect of the temperature of calcination in oxygen on the photoluminescence of the ZnO catalyst, prereduced by degassing at 873 K, showed that when the calcination temperature was increased, the intensity of the UV emission increased markedly at about 600 K, whereas the increase was slight for the visible emission. Kokes (*232*) showed that at 600 K the diffusion of the interstitial zinc ion, Zn^+, under the influence of the electric field caused by the oxygen adsorbed on ZnO becomes rapid enough to produce a new $[Zn^{2+}–O^2]_{surf}$ pair on the surfaces.

It was also found that when the degassing temperature of the ZnO was increased, the λ_{max} of the UV emission was only slightly affected at all, whereas the λ_{max} of the visible emission shifted toward longer wavelengths. Moreover, increasing the degassing temperature of the catalyst hardly changed the intensity of the visible emission at temperature up to 873 K, where upon it then increased slightly. On the other hand, the ultraviolet emission dramatically changes with temperature, increasing up to a maximum at 673 K. Both the wavelength and the intensity of the visible emission only slightly changed as a result of doping with LiO_2. The lifetimes of these emissions were found to depend greatly on the sample pretreatment conditions, such as the degassing temperature and the reduction with H_2. The lifetime of the ZnO catalyst degassed at 673 K was found to be approximately 1.0 ms at 300 K (228).

On the basis of these results, the photoluminescence spectrum of the ZnO catalyst observed at about 380 nm can be associated with the presence of the coordinatively unsaturated surface pair sites $[Zn^{2+}-O^{2-}]_{surf}$ (the subscript "surf" refers to coordinatively unsaturated ions; the subscript "LC" was used in Section IV to refer to low coordinated ions, a similar situation). The O^{2-} ions constituting the $[Zn^{2+}-O^{2-}]_{surf}$ pair sites are neither normal bulk lattice oxygen ions nor adsorbed species; rather, they have the characteristics of an intermediate-like species. A correlation between the photoluminescence at 380 nm and the presence $[Zn^{2+}-O^{2-}]_{surf}$ pair sites is evident from the change in intensity of this emission with temperature. The results suggest that, upon desorption of the O_2^- and/or O^- species, the concentration of the O^{2-}_{surf} species (i.e., the coordinatively unsaturated surface $[Zn^{2+}-O^{2-}]_{surf}$ pair sites) increases up to a degassing temperature of about 600 K. At higher temperatures the concentration of the $[Zn^{2+}-O^{2-}]_{surf}$ pair sites decreases due to a reduction of the surface. Accordingly, the photoluminescence at about 380 nm is inferred to arise from the following charge-transfer process:

$$[Zn^+-O^-]^*_{surf} \xrightarrow{h\nu} [Zn^{2+}-O^{2-}]_{surf} \qquad (28)$$

Since Zn^+ and O^- are regarded as localized photoformed electrons and holes, respectively, the reverse process results in the radiative recombination of these electrons and holes, i.e., the emission of excitons (228).

As shown in Fig. 53, upon introduction of O_2 onto the ZnO catalyst, both emissions markedly decrease in intensity, with the extent of decrease depending on the amount of O_2 added. The intensity recovers upon evacuation of the sample at 300 K, although its recovery is only partial. EPR measurements indicate that the addition of O_2 at 300 K onto the degassed ZnO leads to the appearance of adsorbed O_2^- anion radicals. These results suggest that this reversible quenching is associated with the formation of weakly physisorbed oxygen complexes. The irreversible quenching can be attributed to the formation of O_2^- anion radicals on the ZnO surface, and these species are rather stable and remain on the surface after evacuation at 300 K. They were removed only by evacuation at about 423 K (228). Oster and Yamamoto (233) observed that O_2 suppressed the photoluminescence as well as the photoconductivity of the ZnO, and they attributed the observations to the trapping of electrons by O_2 which also resulted in the formation of O_2^- anion radicals.

On the other hand, the addition of H_2O and/or $CH_3C{\equiv}CH$ led to an increase in the intensity of the photoluminescence, depending on the amounts of adsorbed H_2O and/or $CH_3C\equiv CH$ (*224*). The mechanism of enhancement and quenching of the photoluminescence of the powder semiconductors such as TiO_2 and CdS by added molecules is well explained by the dead-layer model of Meyer *et al.* (*225–227*). According to this model, the formation of the negatively charged adducts through the electron capture causes the quenching of the photoluminescence, whereas the formation of positively charged adducts through hole trapping results in an enhancement of the photoluminescence (*223, 224*).

Photoformed electrons in Pt-loaded metal oxides are easily transferred from the oxides to the Pt particles, which suggests that Pt acts as a scavenger of the photoformed electrons. These results are in good agreement with observations showing that the intensity of the emission decreased when ZnO was loaded with Pt particles, its lifetime being shortened from 3.0 to 0.8 ms at 300 K (*228*).

C. ADSORPTION ON SUPPORTED CATALYSTS

1. *Adsorption of H_2O on Highly Dispersed Anchored Vanadium Oxide Catalysts*

The adsorption of H_2O on the anchored vanadium oxide/SiO_2 catalyst led to a decrease in the photoluminescence intensity of the catalyst without any change in the spectral shape (*33, 34, 63, 69, 115, 116, 119*). The photoluminescence lifetime changed only slightly and remained constant at 5.8 ms at 77 K. The situation is quite different when the quenching is performed with O_2: The intensity of photoluminescence decreased and the lifetime became shorter, depending on the concentration of added O_2. There was no recovery of the photoluminescence intensity after the complete quenching of the emission by evacuation at 298 K for more than 3 h. In related work, EPR spectra of the V^{4+} ion of the catalyst produced by photoreduction by H_2 at 77 K indicate that the addition of H_2O onto the anchored vanadium oxide/SiO_2 led to dramatic changes in the signal attributed to V^{4+} ions, from a typical tetrahedral V^{4+} signal to a typical octahedral signal (*119*).

These results show that the decrease in intensity of the photoluminescence of the catalyst resulting from the addition of H_2O is associated with the reconstruction of the emitting sites, i.e., a change in the tetrahedral coordination of the active surface (V=O) vanadyl species to a nonemitting octahedral coordination by the adsorption of H_2O ligands. Such results obtained by dynamic photoluminescence measurements are in good agreement with results reported by Narayana *et al.* (*234*) and Jonson *et al.* (*235*) showing that the four-coordinated tetrahedral vanadyl species have the ability to coordinate H_2O molecules and change their coordination from tetrahedral to octahedral. The octahedrally coordinated vanadyl species neither luminesce due to efficient radiationless deactivation nor exhibit any photocatalytic reactivity (see Section IV.C.4).

2. *Adsorption of CO on Copper(I)/ZSM-5 Zeolite*

The adsorption of CO molecules is often used to probe surface Cu^+ sites (*107, 236*). After evacuation of Cu(II)zeolite samples at 973 K, the EPR signal assigned to the copper(II) species become weak and can hardly be observed, indicating that the Cu^{2+} ions were reduced to Cu^+ (see Section IV.D.2.a). With the Cu(I)zeolite catalysts prepared in this way, the photoluminescence was observed upon excitation at about 300 nm (Fig. 31). The

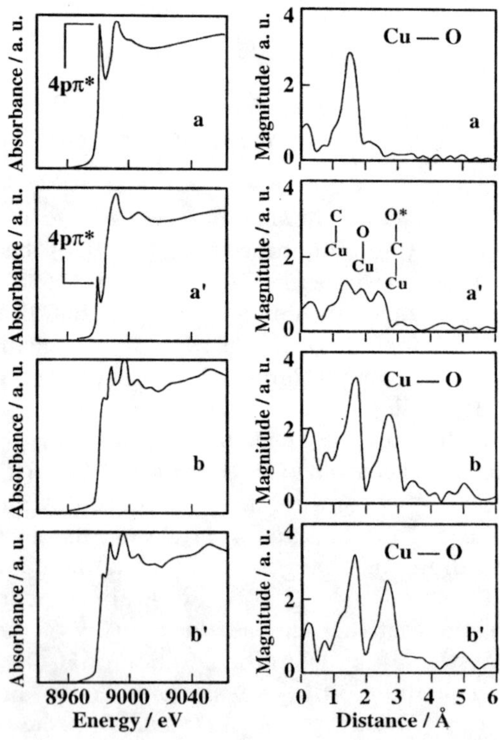

FIG. 54. XANES (left) and EXAFS (right) spectra of the Cu(I)ZSM-5 (a) and Cu(I)Y zeolite catalysts (b) and their respective spectra (a′) and (b′) after the addition of CO to the catalyst. CO pressure, >2660 Pa. Catalysts were prepared by evacuation of the original Cu(II) ZSM-5 and Cu(II)Y zeolite at 973 K [reproduced with permission from Yamashita *et al.* (*180*)].

photoluminescence spectrum observed at about 380–450 nm is attributed to the radiative deactivation of the electronically excited state of the copper(I) ions to their ground state [i.e., $(3d^94s^1 \rightarrow 3d^{10})$], and the photoluminescence band at about 450–550 nm is attributed to the radiative deactivation of the electronically excited state of the (Cu^+-Cu^+) dimer species to their ground state [i.e., $(3d\sigma \rightarrow 3d\sigma^*)$] (*171–181*).

Fig. 54 shows the XANES and EXAFS spectra of the Cu(I)ZSM-5 zeolite after addition of CO (*180*). The addition of CO causes a dramatic decrease in the intensity of the band attributed to the $1s \rightarrow 4p_z$ transition in the XANES of the Cu(I)ZSM-5. In the EXAFS spectra of these catalysts, the Cu–O peak became smaller and shifted to a longer atomic distance (i.e., $1.5 \rightarrow 1.8$ Å), and new peaks attributed to the C atom (Cu–C) and O atom (Cu–C–O) of the adsorbed CO molecule appear due to the addition of CO onto the catalyst. These results indicate that CO molecules adsorb on the isolated Cu^+ ions strongly enough to distort the Cu^+ coordination geometry (*180*).

As shown in Fig. 55, the addition of CO molecules to Cu(I)ZSM-5 leads

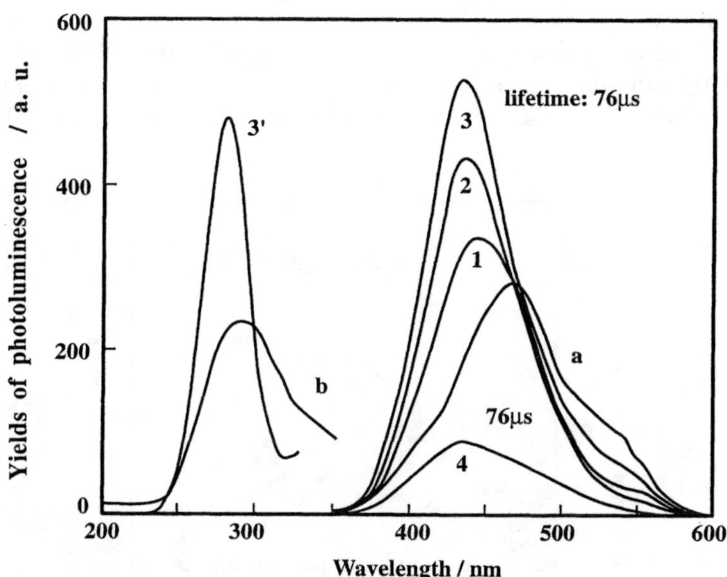

FIG. 55.   Photoluminescence (a) and its excitation spectrum (b) of Cu(I)ZSM-5 catalyst and the effect of CO addition on the photoluminescence (1–4). Catalyst was prepared by evacuation of the original Cu(II)ZSM-5 sample (1.9 wt% as Cu metal) at 973 K. Addition of CO was carried out at 77 K. CO pressure (Pa): 1, 173; 2, 306; 3 and 3′ (excitation spectrum of the photoluminescence spectrum 3), 372; 4, 2660 [reproduced with permission from Yamashita *et al.* (*180*)].

to an efficient quenching of the photoluminescence attributed to the isolated Cu^+ monomer species. Moreover, an increase in the pressure of the added CO leads to the appearance of a new photoluminescence band at about 400–450 nm (*107*), its intensity increasing with the CO pressure. IR spectra also showed that CO adsorption leads to new bands at about 2157 cm^{-1} that can be attributed to the formation of Cu^+–CO complexes. The relative intensities of these new bands observed at 2157 cm^{-1} by IR and at about 420 nm by photoluminescence spectroscopy increase when increasing amounts of CO are added to the catalyst. The good parallel relationship indicates that the newly observed photoluminescence spectrum near 420 nm induced by the addition of CO is linked to the formation of Cu^+–CO complexes (*178, 180*). As shown in Fig. 56, the intensities of the photoluminescence band increase upon addition of CO, passing through a maximum at 2.5×10^{-4} mol of CO g^{-1}. This decrease is attributed to the dynamic quenching of the excited state of the Cu^+–CO complexes by CO molecules. The intensity of the IR band increased when the amount of added CO increased and then leveled off at 2.5×10^{-4} mol of CO g^{-1}. This value corresponds to about 94% of the total amount of Cu^{2+} ions anchored in the zeolite by ion exchange (*178, 180*).

These results indicate that CO molecules selectively adsorb on the isolated Cu^+ monomer species to form stable one-to-one Cu^+–CO complexes, which in turn allows the determination of the exact number of isolated Cu^+

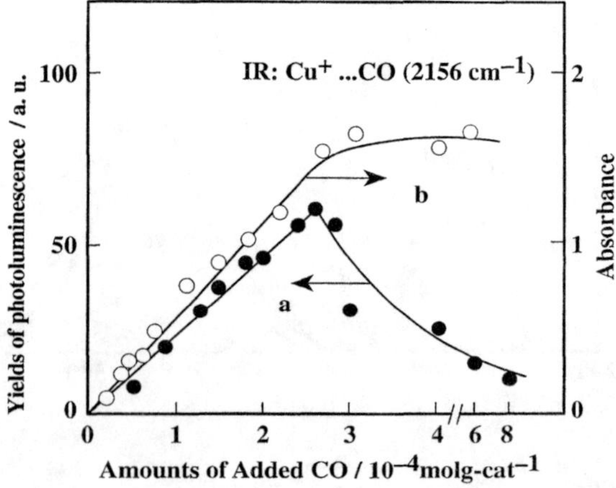

FIG. 56. Effect of CO addition on the intensities of newly appeared photoluminescence spectra at 77 K (a) and of the IR absorption spectra due to Cu^+–CO complexes on Cu(I)ZSM-5 at 295 K (b). Catalyst was prepared by evacuation of Cu(II)ZSM-5 at 973 K [reproduced with permission from Anpo (*178*)].

ions within the zeolite. On the other hand, there were no significant changes in the XANES and EXAFS data of the Cu(I)Y zeolite catalyst—in which the copper species exist mainly as Cu^+–Cu^+ dimer species—after the addition of CO molecules, which shows that CO molecules do not adsorb on the Cu^+–Cu^+ dimer species (*178, 180*).

VII. Applications to Catalysis

A. SUPPORTED CATALYSTS

1. *Molybdenum Oxide*

When a metal oxide is supported on a carrier in low concentrations, its character is dramatically modified, resulting in a change in its catalytic activity and selectivity. The structure of molybdenum oxide supported on SiO_2 (molybdenum oxide/SiO_2) was investigated using XRD, IR, laser Raman, and photoluminescence techniques. At low molybdenum contents, an X-ray-amorphous phase is formed which is characterized by surface molybdates dispersed on SiO_2. The surface concentration of amorphous MoO_3 and the rate of oxidative dehydrogenation of C_2H_5OH show a maximum at a content of about 1 atom% Mo. The rate changes in parallel to the change in the concentration of amorphous MoO_3, i.e., the surface polymolybdate (*138*).

As shown in Fig. 26, molybdenum oxide/SiO_2 catalysts show a characteristic photoluminescence spectrum having a λ_{max} near 480 nm and an excitation band with a λ_{max} near 290 nm. The absorption and photoluminescence are attributed to the charge-transfer processes on the tetrahedrally coordinated molybdenum oxide species on SiO_2 (*126, 127, 133–140*):

$$(Mo^{5+}\!-\!O^-)* \xrightarrow{h\nu} (Mo^{6+}\!=\!O^{2-}). \tag{29}$$

When the molybdenum content was increased, the intensity of the photoluminescence increased, passing through a maximum at approximately 1 atom% of Mo (*138, 140*). The maximum concentration of the tetrahedrally coordinated molybdenum oxide species was found at the same concentration at which the X-ray-amorphous MoO_3 exhibited a maximum concentration. It is therefore most likely that tetrahedrally coordinated molybdenum oxide species are formed on a restricted part of the amorphous MoO_3. The photoluminescence intensity of the catalyst decreased at concentrations higher than 2 atom% Mo, suggesting that the coordination state of the molybdenum oxide species changes from a tetrahedral to an octahedral coordination.

The metathesis reaction of C_3H_6 is catalyzed by molybdenum oxide/ SiO_2. Increasing the molybdenum content leads to an increase in the metathesis rate, which passes through a maximum at 1 atom% Mo and then decreases in a manner similar that observed for oxidative dehydrogenation. However, at Mo loadings higher than 2 atom% these reactions show different features, and the rate becomes zero for the metathesis reaction. From the results of photoluminescence measurements, such phenomena can be associated with the change in concentration of the tetrahedrally coordinated molybdenum oxide species. A good parallel between the concentration of the tetrahedrally coordinated molybdenum oxide species and the rate of the metathesis reaction was observed. It can be concluded that the tetrahedrally coordinated molybdenum oxide species are the active sites for the metathesis reaction.

2. *Copper-Exchanged Zeolites*

Figure 57 shows the activities of various catalysts for the reduction of NO in the presence of NH_3 and O_2 (*237*). Data representing the vanadium oxide supported on TiO_2 anatase (3.8 wt% as V_2O_5) are also shown. The CuZSM-5 catalyst exhibits higher activity than the others in the reduction of NO into N_2 in the presence of NH_3 and O_2, showing similar catalytic behavior as that of the highly active vanadium oxide supported on TiO_2 (*238*). The Cu(II)Vycor and Cu(II)SiO_2 catalysts are two orders of magnitude less active.

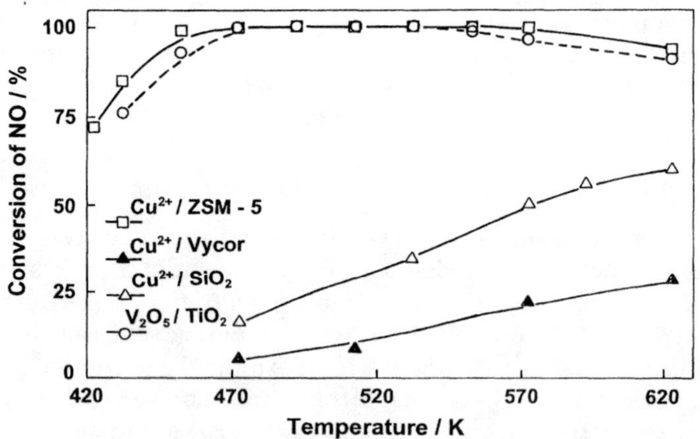

Fig. 57. Conversion of NO into N_2 in the presence of NH_3/O_2 (NO, 612 ppm; NH_3/NO = 1.06; 3% O_2) on Cu(II)ZSM-5, Cu(II)Vycor glass, and Cu(II)SiO_2 and V_2O_5/TiO_2 catalysts, respectively. □, Cu(II)ZSM-5; ▲, Cu(II)Vycor; △, Cu(II)SiO_2; ○, V_2O_5/TiO_2 [reproduced with permission from Centi *et al.* (*237*)].

Figure 57 also indicates that the spontaneous reduction of the Cu(II)ZSM-5 catalyst proceeds by decreasing the partial pressure of O_2 at lower reaction temperatures. The Cu(II)SiO$_2$ and Cu(II)Vycor catalysts are much less active than the Cu(II)ZSM-5 catalyst for the conversion of NO. Furthermore, in these catalysts, O_2 promoted the reduction of NO into N_2 with NH_3.

Only the copper-based catalysts which were degassed at high temperatures and showed very weak EPR signals attributed to the Cu^{2+} ions exhibit a photoluminescence spectrum near 400–500 nm, attributed to the presence of a Cu^+ species of which the absorption band could be observed at about 300 nm (see Section IV.D.2.a) (*105–108, 192–197*). The Cu(I)ZSM-5 catalyst exhibits a major photoluminescence near 440 nm and has a weak shoulder at about 510 nm. On the other hand, the Cu(I)SiO$_2$ and Cu(I)Vycor catalysts show photoluminescence spectra near 520 nm and have very weak shoulders at about 430 nm; and the Cu(I)SiO$_2$ and Cu(I)PVG catalysts exhibit very similar photoluminescence spectra, with the spectrum of the former being more intense.

These results indicate that the isolated copper species on ZSM-5 have an activity for the decomposition reaction of NO different from that of the dimeric or polynuclear copper species, probably because there are different reaction mechanisms (*237*). The results obtained with the Cu(II)ZSM-5 catalyst also suggest that Cu^{2+} ions promote the spontaneous low-temperature dehydroxylation of nearby Brønsted sites or the elimination of lattice oxygen anions which play a vital role in the decomposition of NO. When the dimeric or polynuclear species of Cu^{2+} are present, the spontaneous elimination of the lattice oxygen bridging the two Cu^{2+} sites does not occur at low temperatures; however, this reaction occurs at high temperatures. The activity for the decomposition of NO is nearly zero at about 573 K, but in the presence of O_2 a different reaction mechanism is initiated and this results in the enhancement of NO conversion. Moreover, the presence of stronger Brønsted sites in ZSM-5 can explain why only the CuZSM-5 catalyst exhibits much higher activity for the reduction of NO in NO–NH_3–O_2 reaction systems.

B. BULK CATALYSTS

1. *Selective Dimerization and Hydrogenation of C_2H_4 on ZrO_2 with Coordinatively Unsaturated Surface Sites*

Degassing ZrO_2 at high temperatures leads to the appearance of an abnormal absorption and photoluminescence spectra which could be attributed to the formation of surface sites in low coordination or coordinatively unsaturated surface sites (see Section IV.A.2.b) (*101–104*). Moreover, ZrO_2

catalysts with such surface sites exhibit high catalytic activity for the selective dimerization of C_2H_4 to 2-C_4H_8 (*171*).

The addition of C_2H_4 to ZrO_2 leads to an efficient quenching of the photoluminescence. After quenching, the evacuation of the system at 298 K leads to only a partial recovery of the emission, which is referred to as "reversible quenching." Both the reversible and irreversible quenching of the photoluminescence can be observed as a result of the addition of C_2H_4 to the ZrO_2 catalyst, which suggests the facile interaction between the added C_2H_4 and surface sites in low coordination. Two different interactions were observed; a weak and reversible interaction and a strong and irreversible interaction (*171*). Kondo *et al.* (*239*) showed that there are two different types of interactions between C_2H_4 and the ZrO_2 surfaces: a weak adsorption of species identified as π-C_2H_4 and strong adsorption of species identified as s-π-C_2H_4. These results, obtained by IR and photoluminescence methods for the adsorption of C_2H_4 on the active ZrO_2 catalyst, show good agreement with each other.

Furthermore, the addition of H_2 to the catalyst after quenching of the photoluminescence by the addition of C_2H_4 leads to additional quenching of the photoluminescence, suggesting that H_2 interacts with the remaining unquenched surface sites. In other words, H_2 was able to adsorb onto sites with which C_2H_4 does not interact. After quenching by H_2, evacuation of the system led to only a partial recovery of the emission, suggesting that H_2 adsorbs irreversibly on the ZrO_2.

Table VI indicates that the strongly adsorbed species formed from C_2H_4 selectively contribute to the formation of C_2H_6 (*102*). It is clear that the adsorption of H_2 is much stronger than that of C_2H_4, in good agreement with results showing that quenching of the photoluminescence is more efficient as a result of the addition of H_2 than of C_2H_4. Therefore, as can be seen in Table VI (column 4), the introduction of C_2H_4 at 0.24 and at 12.3 kPa onto ZrO_2 on which H_2 was presorbed and from which excess gaseous H_2 was evacuated leads to the formation of only small amounts of C_2H_6 and C_4H_{10}, respectively. These results indicate that strongly adsorbed species formed from H_2 do not play a role in the hydrogenation reaction of C_2H_4, whereas species formed from H_2 and C_2H_4 that are only weakly adsorbed play significant roles in the hydrogenation and dimerization reactions of C_2H_4 on ZrO_2 with surface sites of low coordination.

2. Catalytic Properties of Li-Doped MgO

In the methane oxidative coupling reaction, a small amount of Li doping of the MgO catalyst leads to an enhancement of the yield of C_2 compounds (mainly C_2H_4 and C_2H_6) as well as O_2 and CH_4 conversions (*99, 100*). The

TABLE VI

Yields (in μmol/g of Catalyst) of Products in Dimerization and Hydrogenation Reactions of C_2H_4 on Active ZrO_2 Catalyst at 298 K

| | Reactants | | | | | | | |
| | C_2H_4 only | | $C_2H_4 + H_2$[a] | | C_2H_4 only + H_2[b] | | $H^2 + C_2H_4$[c] | |
Amount of C_2H_4	Product	Yield	Product	Yield	Product	Yield	Product	Yield
0.24 kPa (6 μmol/g of catalyst)	Not detected		C_2H_6	1.96	C_2H_6	15.6	C_2H_6	0.14
1.23 kPa	2-C_4H_8	1.26	C_2H_6 C_4H_{10}	5.01 21.0	C_4H_{10}	22.0	2-C_4H_8 C_4H_{10}	48.9 <0.1
20.2 kPa	2-C_4H_8	20.0	C_2H_6 C_4H_{10}	16.2 81.5	C_4H_{10}	30.0	No product detected	

[a] H_2 (2.66 kPa) added to the sample onto which C_2H_4 was preadsorbed at the desired pressures. [b] H_2 (2.66 kPa) added to the sample onto which C_2H_4 was preadsorbed and subsequently evacuated. [c] C_2H_4 added to the sample onto which H_2 was preadsorbed and subsequently evacuated.

pure undoped MgO catalyst has a low activity and lower C_2 selectivity. With Li-doped MgO, both yield and selectivity were increased markedly when the amount of Li increased. An optimum activity was observed at about 3 mol% of Li, and at when more Li was added the activity decreased because of a reduction in the catalyst surface area.

Figure 58 shows the photoluminescence spectrum of the undoped MgO degassed at the same temperature as the methane oxidative coupling reaction together with the photoluminescence spectrum of the 3 mol% Li-doped MgO (Fig. 58, 2) and its deconvoluted curves (Fig. 58, 2-a and 2-b). In addition to a characteristic photoluminescence spectrum at around 370 nm, attributed to the surface sites in low coordination on MgO, the Li-doped MgO exhibits a new photoluminescence band at about 350–550 nm with a λ_{max} at about 450 nm (Fig. 58, 2-b). The intensity of this new emission depends on the amount of Li doped. The excitation spectrum corresponding to this new emission is evident at about 260–290 nm (*100, 240*), which suggests that surface sites with a coordination number of four may be associated with this new photoluminescence.

Figure 59 shows the effect of Li doping on the relative intensity of the newly observed photoluminescence. When the amount of Li increases, the intensity of this photoluminescence increases, passing through a maximum at approximately 3 mol% of Li. A parallel relationship between the intensity

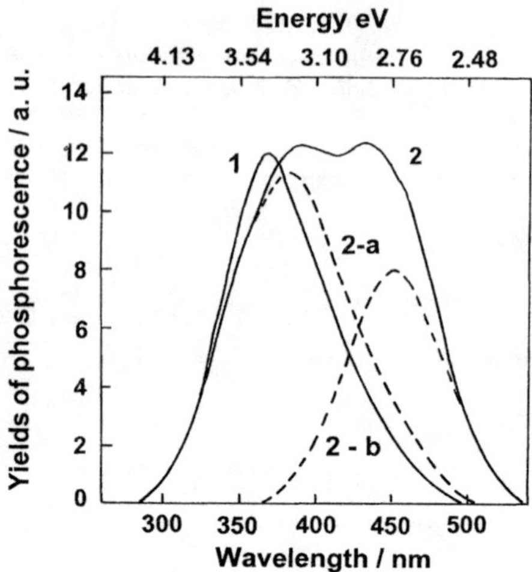

FIG. 58. Photoluminescence spectra of powder MgO (1) and of 3 mol% Li-doped MgO-(II) (2) and the corresponding deconvoluted curves (2-a and 2-b). Excitation wavelength, 240 nm; recording range, 500 mV for line 1 and 200 mV for line 2 [reproduced with permission from Anpo *et al.* (*240*)].

of the new photoluminescence and the activity of the Li-doped MgO catalyst (Fig. 59) (*100, 240*) shows that surface sites with a coordination number of four are formed on the Li-doped MgO catalyst and play a vital role in the oxidative coupling reaction of methane.

The addition of CH_4 at 13.3 Pa onto the Li-doped MgO only slightly quenched the photoluminescence at temperatures between 298 and 573 K. However, the addition of CH_4 to the catalyst at temperatures above 723 K quenched the newly observed photoluminescence, but it did not quench the characteristic emission at about 380 nm. The characteristic photoluminescence of the undoped MgO was quenched with CH_4 only at temperatures higher than 873 K. The addition of O_2 led to the quenching of both the characteristic and the new photoluminescence, even at room temperature (*100, 240*).

These findings show that the newly formed surface sites in low coordination on the Li-doped MgO are more active than those on the undoped catalyst, which explains the high activity of the Li-doped MgO catalyst for methane oxidative coupling. Ito *et al.* (*99*) detected the (Li^+-O^-) or O^- sites on the Li-doped MgO catalyst by EPR and suggested that these sites play a significant role in the formation of CH_3 radicals from CH_4. It is unclear whether the newly produced surface sites in low coordination are directly associated with the existence of such active sites.

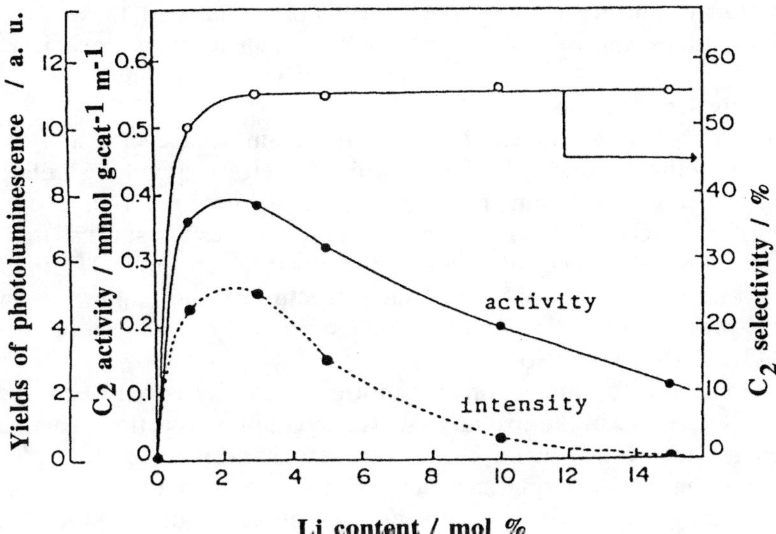

FIG. 59.   Influence of the amount of Li on the activity and selectivity of the CH_4 oxidative coupling reaction and on the intensity of the newly observed photoluminescence spectrum of Li/MgO-(II) at about 420–520 nm. Reaction temperature, 965 K; photoluminescence measurements at 77 K; excitation at 240 nm; selectivity for C_2 compounds [reproduced with permission from Anpo *et al.* (*240*)].

3.   Catalytic Properties of La₂O₃ Methane Coupling Catalysts

La_2O_3 catalysts with different selectivities and catalytic activities for methane coupling were prepared by the decomposition of $La(NO_3)_3 \cdot 5H_2O$ at 923 K (referred to as La_2O_3-LT) for 2 h or at 1073 K (referred to as La_2O_3-HT) for 16 h. The morphologies and specific surface areas are as follows: La_2O_3-LT, thin and monocrystalline plates, BET surface area = 3.8 m^2g^{-1}; La_2O_3-HT, polycrystalline plates with larger thickness, BET surface area = 1.4 m^2g^{-1} (*241*).

The photoluminescence spectra of these catalysts were observed at 77 K, with a λ_{max} near 470 nm when the catalysts were excited at approximately 290 nm after the evacuation at various temperatures. The intensity (i.e., yield) of the photoluminescence depended on the evacuation temperature. The yield increased with increasing evacuation temperature, passing through a maximum at 573 K. The yield was almost the same for the different catalysts, but the shape of the spectrum was influenced by factors such as the presence of other rare earth metal impurities which act as phosphors in the La_2O_3 framework and of surface OH groups. These observations suggest that the photoluminescence should be attributed to the formation of surface sites in low coordination, and that these sites are produced upon the evacuation of La_2O_3 at high temperatures since the original La_2O_3 samples do not exhibit absorption and emission bands in these wavelength regions (*241*).

The La_2O_3-HT catalyst exhibited a photoluminescence spectrum similar in shape to that of La_2O_3-LT, and the two spectra responded similarly to changes in the evacuation temperature, but a lower yield could be observed for La_2O_3-HT than for La_2O_3-LT. These results suggest that the concentration of surface sites in low coordination on La_2O_3-HT is lower than on La_2O_3-LT and that the surface structure of the La_2O_3 is sensitive to the pretreatment temperature applied in the decomposition of the $La(NO_3)_3 \cdot 5H_2O$ precursor.

The lower the decomposition temperature of $La(NO_3)_3 \cdot 5H_2O$, the higher the C_{2+} selectivity in the oxidative methane coupling reaction. The higher selectivity achieved with La_2O_3-LT was associated with the thin platelet shape of the oxide particles, which mainly exhibited the (001) face of the hexagonal La_2O_3 structure. In the decomposition of $La(NO_3)_3 \cdot 5H_2O$ at higher temperatures, the resulting oxide particles showed a three-dimensional structure, and the BET surface area decreased together with a decrease in the C_{2+} selectivity in the oxidative methane coupling reaction.

The addition of CH_4 or O_2 to the catalyst led to quenching of the photoluminescence with almost the same efficiency. After complete quenching of the photoluminescence by CH_4, the evacuation of the system resulted in the original photoluminescence intensity. These results suggest not only that coordinatively unsaturated surface sites are accessible to CH_4 molecules and interact with them even at low temperatures but also that the higher C_{2+} selectivity of the La_2O_3-LT catalyst is linked to the higher concentration of the coordinatively unsaturated surface sites on the surface (*242*).

VIII.   Application To Photocatalysis

A.   PHOTOCATALYSIS ON UNSUPPORTED CATALYSTS

1.   *Photocatalysis on MgO*

Degassed MgO exhibits a photoluminescence spectrum was 340–450 nm when the absorption band attributed to the charge-transfer process described in Eq. (12) is excited at about 240–280nm (*65, 66, 81–83, 96–98*). The addition of *trans*-2-C_4H_8 at 293 K to MgO degassed at 1073 K led to an efficient decrease of the photoluminescence intensity without any changes in shape, with the extent of quenching increasing with an increase in the pressure of *trans*-2-C_4H_8. The addition of 1.16×10^{-6} mol of *trans*-2-$C_4H_8^{-1}$ quenched the photoluminescence by about 66% and the addition of 7.9 × 10^{-6} mol g^{-1} by about 92%. The evacuation of *trans*-2-C_4H_8 at 298K for 30 min after a complete quenching of the emission led to a recovery of most of the emission intensity (*96–98*).

Such quenching behavior suggests that dynamic quenching, whereby quencher molecules interact with the emitting sites in their excited state to give nonradiative deactivation pathways, is the principal cause of the quenching that occurs with butenes. As a result, the quenching efficiency depends on the amount of surface quencher molecules and in turn on the equilibrium pressure of the added molecules (*33, 34*).

UV irradiation of the MgO catalyst in the presence of *trans*-2-C_4H_8 at 273 K led to a marked enhancement in the formation of *cis*-2-C_4H_8 as well as in the formation of 1-C_4H_8. The activity of the MgO catalyst for such photoinduced reactions depends strongly on the evacuation temperature of the MgO. When MgO was degassed at higher temperatures, the yields of the photoinduced isomerization reactions increased linearly with the UV irradiation time, indicating that the reactions proceeded catalytically under UV irradiation at 273 K (*96–98*).

The initial reaction rates of the photocatalytic isomerization determined from the slopes of the yields of the photoformed products plotted against the UV irradiation time were markedly dependent on the evacuation temperature of the MgO. The initial rate of photocatalytic *trans* → *cis* (i.e., geometrical) isomerization increased when the evacuation temperature increased, passing through a maximum at 1173 K. On the other hand, the rate of *trans*-2 → 1-C_4H_8 (i.e., double-bond shift) isomerization increased, passing through a maximum at 773 K, showing again a small increase at a high degassing temperature.

The effect of the MgO evacuation temperature on the intensity and wavelength of the photoluminescence is similar to the effect of the temperature on the initial rate of the photocatalyzed geometrical isomerization on MgO. This result indicates that the concentration of the surface sites in low coordination increases when the degassing temperature is increased up to 1173 K, and then it decreases with further increase in temperature. Simultaneously, when the degassing temperature is increased, the emitting environment involving the four-coordinated surface ions becomes more uniform and homogeneous, with complete uniformity being achieved at about 1073 K. Therefore, the isomerization of butenes proceeds photocatalytically on surface ions in low coordination in a uniform and homogeneous environment under UV irradiation at 273 K (*96–98, 240*). Furthermore, a good agreement between the effect of the evacuation temperature of the MgO sample on the yields of the photoluminescence at about 420 nm (which is associated with the existence of the surface OH^- groups) and the affect on the rate of photocatalyzed double-bond shift isomerization was also observed.

The action spectrum of the photocatalytic isomerization reactions on MgO, defined as the plot of the reaction rate vs the wavelength of the light used, shows a good agreement with the absorption spectrum, i.e., the photoluminescence excitation spectrum of the MgO (96–98, 248). The addition of O_2 or CO to MgO led to the quenching of the photoluminescence. Similarly, the rates of the photocatalytic isomerization reactions on MgO were easily inhibited by the addition of CO, its extent increasing with an increase in CO pressure.

These results indicate that the photocatalytic isomerization reactions on MgO proceed via the same excited state for the catalysis and for the photoluminescence. Thus, the photophysical processes on MgO in the presence of a mixture of 2-C_4H_8 and CO are described as follows (98):

$$(Mg^{2+}-O^{2-})_{LC} \xrightarrow{h\nu} (Mg^{+}-O^{-})^{*}_{LC} \begin{array}{l} \nearrow \text{photocatalyzed reaction } (k_p) \\ \rightarrow \text{radiationless deactivation } (k_d) \\ \searrow \text{deactivation by added CO } (k_q) \end{array} \quad (30)$$

The Stern–Volmer equation (Eq. 24) is obtained for the rate of the photocatalytic isomerization of *trans*-2-C_4H_8 on MgO by using the steady-state treatment: $\Phi_0/\Phi = 1 + \tau k_q [Q]$, where Φ_0 and Φ are the yields of the photocatalytic isomerization of *trans*-2-C_4H_8 in the absence and presence of CO molecules, respectively, and τ, k_q, and $[Q]$ are the lifetime of the excited state of the active sites on the MgO surface, the quenching rate constant, and the concentration of added CO molecules on the surface, respectively. In fact, as shown in Fig. 60, Φ_0/Φ in the presence of CO is a linear function of the CO pressure in the low-pressure region, although there is a deviation from linearity at higher pressures. In the low-pressure region, the dynamic quenching of the excited state of the active sites by CO molecules operates mainly when the quenching efficiency is dependent on the amount of CO on the surface and, in turn, on the equilibrium pressure of the added CO. Deviations from linearity of the Stern–Volmer plots at higher pressures can be attributed to the fact that the number of CO molecules on the surface did not increase linearly with the CO pressure.

Figure 60 also shows the Stern–Volmer plots for the quenching of the photoluminescence intensity. There is good agreement between the Stern–Volmer plots for the photoluminescence intensity and the yields of the photocatalytic isomerization reactions, indicating that both the photocatalytic reaction and the photoluminescence proceed through the same excited state of MgO, i.e., the charge-transfer excited state on the four-coordinated surface sites on MgO (96–98, 240).

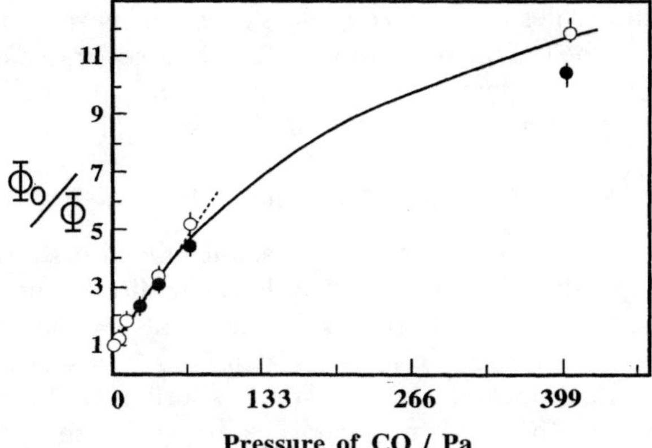

FIG. 60.   Stern–Volmer plots, Φ_0/Φ values at various pressures of CO for the rate of the photocatalytic isomerization of *trans*-2-butene (●) and the photoluminescence yield (○) of the MgO catalyst degassed at 1173 K for 2 h [reproduced with permission from Anpo *et al.* (*98*)].

2.   *Photocatalysis on TiO_2*

The addition of O_2 or N_2O onto TiO_2 leads to the quenching of the photoluminescence of the catalyst observed near 450–550 nm, whereas the addition of various unsaturated hydrocarbons (such as C_3H_6 and C_4H_6) or H_2O or H_2 enhances the photoluminescence intensity (see Section VII.A.1) (*224, 225*). Such quenching and enhancement of the photoluminescence of semiconducting TiO_2 can be attributed to changes in the width of the semiconductor depletion layer, which becomes larger as a result of the formation of negatively charged adduct species through electron capture or smaller as a result of the formation of the positively charged adsorption species through hole trapping on the surface (*226–228*).

UV irradiation of TiO_2 containing a sufficient amount of H_2O in the presence of unsaturated hydrocarbons such as C_3H_6 and C_4H_6 leads to the formation of hydrogenolysis reaction products and oxygen-containing products (*214, 242, 243*). As shown in Fig. 52, the rates of these photocatalytic reactions increase when the ionization potential of the reactants decreases (*224*).

A parallel relationship between the enhancement of the TiO_2 photoluminescence resulting from the addition of unsaturated hydrocarbons and the photocatalytic reaction rates of these unsaturated hydrocarbons with H_2O on the catalyst clearly suggests that these photoinduced surface processes are closely connected, both being crucially dependent on the magnitude of the charge shift to or from the TiO_2 catalyst. In other words, the formation of the adsorbed cationic species which are derived by charge transfer from the reactant molecules into the TiO_2 plays a significant role in determining the efficiency of the charge separation of the photogenerated electrons and holes, which in turn plays a vital role in determining the yields of the photocatalytic reaction and the photoluminescence of the catalyst.

3. *Photocatalysis on Pt-Loaded TiO_2 and ZnO*

The photocatalytic activities of various semiconductors show remarkable differences depending on whether they have small amounts of metal on their surfaces. The large differences in the yields of the photocatalytic hydrogenation reactions of $CH_3C\equiv CH$ with H_2O (wet system) on Pt-free TiO_2 and Pt-loaded powder $TiO_2(Pt/TiO_2)$ as well as on Pt-free ZnO and Pt-loaded powder ZnO (Pt/ZnO) catalysts are shown in Table VII (*224, 244*). The major product of the photocatalytic hydrogenation of $CH_3C\equiv CH$ and C_3H_6 with H_2O on Pt-free TiO_2 and ZnO is C_2H_6. This hydrogenation product is accompanied by the fission of the carbon–carbon bond of $CH_3C\equiv CH$ and C_3H_6, i.e., the hydrogenolysis reaction; CO_2 is formed as the oxidation product.

Such product distributions in photocatalytic hydrogenolysis can be attributed to the closely spaced photogenerated electron–hole pair, i.e., the Ti^{3+}–OH pair species. These electron–hole pairs interact with $CH_3C\equiv CH$ molecules to form radical species such as CH and $CH=C$ (*242–244*). On the other hand, on Pt/TiO_2 and Pt/ZnO, on which a photoelectrochemical

TABLE VII

Comparison of the Rates and Quantum Yields of Photocatalytic Hydrogenation of $CH_3-C\equiv CH$ with H_2O on TiO_2, Pt-Loaded $TiO_2(Pt/TiO_2)$, ZnO, and Pt-Loaded ZnO (Pt/ZnO) Catalysts at 298 K[a]

	Product[b]/rate (10^{-9} mol/m² h)						
Catalyst	CH_4	C_2H_4	C_2H_6	C_3H_6	C_3H_8	CO	CO_2
TiO_2	0.10	0.018	0.57	0.013	0.02	1.03	5.75
Pt/TiO_2	0.17	0.174	0.28	7.57	0.83	1.885	8.72
Quantum yield[c]	—	—	0.004	0.114	—	0.028	0.13
ZnO	0.0017	Trace	0.0094	Trace	Trace	0.034	0.107
Pt/ZnO[d]	0.002	0.0023	0.0021	0.153	0.197	0.040	0.201

Note: From Anpo *et al.* (*224*).

[a] The pressures of $CH_3-C\equiv CH$ and H_2O were 0.4 and 1 kPa, respectively. [b] The photocatalytic reaction proceeded linearly against UV-irradiation time. CH_3CHO, CH_3COCH_3, and $CH_3COC_2H_5$ were detected as minor oxidation products. [c] The quantum yield is the ratio (number of molecules of photoformed product)/(number of incident photons) at the wavelength of 350 ± 10 nm light on Pt/TiO_2 catalyst at 298 K. [d] The photocatalytic reactivity of Pt/ZnO catalyst was too low to allow the quantum yield to be measured.

mechanism is predominant, the major reaction product is C_3H_6, i.e., the hydrogenation of $CH_3C\equiv CH$ gives C_3H_6, whereas the oxidation of $CH_3C\equiv CH$ with the OH radicals produced CO_2 as the product. The rates of the photocatalytic hydrogenation reactions are remarkably enhanced on both TiO_2 and ZnO by the addition of Pt as a cocatalyst (*224*).

The rates of photocatalytic hydrogenolysis on TiO_2 and hydrogenation on Pt/TiO_2 of $CH_3C\equiv CH$ with H_2O are much higher than the rates observed with ZnO and Pt/ZnO, respectively. With TiO_2 and ZnO, the rates of oxidation are much higher than the rates of hydrogenation. However, with Pt/TiO_2 and Pt/ZnO, there is no marked difference between the rates, and this result supports the hypothesis that photoreactions on Pt-loaded TiO_2 and ZnO are much more catalytic than those on Pt-free TiO_2 and ZnO.

TPD spectra obtained for the desorption of H_2O from TiO_2 and ZnO after the adsorption of a monolayer of H_2O at 298 K showed that the amounts of water desorbed from TiO_2 and ZnO in the 273–673 K temperature range are almost the same. These results suggest that differences in the photocatalytic reaction rates do not correlate with the amounts of adsorbed H_2O in the steady state.

ZnO exhibits a high photoluminescence yield when it is excited with band gap light of about 3.2 eV ($= 380$ nm) (*228*). The addition of H_2O led to an increase in the photoluminescence intensity of ZnO, with its extent depending on the amount of adsorbed H_2O. The addition of $CH_3C\equiv CH$ or H_2O to ZnO led to an enhancement of the photoluminescence intensity, with its magnitude being much larger than the enhancement observed for TiO_2. The addition of excess amounts of H_2O led to an enhancement of 1.5–2.0 times in the photoluminescence of ZnO, but on TiO_2 an enhancement of only 1.1 or 1.2 times was observed.

On the other hand, the addition of O_2ZnO led to a decrease in the photoluminescence intensity, with its extent depending on the amount of added O_2. The addition of 0.4–1.33 kPa of O_2 (even excess amounts of O_2) to TiO_2 and ZnO led to a quenching of about 80–95% of their photoluminescence, although the initial photoluminescence yield to TiO_2 is much lower than that for ZnO.

As shown in Fig. 61, these data suggest that the addition of H_2O and/or $CH_3C\equiv CH$ to TiO_2 or ZnO causes a decrease in the extent of structural changes in the bending of the surface band, resulting in an increase in the efficiency of the recombination of the photogenerated electrons and holes and subsequently an enhancement of the photoluminescence—which all have a larger effect on ZnO than on TiO_2 (*223*). On the other hand, the addition of O_2 to these photocatalysts causes a suppression of the recombination efficiency of the photogenerated electrons and holes, i.e., a better and more efficient charge separation resulting in a quenching of the

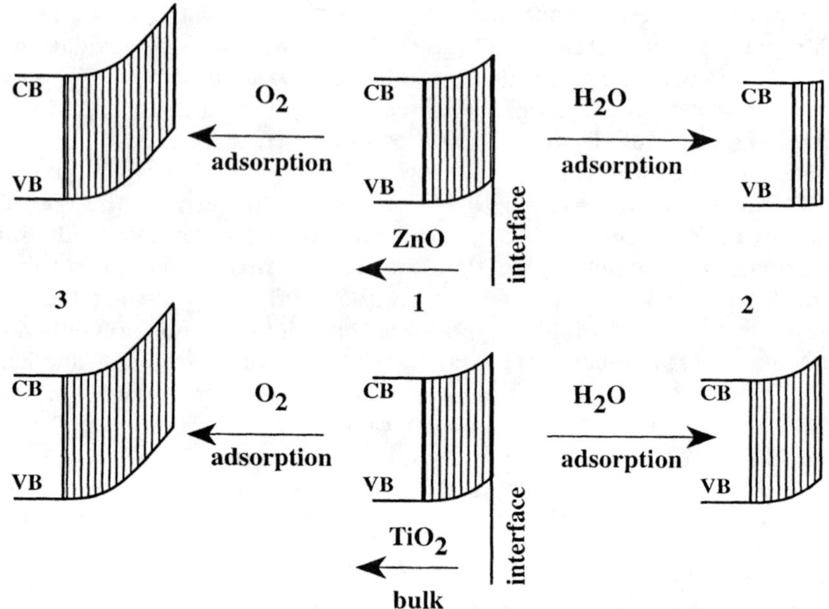

FIG. 61.   Schematic description of the surface band bending of ZnO (top) and TiO$_2$ (bottom). 1, after degassing *in vacuo* (*n*-type semiconductors); 2, after adsorption of H$_2$O (decrease of band bending by formation of positively charged species on the surface); 3, after adsorption of O$_2$ (increase of band bending by the formation of negatively charged species on the surface) [reproduced with permission from Anpo *et al.* (*224*)].

photoluminescence of the same extent for both TiO$_2$ and ZnO (excess amounts of O$_2$ led to an almost complete quenching of the photoluminescence). Thus, the photoluminescence behavior is closely associated with the differences in the electronic structure of the TiO$_2$ and ZnO surfaces since Ti^{4+} and Zn^{2+} ions play significant roles in the formation of surface adduct species.

The characteristics of the photocatalytic properties of TiO$_2$ and ZnO are the same for both Pt-loaded and -unloaded systems, but large differences in the activities have been observed for these catalysts (*223*). Large differences are found in photocatalytic reactions involving the activation of H$_2$O and/or surface OH groups, but only small differences are observed in reactions associated with the activation of O$_2$ in dry systems. Such large differences observed for the photocatalytic activities of TiO$_2$ and ZnO in the presence of H$_2$O can therefore be attributed to the much faster recombination of photogenerated electrons and holes on ZnO than on TiO$_2$. It was also found that the formation of the positively charged surface adduct species results in a much smaller upward band bending on ZnO than on TiO_2. On the other hand, in dry systems, the formation of the negatively charged surface adduct species led to a much larger upward band bending on both TiO_2 and ZnO, which resulted in higher photocatalytic (as well as equal) efficiencies for TiO_2 and ZnO.

4. *Photocatalysis on ZnS*

Preground ZnS exhibits a characteristic photoluminescence spectrum near 420–470 nm with a high efficiency when the absorption band at about 390–410 nm is excited (*245–247*). This photoluminescence spectrum can be attributed to the radiative recombination of the photogenerated holes and electrons at sites associated with sulfur vacancies in ZnS. The intensity of the photoluminescence changes as a result of mechanical grinding. As shown in Fig. 62, the intensity of the photoluminescence decreases when the time of grinding is increased (*245*). Simultaneously, the absorption band of the catalyst gradually shifts toward longer wavelengths, showing a shoulder in these regions. These changes suggest that some extrinsic surface energy levels are newly created in the band structure of the ZnS catalyst caused by mechanical damage due to grinding.

The preground ZnS also exhibited a weak EPR signal with *g* values of

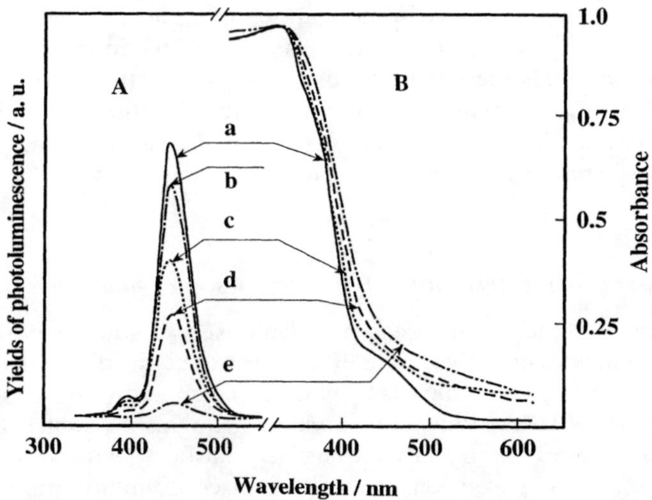

FIG. 62. Photoluminescence spectrum (A), absorption spectra of the ZnS catalyst (B), and the effect of the grinding of ZnS. Number of grindings: a, 0 (original ZnS); b, 50; c, 250; d, 1250; e, 6250. Temperature, 298 K; excitation wavelength at 285 nm [reproduced with permission from Anpo *et al.* (*245*)].

$g_1 = 2.0063$, $g_2 = 2.0021$, and $g_3 = 2,000$ (*245*). In addition to this signal, new EPR signals with g values of 2.016–2.024 were observed with the ground ZnS, and these were attributed to the radical-like sulfur species such as S_n involving S_3 and/or the sulfur cluster species. The concentration of these species increases dramatically with increase in the grinding time. Thus, these spectroscopic data provide clear and direct evidence that extrinsic surface energy states due to radical-like sulfur and/or sulfur cluster species are produced by the damage in the structure due to grinding.

The *trans–cis* isomerization of 2-C_4H_8 proceeds catalytically on the ZnS catalyst under UV irradiation (*245*). The initial rate of this photocatalytic isomerization on preground ZnS was found to be relatively low, but it increased markedly when the grinding time increased. The effect of the grinding of the catalyst on the efficiency of photocatalytic isomerization was the opposite of the effect on the photoluminescence efficiency of the catalyst.

These results indicate that extrinsic surface energy states due to the presence of sulfur cluster species and/or radical-like sulfur species produced on the surfaces by grinding play a decisive role in the trapping of the photoformed holes on the surfaces. Such efficient hole trapping results in the suppression of the photoluminescence arising from the recombination of the photoformed holes and electrons at the sulfur vacancies. Simultaneously, this trapping results in an enhancement of the photocatalytic activity of the catalyst for *trans–cis* isomerization since the trapped holes at the surface sites (i.e., the sulfur radicals) play a vital role in the photocatalytic isomerization of alkenes. It is therefore concluded that extrinsic surface energy states play a significant role in determining the fate of photoproduced holes and electrons; in other words, the efficiencies of the photocatalytic activity and photoluminescence of the catalyst (*245*).

5. Quantum Size Effect on the Photoluminescence and Photocatalysis

A principal objective of research on small semiconductor particles is to understand how to effectively and efficiently utilize the electrons and holes injected into the semiconductors by band-gap excitation for various surface catalytic reactions. A unique feature which distinguishes small semiconductor particles from bulk crystals is their large surfaces, which incorporate a great number of exposed ions with different coordination numbers. Such small particles show various unique and distinct characteristics involving the enhancement of the charge separation and energy transfer in nanometer-size regimes in which a quantum confinement effect can be observed, thus providing many possibilities for innovation that may open new research fields (*242, 243, 248, 249*). As shown in Fig. 63, as the size of the small particle (i.e., cluster) is reduced below a certain critical dimension, its highest occupied molecular orbital–lowest unoccupied molecular orbital energy gap starts to increase, which is often reflected by dramatic changes in the absorption and photoluminescence bands of the cluster. Such a phenomenon is usually referred to as the quantum size effect.

B. PHOTOCATALYSIS ON SUPPORTED CATALYSTS

1. *Photocatalysis on Anchored Vanadium Oxides*

The photoluminescence of the anchored vanadium oxide prepared by the photo-CVD method was quenched in both its intensity and its lifetime by the addition of *trans*-2-C_4H_8, suggesting that the molecules interact with

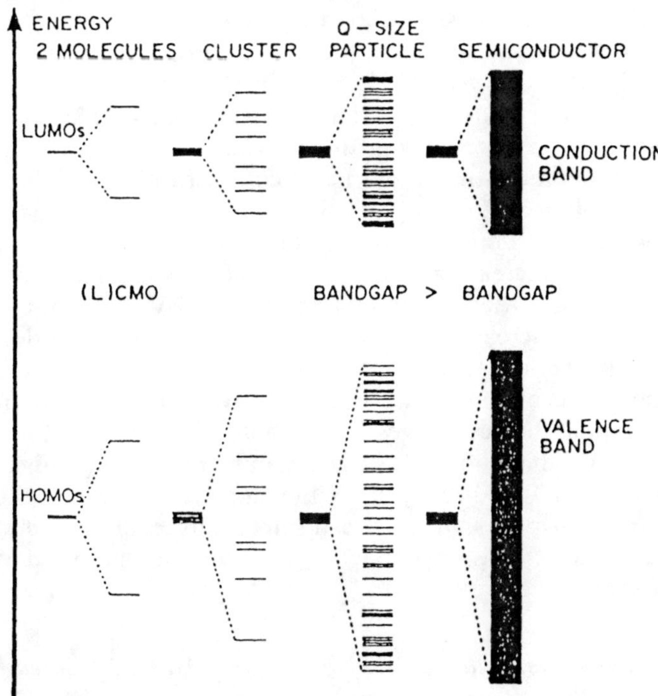

FIG. 63. MO model for particle growth [reproduced with permission from Brus (*248*)].

the surface vanadyl groups (V=O) in their excited states. UV irradiation of the anchored vanadium oxide in the presence of *trans*-2-C_4H_8 led to *trans* → *cis* (geometrical) and 2 → 1 (double-bond shift) isomerization reactions, even at temperatures lower than 273 K (*33, 34, 115–120*). The yields of these photocatalytic reactions increased with the UV irradiation time.

As shown in Fig. 44, the rate of photocatalytic isomerization exhibits a good linear relationship with the yield of the phosphorescence of the anchored vanadium oxide, indicating that the charge-transfer excited triplet state of the surface vanadyl groups plays a vital role in the photocatalytic isomerization reaction. The lifetime of the phosphorescence decreased in the presence of 2-C_4H_8, with the extent depending on the concentration of butene. These data suggest that the interaction of 2-C_4H_8 with the charge-transfer excited triplet of the $(V^{4+}–O^-)^{3*}$ paired state or the O^- species of the paired state results in an opening of the C=C bond of 2-C_4H_8, with the resulting intermediate species participating in a geometrical isomerization. The reaction mechanism is similar to that proposed for the photocatalytic isomerization of *trans*-2-C_4H_8 on TiO_2 (*57, 200, 201, 244*), MgO (*58, 240*), and vanadium oxides supported on SiO_2 (*33, 34, 115, 116, 120*).

The results obtained with vanadium oxide catalysts prepared by the impregnation method show a remarkable contrast with those obtained with anchored vanadium oxide catalysts (*63, 116*). As shown in Fig. 64, the yields of the photocatalytic isomerization as well as the yields of the phosphorescence of the oxide increase with the content of the vanadium ions and then decrease, even when the catalyst contains 0.1 wt% V. When the vanadium content is high, an increase in the efficiency of the radiationless deactivation due to the aggregation of the vanadium oxide species is observed.

Thus, the results obtained with vanadium oxides not only provide useful information about the advantages of the photo-CVD Method for preparing highly dispersed vanadium oxide catalysts and far achieving a high photocatalytic activity but also directly show the significant role that the charge-transfer excited triplet state of the tetrahedrally coordinated vanadium oxide species plays in photocatalytic reactions on supported vanadium oxides (*115, 116*).

2. *Photocatalysis on Titanium–Silicon Binary Oxides Prepared by the Sol-Gel Method*

Titanium oxides anchored to transparent Vycor glass and titanium oxides dispersed into SiO_2 carriers or matrices exhibit extremely high photocatalytic activity due to the high activity of the highly dispersed tetrahedral titanium oxide species (*168, 200, 201*). The sol-gel method was found to be

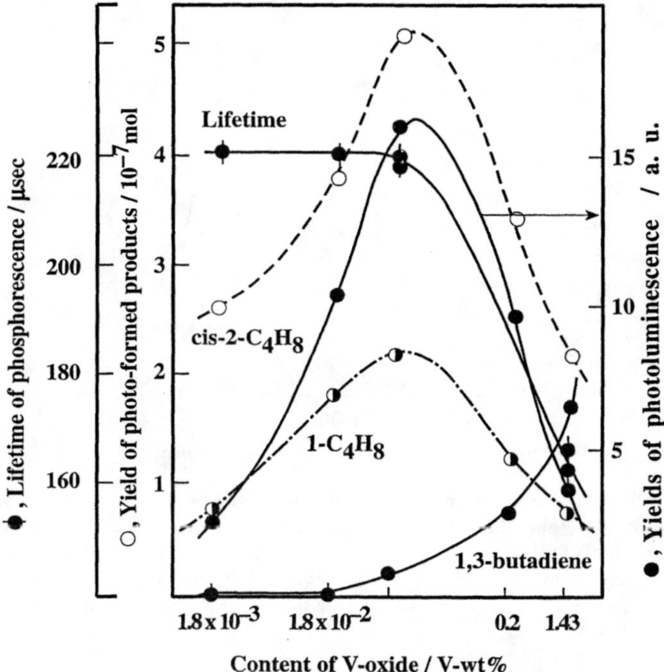

FIG. 64.   Effect of the content of vanadium oxide on the yields of the photoformed products and the phosphorescence spectrum of vanadium oxide catalyst supported on porous Vycor glass (temperature, 290 K) [reproduced with permission from Kubokawa and Anpo (63)].

useful in designing highly efficient and transparent photocatalysts involving highly active tetrahedral titanium oxide species (178).

As shown in Fig. 65, titanium–silicon binary oxides prepared by the sol-gel method exhibit a characteristic photoluminescence spectrum near 480 nm upon excitation at 280 nm. The absorption and photoluminescence spectra are attributed to the charge-transfer processes on the highly dispersed tetrahedral titanium oxide species embedded in the SiO_2 matrices (168, 200, 201). When the titanium content of the oxides was decreased, the intensity of the photoluminescence spectrum increased, and its peak wavelength shifted to shorter wavelengths.

UV irradiation of the titanium–silicon binary oxide catalyst in 1-octanol acetonitrile solutions in the presence of O_2 led to the oxidation of 1-octanol to produce 1-octanal as the main product, whereas no products were detected under dark conditions. The specific photocatalytic reactivities of the titanium oxide moieties in the bindar oxides can be determined since the

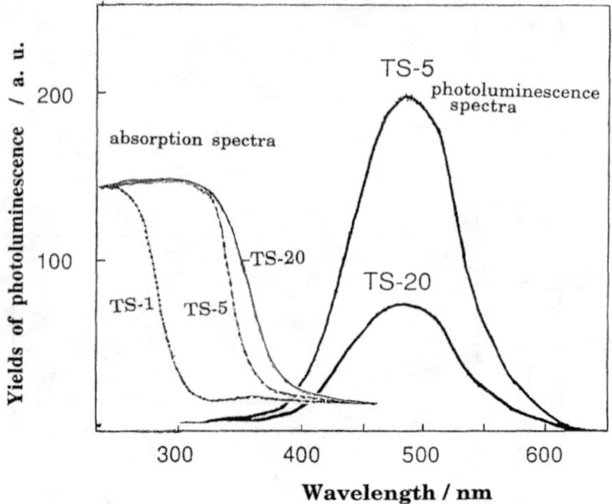

FIG. 65. Photoluminescence spectra of titanium–silicon binary oxides prepared by the sol-gel method at 77 K and the absorption spectra of titanium–silicon binary oxides having various Ti contents. Content of Ti in TiO_2 wt%: TS-1, 1%; TS-5, 5%; TS-20, 20%. The photoluminescence was quenched by the addition of H_2O or CO_2 with different efficiencies (excitation at 280 nm) [reproduced with permission from Anpo and Yamashita (252)].

catalysts involve only the titanium oxide species and SiO_2 which by itself exhibits almost no photocatalytic activity. A marked increase in the photocatalytic activity of the titanium oxide species was observed at lower titanium contents, whereas a slight increase was observed at higher titanium contents. A good relationship was found between the specific photocatalytic activities of the titanium oxide species and the photoluminescence yields of the titanium–silicon binary oxides. These data indicate that the appearance of such high photocatalytic activity for titanium–silicon binary oxides prepared by the sol-gel method is closely associated with the formation of highly dispersed tetrahedral titanium oxide species surrounded by the SiO_2 matrices and with the high activity of the charge-transfer excited state of these species.

3. *Photocatalysis on Titanium Oxide/Zeolite Catalysts*

The reduction of CO_2 with H_2O to produce useful compounds such as CH_3OH and CH_4 active photocatalysts is one of the most desirable objectives of this research field. UV irradiation of titanium oxides included within ZSM-5 zeolite (titanium oxide/Y zeolite) in the presence of CO_2 and H_2O led to the evolution of CO, CH_4, and CH_3OH at 275 K, with the amounts increasing linearly with the UV irrdiation time. The photocatalytic reduction of CO_2 with H_2O to produce CH_3OH and CH_4 proceeds only on the highly active titanium-based oxides prepared within the zeolites, with the yields and distribution of the photoformed products dependent on the kind of zeolite or the environment of the titanium oxide (*178, 250–253*).

As shown in Fig. 66, the ex-titanium oxide/Y Zeolite (titanium oxides prepared within Y zeolite via ion exchange) exhibits a photoluminescence spectrum near 480–500 nm upon excitation at about 280 nm at 77 K. The photoluminescence is attributed to a radiative decay from the charge-transfer excited state of the highly dispersed tetrahedrally coordinated titanium oxide species $(Ti^{3+}-O^-)^*$. As shown in Fig. 67, the addition of CO_2 or H_2O leads to the efficient quenching of the photoluminescence, with its extent depending on the amounts of CO_2 or H_2O added. A good relationship was found between the yields of photoformed CH_4 and the photoluminescence of the catalyst and the titanium content both increasing and then decreasing with a decrease in the titanium content (*251, 252*).

These results indicate that the high activity of the charge-transfer excited state $(Ti^{3+}-O^-)^*$ of the tetrahedral titanium oxide species plays a significant role in the photocatalytic reduction of CO_2 with H_2O on the titanium oxide/zeolite catalysts.

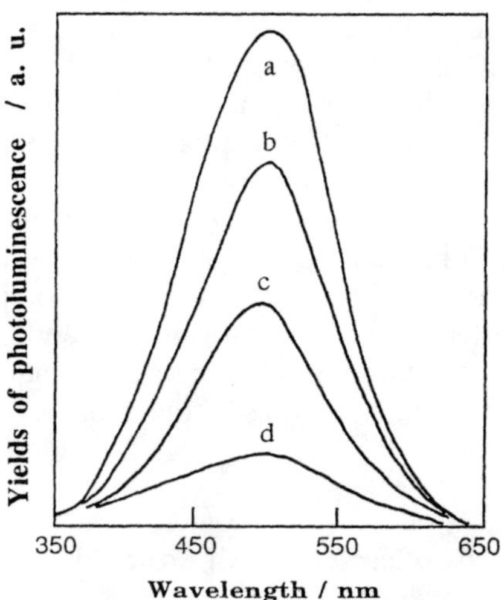

FIG. 66. Photoluminescence spectrum of ex-titanium oxide/Y zeolite catalyst (a) and effect of the addition of CO_2 and H_2O on the photoluminescence spectrum (b–d). Amount of added CO_2: b, 8.5; amount of H_2O: c, 2.9 mmol g^{-1}; measured at 77 K; excitation at 290 nm; emission monitored at 490 nm) [reproduced with permission from Anpo *et al.* (*251*)].

4. *Photocatalysis on Copper(I)ZSM-5 Zeolite*

UV irradiation of the Cu(I)ZSM-5 catalyst even at 275 K in the presence of NO leads to the formation of N_2 and O_2 with a linear relationship between the UV irradiation time and the NO conversion into N_2. This indicates that the direct photocatalytic decomposition of NO proceeds efficiently on the Cu(I)ZSM-5 catalyst (*175, 177–180*).

As shown in Fig. 46 (c), the photocatalytic activity of Cu(I)ZSM-5 for the direct decomposition of NO increases with the evacuation temperature of the Cu(II)ZSM-5, passing through a maximum at 1073 K. As shown in Fig. 46 (b), the yield of the photoluminescence, which is attributed to the presence of the isolated Cu^+ monomer species, also changes dramatically in the same way. As CO molecules selectively adsorb on the isolated Cu^+ monomer species, the number of copper(I) ions on the catalysts can be determined accurately by measurement of the number of CO molecules adsorbed. The number of isolated Cu^+ ions determined using this method is shown in Fig. 46 (c) These values also change with the evacuation temperature of the Cu(II)ZSM-5, with the change being parallel to the yield of the photoluminescence (*175, 177–180*).

Figure 46 shows that a parallel exists between the yield of the photoluminescence assigned to the isolated Cu^+ monomeric species and the yield of the photocatalytic decomposition of NO, indicating the important role that the excited state of the copper(I) species plays in the photocatalytic decomposition of NO into N_2 and O_2.

The direct observation of the $(Cu–NO)^+$ species by IR spectroscopy (*174*) and of its behavior under UV irradiation by EPR measurements (*175, 177*) suggests that a local electron transfer from the excited state of the Cu^+ ion ($3d^9 4s^1$) to the anti-π-bonding orbital of NO and simultaneously an electron transfer from the π-bonding orbital of another NO to the vacant orbital of the Cu^+ ion ($3d^9\ 4s^0$) lead to the decomposition of two NO molecules on the Cu^+ site, selectively forming N_2 and O_2 under UV irradiation, even at 275 K. The data show that new and unique photocatalysis can be achieved within the small cavities of the zeolite.

IX. Application To Chemical Detection

A. CHEMICAL SENSING

The excitation of *n*-CdS at the band gap ($Eg = 2.4$ eV) produces a band edge photoluminescence spectrum with a λ_{max} near 510 nm, which is attributed to the radiative recombination of the photogenerated electrons

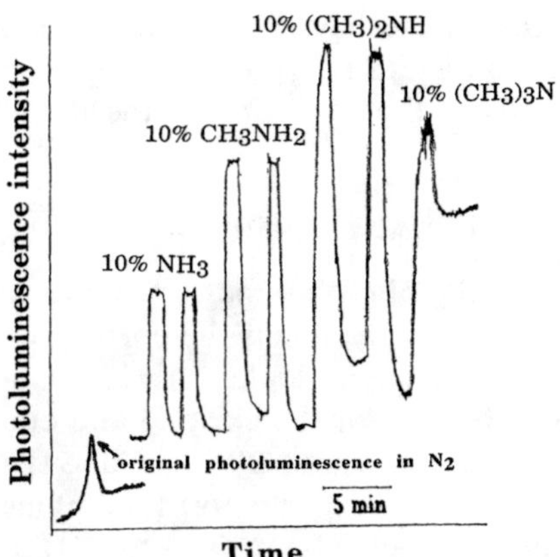

FIG. 67.   Changes in the intensity (yield) of the photoluminescence spectrum at about 505 nm resulting from exposure of an etched n-CdS sample to N_2 (initial response) and the indicated amine. The original photoluminescence spectrum was measured in N_2. Flow rates for all gases were 100 ml min^{-1} at 1 bar pressure; sample excited with 458-nm light [reproduced with permission from Meyer *et al.* (227)].

and holes (*113, 197–198, 225–229*). As shown in Fig. 67, the exposure of this n-CdS to gaseous NH_3, ND_3, $(CH_3)NH_2$, $(CH_3)_2NH$, or $(CH_3)_3N$ causes an enhancement in the intensity of the photoluminescence compared with the photoluminescence intensity in the N_2 atmosphere, and with an exchange of the gas to N_2 the enhanced intensity returns to the original level (*225–227*). The magnitude of the photoluminescence enhancement was found to increase in the order

$$NH_3 = ND_3 < (CH_3)NH_2 < (CH_3)_2NH < (CH_3)_3N,$$

and this is the same order as the intrinsic base strengths of these gases, with the exception of $(CH_3)_3N$. The photoluminescence response time increases with the degree of methyl substitution of these gases. $(CH_3)_2NH$ and $(CH_3)_3N$ require the longest time to reach their maximum enhancement points and also to return to their original levels in the N_2 atmosphere. On the other hand, the exposure of n-CdS to SO_2 leads, to the efficient quenching of the photoluminescence. Thus, enhancements and diminutions of the photoluminescence intensity are observed with the Lewis basic and acidic compounds, respectively (*227*).

These quantitatively reversible enhancements and quenching of the photoluminescence of n-CdS semiconductors by the addition of gaseous bases and acids, respectively, clearly indicate not only that such phenomena are associated with the intrinsic basicity or acidity of these gases but also that the formation of weak adduct species plays a significant role, leading to changes in the depletion width based on the dead-layer model (*227*).

Quantitative analysis of the reduction in the depletion width has been obtained as follows:

$$(\Phi_e)_0/\Phi_e = \exp(-\alpha' \, \Delta D), \tag{31}$$

where $(\Phi_e)_0$ and Φ_e are the photoluminescence intensities in N_2 and in the presence of amines, respectively; ΔD is the corresponding change in dead-layer thickness, i.e., the change in depletion width; and $\alpha' = \alpha + \beta$, where α and β are the absorptivities for the exciting and emitted light. From this treatment, using the values of α and β for n-CdSe (i.e., $9 \times 10_4$ and 5.5×10^3 cm^{-1} at 458 and 510 nm, respectively) the values of the reduction in depletion width are determined to be ~200 Å for NH_3 (a 20% enhancement) and ~400 Å for $(CH_3)_2NH$ (a 40% enhancement), respectively (*225–227*).

Thus, the enhancement and dimunition in the photoluminescence intensity observed with Lewis bases and acids are of great interest from a fundamental and a technological perspective. In the former case, photoluminescence provides a simple *in situ* technique for corresponding physicochemical interaction occurring at the interface with changes in electronic structure of semiconducting catalysts. In the latter case, the technique is readily adapted to optically coupled chemical sensing, even using a remote controlling system (*227*).

B.   Probe of Catalyst Poisoning

Suib *et al.* (*25, 254*) reported the different effects of nickel and vanadium on the catalytic activity and selectivity for the fluid catalytic cracking by a photoluminescence technique and showed that the method is useful in predicting the catalyst deactivation caused by the deposition of metals on surfaces. The activity of the catalyst decreases monotonically with increasing vanadium content. With 1.5 wt% of V, the catalystad lost most of its activity, and with 2.0 wt% of V it became almost completely inactive. Such a deactivation of the catalyst was irreversible, with the extent being closely associated with the surface area covered with vanadium. Moreover, the extent of the deactivation was found to depend on the aging temperature, which was accelerated when aging was carried out under the same conditions normally sized in hydrothermal reactions.

Quenching of the photoluminescence was found for the 2 wt% V-loaded catalyst, and in this system a new band appeared (*25, 254*). Vanadium–naphthenate and the vanadium-loaded catalyst exhibit similar photoluminescence spectra near 350 nm. Moreover, only the vanadium-loaded catalyst exhibits a broad emission at about 550 nm. After aging the vanadium-loaded catalyst, the yield of this additional photoluminescence spectrum dramatically increased at about 550 nm. Such photoluminescence behavior observed with the fluid catalytic cracking catalysts suggests that upon deposition of vanadium, vanadium interacts with the zeolite, destroying the activity.

In contrast, with as much as 3% Ni, the aged catalyst only slightly changes and retains about 95% of its original surface area as well as most of its catalytic activity for cracking. The photoluminescence spectrum of the nickel-loaded catalyst observed near 350 nm changes when nickel–naphthenate is deposited onto it (25, 254). The similarity between the shape and wavelength regions of this photoluminescence spectrum and the spectrum of the LaY zeolite suggests that the photoluminescence is largely due to the presence of the LaY zeolite. Differences in the photoluminescence intensity depend on the nickel coverage on the zeolite. When nickel is not present, the photoluminescence at about 370 nm is observed with a high yield. The deposition of 2% of Ni results in a complete quenching of the photoluminescence of the LaY zeolite, and the subsequent hydrothermal treatment of the catalyst leads to a recovery of some parts of the lost intensity.

X. Relationships to Other Techniques

A. Photoluminescence Related to Other Physical Techniques

We have discussed examples of the various types of systems investigated in both catalysis (mainly by gas chromatography) and photocatalysis (by photoluminescence) in order to provide more insight into the reactivity of electronically excited states and their contribution to the mechanism of catalytic and photocatalytic reactions.

Fluorescence and phosphorescence are emission processes which originate directly or indirectly (see 5 section II.B) from the electronically excited singlet state and triplet state, respectively, produced by charge-transfer processes (Eqs. 1 and 2). Many publications deal with such charge-transfer transitions by diffuse reflectance spectroscopy (DRS) (2–6) showing the link between the latter technique and photoluminescence. It is worthwhile to recall that the emergence of the coordination chemistry of solid-state anions, namely, of surface lattice O^{2-} oxide ions, has almost entirely been based on the results of both photoluminescence and DRS analyses (7, 66). For some catalytic systems, vibrational structures can be detected (see Section IV.B) with an associated vibrational constant, which may be determined directly and independently by IR or Raman spectroscopy, evidencing the relation between these spectroscopies and photoluminescence (33, 34).

Charge-transfer processes may eventually lead to excited triplet states [Eqs. (2) and (3)] usually not observed by EPR because of the short distance between the two paramagnetic centers, leading to line broadening beyond detection. On the other hand, if experiments are performed in the presence of a gas (such as H_2, CO, and hydrocarbons) which may react specifically with the O^- hole center, the remaining paramagnetic center may be the subject of EPR studies (*66, 67*). In some favorable cases, the two radicals (electron and hole) may migrate and react, thus producing several types of paramagnetic centers detectable by EPR. Again, there is the possibility of coupling two types of spectroscopies—EPR and photoluminescence.

B.   A CASE STUDY: IN SITU CHARACTERIZATION OF
Cu(I)ZSM-5 CATALYSTS

Figure 68 shows the XANES and EXAFS spectra of (a) Cu(II)ZSM-5 and (b) Cu(I)ZSM-5 catalysts prepared by the evacuation of the former at 973 K. These spectra exhibit four kinds of bands due to transitions 1s–3d (A), $1s \rightarrow 4p_z$ ($1s \rightarrow 4p\pi^*$) (B), $1s \rightarrow 4p_{x,y}$ ($1s \rightarrow 4p\sigma^*$) (C), and multiple scattering (D) (*179, 180*). The Cu(II)ZSM-5 sample dried at 373 K shows a well-separated weak preedge band due to the $1s \rightarrow 3d$ transition(A) and an intense band due to the $1s \rightarrow 4p$ transition. The band due to the $1s \rightarrow 4p_z$ transition (B) can be observed as a shoulder of the band due to the $1s \rightarrow 4p_{x,y}$ transition (C) accompanied by their shake-up bands (B' and C'). The presence of a band due to the $1s \rightarrow 3d$ transition (A), which is forbidden by the selection rule in the case of perfect octahedral symmetry, and shake-up bands (B' and C') indicates that the Cu(II)zeolite sample contains predominantly Cu(II) ions with slightly distorted symmetries. These results coincide with results of EPR experiments (shape, *g*-tensors, and *A* factors) which indicate the presence of distorted hydrated Cu^{2+} ions in the Cu(II)zeolite sample.

Cu(I)ZSM-5 and Cu(I)mordenite catalysts prepared by the evacuation of the original Cu(II)ZSM-5 and Cu(II)mordenite, respectively, at 973 K exhibited very strong bands due to the $1s \rightarrow 4p_z$ transition (Fig. 68, B) which are clearly separated from a band due to the $1s \rightarrow 4p_{x,y}$ transition (C). The band due to the $1s \rightarrow 4p_z$ transition (B) and without the shake-up band (C') which can be observed with the Cu(I)ZSM-5 and Cu(I)mordenite catalysts is strong enough to identify the copper species, which exist as isolated Cu^+ ions with a square planar geometry (*180*). EPR, photoluminescenc, and UV absorption investigations have also indicated that Cu^{2+} species were reduced to give isolated Cu^+ ions by evacuation of the Cu(II)-zeolite at temperatures higher than 573 K.

Figure 68 also shows the EXAFS spectra of these samples and catalysts.

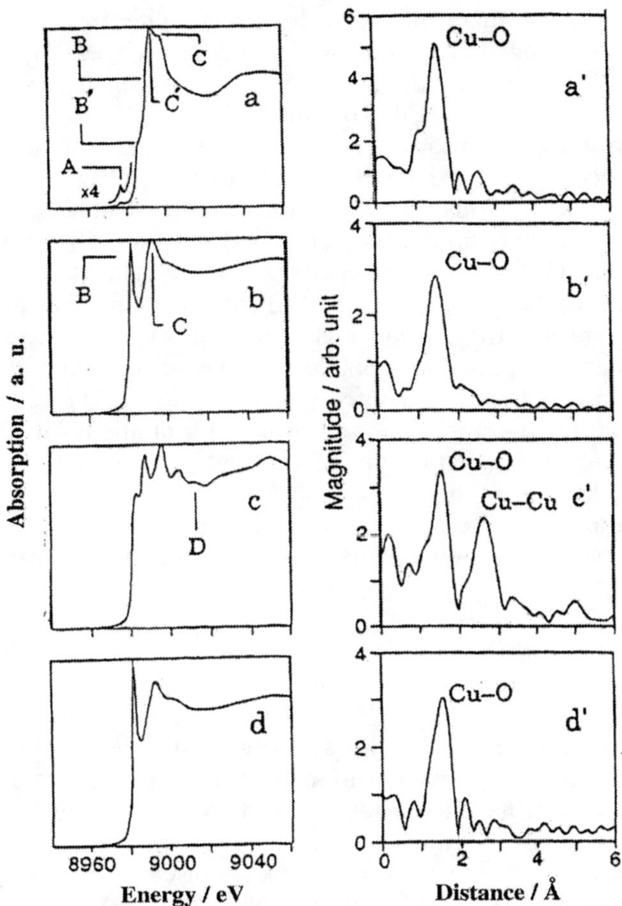

FIG. 68.   XANES (a–d) and EXAFS spectra (a'–d') of Cu(II)ZSM-5 (a, a'), Cu(I)ZSM-5 (b, b'), Cu(I)Y zeolite (c, c'), and Cu(I)mordenite catalysts (d, d'). Cu(I)zeolite catalysts were prepared by evacuation of the original Cu(II)zeolite samples at 973 K [reproduced with permission from Yamashita *et al.* (*180*)].

The Cu(II)ZSM-5 sample exhibits a strong peak at 1.5 Å assigned to the neighboring O atoms (Cu–O). The EPR date indicate the presence of six coordinate isolated Cu^{2+} ions in this sample. The Cu(I)ZSM-5 and Cu(I)mordenite catalysts prepared by evacuation at 973 K also exhibit only Cu–O peaks, indicating the presence of isolated Cu^+ ions (*179, 180*). Comparisons of the intensity of the Cu–O peaks of the Cu(II)ZSM-5 sample containing six-coordinate isolated Cu^{2+} ions and the smaller Cu–O peaks of the Cu(I)ZSM-5 and Cu(I)mordenite indicate that isolated Cu^+ ions exist as a two-coordinate species in the Cu(I)zeolite. Investigations of the isolated two-coordinate Cu^+ ions by EXAFS also suggest the formation of an $O–Cu^+–O$ complex by the addition of CO to the surface species.

After evacuation of the Cu(II)zeolite samples at 973 K, the EPR signal assigned to the Cu^{2+} species became very weak and could hardly be observed, indicating the reduction of Cu^{2+} to Cu^+. With the Cu(I)zeolite catalysts prepared in this way, as shown in Figs. 31 and 32, characteristic photoluminescence became observable at about 400–450 nm upon excitation at about 300 nm. As shown in Fig. 32, the relative intensity of the photoluminescence spectra attributed to the Cu^+ monomer and the (Cu^+–Cu^+) dimer species strongly depends on the type of zeolite. The photoluminescence band at about 430–460 nm is the major component for the Cu(I)ZSM-5 and Cu(I)mordenite. Moreover the Cu(I)Y zeolite exhibits two different intense photoluminescence bands at about 450 and 525 nm.

These results show that on the ZSM-5 and mordenite, most of the Cu^+ ions are isolated Cu^+ monomer species, but on Y zeolite the Cu^+ ions form Cu^+ dimer species. Thus, the photoluminescence observations are in agreement with the results obtained by EXAFS and ESR measurements.

XI. Conclusions

In this review, the basic principles and practical aspects of photoluminescence techniques and their application to adsorption, catalysis, and photocatalysis phenomena have been summarized. We did not discuss studies on adsorption luminescence and/or chemiluminescence in which photons emitted by adsorption or chemical reactions are used to monitor adsorption and/or catalytic reactions on surfaces (*255, 256*).

One of the principal advantages of photoluminescence spectroscopy is its high sensitivity, allowing the investigation of very low concentrations of ions or oxides dispersed on oxide supports, which constitute good models of some supported catalysts. In many cases, photoluminescence has helped to identify or confirm the nature of surface sites, thus contributing to improved understanding of the behavior of oxide-supported transition metal ions of catalytic interest. In contrast to most other spectroscopic techniques, photoluminescence has played a decisive role in the emergence of the coordination chemistry of anions because it opens up the possibility of directly observing surface lattice O^{2-} ions in different positions of coordination.

The importance of supported vanadium oxide in catalysis and photocatalysis is attested to by the richness of information provided by the photoluminescence spectra as well as the associated vibrational fine structure. These catalysts have become some of the best models for detailed case studies.

This review has emphasized the significance of dynamic photoluminescence spectroscopy because of its growing importance and the promising results obtained with it. The method provides insights into the reactivity and dynamics of excited states and their contributions to photocatalytic reactions, especially for determination of the absolute reaction rate constants and the dynamic aspects of intermediate species in photocatalytic reactions (*33, 34*).

In addition to its high sensitivity, which makes this technique very adaptable and effective for dilute systems, photoluminescence also presents advantages in its easy adaptability for *in situ* studies, providing picosecond time resolution. Notwithstanding the importance of this technique, it is only fair to recognize that, in order to have a comprehensive view of a working catalyst, photoluminescence requires the backing of other physical techniques and other spectroscopies, particularly diffuse reflectance, electron paramagnetic resonance, and infrared techniques. Furthermore, if we can also apply laser-induced photoluminescence spectroscopy such as laser-induced fluorescence spectroscopy, it is possible to obtain essential information about reaction intermediate species and their dynamics, especially in working catalysts (*257*).

It is difficult to predict trends, but in our judgment it is likely that future work using photoluminescence will be concentrated in the following areas.

1. Photoluminescence techniques will be applied to a broader range of systems, particularly oxide-supported sulfides (because of their important role in hydrotreating catalysis) as well as unsupported or oxide-supported (oxi)carbides or (oxi)nitrides (because of their growing importance as substitutes for noble metals and because they have metallic and acidic functions). Moreover, improved procedures for preparing catalytic materials will enable the design of tailored oxides with better defined characteristics, such as size, composition, and structure. The accumulation of data concerning the behavior of surface anions will also lead to a more refined view of the coordination chemistry of anions of nontransition elements.

2. The progress in spectrometer technology will allow the use of time-resolved equipment. It is likely that more *in situ* experiments will be performed. These should also facilitate the characterization of excited states and their relevance to catalysis, leading to a better understanding of the chemical nature of photoformed holes and electrons, their reactivities (including their dynamics), and reaction selectivity.

3. The need for a cleaner environment will lead to increased attention to photocatalytic processes and to better use of solar energy. Investigations involving semiconductors with more adaptable band gaps as well as photocatalytic and photoluminescence properties are expected. In fact, photocatalysts developed to operate more efficiently and effectively under visible light and solar beam irradiation are now being introduced as competitive alternatives to conventional systems and technologies while they also constitute new ways to improve our environment (*258–260*).

ACKNOWLEDGMENTS

This work was supported in part by the Grant-in-Aid on Priority Area Research on "Photoreaction Dynamics" (Grant No. 06239110) of the Ministry of Education, Science, Sports, and Culture of Japan. M. Anpo thanks the Université P. et M. Curie for several appointments as Invited Professor since 1986. M. Che thanks the Osaka prefecture University and the Japanese Society for the Promotion of Science for several appointments as visiting professor. The authors also thank the International Joint Project (Research Grant No. 07044162) of the Ministry of Education, Science, Sports, and Culture of Japan for financial support.

REFERENCES

1. Vrieland, E. G., and Selwood, P. W., *J. Catal.* **3**, 359 (1964).
2. Cimino, A., Bosco, R., Indovina, V., and Schiavello, M., *J. Catal.* **5**, 271 (1966).
3. Cimino, A., Schiavello, M., and Stone, F. S., *Discuss. Faraday Soc.* **41**, 350 (1966).
4. Cimino, A., Indovina, V., Pepe, M., and Schiavello, M., *J. Catal.* **14**, 49 (1969).
5. Stone, F. S., *in* "Chemical and Physical Aspects of Catalytic Oxidation" (J. L. Portefaix and F. Figueras, Eds.), p. 457. CNRS, Paris, 1980.
6. Stone, F. S., *J. Solid State Chem.* **12**, 271 (1975).
7. Che, M., Proc. 10th Int. Congr. Catal. Budapest A, 31 (1993).
8. Vol'kenshtein, F. F., Peka, G. P., and Malakhov, V. V., *Kinet. Katal.* **14**, 1052 (1973).
9. Tench, A. J., and Pott, G. T., *Chem. Phys. Lett.* **26**, 590 (1974).
10. Pott, G. T., and Stork, W. H. J., *Catal. Rev.-Sci. Eng.* **12**, 163 (1975).
11. Gritscov, A. M., Shvets, V. A., and Kazansky, V. B., *Chem. Phys. Lett.* **35**, 511 (1975).
12. Kazansky, V. B., *Proc. Int. Congr. Catal. 6th London* **1**, 50 (1976).
13. Coluccia, S., *in* "Adsorption and Catalysis on Oxide Surfaces" (M. Che and G. C. Bond, Eds.), p. 59 and references therein. Elsevier, Amsterdam, 1984.
14. Yun, C., Anpo, M., and Kubokawa, Y., *J. Chem. Soc. Chem. Commun.*, 665 (1977).
15. Kubokawa, Y., Anpo, M., and Yun, C., *Proc. 7th Int. Congr. Catal. Tokyo* **B**, 1170 (1980).
16. Anpo, M., Yun, C., and Kubokawa, Y., *J. Chem. Soc., Faraday Trans I* **76**, 1014 (1980).
17. Anpo, M., Tanahashi, I., and Kubokawa, Y., *J. Phys. Chem.* **84**, 3440 (1980).
18. Texter, J., Strome, D. H., Herman, R. G., and Klier, K., *J. Phys. Chem.* **81**, 333 (1977).
19. Iwasawa, Y., *J. Bull. Chem. Soc. Jpn.* **53**, 3709 (1980).
20. Iwasawa, Y., *Adv. Catal.* **35**, 187 and references therein (1987).
21. Iwamoto, M., Furukawa, H., Matsukami, K., Taknaka, T., and Kagawa, S., *J. Am. Chem. Soc.* **105**, 3719 (1983).
22. Moser, J., and Grätzel, M., *J. Am. Chem. Soc.* **105**, 6547 (1983).
23. Becker, W. G., and Bard, A. J., *J. Phys. Chem.* **87**, 4888 (1983).

24. Weller, H., Koch, U., Gutierrez, M., and Henglein, A., *Berlin Bunsenges. Phys. Chem.* **88,** 649 (1984).
25. Tanguay, J. F., and Suib, S. L., *Catal. Rev.-Sci. Eng.* **29,** 1 (1987).
26. Petras, M., and Wichterlova, B., *J. Phys. Chem.* **96,** 1805 (1992).
27. Hiramoto, M., Hashimoto, K., and Sakata, T., *Chem. Phys. Lett.* **133,** 440 (1987).
28. Nakato, Y., Ohta, A., and Tsubomura, H., *Proc. Electrochem. Soc.* **88**(14), 382 (1988).
29. Barrie, J. D., Dunn, B., Hollingsworth, G., and Zink, J. I., *J. Phys. Chem.* **93,** 3958 (1989).
30. Yoshida, S., Tanaka, T., Okada, M., and Funabiki, T., *J. Chem. Soc. Faraday Trans. 1* **80,** 119 (1984).
31. Tanaka, T., Yoshida, H., Nakatsuka, K., Funabiki, T., and Yoshida, S., *J. Chem. Soc. Faraday Trans.* **88,** 2297 (1992).
32. Idriss, H., and Barteau, M. A., *J. Phys. Chem.* **96,** 3382 (1992).
33. Anpo, M., *in* "Handbook of Heterogeneous Catalysis," (G. Ertl, H. Knözinger, and J. Weitkamp, Eds.), Vol. 2, p. 664. VCH, Weinheim, 1997.
34. Anpo, M., *Shokubai (Catalysts & Catalysis)* **39,** 45 (1997).
35. "Opportunities in Chemistry." National Academy Press, Washington, DC, 1985.
36. Anpo, M., and Matsuura, T. (Eds.), *in* "Photochemistry on Solid Surfaces." Elsevier, Amsterdam, 1989.
37. Anpo, M., and Yamashita, H., *in* "Surface Photochemistry," (M. Anpo, Ed.) p. 117. Wiley, London, 1996.
38. Anpo, M., Negishi, N., and Nishiguchi, H., *Crit. Rev. Surf. Chem.* **3,** 131 (1993).
39. Imbusch, G. F., *in* "Luminescence Spectroscopy," (M. D. Lamb, Ed.), p. 1. Academic Press, New York, 1978.
40. Blasse, G., *Struct. Bond.* **42,** 1 (1980).
41. Blasse, G., *in* "Radiationless Processes" (B. DiBartolo, Ed.), p. 287. Plenum, New York, 1978.
42. Blasse, G., Bleijenberg, K. C., and Powell, R. C., "Luminescence and Energy Transfer," Springer-Verlag, Berlin, 1980.
43. Hurtubise, R. J., "Solid Surface Luminescence Analysis, Theory, Instrumentation, Applications," Dekker, New York, 1981.
44. Vo Dinh, T., "Room Temperature Phosphorimetry for Chemical Analysis," Wiley, New York, 1984.
45. Daniels, F., and Alberty, R. A., "Physical Chemistry," Wiley, New York, 1975.
46. Atkins, P. W., "Physical Chemistry," Oxford Univ. Press, Oxford, 1986.
47. Barrow, G. M., "Physical Chemistry," McGraw-Hill, New York, 1988.
48. Olmsted J., and Williams, G. M., "Chemistry," Mosby, St. Louis, MO, 1994.
49. Calvert, J. G., and Pitts, J. N., Jr., "Photochemistry," New York, 1966.
50. Simons, J. P., "Photochemistry and Spectroscopy," Wiley Interscience, London, 1971.
51. Turro, N. J., "Molecular Photochemistry," Benjamin, New York, 1967.
52. Turro, N. J., "Modern Molecular Photochemistry," Benjamin/Cummings, Menlo Park, CA, 1978.
53. Demas, J. N., "Excited State Lifetime Measurements," Academic Press, New York, 1984.
54. Lewis, G. N., and Kasha, M., *J. Am. Chem. Soc.* **66,** 2100 (1944); **67** 994 (1945); Lewis, G. N., Calvin, M., and Kasha, M., *J. Chem. Phys.* **17,** 804 (1945).
55. Terenin, A. N., *J. Phys. Chem. SSSR* **14,** 1362 (1940).
56. Anpo, M., and Kubokawa, Y., *Res. Chem. Intermed.* **8,** 105 (1987).
57. Anpo, M., *Res. Chem. Intermed.* **9,** 67 (1989).
58. Anpo, M., Yamada, Y., Coluccia, S., Zecchina, A., and Che, M., *J. Chem. Soc. Faraday Trans. 1* **85,** 609 (1989).

59. Che, M., and Giamello, E., *in* "Catalyst Characterization: Physical Techniques for Solid Materials," (B. Imelik, and J. C. Védrine, Eds.). Plenum, New York, 1994.

60. Garrone, E., and Stone, F., *in* "Adsorption and Catalysis on Oxide Surfaces," (M. Che, and G. C. Bond, Eds.), p. 97. Elsevier, Amsterdam, 1985.

61. Che, M., and Giamello, E., *in* "Spectroscopic Characterization of Heterogeneous Catalysts," (J. L. G. Fierro, Ed.) Vol. B, p. 265. Elsevier, Amsterdam, 1990.

62. Che, M., and Giamello, E., *in* "Electron Magnetic Resonance of Disordered Systems," (N. D. Yordanov, Ed.), p. 112. World Scientific, London, 1991.

63. Kubokawa, Y., and Anpo, M., *in* "Adsorption and Catalysis on Oxide Surfaces," (M. Che, and G. C. Bond, Eds.), p. 127. Elsevier, Amsterdam, 1985.

64. Lepetit, C., and Che, M., *J. Mol. Catal.* **100,** 147 (1995).

66. Che, M., and Tench, A. J., *Adv. Catal.* **32,** 1 (1983).

67. Che, M., and Tench, A. J., *Adv. Catal.* **31,** 77 (1982).

68. Tench, A. J., and Nelson, R. L., *Trans. Faraday Soc.* **63,** 2254 (1967).

69. Anpo, M., Sunamoto, M., and Che, M., *J. Phys. Chem.* **93,** 1187 (1989).

70. Che, M., Naccache, C., and Imelik, B., *J. Catal.* **24,** 328 (1972).

71. Cordischi, D., Indovina, V., and Cimino, A., *J. Chem. Soc. Faraday Trans. I* **70,** 2189 (1974).

72. Coluccia, S., Barton, A., and Tench, A. J., *J. Chem. Soc. Faraday Trans. I* **77,** 2203 (1981).

73. Kibblewhite, J. F. J., and Tench, A. J., *J. Chem. Soc. Faraday Trans. I* **70,** 72 (1974).

74. *Natl. Bur. Stand. Tech. Note Ser.* **270**(3), 6 (1968).

75. Whited, R. C., and Walker, W. C., *Phys. Rev. Lett.* **22,** 1428 (1969).

76. Nelson, R. L., and Hale, J. W., *Discuss. Faraday Soc.* **52,** 77 (1971).

76. Coluccia, S., and Tench, A. J., *Proc. 7th Int. Congr. Catal. Tokyo* **B,** 1154 (1981).

77. Zecchina, A., Lofthouse, M. G., and Stone, F. S., *J. Chem. Soc. Faraday Trans. I* **71,** 1476 (1975).

78. Zecchina, A., and Stone, F. S., *J. Chem. Soc. Faraday Trans. I* **72,** 2364 (1976).

79. Garrone, E., Zecchina, A., and Stone, F. S., *Philos. Mag.* **42B,** 683 (1980).

80. Zecchina, A., and Scarano, D., *in* "Adsorption and Catalysis on Oxide Surfaces," (M. Che, and G. C. Bond, Eds.), p. 71. Elsevier, Amsterdam, 1984.

81. Coluccia, S., Deane, A. M., and Tench, A. J., *Proc. 6th Int. Congr. Catal. London* **1,** 171 (1977).

82. Coluccia, S., Deane, A. M., and Tench, A. J., *J. Chem. Soc. Faraday Trans. I* **74,** 2913 (1978).

83. Coluccia, S., Tench, A. J., and Segall, R. L., *J. Chem. Soc. Faraday Trans. I* **75,** 1769 (1979).

84. Levine, J. D., and Mark, P., *Phys. Rev.* **144,** 751 (1966).

85. Coluccia, S., Hemidy, J. F., and Tench, A. J., *J. Chem. Soc. Faraday Trans. I* **74,** 2263 (1978).

86. Nelson, R. L., Tench, A. J., and Hamsworth, B. J., *Trans. Faraday Soc.* **63,** 1427 (1967).

87. Duley, W. W., *Philos. Mag.* **49B,** 159 (1984).

88. Shvets, V. A., Kuznetsov, A. V., Fenin, V. A., and Kazansky, V. B., *J. Chem. Soc. Faraday Trans. I* **81,** 2913 (1985).

89. Nunan, J., Cunningham, J., Deane, A. M., Colbourn, E. A., and Mackrodt, W. C., *in* "Adsorption and Catalysis on Oxide Surfaces," (M. Che, and G. C. Bond, Eds.), p. 83. Elsevier, Amsterdam, 1984.

90. Ito, T., Kuramoto, M., Yoshioka, M., and Tokuda, T., *J. Phys. Chem.* **87,** 4411 (1983).

91. Tench, A. J., *Discuss. Faraday Soc.* **52,** 94 (1971).

92. Stone, F. S., and Zecchina, A., *Proc. 6th Int. Congr. Catal. (London),* **A,** 8 (1976).

93. Zecchina, A., and Stone, F. S., *J. Chem. Soc. Faraday Trans. I* **74,** 2278 (1978).

94. Che, M., *in* "Adsorption and Catalysis on Oxide Surfaces," (M. Che, and G. C. Bond, Eds.), p. 11 and references therein. Elsevier, Amsterdam, 1985.

95. Stone, F. S., *in* "Surface Properties and Catalysis by Non-Metals," (J. P. Bonnelle, B. Delmon, and E. Derouane, Eds.), p. 237. Reidel, Dordrecht, 1983.

96. Anpo, M., Yamada, Y., Kubokawa, Y., Coluccia, S., Zecchina, A., and Che, M., *J. Chem. Soc. Faraday Trans. I* **84,** 751 (1988).

97. Anpo, M., and Yamada, Y., *Mater. Chem. Phys.* **18,** 465 (1988).

98. Anpo, M., Yamada, Y., Kobokawa, Y., Coluccia, S., Zecchina, A., and Che, M., *J. Chem. Soc. Faraday Trans. I* **84,** 751 (1988).

99. Ito, T., Wang, J. X., Lin, C. H., and Lunsford, J., *J. Am. Chem. Soc.* **107,** 5062 (1985).

100. Anpo, M., Sunamoto, M., Doi, T., and Matsuura I., *Chem. Lett.* 701 (1988).

101. Anpo, M., Nomura, T., Kondo, J., Domen, K., Maruya, K., and Onishi, T., *Res. Chem. Intermed.* **13,** 195 (1990).

102. Moon, S. C., Tsuji, K., Nomura, T., and Anpo, M., *Chem. Lett.,* 2241 (1994).

103. Moon, S. C., Fujino, M., Yamashita, H., and Anpo, M., *J. Phys. Chem.* **101,** 369 (1997).

104. Moon, S. C., Tsuji, K., Yamashita, H., and Anpo, M., *Proc. 4th Japan–Korea Symp. (Tokyo),* 17 (1993).

105. Anpo, M., Nomura, T., Kitao, T., Giamello, E., Che, M., and Fox, M. A., *Chem. Lett.,* 889 (1991).

106. Anpo, M., Nomura, T., Kitao, T., Giamello, E., Murphy, D., Che, M., and Fox, M. A., *Res. Chem. Intermed.* **15,** 225 (1991).

107. Strome, D. H., and Klier, K., *J. Phys. Chem.* **84,** 981 (1980).

108. Strome, D. H., and Klier, K., *in* "Adsorption and Catalysis on Oxide Surfaces," (M. Che, and G. C. Bond, Eds.), p. 41. Elsevier, Amsterdam, 1985.

109. Serpone, N., and Pelizzetti, E., (Eds.), "Photocatalysis." Wiley, New York, 1989.

110. Hiramoto, M., Hashimoto, K., and Sakata, T., *Chem. Phys. Lett.* **133,** 440 (1987).

111. Hashimoto, K., Hiramoto, M., Lever, A. B. P., and Sakata, T., *J. Phys. Chem.,* **92,** 1016 (1988).

112. Grätzel, M., (Ed.), "Energy Resources through Photochemistry and Catalysis. Academic Press, New York, 1983.

113. Kuczynski, J., and Thomas, J. K., *J. Phys. Chem.* **89,** 2720 (1985).

114. Gopidas, K. R., Bohorquez, M., and Kamat, P. V., *J. Phys. Chem.* **94,** 6435 (1990).

115. Anpo, M., *J. Surf. Sci. Jpn.* **11,** 39 (1990).

116. Anpo, M., Suzuki, T., Yamada, Y., and Che, M., *Proc. 9th Int. Congr. Catal. (Calgary),* **4,** 1513 (1988).

117. Anpo, M., Tanahashi, I., and Kubokawa, Y., *J. Phys. Chem.* **86,** 1 (1982).

118. Anpo, M., *in* "Photochemical Conversion and Storage of Solar Energy." (E. Pelizzetti, and M. Schiavello, Eds.), p. 307. Kluwer, Dordrecht, 1991.

119. Anpo, M., Sunamoto, M., Fujii, T., Patterson, H., and Che, M., *Res. Chem. Intermed.* **11,** 245 (1989).

120. Anpo, M., Suzuki, T., Yamada, Y., Otsuji, Y., Giamello, E., and Che, M., *in* "New Developments in Selective Oxidation," (G. Centi, and F. Trifiro, Eds.), p. 683. Elsevier, Amsterdam, 1990.

121. Went, G., T., Oyama, T., and Bell, A. T., *J. Phys. Chem.* **94,** 4240 (1990).

122. Cristiani, C., Forzatti, P., and Busca, G. J., *J. Catal.* **116,** 586 (1989).

123. Herzberg, G., "Molecular Spectra and Molecular Structure I, Spectra of Diatomic Molecules," Van Nostrand, New York, 1966.

124. Henderson, J. R., Muramoto, M., and Willett, R. A., *J. Chem. Phys.* **41,** 580 (1964).

125. Patterson, H. H., Cheng, J., Despres, S., Sunamoto, M., and Anpo, M., *J. Phys. Chem.* **95,** 8813 (1991).

126. Kobayashi, H., Yamaguchi, M., Tanaka, T., and Yoshida, S., *J. Chem. Soc. Faraday Trans. I* **81,** 1513 (1985).

127. Anpo, M., Kondo, M., Coluccia, S., Louis, C., and Che, M., *J. Am. Chem. Soc.* **111,** 8791 (1989).

128. Anpo, M., Tanahashi, I., and Kubokawa, Y., *J. Chem. Soc. Faraday Trans. I* **78,** 2121 (1982).

129. Anpo, M., and Kubokawa, Y., *J. Phys. Chem.* **88,** 5556 (1984).

130. Hazenkamp, M. F., on Luminescence of metal ions at oxide surfaces and in zeolites, PhD thesis, 1992.

131. Nag, N. K., and Massoth, F. E., *J. Catal.* **124,** 127 (1990).

132. Kazanskyi, V. B., *Kinet. Katal.* **24,** 1338 (1983).

133. Louis, C., Che, M., and Anpo, M., *Res. Chem. Intermed.* **15,** 81 (1991).

134. Anpo, M., Kondo, M., Kubokawa, Y., Louis, C., Che, M., and Coluccia, S., *Chem. Express* **2,** 61 (1987).

135. Anpo, M., Kondo, M., Kubokawa, Y., Louis, C., Che, M., and Coluccia, S., *J. Lumin.* **40/41,** 829 (1988).

136. Anpo, M., Suzuki, T., Tanaka, F., Yamashita, S., and Kubokawa, Y., *J. Phys. Chem.* **88,** 5778 (1984).

137. Fujii, T., Kuno, K., Suzuki, S., Anpo, M., and Kubokawa, Y., *Physics* **6,** 544 (1985).

138. Ono, T., Anpo, M., and Kubokawa, Y., *J. Phys. Chem.* **90,** 4780 (1986).

139. Anpo, M., Mihara, K., and Kubokawa, Y., *J. Catal.* **97,** 272 (1986).

140. Anpo, M., Kondo, M., Kubokawa, Y., Louis, C., and Che, M., *J. Chem. Soc. Faraday Trans. I* **84,** 2771 (1988).

141. Kazansky, V. B., Pershin, A. N., and Shelimov, B. N., *Proc. 7th Int. Congr. Catal. (Tokyo),* **B,** 1210 (1980).

142. Shelimov, N., Pershin, A. N., and Kazansky, V. B., *J. Catal.* **81,** 426 (1980).

143. Shelimov, B. N., Elev, I. V., and Kazansky, V. B., *J. Catal.* **98,** 70 (1986).

144. Iwasawa, Y., Nakano, Y., and Ogasawara, S., *J. Chem. Soc. Faraday Trans. I* **74,** 2968 (1978).

145. Iwasawa, Y., and Ogasawara, S., *J. Chem. Soc. Faraday Trans. I* **75,** 1465 (1979).

146. Iwasawa, Y., and Ogasawara, S., *Bull. Chem. Soc. Jpn.* **53,** 3709 (1980).

147. Iwasawa, Y., Sato, Y., and Kuroda, H., *J. Catal.* **82,** 289 (1983).

148. Forzatti, P., Attardo, C., and Trifiro, F., *J. Chem. Soc. Chem. Commun.,* 747 (1979).

149. Mahay, A., Kaliaguine, S., and Roberge, P. C., *Can. J. Chem.* **60,** 2719 (1982).

150. Hazenkamp, M. F., and Blasse, G., *J. Phys. Chem.* **96,** 3442 (1992).

151. Hazenkamp, M. F., Strijbosch, A. W. P. M., and Blasse, G., *J. Solid State Chem.* **97,** 115 (1992).

152. Blasse, G., Dirksen, G. J., Hazenkanp, M. F., Verbaere, A., and Oyetola, S., *Eur. J. Solid State Inorg. Chem.* **26,** 497 (1991).

153. Ono, T., Anpo, M., and Kubokawa, Y., *Chem. Express* **3,** 181 (1986).

154. Lammers, M. J. J., and Blasse, G., *Phys. Status Solidi* **A63,** 157 (1981).

155. Amberg, M., Gunter, J. R., Schmalle, H., and Blasse, G., *J. Solid State Chem.* **77,** 162 (1988).

156. Blasse, G., and Schipper, W. J., *Phys. Status Solidi* **A25,** 163 (1974).

157. van Oosterhout, A. B., *Phys. Status Solidi* **A41,** 607 (1977).

158. Anpo, M., Suzuki, T., Kubokawa, Y., Fuhii, T., Kuno, K., and Suzuki, S., *Chem. Express* **1,** 41 (1986).

159. Vuurman, M. A., Wachs, I. E., and Hirt, A. M., *J. Phys. Chem.* **95,** 9928 (1991).

160. Blasse, G., and de Haart, L. G. J., *Mater. Chem. Phys.* **14,** 481 (1985).

161. Blasse, G., *J. Chem. Phys.* **48,** 3108 (1968).

162. Verhaar, H. C. G., Donker, H., Dirksen, G. J., Lammers, M. J. J., Blasse, G., Torardi, C. C., and Brixner, L. H., *J. Solid State Chem.* **60,** 20 (1985).

163. Blasse, G., and Bril, A., *J. Lumin.* **3,** 109 (1970).

164. Blasse, G., and Dirksen, G. J., *Inorg. Chim. Acta* **157,** 141 (1989).

165. Dalhoeven, G. A. M., and Blasse, G., *Chem. Phys. Lett.* **76,** 27 (1980).

166. McDaniel, M. P., and Martin, S. J., *J. Phys. Chem.* **95,** 3289 (1991).

167. Kim, D. S., Tatibouët, J., and Wachs, I. E., *J. Catal.* **136,** 209 (1992).

168. Anpo, M., Aikawa, N., Kubokawa, Y., Che, M., Louis, C., and Giamello, E., *J. Phys. Chem.* **89,** 5017 (1985).

169. Tanaka, T., Yoshida, H., Nakatsuka, K., Funabiki, T., and Yoshida, S., *J. Chem. Soc. Faraday Trans.* **88,** 2297 (1992).

170. Moon, S. C., Yamashita, H., and Anpo, M., *in* "Acid-Base Catalysis II," (H. Hattori, M. Misono, and Y. Ono, Eds.), p. 479. Kodansha, Tokyo, 1994.

171. Moon, S. C., Hieida, T., Yamashita, H., and Anpo, M., *Chem. Lett.,* 447 (1995).

172. Anpo, M., Nomura, T., Kitao, T., Giamello, E., Che, M., and Fox, M. A., *Chem. Lett.,* 889 (1991).

173. Anpo, M., Nomura, T., Kitao, T., Giamello, E., Murphy, D., Che, M., and Fox, M. A., *Res. Chem. Intermed.* **15,** 225 (1991).

174. Giamello, E., Murphy, D., Magnacca, G., Morterra, C., Shioya, Y., Nomura, T., and Anpo, M., *J. Catal.* **136,** 510 (1992).

175. Anpo, M., Nomura, T., Shioya, Y., Che, M., Murphy, D., and Giamello, E., *Proc. 10th Int. Congr. Catal. (Budapest)* **C,** 2155 (1992).

176. Negishi, N., Matsuoka, M., Yamashita, H., and Anpo, M., *J. Phys. Chem.* **97,** 5211 (1993).

177. Yamashita, H., Matsuoka, M., Tsuji, K., Shioya, Y., Giamello, E., Che, M., and Anpo, M., *in* "Science and Technology in Catalysis," p. 227. Kodansha, Tokyo, 1995.

178. Anpo, M., *Solar Energy Mater. Solar Cells* **38,** 221 (1995).

179. Anpo, M., Matsuoka, M., Shioya, Y., Yamashita, H., Giamello, E., Morterra, C., Che, M., Patterson, H. H., Webber, S., Ouellette, S., and Fox, M. A., *J. Phys. Chem.* **98,** 5744 (1994).

180. Yamashita, H., Matsuoka, M., Tsuji, K., Shioya, Y., Anpo, M., and Che, M., *J. Phys. Chem.* **100,** 397 (1996).

181. Shin, K. S., Barrie, J. D., Dunn, B., and Zink, J. I., *J. Am. Chem. Soc.* **112,** 5701 (1990).

182. Jacobs, P. A., Wilde, W., Schoonheydt, R. A., Uytterhoeven, J. B., and Beyer, H., *J. Chem. Soc. Faraday Trans. I* **72,** 1221 (1976).

183. Beer, R., Calzaferri, G., and Kamber, I., *J. Chem. Soc. Chem. Commun.,* 1489 (1991).

184. Dedecek, J., and Wichterlova, B., *J. Phys. Chem.* **98,** 5721 (1994).

185. Wichterlova, B., Dedecek, J., and Vondrova, A., *J. Phys. Chem.* **99,** 1065 (1995).

186. Matsuoka, M., Matsuda, E., Tsuji, K., Yamashita, H., and Anpo, M., *J. Mol. Catal. A Chem.* **107,** 399 (1996).

187. Matsuoka, M., Matsuda, E., Tsuji, K., Yamashita, H., and Anpo, M., *Chem. Lett.,* 375 (1995).

188. Anpo, M., Matsuoka, M., and Yamashita, H., *Catal. Today* **35,** 177 (1997).

189. Anpo, M., Matsuoka, M., Mishima, H., and Yamashita, H., *Res. Chem. Intermed.* **23,** 197 (1997).

190. Arakawa, T., Tanaka, T., Adachi, G., and Shiokawa, J., *J. Chem. Soc. Chem. Commun.,* 453 (1979).

191. Hazenkamp, M. F., and Blasse, G., *Chem. Mater.* **2,** 105 (1990).

192. Nedeljkovic, J. M., Nenadovic, M. T., Micic, O. I., and Nozik, A. J., *J. Phys. Chem.* **90,** 12 (1986).

193. Henglein, A., *Chem. Rev.* **89,** 1861 (1989).

194. Wang, Y., and Herron, N., *J. Phys. Chem.* **92,** 4988 (1988).
195. Bahnemann, D. W., Kormann, C., and Hoffmann, M. R., *J. Phys. Chem.* **91,** 3789 (1987).
196. Kuczynski, J., and Thomas, J. K., *J. Phys. Chem.* **89,** 2720 (1985).
197. Rossetti, R., and Brus, L., *J. Phys. Chem.* **86,** 4470 (1982).
198. Stramel, R. D., Nakamura, T., and Thomas, J. K., *J. Chem. Soc. Faraday Trans. 1* **84,** 1287 (1988).
199. Henglein, A., *Berlin Bunsenges. Phys. Chem.* **86,** 301 (1982).
200. Anpo, M., Nakaya, H., Kodama, S., Kubokawa, Y., Domen, K., and Onishi, T., *J. Phys. Chem.* **90,** 1633 (1986).
201. Anpo, M., Kawamura, T., Kodama, S., Maruya, K., and Onishi, T., *J. Phys. Chem.* **92,** 438 (1988).
202. Anpo, M., Zhang, S. G., and Yamashita, H., *Proc. 11th Int. Congr. Catal. (Baltimore),* **B,** 941 (1996).
203. Zhang, S. G., Ichihashi, Y., Yamashita, H., Tatsumi, T., and Anpo, M., *Chem. Lett.,* 895 (1996).
204. Yamashita, H., Ichihashi, Y., Anpo, M., Hashimoto, M., Louis, C., and Che, M., *J. Phys. Chem.* **100,** 16041(1996).
205. Anpo, M., *Soc. Italiana Fisica Chimi,* IL Nuovo Cimento, **19,** 1641 (1997).
206. Marchese, L., Gianotti, E., Maschmeyer, T., Martra, G., Coluccia, S., and Thomas, J. M., *Soc. Italiana Fisica Chimi,* IL Nuovo Cimenta, **19,** 1707 (1997).
207. Kubokawa, Y., Anpo, M., and Yun, C., *Proc. 5th Soviet–Japan Sem. Catal. Tashkent,* 3 (1979).
208. Anpo, M., Sunamoto, M., Fujii, T., Che, M., and Patterson, H. H., *Res. Chem. Intermed.* **11,** 320 (1989).
209. Anpo, M., Suzuki, T., and Yamada, Y., *Chem. Express* **4,** 447 (1989).
210. Anpo, M., Louis, C., and Che, M., *Chem. Express* **5,** 109 (1990).
211. Anpo, M., Tanahashi, I., and Kubokawa, Y., *J. Chem. Soc., Faraday Trans. 1* **78,** 2121 (1982).
212. Anpo, M., Aikawa, N., Kubokawa, Y., Che, M., and Louis, C., *J. Phys. Chem.* **89,** 5689 (1985).
213. Anpo, M., Aikawa, N., and Kubokawa, Y., *J. Chem. Soc. Chem. Commun.,* 644 (1984).
214. Anpo, M., *Catal. Surv. Jpn.* **1,** 169 (1997).
215. Lin, C. T., Hsu, W. L., Yang, C. L., and El-Sayed, M. A., *J. Phys. Chem.* **91,** 4556 (1987).
216. Fenin, V. A., Shvets, V. A., and Kazansky, V. B., *Kinet. Katal.* **19,** 1041 (1979).
217. Okamoto, S., Nishiguchi, H., and Anpo, M., *Chem. Lett.,* 1009 (1992).
218. Hsu, W. L., and Lin, C. T., *J. Phys. Chem.* **94,** 3780 (1990).
219. Shen, Y. F., Suib, S. L., Deeba, M., and Koermer, G. S., *J. Catal.* **146,** 483 (1994).
220. Anpo, M., Nishiguchi, H., and Fujii, T., *Res. Chem. Intermed.* **13,** 73 (1990).
221. Nishiguchi, H., Yamashita, H., and Anpo, M., *Shokubai (Catalysts & Catalysis)* **35,** 146 (1993).
222. Fujii, T., Murayama, K., Negishi, N., Anpo, M., Winder, E. J., Neu, D. N., and Ellis, A. B., *Bull. Chem. Soc. Jpn.* **66,** 739 (1993).
223. Anpo, M., Tomonari, M., and Fox, M. A., *J. Phys. Chem.* **93,** 7300 (1989).
224. Anpo, M., Chiba, K., Tomonari, M., Coluccia, S., Che, M., and Fox, M. A., *Bull. Chem. Soc. Jpn.* **64,** 543 (1991).
225. Meyer, G. J., Lisensky, G. C., and Ellis, A. B., *J. Am. Chem. Soc.* **110,** 4914 (1988).
226. Meyer, G. J., Lisensky, G. C., and Ellis, A. B., *J. Am. Chem. Soc.* **111,** 5146 (1989).
227. Meyer, G. J., Luebker, E. R. M., Lisensky, G. C., and Ellis, A. B., *in* "Photochemistry on Solid Surfaces," (M. Anpo and T. Matsuura, Eds.), p. 388. Elsevier, Amsterdam, 1989.
228. Anpo, M., and Kubokawa, Y., *J. Phys. Chem.* **88,** 5556 (1984).

229. Nicoll, F. H., *J. Opt. Soc. Am.* **38,** 817 (1948).
230. Heiland, G., Mollow, E., and Stockman, F., *Solid State Phys.* **8,** 191 (1959).
231. Kaminori, O., Yamaguchi, N., and Sato, K., *Bunseki Kagaku* **16,** 1050 (1967).
232. Kokes, R., *J. Phys. Chem.* **66,** 99 (1962).
233. Oster, G., and Yamamoto, M., *J. Appl. Phys.* **37,** 823 (1966).
234. Narayana, M., Narashimhan, C. S., and Kevan, L., *J. Catal.* **79,** 237 (1983).
235. Jonson, B., Rebenstorf, B., Larsson, L. L., and Anderson, S. L. T., *J. Chem. Soc. Faraday Trans. I* **84,** 1897 (1988).
236. Haung, Y., *J. Am. Chem. Soc.* **95,** 6636 (1973).
237. Centi, G., Perathoner, S., Shioya, Y., and Anpo, M., *Res. Chem. Intermed.* **17,** 125 (1992).
238. Boshh, H., and Jansen, F., *Catal. Today* **2,** 369 (1988).
239. Kondo, J., Domen, K., Maruya, K. I., and Onishi, T., *J. Chem. Soc. Faraday Trans.* **86,** 3021 (1990).
240. Anpo, M., Yamada, Y., Doi, T., Matsuura, I., Coluccia, S., Zecchina, A., and Che, M., *in* "Structure and Reactivity of Surfaces," (C. Morterra, A. Zecchina, and G. Costa, Eds.), p. 1. Elsevier, Amsterdam, 1989.
241. Franck, R., Negishi, N., Moon, S. C., Anpo, M., Tatibouët, J. M., and Che, M., *Topics Catal.* **3,** 115 (1996).
242. Anpo, M., Shima, T., Kodama, S., and Kubokawa, Y., *J. Phys. Chem.* **91,** 4305 (1987).
243. Anpo, M., Aikawa, N., Kodama, S., and Kubokawa, Y., *J. Phys. Chem.* **88,** 2569 (1984).
244. Anpo, M., Aikawa, N., and Kubokawa, Y., *J. Phys. Chem.* **88,** 3998 (1984).
245. Anpo, M., Matsumoto, A., and Kodama, S., *J. Chem. Soc. Chem. Commun.,* 1038 (1987).
246. Henglein, A., *J. Phys. Chem.* **86,** 2291 (1982).
247. Matsumura, M., Furukawa, S., Saho, Y., and Tsubomura, H., *J. Phys. Chem.* **89,** 1327 (1985).
248. Brus, L., *J. Phys. Chem.* **90,** 2555 (1986).
249. Chestnoy, N., Harris, T. D., Hull, R., and Brus, L. E., *J. Phys. Chem.* **90,** 3393 (1986).
250. Anpo, M., and Chiba, K., *J. Mol. Catal.* **74,** 207 (1992).
251. Anpo, M., Yamashita, H., Ichihashi, Y., Fujii, Y., and Honda, M., *J. Phys. Chem.* **101,** 2632 (1997).
252. Anpo, M., and Yamashita, H., *in* "Heterogeneous Photocatalysis," (M. Schiavello, Ed.), p. 133. Wiley, London, 1997.
253. Zhang, S. G., Fujii, Y., Yamashita, H., Koyano, K., Tatsumi, T., and Anpo, M., *Chem. Lett.* 659 (1997).
254. Anderson, M. W., Occelli, M. L., and Suib, S. I., *J. Catal.* **118,** 31 (1989).
255. Vol'kenshtein, F. F., Sokolov, V. A., Popov, D. P., and Styrov, V. V., *Kinet. Katal.* **15,** 1250 (1974).
256. Aras, V. M., Breysse, M., Caludel, B., Faure, L., and Guenin, M., *J. Chem. Soc. Faraday Trans. I* **73,** 1039 (1977).
257. Lunsford, J. H., personal communication (1988).
258. Zamaraev, K. I., Khramov, M. I., and Parmon, V. N., *Catal. Rev. Sci. Eng.* **36,** 617 (1994).
259. Anpo, M., Ichihashi, Y., Takeuchi, M., and Yamashita, H., *Res. Chem. Intermed.,* **24,** 143 (1998).
260. Anpo, M., Ichihashi, Y., Takeuchi, M., and Yamashita, H., *Sci. Technol. Catal., (Kodansha),* 305 (1999).

The Surface Science Approach Toward Understanding Automotive Exhaust Conversion Catalysis at the Atomic Level

BERNARD E. NIEUWENHUYS

*Leiden Institute of Chemistry, Gorlaeus Laboratories Leiden University
2300 RA Leiden, The Netherlands*

This review focuses on the reactions and catalysts used for control of emissions of exhaust gases for gasoline-fueled automobiles with emphasis on fundamental understanding of the surface processes. Attention is paid to three-way catalysts, which simultaneously enhance the conversion of CO, hydrocarbons, and nitrogen oxides. The mechanisms of the CO oxidation and nitrogen oxide reactions, the specific differences in behavior of Pt, Pd, and Rh, the effect of alloy formation, and the role of ceria used as additive in three-way catalysts are discussed. Results of surface science studies are compared with results reported for supported catalysts. For CO oxidation, there is excellent agreement between results obtained for single crystal surfaces and supported catalysts. The kinetics of the reactions on pure metals can be understood on the basis of the kinetics parameters obtained from single-crystal studies. For the NO reduction reactions, there is qualitative agreement between results obtained with single-crystal and supported catalysts. The major effects of alloy formation can be understood on the basis of the surface composition.

Abbreviations: AES, Auger electron spectroscopy; E_{des}, activation energy for desorption; ESDIAD, electron-stimulated desorption ion angular dependence; hc, hydrocarbon; HREELS, high-resolution electron energy loss spectroscopy; L–H, Langmuir–Hinshelwood; LEED, low-energy electron diffraction; RAIRS, reflection absorption infrared spectroscopy; SERS, surface-enhanced Raman spectroscopy; SIMS, secondary ion mass spectrometry; TDS, thermal desorption spectroscopy, spectrum, or spectra; TPR(S), temperature-programmed reaction (spectroscopy); TOF, turnover frequency; TWC, three-way catalyst or catalysis; uhv, ultrahigh vacuum; UPS, UV photoelectron spectroscopy; XPS, X-ray photoelectron spectroscopy.

259

I. Introduction

Severe emission limits for motor vehicles were introduced, first in the United States and later in many other countries, starting in the mid-1960s. Meeting the increasingly stringent emission requirements in subsequent years forced the installation in motor vehicles of progressively more advanced emission control devices. The focal point of emission control is the catalytic converter, in which the desired chemical reactions occur. The pollutants carbon monoxide and unburned hydrocarbons (hc) are converted by oxidation into the desired CO_2 and water:

$$2CO + O_2 \rightarrow 2CO_2 \tag{1}$$

$$aC_nH_m + bO_2 \rightarrow cCO_2 + dH_2O, \tag{2}$$

and the pollutant nitrogen oxides are reduced to dinitrogen and CO_2 or H_2O:

$$2NO + 2CO \rightarrow N_2 + 2CO_2$$

$$2NO + H_2 \rightarrow N_2 + H_2O$$

$$aNO + bhc \rightarrow cN_2O + dCO_2 + eH_2O. \tag{3}$$

The first generation of automotive catalysts catalyzed only the two oxidation reactions (Eqs. 1 and 2) and, hence, these are called *two-way* catalysts. At that time much knowledge was available concerning oxidation catalysis, and it was relatively easy to develop the first automotive catalyst. Both noble metal and transition metal oxides were available to oxidize the CO and hc. Many transition metal oxides exhibit good catalytic properties for the oxidation reactions. However, the noble metals have superior properties with significantly lower light-off temperatures (the light-off temperature is the minimum temperature needed to start the reaction) and with better resistance to poisoning by sulfur compounds and other compounds present in the fuel or lubricant. The noble metals Pt and/or Pd were the active components of these catalysts. Small noble metal particles were supported by γ-alumina with high surface area and good thermal stability.

A new generation of catalysts was needed when the exhaust emission regulations in the United States changed, requiring lower levels for NO_x. After many years of research, it was determined that the most effective technology is a three-way catalyst (TWC) system, which simultaneously accelerates the NO_x reduction reactions and the oxidation of CO and hydrocarbons. The principle of TWC is illustrated in Fig. 1, which shows the change of the conversion of the three major pollutants as a function of the air/fuel ratio. The air/fuel ratio is usually expressed as the weight

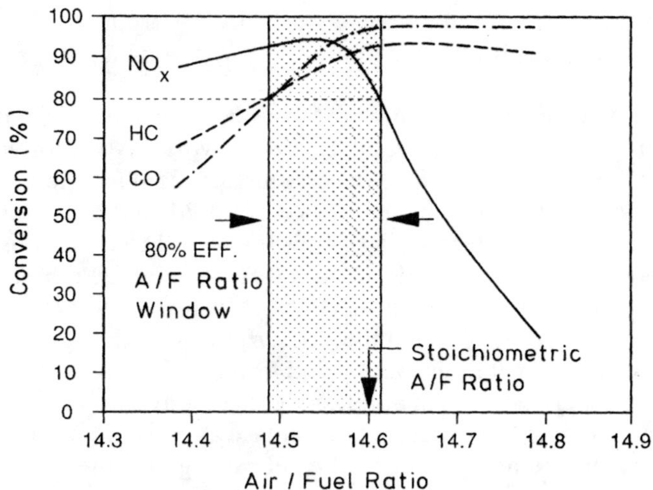

FIG. 1.   Efficiency of a three-way catalyst for the conversions of CO, NO, and hydrocarbons at various A/F ratios.

of air per weight of fuel. It is easily convertable into the equivalence ratio (λ), the value of which is unity at the stoichiometric point when all the fuel is completely converted to CO_2 and water. Obviously, the oxidation reactions (Eqs. 1 and 2) are favored under conditions of excess air ($\lambda > 1$), whereas complete reduction of nitrogen oxides requires reducing conditions ($\lambda < 1$). Fortunately, simultaneous conversion of NO, CO, and hc can occur in a narrow window around the stoichiometric composition $\lambda = 1$, as demonstrated in Fig. 1. Precise control of the air/fuel ratio is required to achieve high conversions of CO, hc, and NO. The composition of the air/fuel mixture introduced to the engine is electronically controlled by a feedback system with an oxygen sensor to monitor the oxygen concentration in the exhaust.

Most of the current converters consist of a flow-through ceramic monolith with its channel walls covered with a high-surface-area γ-Al_2O_3 layer (the washcoat) which contains the active catalyst particles. The monolith is composed of cordierite, a mineral with the composition $2MgO \cdot 2Al_2O_3 \cdot 5SiO_2$. The chemical composition of a modern TWC is quite complex. In addition to alumina, the washcoat contains up to 30 wt% base metal oxide additives, added for many purposes. The most common additives are ceria and lanthana; in many formulations BaO and ZrO_2 are used, and in some converters NiO is present. The major active constituents of the washcoat are the noble metals Pt, Pd, and Rh (typically 1–3 g). Most of the TWC systems in use today are still based on Pt and Rh in a ratio of about 10:1.

In the past 5 years, some of the Pt–Rh TWC formulations have been replaced by Pd-based TWC (the "Pd-only" TWC and Pd–Rh-based TWC), particularly in the United States.

In the past 30 years, automotive catalysis has become the greatest novel application of heterogeneous catalysis in the world. Automotive catalysis is a major application for the precious metals, as illustrated in Table I, which shows that the relative importance of Pt, Pd, and Rh for automotive catalysis changed considerably in the past decade. The current usage of Pd for automotive catalysis exceeds that of Pt. However, the Pd-based converters contain more precious metal than the Pt-based converters.

Many reviews (1–4) give detailed descriptions of the fundamentals of automotive exhaust catalysis. Shelef and Graham (3) gave a broad view of the unique properties of Rh in automotive three-way catalysis. This chapter focuses on fundamental processes taking place at the catalyst surface. Attention is given to the adsorption of the relevant gases, mechanisms of the relevant reactions, specific differences in the surface properties of the various noble metals, effects of alloy formation, and the chemistry of the additives, in particular ceria. The chemistry of the reactions shown in Eqs. (1)–(3) is understood in considerable detail as a result of recent studies using models of the TWC and surface science techniques. Relevant literature of diesel and lean exhaust gas control is also briefly discussed.

It is not the aim of this chapter to review all the available data. The data mentioned and the papers cited here are presented only to the extent that the findings illuminate the discussion.

TABLE I

Western World Demand for Pt, Pd, and Rh[a]

Metal	Year	10^{-6} × total demand, g	10^{-6} × automobile catalyst demand (gross), g	Percentage of total for automobile catalysts
Pt	1987	96	39	40
	1992	119	48	40
	1996	152	57	37
Pd	1987	98	8.4	8
	1992	121	15.2	13
	1996	212	72	34
Rh	1987	9.7	7.0	72
	1992	11.8	9.5	81
	1996	14.8	12.9	87

[a] Data from Platinum 1997 (Johnson Matthey).

II. Adsorption

A. Adsorption of CO, Hydrogen, Oxygen, and NO

The adsorption of diatomic molecules on a metal surface may be considered as competition between molecular and dissociative adsorption:

$$AB_{gas} \rightarrow AB_{ads} \rightarrow A_{ads} + B_{ads}. \tag{4}$$

Dissociative adsorption can occur when the bonds formed between the fragments of the dissociated molecule and the surface are much stronger than the bonds between the molecule and the surface. For example, molecular hydrogen, oxygen, or nitrogen is only weakly adsorbed on transition metals. Oxygen, hydrogen, and nitrogen adatoms, on the other hand, are strongly bound on many metal surfaces. Therefore, dissociative adsorption is often thermodynamically possible, as discussed later. Molecular adsorption of CO and of NO is relatively strong on many metal surfaces. These adsorbates may undergo both dissociative and molecular adsorption on the same surface depending on the experimental conditions.

It is often observed that molecular adsorption prevails at lower temperatures and that dissociative adsorption occurs at higher temperature. This pattern may be caused by kinetics; the activation energy for dissociative adsorption is too high for dissociation at lower temperatures. There may also be a thermodynamic reason. If the number of surface sites at which adsorption can take place is equal for molecular and dissociative adsorption, the surface can accommodate twice as many molecules in the molecular state as it can in the dissociated state. Hence, molecular adsorption will prevail if the heat of dissociative adsorption is not much greater than the heat of molecular adsorption. The entropy change for adsorption is negative and, consequently, at sufficiently high temperatures desorption will occur. In the case considered previously, only half the number of molecules can be adsorbed in the dissociative state as can be adsorbed in the molecular state. As a result, the entropy of the system will be lower for molecular adsorption and dissociation can occur at higher temperatures.

Dissociative adsorption requires a cluster of several free and adjacent metal atoms on the surface. Therefore, often dissociation occurs when the surface coverage is low and molecular adsorption occurs above a certain coverage, provided that both dissociative and molecular adsorption can occur under the experimental conditions considered.

It is assumed that a diatomic molecule adsorbed parallel to the surface is in a transition state for dissociation. The more favorable adsorption complex for molecules such as CO on a group VIIII metal surface is that in which the molecular axis is bonded perpendicular to the surface (or slightly tilted). It has been demonstrated by electron-stimulated desorption ion angular distribution (ESDIAD) that the molecular axis vibrates around the surface normal and that its amplitude increases with increasing temperatures (5) (Fig. 2). The temperature at which dissociation occurs is probably reached when the vibrational amplitude becomes sufficiently large so that a bond between the O atom and the metal surface is formed, resulting in dissociation.

Many of the adsorption data presented in this review are based on thermal desorption spectroscopy (TDS) measurements. This technique yields in a simple way information about the number of binding states and their bond strengths with the surface. For illustration, TD spectra are shown for molecules adsorbed on a Rh filament.

TDS measured for polycrystalline surfaces are often much more complicated than those representing single-crystal surfaces. A polycrystalline surface exhibits all the adsorption sites of the crystal faces of which it is composed. Since these sites are simultaneously present, a TDS represents an average of the spectra of the different surface sites weighted according to the relative concentrations of these sites.

1. Adsorption of Hydrogen

Hydrogen is known to be dissociatively adsorbed on transition metal surfaces, and the initial sticking probability varies from about 0.05 to unity at $T = 100$–300 K, depending on the metal and its surface structure. For example, the initial sticking probability for hydrogen on a Ni(111) surface is only 0.05; on the stepped Ni-8(111) $\times$ (100) surface it is 0.2, and on a Ni(110) surface it is essentially unity (6). In general, the sticking probability is significantly smaller on flat surfaces than on surfaces with high densities of steps, surface defects, etc. It appears that hydrogen prefers to be adsorbed on sites on which it is coordinated with many metal atoms. On (111) surfaces

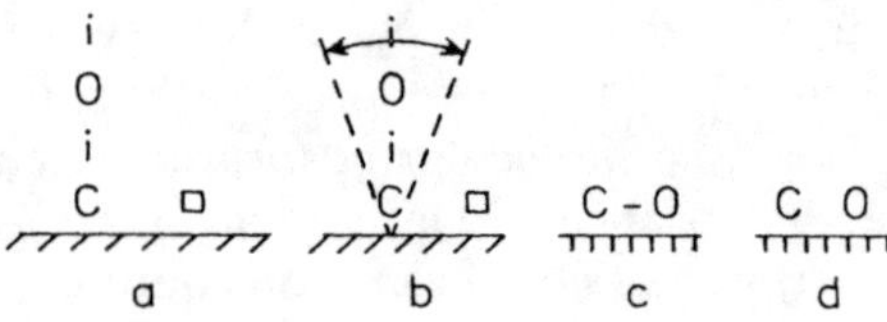

FIG. 2. Molecules such as CO are bonded on most of the group VIII metal surfaces with the molecular axis perpendicular to the surface (a) or slightly tilted. The amplitude of the vibration around the surface normal increases with increasing temperature (b). At a certain temperature, an O–metal bond is formed (c) and dissociation occurs immediately (d).

of the face-centered cubic (FCC) lattice, for example, hydrogen is adsorbed at low coverage on threefold sites. Hydrogen desorbs from the group VIII metals at about 350–400 K; its initial heat of adsorption lies in the range 80–100 kJ mol^{-1} and is the same within 20% on similar surfaces of these metals. More details are given in Nieuwenhuys (7).

2. *Adsorption of Oxygen*

The interaction of oxygen with Pt surfaces has been studied in detail with modern surface analytical techniques (7). Since the TDS of oxygen on most of the other metals of group VIII are qualitatively similar to those for oxygen on Pt, one can expect that essentially the same species occurs on all these metals. At 100 K, oxygen is adsorbed on Pt(111) with a sticking probability of about unity in a molecularly adsorbed state (γ state) with a low heat of adsorption (37 kJ/mol) (8). Electron energy loss spectroscopy (EELS) combined with UV photoelectron spectroscopy (UPS) indicates an essentially single O–O bond with a significant electron transfer from the valence band of Pt into orbitals derived from the π antibonding oxygen levels and with the O–O bond axis parallel to the surface. Heating the surface to temperature above 170 K results in the formation of adsorbed atomic oxygen (β–O) and desorption of some oxygen (γ). The heat of adsorption of β–O decreases rapidly with increasing coverage from 500 to 160 kJ/mol at $\theta = 0.8\ \theta_{max}$. On Pt(111) surfaces, oxygen is adsorbed at 300 K with a low sticking probability ($<$0.1).

A third state is sometimes observed upon heating Pt(111) in the presence of oxygen in the 900 K temperature range (8). This subsurface oxygen, which desorbs at temperatures higher than 1200 K, has a very low reactivity toward CO and hydrogen. There is controversy concerning the nature of this state (7). Many authors believe that this state is a so-called surface oxide (Fig. 3), whereas others attribute its appearance

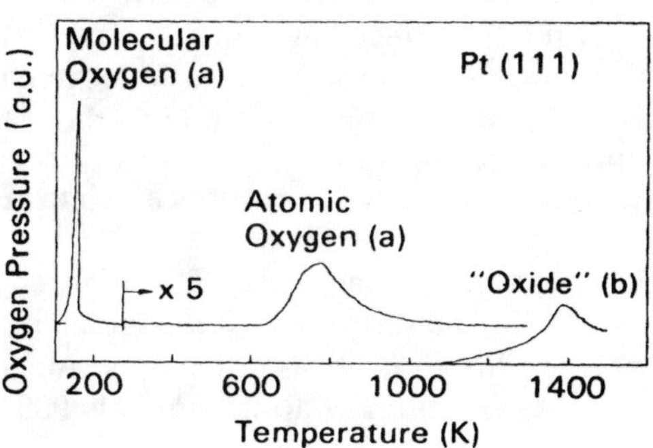

FIG. 3. TDS of oxygen from Pt(111) [reproduced with permission from Gland *et al.* (8)].

to oxide-forming bulk impurities such as Si or Ca and its segregation to the surface during oxygen exposure at high temperature. On polycrystalline surfaces and on open surfaces, the formation of an oxide layer at higher temperature is quite common as is demonstrated by field emission microscopy (7).

In contrast to adsorbed CO, the density of O adatoms at saturation is not determined by the size of the adsorbate; more open overlayer structures are usually formed. The kinetics of oxygen adsorption at ambient temperature varies greatly from metal to metal and from plane to plane. For example, oxygen is adsorbed with a sticking probability near unity on Ni surfaces, whereas on Pt(111) the sticking probability is <0.1. The initial heat of adsorption of β–O is approximately 250 kJ/mol on the group VIII metals, corresponding to a metal–O bond strength of about 370 kJ/mol. Because of the high desorption temperature, the appreciable decrease of the heat of adsorption Q with increasing coverage, and the incorporation of oxygen into the bulk, it is not easy to find a correlation of $Q_{initial}$ with the position of the metal in the periodic table. A careful examination of reliable data suggests that the heat of adsorption increases in the following order: Pt (230 kJ/mol) = Pd < Ir < Rh < Ru < Ni (330 kJ/mol). The variation in the metal–O bond strength is about 55 kJ/mol (or 15%). The order in the heat of adsorption is similar to that in the heat of formation of the oxides. The tendency of the metal to form surface or bulk oxides also increases in this sequence. It can be concluded that, in contrast to hydrogen on group VIII metals, there are significant differences in the metal–O bond strength on these metals.

3. Adsorption of CO

CO on metals is the most extensively studied adsorption system (9). I highlight some of the features that are important for this discussion. TDS of CO on group VIII metal surfaces in general show various peaks, which can be classified into three main groups as shown schematically for CO on polycrystalline Rh in Fig. 4 and corresponding to the following:

1. Low-T γ states desorbing at temperatures much lower then room temperature,
2. α states with a peak maximum in its TDS between 350 and 500 K.
3. Sometimes, depending on the choice of the metal, its surface structure, and the experimental conditions (such as the temperature), additional states (β) are observed with a maximum of approximately 900 K. These β states arise from CO dissociation followed by recombination of C and O at the desorption temperature.

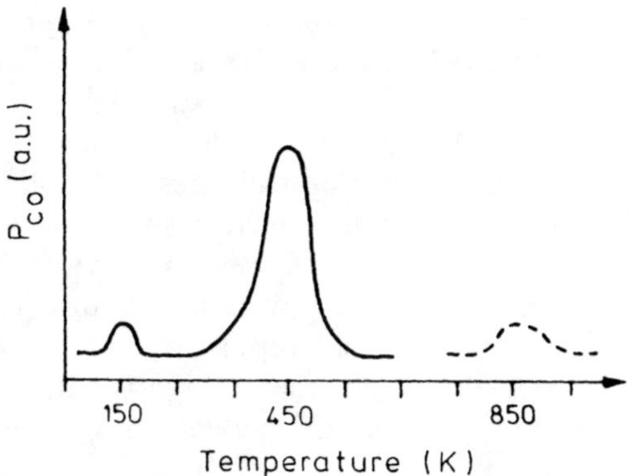

FIG. 4.   TDS of CO from Rh [reproduced with permission from Nieuwenhuys (7)].

At room temperature, adsorption normally takes place in the α state corresponding to molecularly adsorbed CO. Usually two or more α peaks are observed, depending on the surface structure and the CO coverage. These peaks can be rationalized on the basis of distinct species adsorbed on different sites and in terms of lateral interactions.

A model of the α–CO metal bond was first proposed by Blyholder (10). According to this model, the bond is formed by electron transfer from the highest filled molecular orbital of CO (5σ, which is essentially a nonbonding orbital with respect to the C–O bond) to unoccupied metal orbitals, implying a donation of electrons to the metal, accompanied by back-donation of electrons from occupied d orbitals into the lowest unfilled CO molecular orbital (the strongly antibonding 2π orbital). The relative contributions of the σ and the π bonding to the metal–CO bond strength and to the net charge transfer have been discussed in many papers. Sufficient evidence, based on theoretical calculations (11, 12) and inverse photoemission studies (13), indicates that for Pd and Ni metal, π donation to CO is much more important for the CO–metal bond than is the CO-to-metal σ donation. However, the CO–Pt bond has a significant σ-donation component (13). Furthermore, work function measurements indicate that the net electron transfer from metal to CO is much larger for Rh, Pd, and Ni than for Pt (14).

Experimental and theoretical results support the fact that the CO molecules are adsorbed on the densely packed surfaces of Pt, Pd, Rh, and Ir with the C–O axis normal to the surface and with the carbon atom directed to the surface. Experimental evidence is derived from angular resolved UPS, ESDIAD, ion scattering, low-energy electron diffraction (LEED) intensity analysis, and EELS.

LEED, IR, or EELS and TDS have been applied for CO on many single-crystal surfaces. The general features of these are very similar. First, a simple overlayer structure is formed up to medium coverage. For example, on the FCC(111) surfaces, $(\sqrt{3} \times \sqrt{3})R30°$ structures are observed which are fully developed at $\theta = 1/3$. The molecules are then located on most of the densely packed surfaces in identical sites. At higher coverage, the overlayer unit cell is compressed, new surface structures are observed, and a fraction of the CO molecules are forced to move into other sites. At the highest coverage, a close-packed overlayer is formed which is largely determined by the CO–CO mutual repulsion and not by the substrate sites. The maximum density of adsorbed CO molecules is about 1×10^{15} molecules/cm^2. The sites on which CO molecules are bound are reflected in the stretching vibration frequencies v of the CO bond. In the pioneering work of Eischens et al. (15), bands at frequencies below 2000 cm^{-1} were assigned to CO acting as a bridging ligand, $M_2C{=}O$, and bands between 2000 and 2100 cm^{-1} were attributed to linear CO species such as $M{=}C{=}O$. Numerous results have shown that Eischens et al's. interpretation of the bands was essentially correct, and more comprehensive correlations of v with bonding sites have been suggested (16).

A substantial increase in v occurs with increasing surface coverage. Both experimental observations, based on the combined use of TDS, LEED, and EELS/IR, and theoretical predictions suggest that the energy difference of CO bonded in on-top positions or in multifold sites is quite small. The only metal on which bridged (or multifold) sites are the most favorable positions for CO is Pd. On Ir, Rh, and Ru, linearly adsorbed CO is preferred. Also, on Pt(111) on-top positions are preferred, but the energy differences with the other positions are small, resulting in a high degree of disorder at room temperature, as observed by LEED (7). The differences in heat of adsorption on the group VIII metals for the various single-crystal surfaces of a metal are about 20 kJ/mol (or 20%) (7).

Dissociative adsorption of CO has been found on a variety of transition metal surfaces. Broden et al. (17) and Nieuwenhuys (14) correlated the tendency for CO, N_2, and NO to dissociate with the position of the transition metal in the periodic table; the tendency for dissociation increases the further to the left the metal appears in the table, and it decreases from 3d to 5d metals. Furthermore, the borderline for dissociative or molecular adsorption moves to the right in the sequence CO, N_2, NO to O_2, being the same order as the bond strength in the free molecules. There is sufficient evidence for the proposed correlation. For example, W and Mo surfaces dissociate CO easily at room temperature; dissociative adsorption has not been reported for Pt, Ir, and Pd(111) surfaces, and CO dissociation has been reported to occur on Ni, Co, and Ru at elevated temperatures. Benzinger (18) suggested that the state of adsorption (molecular or dissociative) is determined by thermodynamic criteria. In his paper, the heat of dissociative adsorption was estimated from the heats of formation of metal carbides, nitrides, and oxides. In this analysis, a similar correlation as that in Refs. (*14*) and (*17*) relating dissociative adsorption with position in the periodic table was found. The surface structure may have an additional influence on the dissociation. In general, close-packed surfaces are the least reactive and rough surfaces or surfaces with steps or kinks are the most reactive for dissociation. The effect of surface structure on dissociation can be attributed to the variation of, e.g., metal–carbon bond strength with surface structure.

For our purposes, it is relevant to conclude that CO adsorption is predominantly molecular on Pt, Ir, Pd, and Rh and that desorption occurs at about 500 K. Controversy exists in the literature regarding CO dissociation on Rh surfaces (*19*). On supported Rh catalysts and Rh clusters, CO dissociation has been observed. Some authors also reported CO dissociation on certain single-crystal surfaces, whereas other studies indicate no or insignificant dissociation on clean Rh surfaces (*19*).

4. *Adsorption of NO*

Figure 5 presents a TD spectrum for NO adsorbed on a Rh filament at 80 K (*20*), showing desorption peaks of NO, N_2, and O_2. The heat of

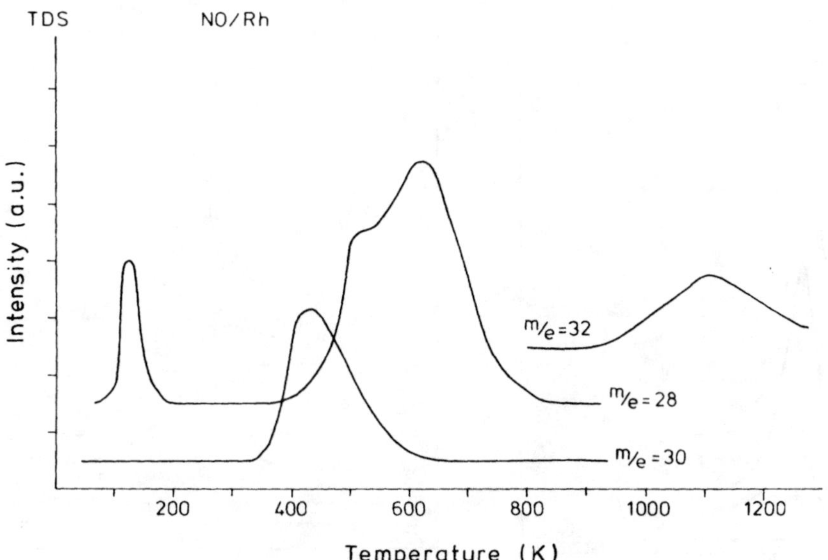

FIG. 5. TDS of NO, N_2, and O_2 following adsorption of NO on a Rh filament at 80 K [reproduced with permission from Hendrickx and Nieuwenhuys (*20*)].

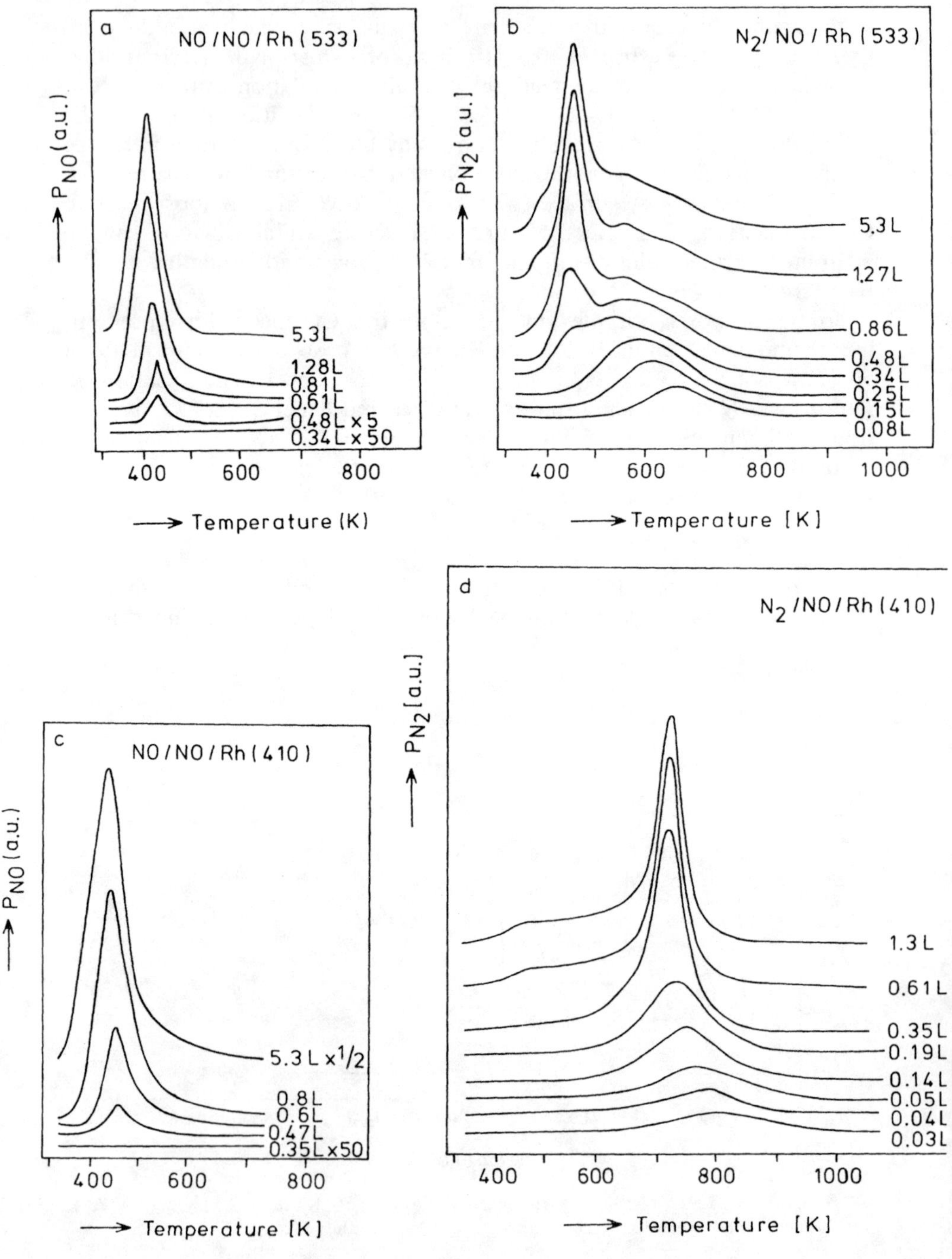
a
NO / NO / Rh (533)
PNO (a.u.)
5.3 L
1.28 L
0.81 L
0.61 L
0.48 L × 5
0.34 L × 50
400 600 800
Temperature (K)

b
N2 / NO / Rh (533)
PN2 (a.u.)
5.3 L
1.27 L
0.86 L
0.48 L
0.34 L
0.25 L
0.15 L
0.08 L
400 600 800 1000
Temperature [K]

c
NO / NO / Rh (410)
PNO (a.u.)
5.3 L × 1/2
0.8 L
0.6 L
0.47 L
0.35 L × 50
400 600 800
Temperature [K]

d
N2 / NO / Rh (410)
PN2 (a.u.)
1.3 L
0.61 L
0.35 L
0.19 L
0.14 L
0.05 L
0.04 L
0.03 L
400 600 800 1000
Temperature [K]

adsorption of molecularly adsorbed NO is on the order of 110 kJ/mol^{-1} for the group VIII metals. The energy barrier for dissociation is much lower for NO than for CO due to the presence of one electron in the antibonding 2π orbital. The temperature at which dissociation starts depends strongly on the NO coverage and is very sensitive to the surface structure. For example, when a NO-covered Rh(331) surface is heated, dissociation starts even at 240 K for a low NO coverage (21). However, a temperature of 350 K is required for a surface that starts fully covered. These results show that some of the molecularly adsorbed NO must desorb in order to create free sites needed for dissociation. Van Hardeveld (22) investigated the adsorption of NO on Rh(111) with secondary ion mass spectrometry (SIMS); a technique that measures the composition of the adsorbate layer and TDS. At temperatures higher than 250 K, a substantial fraction of the NO dissociates during adsorption. At temperature higher than 350 K NO$_{ads}$ is absent; the dissociation is complete. The decrease of the dissociation rate with increasing coverage can be explained by assuming that an ensemble of three or four empty sites is required for NO dissociation (23). The authors argued that a more likely explanation is that lateral interactions between NO$_{ads}$ and O$_{ads}$ or N$_{ads}$ cause the decrease in dissociation rate with increasing coverage.

EELS spectra (7) for NO on Ru(001) at temperatures higher than 180 K are consistent with a model in which NO is adsorbed in bridged or threefold sites (state α_1, ν_{NO} = 1500 cm^{-1}), and in on-top sites (state α_2, ν_{NO} = 1800 cm^{-1}). At low coverage, NO is adsorbed in the more strongly bound α_1 state and, at coverages >1/3 of saturation, the on-top sites become populated. Exposure at 280 K causes dissociation of α_1 NO. ESDIAD studies suggest that the N–O bond is predominantly normal to the surface in both molecular states. Ku et al. (For ref. see (7)) concluded from LEED observations following NO dissociation on a Ru(1010) surface that the N and O adatoms form separate islands on this surface. Many investigations indicate that molecularly adsorbed NO is highly inclined at low coverage, whereas at high coverage, this adsorbate has a perpendicular orientation. Examples are NO on Ni(111), Ni(100), Pt(100), and Rh(100) (24). The tilted species is usually considered to be a precursor for dissociation.

Somorjai (25), Gorte et al. (26), and Masel (27) found that NO dissociation is neglegible on perfect Pt(111) surfaces. Surface defects bind NO

FIG. 6.   TDS of NO on Rh(533) and (410) surfaces. (a) NO from (533), (b) N$_2$ from (533), (c) NO from (410), and (d) N$_2$ from (410), Heating rate = 20 Ks^{-1} [reproduced with permission from Janssen et al. (28)].

more tightly on Pt and induce dissociation. Gorte *et al.* reported that the Pt(100) surface has stronger bonding and much higher decomposition activity than the Pt(110) and (111) surfaces. The major tightly bound state on Pt(100) dissociates to yield 50% N_2 and O_2, whereas the fraction decomposed on the Pt(111) surface is <2%. The stepped Pt(410) surface with (100) terraces is much more reactive in NO dissociation than the (100) surface (27). On Pd(111), the rate of NO dissociation becomes significant at temperatures higher than 500 K (7). In addition to desorption of NO, N_2, and O_2, production of N_2O has been observed on many Pd and Pt surfaces.

On Rh surfaces, only desorption of N_2 and NO is observed in the TDS. Figure 6 shows TDS for NO on two stepped Rh surfaces, Rh(533) [structure 4(111) × (100)] and Rh(410) [structure 4(100) × (100)]. The adsorption of NO and the effect of preadsorbed O and N on the adsorption of NO have been studied on these surfaces and on many other Rh surfaces by Janssen *et al.* (28). The N atoms are markedly more strongly bound on (100) terraces than on (111) terraces. The presence of steps does not affect the thermal stability of N_{ads} on Rh. At higher NO exposures, repulsive N–N and N–O interactions lower the thermal stability of N_{ads}. Following saturation NO exposure, N_2 desorbs in a single state from (100) terraces at 750 K. From (111) terraces several desorption states of nitrogen appear at temperaturs between 450 and 700 K. An important observation is that recombination of N_{ads} and O_{ads} to give NO is more favorable than the 2 $N_{ads} \rightarrow N_2$ reaction when Rh with precovered O_{ads} is exposed to NO.

The literature suggests that on Pd and Pt surfaces adsorption of NO is predominantly molecular at room temperature. Dissociation is observed at elevated temperatures, and the surface structure has a significant influence on the extent of dissociation. In contrast to the adsorption of CO, dissociation of NO can easily be detected on Ir, Rh, and Ru surfaces at room temperature.

B. ADSORPTION OF N_2, H_2O, CO_2, N_2O, AND NH_3

Various techniques have been used to investigate the interaction of N_2, H_2O, CO_2, and NH_3 with clean metal surfaces. The adsorption of N_2O has not been studied in detail.

The interaction of CO_2 with group VIII metals was reviewed by Solymosi (29). At 80 K, the adsorption on a Rh field emitter exhibits an interesting crystal face dependency (30). In addition to molecular adsorption, dissociation occurs at an appreciable rate on the stepped surfaces around (111) and (100) at temperatures higher than 220 K. The field electron microscopy (FEM) patterns suggest that the surface structure of Rh has a striking influence on the ability of the metal to dissociate the CO_2 molecule. From

Pt, CO_2 desorbs at a much lower temperature (90 K, corresponding to a heat of adsorption of about 20 kJ/mol), and dissociation was not detected (*30*).

No evidence was found for significant dissociation of NH_3 on Pt(111) and Ru(001) under low-pressure conditions. However, decomposition of NH_3 occurs at temperatures higher than 600 K in a NH_3 atmosphere (*31*). The effect of the surface structure on the dissociation is large. For Pt the order in reactivity is (210) > (110) > (100) > (111) (*31*).

In the temperature range 80–400 K, nitrogen is only molecularly adsorbed on the group VIII metal surfaces, with the exception of Fe and Co, on which slow dissociation is observed on some faces at temperatures higher than 300 K (*7*). The nature of the adsorption bond is similar to that of the CO–metal bond. The heat of adsorption, however, is significantly lower for N_2 (40 kJ/mol) than for CO (130 kJ/mol) on the same surfaces, as demonstrated by the data in Fig. 7. It can be concluded from the available data that N_2, H_2O, N_2O, and CO_2 molecules leave the surface immediately upon formation.

C. COADSORPTION

To understand the catalytic properties of the various group VIII metal surfaces, the interactions among neighboring adsorbed species and their

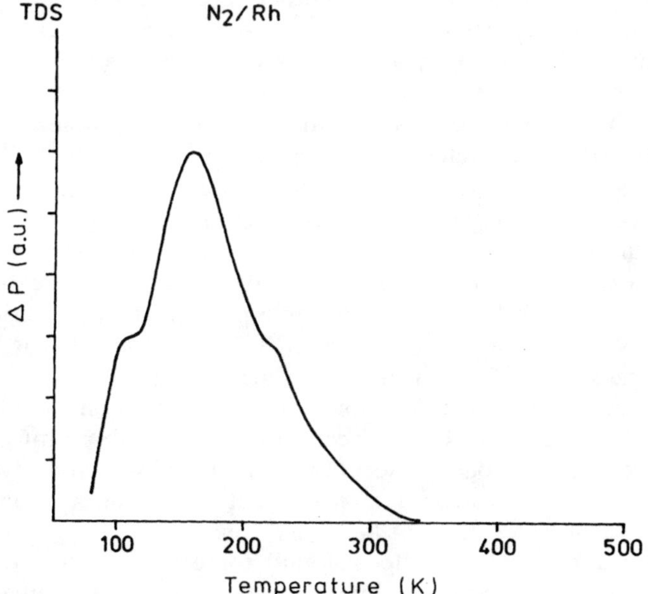

FIG. 7.   TDS of nitrogen from a Rh filament [reproduced with permission from Nieuwenhuys (*7*)].

effects on the surface processes must be known since the surfaces may be covered with different kinds of adsorbates during the reaction.

Conrad *et al.* (*32*) studied in detail the mutual interaction of coadsorbed O and CO on a Pd(111) surface. Some of their relevant results are summarized here. Oxygen adsorption is inhibited by preadsorbed CO. At coverages below $\theta_{CO} = 1/3$, LEED patterns show that O and CO form separate surface domains. However, the behavior is different when O is preadsorbed. CO can be adsorbed on the Pd(111) surface covered with O which is less densely packed than a saturated CO layer. The O adatom islands are then suppressed to domains of a $(\sqrt{3} \times \sqrt{3})R30°$ structure ($\theta = 1/3$), with a much larger local coverage than can be reached with O alone, which orders in a (2×2) structure ($\theta = 0.25$). After further exposure, the LEED patterns suggest the formation of mixed phases of O_{ads} and CO_{ads} (with local coverages of $\theta_O = \theta_{CO} = 0.5$) which are embedded in CO domains. When these mixed phases are present, CO_2 is produced even at temperature lower than room temperature. Coadsorption studies of other noble metal surfaces are consistent with this scenario; preadsorbed CO inhibits the dissociative adsorption of oxygen, whereas CO is adsorbed on a surface covered with O.

Lambert and Comrie (*33*) concluded from TDS on Pt(111) and (110) that gaseous CO displaces molecularly adsorbed NO (α_1) and also causes the conversion of this state to the other more weakly bound state α_2. The α_1 state was reported to be the only one which is reactive with CO. Displacement of molecularly adsorbed NO by gaseous CO has also been observed on Pd surfaces. Thiel *et al.* (*34*) used EELS to observe directly the competition of CO adsorption with one of the two molecular states of NO on Ru(001). The stretching frequencies suggest that (i) CO is linearly bonded to a single Ru atom, whereas NO prefers the bridging or multifold sites, and (ii) at higher coverages, linear sites are also populated. The EEL spectra indicate that conversion from linear to multifold bonded NO occurs during adsorption of CO. Similar observations have been made for CO–NO mixtures on Pd(110) (*35*). For each single-phase system, the bridge site is preferred. When the two are coadsorbed, however, a mixed phase is formed in which the stable CO site is linear. In the presence of a CO–NO atmosphere, Pt surfaces favor CO adsorption, as expected on the basis of the heats of adsorption (*36*). Rh surfaces, however, adsorb both CO and NO (*36*). Blocking of hydrogen adsorption by NO is observed on Rh surfaces (*21*). The occurrence of strong (repulsive or attractive) interactions between coadsorbed H and NO has not been reported.

In conclusion, the main effects found for coadsorption are blocking, island formation, displacement by the component with the higher heat of adsorption, and site conversion. Strong attractive interactions between the two adsorbates do not take place.

III. CO–O$_2$ Reaction

Usually, temperatures of 300–600 K are required for the CO–O$_2$ reaction on the active metal catalysts Pt, Pd, Ir, Rh, Ru, and Ni. CO oxidation catalysts are used mainly for automotive emission control. In addition, air-purification devices for respiratory protection and CO gas sensors commonly employ CO oxidation catalysts. A novel application is in sealed CO$_2$ lasers used for weather monitoring and in other remote-sensing applications. For the development of long-life sealed CO$_2$ lasers, novel catalysts must be developed which are able to oxidize CO at temperatures near ambient. Active catalysts for this purpose are based on gold and a transition metal oxide, e.g., Au–manganese oxide (*37*). The mechanisms of the processes responsible for the high activity are not fully understood. In my opinion, the reaction may occur at the Au–MnO$_x$ interface with CO adsorbed on Au reacting with O on MnO$_x$. An alternative explanation is that O atoms spill over from MnO$_x$ to gold. Gold is not very active in O–O bond scission. However, O adatoms are stable on gold at room temperature. In the presence of CO, the O adatom can easily react with CO due to the low Au–O and Au–CO bond strengths.

Two mechanisms have been suggested for reactions such as oxidation of CO:

1. An Eley–Rideal mechanism in which one of the components reacts in the adsorbed state with the other molecule in the gaseous or in a physically adsorbed state, e.g.,

$$O_{ads} + CO_{gas} \rightarrow CO_2. \tag{5}$$

2. Langmuir–Hinshelwood mechanism in which both components react with each other in the adsorbed state, e.g.,

$$O_{ads} + CO_{ads} \rightarrow CO_2. \tag{6}$$

It has been established (*38*) that the dominant mechanism of the reaction is the Langmuir–Hinshelwood (L–H) mechanism:

$$CO \rightarrow CO_{ads} \tag{7}$$

$$O_2 \rightarrow 2O_{ads} \tag{8}$$

$$O_{ads} + CO_{ads} \rightarrow CO_2. \tag{9}$$

The alternative Eley–Rideal mechanism fails to explain all the experimental results. Direct evidence in favor of the L–H mechanism has been obtained from molecular beam experiments. Figure 8 (*39*) shows a typical example of the variation in rate of CO_2 production with time for a CO beam impinging on an O adlayer on Pd(111) at a constant temperature of 374 K and also the variation of the surface concentrations of O_{ads} and CO_{ads}. The reaction exhibits an induction period and reaches its maximum rate after a few seconds. Obviously, the data of Fig. 8 confirm that both reactants must be adsorbed, in accordance with the L–H mechanism.

The reaction may thus be written as:

$$dp_{CO2}/dt = k\theta_{CO}\theta_O = v\theta_{CO}\theta_O \exp[-E_a(\theta_{CO}, \theta_O)/RT], \qquad (10)$$

where θ_{CO} and θ_O are, respectively, the surface coverages by CO_{ads} and O_{ads}; E_a is the activation energy for the reaction, which may vary with θ_{CO} and θ_O; k is the rate constant and v the frequency factor. Both for Pd(111) and for Pt(111), E_a is equal to 101 kJ/mol at low-coverage of CO and O (*40*). A much lower value of the activation energy (50 kJ) was found when a Pt(111) surface saturated with O was exposed to a CO molecular beam. This decrease in E_a was attributed to a decrease in heats of adsorption of CO and O due to repulsive interactions.

At higher O coverages, deviations from Eq. (10) can be expected because

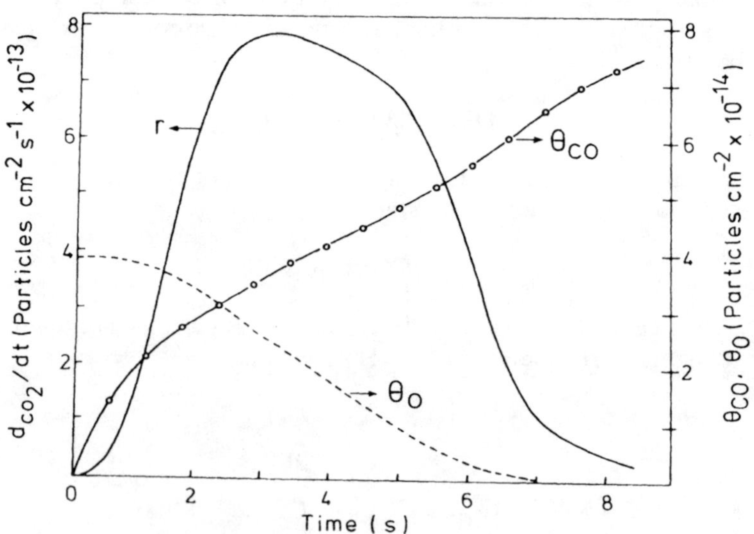

FIG. 8. Rates of CO_2 formation on Pd(111) and O and CO coverages as a function of time at 374 K [reproduced with permission from Engel and Ertl (*39*)].

the O atoms form islands on the surface due to attractive interactions (*41*). Sufficient evidence exists that the reaction proceeds at the boundaries of O and CO islands (*42*), under both low- and high-pressure conditions and on Pt single-crystal surfaces as well as supported catalysts (*43*).

Recently, very interesting results were reported by Wintterlin *et al.* (*44*). The reaction was imaged on Pt(111) by a variable-temperature scanning tunneling microscopy, (STM). Small O islands with (2 × 2) structure and, at high CO coverage, CO adatoms were resolved. The results show clear evidence of separate CO and O domains with the reaction taking place at the island boundaries. The activation energy for CO_2 formation found from these microscopic measurements was equal to the value reported by Campbell *et al.* (*40*) using macroscopic measurements.

Figure 9 shows the rate of CO_2 production catalyzed by the (111), (100), (410), and (210) surfaces of a $Pt_{0.25}-Rh_{0.75}$ single crystal as a function of temperature in the presence of a 2:1 mixture of CO and O_2 at a total pressure of 2×10^{-7} mbar (*45*). It shows that the reaction rate increases rapidly between 400 and 500 K to a temperature T_m at which a maximum is reached and beyond which it gradually decreases. The temperature at which the maximum occurs increases with increasing CO pressure. Similar behavior was found for many Pt, Pd, Ir, Rh, and Ru surfaces. X-ray photoelectron spectroscopy (XPS) measurements show that the maximum rate

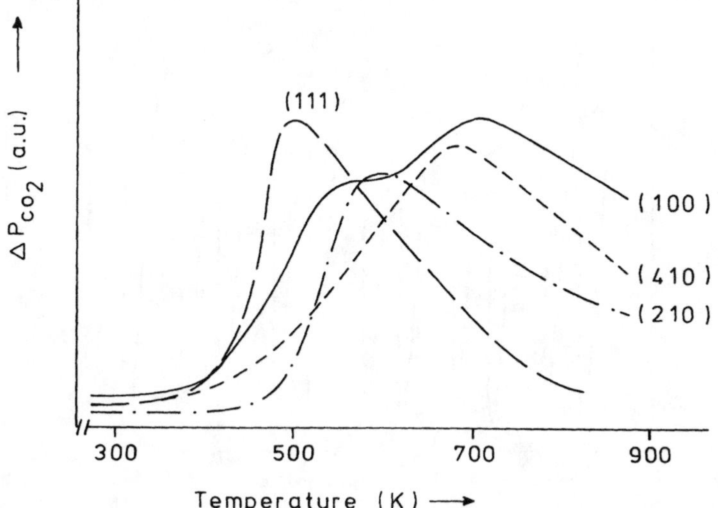

FIG. 9.   Steady-state rates of CO_2 production for the CO + O_2 reaction on the (111), (100), (410), and (210) surfaces of a $Pt_{0.25}-Rh_{0.75}$ single crystal [reproduced with permission from Siera *et al.* (*45*)].

is reached at a temperature at which the CO coverage is already small (7). These and other observations indicate that adsorbed CO acts as an inhibitor for O adsorption and, hence, for the catalytic reaction. The value of T_m is determined by the sticking probability, the enthalpy of adsorption of oxygen, and the enthalpy of adsorption of CO. The relatively small CO inhibition of the reaction on the (111) surface is consistent with the low heat of adsorption of CO on that surface.

Oscillations in the rate of CO_2 production have been observed for many supported metal catalysts and single-crystal surfaces. Similar oscillations have been observed for most of the reactions discussed in this review. An example is shown in Fig. 10 for the NO + H_2 reaction on Pt(100) (46).

Several models have been proposed to explain these oscillatory rates. Sales *et al.* (47) associated the oscillation with a slow and reversible modification of the catalyst surface; slow oxidation and reduction of the metal surface induces transitions between the two branches.

Ertl (38) reported in detail the oscillations on the (100) and (110) surfaces of Pt. The clean (100) and (110) surfaces reconstruct, i.e., the atomic arrangement in the topmost layer is not that of the corresponding bulk plane. However, the reconstruction is lifted by adsorption of CO and NO when

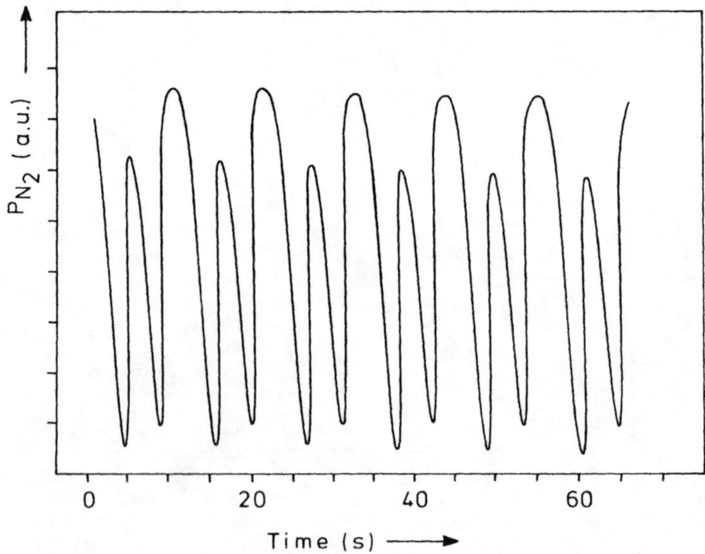

FIG. 10.   Oscillatory behavior of the NO + H_2 reaction on Pt(100) [reproduced with permission from Cobden *et al.* (46)].

the coverage exceeds a certain critical value. The driving force is the higher heat of adsorption of CO on the nonreconstructed surface, which overcompensates the energy required for altering the surface structure. The sticking probability of oxygen is negligibly small on the reconstructed surface, whereas it is much higher on the nonreconstructed surface.

Ertl and coworkers (38) proposed the following model for oscillations: The surface structure changes to the nonreconstructed surface due to CO adsorption. Oxygen can then be adsorbed due to its relatively high sticking probability on this surface. The adsorbed O reacts with adsorbed CO molecules to yield CO_2 that is immediately released into the gas phase. As a result, more sites are created for oxygen adsorption and the reaction rate increases. The CO coverage decreases because of its rapid reaction with O. At the critical CO coverage, the surface structure transforms back into the reconstructed phase. Oxygen is not adsorbed anymore and, hence, the reaction rate decreases. The CO coverage then increases and, beyond the critical coverage, the surface structure transforms again to the nonreconstructed phase. This transformation completes the cycle of the oscillation. Recent studies suggest that rate oscillations under isothermal conditions can also occur on surfaces for which the reconstruction model does not apply. A more general model was proposed involving the key role of vacant sites on the surface required for dissociation of molecules such as NO and O_2 (48). For recent reviews concerning the oscillatory behavior of surface reactions, see Refs. (48, 49).

New regulations in the United States and Europe mandate that automotive emissions must decrease substantially from current levels. As a result, there is a strong incentive to develop improved TWC with better oxidation activity at low temperatures since most of the hydrocarbons and CO are emitted immediately following cold starts of engines. As previously mentioned, the addition of transition metal oxides can have a beneficial effect on the performance of Au catalysts in CO oxidation. Combinations of Pt or Pd with transition metal oxides are also active in CO oxidation at low temperatures (50). Figure 11 shows examples of the reaction over Pt/MO/ SiO_2 catalysts.

Preoxidation does not significantly affect the temperature of 50% conversion ($T_{50\%}$) on Pt/SiO_2. Oxidation increases the $T_{50\%}$ on Pt/CoO$_x$/SiO_2. However, $T_{50\%}$ is still much lower for Pt/MO catalysts than for Pt/SiO_2. Similar effects were found for other CO/O_2 ratios. These experiments were continued for many hours. The $T_{50\%}$ of a Pt/CoO$_x$/SiO_2 catalyst shifts to a value between $T_{50\%}$ for the prereduced catalyst and the $T_{50\%}$ for the preoxidized catalyst, depending on the CO/O_2 ratio. Pt/CoO$_x$/SiO_2 is the most active of the catalysts investigated in our laboratory. At room temperature, CO is oxidized to CO_2. The value of $T_{50\%}$ are 100–200 K lower than those

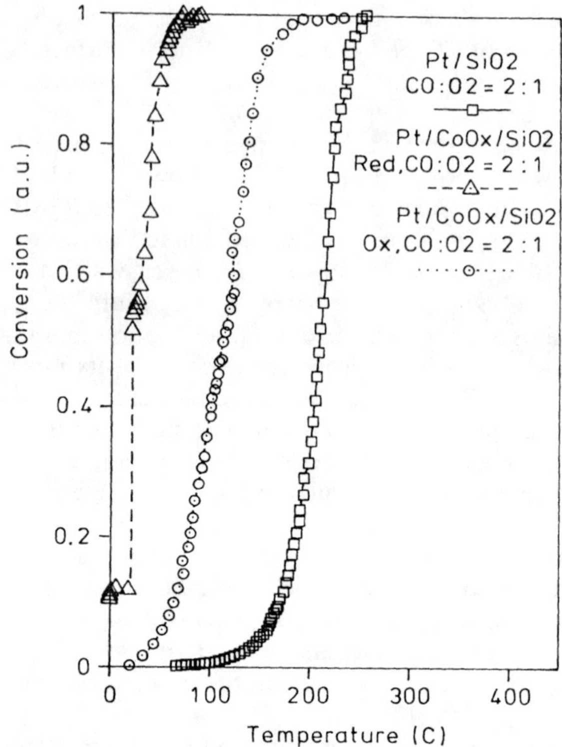

FIG. 11.   CO conversion at various temperatures during CO oxidation with $CO:O_2 = 2:1$ catalyzed by Pt/SiO_2 and by $Pt/CoO_x/SiO_2$ after a reductive and an oxidative pretreatment [reproduced with permission from Mergler *et al.* (*50*)].

for Pt/SiO_2, depending on the CO/O_2 ratio. CoO_x/SiO_2 does not exhibit significant activity at temperatures below 498 K. The Pt/MO catalysts also exhibit improved performance in NO reduction reactions (*51–53*).

IR (*53*) and temporal analysis of products (TAP) (*54*) have been used to investigate the origin of the improved performance of the Pt/MO catalysts in CO oxidation. The TAP experiments shown in Fig. 12 demonstrate that the high activity of $Pt/CoO_x/SiO_2$ in CO oxidation is related to the absence of CO inhibition effects at low temperatures. On the basis of these results it was proposed that CoO_x is the supplier of O, which reacts with CO adsorbed on Pt. It is likely that the reaction takes place at the $Pt–CoO_x$ interface.

An interesting observation is that CO_2 can form on Pt(111) at the low temperature of 160 K upon heating the surface which is covered with

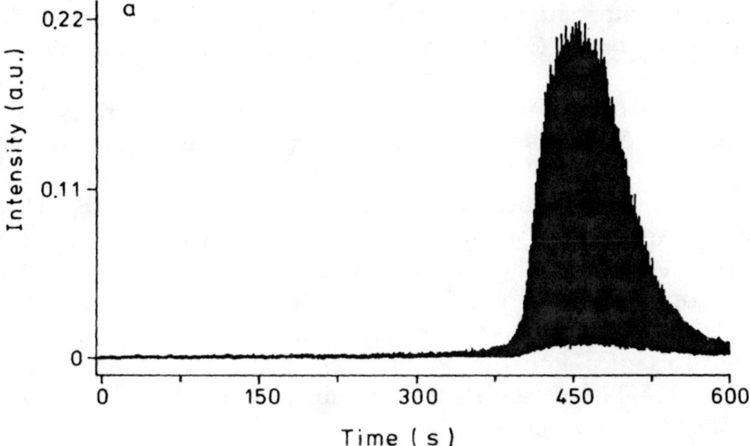

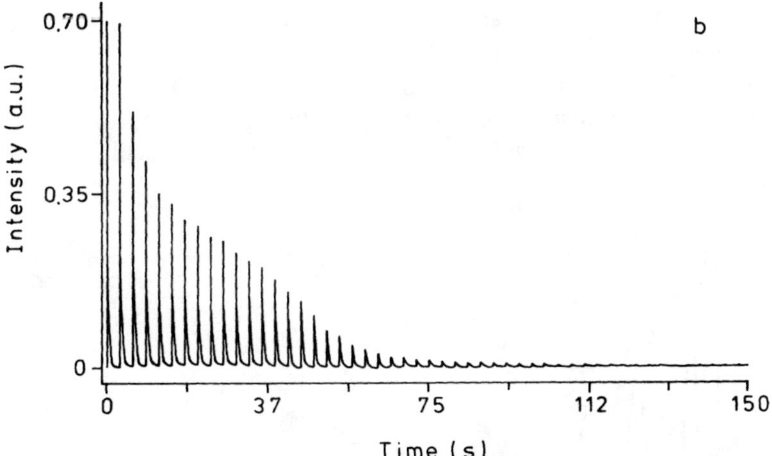

FIG. 12.   Multipulse experiments with Pt/SiO_2 and $Pt/CoO_x/SiO_2$. CO_2 signal intensity: (a) CO-precovered Pt catalyst at 327 K during O_2 pulses; (b) a CO-precovered Pt–CoO_x catalyst at 313 K during O_2 pulses [reproduced with permission from Mergler *et al.* (*54*)].

molecularly adsorbed O_2 and CO (*55*). The authors showed that both O atoms in $O_{2,ads}$ react to give CO_2:

$$O_{2,ads} + 2CO_{ads} \rightarrow 2CO_2. \tag{11}$$

The CO_2 formation temperature coincides with the temperature at which $O_{2,ads}$ dissociates. Therefore, the origin of CO_2 formation at this low temperature may be attributed to a reaction of the recently discovered "hot" O atoms with adsorbed CO. Recent articles have discussed chemical energy which is released during dissociation and subsequent formation of the strong metal–O bonds and is transformed in part into kinetic energy. This excess kinetic energy could cause motion of the adsorbed particles or could induce chemical reaction. STM is an obvious technique to use to characterize the distance that the adsorbed atoms can travel across the surface. This technique was applied by Wintterlin *et al.* (56) to investigate the diffusion distance of adsorbed O atoms when formed on Pt(111) by O_2 dissociation. The authors found evidence for the existence of hot O atoms. However, the O atoms created by dissociation appear in pairs, with an average distance of only two lattice atoms, which is much smaller than that found by the same group (57) for dissociation of O_2 molecules on Al(111). (On Al(111), the distance exceeds 80 Å).

The existence of hot O adatoms was also believed to be responsible for the desorption of CO_2 at 140 K from the Pt(100) hex surface partly covered with CO_{ads} and $O_{2,ads}$. Fadeev *et al.* (58) used TPR and high-resolution EELS (HREELS) to investigate CO oxidation on Pt(100). Some of the results are summarized in Fig. 13.

At 90 K, O_2 is adsorbed both on the reconstructed Pt(100) (hex phase) and on the nonreconstructed Pt(100) (1 × 1) surface as peroxide O_2, with the O–O bond axis parallel to the surface. This molecularly adsorbed

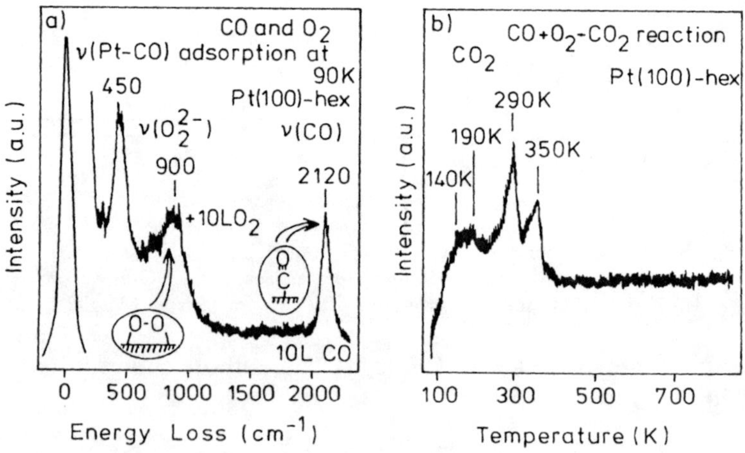

FIG. 13. Low-temperature CO and O_2 adsorption (a) and CO_2 desorption as a result of $CO_{ads} + O_{2,ads}$ reaction (b) on Pt(100)-hex surface [adopted with permission from Fadeev *et al.* (58)].

state desorbs at temperatures near 140 K from the hex phase, without dissociation. However, on the (1×1) surface, partial dissociation occurs simultaneously with dissociation at 160 K. Interaction of O_{ads} with CO_{ads} results in the formation of two CO_2 peaks, at 290 and 350 K. In addition to these CO_2 formation peaks, two low-temperature CO_2 peaks were observed at 140 and 190 K.

It has been reported that the CO_2 molecules produced on Pt surfaces can eject into the gas phase with a translational, vibrational, and rotational energy in excess of that expected from the surface temperature (*59, 60*). The group of King (*61*) applied the technique of single-crystal adsorption microcalorimetry to the investigation of CO oxidation on Pt(110). It was found that when CO is dosed onto a saturated O overlayer, the product CO_2 molecules have an additional 9 ± 17 kJ mol^{-1} energy in excess of that expected for thermally accommodated molecules. However, when O_2 is dosed onto a CO overlayer, the product CO_2 molecules have an excess energy of 52 ± 21 kJ mol^{-1}. It was suggested that these highly excited CO_2 molecules are formed by reaction of CO_{ads} with hot O adatoms produced by the O_2 dissociation process.

Many of the results discussed in this paper were obtained on idealized models of real catalysts, usually single-crystal surfaces investigated at low pressure. The obvious advantage of this approach is that these surfaces can be characterized on the molecular/atomic level by use of surface science techniques. Can we extrapolate these results to the behavior of supported catalysts with high surface areas at elevated pressures? In this context, the two main questions are (i) What is the effect of the pressure gap? and (ii) What is the effect of the structure gap? In many studies, both questions have been addressed because of their general importance to the understanding of catalysis with the aid of idealized models based on the ultrahigh vacuum surface science approach. Fortunately, sufficient information is available for the CO oxidation on both real and model catalysts to discuss these topics in detail for this reaction. In modern surface science equipment, the kinetics of the reaction can be followed in a microcatalytic reactor coupled to (but isolated by a valve from) the ultrahigh vacuum chamber. The pressure gap in the kinetics is then eliminated, and the structure effect can be examined. The chemical and physical state of the model catalyst is analyzed in the ultrahigh vacuum chamber by transfer under vacuum from the reactor to the analysis chamber.

The CO oxidation is usually considered to be a typical example of a structure-insensitive catalytic reaction. For example, the catalytic activity is almost equal on Pt/Al_2O_3 catalysts with widely varying particle sizes at high CO concentrations (*7*). However, the reaction becomes structure sensitive in excess oxygen. The observed temperature dependences of the steady-state rates are almost similar on Pd(111), (100), (110), and (210) and polycrystalline Pd surfaces (*38*). Boudart (*62*) concluded that under similar experimental conditions the rate per Pd atom is equal on small Pd clusters (~5 nm) and Pd(111). Hence, data observed with a number of both single-crystal and supported metal catalysts indicate that the reaction is essentially surface insensitive.

However, Goodman *et al.* (*63, 64*) concluded that the CO oxidation on Pd is affected by the surface structure in a subtle manner. These authors combined kinetic and reflection absorption infrared spectroscopy (RAIRS) studies of CO oxidation on Pd(111) and (100) surfaces in the pressure range up to 10 mbar. RAIRS was used to follow the CO coverage during reaction. In the low-temperature range studied with first-order oxygen and negative first-order CO partial pressure dependencies of the rate, the apparent activation energy equals the heat of CO adsorption. The variation in heat of CO adsorption with coverage differs for Pd(111), (110), and (100). For example, at $\theta_{CO} = 0.50$, the heats of adsorption are 71, 139, and 122 kJ/mol for the (111), (110), and (100) surfaces, respectively. The authors argued that the reported apparent structure insensitivity of the reaction on Pd catalysts may be due to the varying CO coverages of the different surfaces under similar reaction conditions. Clearly, Fig. 9 illustrates that the reaction may be very sensitive to the surface structure under certain experimental conditions. A reactive mixture of CO and O_2 could also modify the catalyst particle shape in such a way that one type of surface structure dominates, leading to structure insensitivity. Another explanation could be that only the activity of one type of site is measured under certain experimental conditions. Results of Ramsier and Yates (*5*) are consistent with this model. They found from temperature programmed reaction experiments that the CO + O_2 reaction takes place at lower temperatures on the terraces than on the steps of the Pt(211) surface. However, at higher temperatures the high diffusion rate of O adatoms over the surface obscured the differences of step and terrace sites.

Detailed information is also available for the CO oxidation reaction on Rh. Peden (*65*), Peden *et al.* (*66*), and Oh *et al.* (*67*) compared the turnover rates of CO oxidation on Rh(111) and Rh(100) surfaces with those obtained on supported Rh catalysts with high surface areas. Figure 14 is a summary of some of the main results obtained for a CO and O_2 pressure of about 10^{-2} bar. Obviously, there is no intrinsic sensitivity of the reaction to the surface structure under these experimental conditions. Furthermore, Fig. 14 demonstrates that the single-crystal surfaces are in fact very good models for the practical catalysts.

Su *et al.* (*68*) investigated the reaction and the nature of CO adsorbed on Pt(111) at pressures up to 1 bar, characterizing the surface with sum

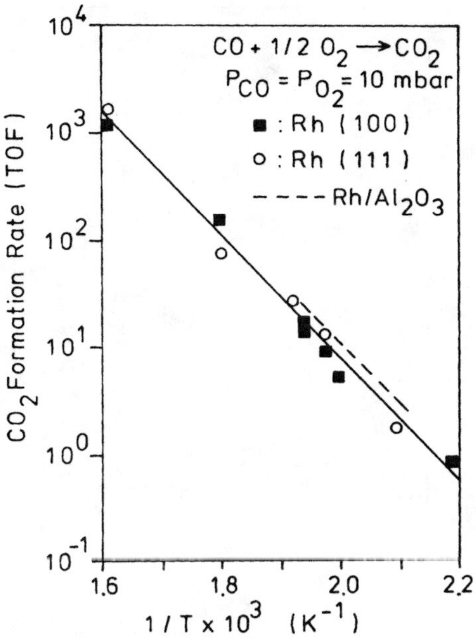

FIG. 14.   Arrhenius plots of CO oxidation reaction rates (turnover frequencies) on Rh showing a comparison between two [(111) and (100)] Rh single-crystal surfaces and supported Rh catalysts. $P_{O_2} = P_{CO} = 10^{-2}$ bar. The activation energy is ca. 126 kJ mol^{-1} in all cases [adopted with permission from Oh et al. (67)].

frequency generation used as vibrational spectroscopy. Upon raising the CO pressure beyond 10 mbar, they observed that the intensity of linear CO (~2100 cm^{-1}) decreased and a new peak at 2045 cm^{-1} developed. At 1 bar the spectrum was dominated by the new feature, which was attributed to metal carbonyl clusters with a CO/Pt ratio of >1. Interestingly, the spectra were completely reversible with variation of the CO pressure. The spectra at 10^{-7} mbar of CO were similar to those obtained after exposing the Pt(111) sample to 1 bar and then back to 10^{-7} mbar (mainly, linearly adsorbed, atop, CO at 2100 cm^{-1}). The CO–O_2 reaction was investigated with a batch reactor with initially 130 mbar O_2 and 50 mbar CO in the temperature range 540–850 K. The results are summarized in Fig. 15.

The vibrational spectra at temperatures lower than 600 K are dominated by CO adsorbed on atop sites. At higher temperatures, atop CO is absent. The three bands observed were assigned to metal carbonyl clusters (2050 cm^{-1}), CO on oxidized Pt (2130 cm^{-1}), and CO_2 (2240 cm^{-1}). The rate of CO oxidation at T > 600 K increased linearly with the intensity of the 2050

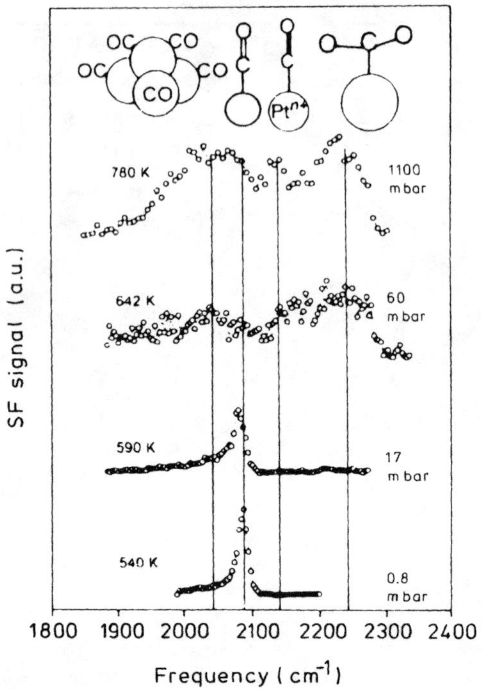

FIG. 15. *In situ* sum frequency generation spectra of CO oxidation on Pt(111) at different temperatures. The corresponding turnover rate (TOR) is also shown. The initial reaction conditions were 130 mbar O_2–50 mbar CO [adopted with permission from Su *et al.* (*68*)].

cm^{-1} peak. It was suggested that the metal carbonyl clusters are the reaction intermediates. The activation energy of the reaction decreases from 176 kJ/mol^{-1} at temperatures below 600 K (with a negative order of reaction in P_{CO}) to 59 $kJ\ mol^{-1}$ at higher temperatures (with a positive reaction order in P_{co}). Su *et al.* (*68*) concluded that the reaction mechanism changed at 600 K. I believe that the observed change in apparent activation energy is in fact consistent with those reported earlier for low-pressure measurements. The E_{act} for the reaction (59 $kJ\ mol^{-1}$) is within experimental accuracy equal to that found by Campbell *et al.* (*40*) for the CO + O_2 reaction on Pt(111) at low pressure, starting with an O-covered surface. STM studies show significant reconstruction of Pt(110) in both CO and O_2, whereas the surface structure of Pt(111) exhibits only minor changes (*68*).

As discussed previously, at relatively low temperatures the surfaces of Pt, Rh, and Pd are predominantly covered by adsorbed CO. At higher temperatures and/or under strongly oxidizing conditions, the metal surface can become oxidized. CO oxidation rates on the surfaces of oxidized Pd and Rh are much lower than those on reduced surfaces. This effect is large for Rh (65). However, Pt showed no significant deactivation even under severely oxidizing conditions. Bowker *et al.* (69) reported that the reaction is sensitive to the surface structure at high temperatures when Rh surfaces are covered by O. Recent STM results demonstrated the high mobility of surface metal atoms as manifested in adsorbate-induced surface reconstructions, in particular by oxygen. CO can also induce changes in surface structure and even in morphology of the metal particles. This effect was demonstrated with the field ion microscope. An original hemispherial Rh tip becomes faceted by reaction with 10^{-4} Pa CO at 420 K. The resulting tip shape is that of a polyhedron with large (100) and (111) facets (70).

IV. Reduction of NO by CO and H_2

In Section II.A, it was shown that the decomposition of NO proceeds on Pt, Pd, and Rh surfaces. However, it is not an efficient process and probably contributes only slightly to the NO removal. Exhaust gas contains the reducing gases CO, H_2, and various hydrocarbons, with H_2 produced by the water gas shift reaction

$$CO + H_2O \leftrightarrows H_2 + CO_2 \tag{12}$$

and by cracking of hydrocarbons. It is assumed in many papers that CO is mainly responsible for the NO conversion (71). The role of hc in conversion of NO has largely been neglected. Recent work by van den Brink and McDonald (72) suggests that CO mainly contributes to the conversion of O_2, whereas hcs have a significant role in NO conversion. Hydrogen, the concentration of which is about one-third that of CO in the exhaust, is a more efficient reducing gas for NO than CO at low temperatures (73).

Unfortunately, due to the presence of various gases in the exhaust, dinitrogen is just one of a variety of reaction products that can be formed in the catalytic converters used in automobiles. The main undesirable reaction products are N_2O and NH_3, which are formed especially under reducing conditions. Some of the processes that have been proposed involve adsorbed isocyanate (NCO) and adsorbed HNCO as intermediates. The formation of NCO_{ads} complexes has been observed on supported metals by IR spectroscopy. It has been established that the presence and the stability of the NCO groups depend on the support (74). It is most likely that the isocyanate species resides mainly on the support and its role is merely that of a spectator.

The reduction of NO_x by hc on noble metal surfaces has not been investigated in detail using the surface science approach. The recent finding that NO reduction by hc significantly contributes to NO conversion is likely to stimulate fundamental investigations of NO–hc interactions on metal surfaces. IR studies by Bamwenda *et al.* (*75*) point to the presence of NCO and CN species on Rh/Al_2O_3 during the NO–propene reaction. Van Hardeveld *et al.* (*76*) investigated the reaction of C_2H_4 and NO on Rh(111) by temperature-programmed reaction spectroscopy (TPR) and SIMS. No indication was found of a direct reaction between the molecular species. The first steps are NO dissociation and ethylene decomposition. The dominant reaction products are, as expected, H_2O, CO_2, and N_2. At low NO–C_2H_4 ratios (<3), significant amounts of H_2, CO, and HCN are produced. Surface cyanide is formed by reaction of C_{ads} with N_{ads} in the absence of O_{ads}. Depending on the availability of hydrogen, CN_{ads} can be hydrogenated to give HCN that desorbs at the reaction temperature. (In earlier papers on automotive catalysis, the possible formation of HCN was proposed, but it has not been detected in the exhaust gas.)

Kobylinski and Taylor (*73*) studied the NO + CO and NO + H_2 reactions on supported noble metals and found that the activity for the first reaction increases in the order Pt < Pd < Rh < Ru and that for the second reaction Ru < Rh < Pt < Pd. The first reaction is slower than the second; only for Ru was the order reversed. Ru is an excellent catalyst for the NO reduction with a minimum of NH_3 production. However, Ru forms volatile oxides under operating conditions resulting in an unacceptable catalyst loss. The most efficient catalyst appears to be Rh (*1*).

Many techniques have been applied to examine the reaction pathways of the NO–CO and NO–H_2 reactions and to elucidate the reaction mechanisms. In this section some of the relevant results are discussed with emphasis on the reaction mechanism. Both the NO–CO and NO–H_2 reactions are discussed here since it is likely that the mechanisms of N_2 and N_2O formation are independent of the type of reducing agent. Note that CO dissociation is not considered to be involved in the mechanism. However, dissociative adsorption of NO into N and O adatoms is an important process on the relevant metals, as discussed in Section II.A. Possible mechanisms of N_2, N_2O, and NH_3 formation can then be evaluated on the basis of the following hypothetical mechanisms involving all the possible elementary steps in which NO_{ads}, N_{ads}, O_{ads}, CO_{ads}, and H_{ads} can participate:

N_2 formation

I. Without direct NO dissociation on the surface, the reduction may proceed via a bimolecular reaction between two molecularly adsorbed NO molecules:

$$2NO \rightarrow N_2' + 2O_{ads} \tag{13a}$$

or via a bimolecular reaction of NO_{ads} with CO_{ads}:

$$2NO_{ads} + 2CO_{ads} \rightarrow N_2 + 2CO_2. \tag{13b}$$

For hydrogen as reducing gas, a similar mechanism has been proposed via

$$NO_{ads} + H_{ads} \rightarrow HNO_{ads}, \tag{13c}$$

followed by $HNO_{ads} + H_{ads} \rightarrow N_{ads} + H_2O$ and combination of $2N_{ads}$ to N_2.

II. Dissociation of NO_{ads} followed by combination of 2N adatoms:

$$2N_{ads} \rightarrow N_2'. \tag{14}$$

III. Dissociation of NO_{ads} followed by reaction of N_{ads} with NO_{ads}:

$$N_{ads} + NO_{ads} \rightarrow N_2' + O_{ads}, \tag{15}$$

with N_2O_{ads} as an intermediate.

The oxygen adatoms formed by I–III may then react with H_{ads} or CO_{ads} as they do in the $H_2 + O_2$ and $CO + O_2$ reactions, although the presence of N_{ads} and NO_{ads} may modify the activation energies for these reactions due to lateral interactions. The N_2, CO_2, and H_2O desorb as soon as they are formed.

IV. A fourth proposed mechanism proceeds as follows:

$$2NO_{ads} + 2\,O_{ads} \rightarrow 2NO_{2,ads} \leftrightarrows N_2O_{4,ads} \tag{16}$$

$$N_2O_{4,ads} \rightarrow N_2' + 4\,O_{ads}. \tag{17}$$

NH_3 formation

V. Hydrogenation of molecularly adsorbed NO with HNO_{ads} as an intermediate:

$$NO_{ads} + 3H_{ads} \rightarrow NH_3 + O_{ads}. \tag{18}$$

VI. Hydrogenation of N_{ads} formed by NO_{ads} dissociation:

$$N_{ads} + H_{ads} \rightarrow NH_{ads} \tag{19}$$

$$NH_{ads} + H_{ads} \rightarrow NH_{2,ads} \tag{20}$$

$$NH_2 + H_{ads} \rightarrow NH_{3,ads} \tag{21}$$

$$NH_{3,ads} \rightarrow NH_{3,gas}'. \tag{22}$$

N_2O formation

VII. Reaction of N_{ads} with NO_{ads}:

$$NO_{ads} + N_{ads} \rightarrow N_2O'. \tag{23}$$

VIII. Via reaction between two molecularly adsorbed NO molecules:

$$2NO_{ads} \rightarrow N_2O' + O_{ads}. \tag{24}$$

IX. Via dissociation of NO_{ads} and reactions between two N_{ads} and O_{ads}:

$$2N_{ads} + O_{ads} \rightarrow N_2O'. \tag{25}$$

The steps listed previously are the possible steps at very low NO conversions. At higher NO conversions, reactions of N_2, NH_3, N_2O, and H_2O with each other and the reactants should also be considered. In particular, decomposition of NH_3 and N_2O and reduction of N_2O may contribute to N_2 formation.

X.

$$N_2O \rightarrow N_2O_{ads} \rightarrow N_2' + O_{ads}. \tag{26}$$

XI.

$$NH_3 \rightarrow NH_{3,ads} \tag{27}$$

$$2NH_{3,ads} \rightarrow N_2 + 6H_{ads}, \tag{28}$$

via $NH_{2,ads}$, NH_{ads}, N_{ads}, and recombination of $2N_{ads}$. In addition to the previous steps, O_2 and NO formation via

XII.

$$2O_{ads} \rightarrow O_2 \tag{29}$$

and

XIII.

$$N_{ads} + O_{ads} \rightarrow NO_{ads} \tag{30}$$

have been documented in the literature.

In the presence of both H_2 and CO as reducing agents the situation will be even more complicated due to competition between CO and H_2 and, possibly, formation of specific products and intermediates such as HCN. This topic was addressed briefly at the beginning of this section.

Ample evidence exists for the major role of process II in N_2 formation on the noble metal surfaces, with the possible exception of Pt(111), as discussed later. Figure 5 shows that TDS of NO from a Rh filament following a saturation exposure to NO exhibits two distinct N_2 peaks: β-N_2, with a maximum at about 625 K, and α-N_2, with a maximum at about 500 K. Similar behavior has been reported for Rh single-crystal surfaces, although the TDS of nitrogen from Rh(100)-like surfaces differ, significantly from those of the Rh(111)-like surfaces (28), as has been illustrated by the data of Fig. 6. At low NO exposures, no NO desorbs, and N_2 is evolved only in the β peak, which exhibits second-order desorption. Following higher NO exposures, α-N_2 desorption is observed at those NO coverages at which NO desorption is also observed, with a peak maximum only slightly lower than the α-N_2 peak. On the basis of these results, many authors assigned the β peak to N–N recombination (process II) and the α peak to process III. N_2O formation has been found for some supported Rh catalysts at the temperature regime of the α-N_2 and NO desorption peaks (77). This correlation between α-N_2, N_2O, and NO desorption may point to a common N_2O-like surface intermediate for α-N_2 and N_2O.

Lambert and Comrie (33) investigated the CO + NO reaction on Pt(111) and (110) surfaces and concluded that the reaction proceeds by a L–H mechanism between O_{ads} and molecular CO_{ads}. NO dissociation is also the prime step in the NO + H_2 reaction on a Pt foil at a pressure of 10^{-7} mbar (7). NH_3 was found to be the major product at temperatures lower than 600 K and N_2 was the major product at temperatures higher than 600 K when the NO/H_2 ratio was ~1/5.

Siera et al. (45) investigated the CO + NO reaction on the same surfaces and under the same experimental conditions as described for the CO + O_2 reaction. At low temperatures (<500 K), the reaction rate decreases in the order Pt–Rh(111) > (100) > (410) > (210). This is the same order as was observed for the CO + O_2 reaction, and it corresponds to the order in enthalpy of adsorption of CO, which increases from the (111) to the (210) surface. NO dissociation decreases in the order (410) ~ (210) > (100) > (111). For both reactions, the rate at low temperatures is controlled by CO desorption and not by NO dissociation activity.

Similar effects were found for the NO reduction by hydrogen on Rh single-crystal surfaces; Wolf et al. (21) studied the NO + H_2 reaction on several Rh surfaces by means of a scanning field emission probe-hole microscope. The temperature at which the reaction starts is strongly dependent on the NO coverage and is sensitive to the surface structure. At low NO precoverages and with hydrogen in the gas phase, the reaction starts already at 240 K on Rh(331), a surface with (111) terraces and (111)-like steps. On Rh(211), a surface with (111) terraces and (100) steps, the reaction starts at a significantly higher temperature. On Rh(321), a surface with (111) terraces and both (100)- and (111)-type steps, the reaction features characteristic of both (111) and (100) steps are observed. During reaction, an electropositive adsorbate, assigned as $NH_{x,ads}$, is formed that dissociates at a temperature of approximately 300 K. At high NO precoverages, NO desorption is the initiation process; vacancies required for NO dissociation and hydrogen adsorption are then created. Under these conditions, surface structural effects are related to differences in enthalpy of adsorption of NO. However, hydrogen has a beneficial effect on the initiation temperature; increases of hydrogen pressure result in decreases of the onset temperature of the reaction. This effect can be explained in terms of hydrogen-assisted NO desorption, i.e., displacement of NO_{ads} by hydrogen. It is not a direct hydrogen-assisted NO dissociation, for example, by formation of HNO_{ads}. At low NO precoverages, the influence of the surface structure is related to the intrinsic differences in NO dissociation reactivity of the various surfaces. The observed onset temperatures of reaction did not differ significantly from the NO dissociation temperatures. The authors concluded that hydrogen-assisted NO dissociation proposed by Chin and Bell (77) does not seem to be important under their conditions (21).

Recently, spectroscopic techniques and the use of isotopically labeled N have provided a wealth of novel information about the mechanisms of N_2, NH_3, and N_2O formation. It is illustrative to discuss the detailed studies reported by Hirano *et al.* (78–80) concerning the mechanisms of the NO + H_2 reactions in light of recent studies. The surface investigated in most detail by Hirano *et al.* was a Pt–Rh(100) alloy surface (75% Rh and 25% Pt in the bulk).

The (1 × 1) structure, representative of the clean Pt–Rh(100) surface, changes into the c(2 × 2) surface structure after exposure to a reaction mixture of 1×10^{-7} mbar NO and 5×10^{-8} mbar H_2. Analysis of the c(2 × 2) surface by means of TDS, auger electron spectroscopy (AES), and HREELS indicated that atomic nitrogen was the main species present. For comparison, similar experiments were carried out on the Rh(100), Pd(100), and Pt(100) surfaces (78–80). Figure 16 illustrates some of the results.

Accumulated N atoms ordered in the c(2 × 2) surface structure were found on both Pt–Rh(100) and Rh(100) but not on Pt(100). In the presence of H_2 or D_2 at a crystal temperature of 400 K, a nitrogen–hydrogen vibration was found. Its intensity depends strongly on the partial pressure of hydrogen. Evacuation of H_2 and D_2 results in a quick loss of the signal. On the basis of these results, it was concluded that hydrogenation of N can occur via the reaction

$$N_{ads} + xH_{ads} \leftrightarrows NH_{x,ads}. \tag{31}$$

The hydrogenation of atomic nitrogen is largely reversible. By AES it was established that the N/Rh signal ratio was decreased only slightly

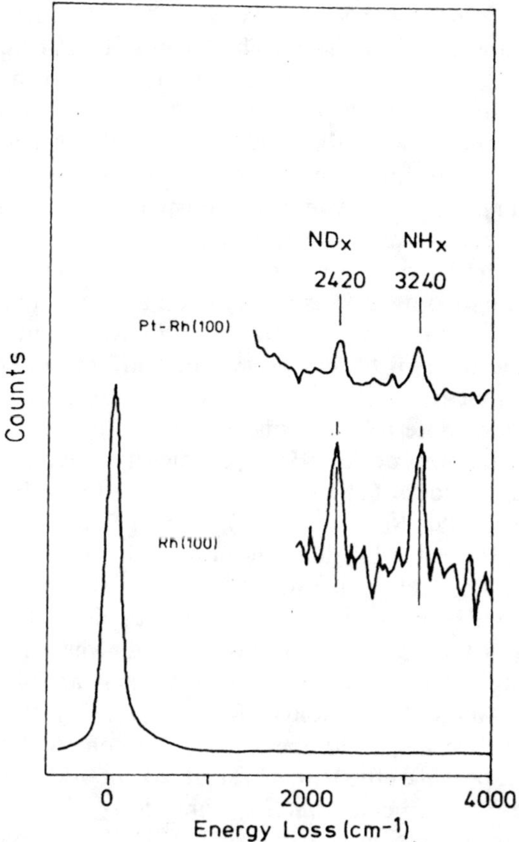

FIG. 16.   *In situ* EEL spectra of c(2 × 2)–N on the Rh(100) and Pt–Rh(100) surface in equilibrium with a hydrogen–deuterium mixture at 10^{-7} mbar and a temperature of 400 K. The intensity of the N–H signal is 1.6 times larger on Rh(100) than on Pt–Rh(100) [reproduced with permission from Nieuwenhuys *et al.* (*80*)].

during the hydrogen treatment. This result indicates that complete hydrogenation of the atomic nitrogen to NH_3 can occur but with a very low rate and only under these experimental conditions. The intensity of the N–H EELS signal varies with the square root of the hydrogen pressure:

$$I_{N-H} = AP_{H_2}^{1/2}. \tag{32}$$

Furthermore, N–H scissor vibrations were absent from the EELS spectra. On the basis of these observations it was concluded that the dominant NH_x adsorption complex is NH_{ads} and not $NH_{2,ads}$ or $NH_{3,ads}$.

Attempts to form a N layer on the Pt(100) surface by means of the NO–H_2 reaction were unsuccessful. The NO and H_2 partial pressures were varied from 10^{-8} to 10 mbar and the temperature from 400 to 600 K. Formation of a nitrogen overlayer was not observed under these conditions.

Useful information concerning the reactivity of adsorbed species is obtained by TPR. In the presence of a NO + hydrogen (H_2 or D_2) flow ($\sim 2 \times 10^{-7}$ mbar), the greatest ammonia formation was observed at approximately 500 K (heating rate, ~ 3 K s^{-1}) (78–80). However, if the surface was covered with ^{15}N prior to exposure to the ^{14}NO + hydrogen flow, ammonia resulting from ^{15}N was found even at 435 K (78). These results show that N atoms on the surface can react with hydrogen to give ammonia at 435 K. On the basis of these observations and the HREELS results, it was concluded that hydrogenation of atomic nitrogen via NH_{ads} is most likely the reaction route to ammonia formation (process VI).

Evidence for the presence of $NH_{x,ads}$ species has been obtained for many surfaces, including those of Pt–Rh (78, 80, 81), Pt (82–84), Rh (70–81, 85), Pd (81), and Ru (86). Sun et al. (84) investigated the stability of NH_x species on Pt(111) formed by electron-induced dissociation of $NH_{3,ads}$ using HREELS and XPS. $NH_{2,ads}$ formed following electron irradiation at 100 K. Upon heating, many of the $NH_{2,ads}$ species are rehydrogenated and desorb as NH_3 at temperatures of about 200 K, whereas a small fraction dehydrogenates, forming NH_{ads} at temperatures near 300 K. At temperatures exceeding 400 K, NH_{ads} dehydrogenates, leaving N_{ads} on the surface, which then desorbs as N_2 at temperatures between 500 and 700 K.

Recent SIMS spectra of the Rh(111) surfaces obtained during hydrogenation of atomic N_{ads} indicate that N_{ads} and $NH_{2,ads}$ (not NH_{ads}) are the predominant surface intermediates on Rh(111) (85). According to the authors, the third hydrogenation step in process VI, i.e., the hydrogenation of $NH_{2,ads}$ to give $NH_{3,ads}$, is rate limiting for NH_3 formation, with an activation energy of 69 kJ/mol.

Because of the interest in ruthenium as a potential catalyst for ammonia synthesis from N_2 and H_2, the hydrogenation of N_{ads} on Ru surfaces has been investigated (86). Both NH_{ads} and $NH_{2,ads}$ were found, in addition to $NH_{3,ads}$. Dietrich et al. (86) reported that the thermal stability of NH_{ads} is the highest of the three $NH_{x,ads}$ species ($x = 1$, 2, and 3) on Ru(0001) and on Ru(1121) at temperatures up to 400–450 K. On Ru (1010), however, the thermal stability of $NH_{2,ads}$ is higher than that of NH_{ads} and much higher than those on other Ru surfaces. The reason for the enhanced thermal stability of $NH_{2,ads}$ on Ru(1010) is not clear. It was also found that coadsorbed N has a positive effect on the thermal stability of NH_{ads} on Ru(0001), whereas NH_{ads} is stable at temperatures up to 400 K, and NH_{ads} in the N/NH coadsorbate is stable at temperatures up to 460 K.

Yamada and Tanaka (*81*) found that the N c(2 × 2) structure on the (100) surfaces of Pd, Rh, and Pt–Rh does not change significantly upon the formation of NH_{ads} in the presence of hydrogen. This observation was interpreted on the basis of formation of NH_{ads} at the borders of c(2 × 2) islands during exposure to hydrogen.

An interesting effect of the surface structure was also found for Pt by Zemlyanov *et al.* (*82, 83*). HREELS and TDS confirmed that NO desorbs completely from Pt(111) without dissociation. However, N_2 and NH_3 formation were found upon heating the NO-covered Pt(111) surface in a hydrogen atmosphere. HREELS results were interpreted in terms of formation of HNO_{ads}. It was proposed that NH_3 and N_2 are formed without direct dissociation of NO:

$$HNO_{ads} + H_{ads} \rightarrow N_{ads} + H_2O', \tag{33}$$

followed by reaction II and hydrogenation of N_{ads} to NH_3. On Pt(100), N_{ads} formed by NO dissociation at 300 K reacts readily with hydrogen to give $NH_{2,ads}$ on the Pt(100) (1 × 1) surface and to give NH_{ads} on the reconstructed Pt(100)-hex surface. It was proposed that the NH_{ads} and $NH_{2,ads}$ species are the intermediates for NH_3 formation. However, NH_3 formation was not observed under the experimental conditions; the only product observed was dinitrogen.

The presence of HNO_{ads} was also proposed by Williams *et al.* (*87*) [on the basis of surface-enhanced Raman spectra (SERS)] as an intermediate for the NO + H_2 reaction on polycrystalline Rh films. The authors demonstrated that SERS can be used to characterize adsorbed species for CO + NO and NO + H_2 reaction (*87, 88*) on ultrathin Rh films deposited on a roughened gold substrate. (The Rh surface contained carbonaceous contaminants and possibly gold atoms.) This type of measurement is interesting because it yields vibrational data *in situ,* even at high pressures. In the presence of a NO + N_2 flow (ratio NO/H_2 = 1, total pressure-1 bar), a band characteristic of N_{ads} is present, and at temperatures from 420 to about 620 K, a band attributed to HNO_{ads} is present. The feature observed was attributed to HNO_{ads} because the surface species was formed only from NO_{ads} and hydrogen and not from N_{ads} and hydrogen. The authors proposed an additional pathway for NO dissociation, illustrated by Eq. (33).

This proposed process may explain the hydrogen-assisted NO dissociation first proposed by Hecker and Bell (*89*). Investigations of Rh single-crystal surfaces have also indicated that the presence of hydrogen may have a beneficial effect on NO dissociation (*21, 90*). It was also proposed that this alternative process for NO dissociation may be fast under conditions of slow direct decomposition because of the absence of free adjacent sites needed for NO dissociation (*87*). It was argued that a high coverage of

H_{ads} may block the vacant sites and slow the direct dissociation. [Blocking of sites needed for NO dissociation by hydrogen is an unlikely process, in contrast to blocking by NO, CO, N, or O. Furthermore, the hydrogen-assisted NO dissociation can also be attributed to the removal of O_{ads} and/or N_{ads} by hydrogen (21).]

The NO + H_2 reaction catalyzed by Rh(533), which is composed of four-atom-wide (111) terraces separated by (100) steps, was investigated *in situ* by Cobden *et al.* (91) in the 10^{-6} mbar pressure range by using fast XPS and mass spectrometry. The emphasis of this study was on understanding the oscillatory behavior of the NO + H_2 reaction on Rh surfaces. Some of the results are also relevant to the current discussion. It was shown that NO_{ads} is not present on the surface in the temperature range in which N_2 and NH_3 formation is observed. Two different N_{ads} species were observed in addition to O_{ads}. One N_{ads} species (NI) is probably N_{ads} on the (111) terraces, and the other (NII) is either N_{ads} adsorbed on the steps or $NH_{x,ads}$. O_{ads} plays an important role in controlling the surface reactivity. O_{ads} present in small amounts destabilizes the NI species, and at high O_{ads} coverage it favors the formation of NII.

The results summarized previously show that the surface structure and the presence of coadsorbed species have a large effect on the relative stability of NH_{ads} and $NH_{2,ads}$. However, more detailed studies are required to understand the relative stabilities of NH and NH_2 on noble metal surfaces. In conclusion, there is ample evidence that process VI is a major mechanism for NH_3 formation. The intermediates NH_x have been identified. The only possible exception may be Pt(111).

As stated in Section II.A, NO dissociation is negligible on Pt(111) at low pressure. However, Pt(100) is active in NO bond breaking. Polycrystalline Pt surfaces have low activity for NO dissociation. The group of Schmidt (92, 93) investigated the NO + CO reaction on polycrystalline and single-crystal Pt surfaces. At low pressures, Pt(111) is unreactive for the reaction. The kinetics of the reaction was investigated on polycrystalline Pt at temperatures from 300 to 1200 K and pressures from 10^{-8} to 1 mbar. The authors proposed that the reaction is a true bimolecular reaction between NO_{ads} and CO_{ads} (Eq. 13b) rather than NO decomposition with CO scavenging of O_{ads} (mechanism II). For the reactive Pt(100) surface, the reaction of NO + H_2 was very similar to that of NO + CO, and mechanism II qualitatively explains the observations.

The possible relevance of mechanisms I and IV to the formation of N_2 in TWC has not been addressed, except for polycrystalline Pt. Mechanism IV has only been added for the sake of completeness since it has not been proposed for NO reduction catalyzed by TWC. This and related mechanisms have been proposed for the selective catalytic reduction of NO_x by hydrocarbons (*94–96*) and hydrogen (*97*) in the presence of O_2 and catalyzed by various zeolites containing metal ions in their cavities. The rate of N_2 formation is enhanced in the presence of excess O_2. In several proposed mechanisms, $NO_{y,ads}$ (with $y > 2$) plays a key role. These mechanisms, although probably not important for TWC, may play a role when dissociation of NO cannot occur due to the absence of adjacent metal atoms required for NO dissociation. Shelef and Graham (*3*) proposed that the high selectivity for N_2 production that distinguishes Rh from Pt and Pd catalysts may be rooted in the promotion of N pairing in $2NO_{ads}$ molecules before the N–O bond is broken. These authors noted that IR spectra of NO on well-dispersed Rh/Al_2O_3 catalysts point to the presence of dinitrosyl $(NO)_{2,ads}$, a species that is not found on Pt and Pd/Al_2O_3 catalysts. These adsorbed dinitrosyl species could represent a locus for the event of pairing of the nitrogen atoms. It is probable that $(NO)_{2,ads}$ is formed on ionic Rh sites. No experimental evidence supporting this mechanism is available for TWC.

The formation rates of N_2 and N_2O on Pt–Rh(100) as a function of crystal temperature are shown in Fig. 17a for a mixture of 1.2×10^{-7} mbar NO and 3.6×10^{-7} mbar H_2 under steady-state conditions (*78–80*). The N_2 and NH_3 formation rates increase rapidly as temperature rises from 400 K until the maximum rates are obtained at approximately 600 K for N_2 and at the significantly lower temperature of about 525 K for NH_3. Nitrous oxide (N_2O) is only a minor product under these conditions. N_2 production remains high and almost constant until a temperature of about 1000 K is reached, indicating that the reaction rate in this temperature range may be controlled by the collision frequency of NO on the surface. At higher temperatures, the formation rate drops. For reaction mixtures with excess hydrogen, the N_2 formation rate is almost independent of the hydrogen pressure (zero order) at temperatures from 600 to 800 K. The ammonia production is approximately first order in hydrogen over the whole temperature range of 400 to 800 K. N_2O formation is observed only at relatively high NO pressures, and it is only slightly affected by changing the hydrogen pressure. Under the conditions of these experiments ($P_{NO} < 5 \times 10^{-6}$ mbar), the N_2 formation remains at least a factor of 100 larger than the N_2O production rate.

In another series of experiments of H_2 in the gas flow was replaced by NH_3 and the ^{14}NO by ^{15}NO to allow examination of the distribution of the two types of N atoms over the N-containing products. Figure 17b shows the variation of dinitrogen formation rates with increasing temperature in the presence of a flow of 1.2×10^{-7} mbar of ^{15}NO and 3.0×10^{-7} mbar of $^{14}NH_3$. In the low-temperature range ($T < 800$ K), the rate of $^{14}N^{15}N$ formation is higher than the rate of $^{15}N^{15}N$ formation, which in turn is much

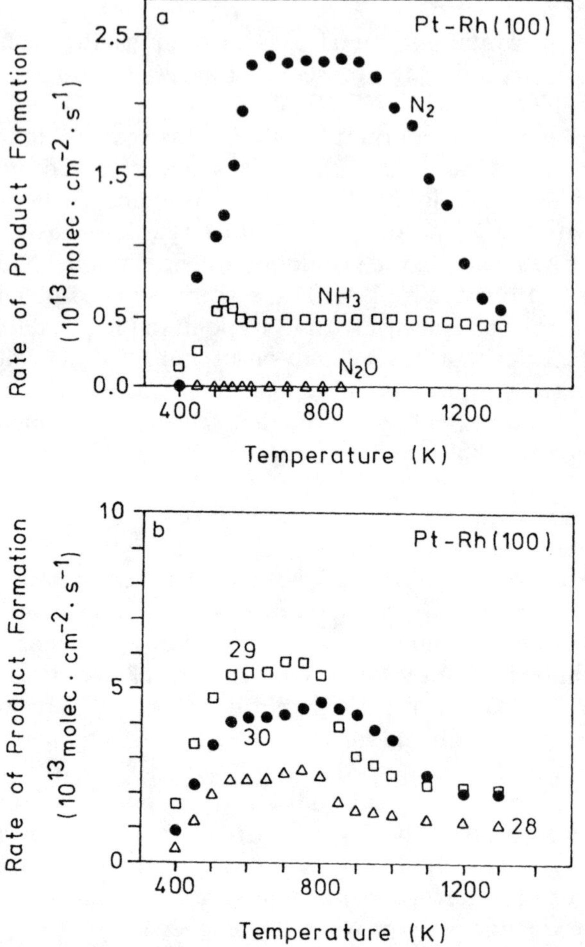

FIG. 17. (a) Steady-state formation of N_2 ($\bullet$), NH_3 ($\square$), and N_2 ($\triangle$) on the Pt–Rh(100) surface in the presence of 1.2×10^{-7} mbar NO and 3.6×10^{-7} mbar H_2. (b) Steady-state formation rates of N_2 on Pt–Rh(100) with amu 28 ($\triangle$), 29 ($\square$), and 30 ($\bullet$) in the presence of 1.2×10^{-7} mbar ^{15}NO and 3.0×10^{-7} mbar ^{14}NH$_3$ [reproduced with permission from Hirano *et al.* (*78*)].

higher than the rate of $^{14}N^{14}N$ formation. [The ^{14}N ($^{14}NH_3$) concentration in the gas phase was a factor of 2.5 larger than the ^{15}N (^{15}NO) concentration.] The relatively low production of $^{14}N^{14}N$ is in qualitative agreement with results reported by Otto *et al.* (*98*) for the NO + NH$_3$ reaction catalyzed by supported Pt. Using N isotopes, these authors showed that N_2 formed on Pt exclusively from NH_3 is only a minor path. Under the conditions used in the experiments shown in Fig. 17b, N_2 is formed on Pt–Rh(100) exclusively from NH_3, albeit at a much lower rate than from NO + NH_3 and from NO only. At temperatures higher than 700 K, the rate of $^{14}N^{15}N$ formation decreases rapidly with increasing temperature, and the $^{15}N^{15}N$ production increases. The $^{14}N^{14}N$ production shows the same temperature dependence as the $^{14}N^{15}N$ production.

The temperature dependence of the N_2 formation is the same for the $NH_3 + {}^{15}NO$ and the H_2 + NO reactions, suggesting that the same processes are involved. The N-combination reactions can occur over the whole temperature range, as demonstrated by the rate of $^{14}N_2$ formation. The production of $^{14}N_2$ is much smaller than the production of $^{14}N^{15}N$ and $^{15}N_2$, whereas the $^{14}NH_3$ concentration in the gas phase is a factor of 2.5 greater than that of ^{15}NO. This comparison may indicate that NH_3 is a slower producer of N_{ads} than is NO. However, this argument is in contradiction with the high rate of $^{14}N^{15}N$ formation. It was suggested that the relatively high rate of production of $^{14}N^{15}N$ may point to a large contribution of the reaction of N_{ads} with NO_{ads} (process III) to the N_2 formation in the lower temperature range ($T < 700$ K). At 1100 K, the rate of $^{14}N_2$ formation is a factor of 1.5 less than at 600 K; the rate of $^{14}N^{15}N$ formation is a factor of 2.5 less and the rate of $^{15}N_2$ formation only slightly lower. On the basis of this result, it was concluded that in the high-temperature range (1100 K), the importance of the N-combination reaction is much greater than in the low-temperature range and that ammonia becomes a less effective N producer than NO.

The same authors investigated the NO + H_2 reaction catalyzed by Pt–Rh(100) in the 10-mbar pressure range (*78–80*). Figure 18 shows some results obtained at 550 K.

At a temperature less than 500 K and a NO/H_2 ratio of 1/5, most of the NO reacts at low conversions to give N_2, and the formation of NH_3 is greater than that of N_2O. However, as the temperature increases above 500 K, the NH_3 production exceeds the N_2 production, and N_2O formation essentially ceases. The rate of NO conversion is much lower for a NO/H_2 ratio of unity than for the ratio of 1/5. N_2 production is only slightly affected by lowering of the H_2 pressure. However, NH_3 formation decreases drastically as a result of lowering of the hydrogen pressure. Consequently, the selectivity toward dinitrogen is much improved by lowering of the H_2/NO ratio. The hydrogen pressure was varied from 1.2 mbar to 24 mbar at a constant NO pressure of 1.2 mbar. It appeared that the reaction order in hydrogen is essentially zero for formation of N_2 and N_2O and about unity for formation of NH_3 under these conditions. It is emphasized that NO, NH_3, and N_2O decomposition and the NO + H_2 and NH_3 + NO

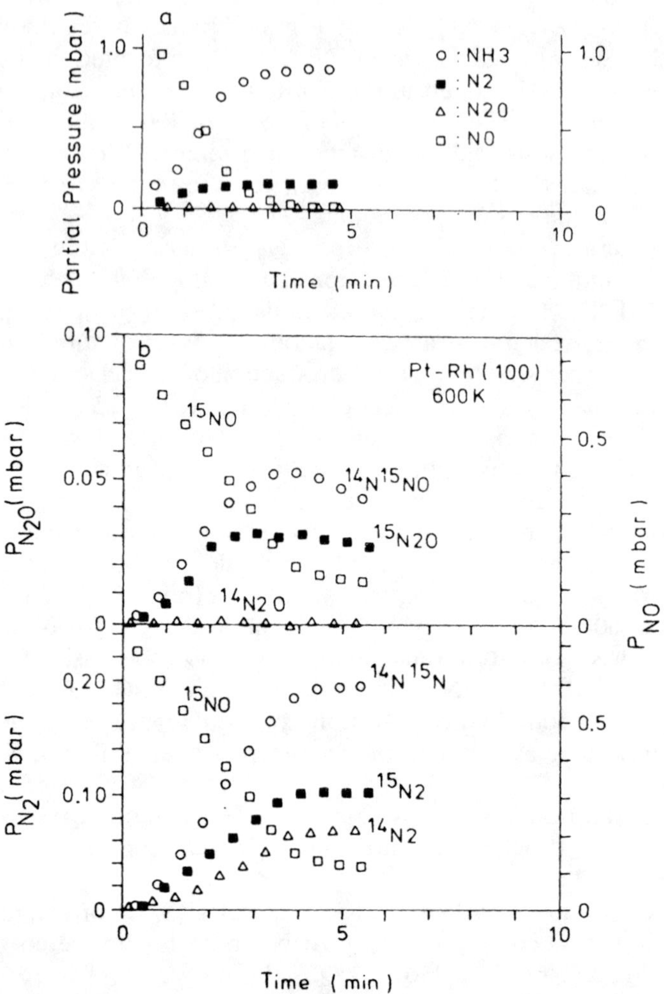

FIG. 18.   (a) Formation of N_2, NH_3, and N_2O on Pt–Rh(100) at 550 K at a NO/H_2 ratio of 1/5 with total pressure of 3 mbar. (b) The observed reaction products for the ^{15}NO + $^{14}NH_3$ reaction on Pt–Rh(100) as a function of the reaction time at 600 K with a $^{15}NO/^{14}NH_3$ ratio of 1 and total pressure of 3 mbar [reproduced with permission from Hirano *et al.* (78)].

reactions can occur simultaneously. Hence, useful information concerning one specific reaction is obtained only at the beginning of the reaction when the conversion is still low.

The most striking differences between the low- and higher pressure regimes are the following: (i) The relatively high selectivity to N_2O in the higher pressure regime and (ii) the relatively high selectivity to ammonia. The first observation is completely in line with the expectations based on the mechanism for N_2O formation via process VII, requiring a high concentration of molecularly adsorbed NO. The dissociation of NO is inhibited by high concentrations of adsorbed NO, as discussed previously. It has also been reported that during the NO–hydrogen reaction a silica-supported Rh catalyst is largely covered with NO. On the other hand, in the low-pressure regime, Rh surfaces are covered with N_{ads}. As the temperature increases, the concentration of NO_{ads} decreases as a result of desorption and the selectivity to N_2O is expected to decrease, as observed.

The same argument can be used to interpret the relatively high selectivity to NH_3. A high concentration of N_{ads} favors a high selectivity to N_2 via process II. A low concentration of N_{ads} favors the formation of NH_3 (process VI). Obviously, the N_{ads} concentration at 600 K in the 10-mbar pressure regime is lower than that in the 10^{-7} mbar regime. It has been stated that the selectivity for NH_3 formation on supported metals such as Pt and Rh increases with increasing temperature up to a maximum T_m, beyond which it decreases; for Pd and Pt, T_m is about 600 K. The maximum temperature used in one investigation (78) was 600 K and, hence, the increasing selectivity to NH_3 observed when the temperature increases is in agreement with these data. The ^{15}NO + $^{14}NH_3$ reaction was also investigated in the pressure range of 10 mbar (78). Some results are shown in Fig. 18b for a $^{15}NO/^{14}NH_3$ ratio of 1 at 600 K. $^{14}N_2O$ is not formed, whereas $^{14}N^{15}NO$ and $^{15}N_2O$ are both formed, showing that N_2O is formed only via process VII. At 600 K the formation of $^{14}N^{15}NO$ is faster than that of $^{15}N_2O$, which suggests that NH_3 is a better producer of N than NO at this temperature. $^{14}N_2$ formation is not detected at 550 K, whereas at 600 K the rate of $^{14}N_2$ formation is much less than that of $^{14}N^{15}N$ and $^{15}N_2$. On the basis of this observation it was suggested that process III is the dominant path for N_2 production in the low-temperature regime.

The molecular picture that the authors proposed from their results is the following (78–80): At low temperatures ($T < 450$ K), the majority of the adsorbed species are NO, and N is only a minor species. Adsorbed hydrogen is very mobile and can easily react with either O or N, provided it can find a vacant site in the neighborhood of an adsorbed O or N atom. The concentration of vacant sites is low in this temperature range. As the temperature increases, the concentration of NO becomes lower, and the concentration of N_{ads} increases. When the NO_{ads} concentration becomes much lower than the N_{ads} concentration, the rate of process VII, and hence the selectivity to N_2O, becomes low. For N_2 formation via process III, high concentrations of both N_{ads} and NO_{ads} are required; for N_2 formation via process II, a high concentration of N_{ads} is required. Consequently, process

III may contribute to the N_2 formation in the lower temperature range, and process II dominates in the higher temperature range. The selectivity to NH_3 increases with increasing temperature in the temperature range 300–600 K, again suggesting that the dominant mechanism of NH_3 formation is via process VI and not via hydrogenation of NO_{ads}. The selectivity of NH_3 will be the greatest at a temperature at which both the N_{ads} and H_{ads} concentrations are sufficiently high. The strong positive reaction order in hydrogen pressure shows that the concentration of H_{ads} is the limiting factor under the experimental conditions. At some very high temperature, the reaction rates of both NH_3 and N_2 formation become low because of the low N_{ads} and H_{ads} concentrations. Obviously, this temperature depends strongly on the absolute NO and H_2 pressures and on the metal surface. Rh, with its much stronger metal–N bond strength, will exhibit a much better selectivity toward N_2 than Pt, which has a weaker metal–N bond strength. The observed decrease in selectivity toward NH_3 as temperature becomes very high can be attributed to a combination of limited availability of H at high temperature and the high rate of NH_3 decomposition.

In contrast to the apparent structure insensitivity of the $CO–O_2$ reaction catalyzed by Rh and reported by Peden (*65*), Peden *et al.* (*66*), and Oh *et al.* (*67*), these researchers found substantially different kinetic behavior for the NO + CO reaction on Rh(111), Rh(100), and supported Rh/Al_2O_3 (*66*, *67*), as illustrated in the Arrhenius plots of Fig. 19. The corresponding activation energies are 124 kJ/mol for Rh(111) and 101 kJ/mol for Rh(100). The authors postulated that the structure sensitivity of the CO + NO reaction catalyzed by Rh is due to different rates of NO dissociation. On Rh(111), the rate would be limited by the formation of N_2 from process II, and the surface would be largely covered with N_{ads}. On the supported catalyst, the rate-limiting step was judged to be NO dissociation.

STM is an attractive technique for elucidating reaction mechanisms since, in addition to the observation of the surface structure, it allows imaging of adsorbed atoms and molecules under some conditions. Many recent examples indeed show that STM can provide useful additional information (*44*, *99*, *100*). Leibsle *et al.* (100) observed a one-dimensional CO + O_2 reaction on Rh(110). Xu and Ng (*99*) imaged NO_{ads} molecules coadsorbed with O or N atoms on Rh(111); NO_{ads} molecules form ordered islands with (4 × 4) structure. Repulsive interactions between NO and O or N lead to segregation of NO and O or N islands. The results also suggest the formation of Rh–N(1 × 2) added row reconstruction through NO dissociation.

Some confusion exists in the literature concerning N_2O formation as a product of the CO + NO reaction. In the papers of Oh *et al.* (*67*) and Peden *et al.* (*66*), N_2O formation was not reported for the CO + NO reaction on Rh single-crystal surfaces. However, Hecker and Bell (*101*)

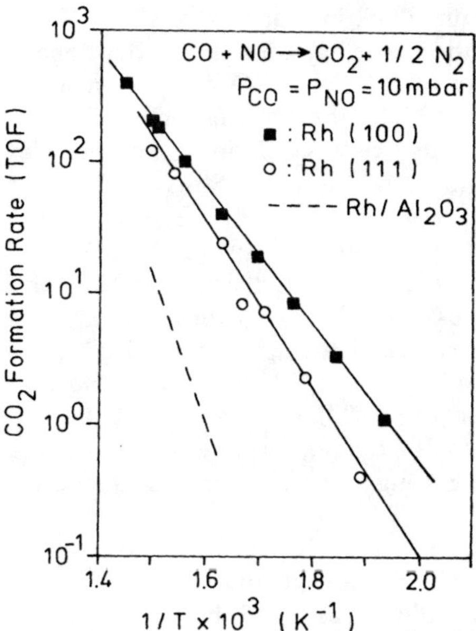

FIG. 19.   Comparison of the specific rates of the NO–CO reaction on Rh(111), Rh(100), and Rh/Al$_2$O$_3$ at $P_{CO} = P_{NO} = 10^{-2}$ bar [adapted with permission from Oh et $al.$ (67)].

reported that N$_2$O is a major N-containing product of the CO + NO reaction catalyzed by Rh/SiO$_2$. Belton and Schmieg (102) studied the reaction on a Rh(111) catalyst and found that N$_2$O is formed with a selectivity of 70% from a mixture of 1.06 kPa of NO and 1.06 kPa of CO at temperatures between 525 and 675 K. Both Rh(111) and Rh/SiO$_2$ give similar product distributions, with more N$_2$O than N$_2$ and an activation energy of 139 kJ at temperatures higher than 480 K.

To what extent does reaction III contribute to N$_2$ formation? Whereas Hirano et $al.$ (78–80) concluded from that reaction III may make a significant contribution at low temperatures, Belton et $al.$ (103) concluded that there is no experimental evidence for the reaction step

$$NO_{ads} + N_{ads} \rightarrow N_2O_{ads} \rightarrow N_2 + O_{ads}, \qquad (34)$$

and they recommended that in future NO reduction mechanisms this step should be omitted. Belton et $al.$ prepared a Rh(111) surface covered with 0.4 monolayers of ^{15}N by electron beam dissociation of ^{15}NO$_{ads}$ and saturated this surface with ^{14}NO at 200 K. TDS did not reveal ^{15}NO, showing that scrambling of ^{15}N and ^{14}NO did not occur during their experiments. In addition, all three N$_2$ isotopic mixtures (masses 28, 29, and 30) have been observed in the TDS, indicating that the α-N_2 feature is also due to N atom recombination. The apparent first-order behavior of the α-N_2 peak and the correlation of α-N_2 and NO desorption were attributed to rate-limiting formation of N_2 by the dissociation of NO_{ads} at high NO coverage (NO_{ads} inhibition of its own dissociation). Another view was presented by Borg *et al.* (*23*) and by Makeev and Slinko (*104*): Repulsive interactions in the adlayer are required to produce α-N_2. Bugyi and Solymosi (*105*) prepared a N overlayer with high coverage on Rh(111) from atomic N in the gas phase by means of a discharge tube and observed α-N_2 and β-N_2 desorption states very similar to those observed after NO exposure. Apparently, the presence of NO_{ads} is not required to produce α-N_2 in TDS. Makeev and Slinko (*104*) were able to simulate the TDS, reported by Root *et al.* (*106*) and Borg *et al.* (*23*), of N_2 and NO from Rh(111) following NO exposure on the basis of repulsive interaction between N_{ads} and NO_{ads}, together with the strong inhibition of NO dissociation by NO_{ads} N_{ads}, and O_{ads}.

The NO + CO reaction on Rh(111) was investigated by Permana *et al.* (*107*) by RAIRS. The measurements were performed with NO and CO pressures in the 10-mbar range and, hence, in the same range as in the automotive exhaust gas contacting the TWC. IR spectra taken under reaction conditions showed only atop CO and multiply bonded (either twofold or threefold) NO. Changes in the surface coverages of NO_{ads} and CO_{ads} correlated well with the observed changes in N_2O selectivity; at temperatures below 635 K, NO_{ads} dominates and N_2O formation is favored. At temperatures above 635 K, at which N_2 formation is preferred, CO is the majority surface species. Obviously, these RAIRS data support the model according to which N_2O and N_2 are formed by parallel pathways by reaction of N_{ads} with either NO_{ads} or N_{ads}, respectively. By adding $^{15}N_2O_{gas}$ to the ^{14}NO and ^{12}CO reactant mixture it was shown that $^{15}N_2O$ was not consumed during the reaction. Therefore, readsorption of N_2O is not an important path to produce N_2 under the conditions used in Permana *et al.* (*107*) (650 K and low conversion). However, the same authors also showed that N_2O is readily converted to N_2 at temperatures higher than 700 K (*102*). The CO + N_2O reaction runs only after complete NO conversion at temperatures higher than 700 K.

The mechanism and kinetics of the NO + CO reaction on Rh(111) have been discussed in detail by Zhdanov and Kasemo (*108*). They showed that simulations based on surface science data obtained at low pressures reproduce the scale of the reaction rate at the pressure regime of interest for the TWC but fail to predict accurately the apparent activation energy and reaction orders.

Can we conclude that process III does not at all contribute to N_2 formation? In this context it is relevant to review recent work of Ikai *et al.* (*109–111*) on N_2 desorption from Pd(110). They showed that the behavior of Pd(110) toward NO adsorption is very different from that of Pd(100). TDS experiments (with an NO-covered surface rapidly heated in uhv) characterizing NO adsorbed on Pd(110) showed NO, N_2, and N_2O peaks, with a maximum at about 490 K. In addition to the 490-K peak, a second NO desorption peak was observed at 370 K. The authors also measured the spatial distribution of the molecules desorbing during TDS and TPR. This technique can provide additional information about the nature of the chemical processes involved. N_{ads} can accumulate on Pd(100) during NO exposure at higher temperatures but not on Pd(110). N_{ads} accumulation on Pd(110) was obtained by exposing the surface to N^+/N_2^+ ions. The resulting N_2 TDS are distinctly different from the N_2 TDS observed following NO adsorption. In addition, its spatial distribution is very different from the off-normal desorption found for N_2 formed from NO adsorption and also from N_2 formed by the $NO + H_2$ reaction. In the same way, a Pd(110) surface precovered with ^{15}N was prepared. Exposure to ^{14}NO resulted in $^{15}N^{15}N$ peaks at 460 and 520 K. The spatial distribution is very different from that of $^{15}N_2$ at 460 K. Ikai (*111*) suggested that reaction III may be responsible for the off-normal desorption of N_2 at 490 K.

On the basis of the results discussed previously, it is concluded that the formation of N_2 on Rh(111) can be understood solely in terms of N_{ads} combination. However, it is premature to conclude that process III does not play any role in the N_2 formation on other metal surfaces.

The CO + NO reaction on various kinds of Pd catalysts was investigated by Rainer *et al.* (*112*). Kinetics data for Pd/Al_2O_3 and "planar model" $Pd/Al_2O_3/Ta(110)$ catalysts were compared with those for CO + NO on Pd(111), (100), and (110) surfaces. Figure 20 is a comparison of the Arrhenius plots measured at partial pressures of about 1 mbar in each reactant for the model catalysts and at a CO pressure of 5.9 and a NO pressure of 6.8 mbar (in a helium carrier) for the Pd/Al_2O_3 powder catalyst.

The reported apparent activation energies are 67 kJ/mol for Pd(111) and (100), 71 kJ/mol for Pd(110), and 155 kJ/mol for the Pd/Al_2O_3 powder catalyst. The authors concluded from the comparison of rates per surface Pd atom (TOF) at 560 K that the supported Pd/Al_2O_3 powder catalysts exhibit a pronounced particle size effect, with an increase in activity with increasing particle size. The Pd(111) surface is more active than the (100) and (110) surfaces. The authors argued that the smaller Pd particles with their higher step/edge/defect densities have more in common with the open single-crystal faces, whereas the larger particles have more Pd(111) character. Pd(100) is more effective at dissociating NO_{ads} than Pd(111). As a result, Pd(100) yields a higher ratio of N_2/N_2O than Pd(111). TDS of N_2

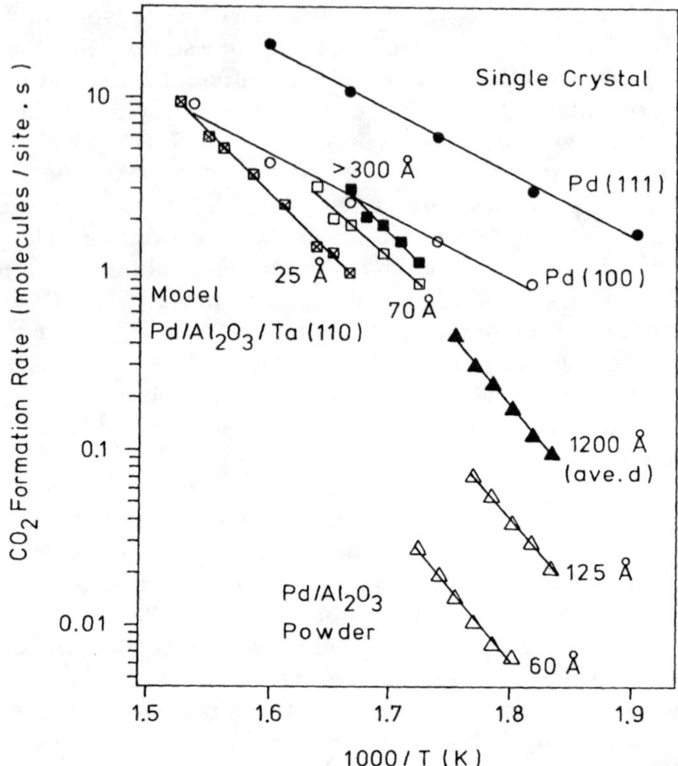

FIG. 20. Arrhenius plots for the rate of the NO–CO reaction on various kinds of Pd catalysts. The planar model and single-crystal surface data were taken in a batch reactor (1 mbar of each reactant) and the powder catalyst in a flow reactor (P_{CO} = 5.9 mbar, P_{NO} = 6.8 mbar) [adopted with permission from Rainer *et al.* (*112*)].

following NO exposure and TPR of CO + NO show essentially two peak maxima for N_2 production: α at 450 K and β between 530 and 590 K for Pd(111) and between 575 and 625 K for Pd(100). The desorption temperature for β-N is higher than the reaction temperature employed in these experiments. The Pd(100) surface, more active for NO dissociation than Pd(111), exhibits the lower activity for the reaction. This correlation implies that the removal of β-N is more important than NO dissociation in determining the reaction kinetics. Obviously, the dissociation of NO is not the rate-limiting step on Pd. The origin of the huge difference in activity between the model catalysts and the powder catalyst is not clear.

V. Effects of Alloy Formation

In most of the early papers, the formation of alloy particles was considered to be undesirable. For example, Kummer (*113*) mentioned that each noble metal is used for a specific purpose which can be served by keeping it separate from other metal components. However, according to Nieuwenhuys *et al.* (*114–116*), alloy formation cannot be avoided at the extreme conditions under which the catalyst must operate. Many studies have shown that nearly all the noble metal particles are alloyed in real vehicle-aged catalysts (*117, 118*). Many possible detrimental effects of alloy formation have been proposed (*113*), including reduction of the effectiveness of Pt for alkane oxidation because of Rh surface segregation. In addition, Pt alloy formation may result in a loss of the availability of Rh or Pd to the exhaust gas since Pt readily sinters to large crystallites at temperatures higher than 870 K, whereas Rh and Pd alone can remain dispersed under usual operation conditions (*113*). Alloy formation in general may also result in a distinctly different behavior, relative to those of the pure component catalysts (*119*). Therefore, it is of particular interest to review the available data concerning the effects of alloy formation on the reactions of importance in automotive catalysis.

The first step in understanding the results of activity measurements with alloy catalysts is understanding knowledge concerning the surface composition. Under vacuum, Pd segregates to the surface of Pt–Pd (*120*) and Rh–Pd alloys (*121*). The surface composition of Pt–Rh alloys has been investigated by several groups with techniques such as AES, ion-scattering spectroscopy, work function measurements, atom-probe FIM, and STM (*114–116, 119, 122–131*). Most of the results point to a strong Pt surface segregation of clean Pt–Rh crystals under vacuum. The Pt enrichment in the top layer is accompanied by Pt depletion in the second layer (*123, 124, 126*). Calculations of LeGrand and Treglia (*126*) and Schoeb *et al.* (*127*) show that the experimentally found Pt surface enrichment can be understood on the basis of a difference in surface energy. Pt surface enrichment should be expected, whatever the bulk composition and equilibrium temperature.

According to van Delft *et al.* (*114, 115*), the surface composition of Pt–Rh alloys may be extremely sensitive to the presence of adsorbate atoms due to the very small differences in factors resulting in Pt surface segregation of the clean surfaces. Experiment show that the surface composition may easily change in the presence of other elements. Carbon monoxide and hydrogen do not exert a measurable influence on the surface composition (*114, 115*). For oxygen, however, a strong oxygen-induced Rh surface segregation was found that was caused by the much stronger Rh–O bond strength relative to that of Pt–O (*114, 115, 124*). Similar results were obtained for

NO adsorption. NO dissociates on the surface, resulting in oxygen-induced Rh surface segregation. The presence of sulfur also causes Rh surface segregation (123).

In view of the relative thermal stabilities of the oxides (Rh_2O_3 > PdO $\gg$ PtO_2), it is not surprising that under oxidizing conditions and at temperatures higher than 870 K, metallic Pt particles exist in combination with Rh oxide (132, 133) and that PdO crystallites separate from Pt crystallites in Pd–Pt alloys in the temperature range 670–870 K (134). A more complicated behavior was found for Pd–Rh (135).

Beck et al. (124) reported that the $Pt_{0.10}$–$Rh_{0.90}$ (111) single-crystal surface, which has a surface composition of about 30% Pt in vacuum, remains Pt rich, even under 50 mbar of hydrogen at temperatures typical of automotive catalytic converter operation 770–870 K. A few studies of supported Pt–Rh/ Al_2O_3 catalysts have been pursued using NMR spectroscopy. Wang et al. (136) used ^{13}C NMR of adsorbed CO as well as ^{195}Pt NMR and concluded that the surface is slightly enriched in Rh in the presence of CO. Savargaonkar et al. (137) reported the use of ^{1}H NMR spectroscopy to determine the surface composition of Pt–Rh/Al_2O_3 catalysts in the presence of hydrogen. The authors concluded that their catalysts are slightly enriched in rhodium relative to the adsorbate-free catalysts, which are enriched in Pt. Simulations indicated that the heat of adsorption of hydrogen must be 13 kJ/mol higher on Rh than on Pt to achieve the reported Rh surface segregation. It should be noted that the interpretation of the observed Knight shifts in terms of surface composition is not straightforward. Furthermore, literature data characterizing hydrogen adsorption on pure well-defined single-crystal surfaces suggest that the initial heat of adsorption of hydrogen on Rh (80 kJ/mol) is similar to that on Pt (within 10%) (138). In my opinion, there is no doubt of the correctness of the earlier conclusion: Pt–Rh surfaces are Pt enriched under reducing conditions and Rh enriched under oxidizing conditions.

The most direct determination of both the surface structure and the composition of Pt–Rh alloy surfaces was obtained with STM (128, 129). The (100) surfaces of $Pt_{0.25}$–$Rh_{0.75}$ and $Pt_{0.50}$–$Rh_{0.50}$ single crystals were imaged with atomic resolution and with discrimination of the Pt and Rh atoms. The STM image shown in Fig. 21 demonstrates that there is a limited tendency for Rh–Rh and Pt–Pt clustering on the surface, and Pt–Rh ordering is absent. Interestingly, Pt preferentially populates the step edges. Oxygen causes a large reconstruction of the surface (130, 131). For example, Fig. 22 shows the STM image of the (100) surface of $Pt_{0.25}$–$Rh_{0.75}$ exposed to O_2 at 770 K and subsequently cooled to room temperature in oxygen. The STM image points to the formation of linear Rh–O chains.

Adsorption and reactivity investigations of Pt–Rh alloy surfaces were

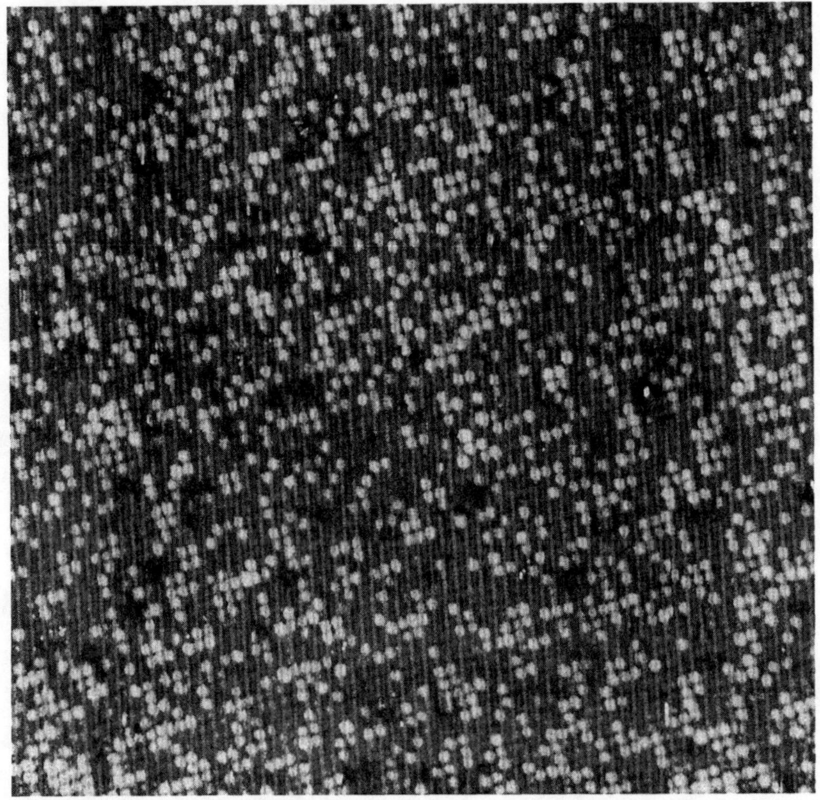

FIG. 21. STM image of $Pt_{0.5}$–$Rh_{0.5}(100)$; image size, 200 × 200 Å. Bright spots correspond to Rh atoms [reproduced with permission from Wouda et al. (*129*)].

carried out in a number of laboratories, particularly to examine the effect of alloy formation on the reactions of importance in automotive catalysis: oxidation of CO, oxidation of hydrocarbons, and reduction of NO (*45, 79, 80, 114–116, 139–147*). NO bond breaking is usually considered the first step in the reduction of NO by CO and hydrogen and, therefore, many papers deal with studies of NO adsorption and dissociation (see Section II.A.4). NO adsorption is also a sensitive probe for examining the possible effect of alloy formation since there are large differences in the behavior of Pt and Rh toward NO interaction. The extent of dissociation of NO on Pt is sensitive to the surface structure. Rh has a much greater reactivity for NO bond breaking, and the effect of the surface structure is smaller.

At temperatures of about 210 K, complete dissociation of NO occurs on Rh(100) and on Pt–Rh(100) alloy surfaces at low NO coverages, and partial

FIG. 22. STM image of $Pt_{0.5}-Rh_{0.5}(100)$ exposed to O_2 at 770 K and subsequently cooled to room temperature in oxygen [reproduced with permission from Wouda *et al.* (*131*)].

dissociation occurs during heating following saturation. The extent of N_2 desorption is a measure of the activity of a surface in NO bond breaking. On the (100) surfaces of Pt, Rh, and $Pt_{0.25}-Rh_{0.75}$, NO desorption occurs at about 450 K. Figure 23 illustrates the effect of alloy formation on NO dissociation under TDS conditions. It shows the relative amount of NO desorbing as NO as a fraction of the total coverage following exposure at a temperature of 210 K or lower. On Rh(100) and on Pt–Rh(100) alloy surfaces, complete dissociation occurs at low NO coverages and partial dissociation occurs during heating following saturation (*141*).

In contrast to the behavior on Rh(100) and Pt–Rh(100) surfaces, the fraction of NO decomposing on Pt(100) does not change dramatically with coverage. The behavior of the Pt–Rh(100) surface resembles that of the pure Rh(100) surface at low NO coverage and that of pure Pt(100) at high coverage. This pattern illustrates that Rh atoms on the surface are very effective in NO dissociation and that NO dissociation occurs mainly on Rh sites. These results also indicate that mixed Pt–Rh sites are not very reactive in NO bond breaking (*141*).

NO dissociation is sensitive to the surface structure of Rh and, in particu-

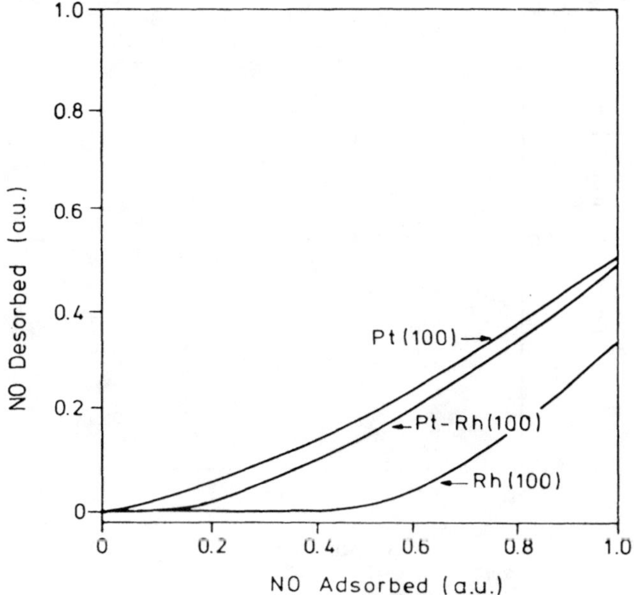

FIG. 23. Molecular NO desorption from Pt–Rh, Pt, and Rh(100). The abscissa is NO adsorbed as a fraction of saturation NO coverage [reproduced with permission from Siera *et al.* (*141*); Pt and Rh(100) data from Root *et al.* (*106*) and Gorte and Schmidt (*148*)].

lar, to that of Pt. For example, it has been reported that the Pt(111) surface cannot break the NO bond, whereas the Pt(410) surface is very reactive in NO bond breaking (*27*). XPS has been used to investigate the NO dissociation on Pt–Rh single-crystal surfaces (*114*); the results are summarized in Fig. 24. This figure shows the following: (i) The dissociation activity is sensitive to the surface structure. The (410) and (210) surfaces are more reactive in NO bond breaking than the (321) surface, the activity of which is larger than that of the (111) surface. The effect of the surface structure is greater for Pt-rich than for Rh-rich surfaces; and (ii) the dissociation reactivity is higher for Rh-rich than for Pt-rich surfaces.

Fisher *et al.* (*143, 144*) investigated NO adsorption and reduction on the (111) surface of a $Pt_{0.10}–Rh_{0.90}$ single crystal with a surface composition of about 30% Pt in vacuum. The presence of 10 atom% Pt in the bulk significantly reduces the ability of the surface to dissociate NO. The activation energy of NO dissociation was reported to be intermediate between those of the reaction on Rh(111) and on Pt(111).

An interesting behavior was found for the Pt–Rh(100) surface after exposure to NO at 500 K (*139*). Initially, a c(2 × 2) surface structure is

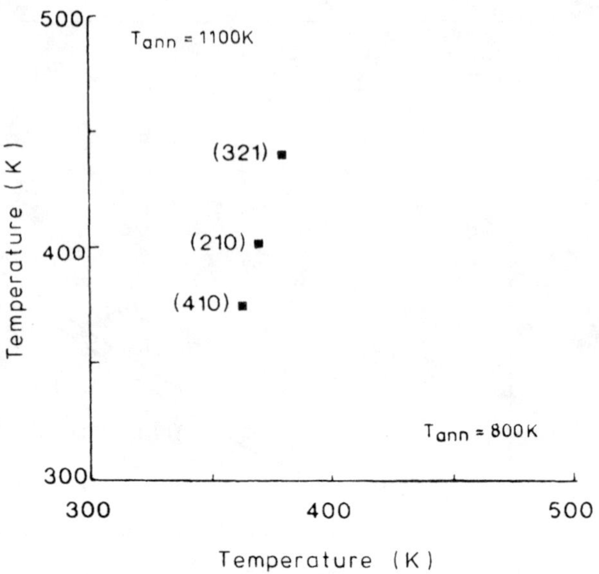

FIG. 24.   The temperature at which a dissociation percentage of 25% is obtained for Pt-rich Pt–Rh alloy surfaces versus the same parameter for Rh-rich Pt–Rh alloy surfaces [reproduced with permission from Wolf *et al.* (*114*)].

formed which, via a combination of c(2 × 2) and (3 × 1), changes slowly into a (3 × 1) surface structure. The time required for formation of the (3 × 1) structure depends strongly on the initial Rh surface concentration. On Rh-rich surfaces, the (3 × 1) structure is formed much faster than on an originally Pt-rich surface. The O formed by NO dissociation slowly replaces the N adatoms also formed by NO dissociation. The O atoms extract Rh atoms to the surface. Since the second layer is Rh enriched, the slow formation of the (3 × 1) structure is related to an exchange of Rh and Pt atoms between the second and the top layer.

Similar observations were made by Tanaka *et al.* (*146, 147*) for Pt–Rh surfaces prepared by electrochemical deposition of Pt on Rh(100) or Rh on Pt(100). All these studies demonstrate that the first two layers of Pt–Rh samples are very flexible; both composition and structure change easily with changing experimental conditions caused by exchange of atoms between the first and second layer. A pure Rh(100) surface is not very reactive for NO reduction by hydrogen at temperatures <650 K. A Pt-enriched Pt–Rh(100) surface shows reactivity at temperatures higher than 650 K. However, on a Rh-enriched Pt–Rh(100) surface with (3 × 1) structure NO reacts with hydrogen even at 500 K. The authors suggested that a Pt–Rh(100) hybrid surface with (3×1) structure is the active catalyst which can be formed during the catalytic process (*146, 147*).

In contrast to the activity of Pt(100), the activity of the Pt(110) surface is very low for the NO + H_2 reaction (*149*). Deposition of a submonolayer of Rh on this surface results in a drastic enhancement of the activity, which is also higher than that of pure Rh(110). The resulting p(1×2) Pt–Rh(110) catalyst exhibits an almost equally high activity as p(3×1) Pt–Rh(100). It was proposed that a specific Pt–RhO–Pt arrangement is needed for high activity (*149*).

The previous results show that Rh sites are those with the higher activity for NO dissociation. What do we know about CO adsorption? Does CO prefer Rh or Pt sites? These questions were addressed by Rutten *et al.* (*145*) in an investigation of CO adsorption on a $Pt_{0.25}$– $Rh_{0.75}$ (111) surface with RAIRS, TDS, and work function measurements. It was found that CO shows a pronounced preference for Rh sites at both 100 and 300 K up to a coverage of $\theta = 0.27$. Adsorption was found to be predominantly on atop sites, and only weak bridge bands were detected. Unfortunately, information concerning coadsorption of CO and NO on Pt–Rh single-crystal surfaces is missing. However, IR spectra of NO–CO mixtures are available for Pt–Rh supported on silica. Heezen *et al.* (*150*) concluded that for CO–NO mixtures the NO molecules are primarily adsorbed on Rh and the CO molecules on Pt atoms.

The $CO-O_2$ and CO–NO reactions have been investigated in the presence of several surfaces of $Pt_{0.25}-Rh_{0.75}$ and $Pt_{0.10}-Rh_{0.90}$ single crystals (*45, 119, 142–144*). The results were compared with data for pure Pt and Rh surfaces to examine the effect of alloy formation. This comparison is of interest for understanding the synergistic effects reported by many groups for the CO oxidation catalyzed by supported Pt–Rh (*151–153*). Siera *et al.* (*142*) argued that beneficial effects of Pt–Rh alloy formation might be related to the strong preference of oxygen for the Rh sites on the surface. This effect might result in a better mixing of CO and O on the surface (with the O bound to Rh and CO adsorbed on Pt) and, hence, in a faster reaction at lower temperature. Similar effects might be expected for CO + NO (*142*) because CO has a slight preference for Pt and NO for Rh. However, Siera *et al.* (*45, 80, 142*) did not find any indication of a synergistic effect due to alloy formation. This result is consistent with those obtained in the same laboratory for supported Pt–Rh alloy catalysts (*80, 114, 150*).

Large differences were found by Siera *et al.* (*45*) in the steady-state reaction rates of both reactions catalyzed by four $Pt_{0.25}-Rh_{0.75}$ alloy surfaces in the low-pressure range. An example is shown in Fig. 25 for the NO + CO reaction and in Fig. 9 for the CO + O_2 reaction.

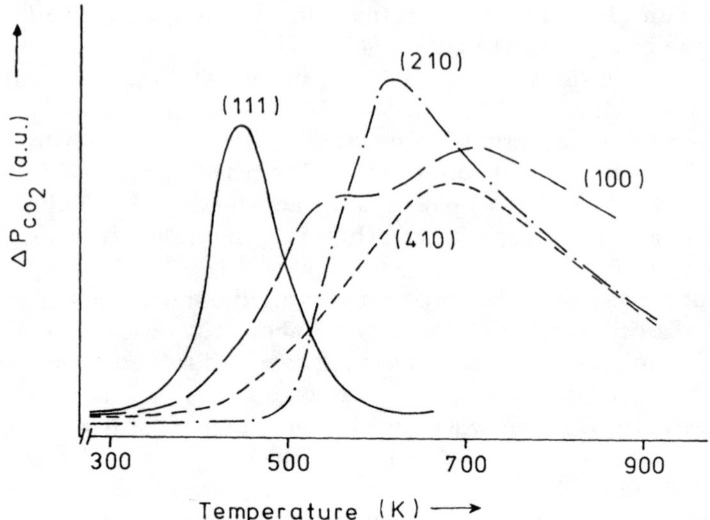

FIG. 25. Steady-state reaction rates of CO_2 production from CO + NO on Pt–Rh(111), (100), (410), and (210) under stoichiometric conditions [reproduced with permission from Siera *et al.* (45)].

This large effect of the surface structure is a remarkable observation for the CO + O_2 reaction since this reaction is usually considered to be a typical example of a reaction that is insensitive to the surface structure. The influence of the surface structure can be explained on the basis of the interaction of these surfaces with CO, NO, and O_2. A large CO inhibition occurs in the low temperature range (<400 K), especially for the CO + O_2 reaction. This effect is larger for the (210) surface than for the (111) surface due to the higher heat of adsorption of CO on the open (210) surface. The reaction order in CO becomes positive in the higher temperature range (550 K). A positive reaction order was found for oxygen in the whole temperature range, just as for pure Pt(111). However, for pure Rh(111) a negative reaction order in oxygen was found in the higher temperature range (154). This observation suggests that the Rh atoms in the Pt–Rh alloy are more difficult to oxidize in the presence of Pt atoms. Similar observations were made for the CO + NO reaction (45). In the low-temperature regime, the order in CO is negative. However, the reaction starts at significantly lower temperature, especially on the (111) surface, which is the least reactive surface for the N–O bond breaking. Hence, N–O dissociation is not the rate-determining step at low temperatures. For pure Rh(100) a reaction order of zero was reported by Oh *et al.* (155). Oh *et al.*

concluded that the Rh(100) surface is predominantly covered by NO. Pt–Rh alloy surfaces, on the other hand, are covered with both NO and CO at low temperatures. The CO molecules reside predominantly on the Pt atoms and the NO molecules on the Rh atoms in the alloy surface (*150*). Hence, it is likely that the Pt atoms in the alloy surface are covered with CO at low temperatures, causing the negative dependence of reaction rate on CO pressure.

Ng *et al.* (*144*) examined the NO–CO, CO–O_2, and CO–N_2O reactions on a $Pt_{0.10}$–$Rh_{0.90}$ (111) surface. The NO–CO activity of this alloy surface is similar to that of Rh(111) at temperatures from 573 to 648 K in that the two surfaces are represented by the same activation energy, reaction orders, and selectivity. The turnover frequencies are slightly lower than those for Rh(111) when compared on a per surface atom basis; however, the rates per surface Rh atom are virtually unchanged. The authors suggested that the primary effect of Pt is to dilute the Rh surface concentration.

Hirano *et al.* (*78–80*) investigated the reaction of NO with hydrogen on the (111), (100), and (410) surfaces of a $Pt_{0.25}$–$Rh_{0.75}$ single crystal and, for comparison, the pure Pt(100) and Rh(100) surfaces. Both the activity and the selectivity depend strongly on the surface structure and composition, as shown in Fig. 26.

In the low-temperature range ($\sim$ 520 K), the activity of the (100) surfaces decreases in the order Pt(100) > Pt–Rh(100) > Rh(100). The activity of pure Rh(100) is drastically enhanced by alloying with Pt. For the alloy surfaces, the order in activity at 520 K is (100) > (410) > (111). The order of intrinsic reactivity for NO bond breaking is Pt–Rh(410) > Rh(100) > Pt–Rh(100) > Pt(100) > Pt–Rh(111). Hence, these results indicate that NO bond scission is not the rate-determining step of the reaction. The selectivity toward N_2 decreases in the order Rh(100) $\sim$ Pt–Rh(410) > Pt(100) > Pt–Rh(111) at 520 K. It is well established that many vacant metal sites are required before NO dissociation can occur. If the surface is largely covered with molecularly adsorbed NO or with N adatoms, the reaction is inhibited by blocking of active sites. The conversion of NO at temperatures below 600 K is low for the Rh(100) surface due to the presence of strongly bound N atoms. The high selectivity of this surface is also caused by the high concentration of N adatoms. On Pt(100), the Pt–N bond strength is much weaker, resulting in a much lower concentration of N adatoms at 520 K. Free Pt sites remain available and, consequently, the reaction is fast but with a much lower selectivity toward N_2. The Pt–Rh(100) alloy surface shows a behavior intermediate between those of pure Pt and Rh(100).

The CO oxidation and NO reduction reactions have also been investigated in the presence of supported Pt–Rh catalysts. The results seem confusing: Whereas Oh and Carpenter (*151*) and Lyman *et al.* (*153, 156*) reported

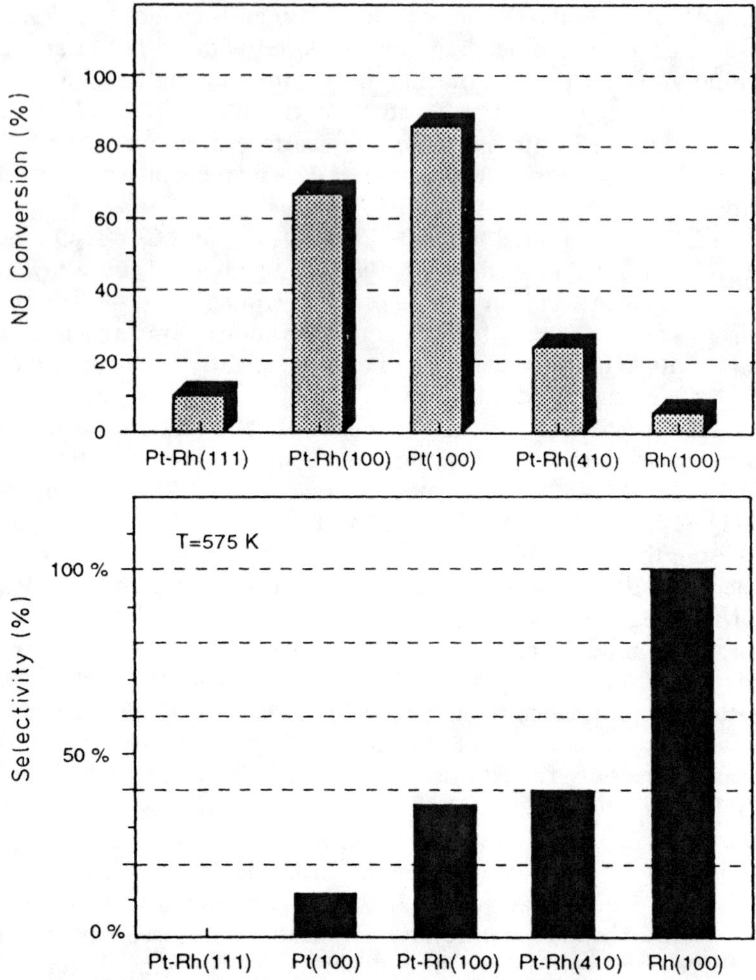

FIG. 26. Relative activity (NO conversion after 3 min at 520 K) and selectivity for N_2 (at a conversion of 10%) for the reduction of NO by hydrogen on Pt, Rh, and Pt–Rh alloy single-crystal surfaces [reproduced with permission from Hirano *et al.* (*79*)].

a better performance of some bimetallic Pt–Rh/Al_2O_3 catalysts than those of individual Pt or Rh, results of other groups did not support the evidence for a synergistic effect (*80, 114, 157, 158*). For illustration, the results for the NO reduction reactions reported by Nieuwenhuys *et al.* (*80, 114, 157*) are summarized in Fig. 27, which shows the temperature required to achieve a constant turnover frequency versus the bulk catalyst composition. Both

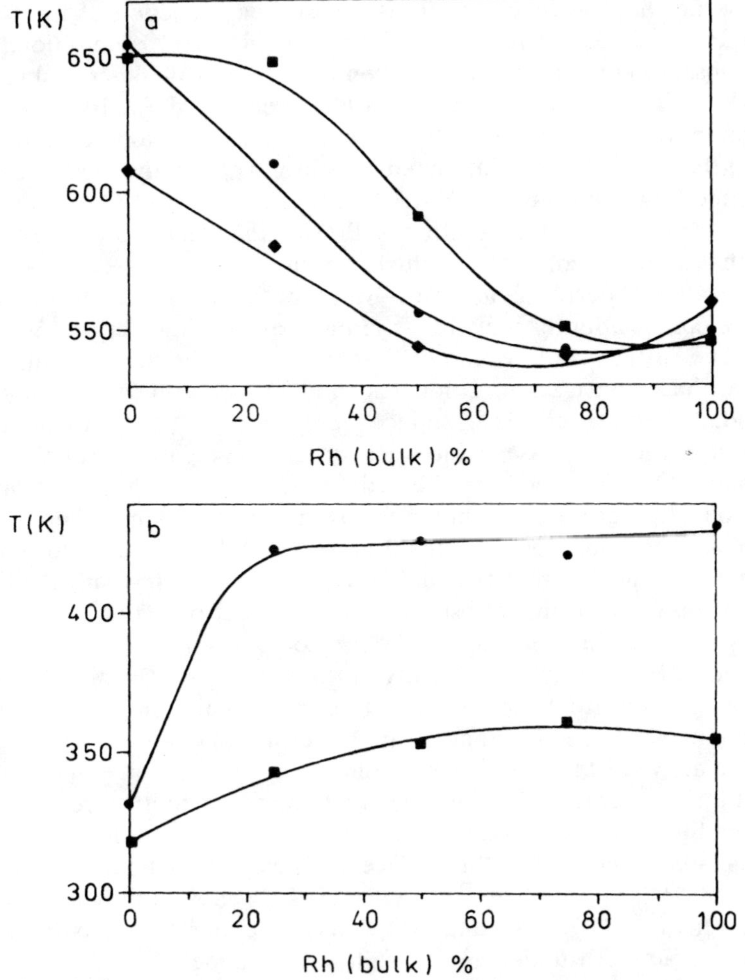

FIG. 27. (a) Temperature required for a constant turnover frequency of 0.05 s^{-1} for the NO + CO reaction on Pt–Rh on SiO$_2$ catalysts as a function of the bulk composition for three NO/CO ratios. ■, NO/CO = 1/4; ●, NO/CO = 1; ◆, NO/CO = 4. (b) Temperature required for a constant turnover frequency of 0.005 s^{-1} for the NO + H$_2$ reaction on Pt–Rh on SiO$_2$ catalysts as a function of the bulk composition for two NO/H$_2$ ratios. ●, NO/H$_2$ = 1; ■, NO/H$_2$ = 1/3 [reproduced with permission from Nieuwenhuys *et al.* (*80*)].

the activity and the selectivity of the Pt–Rh alloy catalysts are between those of the constituent metal catalysts. Rh/SiO$_2$ is more active in the NO + CO reaction than Pt/SiO$_2$. For this reaction the effect of changing feed composition is small for Rh and large for Pt. In uhv studies it was also found that the reaction rate is a very weak function of the CO/NO ratio. For Pt/SiO_2 the results point to a significant CO inhibition, as has been observed under low-pressure conditions on Pt single-crystal surfaces. Only in a large excess of NO can a CO precovered Rh(100) surface be saturated with NO (*159*). On Rh the reaction occurs at temperatures much higher than the NO dissociation temperature. Apparently, the reaction rate is limited by the presence of N on the Rh catalyst. The $NO + H_2$ reaction starts at a much lower temperature on Pt/SiO_2 than on Rh/SiO_2, an observation that is consistent with the single-crystal work, discussed previously.

The catalytic performance of Pt–Rh alloys has been correlated with the surface composition expected on the basis of the single-crystal work (*80, 114*). The surface composition of Pt–Rh alloys varies strongly with conditions such as the gas composition. Clean Pt–Rh alloy surfaces are enriched with Pt. Adsorbates can easily induce segregation of Pt or Rh to the surface. Oxygen in the gas phase induces Rh surface segregation. For the $NO + CO$ and $NO + H_2$ reactions, the activity of the $Pt_{0.5}-Rh_{0.5}$ alloy catalysts resembles that of pure Rh under net oxidizing conditions, whereas under net reducing conditions its activity is between those of Pt and Rh. This comparison suggests that the surface composition varies with the experimental conditions from almost pure Rh under net oxidizing conditions to, perhaps, a bulk-like or Pt-rich surface composition under net reducing conditions. However, the selectivity of this catalyst for the $NO + H_2$ reactions is different from those of both Pt and Rh. This comparison indicates that this catalyst contains Pt atoms in the surface that influence the selectivity. The activity of the $Pt_{0.75}-Rh_{0.25}$ alloy catalyst is almost equal to that of pure Pt under net reducing conditions, whereas its activity is between those of Pt and Rh under stoichiometric and net oxidizing conditions. Again, this comparison suggests that the surface composition changes with the feed composition. The properties of the $Pt_{0.25}-Rh_{0.75}$ alloy catalysts are notable. For both the $NO + CO$ and $NO + H_2$ reactions the activity is equal to that of pure Rh under net oxidizing, stoichiometric, and net reducing conditions. Under net oxidizing conditions, a behavior such as that of pure Rh may indeed be expected. However, under net reducing conditions the presence of both Pt and Rh atoms should be expected on the surface. The relatively high activity of this catalyst for the $NO + CO$ reaction might be caused by a beneficial effect of the presence of both Rh and Pt, i.e., the low CO inhibition on Rh sites and a beneficial effect of Pt on, for example, the amount of N or NO on the surface. For the $NO + H_2$ reaction, however, the presence of Pt atoms does not seem to diminish the inhibition effect of NO, probably because of the lower reaction temperatures.

The synergistic effect reported by Lyman *et al.* (*153, 156*) was found only for a Pt-rich catalyst with a very homogeneous composition of the particles.

Bimetallic catalysts with a bimodal particle composition distribution (Rh-rich and Pt-rich particles) did not show synergistic performance. Cai *et al.* (*153*) suggested that the absence of synergism in the observations of Nieuwenhuys *et al.* (*80, 114, 157*) may be attributed to the presence of bimodal particle composition distribution or to the low operating temperature regime used by Nieuwenhuys *et al.* due to the high activity of their catalysts. It is emphasized that the synergism observed by Cai *et al.* (*153*) is very small, smaller than 9 K expressed in terms of the temperature needed to achieve a conversion of 50%. In addition, the effect was not observed for the Pt–Rh catalysts with other compositions than 96% Pt, including a very Pt-rich $Pt_{0.99}$–$Rh_{0.01}$ catalyst. The synergistic effect found for the $Pt_{0.96}$–$Rh_{0.04}$ catalyst was explained by a beneficial effect of Pt on Rh reduction and by a higher probability of having both reactant species on neighboring sites.

In conclusion, the catalytic properties of supported Pt–Rh alloys are strongly dependent on the gas-phase composition, the bulk composition of the alloy particles, and the reaction temperature. A combination of factors determines the activities of the different catalysts. Under oxidizing conditions at sufficiently high temperatures, Rh segregation to the surface occurs. This is reflected in the catalytic activity of the Pt–Rh alloys which can vary between that of pure Pt and that of pure Rh. Large synergistic effects caused by alloying of Pt with Rh have not been observed. Differences in activity and selectivity for catalysts with different Pt–Rh compositions can be explained in terms of specific properties of the pure metals toward the adsorbates. Although most results show that the behavior of the alloy catalysts varies between those of Pt and Rh, care must be taken when interpreting the results in terms of absolute surface concentrations of Pt and Rh. The activity of an alloy catalyst depends on the surface composition of the metal particles. However, the relationship between activity, selectivity, and surface composition may be complicated for many reasons. The catalytic behavior varies with the concentration and the size of the various ensembles of atoms on the surface. All reaction steps (molecular adsorption, dissociation, and reaction between the species on the surface) may depend in different ways on the distribution of atoms over the surfaces. Furthermore, the small effect of alloying on the binding energies of adsorbed atoms and molecules may play a role in the catalytic performance of the alloys.

VI. Effects of the Additives Cerium and Lanthanum Oxides

It was noted in the introduction that the washcoat of the automobile catalysts contains several other oxides, mainly cerium and lanthanum oxide.

These rare earth oxides make an important contribution to catalyst performance and durability and have multiple functions (*160*). The ceria content of current three-way catalyst formulations is about 25 wt% (as cerium) of the washcoat. Ceria was originally added to increase the stability and dispersion of the noble metal (*161*). Ceria and, in particular, lathana are thought to stabilize the γ-Al_2O_3 support by inhibiting its structure change from the γ to the α modification (*162–164*). An important difference between the two rare earth oxides is the facile change of the oxidation state of Ce, whereas La is valence invariant (La_2O_3) under operation conditions. Cerium oxide can be partly reduced from CeO_2 to give oxygen-deficient CeO_{2-x} (x is near one-half) and reoxidized to CeO_2 under oxidizing conditions. Ceria is a chemically active component, as an oxygen buffer component (*165*), as a promoter for the noble metal catalyst, and as a catalyst for the water–gas shift reaction (*166*)

$$CO + H_2O \leftrightarrows CO_2 + H_2. \tag{35}$$

This reaction leads to an additional increase in CO conversion and to the formation of hydrogen, which has a beneficial effect on NO reduction. The oxygen storage capacity of ceria results in a widening of the effective air/fuel ratio at which the reduction and oxidation reactions can operate during the oscillatory cycle. Under fuel-rich (reducing) conditions, the stored oxygen is released and is available for the oxidation of CO and hc; oxygen is stored during fuel-lean (oxidizing) conditions, thereby enhancing NO reduction to N_2.

Recently, direct promoter effects of ceria on the catalytic properties of noble metals have been reported (*167, 168*). Addition of ceria to Rh/Al_2O_3 was found to improve NO reduction at low temperatures, with a decrease in the apparent activation energy for the CO + NO reaction (*168*). A large effect of the particle size of ceria was found for NO reduction catalyzed by a $Pt/ceria/Al_2O_3$ (*169*). The authors proposed that the Pt–ceria interaction increases by reduction of the ceria particle size. Although these promoter effects of ceria on the noble metals are well documented, the origin of the effects is not fully understood. The few surface science studies directed at understanding these promoter effects provide some novel information. Zafiris and Gorte (*170*) examined the structure and adsorption properties of Pt and Rh deposited on amorphous ceria films. They reported that ceria-supported Pt exhibits adsorption properties very similar to those of Al_2O_3-supported Pt for CO, NO, and H_2. In contrast, large effects were found for Rh on ceria; adsorption of CO and of NO is affected by ceria, and CO adsorption results in a CO_2 formation peak during TDS. It was suggested that O can migrate from ceria to the Rh surface at temperatures near room temperature.

Schmidt and Krause (*171*) and Schmidt and Schwartz (*172*) observed dramatic changes in the microstructures of Rh supported on thin films of SiO_2 and Al_2O_3 during heating in an atmosphere of NO, CO, or a mixture of NO + CO. Rh on silica is volatilized even by treatment in NO + CO at 570 K, but the addition of ceria retards this effect. Heating Rh on alumina in NO + CO had much less effect. However, heating in NO alone resulted in redispersion of Rh into small (20-Å) particles on the Al_2O_3. It was speculated that a mobile nitrosyl complex is formed during heating in NO, resulting in redispersion, and that volatilization occurs via formation of metal carbonyls.

Hardacre *et al.* (*173, 174*) investigated the properties, structure, and composition of cerium oxide films prepared by cerium deposition on Pt(111), finding that the activity for CO oxidation is enhanced on Pt(111) that is partially covered by ceria. It was suggested that new sites at the Pt–oxide interface become available for reaction. A remarkable observation is the high activity for CO oxidation when the Pt(111) sample is fully encapsulated by ceria (Pt was undetectable by XPS and AES). It was proposed that an ultrathin, disordered ceria film becomes the active catalyst. It was also demonstrated by XPS and AES that Pt dramatically increases the reducibility of cerium oxide that is in intimate contact with Pt. This result suggests that intimate contact between the noble metal and oxide phases is indeed crucial to facile oxygen release from ceria. High-resolution electron microscopy demonstrated the presence of direct contact between ceria and noble metal for supported Pt–Rh catalysts (*175*). Hardacre *et al.* (*173, 174*) related the catalytic activity of the ceria phase to partially reduced cerium oxide.

In conclusion, there is overwhelming evidence for the beneficial effect of cerium oxide on the activity of the noble metal catalyst. However, the nature of the "promoter" effect of ceria is not fully understood. Most likely, the noble metal–cerium oxide interface is of crucial importance for some of the effects observed. More studies with model systems are needed for a better understanding of the promoter effects of ceria.

VII. Summary, Assessment, and Forecast

The automotive TWC is one of the major achievements of modern research in heterogeneous catalysis. There have been major efforts to elucidate the fundamental reaction pathways and the catalyst characteristics that account for the success of the TWC. The chemistry involved is understood in considerable detail as a result of work with idealized models of the TWC. The mechanisms of the CO oxidation and NO reduction reactions with

CO, hydrogen, and hc have been discussed in detail in this review. In particular, the application of various modern surface science techniques has advanced our understanding of the principles of the reactions on the atomic scale and the relationship of surface composition and structure to catalytic properties.

Surface science offers many opportunities in catalysis research because a variety of techniques are available to characterize in detail the composition and structure of the catalyst surface and to identify the adsorbed species. A frequent criticism of the surface science approach is that it is far removed from real catalysis since most of the surface science techniques can only be applied at low pressures and with "model" catalysts, often single-crystal surfaces. The so-called pressure gap has been bridged by combining, in the same apparatus, the techniques needed for surface analysis and characterization with the ability to measure reaction rates at elevated pressures. In addition, many techniques can also be applied *in situ* at elevated pressures.

In this review, literature data concerning CO oxidation and NO reduction on model catalysts have been reviewed and compared with those reported for supported catalysts. The major differences in behavior of the three noble metals—Pt, Pd, and Rh—used in TWC have been assessed. It is concluded that the major mechanisms are reactions of the L–H type between O_{ads}, CO_{ads}, and the dissociation products of NO, viz., N_{ads} and O_{ads}, with N_2 formed by combination of 2 N_{ads}, NH_3 by hydrogenation of N_{ads}, and N_2O by reaction between N_{ads} and NO_{ads}. Although other mechanisms have been proposed and their possible existence cannot be ruled out, the effects of the surface composition and structure, the specific differences in behavior of Pt, Pd, and Rh, the effect of changes in temperature, and variations in partial pressures can be fully understood on the basis of these reaction pathways.

The effects of alloy formation and the chemistry of the additive cerium oxide, although less well understood, have also been evaluated in detail. It was shown that the catalytic properties of Pt–Rh alloy catalysts are strongly dependent on the gas-phase composition and the bulk composition of the alloy particles. Many factors determine the activities of these catalysts. Under oxidizing conditions, Rh segregation to the surface occurs. This is reflected in the catalytic activity of the Pt–Rh alloys, which can vary between that of pure Pt and that of pure Rh. No large synergistic effects resulting from the alloying of Pt with Rh have been observed. The differences in activity observed for the different catalysts can be explained in terms of the specific properties of the pure metals toward the adsorbates. The activity and the selectivity of the Pt, Rh, and Pt–Rh surfaces depend on the surface structure and composition. The selectivity is determined by the relative concentrations of CO, NO, N, and H adsorbed on the various surfaces.

In fresh automotive exhaust catalysts, the precious metals are well distributed in the form of small crystallites with sizes between 0.5 and 5 nm. During aging, a severe loss of Pt surface area occurs due to sintering. Rh, however, remains in a highly dispersed state because it is partly in an oxidized state, whereas the Pt particles remain metallic. The morphology and composition of the catalyst particles change continuously during operation. Pt–Rh alloy particles are formed under reducing conditions. Under oxidizing conditions dealloying occurs, resulting in rhodium oxide particles being separated from metallic Pt and Pt–Rh alloy particles. During reaction, cerium switches between oxidation states leading to changes in the noble metal–Al_2O_3 (especially Rh) interactions. Ce-noble metal and Ce-support interactions can occur. Hence, a continuous restructuring and modification of catalytic behavior can occur. All these effects make the automotive TWC one of the most dynamic catalysts in use.

Interest in the control of emissions from automobiles will continue to grow because of increasingly stringent legislation worldwide for emission of hc, CO, NO_x, and particulates. Major new developments include (i) the introduction of a small catalytic converter, combined with the main converter containing the TWC, close to the engine manifold, enabling quicker light-off and therefore overall better CO and hc conversion and (ii) the development of catalytic converters for lean-burn gasoline-fueled engines and for diesel engines. Lean-burn versions of the four-stroke engine have been in development for many years. From the perspective of fuel economy and, hence, CO_2 emission, the use of lean-burn engines is desirable. The difficulty is in achieving NO_x conversion under such oxidizing conditions. The most thoroughly investigated catalysts are zeolites containing metal ions, in particular, Cu/ZSM5, Co/ZSM5, and other metal–zeolite systems. Unfortunately, these zeolites are deactivated at high temperatures in the presence of water (because of destruction of the zeolite structure), which is an inevitable component of automotive emissions. Although some reports suggest that many metal-containing zeolites, such as Fe/ZSM5 (*176, 177*) and cerium-exchanged mordenite (*178*), are stable in water vapor, it is not certain that a zeolite-based catalyst that meets the legislation standards for emission and catalyst lifetime will be developed. New technologies that are developing include NO_x catalysts that can reduce NO_x under oxidizing conditions, adsorbers/absorbers for NO_x and hc, particulate traps that are regenerated by catalytic oxidation using Pt, and novel catalysts that can convert CO immediately after the cold start.

REFERENCES

1. Taylor, K. C., *Catal. Sci. Technol.* **5,** 119 (1984).
2. Taylor, K. C., *Catal. Rev. Sci. Eng.* **35,** 457 (1993).

3. Shelef, M., and Graham, G. W., *Catal. Rev. Sci. Eng.* **36**, 433 (1994).
4. Bowker, M., and Joyner, R. W., *Insights Spec. Inorg. Chem.* 145 (1995).
5. Ramsier, R. D., and Yates, J. T., *Surf. Sci. Rep.* **12**, 243 (1991).
6. Winkler, A., and Rendulic, K. D., *Surf. Sci.* **118**, 19 (1982).
7. Nieuwenhuys, B. E., *Surf. Sci.* **126**, 307 (1983); *NATO-ASI Ser.* C **398**, 155 (1993) and references therein.
8. Gland, J. L., Sexton, B. A., and Fisher, G. B., *Surf. Sci.* **95**, 587 (1980).
9. Campuzano, J. P., *Chem. Phys. Solid State Surf. Sci. Catal.* **3A**, 389 (1991).
10. Blyholder, G., *J. Phys. Chem.* **68**, 2772 (1964).
11. Bagus, P. S., Nelin, C. J., and Bauschlicher, C. W., *J. Vac. Sci. Technol.* **A2**, 905 (1984).
12. Muller, W., and Bagus, P. S., *J. Vac. Sci. Technol.* **A3**, 1623 (1985).
13. Rangelov, G., Memmel, N., Bertel, E., and Dose, V., *Surf. Sci.* **251**, 965 (1991).
14. Nieuwenhuys, B. E., *Surf. Sci.* **105**, 505 (1981).
15. Eischens, R. P., Pliskin, W. A., and Francis, S. A., *J. Chem. Phys.* **22**, 1786 (1954).
16. Sheppard, N., and Nguyen, T. T., *in* "Advances in IR and Raman Spectroscopy" (R. E. Hester and R. J. H. Clark, Eds.), Vol. 5. Heyden, London, 1978.
17. Broden, G., Rhodin, T. N., Brucker, C., Bendow, R., and Hurych, Z., *Surf. Sci.* **59**, 593 (1976).
18. Benzinger, J. B., *Appl. Surf. Sci.* **6**, 105 (1980).
19. Gorodetski, V. V., and Nieuwenhuys B. E., *Surf. Sci.* **105**, 299 (1981).
20. Hendrickx, H. A. C. M., and Nieuwenhuys, B. E., *Surf. Sci.* **175**, 185 (1986).
21. Wolf, R. M., Bakker, J. W., and Nieuwenhuys B. E., *Surf. Sci.* **246**, 135 (1991).
22. van Hardeveld, R. M., Thesis, Technology University of Eindhoven, 1997.
23. Borg, H. J., Reijerse, J. F. C. J. M., van Santen, R. A., and Niemantsverdriet, J. W., *J. Chem. Phys.* **101**, 10051 (1994).
24. Sandell, A., Nilsson, A., and Martensson, N., *Surf. Sci.* **241**, L1 (1991).
25. Somorjai, G. A., *Surf. Sci.* **107**, 573 (1981).
26. Gorte, R. J., Schmidt, L. D., and Gland J. L., *Surf. Sci.* **109**, 367 (1981).
27. Masel, R. I., *Catal Rev.-Sci. Eng.* **28**, 335 (1986).
28. Janssen, N. M. H., Cholach, A. R., Ikai, M., Tanaka, K., and Nieuwenhuys, B. E., *Surf. Sci.* **382**, 201 (1997).
29. Solymosi, F., *J. Mol. Catal.* **65**, 337 (1991).
30. van Tol, M. F. H., Gielbert, A., Wolf, R. M., Lie, A. B. K., and Nieuwenhuys, B. E., *Surf. Sci.* **287/288**, 201 (1993).
31. Loffler, D. G., and Schmidt, L. D., *Surf. Sci.* **59**, 195 (1976).
32. Conrad, H., Ertl, G., and Küppers, J., *Surf. Sci.* **76**, 323 (1978).
33. Lambert, R. M., and Comrie, C. M., *Surf. Sci.* **46**, 61 (1974).
34. Thiel, P. A., Weinberg, W. H., and Yates, J. T., *J. Chem. Phys.* **71**, 1643 (1979).
35. King, D. A., *Faraday Disc. Chem. Soc.* **87**, 303 (1989).
36. van Slooten, R. F., and Nieuwenhuys, B. E., *J. Catal.* **122**, 429 (1990).
37. Gardner, S. D., Hofland, G. B., Upchurch, B. T., Schryer, D. R., Kielin, E. J., and Schryer, J., *J. Catal.* **129**, 114 (1991).
38. Ertl, G., *Adv. Catal.* **37**, 213 (1990).
39. Engel, T., and Ertl, G., *J. Chem. Phys.* **69**, 1267 (1978).
40. Campbell, C. I., Ertl, G., Kuipers, H., and Segner, J., *J. Chem. Phys.* **73**, 5862 (1980).
41. Gland, J. L., Sexton, B. A., and Fisher, G. B., *Surf. Sci.* **95**, 587 (1980).
42. Su, C. X., Cremer, P. S., Shen, Y. R., and Somorjai, G. A., *J. Am. Chem. Soc.* **119**, 3994 (1997).
43. Li, Y., Boecker, D., and Gonzalez, R. D., *J. Catal.* **110**, 319 (1988).

44. Wintterlin, J., Völkening, S., Janssens, T. V. W., Zambelli, T., and Ertl, G., *Science* **278**, 1931 (1997).
45. Siera, J., Rutten, F., and Nieuwenhuys, B. E., *Catal. Today* **10**, 353 (1991).
46. Cobden, P. D., Siera, J., and Nieuwenhuys, B. E., *J. Vac. Sci. Technol.* **A10**, 2487 (1992).
47. Sales, B. C., Turner, J. E., and Maple, M. B., *Surf. Sci.* **114**, 381 (1982).
48. Janssen, N. M. H., Cobden, P. D., and Nieuwenhuys, B. E., *J. Phys. Condens. Matters* **9**, 1889 (1997).
49. Imbihl, R., and Ertl, G., *Chem. Rev.* **95**, 697 (1995).
50. Mergler, Y. J., van Aalst, A., van Delft, J., and Nieuwenhuys, B. E., *Appl. Catal.* **B10**, 245 (1996).
51. Mergler, Y. J., and van Aalst, A., *ACS Symp. Ser.,* **587**, 196 (1995).
52. Mergler, Y. J., and Nieuwenhuys, B. E., *Appl. Catal.* **B12**, 95 (1997).
53. Mergler, Y. J., van Aalst A., van Delft, J., and Nieuwenhuys, B. E., *J. Catal.* **161**, 310 (1996).
54. Mergler, Y. J., Hoebink, J., and Nieuwenhuys, B. E., *J. Catal.* **167**, 305 (1997).
55. Matsushima T., *Surf. Sci.* **127**, 403 (1983).
56. Wintterlin, J., Schuster, R., and Ertl, G., *Phys. Rev. Lett.* **77**, 123 (1996).
57. Brune, H., Wintterlin, J., Trost, J., Ertl, G., Weichers, J., and Behm, R. J., *J. Chem. Phys.* **99**, 2128 (1993).
58. Fadeev, E., Gorodetskii, V., and Smirnov, M., *in* "Proceedings of the Boreskov Memorial Symposium" (1997) Novosibirsk, Russia.
59. Watanabe, K., Uetsuka, H., Ohnuma, H., and Kunimori, K., *Catal. Lett.* **47**, 17 (1997).
60. Mullins, C. B., Rettner, C. T., and Auerbach, D. J., *J. Chem. Phys.* **95**, 8649 (1991).
61. Wartnaby, C. E., Stuck, A., Yeo, Y. Y., and King, D. A., *J. Chem. Phys.* **102**, 1855 (1995).
62. Boudart, M., *J. Mol. Catal. A* **120**, 271 (1997) (and references therein).
63. Szanyi, J., and Goodman, D. W., *J. Phys. Chem.* **98**, 2972 (1994).
64. Szanyi, J., Kuhn, W. K., and Goodman, D. W., *J. Phys. Chem.* **98**, 2978 (1994).
65. Peden, C. H. F., *ACS Symp. Ser.* **482**, 143 (1992).
66. Peden, CH. F., Goodman, D. W., Blair, D. S., Berlowitz, P. J., Fisher, G. B., and Oh, S. H., *J. Phys. Chem.* **92**, 1563 (1988).
67. Oh, S. H., Fisher, G. B., Carpenter, J. E., and Goodman, D. W., *J. Catal.* **100**, 360 (1986).
68. Su, X., Jensen, J., Yang, M. X., Sameron, M. B., Shen, Y. R., and Somorjai, G. A., *Faraday Disc.* **105**, 263 (1996).
69. Bowker, M., Guo, Q., and Joyner, R. W., *Catal. Lett.* **18**, 119 (1993).
70. Kruse, N., and Gaussmann, A., *Surf. Sci.* **266**, 51 (1992).
71. Schwartz, S. B., Fisher, G. B., and Schmidt, L. D., *J. Phys. Chem.* **92**, 389 (1988).
72. van den Brink, P. J., and McDonald, C. R., SAE Tech. Paper No. 952399 (1995); *Appl. Catal. B* **6**, L97 (1995).
73. Kobylinski, T. P., and Taylor, B. W., *J. Catal.* **33**, 376 (1974).
74. Solymosi, F., Völgyesi, L., and Sárkány, J., *J. Catal.* **53**, 336 (1978).
75. Bamwenda, G. R., Obuchi, A., Ogata, A., and Mizuno, K., *Chem. Lett.* **2109**, (1994).
76. van Hardeveld, R. M., Schmidt, A. J. G. W., van Santen, R. A., and Niemantsverdriet, J. W., *J. Vac. Sci. Technol.* **A15**, 1558 (1997).
77. Chin, A., and Bell, A. T., *J. Phys. Chem.* **87**, 3700 (1983).
78. Hirano, H., Yamada, T., Tanaka, K., Siera, J., Cobden, P. D., and Nieuwenhuys, B. E., *Surf. Sci.* **262**, 97 (1992).
79. Hirano, H., Yamada, T., Tanaka, K., Siera, J., and Nieuwenhuys, B. E., *Stud. Surf. Sci. Catal.* **75**, 345 (1993).
80. Nieuwenhuys, B. E., Siera, J., Tanaka, K., and Hirano, H., *ACS Symp. Ser.* **552**, 114 (1994).
81. Yamada, T., and Tanaka, K., *J. Am. Chem. Soc.* **113**, 1173 (1991).

82. Zemlyanov, D. Y., Smirnov, M. Y., Gorodetskii, V. V., Block, J. H., *Surf. Sci.* **329,** 61 (1995).

83. Zemlyanov, D. Y., Smirnov, M. Y., and Gorodetskii, V. V., *Surf. Sci.,* in press.

84. Sun, Y. M., Sloan, S., Ihm, H., and White, J. M., *J. Vac. Sci. Technol.* **A14,** 1516 (1996).

85. van Hardeveld, R. M., van Santen, R. A., and Niemantsverdriet, J. W., *J. Phys. Chem.* **101,** 998 (1997).

86. Dietrich, H., Jacobi, K., and Ertl, G., *Surf. Sci.* **377–379,** 308 (1997); *J. Chem. Phys.* **106,** 9313 (1997).

87. Williams, C. T., Tolia, A. A., Weaver, M. J., and Takoudis, C. G., *Chem. Eng. Sci.* **51,** 1673 (1996).

88. Tolia, A. A., Williams, C. T., Takoudis, C. G., and Weaver, M. J., *J. Phys. Chem.* **99,** 4599 (1995).

89. Hecker, W. C., and Bell, A. T., *J. Catal.* **92,** 247 (1985).

90. Kiskinova, M., Lizzit, S., Comelli, G., Paolucci, G., and Rosei, R., *Appl. Surf. Sci.* **64,** 185 (1993).

91. Cobden, P. D., Nieuwenhuys, B. E., Esch, F., Baraldi, A., Comelli, G., Lizzit, S., and Kiskinova, M, *J. Vac. Technol. A,* **16,** 1014 (1999).

92. Klein, R. L., Schwartz, S., and Schmidt, L. D., *J. Phys. Chem.* **89,** 4908 (1985).

93. Lesley, M. W., and Schmidt, L. D., *Surf. Sci.* **155,** 215 (1985).

94. Li, Y., Slager, T. L., and Armor, J. N., *J. Catal.* **150,** 388 (1994).

95. Witzel, F., Sill, G. A., and Hall, W. K., *J. Catal.* **149,** 229 (1994).

96. Adelman, B. J., Beutel, T., Lei, G. D., and Sachtler, W. M. H., *Appl. Catal.* **B11,** L1 (1996).

97. Salama, T. M., Ohnishi, R., and Ichikawa, M., *Chem. Commun.* **105,** (1997).

98. Otto, K., Shelef, M., and Kummer, J. T., *J. Phys. Chem.* **74,** 2690 (1970).

99. Xu, H., and Ng, K. Y. S., *Appl. Phys. Lett.* **68,** 496 (1996); *Surf. Sci.* **365,** 779 (1996).

100. Leibsle, F. M., Murray, P. W., Francis, S. M., Thornton, G. T., and Bowker, M., *Nature* **363,** 706 (1993).

101. Hecker, W. C., and Bell, A. T., *J. Catal.* **84,** 200 (1983); **85,** 389 (1984).

102. Belton, D. N., and Schmieg, S. J., *J. Catal.* **144,** 9 (1993); **138,** 70 (1992).

103. Belton, D. N., DiMaggio, C. L., Schmieg, S. J., and Ng, K. Y. S., *J. Catal.* **157,** 559 (1995).

104. Makeev, A. G., and Slinko, M. M., *Surf. Sci.* **359,** L467 (1996).

105. Bugyi, L., and Solymosi, F., *Surf. Sci.* **188,** 475 (1987).

106. Root, T. W., Schmidt, L. D., and Fisher, G. B., *Surf. Sci.* **134,** 30 (1983).

107. Permana, H., Ng, K. Y. S., Peden, C. H. F., Schmieg, S. J., Lambert, D. K., and Belton, D. N., *J. Catal.* **164,** 194 (1996).

108. Zhdanov, V. P., and Kasemo, B., *Surf. Sci. Rep.* **29,** 31 (1997).

109. Ikai, M., He, H., Borroni-Bird, C. E., Hirano, H., and Tanaka, K., *Surf. Sci.* **315,** L973 (1994).

110. Ikai, M., and Tanaka, K., *Surf. Sci.* **357/358,** 781 (1996).

111. Ikai, M., Thesis, University of Tokyo, 1996.

112. Rainer, D. R., Vesecky, S. M., Koranne, M., Oh, W. S., and Goodman, D. W., *J. Catal.* **167,** 234 (1997).

113. Kummer, J. T., *J. Phys. Chem.* **90,** 4747 (1986).

114. Wolf, R. M., Siera, J., van Delft, F. C. M. J. M., and Nieuwenhuys, B. E., *Faraday Disc. Chem. Soc.* **87,** 275 (1989).

115. van Delft, F. C. M. J. M., Nieuwenhuys, B. E., Siera, J., and Wolf, R. M., *Iron Steel Inst. Jpn. Int.* **29,** 550 (1989).

116. Nieuwenhuys, B. E., *Surf. Rev. Lett.* **3,** 1869 (1996).

117. Kim, S., and d'Aniello, M. J., Jr., *Appl. Catal.* **56,** 23, 45 (1989).

118. Maire, F., Capelle, M., Meunier, G., Beziau, J. F., Bazin, D., Dexpert, H., Garin, F., Schmitt, J. L., and Maire, G., *Stud. Surf. Sci. Catal.* **96,** 749 (1995).
119. Nieuwenhuys, B. E., *in* "The Chemical Physics of Solid Surfaces" (D. A. King and D. P. Woodruff, Eds.), Vol. 6, p. 185. Elsevier, Amsterdam, 1993.
120. Kuijers, F. J., Tieman, B., and Ponec, V., *Surf. Sci.* **75,** 657 (1978).
121. Batirev, I. G., and Leiro, J. A., *J. Electr. Spectr. Rel. Phen.* **71,** 79 (1995).
122. Siera, J., van Delft, F. C. M. J. M., van Langeveld, A. D., and Nieuwenhuys, B. E., *Surf. Sci.* **264,** 435 (1992).
123. Tsong, T. T., *Surf. Sci. Rep.* **8,** 127 (1988).
124. Beck, D. D., DiMaggio, C. L., and Fisher, G. B., *Surf. Sci.* **297,** 293, 303 (1993).
125. Altenstaedt, W., and Leisch, M., *Appl. Surf. Sci.* **94/95,** 403 (1996).
126. LeGrand, B., and Treglia, G., *Surf. Sci.* **236,** 398 (1990).
127. Schoeb, A. M., Raeker, T. J., Yang, L., Wu, X., King, T. S., and De Pristo, A. E., *Surf. Sci. Lett.* **278,** L125 (1992).
128. Matsumoto, Y., Okawa, Y., Fujita, T., and Tanaka K., *Surf. Sci.* **355,** 109 (1996).
129. Wouda, P. T., Nieuwenhuys, B. E., Schmid, M., and Varga, P., *Surf. Sci.* **359,** 310 (1996).
130. Matsumoto, Y., Aibara, Y., Mukai, K., Moriwaki, K., Okawa, Y., Nieuwenhuys, B. E., and Tanaka, K., *Surf. Sci.* **377–379,** 32 (1997).
131. Wouda, P. T., Schmid, M., Hebenstreit, W., and Varga, P., *Surf. Sci.* **388,** 63 (1997).
132. Zhu, Y., and Schmidt, L., *Surf. Sci.* **129,** 107 (1983).
133. Wang, T., and Schmidt, L., *J. Catal.* **70,** 187 (1981).
134. Chen, M., and Schmidt, L., *J. Catal.* **56,** 198 (1979).
135. Graham, G. W., Potter, T., Baird, R. J., Ghandi, H. S., and Shelef, M., *J. Vac. Sci. Technol.* **A4,** 1613 (1986).
136. Wang, Z., Ansermet, J. P., Slichter, C. P., and Sinfelt J. H., *J. Chem. Soc. Faraday Trans I* **84,** 3785 (1988).
137. Savargaonkar, N., Khanra, B. C., Pruski, M., and King, T. S., *J. Catal.* **162,** 277 (1996).
138. Gorodetski, V. V., Nieuwenhuys, B. E., Sachtler, W. M. H., and Boreskov, G. K., *Surf. Sci.* **108,** 225 (1981).
139. Yamada, T., Hirano, H., Tanaka, K., Siera, J., and Nieuwenhuys, B. E., *Surf. Sci.* **226,** 1 (1990).
140. Tanaka, K. I., Yamada, T., and Nieuwenhuys, B. E., *Surf. Sci.* **242,** 503 (1991).
141. Siera, J., Koster van Groos, M. J., and Nieuwenhuys, B. E., *Recl. Trav. Chim. Pays–Bas.* **109,** 127 (1990).
142. Siera, J., van Silfhout, R., Rutten, F. J. M., and Nieuwenhuys, B. E., *Stud. Surf. Sci. Catal.* **71,** 395 (1991).
143. Fisher, G. B., Dimaggio, C. L., and Beck, D. D., *Stud. Surf. Sci. Catal.* **75,** 383 (1993).
144. Ng, K. Y. S., Belton, D. N., Schmieg, S. J., and Fisher, G. B., *J. Catal.* **146,** 394 (1994).
145. Rutten, F. J. M., Nieuwenhuys, B. E., McCoustra, M. R. S., Hollins, P., and Chesters, M. A., *J. Vac. Sci. Technol.* **A15,** 11619 (1997).
146. Taniguchi, M., Kuzembaev, E. K., and Tanaka, K., *Surf. Sci.* **290,** L711 (1993).
147. Tamura, H., Sasahara, A., and Tanaka, K., *Surf. Sci.* **303,** L379 (1994).
148. Gorte, R. J., and Schmidt, L. D., *Surf. Sci.* **111,** 260 (1981).
149. Sasahara, A., Tamura, H., and Tanaka, K., *Catal. Lett.* **28,** 161 (1994); *J. Phys. Chem.* **100,** 15229 (1996); *J. Phys. Chem. B* **101,** 1186 (1997).
150. Heezen, L., Kilian, V. N., van Slooten, R. F., Wolf, R. M., and Nieuwenhuys, B. E., *Stud. Surf. Sci. Catal.* **71,** 381 (1991).
151. Oh, S. H., and Carpenter, J. E., *J. Catal.* **98,** 178 (1986).
152. Tzou, M. S., Asakura, K., Yamazaki, Y., and Kuroda, H., *Catal. Lett.* **11,** 33 (1991).
153. Cai, Y., Stenger, H. G., Jr., and Lyman, C. E., *J. Catal.* **161,** 123 (1996).

154. Schwartz, S. B., Schmidt, L. D., and Fisher, G. B., *J. Phys. Chem.* **90,** 6194 (1986).
155. Oh, S. H., Fisher, G. B., Carpenter, J. E., and Goodman, D. W., *J. Catal.* **100,** 360 (1986).
156. Lakis, L. E., Cai, Y., Stengerrand, H. G., and Lyman, C. E., *J. Catal.* **154,** 276 (1995).
157. van den Bosch-Driebergen, A. G., Kieboom, M. N. H., van Dreumel, A., Wolf, R. M., van Delft, C. M. J. M., and Nieuwenhuys, B. E., *Catal. Lett.* **2,** 83, 235 (1989).
158. Anderson, J. A., *J. Catal.* **142,** 153 (1993).
159. Schwartz, S. B., and Schmidt, L. D., *Surf. Sci.* **206,** 169 (1988).
160. Harrison, B., Diwell, A. F., and Hallett, C., *Plat. Met. Rev.* **32,** 73 (1988).
161. Summers, J. C., and Ausen, S. A., *J. Catal.* **58,** 131 (1979).
162. Ozawa, M., and Kimura, M., *J. Mater. Sci. Lett.* **291** (1990).
163. Bettman, M., Chase, R. E., Otto, K., and Weber, W. H., *J. Catal.* **117,** 447 (1989).
164. Oudet, F., Courtine, P., and Vejux, A., *J. Catal.* **114,** 112 (1988).
165. Yao, H. C., and Yao, Y. F. J., *J. Catal.* **86,** 254 (1984).
166. Kim, G., *Ind. Eng. Chem. Prod. Res. Dev.* **21,** 267 (1982).
167. Loof, P., Kasemo, B., Anderson, S., and Frestad, A., *J. Catal.* **130,** 181 (1991).
168. Oh, S. H., *J. Catal.* **124,** 477 (1990).
169. Nunan, J. G., Robota, H. J., Cohn, M. J., and Bradley, S. H., *J. Catal.* **133,** 309 (1992).
170. Zafiris, G. S., and Gorte, R. J., *Surf. Sci.* **276,** 86 (1992); *J. Catal.* **139,** 561 (1993).
171. Krause, R. S., and Schmidt, L. D., *J. Catal.* **140,** 424 (1993).
172. Schwartz, J. M., and Schmidt, L. D., *J. Catal.* **148,** 22 (1994).
173. Hardacre, C., Ormerod, R. M., and Lambert, R. M., *J. Phys. Chem.* **98,** 10901 (1994).
174. Hardacre, C., Roe, G. M., and Lambert, R. M., *Surf. Sci.* **326,** 1 (1995).
175. Benaissa, M., Pham-Huu, C., Werckman, J., Crouzet, C., and Ledoux, M. J., *Catal. Today* **23,** 263 (1995).
176. Feng, X., and Hall, W. K., *J. Catal.* **166,** 368 (1997).
177. Chen, H., and Sachtler, W. M. H., *Catal. Today,* **42,** 73 (1998).
178. Ito, E., Mergler, Y. J., Nieuwenhuys, B. E., Calis, H. P. A., van Bekkum, H., and van der Bleek, C. M., *J. Chem. Soc. Faraday Trans.* **92,** 1799 (1996).

Experiments and Processes in the Transient Regime for Heterogeneous Catalysis

CARROLL O. BENNETT

Laboratoire de Réactivité de Surface
Université Pierre et Marie Curie
75252 Paris Cedex 05

Experiments in the transient regime are now widely used in mechanistic studies of heterogeneous catalysis. They permit the quantitative determination of the surface composition during catalysis and also give important information on the sequence of steps that underlie the observed global reaction. In particular, by computer-based modeling and simulation, the forward and backward rate constants of the steps can be estimated so that the behavior of the reacting system can be simulated by numerical methods over a wide range of variables. Results from transient experiments lead to reliable models for reactor design, and they are essential for reactors operating in the transient regime, a situation becoming increasingly common. In this review, I consider the general principles of common experiments in the transient regime: step response, pulse response, frequency response, and temperature-programmed response. The use of feeds of stable isotopes is widespread in transient experiments, greatly increasing the power of these methods. All these methods can be applied at atmospheric pressure and higher, and also under surface science conditions. Following a discussion of general principles, many examples have been chosen from the literature for more detailed case studies. Most of the reactions chosen have a practical application since the design of large-scale reactors requires quantitative information that is best obtained by transient experiments. Among the examples treated are those involving methane or carbon monoxide as reactants, and attention is also devoted to the reduction of the concentration of pollutants in exhaust gases. The response of the laboratory reactor is manifested by the concentration of the various components in the gas leaving the reactor as a function of time. Mass spectrometry is well suited for these measurements, but it is extremely useful to also measure the response of the surface phase composition of the catalyst. This measurement is usually made by infrared spectroscopy (including diffuse reflection from powders), but interesting results have also been obtained by Raman spectroscopy, X-ray spectroscopy, Mössbauer spectroscopy, and scanning tunneling microscopy. These are all applicable at atmospheric pressure; under vacuum conditions, there are many other techniques available. In the search for better catalysts or for a better understanding of structure–performance relations, the determination of the rate parameters of the elementary steps of a reaction is of great utility.

329

I. Introduction

The title of this review can be shortened to a consideration of the *transient method*, as in two earlier reviews (*1, 2*) in 1976 and 1982. In the simplest version of the method, the composition of a stream in steady flow to an open heterogeneous catalytic reactor is perturbed (step function, pulse, sine or square wave, etc.) and the response of the system to the signal is measured, as manifested by the composition of the outlet stream as a function of time. Of particular interest is the measurement by appropriate spectroscopies of the response during the transient period of the intermediates adsorbed on the surface of the catalyst.

Currently, the behavior of a catalytic reaction is most usefully described as the result of many elementary steps, and the goal of the experiments is to identify the sequence of steps underlying the global reaction(s) and to measure the forward and backward rate parameters for the steps that are kinetically significant. Much can be learned by a qualitative consideration of the results of such experiments, but I begin this discussion of the method by describing various quantitative models of the kinetics that lead to the determination of the parameters for the elementary steps.

In considering elementary steps it is usually assumed that the kinetics of each step follows a mass action law: for example, the forward reaction of two surface species may be represented by $r = k\Theta_1\Theta_2$, where k is a function of temperature only and Θ_1 and Θ_2 are coverages ($0 \leq \Theta_j \leq 1$) of two reactants 1 and 2. Until now, identifying the most abundant surface intermediates and the rate parameters for the forward and backward rate constants for the steps that influence the overall reaction rate has been a sufficiently arduous task. However, it is known that k must often be a function of Θ_1 and/or Θ_2 because of lateral interactions among the adsorbed species. Also, it is known that the above surface reaction may take place only at the perimeters of islands of 1 and 2 so that the appropriate concentration measure may be $(\Theta_j)^{0.5}$, or more generally $(\Theta_j)^n$, and n may vary with coverage. In addition, the structure of the underlying catalytic surface may vary with the coverage of adsorbed intermediates. Unless otherwise specified, it is assumed that the kinetics of an elementary step follow the simple mass action law.

The appropriate reaction velocities are preferably expressed as *turnover rates* (TOR) (*1*), and these are the rates that we hope to extract from (transient) experiments, the parameters of which should be in accord with statistical and quantum mechanics. With modern methods for the characterization and preparation of catalysts (*3*), it is possible to make reproducible materials and to propose kinetic parameters for certain elementary steps (*4*). It was suggested (*1*) that it might eventually be possible to predict a catalytic reaction rate based on the parameters describing the elementary steps, supposedly found from the organization of previous results and not obtained by direct experimentation. Progress in this direction may be made through the paradigm of microkinetics (5).

The rate used here is what has been called the nominal turnover rate (4), moles reacted per second per mole of sites, taken as surface atoms for metal catalysts. These sites are measured, for example, by hydrogen chemisorption, electron microscopy, X-ray diffraction, and magnetic methods. Of course, only a fraction of these sites may be active, but this fraction has to be learned by kinetic experiments and is subject to change as new kinetic results become available (6). This fraction seems to be known for ammonia synthesis over iron and can be measured by nitrogen chemisorption.

In addition to experiments done in the laboratory with the aim of understanding catalytic behavior, it is interesting to consider the operation of large-scale reactors in the transient regime, in this case with the aim of improving their operating characteristics such as the yield of desired product. The simulation and design of such reactors is governed by the same partial differential equations as those needed to describe laboratory reactors. Thus, the rate parameters of the elementary steps are needed, unless the characteristic time of the imposed signal (typically cyclic feed) is so long that the chemical reaction can be represented by its steady-state rate equation.

In Section II we consider the general principles that underlie the great variety of studies that have been carried out by using the transient method. In Section III the principles will be illustrated by recent studies that use the methods discussed in Section II to investigate the kinetics of interesting catalytic systems. In both sections an attempt is made to describe the essential features that make the various topics interesting. Articles involving the transient regime have become so numerous in recent years that it is not possible to refer to all such publications in this review.

II. General Principles

A. Governing Equations and Experimental Methods with Step and Pulse-Forcing Functions

I begin the discussion for laboratory reactors by presenting the equations for an ideal plug-flow reactor. To avoid concentration and temperature gradients inside porous particles, the particle diameter must be no more than a few tenths of a millimeter for a typical case. Thus, the gas volume inside the pores is simply added to that of the bulk gas phase in the volume element considered. At atmospheric pressure an inert carrier gas is often used, and the conversion of reactant is not more than a few percent. The resulting product concentrations leaving the reactor are easily measured by a simple quadrupole mass spectrometer. The low conversion and associated conditions ensure that the molar or volumetric flow does not change significantly across the reactor. The basic flow diagram has not changed much since 1982 (2). The control valves should be operated choked, and care should be taken that the pressure excursions generated by switching valves to create steps or pulses are minimized. However, it is no longer necessary to prepare and store gas mixtures (sometimes dangerous) in order to avoid introducing spurious concentration changes at the moment of switching. Fast response mass flow controllers are available so that mixtures can safely be made by mixing the components during the experiment. The whole setup can be computer controlled so that reaction mixtures corresponding to different total conversions can be studied in a differential reactor. These conditions usually ensure the absence of temperature gradients in the reactor.

For the isothermal tubular plug-flow reactor (PFR) discussed previously, the mass balance for the G gaseous components is

$$\frac{\partial c_j}{\partial t} = L_V \sum_{i=1}^{R} r_i \alpha_{ij} - \frac{\Delta z}{\tau} \frac{\partial c_j}{\partial z}; \qquad j = 1, \cdots, G, \tag{1}$$

where c_j is the concentration (mol/ml) of the jth gaseous component, including inerts; t is the residence time V/q; V is the volume of gas in the bed and in the pores of the catalyst (ml), q is the steady flow (ml/s); L_V is the total moles of surface sites per milliliter of gas; r_i is the turnover rate of the ith elementary step; and α_{ij} is the matrix of stoichiometric coefficients, including all the species appearing in the elementary steps. For the surface species, including the vacant sites, we have S species, and

$$\frac{\partial \Theta_j}{\partial t} = \sum_{i=1}^{R} r_i \alpha_{ij}; \qquad j = G + 1, \cdots, S, \tag{2}$$

where

$$\sum_{j=1}^{G} c_j = c = p/RT; \qquad \sum_{j=G+1}^{G+S} \Theta_j = 1, \tag{3}$$

and G is the number of gaseous species, including inert gases. The coupled Eqs. (1)–(3) must be solved simultaneously. The initial and boundary conditions must be specified, of course; we consider at first as forcing function a step up or down of concentration in the continuous steady feed to the reactor. Other possibilities will be discussed later. Notice that not only is $c_j = c_j(t, z)$ but also $\Theta_j = \Theta_j(t, z)$, and it is assumed that the adsorbed species do not migrate on the catalyst in the z direction.

To ensure the absence of axial dispersion [not included in Eq. (1)], the reactor length, Δz, should be at least 50 particle-diameters long, typically about 1 cm for the small particles needed to avoid intraparticle diffusion effects. The bed diameter can be about 4 or 5 mm.

An ideal PFR is experimentally simple, but its behavior is governed by partial differential equations. For a trial set of kinetic parameters for the elementary steps, it is necessary to simulate the reactor and then adjust the parameters to obtain the best fit to the experimental data obtained from the experiments in the transient regime. The analytical solution is so complicated that only a simplified sequence of steps can be considered. Of course, interesting qualitative deductions can be made from the experimental response to an inlet step function.

An analytic solution to Eqs. (1)–(3) has been presented by Fiolikatis *et al.* (7); it is related to wavefront analysis (8) and has been discussed in a review paper (9). However, until recently, these authors were among the few who had actually done experiments in the transient regime in a nondifferential PFR (10); the effort of dealing with partial differential equations favors the use of a continuous stirred task reactor (CSTR), which can be simulated via ordinary differential equations. Nevertheless, the increasing availability of powerful computers and efficient, easy-to-use algorithms is leading researchers to choose the experimentally simple PFR. Nibbelke *et al.* (11) used an integral PFR to study the oxidative coupling of methane in the transient regime; their results will be discussed later along with some others based on the simulation of a PFR in the transient regime. Their simulation is accomplished by numerical methods alone.

For a CSTR the mass balance for the gas phase is the following ordinary differential equation:

$$\frac{dc_j}{dt} = L_v \sum_{i=1}^{R} r_i \alpha_{ij} - \frac{1}{\tau}(c_j - c_{jf}); \qquad j = 1, \cdots, G. \tag{4}$$

Equation (2) (as an ordinary differential equation) and Eq. (3) apply now with Eq. (4). As already implied, a laboratory well-mixed reactor for heterogeneous catalysis is more difficult to realize than a PFR. Many versions have been used (12), and Froment and Bischoff (13) illustrate reactors with external recycle, with internal recycle (1, 14), and with an internal spinning basket (15). When using these reactors for experiments in the transient regime, it is important to keep to a minimum the volume outside the bed of catalyst. Internal recycle reactors involve bearings exposed to hot reactive gases and require a magnetic drive system for leak-proof operation. External recycling requires a separate pump, often operated at a lower temperature than that of the reactor. The unavoidable volume of the pump means that a relatively large catalyst sample is needed to compensate for the effect of the dead volume of the pump and lines. The spinning-basket reactor also has a relatively large dead volume.

As suggested by Zielinski (*16*), a "self-mixed" flow can be achieved in a very shallow bed (a few particles deep) supported on a fine frit. Stockwell *et al.* (*17*) studied methanation in the transient regime in such a reactor. The gas is introduced into the space above the particles by jets so that this volume is well mixed by eddies; the frit, which offers a relatively high pressure drop, ensures that there is little channeling in the bed. The reactor can be made of stainless steel, glass, or quartz. When the slight S shape of the experimental forcing step function (*F*, Fig. 1) between inert gases is taken into account, the reactor responds as a CSTR, with a curvature that is upwardly convex from the beginning of the response; the response time is less than 1 s. The slight S shape of the observed mixing curve (*M*, Fig. 1) is a result of the imperfect forcing function and not from a lack of complete mixing in the bed.

If the flow rate across a bed is sufficiently high, the conversion can be

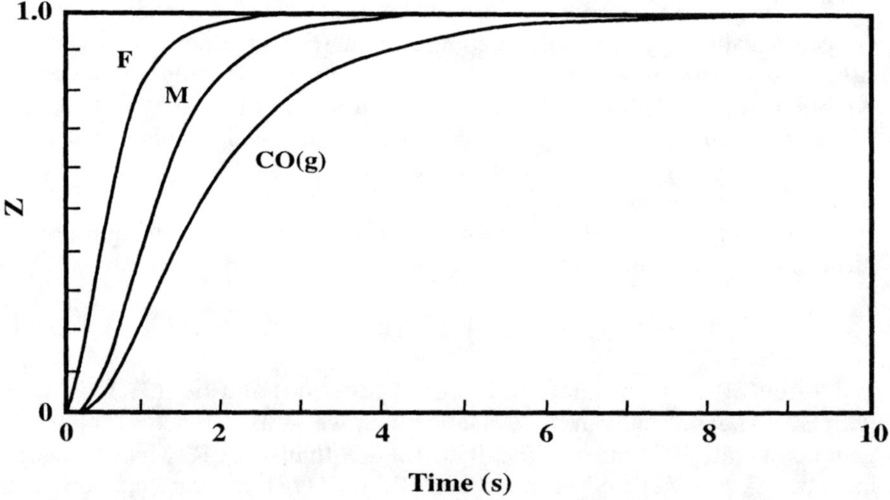

Fig. 1. Response of the output of a CSTR caused by a switch from inert to inert + CO. *F* (forcing function) is the measured response to the switch with flow through a low-volume bypass around the reactor. *M* (mixing function) is the response to the switch with flow through the reactor filled with inert particles. When the same particles carry an active metal (e.g., Ni), CO is adsorbed, and the response is given by the curve labeled CO_g (after *17*).

kept as low as desired so that the concentration of reactant(s) decreases only slightly from $z = 0$ to $z = \Delta z$. In this differential reactor, $\partial c_j/\partial z$ can be approximated by $(c_j - c_{jf})/\Delta z$, and Eq. (1) becomes identical to Eq. (4), with r_{ij} evaluated at $(c_j + c_{jf})/2$. Such a reactor has been used in the extensive and interesting work of Kobayashi and collaborators, first reviewed in 1974 (*18*); a current version of the method has recently been described (*19*). The isothermal PFR is of the order of 40 cm long with a gas flow rate of about 160 ml/min, and it operates with a conversion of <10% for the relatively slow reactions studied. After the wavefront reaches the outlet of the reactor, the response curves have the upwardly convex shapes associated with the CSTR discussed previously. There must be no chromatographic effects between the gases being used because this would require representing the system by Eq. (1) rather than Eq. (4). There is almost no dead volume in such a reactor. Note that this reactor may require as much as 200 g of catalyst, whereas the self-mixed reactor (*17*) requires about 20 mg, depending of course on the reaction rate of the system being studied. The response times are also about two orders of magnitude smaller in this reactor.

It is of interest that similar ideas have been applied in the conception of a flow system for measuring the diffusion coefficients for gases in porous or microporous solids. Ruthven and Eic (*20, 21*) use a "zero-length column" (ZLC) to suppress concentration gradients along the bed in the gas phase. As in the differential reactor described previously, a high gas flow rate is used so that the fixed bed acts as if it were very short. A preadsorbed adsorbate is removed by an inert gas stream. The diffusion inside the solid is very close to the classical solution for zero concentration on the surface, but the small concentration actually present in the gas leaving the bed (column) can be measured accurately.

Kobayashi (*22*) performed computer simulations via Eq. (4) as applied to his differential fixed-bed reactor. The model gas-phase reaction $X \rightarrow Y$ is considered to pass in series through elementary steps to adsorbed X, an adsorbed intermediate *in*, adsorbed Y, to give finally Y. The forward and backward rate parameters were adjusted to simulate various mechanisms with their rate-determining steps. The shapes of the response curves for Y for typical mechanisms are classified as instantaneous, monotonic, overshoot, S shaped, false start, and complex. This paper is a good source of ideas for the interpretation of transient responses. These ideas are illustrated by application to the oxidation of ethylene over a silver catalyst (*23*). The response curves last more than 100 min because the temperature is only 91°C and the bed contains 261 g of catalyst; the flow rate is 160 ml/min.

The decomposition of N_2O over NiO/SiO_2 has been studied experimentally (*24*) in an internally mixed CSTR, and it was proposed that the reaction can be represented by the following two elementary steps (* denotes adsorption sites):

$$N_2O + * \rightarrow N_2 + O^* \qquad (\times 2) \tag{5}$$

$$2O^* \rightleftharpoons O_2 + 2^*, \tag{6}$$

with neither step rate controlling. The parameters found by Yang *et al.* (*24*) have been used to model the response of a step-function input of N_2O to a CSTR and a PFR sized so that both lead to the same steady-state N_2 production rate (*1*). The partial differential equations for the PFR were solved numerically by considering it to be equivalent to 40 mixed reactors in series. The essence of the kinetics, shown by the overshoot in the N_2 response, is more evident in the simulated PFR having the same conversion as the experimental CSTR.

Salmi (*25*) set up equations needed to simulate the transient response of both the PFR and the CSTR. The balance equations and the generalized equations for the rates of the elementary steps are compactly expressed in vector and matrix notation. Details of the computational algorithms are discussed, and they are applied to the N_2O decomposition (Eqs. 5 and 6). In another paper (*26*) these equations are used to simulate (for both PFRs and CSTRs) the responses of systems following many mechanisms: Eley–Rideal, Langmuir–Hinshelwood, a combination of the two, with and without dissociative adsorption, etc. These curves can be added to those of Kobayashi (*22*), to expand the general view of how various systems respond.

The solution of the partial differential equations needed to simulate a PFR has been discussed by van der Linde *et al.* (*27*). Numerical methods are considered preferable to analytic methods and are described in detail, along with statistically correct search procedures for extracting the kinetic parameters from transient data.

An important advantage of using a step function perturbation of the feed is that the catalytic system can move from the steady state before the change to another steady state after the transient period. The transient period is usually short compared to the time needed to change the activity of the catalyst by coking, aging, or poisoning. Of course, for a reaction such as catalytic cracking, the time scales may overlap. Unless specifically mentioned, the intrinsic activity of the catalyst is considered constant during the transient periods discussed here.

It is perhaps easier to introduce a step function into the flow rate to a reactor than to generate a step function in the feed concentration. A recent interesting example (*28*) involves the disproportionation of ethylbenzene to benzene and diethylbenzenes over an HY zeolite at approximately 500 K. When the flow q is decreased from 120 to 30 ml/min, there is an immediate overshoot in the yield of benzene, but the yield of diethylbenzenes rises quite slowly; the two yields approach each other as time increases. This result is explained by product inhibition by the ethylbenzenes, more strongly adsorbed than benzene. In another experiment, the steady-state yield in a diluted bed (similar to a PFR) is higher than that in a shallow bed containing the same amount of HY zeolite (similar to a CSTR), in accord with reactor engineering principles. However, if one wants to use this type of experiment for the quantitative determination of the parameters for the elementary steps via Eqs. (1)–(4), the mathematics is encumbered by the fact that q occurs in the differential equations as a coefficient and not as an independent variable such as c_j.

Before continuing the general discussion of methods, I pause to consider a simple adsorption measured by using a step input of A in an inert carrier flowing to a CSTR. The experiment is described via Fig. 1, in which Z is c_A/c_{Af} and τ is the residence time V/q. By using a short, low-volume bypass around the CSTR, one can measure the response caused by a switch from pure carrier gas to one containing a few percent of Ar. This has been called the forcing function (17). If the response of the adsorbing or reacting system is slow enough, it is sufficient to assume that the curve F in Fig. 1 is a perfect step function. It is clear that it is important to model the forcing function for a fast-responding system, as shown (17). The same signal after passing through the self-mixed CSTR gives curve M, the mixing curve. With the proper boundary conditions, from Eq. (4) it can be shown that the gas volume in the reactor is proportional to the area between curves F and M. Small particles must be used so that there are no concentration gradients in the gas inside the porous catalyst particles. If a gas such as CO is used instead of Ar, the quantity of gas adsorbed is proportional to the area between M and CO_g. For this simple adsorption, the steady state at the end of the response corresponds to equilibrium at the concentration c_{Af} chosen.

If a high flow rate is used for the previous experiment, the response may occur so quickly that gradientless operation becomes impossible. At the other extreme, it may be possible to find a flow rate small enough so that the response curve CO_g is no longer a function of τ. Each point on CO_g corresponds to quasi-equilibrium with the catalyst surface at that particular point. Between these two extremes the position of CO_g is determined by the rate constants for adsorption and for desorption, k_a and k_d.

It is instructive to consider a simple model reaction, $A \rightarrow P$, passing through the elementary steps in a CSTR:

$$A + * \rightarrow A*; \qquad i = 1 \tag{7}$$

$$A* \rightarrow P + *; \qquad i = 2. \tag{8}$$

Equation (4) is written in a more general form as

$$\frac{dc_j}{dt} = L_V \sum_{i=1}^{R} r_i \alpha_{ij} - \frac{q}{V} c_j + \frac{q_f}{V} c_{jf}; \qquad j = 1, \cdot\cdot\cdot, G, \tag{9}$$

in which the flow out (q) may differ from the flow in (q_f). Equation (9) is combined with Eq. (3) and Eqs. (7) and (8) to give

$$q = q_f + (r_2 - r_1)VL_V/c. \tag{10}$$

For the simple example considered here (Eqs. 7 and 8), the rates of the elementary steps are

$$r_1 = k_1 c_A \Theta^* \tag{11}$$

and

$$r_2 = k_2 \Theta_A, \tag{12}$$

where k_1 is in cm^3 mol^{-1} s^{-1} and k_2 is in s^{-1}. If step r_1 were reversible, one would need to write

$$r_1 = r_{+1} - r_{-1}, \tag{13}$$

where

$$r_{+1} = k_{+1} c_A \Theta^* \tag{14}$$

and

$$r_{-1} = k_{-1} \Theta_A .$$

If the unidirectional rates of step 1 become large enough, the step is in pseudoequilibrium and we can measure only the ratio of the k's, and

$$k_{-1}/k_{+1} = K_1 = \Theta_A/\Theta^* C_A \tag{15}$$

even though the global reaction is at steady state and not in equilibrium. Step 2 is the rate-determining step. We know that for Eqs. (7) and (8) or for the more complicated sequence, Eqs. (5) and (6), an algebraic solution for the steady-state rate equation can be found (1). However, for more complicated sequences, after having found the $k_{\pm i}$ from suitable experiments, one can determine how the global rate behaves with respect to conversion by numerically solving the appropriate equations such as Eq. (9) as time increases and dc_j/dt approaches zero. Precautions must be taken to ensure that the system of equations does not tend toward a physically impossible steady state.

Returning to Eqs. (7) and (8) as written, we consider the usual transient period after a step increase of A. The steady flow q_f of inert gas to the reactor is quickly replaced by the same total flow of inert plus A. Thus, A begins to be stored as A^* at a rate r_1 and the rate of appearance (r_2) of P is smaller than r_1; $q_f > q$. When steady state is reached we must have $r_1 = r_2$. Using a dilute mixture of A means that the difference between q and q_f can be neglected during the transient period; $r_1 \neq r_2$ but VL_v/c is chosen small. Of course, if there is a change in moles for the global reaction [as for Eqs. (5) and (6)], $r_2 - r_1$ takes on an appropriate value at steady state that is a simple multiple of r_1. However, as already mentioned, for a feed of 10% A and a conversion of 2%, for example, it is sufficiently precise to consider that $q = q_f$.

The coverage of A at steady state, Θ_{As} can be found by the integration of Eqs. (2) and (4) subject to Eqs. (3), (7), and (8). Tamaru (29) referred to this aspect of the method as "adsorption measurements during surface catalysis." The result depends only on the material balances. However, rather than only integrating, one can compare the experimental results during the transient period to the curves simulated by a trial sequence of steps, optimizing the values of the kinetic parameters as discussed previously and as most recently described by van der Linde *et al.* (27).

Although the step function described previously has been taken as from $c_A = 0$ to c_{Af}, one can start from a steady state at relatively high concentration ($c_A = c_{Ao}$ before the step up or down), corresponding to measurements at a high global conversion. This may be important since the steady-state coverage of intermediates changes with conversion. This is supposed to be taken care of through the use of the proper sequence of steps, but because of surface heterogeneities or lateral interactions (non-Langmuir-like adsorption) between adsorbed species the values of some surface kinetic parameters may vary with coverage, i.e., conversion.

If $\Delta c_j(t)$ is defined as $(c_j - c_{js})$, where c_{js} is the eventual steady-state value of c_j, it is possible to choose conditions so that $|\Delta c_j| \ll c_j$ during the transient, and thus linearize the governing equations (30), leading eventually to the interpretation of experimental data in terms of moments. This procedure has apparently not been put into practice; in subsequent work it seemed preferable to use more direct numerical methods via the computer (1, 24, 27). Moments in connection with experiments using a truly harmonic input perturbation and also for isotopic tracing, involving truly first-order processes, will be discussed.

B.   Experiments at High or Low Pressure: Real Catalysts *versus* Well-Defined Surfaces

We return to the consideration of some fundamental ideas. In addition to the obvious goal of understanding what is occurring, it is good to identify the sequence of steps with their kinetic parameters so as to generate ideas for improved catalysts. We should concentrate our attention on the "slow" steps. In such studies one often deals with model catalysts or catalytic materials—that is, materials that do not necessarily have sufficient resistance to aging or physical attrition, or that are too expensive, to be used as commercial catalysts. It is hoped that better catalysts can be obtained through improved understanding.

However, one can also study real catalysts, leading to a mathematical representation of their behavior by a single model for an interesting range of temperatures, concentrations, and conversions. The sequence of steps can include reactions leading to both desired and undesired products so that selectivity can also be modeled. The data should not be falsified by transport effects, leading to what is called in reactor engineering intrinsic rates. For example, in designing, a large-scale fixed-bed reactor, large pellets may be used to reduce pressure drop, and the economically optimal reactor often presents large gradients of temperature and concentration. By crushing and sieving the pellets, a solid is obtained suitable for use in the laboratory with the ideal reactors previously discussed. This procedure is sometimes complicated by the presence of a nonuniform distribution of metal in a supported catalyst.

Laboratory results lead to an expression for the intrinsic rates for a network of reactions, $r_i = f_i(c_j, T, p)$, and these rates appear in the mass and energy balances for the large-scale, nonideal reactor, for the gas in the pellet, for the adsorbed phase, and for the gas phase outside the pellets. Also needed are values of the effective diffusion coefficient in the pellets, the longitudinal dispersion coefficient in the bed, the interphase mass transfer coefficient, and the appropriate heat transfer coefficients and thermal conductivities for use in the energy balances. The effect of temperature on all these quantities must also be known. Through numerical procedures, this problem can be solved so as to simulate the behavior of the real reactor in the steady state or during a transient. Given the great expense of experiments on increasingly larger pilot plants, the correct simulation is extremely valuable. Naturally, a pilot plant of a certain size may be necessary to test the reliability of the simulation and to use real feeds that may contain small amounts of poisons that were not present in the laboratory studies. Sufficient product may also be obtained so that its purity and other properties can be ascertained. These involved procedures are part of the scientific field of reaction engineering, represented by a vast literature (13).

As can be seen from the preceding discussion, there is interest in using real, high-area catalysts for the kinetic experiments on a laboratory scale. However, we want to identify the active surface sites and how they interact with adsorbed species. Since real catalysts present a distribution of particle sizes for a simple oxide catalyst ($d = 0.5$ mm) or a distribution of metal particle sizes for a metal on an "inert" support ($d = 5$ nm), the nature of the surface sites is not well defined. For this reason, it is interesting to perform surface science studies on well-characterized surfaces, in particular the various faces (including those obtained by cuts exposing atoms at edges, kinks, etc.) of single crystals. However, the surface exposed is of the order of 0.1 cm^2/g of metal (low-area catalysts), whereas it is approximately 10^5 cm^2/g of metal for a supported metal particle (*high-area* catalyst). In order to avoid rapid poisoning by traces of molecules such as H_2S and CO in a gas, it is convenient to work at low pressure to keep the surface of the crystal sufficiently clean for a reasonable time. Since these experiments are aimed at fundamental understanding, often a simple adsorption of a strongly adsorbed gas such as N_2 on Fe or CO on Pt is studied (an elementary step). The rate of adsorption is so high that at atmospheric pressure the process is immeasurably fast. However, from the kinetic theory of gases it is known that the rate of collision of a given molecule with a surface is proportional to its partial pressure. Thus, at a typical pressure (10^{-9} Pa) the rate of collision becomes slow enough to measure conveniently. The net rate of adsorption is $(d\Theta/dt) = r = r_a - r_d$, and r_a is the collision rate multiplied by the sticking probability. Since the sticking probability, the frequency factor, and the activation energy of desorption are functions of the surface structure only, these values may be useful for the estimation of rates at normal pressure (*5*).

It is also possible to simulate supported metal catalysts by the vapor deposition of metal on a flat surface of silica, alumina, etc. The particle size distribution can be closely controlled and the results verified by various electron spectroscopies, for example (*31*). For the reverse situation of a flat metal surface decorated by oxide particles, one can simulate catalysts in the strong metal-support interaction state (*32*).

Kinetics can also be studied at surface science conditions. Feed can be leaked at a constant rate into the chamber containing the crystal face, and the gas is removed at a constant rate by the pumps. The composition of the chamber gas can be continuously monitored by mass spectrometry. The pressure in the reaction chamber is low enough to ensure Knudsen flow: The gaseous molecules collide almost exclusively with the exposed solid surfaces, and the system behaves as a perfectly mixed flow reactor (CSTR). Experiments in the transient regime with various forcing functions can be performed, and response times can be orders of magnitude smaller than those at atmospheric pressure. The catalytic oxidation of CO on Pt(110) was one of the first studies of this type (*33*).

It would clearly be advantageous to use the time and concentration resolution of experiments at low pressure and surface science conditions with high-area catalysts. Gleaves *et al.* (*34*) developed a method for doing this using their reactor system called "temporal analysis of products" (TAP). A small fixed-bed reactor is placed in a vacuum system in such a way that the flux of molecules leaving the reactor is efficiently directed to the ionization chamber of a mass spectrometer. Starting with the background pressure in the bed, specially designed fast-action valves can introduce very short pulses of gas into the entrance of the bed. There is no carrier gas, and if the pressure is low enough, the molecules move in the bed by Knudsen diffusion. The diffusivity is relatively small because of the short distances between particles. After the pulse, nothing more enters the bed so that at $z = 0$, $\partial c_j/\partial z = 0$. There is no bulk (viscous) flow, and each molecule diffuses toward the low-pressure end of the bed according to its own Knudsen diffusion coefficient, colliding only with the solid surfaces. The particles must be small enough so that diffusion resistance inside the particles can be neglected. The boundary condition at $z = L$ is $c_j = 0$. This is consistent with the fact that at $z = L$, $\partial c_j/\partial z \neq 0$, and the flux of j is not zero; this flux is measured by the mass spectrometer.

For the TAP reactor, Eq. (1) becomes

$$\frac{\partial c_j}{\partial t} = L_v \sum_{i=1}^{R} r_i \alpha_{ij} + D_{Kj} \frac{\partial^2 c_j}{\partial z^2}; \qquad j = 1, \cdots, G, \tag{16}$$

and Eqs. (2) and (3) still apply. When a pulse of gas for which all r_i's are zero is introduced, Eq. (16) [unlike Eq. (1)] can be solved by the method of separation of the variables, and a comparison of the measured outlet pulse with the analytical solution of Eq. (16) permits the determination of D_{Kj}. These are effective diffusion coefficients, functions of the porous properties of the bed. When various nonreacting gases are used, their diffusivities are inversely proportional to the square root of their molecular weights, in accordance with the kinetic theory of gases (*34*).

The results to be expected from a typical TAP experiment have been simulated (*35*) for a simple irreversible adsorption and are shown in Fig. 2. If there is no adsorption, the pulse at the reactor outlet is represented by curve A. For $k_a > 0$, some of the inlet pulse (N_{pA} moles of A) remains on the catalyst, and the amount is proportional to the difference between the areas under curve A and curve B, for example. Figure 3 shows what happens with reversible adsorption. For fast adsorption and slow desorption, there may be two peaks as shown by curve C. After a sufficient length of time, all the gas that was initially adsorbed will have left in the exit peak. Gleaves *et al.* (*35*) also show that for values of k_a and k_d that are sufficiently high so that the gas and adsorbed phases are everywhere in equilibrium, the response curve has the shape of curve A but the peak height is reduced to $1.85/(1 + K_{eq})$.

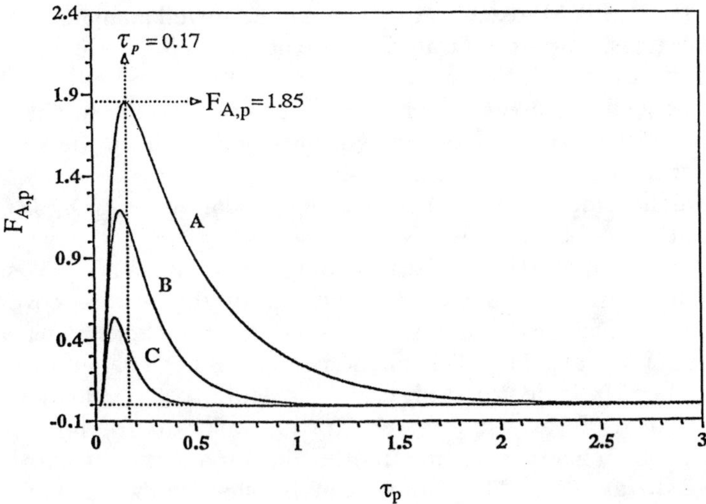

FIG. 2.   Response of the TAP reactor to an inlet pulse of a gas that is irreversibly adsorbed (or reacted) with a dimensionless rate constant k_a. τ_p is the dimensionless time, and $F_{A,p}$ is the dimensionless flow rate. The model takes into account the number of molecules in the pulse $N_{p,A}$, the effective Knudsen diffusion coefficient D_{eA}, the number of surface sites, and the dimensions of the reactor (after 35). A, $k_a = 0$, standard diffusion curve; B, $k_a = 3$; C, $k_a = 10$.

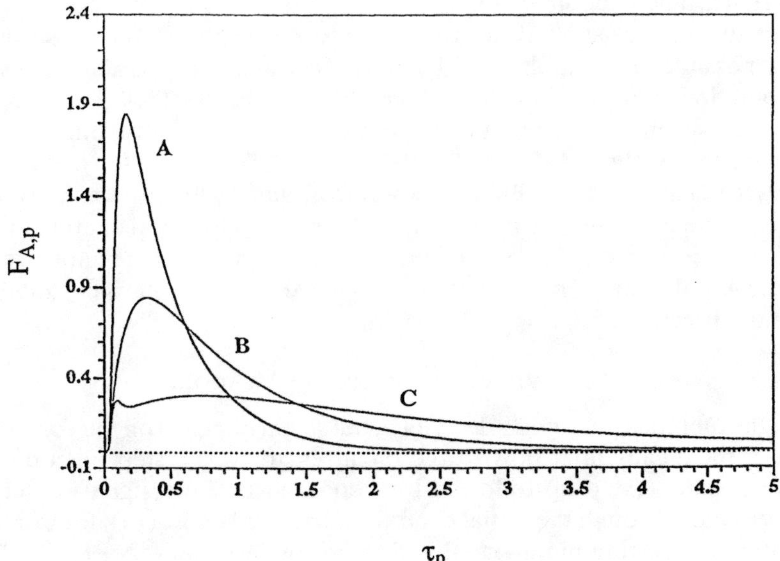

FIG. 3.   TAP response for a reversibly adsorbed gas (after 35). A, $k_a = 0$; B, $k_a = 20$, $k_d = 20$; C, $k_a = 20$, $k_d = 5$.

Although the TAP reactor equations can be solved analytically for some simple systems, more complicated reation networks require numerical solutions. Gleaves *et al.* (*35*) show that useful relations can be found between quantities such as conversion and residence time and the moments of the response of the TAP reactor, analytically obtained from the solutions of the linear system described previously.

Although some reactions such as the oxidation of CO (*33*) can be studied at low pressure, others, such as the hydrogenolysis of cyclopropane on Pt (*36*), cannot. The surface coverage of reactants is not sufficient to give a measurable rate of reaction and/or the surface coverages at atmospheric reaction conditions are very much higher than those that can be obtained at UHV. This phenomenon has been called the pressure gap problem. In order to study such a reaction on a well-characterized surface, it is useful to use a system, for example, that permits the single crystal to be enclosed in or transferred to a small reactor without exposure to air (*36*). The sample can be characterized under vacuum conditions before and after reaction. The reaction mixture is admitted to the batch reactor containing the low-area sample, and the conversion is followed with time in a conventional way. The dc_j/dt are so small that the system passes through a series of steady states. An interesting discussion and review of the literature on turnover rates measured at both low- and high- ("real-world") pressure conditions is provided by Rodriguez and Goodman (*37*).

For the TAP reactor, these same effects of pressure also exist so that it is not possible to extrapolate the kinetics of certain reactions at low pressure to conditions that would exist at high pressure. In the TAP-2 system (*36*) there is a second reactor that can operate at higher pressure, but of course the essence of the TAP reactor is that it makes it possible to deal with catalyst particles, not well-defined surfaces, and at low pressure so as to increase the possibility of measuring fast kinetics. Operated at atmospheric or higher pressures, the results available from the TAP reactor are not fundamentally different from those produced by other transient methods already discussed.

C. Frequency Response Methods

In the methods previously described, the signal imposed on the concentration of the steady inlet flow to the open reactor was a step function or a pulse function (step up followed by step down). These signals enter the analysis only through the initial conditions imposed on Eq. (4), for example. Another interesting input signal is the sinusoidal variation of c_{jf},

$$c_{jf}(t) = c_{jfo}(1 + A_{jf}\sin(\omega t)). \tag{17}$$

Other feed components are varied so as to keep c = constant. The amplitude A_{jf} is often chosen as a small fraction ($\sim$2%) of the average (original steady-state) concentration c_{jfo} so that τ is almost constant and the governing equations form a linear system that permits the use of the classical methods of harmonic analysis. A square wave, easier to generate experimentally, can also be used as an approximate input signal. For the constant-flow system, as t increases from zero, the response c_j passes through a transient stage before reaching the steady periodic response at t_s, given by

$$c_j(t) = c_{jo}(1 + A_j \sin(\omega t + \phi_j)), \text{ for } t > t_s, \tag{18}$$

where A_j is the amplitude and ϕ_j is the phase lag of this response. These quantities are functions of ω. If ω is low enough, the system merely passes through a series of steady states corresponding to the successive values of c_{jfo} so that $\phi_j = 0$. As ω increases to such high values that the system cannot respond, A_j approaches zero and ϕ_j is irrelevant. At frequencies between these extremes, the curves of A_j and ϕ_j (or characteristic functions derived from them) as functions of ω permit the evaluation of parameters of a model for the system. It is interesting to note that a step input signal activates successively a wide range of frequencies, whereas the frequency response method permits the study of the response at a chosen value of ω. For instance, two peaks may occur in the curve $\phi_j(\omega)$, corresponding to a slow and a fast step in a sequence. An early application of the method, via chopping of a molecular beam, was presented by Jones *et al.* (*38*).

Following the pioneering work of Naphtali and Polinski (*39*), Yasuda (*40*) extensively developed the frequency-response method, beginning with applications to the adsorption and desorption of a single gas on a solid adsorbant. He later was able to add the measurement of diffusion coefficients inside the porous adsorbents (*41*). The method is particularly attractive for use with zeolites. The effective diffusion coefficient through the microporous structure is exceedingly small (about 10^{-7} cm^2 s^{-1} vs about 10^{-2} cm^2 s^{-1} for typical oxide supports), but the crystallites typically have a diameter on the order of only 1 mm. Thus, the response time is often too short for measurement by a simple response to a step function in a flow system (τ = 1 s). To avoid this slow response, this version of the frequency-response method has been used in a batch (closed) system in which the total volume is varied sinusoidally by a bellows. The interaction of the gas with the solid leads to a pressure response that exhibits an amplitude ratio and a phase lag that can be used to estimate the rates of adsorption, desorption, and diffusion. Since the diffusion coefficient is often a function of concentration, an advantage of the frequency-response method is that the concentration inside the particles is almost constant. With a step input, one measures the average diffusivity, and it may change with time as the concentration inside the particles changes. Note that the ZLC method already mentioned (*20*) can be used with isotopic tracers without changing the chemical composition so that the resulting self-diffusion coefficients can be measured for known concentrations (*42*). Reviews of the occasionally conflicting results obtained on diffusion in porous and microporous solids by various methods have appeared recently (*43, 44*). A detailed and up-to-date review, "Molecular Mobility Measurement of Hydrocarbons in Zeolites by NMR Techniques" (*45*), focuses on the measurement of self-diffusion coefficients by pulsed-field NMR.

Yasuda and coworkers (*46, 47*) extended the use of the frequency-response method to heterogeneous catalytic reactions. The input remains the sinusoidal variation of the volume of the reactor, but with a continuous flow of reactants and measurement by mass spectrometer of the response of the concentrations of the products. Yasuda recently reviewed all his work (*48*).

Regarding adsorption and diffusion without reaction, Jordi and Do (*49*) simulated the expected results for the frequency response by completely numerical methods, with no need for linearization. In a later study, they used a linearized model coupled with analytic solutions for the diffusion inside the particles, which also took into account transport in both macropores and micropores (*50*). The mathematical details are clearly presented in these papers.

Infrared spectroscopy is well suited to *in situ* studies so that much information can be obtained on the structure and concentration of adsorbed intermediates. Of interest in the current context is the measurement of the concentration of such intermediates as they change with time in response to the input signals discussed here. Li *et al.* (*51*) imposed a harmonic signal for the concentration of CO in He on the steady input flow to a suitable cell and measured the response of the band for adsorbed CO at 2080 cm^{-1} at 343 and 429 K at atmospheric pressure. The results are interesting, but the equilibrium constant $K = k_a/k_d$ is so high that it is difficult to determine the parameters for this particular case. CO is "irreversibly" adsorbed, and only the "reversible" part is involved in these experiments. A thorough study would require the use of an excessively large span of temperatures and CO partial pressures. In addition, the measured resonant frequency is so high that transport effects may be involved.

Many similar studies have been done at surface science conditions using step-function concentration signals. For example, Takagi *et al.* (*52*) studied the adsorption and desorption of CO on Ni(100) via infrared reflection absorption spectroscopy (IRAS). In one case, CO dosing was turned on for 214 s and then turned off. The response of the spectrum of the surface intermediate was followed by IRAS. Of course, under these high-vacuum conditions, other spectroscopies such as high-resolution electron energy loss spectroscopy can be used (*53*).

Methods based on the variation of the reactor volume have been discussed previously, but most work involving a periodic input has been carried out by varying the concentration of the steady inlet flow to a reactor, as in Eq. (17). The goal may be to find the parameters of the sequence of steps and improve understanding of the kinetics via the sequence of elementary steps. However, periodic operation may lead to improved selectivity with the same catalyst as is normally used, leading to important practical results, described by Sylveston *et al.* (*54*). Conversion may also be improved, but it may often be preferable to achieve this goal by less complicated methods, such as increasing the length of a reactor.

The inlet signal $c_{jf}(t)$ need not be a simple harmonic, as given by Eq. (17). One can step up component j in the fed by a quantity A_{j1} with respect to c_{jfo} for a time t_1, and then step it down by a different amount A_{j2} for a time t_2, forming an asymmetric periodic sequence. In addition, the amplitudes of the steps need not be limited to a small fraction of the average value. This procedure may also lead to appreciable temperature oscillations. The simulation of the result must be performed by numerical analysis, taking into account the optimized sequence of steps. The parameters of the periodic signal can be varied (by simulation, but usually by experiment) so as to give the best yield. The average surface composition for the best result is supposedly one that is not accessible through steady-state operation. A table of publications on this subject was published in 1995 (*55*).

An early investigation of the effect of concentration cycling on the flow to an approximately isothermal CSTR was published by Cutlip (*56*). For the oxidation of CO over Pt/Al_2O_3 the average rate with the cycled feed was markedly increased over that for a steady feed. The inlet feeds were cycled between 2% CO and 3% O_2 in argon for periods up to 3 min, with varying fractions of the time spent on CO. The behavior of the system could be explained qualitatively in terms of a high rate at low CO coverage and a low rate at high CO coverage. Lynch (*57*) used the elementary step model

$$CO + * \rightleftharpoons CO*$$

$$O_2 + 2* \rightleftharpoons 2O*$$

$$CO* + O* \rightarrow CO_2 + 2* \tag{19}$$

to simulate the behavior of such a system, and he was able to reproduce the results of Cutlip qualitatively. When the period is long, the system is in a quasi-steady state. Starting with a CO-covered surface, a feed of oxygen gradually reacts with surface CO and then covers the surface; there is a peak of CO_2 production. Upon switching to CO in the feed, there is another peak of CO_2, and the surface is covered with CO. The average rate for this process is higher than that at steady state corresponding to a CO-covered surface, for example. As the frequency of the switches is increased, a maximum rate may be found at a resonant frequency, but at even higher frequencies the rate approaches the (relaxed) steady state corresponding to average composition of the reactants fed. The system is no longer able to sense the rapid variations in the feed concentration. It should be noted that in many simulations the adsorption of oxygen is taken as irreversible.

Kaul *et al.* (*58*) also used the elementary steps of Eq. (19) to model their results for CO oxidation over Pt/SiO_2, for which they used the experimental techniques of transient Fourier-transform infrared (FTIR), temperature-programmed reaction, and concentration-programmed reaction (*59*). They later applied the same methods to the CO oxidation over Rh/SiO_2. In the numerical calculations many parameters were taken from surface science results, and the agreement between experiment and simulation is good (*60*) when spatial nonuniformities are not present.

It must be borne in mind that the imposed signals discussed previously have characteristic times of less than a few minutes. To be effective they must be on a time scale similar to that of the controlling steps of the reaction. To date, it has not been practical to use such rapid concentration switches on the scale of an industrial reactor. However, a procedure called reverse flow operation has been applied to a few commercial-scale processes (*61*). Here, the flow to a simple packed-bed reactor is switched from what had been the entrance to the reactor, E, without changing the concentration, to what had been the outlet, O. The effects obtained depend on the energy balance; the amount of energy stored in the solid particles is equivalent to that carried in by the flowing reactive gas over a period of hours so that the cycling times can be on the order of hours rather than minutes. The instantaneous reaction rate $r(c_j, T, t)$ is approximately at steady state, and there are no surface phase delays arising from storage effects. The usual mass and energy balances of reaction engineering can be applied to analyze the system response.

For an exothermic reaction the gas now entering at O will be heated as it flows over the hot bed, and the bed is cooled. Reaction then occurs as the E end is approached, and the gas leaves with the appropriate conversion at E. The O end of the reactor now has an axial temperature gradient resulting from the passage of the cool gas. At an appropriate moment the reactive gas is switched back to E. The feed is now heated by the hot catalyst and reacts, but further down the bed it encounters the cold catalyst layers, where its conversion continues to increase for a reversible exothermic reaction such as the oxidation of SO_2. Before the temperature at O

becomes too low, the gas is switched again, and so on. The resulting process does not need the reactant product heat exchanger, and no expensive heat transfer configuration is needed inside the reactor or between reactor beds.

A discussion of transient effects in reactors would not be complete without mention of the wrong-way behavior found in a packed-bed reactor for an exothermic reaction. This effect can also be described by the usual balances of reaction engineering, but with the reaction at quasi-steady state (*62*). Starting with the packed-bed reactor at steady state, the temperature of the inlet gas is suddenly reduced. The gas is now too cool to react in the first part of the bed, and it only reacts and heats up some distance down the reactor. However, the cold zone at the entrance to the bed becomes progressively longer, until at some point the fresh gas encounters a zone in which the catalyst is sufficiently hot so that the temperature becomes higher than its former steady-state level before returning to the original level near the reactor exit. As time passes the hot spot travels out of the bed and the outlet temperature decreases so that the temperature profile over the whole bed is lower than that of the original. This creation of a hot spot may be dangerous or at least damage the catalyst.

D.   Spontaneous Oscillating Reactions

1.   *Studies on Supported Metals*

Schüth *et al.* (*63*) published a thorough review of heterogeneous catalytic reactions that exhibit spontaneous oscillations; the review includes 344 references, including those to earlier reviews. Using the oxidation of CO over supported Pt as an example, there may be two steady-state rates for a given set of conditions: the low rate corresponding to a high coverage of CO (low coverage of oxygen), and the high rate corresponding to a low coverage of CO (high coverage of oxygen). By increasing the temperature at constant inlet flow one can follow the low-rate branch until the rate suddenly jumps to the high-rate branch. Decreasing the temperature causes the rate to decrease along the upper branch until it finally jumps back down to the lower branch; there is hysteresis. An oscillating inlet forcing function produces oscillations in the outlet concentration. However, Hugo (*64*) and Wicke *et al.* (*65*) found (in approximately 1970) that certain steady inlet conditions led to self-sustained oscillations in the outlet gas concentration. Schüth *et al.* present a table of about 90 references to spontaneously oscillating systems in heterogeneous catalysis; among previous reviews are those by Sheintuch and Schmitz (*66*) and Razon and Schmitz (*67*).

Of the daunting number of models proposed to explain the observed oscillations at atmospheric pressure, I mention two here. Eigenberger (*68*) proposed the idea of a buffer state on the surface, arising from the slow, reversible adsorption of one of the reactants in a nonreactive form. This accumulation of buffer occurs during the high-rate part of the cycle, as the coverage of key reactant decreases. After the rate has decreased sufficiently the reactant starts to accumulate on the surface again. Now the buffer desorbs, and the reaction rate increases as the coverage of reactant increases to the starting point of the cycle.

Ivanov *et al.* (*69*) worked with a model involving a coverage-dependent activation energy so that the rate constant for the surface reaction is an exponential function of a reactant coverage. Predictions obtained by the numerical application of both this model and that involving a buffer provide fairly good representations of the experimentally measured oscillations (*63*).

In these models, the catalyst surface has been assumed to be spatially uniform as its surface composition varies with time. There is synchronization. However, measurements obtained by infrared thermographic imaging by Pawlicki and Schmitz (*70*) indicate that there are large temperature inhomogeneities over a catalyst disk on which spontaneous oscillations are occurring. A hot spot changes in intensity and moves back and forth on the disk as the oscillations proceed. There is coupling between the energy balance and the component balances; concentration and temperature gradients may exist for the solid, the surface, and the gas phase. For a disk of supported catalyst it is difficult to model the heat-transfer problem involved.

Experimental measurements of the degree of synchronization have been made by Onken and Wolf (*71*) and Onken and Wicke (*72*). A disk of SiO_2 with a single millimeter-size spot of Pt produces coherent synchronized oscillations of the temperature at the spot and the CO_2 concentration in the gas. Two separate spots give different temperature oscillations, and the CO_2 concentration seems to be what would be expected from the combination of the independent behaviors of the two spots. A complete disk of supported catalyst produces a chaotic variation of the two variables. A recent study by Qin and Wolf (*73*) provides clear maps of the CO coverage and the temperature on a disk as a function of time during the oscillations. These authors (*74*) later showed that imposing a high-frequency cycled feed (a vibrating feed) to the reactor at conditions for which there would otherwise be self-sustained oscillations damps these effects. The system approaches a kind of relaxed steady state, although small limit cycles persist.

2. *Surface Science Studies*

As is often the case, studies on well-defined surfaces, such as single-crystal faces, at appropriate vacuum conditions might be expected to shed light on the fundamentals of oscillating systems because (i) temperature gradients are absent, (ii) high sensitivity and precision of measurement of gas-phase concentrations are attainable because of the absence of carrier gas, and (iii) the crystal surface can be characterized *in situ* by techniques not available for use at atmospheric pressure. The results of such studies for the oxidation of CO, mostly for steady state, have been reviewed by Engel and Ertl (75), and the first work on autonomous oscillations for this reaction on Pt(100) was published by Ertl *et al.* in 1982 (76). The results of extensive work in this field are the subject of reviews by Ertl (77, 78). Here, I provide only an outline of this complex and fascinating subject.

It has been found (79–81) that there are no oscillations on a Pt(111) surface under any conditions, whereas for conditions of high surface coverage of CO, for which the rate is limited by the rate of adsorption of oxygen, many authors (77) have found oscillations on Pt(100), (110), and (210). Ertl and associates explained this behavior in terms of the structure of the surfaces exposed to the reacting gases. The surface atoms of the clean, closely packed (111) crystal do not differ in structure from bulk atoms, and their positions are not altered by adsorption of CO or O_2. However, for the more open (100) face, the environment of the Pt atoms on the clean surface is altered from that of the bulk atoms; the surface is reconstructed. It has a "hex" or 1×2 configuration rather than the $c(2 \times 2)$ (or 1×1) structure that would exist if the surface atoms had retained the positions they occupied in the bulk, before the outer Pt atoms were removed. In addition, as the surface coverage of CO is increased above about 0.2, the atoms under them tend to revert to the unreconstructed structure. CO can adsorb on either surface phase, but O_2 adsorbs dissociatively only on the (2×2) phase and hardly at all on the hex phase. This general behavior of the adsorbing species was determined by experiments that were mostly carried out before those on oscillating reactions (75, 77).

I now discuss the sequence of events that lead to oscillations, starting from a sufficiently high coverage of CO. There is enough CO on the surface to create patches of bare (2×2) areas. O_2 adsorbs on these places and rapidly reacts with adjacent CO_{ads} to release CO_2. Now, as the surface is depleted of CO_{ads}, a coverage of CO is reached below which the (2×2) surface switches over to the hex form so that O_2 is no longer adsorbed and CO_2 production virtually stops. Now CO can accumulate on the hex surface, with little production of CO_2 because the O_2 has no place to adsorb. When the CO coverage increases to a critical value, the hex structure of the surface Pt changes over to the (2×2) structure, O_{ads} starts to form, and the CO_2 production increases again.

Cox *et al.* (82) used low-energy electron diffraction (LEED) to probe the surface structure of the Pt(100) crystal during the oscillations of the CO oxidation at 500 K. Figure 4 shows the CO_2 formation rate, the LEED

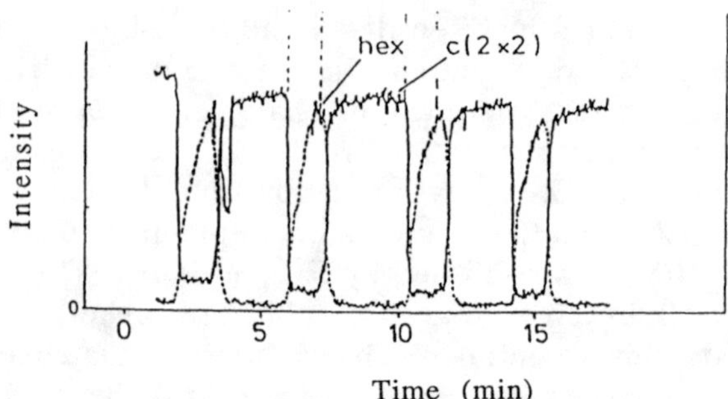

FIG. 4. Sustained oscillations that occur during the oxidation of CO over Pt(100) at 500 K, $P_{O_2} = 5 \times 10^{-2}$ Pa, $P_{CO} = 5 \times 10^{-2}$ Pa. The dashed vertical lines that coincide with the sudden decrease in the c(2 × 2) signal occur at the maximum work function (oxygen coverage), which also corresponds to the maximum production rate of CO_2. The next vertical line corresponds to the maximum in the hex signal, and the simultaneous minima for the oxygen surface coverage and the CO_2 production rate (after 82).

signal from the hex surface, and the LEED signal from the (2 × 2) surface as a function of time. The process of the preceding paragraph can be followed on this graph. The results have been simulated successfully by a model expressed in terms of four ordinary differential equations; it is assumed that there is synchronization over the entire surface. However, later experiments revealed that there are spatiotemporal variations in the surface structure and coverages of CO_{ads} and O_{ads} as waves travel across the surface (81, 83). The representation of these effects has been achieved (84) by the solution of partial differential equations with the inclusion of surface diffusion, leading to gradients of surface concentrations. The gas phase is always well mixed by rapid Knudsen diffusion, and the temperature of the metal is constant. More details, with results for other metals, crystal faces, and reactions, are discussed by Ertl (77).

In a recent study, photoemission electron microscopy (85) was used to reveal remarkable patterns of spaciotemporal variations, as shown in Fig. 5 (86). These patterns are similar to those observed with a homogeneous solution in which the Belousov–Zhabotinsky reaction (97) is occurring.

As pointed out by Qin and Wolf (74), the window in which oscillations occur at UHV conditions does not seem to be accessible at atmospheric conditions with supported metal catalysts. It is not evident that oscillations can be explained via a consideration of the changing surface structures, which cannot be measured. For the supported catalysts there are gradients of gas-phase temperature, solid temperature, gas composition, and surface composition so that a successful simulation via the solution of the coupled

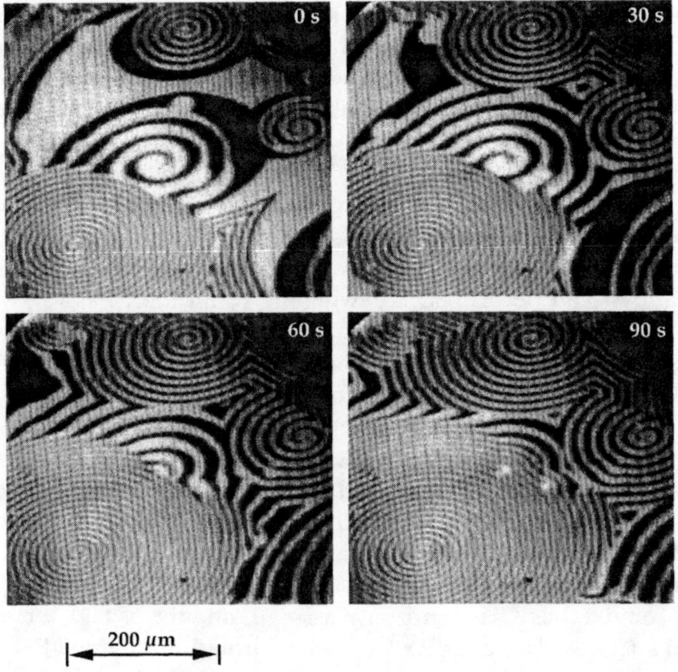

FIG. 5.   Photoemission electron microscopy images of a Pt(110) surface during CO oxidation. The interval between successive images is 30 s (after *86*).

partial differential equations does not seem to have been accomplished to date.

E.   TEMPERATURE-PROGRAMMED METHODS

1.   *General*

To this point, the discussion has been concerned with isothermal experiments, and the transient response has been caused by an input concentration signal, except for the spontaneous oscillations. Following the adsorption described in Fig. 1, a return to pure carrier gas causes a desorption of adsorbed species that are labile at the temperature involved. For instance, at room temperature it would be found that only part of the CO desorbs; of course, this can be measured so that we know the quantity of strongly and weakly adsorbed CO. However, the strongly adsorbed CO can be desorbed in a stream of carrier gas by increasing the reactor temperature and observing the desorbed peaks (temperature-programmed desorption; TPD). Since different forms of an adsorbed species have different ΔH's of desorption, the corresponding peaks appear at different temperatures. This simple experiment is often used to give a fingerprint of the adsorbed species. In the past, the peaks were detected by a thermal conductivity cell. However, for many cases it is essential to use mass spectrometry and estimate quantitatively the amount of each desorbed species. For instance, when CO adsorbed on Ru/SiO_2 is subjected to a TPD in He, the results shown in Fig. 6 are obtained (*88*). The Boudouard reaction, producing CO_2, occurs in addition to the simple desorption of CO.

For a reaction occurring at steady state the composition of the surface of the catalyst can be probed by switching to an inert gas to produce an isothermal desorption, followed by a TPD. Alternatively, various reactive gases can be used to titrate the surface. For example, during methanation the feed of CO/H_2 can be replaced by H_2 alone to produce an isothermal hydrogenation of the surface species. This procedure is sometimes called washing or etching. After the peaks produced isothermally have passed, one can then program the temperature to react off of refractive surface species, typically similar to graphitic carbon. This is a temperature-programmed reaction (TPRx). As an example of a simple TPRx, Fig. 7 shows the result of treating CO adsorbed on Ru/SiO_2 with H_2 (*88*). The CO_{ads} is removed as CH_4, and only a small amount of CO_{ads} is desorbed directly as CO_g. Also, a TPRx in H_2 performed after the TPD of Fig. 6 gives a quantity of CH_4 equivalent to the carbon deposited during the production of CO_2 observed during the TPD.

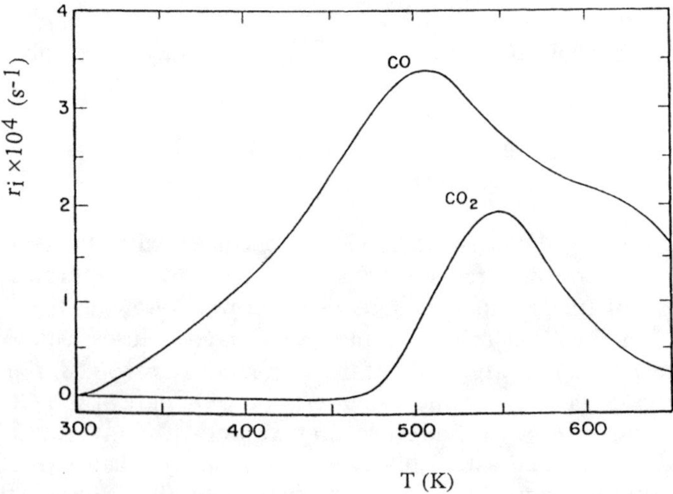

FIG. 6.   TPD obtained by mass spectrometry after the adsorption of CO at 298 K ($\theta <$ 0.35) on Ru/SiO_2 (after *88*).

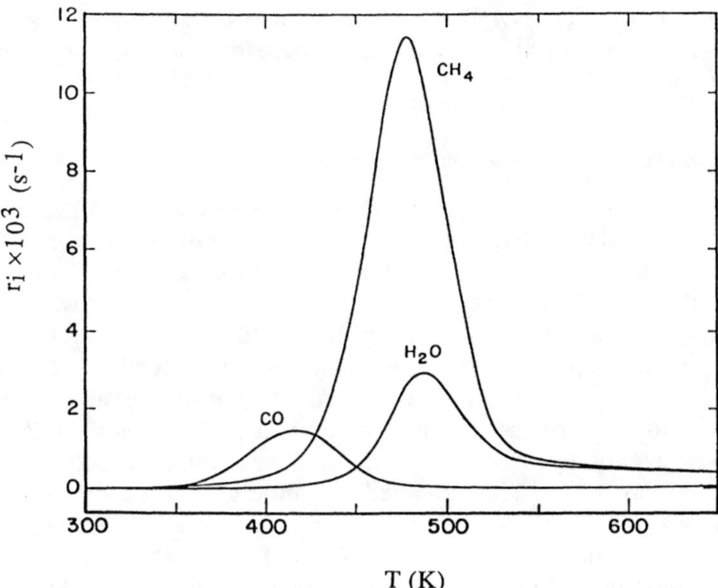

FIG. 7. TPRx obtained by reaction with H_2 during a temperature ramp after adsorption of CO on Ru/SiO$_2$ to saturation at 298 K (after *88*).

When the reactant is hydrogen, a temperature-programmed reduction (TPR) occurs. This method is often applied to the measurement of the reducible components of a solid catalyst. It may also be appropriate to perform a temperature-programmed oxidation (TPO). Temperature-programmed methods were reviewed in 1983 by Falconer and Schwartz (*89*) and again in 1990 by Schwarz and Falconer (*90*), who concentrated on supported Ni catalysts and included experiments involving isotopes, with 151 references. Among the many interesting experiments described in this review (*90*) is the TPRx of CO in H_2 preadsorbed on Ni/Al$_2$O$_3$ at 300 K. At a heating rate of 0.7 K/s, two peaks of methane are observed, probably arising from CO on Ni and from CO on the support. However, at a heating rate of 0.07 K/s, only one peak is observed. Perhaps during this slow desorption the CO on the support keeps the CO concentration high on the Ni so that the two sources of CO are not distinguished. The explanation of this experiment might be clarified by a model based on elementary steps and the appropriate material balances.

In surface science studies it is common to preadsorb the reactants at a temperature sufficiently low so that they do not react with each other and then to program the temperature to produce a temperature-programmed surface reaction (TPSR). This technique has led to a remarkably detailed understanding of certain reactions. An interesting personal account of the development of the method has been written by Madix (*91*).

2. Temperature-Programmed Desorption

A consideration of a simple TPD will illustrate some of the principles that apply to all transient experiments. The quantitative interpretation of a TPD originated in surface science work, pioneered by Redhead (*92*); such studies are also referred to as thermal desorption spectroscopy (TDS). The experiments can now be interpreted by Eqs. (9) and Eq. (2). Here, q is the pumping speed, V is the volume of the perfectly mixed vacuum chamber, and c_{jf} is zero. It is customary to consider that $r_i = r_d$; r_a is supposed to be zero. By suitable numerical procedures $r_d(\theta, T)$ can be found and eventually fitted to an Arrhenius-type equation and the quantities $E(\theta)$ and $\nu(\theta)$ found. This is called a complete method by de Jong and Niemantsverdriet (*93*) in an interesting article that considers the merits of various approximate (and easier) methods of treating the data. Sometimes θ can be measured directly by the work function, by Auger electron spectroscopy, etc. These experiments produce strictly kinetic data, but if adsorption is not activated, $E = \Delta H_{ads}$. However, equilibrium measurements can be made at low pressure as a function of pressure and temperature, for instance, for CO on Ni(111) (*94*). Many powerful methods of surface science involve phenomena occurring in a transient regime. For instance, Block *et al.* (*95*) reviewed methods that depend on the use of high electric fields, such as field electron microscopy, pulsed-field desorption mass spectrometry, and field desorption spectroscopy.

A TPD run on a supported metal by using a carrier gas at atmospheric pressure furnishes data for the mixture of surface sites present. Usually, such studies give only a qualitative idea of the different species present on the various sites. However, simultaneous measurement of the IR spectra of the adsorbed species may furnish valuable information (*88*). In addition, with a carrier gas it is extremely difficult to get kinetic data that are not altered by transport effects. Furthermore, the adsorption–desorption step is not unidirectional; the reverse reaction (adsorption) may occur during the TPD. This effect is called readsorption in surface science studies in which it is unusual, but it is usually present in a TPD obtained with a carrier gas.

Rieck and Bell (*96*) performed modeling studies for typical experimental TPD conditions. To understand the influence of the experimental parameters, it is helpful to refer to the dimensionless groups proposed by Demmin and Gorte (*97*). For example, to evaluate the effect of intraparticle diffusion, the dimensionless parameter $Go = q\rho_p d_p^2/12WD_e$ should be small (<1), where q is flow rate in cm^3/s, ρ_p is the density of the particle in g/cm^3, d_p is the diameter of the particle in centimeters, W is mass in g, and D_e is the diffusion coefficient in cm^2/s.

Figure 8 shows a result for the simulation of a TPD for typical parameters (96). For a flow rate of 5.0 cm^3/s, the positions of the response curves are considerably altered in passing from 0.04-cm particles to 0.2-cm particles or for the large particles in passing from a flow rate of 5.0 cm^3/s to 0.5 cm^3/s. However, for the low flow rate, the curves are little altered by changing the particle size. In this case, the changes in concentration induced by the flow ($q\rho_p/W$) are small compared with the rates of diffusion, represented by D_e/d_p^2. However, this TPD is so "slow" that the rate of adsorption is similar to the rate of desorption. The gas and surface phases are in quasi-equilibrium. This type of TPD has been performed for the desorption of H_2 held on Rh/Al_2O_3; the results are shown in Fig. 9 (98). It is shown that the rate of adsorption is much greater than the net rate of desorption, which permits the use of the data of Fig. 9 to find the thermodynamic functions $\Delta H^\circ(\Theta)$ and $\Delta S^\circ(\Theta)$. $\Delta H^\circ(\Theta)$ decreases from a maximum at low coverage to a minimum at high coverage because of lateral repulsions among the adsorbed species. Clearly, no kinetic data can be obtained. In general, if the process is speeded up, the analysis of the response must take

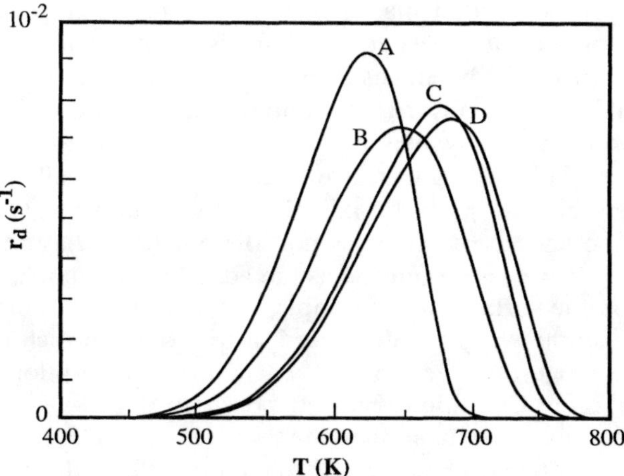

FIG. 8.   Simulation of TPD spectra for CO desorption (after 96). A, d_p = 0.04 cm, q = 5.0 cm^3/s; B, d_p = 0.2 cm, q = 5.0 cm^3/s; C, d_p = 0.04 cm, q = 0.5 cm^3/s; D, d_p = 0.2 cm, q = 0.5 cm^3/s.

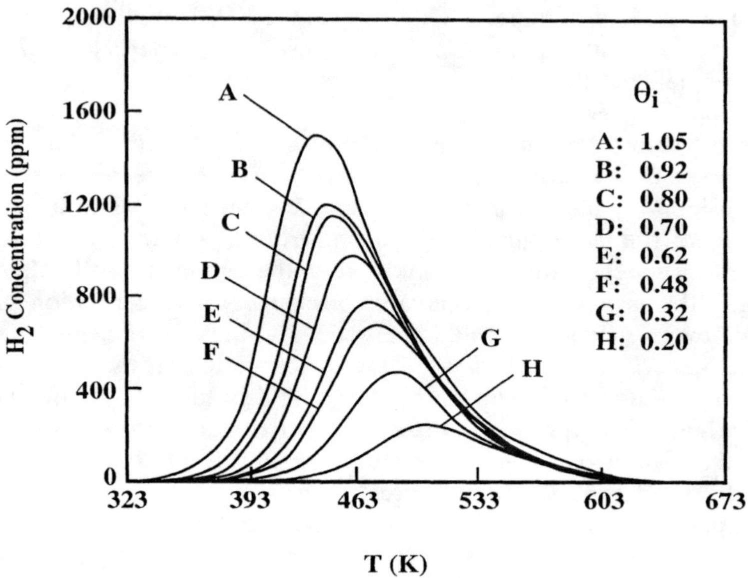

FIG. 9. TPD curves for H_2 adsorbed on Rh/Al_2O_3 at various initial surface coverages, θ_i (after *98*).

into account the effects of the transfer phenomena, as discussed previously (*96, 97*).

Another procedure that makes use of an experiment in the transient regime performed at quasi-equilibrium is high-resolution adsorption (HRADS), carried out by means of an Omnisorb apparatus. Measurement of the rate of adsorption (far from equilibrium) is a classical procedure in surface science, as already mentioned. To measure the adsorption isotherm of a powdered sample (high surface area), starting typically at 10^{-7} bar, a constant flow of gas (q_f) to be adsorbed is sent into the vacuum vessel containing the adsorbent. There is no flow out ($q = 0$), and the varying pressure ($p = c = c_j$) is monitored as the adsorbate accumulates in the gas space and on the surface of the sample. As the ratio of the inlet gas flow rate to the sample weight is decreased, a rate below which the apparent isotherm no longer changes can be reached; the adsorption proceeds at quasi-equilibrium. Conditions can be chosen so that the acquisition time for the isotherm is reasonable (less than 1 h). In the TDS discussed earlier, $q_f = 0$, and q is approximately constant. In the TPD with carrier gas, $q_f \cong q$, $c_{jf} = 0$, and c_j (p = const.) is monitored. An interesting example of the use of HRADS with zeolites is given by Hathaway and Davis (*99*).

F. Isotopic Tracers in the Transient Regime

1. *Steady-State Tracing*

Steady-state tracing consists of following the composition of the outlet of a reactor initially at steady state when one of the reactants is suddenly replaced by the same species but with one of its atoms replaced by one of its stable isotopes. Although radioactive isotopes may be employed, the use of a simple mass spectrometer makes it possible to follow a great variety of species with the same equipment and without the precautions necessary when dealing with ionizing radiation. We begin the discussion by considering the adsorption of ^{12}CO on an Ni/Al_2O_3 catalyst, in terms of Fig. 1. If ^{13}CO had been adsorbed instead, its response curve would be the same as that for ^{12}CO. During the adsorption the surface coverage increases from zero to that corresponding to saturation at c_{COf}. At this coverage the repulsive forces between the adsorbed molecules have reached a maximum, corresponding to a minimum in the activation energy for desorption. Now the ^{12}CO in the feed is replaced by ^{13}CO. At all times the concentrations of $^{12}CO + {}^{13}CO$ in the gas leaving the reactor are equal to the original feed concentration of ^{12}CO. The concentration of the surface phase will also change from $^{12}CO_{ads}$ to $^{13}CO_{ads}$. Any kinetic isotope effect is neglected but of course would need to be considered for a light molecule such as H_2. Notice that this exchange process occurs at a constant saturation coverage so that the response of the ^{13}CO would not be the same as that for the ^{12}CO in Fig. 1 which was obtained as Θ_{CO} increased.

Another important fact is that the quantity of CO exchanged on the surface is the same as that initially adsorbed. However, if the initial ^{12}CO is desorbed by a switch to He, after a long time there will be a residual coverage of ^{12}CO. At this coverage the activation energy of desorption has become so high that the remaining CO is "irreversibly" adsorbed. All this is done at constant temperature, 220°C in Ref. (*17*). The rest of the CO could be removed and measured by a TPD or a TPRx in H_2.

Experiments and calculations have shown that the CO exchange process over Ni at 220°C is immeasurably rapid so that the fraction of labeled CO in the surface phase is the same as that in the gas phase (*17*). The response curves can be calculated by using Eq. (4) with j representing one of the isotopes, all the r_i's equal to zero, and $\tau_{co} = N/R$, where N is the total moles of both kinds of CO in both phases and R is the molar feed rate of the appropriate isotope. Since we know R, τ_{CO}, and the gas volume, we can calculate the surface coverage of CO from this experiment. The response time τ_{CO} is measured for this first-order mixing process, but no information on the kinetics of chemisorption is available from this experiment.

If the same concentration of CO is used but with H_2 instead of He, there will be steady-state methanation over the Ni catalyst. This rate is measured by the rate of production of CH_4, essentially the only hydrocarbon produced. Now the same isotopic tracing experiment can be done to determine the coverage of CO present on the surface during the reaction. A correction for the CO reacted must be included, although it may be small in a differential reactor (*17*).

However, the fraction of the $^{13}CH_4$ isotope in the methane produced does not follow that of the ^{13}CO but increases at a much slower rate, as shown in Fig. 10 for the same conditions as in Fig. 1. There is a reservoir of adsorbed intermediate (CH_x) through which the ^{13}C passes irreversibly, and the residence time is $\tau_{CH_x} = N_{CH_x}/R_{CH_4}$ or Θ_{CH_x}/TOR_{CH_4} ($\tau = \Theta/r$ in simplified notation). The intermediate is hydrogenated via surface H to CH_4, which does not accumulate on the surface. The process through which the carbon passes can be shown by

$$CO_g \underset{1}{\rightleftharpoons} CO_{ads} \underset{2}{\rightarrow} CH_x \underset{3}{\rightarrow} CH_{4g}.$$

Step 1 is in quasi-equilibrium, and steps 2 and 3 control the rate of appearance of CH_4. The measured response times τ_{CO} and τ_{CH_x} give the surface coverages present during reaction. The value of Θ is proportional to the area between the curves for ^{13}CO and $^{13}CH_4$. This analysis is based on the assumption that each *pool* of surface species behaves like a CSTR. This is

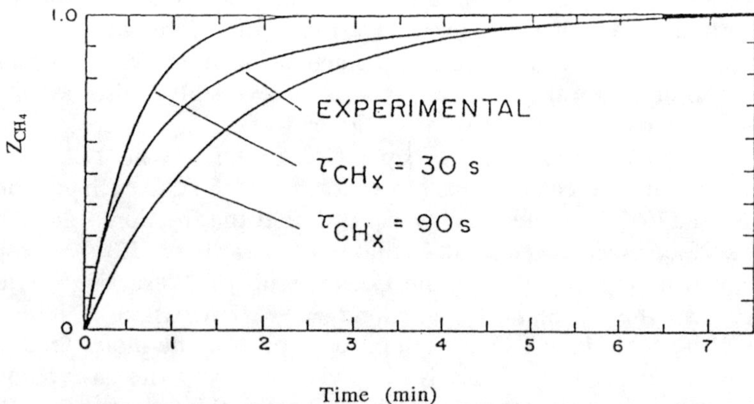

FIG. 10. Steady-state tracing result for $^{13}CH_4$ after a switch from $^{12}CO/H_2$ to $^{13}CO/H_2$ at 250°C during the methanation of CO over Ni/Al₂O₃ (after *17*).

strictly true for the quasi-equilibrium ^{13}CO response of Fig. 1, but for the $^{13}CH_4$ response the curve produced is not fitted by a simple exponential. It can perhaps be modeled by two or more pools in parallel of different size and reactivity; each pool has a τ that can be estimated from fitting the observed response, but we can measure neither the capacity (Θ) nor the rate for an individual pool but only their ratio.

The ratio Θ/r is often called k, but this symbol corresponds to a reaction rate constant only if $r = k\Theta$; then k is simply $1/\tau$. Although the rate-determining step for methanation may be the hydrogenation of CH_x for iron where almost no CO is detectable on the surface, for Ni, Ru, and Rh there is a reservoir of CO on the surface and the speed of step 2 influences the overall rate.

Rather than modeling via a set of finite pools, De Pontes *et al.* (*100*) consider that there may exist a distribution function for k, $f(k)$, i.e., there are many pools in parallel. They present an interesting mathematical development through which it is possible to calculate the curve $f(k)$ by fitting the observed response for $^{13}CH_4$. However, the r and θ corresponding to each k are not estimated, and step 3 is considered to be rate determining. Hoost and Goodwin (*101*) presented a different method for obtaining $f(k)$.

Goodwin and associates have published many interesting studies, predominately making use of steady-state isotopic-transient kinetic analysis (SSITKA), the basic ideas of which were outlined previously. The work was reviewed by Shannon and Goodwin (*102*), and the method has been applied to many systems in heterogeneous catalysis. The linear nature of the equations describing such experiments was pointed out by Le Cardinal *et al.* in 1977 (*103*). Recently, methods have been described (*104*) that take advantage of the mathematical power of linear modeling for the results of SSITKA experiments: Laplace transforms, transfer functions, moments, etc. Such methods permit the correct analysis of the differential PFRs that are convenient to use in the laboratory.

Happel (*105*) began using isotopic tracers in the transient regime in approximately 1970, and the state of the art in 1986 is published in his book. Happel began working on advanced methods for the interpretation of the data with Lecourtier and Walter in approximately 1985, including an analysis of the role of PFRs (*106*). The procedures developed have been applied to the study of a reversible reaction, the dehydrogenation of isobutane to isobutene (*107*) over chromia. In this work transient isotopic tracing is used in a straightforward manner to give results for the individual forward and reverse rates of a heterogeneous catalytic reaction occurring at steady state. This represents an important advance, so I discuss this matter in detail.

Figures 11 and 12 are taken from the work of Kao *et al.* (*107*) referred

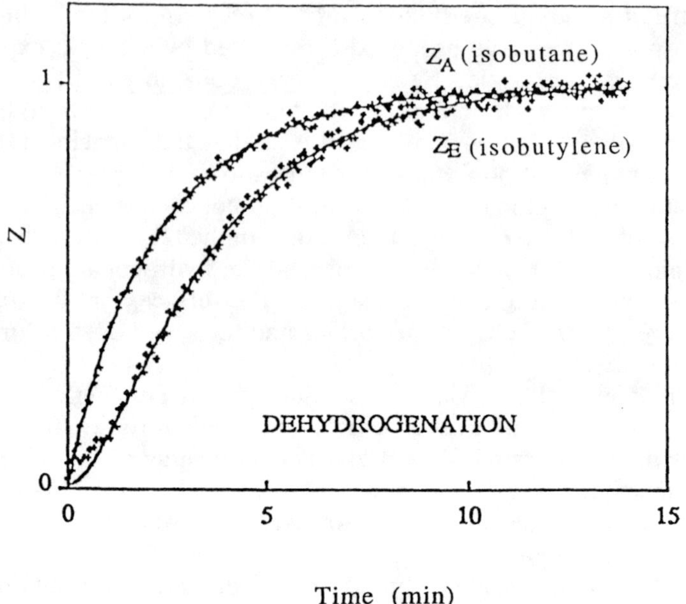

Time (min)

FIG. 11.   Responses of the fractions of ^{13}C-labeled fractions of isobutane (Z_A) and isobutene (Z_E) to a switch in the feed of pure isobutane to its ^{13}C equivalent (after *107*).

to previously. At time zero for Fig. 11 the feed of pure isobutane was replaced, in a mixed flow reactor (CSTR), by its ^{13}C equivalent, and the fraction (Z_A or Z_E, in isobutane or isobutene, respectively) of tracer in the outflow is plotted vs time. Figure 12 shows a similar experiment, but the feed is a mixture of 76.2% i-C_4H_{10} and 23.8% i-C_4H_8; only the i-butane is traced so that the ^{12}C introduced with the i-butene must appear in the products. Up to four pools in series have been considered as models, but a good fit is obtained by the following two-pool model:

$$C_4H_{10,g} + C_4H_{9,ads} \overset{r+}{\underset{r-}{\rightleftharpoons}} C_4H_{8,ads} + C_4H_{8,g}. \tag{20}$$
$$\text{pool1} \qquad\qquad\qquad \text{pool2}$$

This scheme means that the adsorption and half-dehydrogenation are fast (in quasi-equilibrium) so that the left side of the equation represents one pool. The reversible but nonequilibrium step then leads to the second pool; the adsorption–desorption of i-butene is also in quasi-equilibrium. Note that the reactants and products are both in the same physical well-mixed gas phase, and the surface species are on the same surface phase; this will be taken into account in the balance equations that follow.

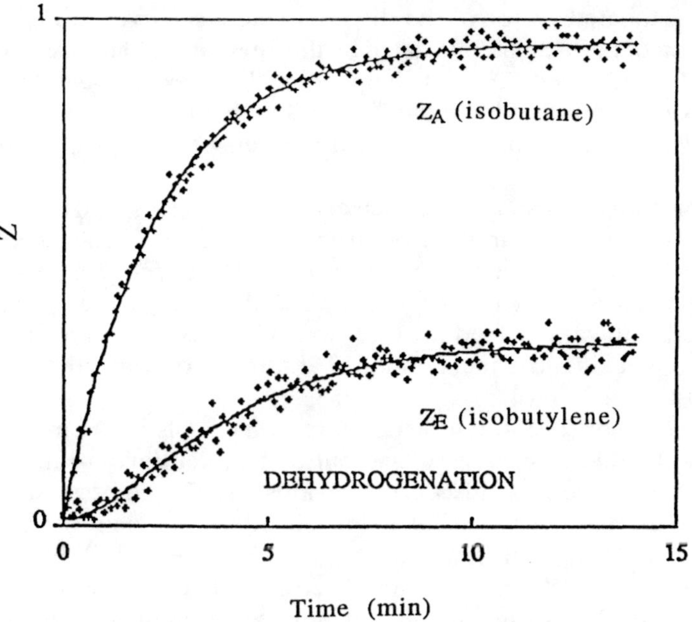

Time (min)

FIG. 12.   Responses of the fractions Z_A and Z_E to a switch of isobutane to its ^{13}C equivalent but in a feed mixture containing 76.2% isobutane and 23.8% isobutene (after *107*).

For the overall steady-state reaction,

$$r_+ - r_- = r = (1/V)(q_f c_{Af} - q c_A) = (1/V)(q c_E - q_f c_{Ef}). \qquad (21)$$

The notation is that of Eqs. (2)–(4). Thus, from the usual steady-state experiment we measure r; we must use the tracing experiment to find r_+ and r_-. For the conditions used (*107*), the inlet gas is more than 95% He so that to a good approximation $q_f = q$ and q/V can be set equal to $1/\tau$; however, with the numerical methods used (*107*) it is not necessary to make any simplifications. The H_2 balance can also be ignored. We now apply the differential equations (Eqs. 2–4) to each pool, with the following results for balances on the marked molecules:

Pool 1: $\quad (c_A + \Theta_9 L_v)(dZ_A/dt) = L_v(r_- Z_E - r_+ Z_A)$
$$+ (1/\tau)(c_{Af} Z_{Af} - c_A Z_A) \qquad (22)$$

Pool 2: $\quad (c_E + \Theta_8 L_v)(dZ_E/dt = L_v(r_+ Z_A - r_- Z_E)$
$$+ (1/\tau)(c_{Ef} Z_{Ef} - c_E Z_E). \qquad (23)$$

The unknown parameters of this system are Θ_9, Θ_8, and r_+; once a value of r_+ is chosen, r_- is calculated from the measured constant r by Eq. (21).

For an estimated vector of the three parameters, curves of Z_A and Z_E as a function of time can be calculated by the integration (if necessary, numerical) of Eqs. (22) and (23) and compared with the experimental results. An objective function can be minimized by suitable-least-squares calculations as the vector of parameters is varied according to the proper procedures (27, 106, 107).

It would be interesting to do such experiments over a range of conditions that lead to a wide variation of surface coverages. Then one could find experimentally a function such as $r_+ = k(\Theta_9, \Theta_8, c_A, c_E, T)$. This approach was applied in a recent article (108) in which the steady-state tracing method was used with various feed mixtures to study methanation over supported nickel so that kinetics could be studied at surface coverages that correspond to realistic conditions.

For some conditions, the inclusion of other pools in the model may be important, and this can be carried out in a straightforward manner using the numerical methods discussed previously (107). The overall rates and the forward and backward rates are related by the identities formulated by Temkin (105, 109). One can hope to represent the kinetics over a wide range by such an elementary-step model. This way of treating the experimental data seems clearer and more powerful than linking the analysis to a parameter such as the response time obtained from certain isotopic experiments.

It is also interesting to determine what can be learned from a tracing experiment in which the feed is switched from CO/H_2 to CO/D_2 in a perfect step function. If the D_2 were incorporated into the product methane by following the same path through the reaction as the original H_2, it would be expected that the first traced methane peak of importance would correspond to CD_xH_{4-x} if the surface intermediate in the series of steps is CH_x. For Ni/Al_2O_3, an experiment (17) shows that CH_4 in the product is almost immediately replaced by CD_4. This could mean that $x = 0$, but it could also mean that the D very rapidly exchanges with all the H in the surface species; the resulting peak of H_2 would be invisible in the experiment as performed (17). Note that the usual assumption of similar chemical properties between isotopes does not apply for D_2 and H_2. There may be a kinetic isotope effect.

Happel et al. (109) performed the previous experiment for methanation over molybdenum sulfide catalysts. Here, it seems that exchange for almost all the H_{ads} is rapid. The gas dead volume is large so that five residence times amount to about 30 min. Thus, during the changeover from H_2 to D_2, peaks for all the intermediate CD_xH_{4-x} species appear successively in the gas phase, according to the first-order differential equations involved (109). A large amount of H is held by the working catalyst, corresponding to the formula $H_{0.074}MoS_2$. This hydrogen-rich solid is linked to many of the desirable properties of the catalyst, such as resistance to sulfur poisoning and lack of deactivation by carbon at low H_2/CO ratios.

2. *Adsorption-Assisted Desorption*

Tamaru, who has done much seminal work on heterogeneous catalysis in the transient regime (*1, 29*), has written a review of recent developments (*110*) in which some interesting results involving isotopic tracing are discussed. From the first article (*111*) in a series treating the adsorption of CO on noble metals, we discuss the adsorption of CO at 2×10^{-6} Pa and 380 K on a Pd polycrystalline surface held in a vacuum chamber that behaves like a CSTR when a leak valve is opened so that $C^{18}O$ flows through the system and is removed at a constant pumping speed. When the valve is opened the pressure rises from its background level and increases to a steady-state value (approximately at equilibrium with the CO adsorbed on the Pd) within about 2 min at a typical surface coverage of about 0.8. The surface coverage at a time chosen during the adsorption is found by closing the inlet valve and then doing a TPD via mass spectrometry. This is a classical procedure in surface science. The interesting part of the experiment consists of changing the feed gas from $C^{18}O$ to $C^{16}O$ at a coverage of, for example, 0.5 ($\Theta^{18} = 0.5$), and then continuing the adsorption toward saturation ($\Theta^T = \Theta^{18} + \Theta^{16} = 0.8$). The response time of the gas phase is small enough so that its changes in composition appear to be instantaneous. The *expected* result is shown in Fig. 13 (*112*).

In Fig. 13, Θ^T increases linearly at low coverage, but as saturation is

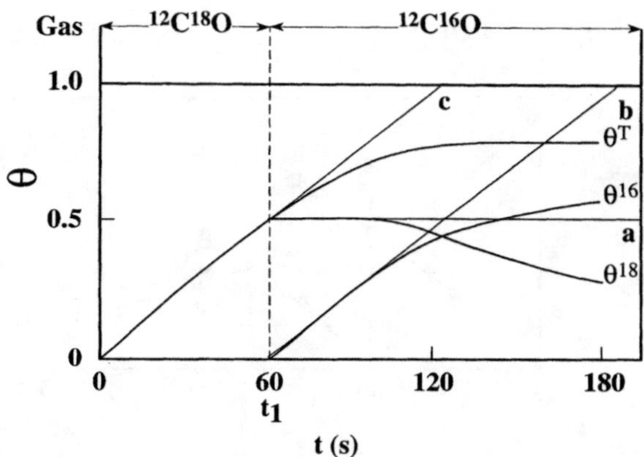

FIG. 13.   Replacement of $^{12}C^{18}O$ by $^{12}C^{16}O$ after 1 min of adsorption in a CSTR [based on the model of Zhadanov (*112*)]. Slopes a–c are defined in the text.

approached the repulsive forces between adsorbed CO molecules cause an increase in the rate of desorption. At saturation $r_a = r_d$ and Θ^{18} goes to zero. At the instant of switch (t_1) there should be no desorption of $C^{18}O$, and the slope $d\Theta^{16}/dt$ (curve b) should equal $d\Theta^T/dt$ (curve c); the slope $d\Theta^{18}/dt = 0$ (curve a).

The results of Yamada *et al.*, (*111*) experiment are shown in Fig. 14; there *is* desorption of $C^{18}O$ at the switch. There must be two different reservoirs of CO on the surface: one for the usual strongly adsorbed CO and the other for CO that can desorb, probably in a precursor state. A material balance shows that at t_1, $d\Theta^{16}/dt + d\Theta^{18}/dt = d\Theta^T/dt$; slopes b and c are not equal. This segregation of the two kinds of CO is a transient effect. If one stops the adsorption and performs a TPD, the $C^{18}O$ and the $C^{16}O$ should become indistinguishable in their behavior. This complicated matter has been considered by Yates and Goodman (*113*), Lombardo and Bell (*114*), and Tamaru and colleagues (*110*). Again, the explanation of transient and isotopic experiments requires a more detailed understanding of the processes involved than that needed merely to explain steady-state experiments.

Tamaru (*110*) also discusses examples from heterogeneous catalysis in which reaction rates of one step seem to be influenced by the adsorption of other components. For example, Nishimura *et al.* (*115*) studied the dehydrogenation of ethanol to acetaldehyde and hydrogen over a specially prepared Nb/SiO_2 catalyst at 523 K. The studies were done in a recirculating closed (batch) reactor. The rate is about constant as time increases, and the IR spectrum of an adsorbed intermediate remains constant. A sudden evacuation of the gas-phase ethanol stops the reaction but does not affect

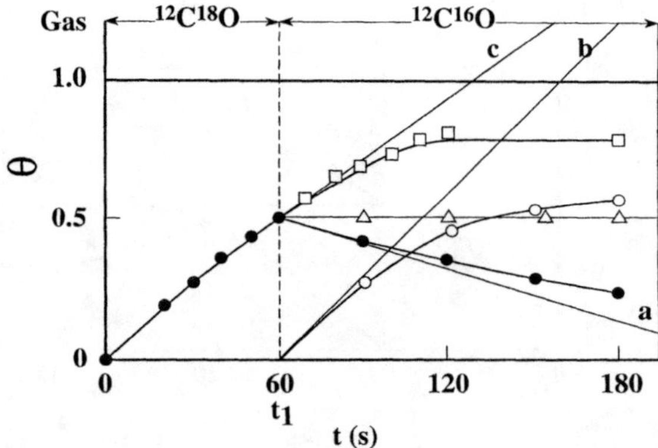

FIG. 14.　Experimental results for the experiment of Fig. 13 that were obtained by Yamada *et al.* (*111*). Slope a is not zero, and slope b does not equal slope c.

the intermediate. When C_2H_5OH is reintroduced, the reaction begins again at the same rate. Dehydrogenation seems to follow a path that does not involve the observed surface intermediate. If, in the absence of ethanol, the intermediate is heated by temperature programming, peaks of C_2H_4 and $(C_2H_5)_2O$ are observed at about 673 K. The decomposition of the intermediate leads to the products of *dehydration*.

G. Summary

From the preceding discussion it is clear that our understanding of heterogeneous catalysis is enhanced by doing experiments in the transient regime, and that during these experiments we should measure as many effects as possible so as to follow the composition of the gas phase and the surface phase during the experiments. The perturbing signal can be sent via the inlet gas composition (step, pulse, and cycled concentration) or via the reactor temperature (TPD, TPRx, TPO, TPR, TPSR, etc.). Additional information is obtained by using isotopes. The measurement of the surface phase at atmospheric pressure during the transients is often difficult, but it is possible to make *in situ* measurements via IR, Raman, and X-ray absorption spectroscopies. A striking example of the results that can be obtained by time-resolved X-ray absorption spectroscopy was obtained by Coulston *et al.* (*116*) for the partial oxidation of *n*-butane on vanadium phosphate catalysts. Other X-ray methods such as Mössbauer effect spectroscopy (*117*) have been used. Interesting results have also been obtained by studies involving positron emitters (*118, 119*), which permit following the concentration of a tracer as a function of position in a packed-bed reactor as the profile changes with time in the transient regime. At surface science conditions some electron spectroscopies may also be applied; for example, the work function can be measured at steady state or during transients to deduce the oxygen coverage on Pt during CO oxidation (*78*). Recently, scanning tunneling microscopy (STM) and photoelectron emission microscopy have produced remarkable results (*78, 120*) showing the mobility of metal surface atoms in both the apparent steady state and the transient regime.

Other interesting results have been described by Bowker (*121*), in particular those obtained by STM. During the (partial) oxidation of methanol over Cu(110) at 353 K the surface was observed by STM, and the rate of adsorption (and reaction to formaldehyde) was measured. The clean surface is first exposed to oxygen to produce a partial monolayer of oxygen. The oxygen atoms form long islands. When exposed to methanol above 400 K, methoxy formation occurs only at the ends of the islands, which grow shorter as time proceeds and formaldehyde and water desorb. The sticking probability of the impinging methanol (proportional to the reaction rate) has been measured as a function of time during this experiment in the transient regime. During this process, the coverage of oxygen diminishes until the reaction stops (Fig. 15). It is clear from Fig. 15 that the reaction rate at a given low coverage (ca. 0.125) is much higher for the system that started out at an oxygen coverage of 0.25 than for the one that started out at a coverage of 0.5. The configuration of the oxygen islands at the same coverage depends on how much oxygen originally covered the surface. A usual equation, such as $r = k\Theta_{\text{oxygen}}\Theta_{\text{methanol}}$, is not adequate. The rate equation must take into account the detailed geometry of the surface, perhaps through Monte Carlo methods. As discussed by Bowker (*121*), it is not possible to perform similar experiments at pressures and temperatures typical for the steady-state production of methanol. Similar preferential reactions at the ends of surface oxygen strips have been observed by STM for CO on Cu(110) at 400 K (*122*) and for NH_3 on Ni(110) at room temperature (*123*). A thorough treatment of this subject is not attempted here.

In Section III, I discuss many articles published mostly during the past 10 years to show the progress that has been made via transient methods in understanding many interesting catalytic systems. For each case some background is presented, but the emphasis is on the applications of the transient method. Some of these articles (and some already cited) can be

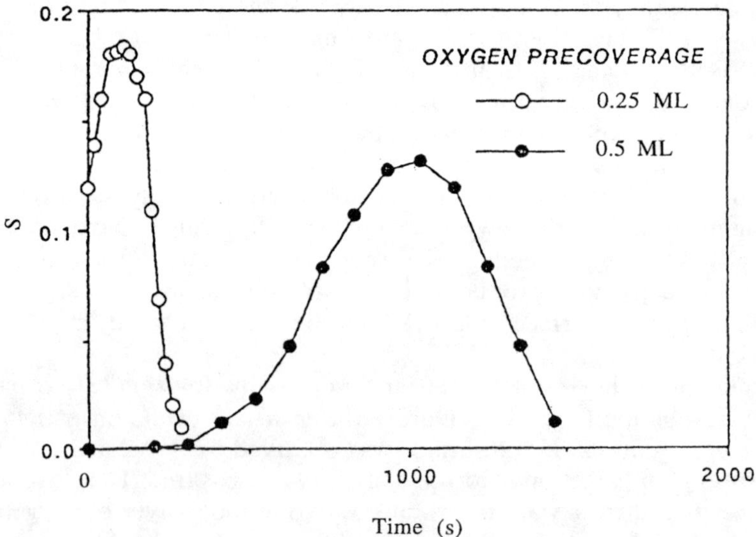

FIG. 15. The sticking probability of methanol on Cu(110) as a function of time for two different initial coverages of oxygen (after *121*).

found in Refs. (*124*)–(*128*). A review concentrating on applications has been written by Mills and Lerou (*129*).

III. Case Studies

A. Oxidative Coupling of Methane

In recent years methane has become plentiful compared to other hydrocarbons, and its resulting low price has been a driving force for research on processes to convert methane directly into higher hydrocarbons or synthesis gas (CO/H_2). For example, research on the reaction

$$2\,CH_4 + 1/2\,O_2 \rightarrow C_2H_6 + H_2O \tag{24}$$

has been pursued by many groups, especially in the early 1990s. The reaction takes place over various oxide catalysts at atmospheric pressure and about 800°C, usually with less than the stoichiometric proportion of oxygen. Transient methods have proven quite effective for understanding the kinetics and mechanism of this reaction, and this subject was reviewed in 1997 by Schuurman and Mirodatos (*130*) and Efstathiou and Verykios (*131*).

Lunsford and coworkers (*132*), in a seminal paper, showed that a Li-doped MgO catalyst showed good selectivity for oxidative coupling of methane (OCM). By ESR measurements in the gas phase leaving the bed of their reactor, they found that CH_3 radicals leaving the bed rapidly combined to form C_2H_6 in accord with existing knowledge of the rates of reactions among radicals and molecules. These rates had been extensively studied in work leading to understanding combustion and the thermal cracking of small hydrocarbons. The homolytic splitting of methane can be represented by the steps

$$CH_4 + O^* \rightarrow CH_3 + OH^* \qquad (\times 2) \tag{25}$$

$$2\,OH^* \rightarrow H_2O + O^* + {}^* \tag{26}$$

$$\tfrac{1}{2}O_2 + O^* + {}^* \rightarrow 2O^*, \tag{27}$$

where the first step controls, leading to an Eley–Rideal process. As the radicals leave the surface, they rapidly combine in the gas phase to form ethane,

$$2CH_3 \rightarrow C_2H_6. \tag{28}$$

An early confirmation of Eq. (28) was published by Mims *et al.* (*133*). Steady-state OCM was performed with a mixture of CH_4 and CD_4, and practically all the ethane formed was limited to CH_3CH_3, CH_3CD_3, or CD_3CD_3.

Ekstrom and Lapszewicz (*134*) were apparently the first to use isotopic methods to study the kinetics of OCM, and they reported considerable holdup of methane on Sm_2O_3 via a switch from $^{12}CH_4$ to $^{13}CH_4$ in the continuous reactor feed. However, Peil *et al.* (*135*) and subsequent investigators found practically no methane adsorption, in agreement with Eq. (25). Their SSITKA results at 600°C are shown in Fig. 16 (*135*). The methane response closely follows that of the argon tracer, whereas that of CO_2 shows a delay related to the storage of intermediates on the surface leading to undesired products, especially CO_2. Thus, methane also reacts with O* to form CO and CO_2 by other surface reactions, and no catalyst has been found to give a yield of more than about 25%, not high enough for an economic exploitation of the process. The well-documented kinetic properties of radicals in the gas phase argue against the direct oxidation of methane in the gas phase.

In order to understand the oxygen pathways during the reaction, SSITKA experiments involving $^{18}O_2$ (*136*) have also proven useful. Some results

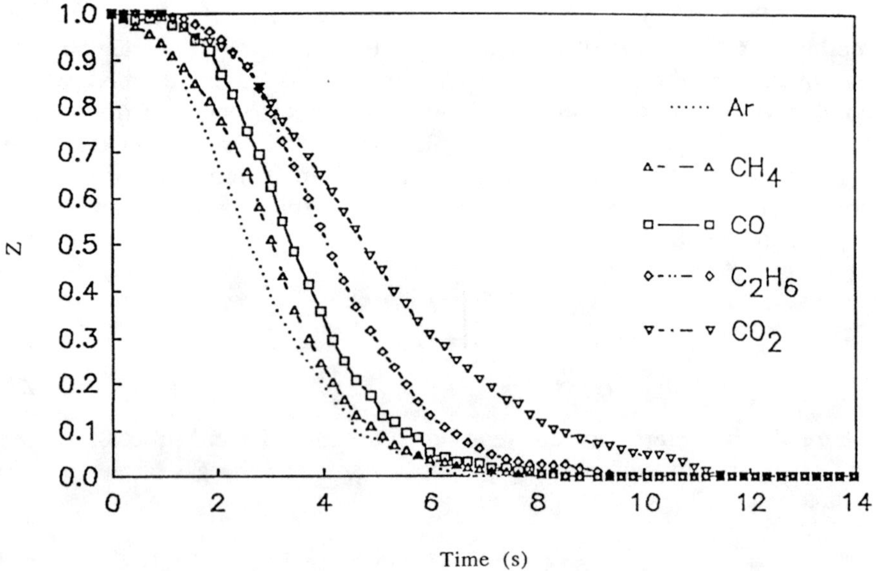

FIG. 16.   Carbon tracing SSITKA results after a switch from $^{12}CH_4$ to $^{13}CH_4$ during the steady-state coupling of methane over Sm_2O_3 at 600°C (after *135*).

over Li/MgO at 600°C are shown in Fig. 17, in which the response to the switch $^{12}CH_4/O_2 \rightarrow {}^{13}CH_4/O_2$ is plotted. There is no surface reservoir of CH_4, but there is about a monolayer of carbon-containing intermediate that leads to CO and CO_2. The interpretation of this *tracing* experiment is straightforward because carbon does not exchange with anything on the surface. Such is not the case for oxygen; a switch from $^{16}O_2$ to $^{18}O_2$ at 600°C with no methane produces $^{16}O^{18}O$ and $^{16}O_2$ in quantities much larger than those corresponding to a monolayer. This result means that lattice oxygen of the solid catalyst is *exchanging* with surface oxygen. If the previous isotopic switch is made during OCM, the oxygen changes over more quickly to $^{18}O_2$ than in the absence of reaction. However, $C^{16}O_2$ and $C^{18}O^{16}O$ are produced as in Fig. 18, showing that these undesired products arise from a considerable reservoir of oxygen from the catalyst. In a related experiment, Peil *et al.* (*137*) show for Sm_2O_3 that the nonisotopic switch $CH_4/O_2/He \rightarrow CH_4/He$ gives an overshoot in the ethane production rate, followed by a continued decreasing production arising from the reaction of methane with oxygen originating in the bulk of the oxide.

These results can be qualitatively explained by adding to Eqs. (25)–(28) the following steps;

$$O^* \rightleftharpoons O_{bulk} + * \tag{29}$$

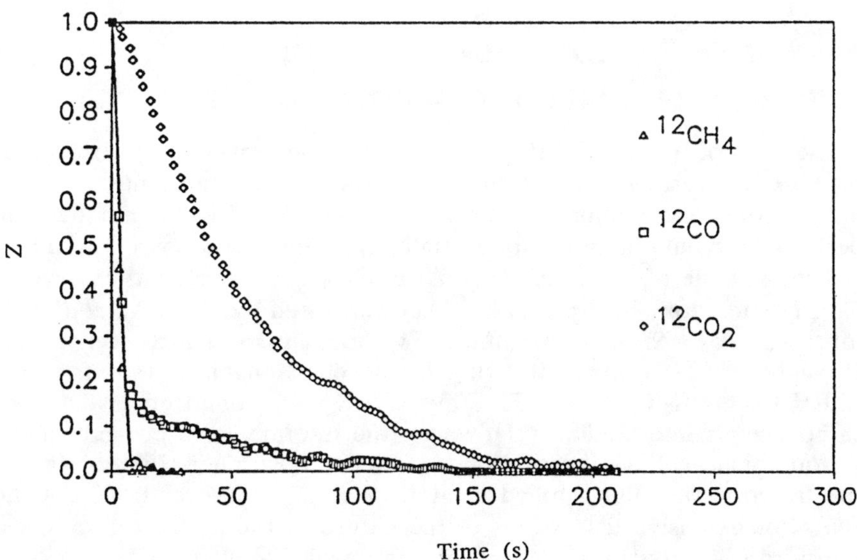

FIG. 17.   The same tracing experiment as shown in Fig. 16 but for steady-state coupling over Li/MgO at 600°C (after *136*).

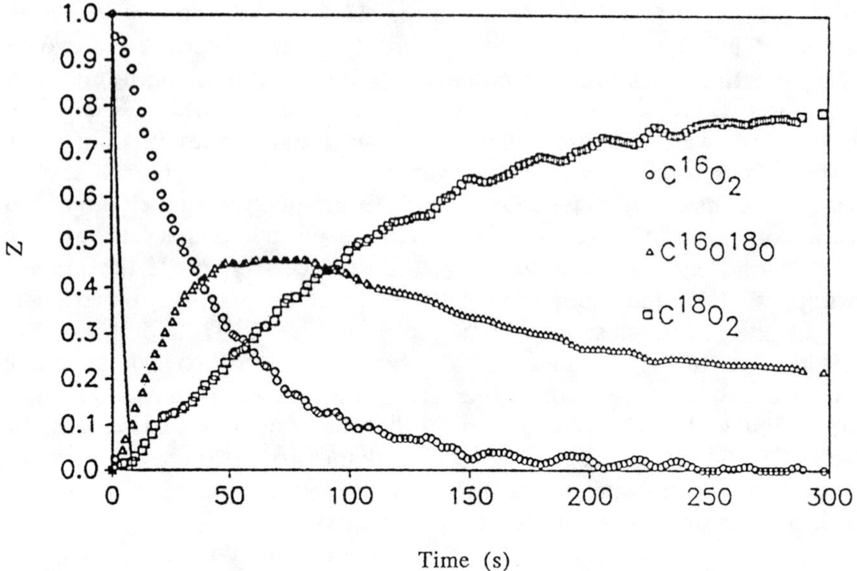

FIG. 18.   Oxygen tracing SSITKA after the switch $CH_4/^{16}O_2$ to $CH_4/^{18}O_2$ showing the large reservoir of ^{16}O that participates in the production of CO_2 (after *136*).

$$2O^* + CH_4 \rightarrow CH_3O^* + OH^* \tag{30}$$

$$CH_3O + 4O^* \rightarrow CO_2 + 3OH^*. \tag{31}$$

These are not necessarily elementary steps. The movement of oxygen in the bulk is governed by diffusion, and the last two equations must be made up of more "elementary" processes. The nature of O^* is not defined; perhaps O^- would be more appropriate, as originally proposed in the heterolytic scheme of Ito *et al.* (*132*). Also, O_{bulk} is probably lattice oxygen, O^{2-}. The methoxy in Eq. (31) has been suggested by Nibbelke *et al.* (*11*) for reaction over Sn/Li/MgO, although without any spectroscopic evidence. Lacombe *et al.* (*138*) were able to use diffuse reflection infrared spectroscopy (DRIFT) during OCM at 750°C over La_2O_3 and found no evidence of carbonaceous intermediates. However, the intermediates present on the various oxide catalysts vary from one solid to another. *In situ* Raman spectroscopy was also applied at atmospheric pressure and 800°C (*139, 140*). For extensive discussions of the nature of the surface species, see Lunsford (*141*), Lacombe *et al.* (*142*), and Nibbelke *et al.* (*11*).

Other studies in the transient regime have been made for Sm_2O_3 (*137, 143–145*), Li/MgO (*11, 145, 146*), La_2O_3 and Sr/La_2O_3 (*138, 142, 147, 148*), and Li/TiO$_2$ (*149–151*). The comparison of the catalysts between studies is difficult, especially since temperatures ranging from 600 to 900°C were used. The general field of OCM has been reviewed by Lunsford (*141*) and in a book edited by Wolf (*152*).

Following the previous introduction to OCM, I turn to some recent studies. Nibbelke *et al.* (*11*) used principally lined-out Sn/Li/MgO in their transient isotopic experiments at atmospheric pressure and 750°C. The CH$_4$/O$_2$ ratio was 4, with 80% helium dilution. In their small PFR the conversion of methane was 24% and that of oxygen 85%, leading to a selectivity of 31% to ethane and 23% to ethylene. The yield of C$_2$ is thus about 13%. At this high conversion, the labeled (transient) species in the reactor must be modeled by a balance like that shown in Eqs. (1)–(3), a set of partial differential equations. Note that the reactor has many layers of particles, ensuring plug-flow behavior. For the sum of the labeled and unlabeled components in the gas phase, the system is in steady state, with gradients of concentration along the length of the reactor. The equations have been integrated by numerical procedures, and the vector of parameters has been optimized to give the best value of an objective function. The rate constants of the elementary steps have been used to reproduce the experimental results, as shown for example in Fig. 19. This and other graphs

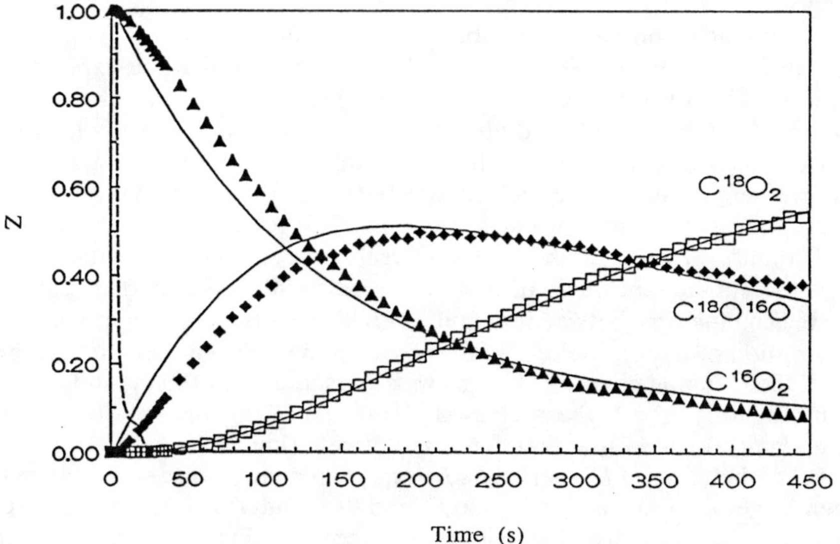

FIG. 19.   Oxygen tracing results obtained in a plug-flow reactor for OCM at 750°C over Sn/Li/MgO, with 85% conversion of the oxygen fed. The continuous lines are calculated from a model based on elementary steps such as Eqs. (25)–(31) (after *11*).

of the results are similar to Figs. 16–18 (*135, 136*). The general procedure is similar to that used by Kao *et al.* (*107*), but these workers used Eqs. (2)–(4), appropriate for their CSTR.

Equations (25)–(31) are basically those used by Nibbelke *et al.* (*11*) to model their results. However, Ekstrom *et al.* (*153*) added $^{13}C_2H_6$ and $^{13}C_2H_4$ to the feed to an OCM reactor and found that at steady state a large amount of the $^{13}CO_2$ is produced, indicating that CO_2 does not come from Eqs. (30) and (31) alone. If the gas-phase oxidation of C_{2+} is ruled out, they must react via the surface, perhaps by an equation such as

$$C_2H_6 + 2O^* \rightarrow 2CH_3O^* \tag{32}$$

and then oxidation to CO_2 by Eq. (31). In this way the selectivity of methane to ethane decreases as conversion increases, as shown for La_2O_3 catalysts by Le Van *et al.* (*154*). In accord with the previous discussion, Colussi and Amorebieta (*155*) propose that the maximum possible yield of C_{2+} via a purely heterogeneous mechanism is about 22%.

It has been proposed that the more effective catalysts are those that show higher mobility of lattice oxygen (*147*). Kalenik and Wolf (*148*) demonstrated this effect by using temperature-programmed isotopic exchange, as illustrated in Fig. 20. Exchange of oxygen in La_2O_3 is facilitated by strontium promotion, and the Sr/La_2O_3 catalyst also shows better performance for OCM.

Many studies on OCM have been made by using the TAP system (*34, 35*) already discussed. For example, I discuss a pump–probe experiment over Sm_2O_3 at 650°C done as follows (*143*). The pump is a pulse of 95% O_2 and 5% Kr; it is followed after an interval Δt by the probe, which is a pulse of 95% CH_4 and 5% Kr. The pulses are made sufficiently long so that the pressure in the pulses is high enough so that there are abundant collisions among the gas-phase molecules, ensuring the production of ethane from CH_3 radicals. Figure 21 shows that oxygen is retained on the catalyst and that the subsequent pulse of methane reacts with this adsorbed oxygen. Increasing the time between the pulses gives more time for the dissociative adsorption of oxygen before it is exposed to the pulse of methane. Therefore, the resulting ethane pulses grow as Δt increases to 100 ms and remain constant up to the highest Δt used (3000 ms). The work of Buyevskaya *et al.* (*144*) for Sm_2O_3 at 800°C is qualitatively similar.

Lacombe *et al.* (*142*) performed oxygen–methane pulse–probe experiments over a La_2O_3 catalyst at 750°C, and the results are completely different from those discussed previously, as shown in Fig. 22. The maximum disappearance of methane occurs over this catalyst for $\Delta t = 0$ and decreases to zero as the interval between the pulses increases. Conversions to desired products and to CO decrease together. CO_2 is retained on the catalyst, but

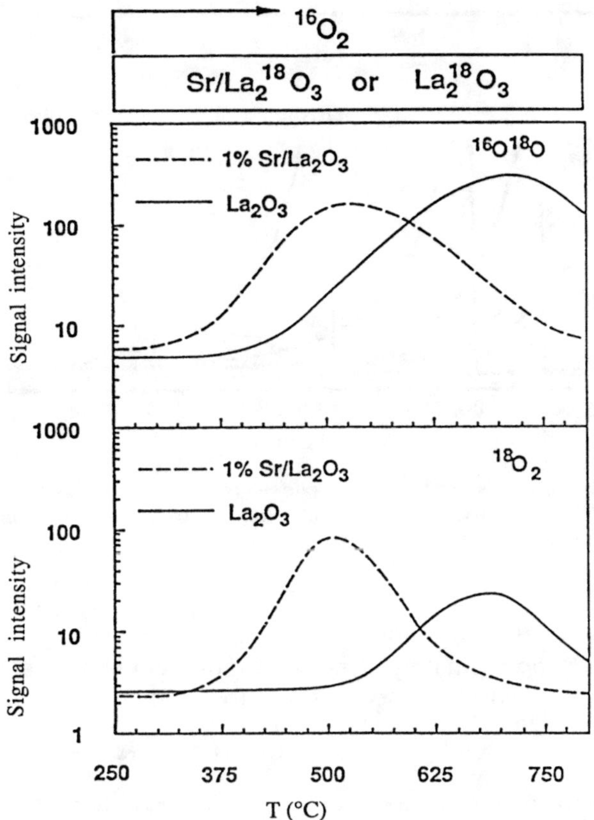

FIG. 20.   Temperature-programmed isotope exchange over two lanthana-based catalysts for OCM (after *148*).

O_2 is not. Oxygen is not stored on lanthana at 750°C (*142*), but it is stored on samaria at 650°C (*143*).

The results of Nibbelke *et al.* (*11*) were confirmed by a later TAP study by Mallens *et al.* (*146*).

B.   PARTIAL OXIDATION OF METHANE TO SYNTHESIS GAS

When methane is oxidized over a noble metal instead of a metal oxide, it is possible to produce synthesis gas directly. As early as 1946, Prettre *et al.* (*156*) showed that the reaction

$$CH_4 + \tfrac{1}{2}O_2 \rightarrow CO + 2H_2, \ \Delta H = -8.5 \text{ kcal/mol} \tag{33}$$

produces synthesis gas in good selectivity when a 2/1 mixture of methane and oxygen is passed over a Ni catalyst at 700–900°C at a space velocity

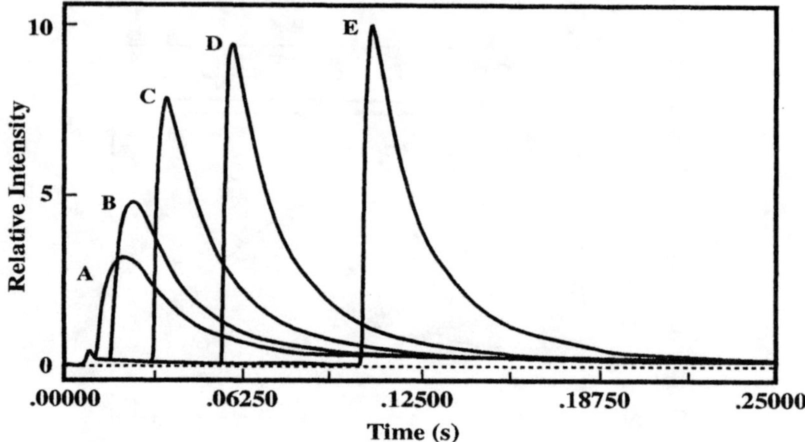

FIG. 21. Ethane pulses produced by a pump–probe experiment in a TAP system over Sm_2O_3 at 650°C. The interval between a inlet pulse of oxygen and the following pulse of methane is as follows: A, 0.005 s; B, 0.010 s; C, 0.025 s; D, 0.050 s; E, 0.100 s (after *143*).

leading to complete consumption of methane and oxygen. Thermodynamic equilibrium is approximately attained at these temperatures, and Eq. (33) is followed quite closely, instead of that for total combustion,

$$CH_4 + 2O_2 \rightarrow CO_2 + 2H_2O, \Delta H = -192 \text{ kcal/mol.} \tag{34}$$

This extremely exothermic reaction is successively suppressed in favor of Eq. (33) as the temperature increases. It should be possible to use a simple autothermal reactor for the slightly exothermic reaction of Eq. (33). Proper design and an appropriate catalyst are required to avoid excessive reaction via Eq. (34) at the entrance to the reactor, causing a hot spot. The ratio $H_2/CO = 2$ is more convenient for downstream processes than that obtained by steam reforming:

$$CH_4 + H_2O \rightarrow CO + 3H_2, \Delta H = 49 \text{ kcal/mol.} \tag{35}$$

This endothermic reaction requires an expensive fired heater as reactor and several subsequent steps if it is desired to lower the H_2/CO ratio below 3. A review of this subject forms the introduction to an interesting article by Mallens *et al.* (*157*).

Hickman and Schmidt (*158, 159*) thoroughly studied Eq. (33) over Pt- and Pt–Rh-coated monoliths and gauzes. Best results are obtained at about 1100°C with residence times of 10^{-4} to 10^{-2} s. High flow rates minimize mass transfer effects. Since the gas at the exit of the reactor is almost in

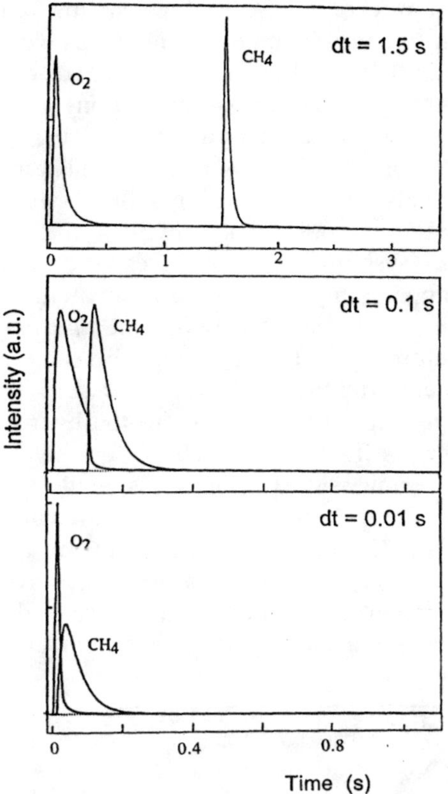

FIG. 22. The same type of pump–probe experiments as discussed in Fig. 21 but over a La$_2$O$_3$ catalyst at 750°C. The unreacted pulses leaving the reactor are shown (after *142*).

equilibrium, it is not possible to measure kinetics under the conditions used for practical application. However, a model has been devised based on literature data on elementary steps, largely from surface science, and the experimental results have been well simulated (*159*). Nevertheless, it is important to have an experimental basis for understanding the kinetics, in particular the competition between Eqs. (33) and (34). For instance, Prettre *et al.* (*156*) suggested that Eq. (34) occurs first (indirect route), until the oxygen is used up, and then the remaining methane reacts with H$_2$O and CO$_2$ via steam reforming and CO$_2$ reforming. On the other hand, there is good reason to believe that Eq. (33) represents the primary reaction (direct route) (*158*). Simulations indicate that CH$_4$ dissociates to form surface carbon and gaseous hydrogen, and then reaction occurs between surface carbon and oxygen to form CO (*159*). Experiments in the transient regime have shed light on this problem, and some of this work will be discussed.

Mallens *et al.* (*157*) studied the reaction system over rhodium sponge (99.9%, BET 0.25 m²/g) and over platinum sponge (99.9%, BET 0.05 m²/g). Results on Pt alone were also published (*160*). Figure 23 shows the TAP result at 700°C for a 2/1 CH_4/O_2 pulse over a Rh sponge that contained 0.4% Rh_2O_3, the steady-state composition of the catalyst. For the Pt catalyst, the steady-state catalyst consists of 0.9% PtO_2. Recall that one of the features of the TAP system is that the pulses used can be small enough so as not to affect appreciably the composition of the catalyst during the passage of the pulse. The starting point and the maximum for the H_2 response occur before those for H_2O, and the same is true for CO and CO_2, qualitatively confirming that Eq. (33) represents the primary reaction. The response curves are all normalized (to the hydrogen response), as is frequently the case for TAP results. The actual height of the CH_4 pulse (shown x25; note the noisy signal in Fig. 23) is small since little CH_4 survives as it reacts (irreversibly) during passage over the catalyst, and the actual heights of the CO_2 (x12) and H_2O (x10) peaks are also small, compared to those for CO and H_2. The authors used data from the literature to simulate the adsorption, desorption, and Knudsen diffusion of the four individual gases, each in a single-pulse experiment. In contrast to the data

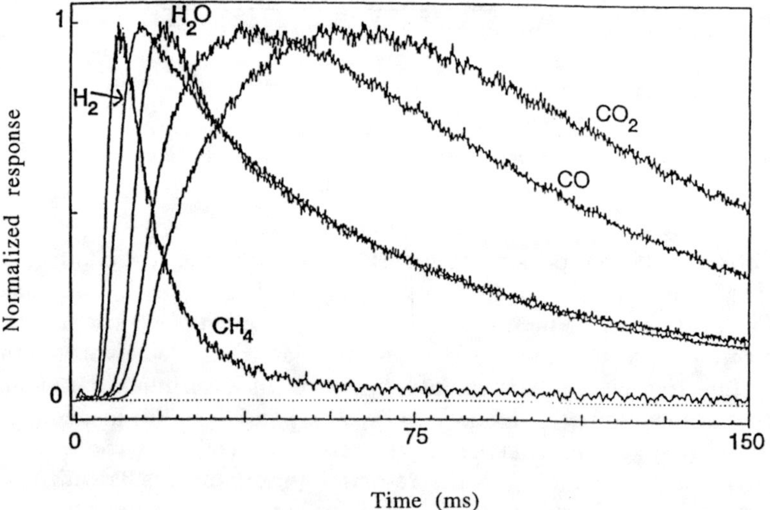

Fig. 23.   Responses in the TAP system for the partial oxidation of CH_4 to synthesis gas over a Rh sponge at 700°C. The inlet pulse is 2/1 CH_4/O_2. See text for the enhancement factors of the responses shown (after *157*).

in Fig. 23, the CO_2 peak occurs before that of CO, and water appears before H_2. The simulation of the data of Fig. 23, involving the elementary steps in the reacting system, would be much more difficult.

The sequence of steps proposed (157) is

$$CH_4 + 5* \rightarrow C* + 4H* \qquad \times 2 \qquad (36)$$

$$C* + Rh_xO \rightarrow Rh_x^\circ + CO + * \qquad \times 2 \qquad (37)$$

$$O_2 + 2Rh_x^\circ \rightarrow 2Rh_xO \qquad \times 1 \qquad (38)$$

$$2H* \rightarrow H_2 + 2* \qquad \times 4. \qquad (39)$$

It was also found that oxygen interacts with Rh to form some O* but continued exposure at 700°C eventually leads to complete oxidation to Rh_2O_3. Experiments with ^{18}O show that over the selective (very little oxidized) catalyst the oxygen acts through Rh_2O_3, as shown in Eqs. (37) and (38). Chemisorbed oxygen, present on a more oxidized surface, leads to more CO_2 and H_2O. Continuous-flow, steady-state experiments show that at 800°C about 90% selectivity to H_2 and CO is obtained over Pt. Rhodium exhibits a better performance; at about 700°C selectivities to the desired products are essentially 100%. Decreasing the CH_4/O_2 ratio below 2 sharply increases the production of CO_2 and H_2O in this temperature range. Good results also require that the catalyst be close to completely reduced, as it is at steady state. If 2/1 CH_4/O_2 is passed over a catalyst rich in Rh_2O_3, production of CO_2 and water is favored until the catalyst is properly reduced. Thus, in a PFR Eq. (34) may be favored near the entrance, and methane may then react with CO_2 and water further down the reactor.

Wang *et al.* (161) studied the role of the support in the partial oxidation of methane on 0.05% Rh/Al_2O_3 by experiments performed in a TAP reactor system at approximately 600°C. They give results only in arbitrary units, normalized with no enhancement factor furnished, so that it is difficult to interpret their graphs. Much of the work involves the effect of water or OH on the support and inverse spillover toward the Rh. This effect seems to lead to reduced selectivities toward CO and hydrogen and greatly complicates the analysis of the results. Since such good results are obtained with unsupported metal sponges or gauzes, this work is difficult to compare with those already discussed.

The partial oxidation of methane has also been studied between 550 and 750°C over Ru/TiO_2 (162). The importance of the indirect route has been shown at 550°C by an appropriate steady-state tracing experiment.

A TAP study of the reaction over 1% Rh/Al_2O_3 at 740°C (163) gives results similar to those discussed previously (161). Only normalized responses are shown. The role of water is important, and a companion *in situ* DRIFT study (*164*) shows that OH on the support participates in the network of reactions at 700°C.

C.   Reforming of Methane by Carbon Dioxide

The so-called dry reforming of methane occurs at 600–800°C according to the reaction

$$CH_4 + CO_2 \rightarrow 2CO + 2H_2, \Delta H = 59.0 \text{ kcal/mol.} \tag{40}$$

Here, $CO/H_2 = 1/1$, in contrast to 1/2 for Eq. (33) and 1/3 for Eq. (35), and the H/C ratio of 2/1 may be attractive as a feed to the Fischer–Tropsch synthesis for higher hydrocarbons. However, the endothermic Eq. (40) leads to the use of a fired heater, as for ordinary steam reforming Eq. (35). The fact that CO_2 is a greenhouse gas is not important because the quantities involved are tiny in comparison to other sources. CO_2 reforming has been much studied in recent years, and transient studies have been useful in unraveling its mechanism.

Fischer and Tropsch (*165*) in 1928 used nickel- and cobalt-based catalysts for CO_2 reforming, and the Calcor process (*166*) is an industrial version of the reaction of Eq. (40). For fundamental studies Ni/SiO_2 can be considered as a reference catalyst, and its deactivation behavior has been studied (*167*), followed later by a mechanistic investigation via transient methods (*168*). However, Zhang and Verykios (*169*) found that Ni/La_2O_3 has remarkable resistance to deactivation based on comparisons with Ni supported on Al_2O_3 and CaO. In subsequent papers (*170–173*) it is shown that the superior behavior of the catalyst is related to the formation of $La_2O_2CO_3$ and formates by reaction with the support and that the latter participate in the catalysis, probably through sites at the boundaries of the metal and the support; these sites may arise because of decoration of the Ni by the oxide phases. This decoration would inhibit the formation of a blocking graphite layer on the Ni surface.

The mechanism of the processes has been studied for Ni/SiO_2 (*130, 168*) and for Ni/La_2O_3 (*174*) by Mirodatos and coworkers using many interesting experiments in the transient regime. For the silica-supported catalyst, the following elementary steps are proposed (*174*):

$$CH_4 + Ni \rightleftharpoons NiC + 2H_2 \tag{41}$$

$$CO_2 + Ni \rightleftharpoons CO + NiO \tag{42}$$

$$NiC + NiO \rightarrow CO + 2Ni. \tag{43}$$

Equation (43) is the rate-controlling step, in accord with the observation that the overall reaction rate is not appreciably altered by a switch from

CH_4 to CD_4 in the feed. There is no kinetic isotope effect. The first two steps are fast and in quasi-equilibrium. A switch from He to $CH_4/CO_2/Ar/$ He over a fresh catalyst shows that CO_2 and CH_4 rise less rapidly than Ar, and CO and H_2 rise more rapidly (there is even an overshoot in H_2), in accord with Eqs. (41) and (42), as C and O build up on the surface. After 10 min of reaction at 700°C, quenching the reaction followed by a TPO produces peaks of CO_2 equivalent to about a monolayer of C, probably present in the form of a carbide-like surface species. *In situ* DRIFT measurements do not reveal bands of any absorbed CO, OH, or CH_x species.

In addition, when the steady feed is changed to CO_2/He alone, large amounts of CO are produced according to Eq. (42). A similar switch, but to CH_4/He alone, produces H_2 according to Eq. (41). Figure 24(a) confirms the reversible nature of Eqs. (41) and (42). The ^{13}C introduced ends up not only as ^{13}CO and $^{13}CH_4$ but also as $^{13}CO_2$. Figure 24(b) can be used to calculate the surface coverage of the traced species, according to the areas between the various responses. Kroll *et al.* (*168*) give quantitative details and other data on the characterization of the catalyst, but space does not permit a complete discussion here. The results are summarized by Schuurman and Mirodatos (*130*).

For the lanthana-supported catalyst, the following elementary steps are proposed (*174*):

$$CH_4 + Ni \rightarrow NiC + 2H_2 \tag{44}$$

$$CO_2 + La_2O_3 \rightarrow La_2O_2CO_3 \tag{45}$$

$$La_2O_2CO_3 + Ni \rightarrow CO + NiO + La_2O_3 \tag{46}$$

$$NiC + NiO \rightarrow CO + 2Ni. \tag{47}$$

For this system, the step shown in Eq. (44) is no longer in quasi-equilibrium, and there is a strong kinetic isotope effect related to the difficult breaking of the CH bonds (*172, 173*); this step has thus been proposed as the rate-limiting process. This result is consistent with the proposition that the Ni surface is decorated by the support. *In situ* DRIFT experiments indicate that a switch from $^{13}CO_2/Ar/He$ to $^{12}CO_2/Ar/He$ at 800°C leads to a shift in the frequencies of the carbonate bands associated with the support, as expected from Eq. (45). During reaction at 800°C, a SSITKA experiment consisting of a switch from $^{13}CO_2/CH_4/Ar/He$ to $^{12}CO_2/CH_4/Ar$ leads to a long tail in the production of ^{13}CO by Eq. (46), permitting the estimation of the relatively large number of monolayers (based on the metal surface) of carbonate present during the steady-state reaction.

The difference in the behavior of Ni/SiO_2 and that of Ni/La_2O_3 has been shown by TAP experiments performed in the pump–probe mode at 600°C

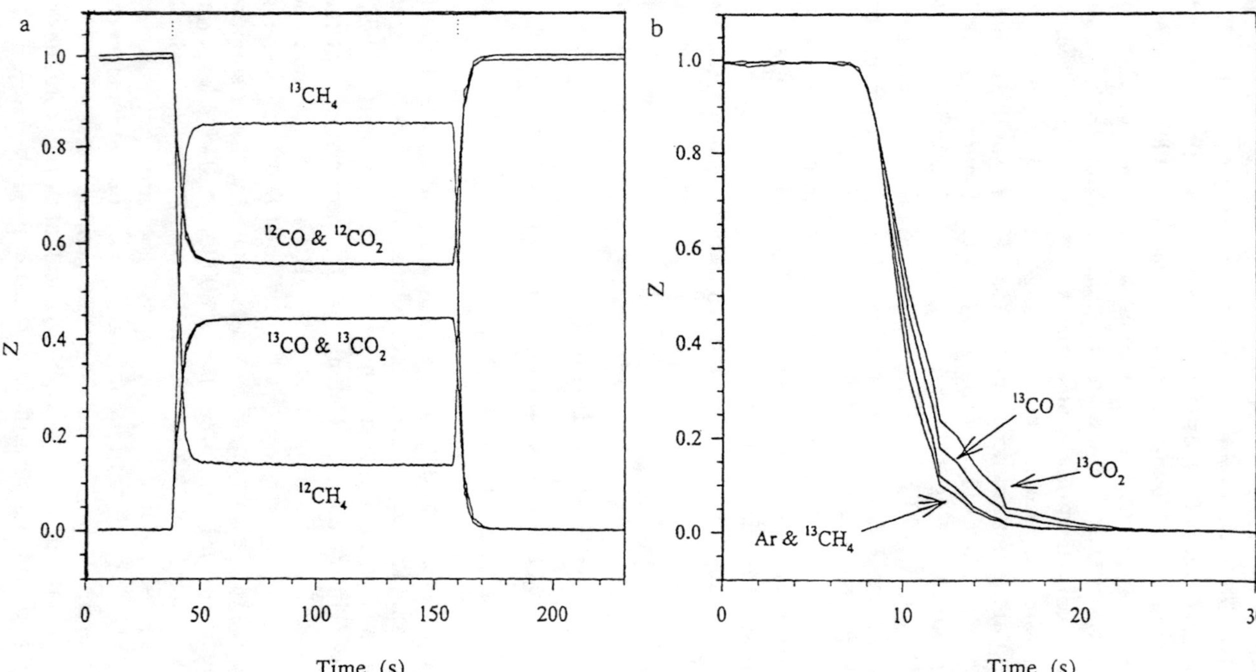

FIG. 24.   (a) Normalized concentration responses obtained after the SSITKA experiment $^{12}CH_4/CO_2/He \rightarrow {}^{13}CH_4/CO_2/He/Ar \rightarrow {}^{12}CH_4/CO_2/He$ at 700°C over Ni/SiO$_2$. (b) Normalized concentrations for the switch $^{13}CH_4/CO_2/He \rightarrow {}^{12}CH_4/CO_2/He/Ar$ (the right-hand part of a for the measurement of the amounts of active intermediates) (after *168*).

(*174*). Figure 25 shows the dramatically different results obtained with these two catalysts. A pulse of $^{12}CH_4/Ar$ is followed 0.5 s later by a pulse of $^{13}CO_2$.

Considering first Fig. 25a, the methane pulse is quickly consumed and a large H_2 peak is produced, according to Eq. (41). The following pulse of $^{13}CO_2$ immediately reacts to form ^{13}CO by Eq. (42), and the NiO formed then can react by Eq. (43) with the $Ni^{12}C$ accumulated in the first pulse, giving a ^{12}CO peak with a long tail.

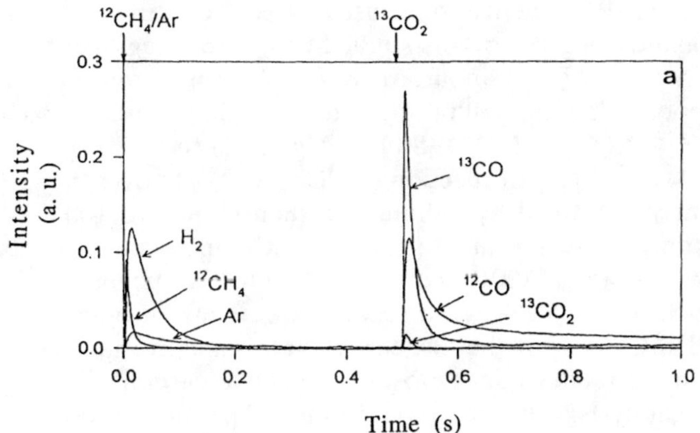

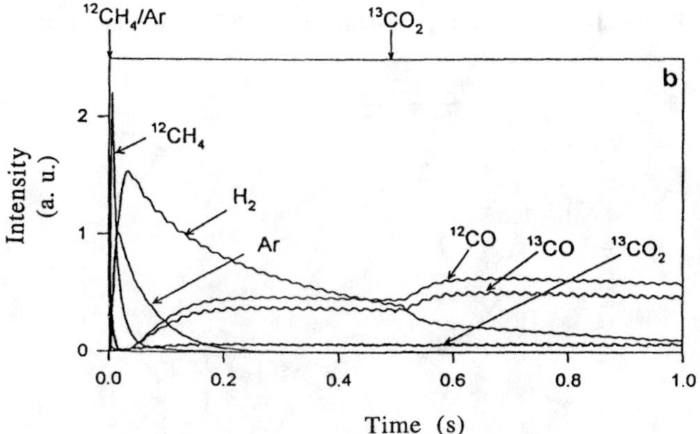

FIG. 25. TAP pump-probe experiments investigating the CO_2 reforming of methane at 600°C. A pulse of $^{12}CH_4$ is followed 0.5 s later by a pulse of $^{13}CO_2$ (after *173*). (a) Responses over Ni/SiO$_2$; (b) responses over Ni/La$_2$O$_3$.

Figure 25b is difficult to interpret because of the arbitrary scales for the intensities. Assuming that the Ar peaks in the first responses of Figs. 25a and 25b should have equal peak heights, the H_2 peak height in Fig. 25b is much lower than that in Fig. 25a, and it clearly has a long tail. This result supports the idea that over Ni/La_2O_3 Eq. (44) is the rate-limiting step. The response to the $^{13}CO_2$ pulse shows that there is a large reservoir for the storage and exchange of CO_2 and CO, associated with the $La_2O_2CO_3$, and which is absent for Ni/SiO_2.

Regarding other metals, it has been found that on noble metals and particularly on Rh deactivation caused by surface carbon is less than that on Ni. Solymosi and coworkers studied CO_2 reforming over supported Rh extensively (*175, 176*). Tsipouriari *et al.* (*177*) compared the performance of Rh supported on many supports and found that activity decreases in the order YSZ > Al_2O_3 > TiO_2 > SiO_2 ≫ MgO, where YSZ is yttria-stabilized zirconia. After 10 min of reaction of $^{13}CH_4/CO_2/He$ over 0.5% Rh/Al_2O_3, the system was purged by helium and then cooled to 100°C. A TPO in O_2/He then produces a small peak of $^{13}CO_2$ and a much larger peak of $^{12}CO_2$, both at about 120°C, corresponding to the oxidation of the active carbon surface species. Then a second large peak of only $^{12}CO_2$ appears at about 330°C, showing that the carbon in this inactive species comes from the CO_2 in the feed, as does most of the active carbon.

The Rh catalysts on the supports mentioned previously have been further characterized and their performance in the CO_2 reforming of methane has been measured (*178, 179*). The best catalysts in terms of both activity and stability (50 h) are those supported on YSZ, SiO_2, and Al_2O_3. FTIR spectra obtained during the first 30 min on stream for the Rh/Al_2O_3 catalyst at 500°C are shown in Fig. 26. The formate at 1590 cm^{-1} accumulates but is

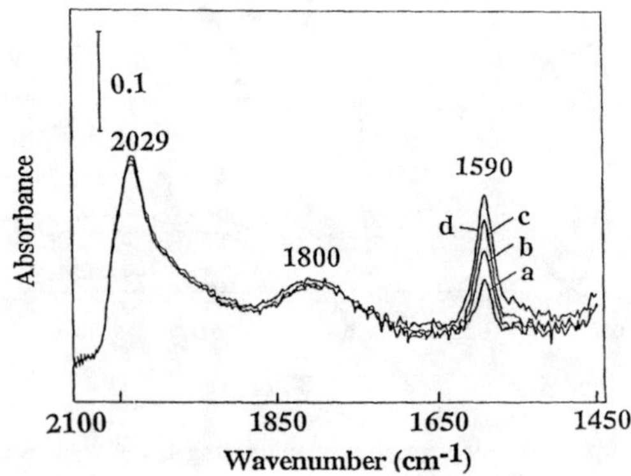

FIG. 26.   FTIR spectra after various reaction times for CH_4/CO_2 over Rh/Al_2O_3 at 500°C. a, 4 min; b, 6 min; c, 10 min; d, 30 min (after *179*).

a spectator species. Linear and bridged CO are observed. Various transient and transient isotopic experiments led to the determination of the active carbon surface coverage as about 0.2, consisting of both C and CO. For Rh/YZS, the corresponding figure is $\Theta_c < 0.02$, but both catalysts show good performance. The lattice oxygen of this catalyst participates in the reforming reaction and may help prevent deactivation by inactive carbon species.

In another study (*180*) of Rh/Al_2O_3 at 650°C, the individual step Eq. (41) was studied and the resulting surface carbon (graphitic in nature) characterized by TPO and TPRx. Treatment of the catalyst by CO_2/He in the absence of CH_4 produces two types of surface carbons, the principal one being carbidic in nature. Finally, after reacting in $CH_4/CO_2/He$, there are three types of surface carbon present, with most of the carbon arising from the CO_2 in the feed. A TAP investigation of CO_2 reforming over Pt/ZrO_2 at 700°C has been carried out (*182*). The qualitative results obtained are in accord with those of the studies on Rh discussed previously.

D.   METHANATION AND FISCHER–TROPSCH REACTIONS

In the years following World War II it appeared that it might be economical to convert synthesis gas to methane by the reverse of Eq. (35), the methanation reaction:

$$CO + 3H_2 \rightarrow CH_4 + H_2O, \Delta H = -49 \text{ kcal/mol.} \tag{48}$$

Although this reaction is now not economically attractive, it is the first step in the Fischer–Tropsch process, in which the carbonaceous fragments that are produced on the way to methane via Eq. (48) are oligomerized to useful higher hydrocarbons. This situation led to many fundamental studies of Eq. (48); for high H_2/CO ratios (often 9) and atmospheric pressure, mostly methane is produced, even over a metal such as iron; lower ratios and higher pressures lead to higher hydrocarbons. Nickel is a typical metal catalyst leading to CH_4, and iron, ruthenium, or cobalt typically lead to higher hydrocarbons. We first consider some studies in the transient regime that have shed light on the sequence of steps underlying the methanation reaction.

Some aspects of methanation were discussed in Section II, and some transient results were discussed in connection with Figs. 1 and 10 for Ni/Al_2O_3. We now discuss some data for a 10% Fe/Al_2O_3 catalyst. After a certain time on stream in a 9/1 H_2/CO mixture at 285°C, the feed is changed to He and the reactor cooled to the desired constant hydrogenation temperature. The He is then replaced by H_2, and the methane concentration arising from reaction with the surface carbonaceous species is measured as a function of time. After 5 min on stream the curves shown in Fig. 27 are observed, and after 30 s those shown in Fig. 28 are observed (*183*). Based also on extensive earlier studies (*184, 185*) it is proposed that only one surface species is present after 30 s, whereas after 5 min a second, less active species is also detected. At longer times, a third peak appears, attributed to bulk carbides, identified by Mössbauer effect spectroscopy (*186*).

Bianchi and Gass (*183*) modeled the kinetics of the appearance of a peak corresponding to a given intermediate according to the following scheme:

$$H_2 \rightleftharpoons 2H_{ads}$$

$$C_{ads} + H_{ads} \rightarrow CH_{ads}$$

$$CH_{ads} + H_{ads} \rightarrow CH_{2ads}$$

$$CH_{2ads} + H_{ads} \rightarrow CH_{3ads}$$

$$CH_{3ads} + H_{ads} \rightarrow CH_4. \tag{49}$$

They reason that if, for example, CH_{ads} is the formula of the observed intermediate, then the following steps must be fast and kinetically irrelevant. In addition, if both the first two hydrogenation steps are significant, they must have approximately the same reaction rate constants k; if this were not so, there would be no delayed peak formation as observed in Fig. 27,

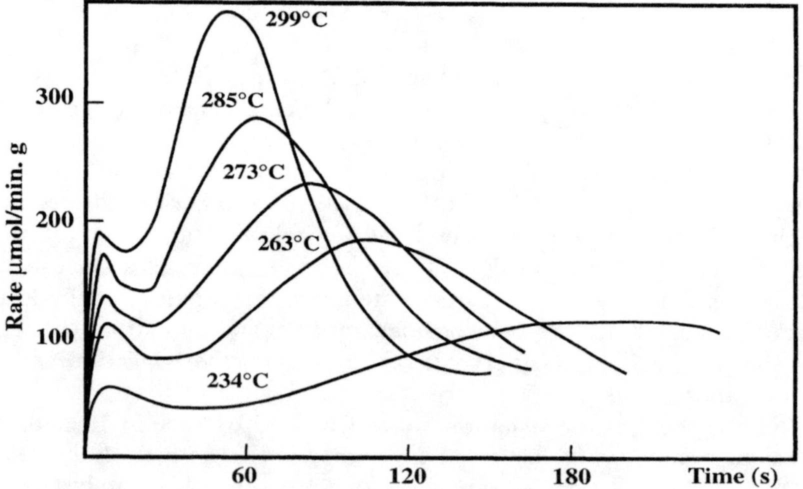

Fig. 27.   Methane produced by the isothermal hydrogenation at the temperatures indicated of the surface species formed after 5 min of CO/H_2 reaction over Fe/Al_2O_3 at 285°C and 1 bar (after *183*).

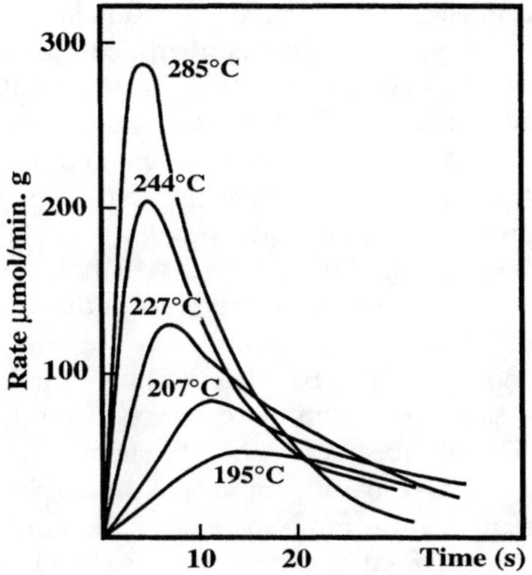

FIG. 28.   The same procedure as in Fig. 27 but after 30 s on stream in H_2/CO.

and the system would behave as if there were only one species on the surface. The model is based on a constant value of H_{ads} so that it is possible to represent the hydrogenation by first-order reactions in series. The initial rise in methane formation rate results merely from the short time required for the hydrogen concentration to rise to saturation and for the experimental measuring system (mass spectrometer with its continuous inlet) to respond. After this initial maximum, if there is only one controlling step the rate should decrease with time according to $\exp(-kt/\Theta_H)$, where Θ_H is the constant surface coverage of hydrogen; the maximum is at $t_m = 0$. If there are two steps in series with the same k, the model shows that a rate maximum should be delayed until the time $t_m = 1/k\Theta_H$. The initial rate is zero. The relations for more steps in series are also developed (183). Note that if we consider that the first species reacts according to a simple one-step first-order reaction with a rate constant k_1, and the second with a smaller rate constant k_2, the second species would merely add a tail to the first peak, without forming the observed second peak.

For the curves of Fig. 27, it is shown (183) that the best agreement with experimentally observed peak maxima and areas for the second species is obtained by the model based on three or four steps in series. The values found for $\ln t_m$ give a straight line when plotted against $1/T$, as required by the model, and the activation energy of the hydrogenation can be obtained from such a curve. After 30 s of reaction, only the first species is present, as shown in Fig. 28. The model for this peak is best fit with two steps in series so that the most abundant surface species is CH_{ads}.

The method has also been applied (183) to the results of Winslow and Bell (187) for similar curves over Ru/SiO_2. The existence of three types of carbonaceous intermediates on Ru has been confirmed by Duncan *et al.* (188) by ^{13}C nuclear magnetic resonance spectroscopy. Bianchi and Gass (189) have also applied their method based on Eq. (48) to the temperature-programmed hydrogenation of the same adsorbed carbonaceous intermediates, but the results are not as easy to interprete as those from the isothermal hydrogenations, for which the peaks are better separated.

The H_2/CO reaction has also been studied by various transient isotopic methods over the same iron/alumina catalyst. After 1.5 h of reaction of 10% CO/H_2 at 285°C, the feed is switched to He and then to H_2. The curve $CH_{4\text{-s}}$ of Fig. 29 represents the usual three peaks that arise from this titration, including a final temperature ramp to remove the most refractory surface carbon. Now, when the feed is changed to $H_2/^{13}CO$ before the titration with H_2, the $^{13}CH_4$ produced after various times in $^{13}CO/H_2$ is shown (190). Even after 30 min of exposure, only 86% of the first peak (CH) has been changed to the ^{13}C form, meaning that the rest of the surface and bulk carbonaceous species are spectator species. It is of course also interesting to measure the rate of formation of $^{13}CH_4$ immediately after the switch

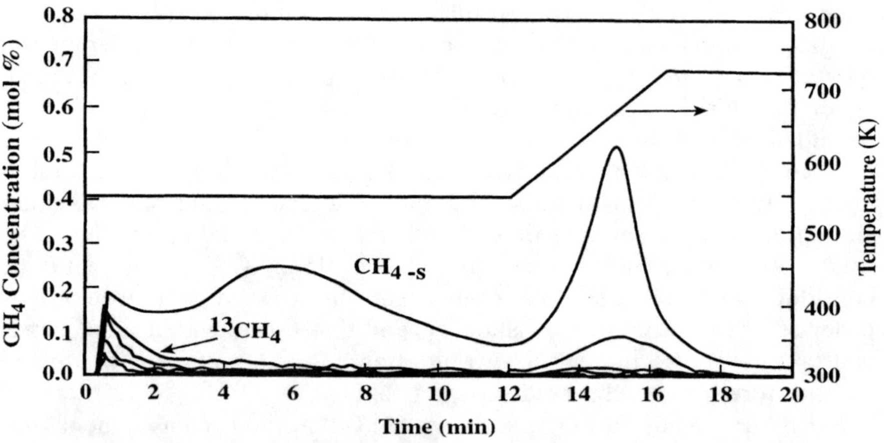

FIG. 29. Titration of the surface of the Fe/Al_2O_3 catalyst by H_2 after reaction in $^{12}CO/H_2$ for 1.5 h, giving a methane production according to the curve $CH_{4\text{-s}}$. Also shown are the curves for $^{13}CH_4$ produced after the feed is changed to $^{13}CO/H_2$ for the times 25 s, 1 min, 5 min, and 30 min at the end of the 1.5-h period. The peaks of $^{13}CH_4$ grow with time in $^{13}CO/H_2$ as shown (after 190).

from $^{12}CO/H_2$ to $^{13}CO/H_2$, the steady-state tracing or SSITKA experiment. This procedure also shows that only a small part of the surface carbon is active (*190*). Over iron, there is negligible molecular CO on the surface, and the rate-determining step is the hydrogenation of the surface CH species.

Winslow and Bell (*191*) carried out a thorough study of methanation over Ru. The unsupported metal was used so that D_2 tracing could be used without troublesome exchange of D with the OH groups on a typical oxide support. In contrast to iron, the working Ru catalyst is highly covered with undissociated CO, which can be observed (for supported catalysts) by IR spectroscopy. The behaviors of the active and less active surface carbonaceous species are similar to those found for iron; however, unlike Fe, Ru does not form a bulk carbide. This article (*191*), in association with those already cited (*187, 188*) and the IR studies of Tamaru and associates (*192, 193*), gives a clear picture of the CO/H_2 reaction over ruthenium, obtained largely by experiments in the transient regime.

Many studies have been done on the CO/H_2 reaction over rhodium. In particular, SSITKA experiments have been performed at 220°C over 5% Rh/Al_2O_3 for a steady conversion of 1.5% (*194*). After the change from $^{12}CO/H_2$ to $^{13}CO/H_2$, the shape of the ^{13}CO response curve does not agree with that predicted (*17*) for a fast, continuously equilibrated exchange with the surface ^{12}CO. For instance, over Ni the reverse exchange rate divided by the net forward rate (b) can be assumed to be infinite (*17*). Using the data over Rh, the data are best fit with $\beta = 2.2$ (*194*). From these data and the response of the $^{13}CH_4$ curve, it is determined that the coverage of the active carbon, C_α, is only about 0.03. From a series of experiments, it has been possible to estimate the surface coverages of C_α, C_β, CO, and a formate species, COOH, the latter present on the support.

An additional study (*195*), of a similar catalyst (1% Rh/Al_2O_3) included transient FTIR studies, one of which is shown in Fig. 30. The reaction of 9/1 H_2/CO at 220°C for 1 h gives spectrum **a** in Fig. 30A. The peaks are assigned as follows: 2046 cm^{-1}, linear CO; 1837, bridged CO; 1592, 1392, and 1378, formate bands; and 1460, ionic carbonate band. Within 40 s the CO bands form, and the bridged band increases until that shown in Fig. 30 is reached at 1 h. The other bands increase slowly and continuously and continue to do so at 1 h. At this time the reaction is stopped, and an isothermal hydrogenation is performed, leading to spectra **b–e** of Fig. 30. After 320 s the CO has disappeared, but the formate and carbonate bands are little affected. Figure 30-B shows the decrease of the bands as the temperature of hydrogenation is programmed up to 370°C. Most of these spectator species, which form on the support, are removed by this treatment. Spectra obtained in the 2700–3100 cm^{-1} region confirm the behavior deduced from the 1150- to 2250-cm^{-1} region. The IR experiments do not

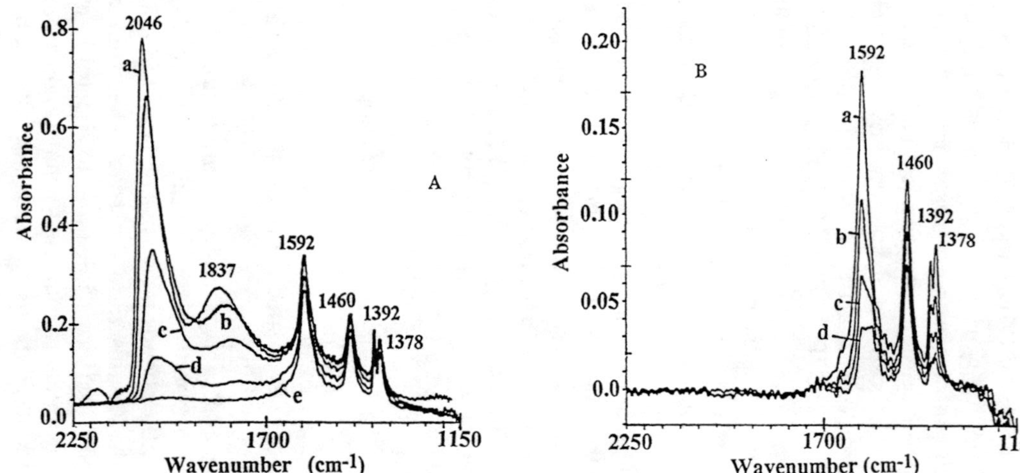

FIG. 30. CO/H_2 reaction over Rh/Al_2O_3 for 1 h at 220°C. (A) The infrared spectra of the adsorbed species are shown for the isothermal hydrogenation after various times in H_2: a, $t = 0$; b, $t = 80$ s; c, $t = 160$ s; d. $t = 240$ s; e, $t = 320$ s. (B) The hydrogenation is continued during temperature programming, and the spectra shown correspond to the following temperatures: a, $T = 265$°C; b, $T = 330$°C; c, $T = 350$°C; d, $T = 370$°C (after *195*).

detect surface C_α, C_β, or C_xH_y; the coverages of these species on the working catalyst are determined by transient isotopic studies, as also performed previously (*194*).

It is interesting that in TPRx studies with an initial state of adsorbed $CO + H_2$ on Ni/Al_2O_3 at 112°C, this leads to the formation of methoxy groups on the support (*196*). From a series of ingenious experiments involving isotopes, TPRx, interrupted TPRx, and TPD, Falconer and coworkers constructed a coherent explanation of their results. However, for the steady-state CO/H_2 reaction at 220°C or higher, over supported Ni, Fe, Ru, and Rh, the adsorbed species on the alumina support are not in the reaction pathway toward methane (*17, 190, 191, 195*).

Chuang and coworkers (*197*) published many interesting studies involving experiments in the transient regime; here, we consider a study of the CO/H_2 reaction over 3% Rh/SiO_2 in the range 180–300°C. Responses have been measured by both mass and IR spectroscopy. Steady-state tracing via ^{13}CO has been used, and the results are quantitatively analyzed as already described in connection with Happel *et al.*'s work (*107*) (Eqs. 22 and 23). Two parallel pools of intermediates, $C^\alpha H_x$ and $C^\beta H_x$, are suggested to react irreversibly to CH_4. CO adsorbs reversibly, as previously discussed. The rates of the elementary steps and the surface coverages of the intermediates are calculated; the results agree with the more qualitative approach used for Rh/Al_2O_3 (*195*). The parallel pools of CH_x have also been treated according to the method of De Pontes *et al.* (*100*). It should be noted that the rapid response of CO and the slow response of CH_4 in the SSITKA experiments does not mean that the hydrogenation step is rate determining. The ^{13}C passes through each pool at the same net steady rate; the reversible nature of the CO adsorption (fast, almost equal unidirectional rates) means that this pool responds rapidly compared to the pools of the surface carbons, through which the ^{13}C passes at the slow unidirectional rate of the overall reaction. At the same time, Θ_{CO} is about 0.75 and $\Theta_{(c^\alpha + c^\beta)}$ is only about 0.06 (*197*). The mathematical model used does not need the assumption of any rate-determining step. Interesting information has been obtained by pulsing CO into a steady hydrogen stream (*197*).

The steady-state tracing procedure can be extended to the analysis of the ^{13}C content of higher hydrocarbons than methane for the Fischer–Tropsch reaction over Fe/Al_2O_3 (*190*) and Ni/Al_2O_3 (*198*). After the switch from $^{12}CO/H_2$ to $^{13}CO/H_2$, samples taken at successive times are stored in a multiple-loop sampling valve. After this procedure, the content of each loop is separately passed through a gas chromatograph to separate the various hydrocarbons into peaks. Use of a flame ionization detector gives the steady composition of the gas. The effluent from the column is then passed through a hydrogenolysis reactor to convert all the peaks to methane.

Then the gas goes to the mass spectrometer, in which the ^{13}C fraction in each methane peak is measured; this fraction is also that of the original hydrocarbon represented by the peak. Instead of treatment by hydrogenolysis, the peaks can be sent through a combustor to convert all the products to CO_2, which is then analyzed for ^{13}C content in a mass spectrometer (*199*); this method has been applied to reaction over 3.3% Ru/TiO_2 at about 200°C and 1 bar.

Krishna and Bell (*199*) described the results of their steady-state tracing experiments by the model shown in Fig. 31. The scheme is in accord with the Anderson–Schulz–Flory distribution of products, based on chain growth by the successive addition of monomers $C_{m,s}$ to chain fragments $C_{n,s}$. $C_{m,s}$, is different from $C_{1,s}$. It is assumed that the probability of chain growth α is not a function of n, where

$$\alpha = k_p\Theta_m\Theta_n/(k_p\Theta_m\Theta_n + k_t\Theta_n);$$

that is, the ratio of the rate of propagation (chain growth) to the rate of propagation plus the rate of termination, where the latter is the rate of production of gaseous polymers. In relating these rates to Fig. 31 it must be borne in mind that all the species are mixed together on the surface. For example, Θ_m is present with Θ_n in each box. Appropriate material balances can then be applied to each pertinent species, according to Eqs. (2)–(4) and as also discussed in connection with Eqs. (21)–(23) for balances on isotopes. The work has been done at various steady states, obtained by varying the temperature and the D_2/CO ratio. Estimates of α, k_i, k_p, and k_t have been made at these various steady states. It was found that k_i is independent of temperature T; for k_p, E_p is 8 kcal/mol, and for k_t, E_t is 20 kcal/mol. These activation energies are consistent with the observed

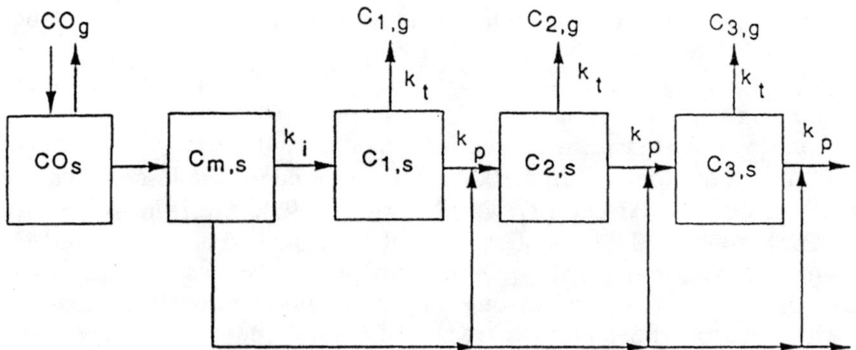

FIG. 31. Fischer–Tropsch synthesis over Ru/TiO_2. Model for chain growth (after *199*).

decrease in α with increasing temperature. Examples of the steady-state coverages are given in Fig. 32. The coverage of the sum of the growing chains is much smaller than that of monomer. For iron, a more qualitative analysis suggests that termination is so fast that the coverage of the growing chains approaches zero (*190*).

Goodwin and coworkers studied H_2/CO reactions over many different catalysts. Steady-state and SSITKA experiments occupy a prominent place in their studies, which provide a link between various catalyst formulations and the surface coverages of active intermediates during reaction, well defined by the isotopic tracing experiments involved. Some of the catalytic systems studied are Ru/SiO_2 (*200*), Cu-modified Ru/SiO_2 (*201*), La^{3+} promotion of Co/SiO_2 (*202, 203*), Ru promotion of Co/Al_2O_3 (*204*), and Zr promotion of Co/SiO_2 (*205*).

It is also possible to react CO_2 with H_2 to form hydrocarbons (mostly

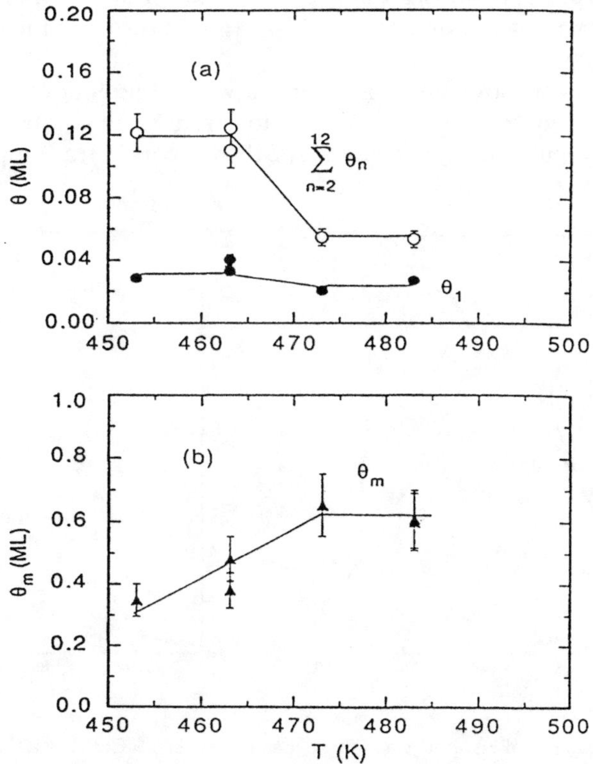

FIG. 32. Variation with temperature of the surface coverage of intermediates during Fischer–Tropsch reaction over Ru/TiO_2. $D_2/CO = 3$ (after *199*).

CH_4); this is an interesting reaction because of the ready availability of CO_2. This process has been little studied by transient methods, but Renken and coworkers reported interesting results using both mass and IR spectroscopy (*206–208*).

E. Oxidation of Carbon Monoxide

This reaction was discussed in Section I; it is one of the most frequently studied in heterogeneous catalysis and reaction engineering. Figure 33 (*59*) shows the TPRx of this reaction over Pt/SiO_2, and it can be understood in terms of the sequence of steps of Eq. (19). Starting from a surface covered with CO, sites for the dissociative adsorption of O_2 are rare; however, as the temperature increases, oxygen can adsorb and react to produce CO_2 at increasing rates, producing an overshoot in the surface temperature. During the subsequent decrease in reactor temperature, CO removes O from the surface, producing a new peak in the surface temperature before the whole process is quenched at low temperature for the CO-covered surface.

Figure 34 (*59*) shows an experiment in which the inlet concentration of CO is increased from zero and then lowered again at 200°C and for a constant oxygen flow. This process is called a concentration-programmed

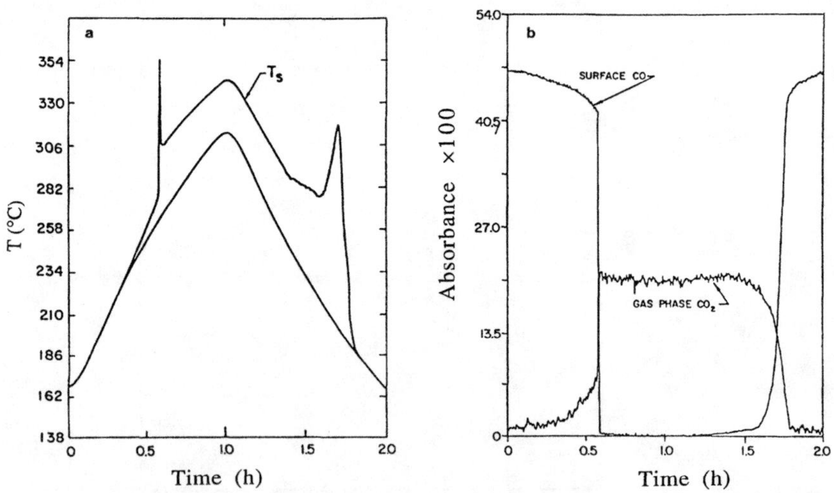

FIG. 33. Reaction of an approximately stoichiometric 3.85% CO/2.14% O_2 (diluted in N_2) mixture over Pt/SiO_2. (a) Reactor temperature ramps up and down as shown lead to the surface temperature profile shown, with peaks at ignition and extinction. (b) Response (via FTIR) of surface CO and gaseous CO_2 concentration for the same temperature ramp (after *59*).

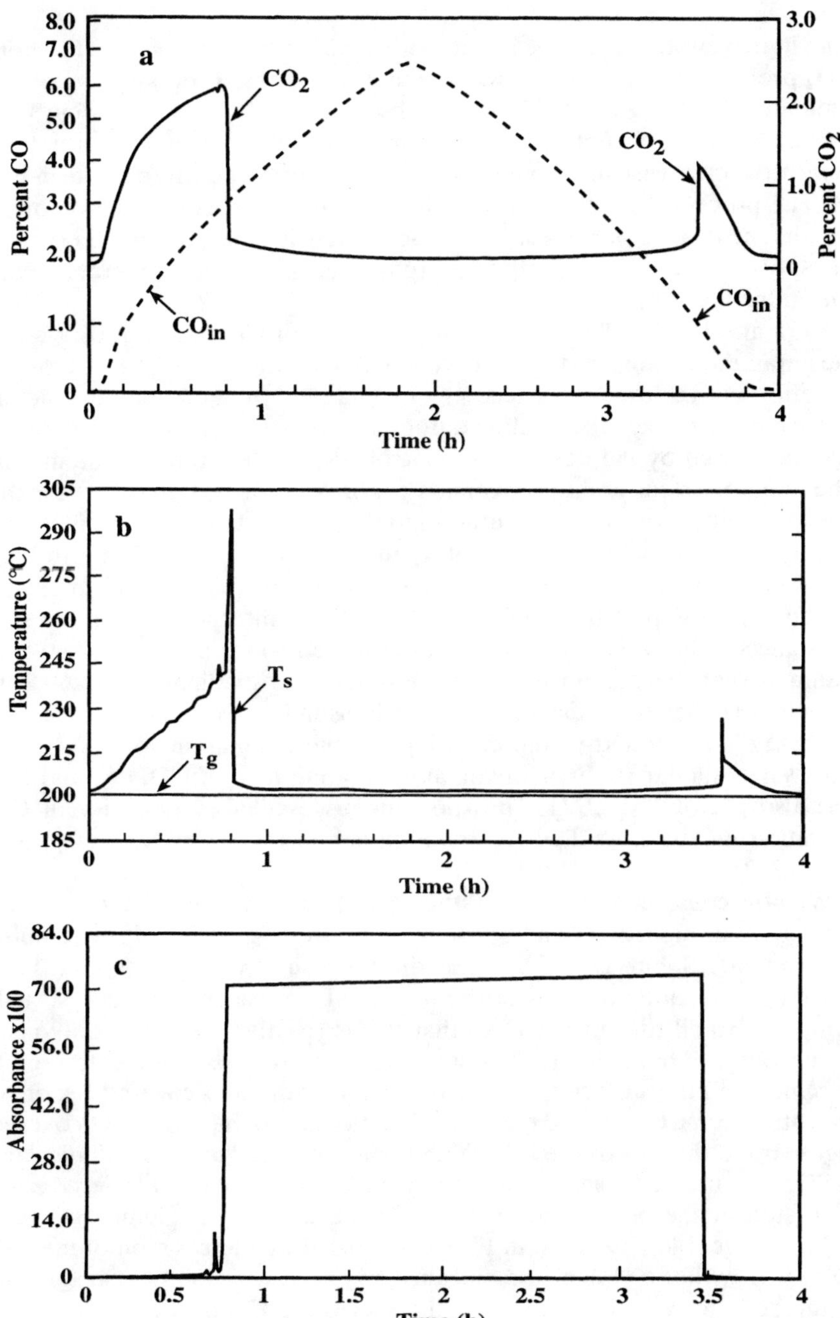

FIG. 34. Response of the same system shown in Fig. 33 to a ramp of CO concentration in about 2.0% O_2 from zero to about 6.0% CO (220°C). (a) CO_2 response (note overshoots); (b) surface temperature response; (c) surface CO response (after *59*).

reaction. Now there are peaks at both ignition and quenching for both CO_2 production rate and surface temperature. These processes have been simulated by Kaul *et al.* (*58*). The absence of CO_2 peaks in Fig. 33 is explained by the abrupt decrease in surface CO concentration as the temperature increases, even though there are similar temperature peaks in Fig. 34. The process is very complicated and it is remarkable that the reported agreement can be achieved. The model used is based on a CSTR, and transport effects are limited to interphase gradients of temperature and concentration.

Herz and Marin (*209*) studied the oxidation of CO over Pt/Al_2O_3, and the parameters found by them have been used by Lie *et al.* (*210*) to simulate the effects of the oscillatory feeding of CO and O_2 to a monolithic reactor. In a catalytic reactor for treating automobile exhaust gas, such oscillations may be caused by the device that controls the air/fuel ratio to the motor. The apparent steady-state conversion (improved for certain cases) for the reactor is a function of the amplitude and frequency of the input oscillations. Ideally, the control device, the motor, and the monolith should be studied together to obtain the best operating conditions. The reactor model is based on laminar flow past the catalytic walls, with an interphase mass transfer resistance at the wall, between the gas and the surface of the catalyst. It is assumed that there is no interphase temperature gradient, although it is known (*13*) that the existence of a Δc is usually accompanied by a ΔT. Large ΔT's are found in the experiments and simulations of Kaul *et al.* (*58, 59*). A similar study involving an oscillating feed of CO, O_2, and NO was also performed (*211*). The spontaneous oscillatory behavior of CO oxidation has been studied for rhodium supported on SiO_2, Al_2O_3, and TiO_2 (*212*).

We now consider a few recent studies that have been made by using the TAP system. Figures 35 and 36 show an interesting series of single-pulse experiments. Nijhuis *et al.* (*213*) explain this result by proposing that there is segregation along the axis of the reactor. The first pulse of CO is small compared to all the surface O so that this is (partly?) removed only near the entrance of the reactor. Then the next O_2 pulse is (by choice) too small to remove all the surface CO. The bed is left with an O-covered region at the entrance, a CO-covered region farther along, and the original O-covered region up to the exit of the bed. With further pulsing, the CO will displace the O from the bed, except that after each O_2 pulse a new O-covered zone is created at the bed entrance. The CO_2 peak at the beginning changes little. The double CO_2 peak in Fig. 35 is caused by the reaction of the CO pulse first with the O near the entrance and then with the O near the end of the bed, not yet all removed by the successive CO pulses.

The existence of the CO_2 peak at 1 s in Fig. 36 requires explanation.

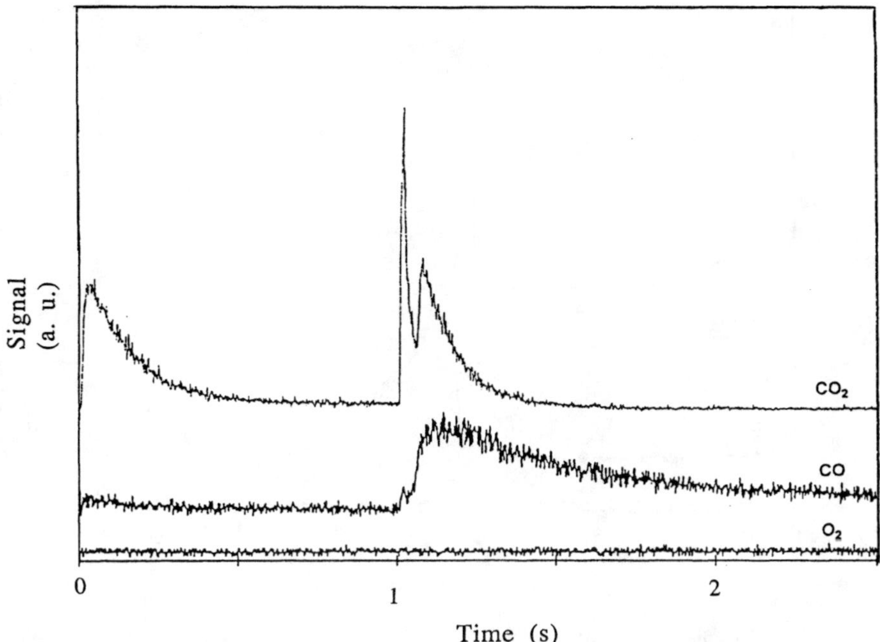

FIG. 35.   CO oxidation over Pt sponge at 300°C in a TAP system. Initially, the 291 mg of Pt are completely covered with oxygen. Successive pulses of 2.3×10^{16} molecules of O_2 and then 1 s later 1.0×10^{17} molecules of CO are passed over the bed. After the fifth set, the responses shown in the figure are measured. At 0 s a pulse of O_2, completely consumed, leads to the first peak of CO_2. At 1 s the larger pulse of CO is injected, and it more slowly reacts to form the double peak of CO_2 shown. Then, after 3.5 s another pulse of O_2 is introduced (after *213*).

Since the bed is mostly covered with CO, one would expect that there would be no O left on the surface after the first peak of Fig. 36. Therefore, where does the oxygen come from to form the second peak at 1 s? Apparently O and CO can coexist on the surface; one explanation is that they may exist as separate islands, as most recently proposed by Hoebink *et al.* (*214*). Note that all the action for Fig. 36 occurs at the bed entrance; downstream the surface is saturated with CO. The model (*213*) is supported by some experiments involving $^{18}O_2$ and by a simulation of the results by a model based on the previous discussion and the classical sequence of steps. Multipulse experiments of CO over an O-covered surface and of O_2 over a CO-covered surface have also been performed. It would be interesting to know the effect of pulsing frequency on the results. No additional

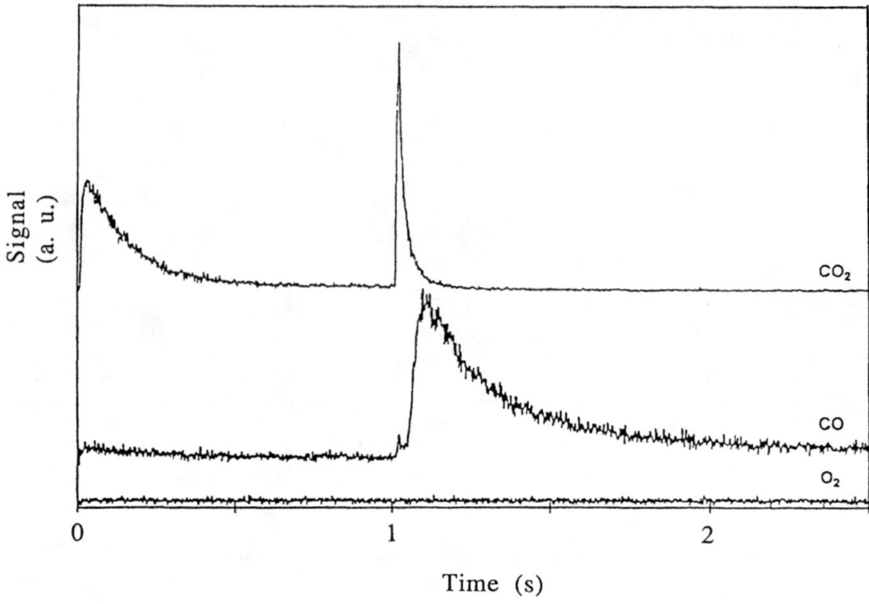

FIG. 36.   This figure is a continuation of Fig. 35. After many pulses, the responses become identical as shown (a pseudo-steady state). Because of the excess CO fed, there is now no O on the surface at $t = 0$ s; the O_2 pulse at this time creates a CO_2 peak but apparently leaves some O on the Pt. At 1 s the CO pulse reacts with the residual O to form the narrow CO_2 peak shown and resaturates the bed with CO (after *213*).

insight seems to be obtained in comparison with previous results (*33, 75, 77, 215*).

TAP multipulse experiments have been done over polycrystalline Pt at 72°C; O_2 is pulsed over a surface initially covered with CO (*214*). Figure 37 shows the usual induction period and then two well-separated peaks of CO_2. Nijhuis *et al.* (*213*) did not find a double peak, but their pulse size was 20 times larger. It is difficult to explain Fig. 37, but a segregation of CO and O into islands on the surface has been used in a complicated model to try to reproduce some of the results (*214*). However, the model was tested only on single-pulse data. In fact, the attractive feature of the TAP reactor is the use of single pulses, producing results that can be compared with those of models involving the kinetics of rapid elementary steps (*35*). Multipulse operation seems to be treated as a kind of titration, and no kinetics are available from the ensemble of the pulses. Indeed, the series of maxima seems to be taken to be the development of concentration with time (*213, 215*), or in some cases the total accumulated amounts are

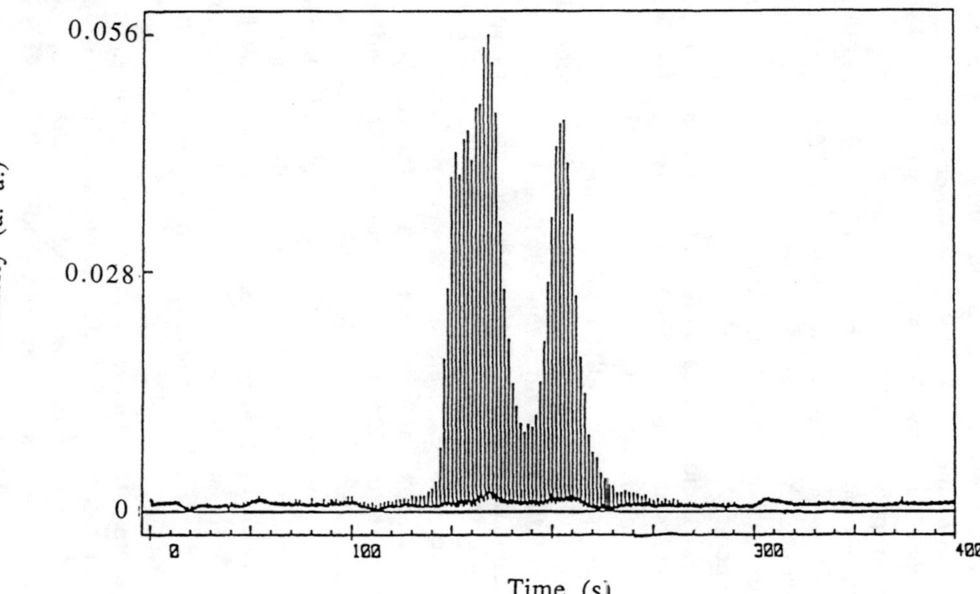

FIG. 37.   TAP experiments. Peaks of CO_2 formed over polycrystalline Pt at 73°C caused by multiple pulses of O_2 over an initial CO-covered surface (after 214).

computed from the sum of the areas under the individual peaks. Recall the discussion in Section II about the effect of an oscillating feed on a reactor output. In the multipulse TAP reactor, it is even an almost on–off feed. Bonzel and Ku (*33*) were able to study a similar reaction system at low pressure (high resolution in time), as were Bennett *et al.* (*216*) at atmospheric pressure (lower resolution in time), by using simple step-forcing functions combined with convincing material balances. Note that the cumulative peak shape by TAP is produced over a time that is easily available to step-response experiments at atmospheric pressure.

The oxidation of carbon monoxide has been studied by both the usual step-response and isotopic experiments and by the TAP system (*217*). The general conclusion is that the fast response of the TAP system did not produce any additional mechanistic information to that obtained from step-response experiments. A number of the points discussed in previous paragraphs are mentioned, and it is suggested that the final pattern of multipulse response experiments be termed a pseudo-steady state. A factor not mentioned is that transient IR experiments are valuable with the step-response method but not compatible with the TAP system.

The oxidation of CO over nonnoble metals would be economically attractive if an active and durable catalyst could be found. This subject has been pursued at Delft, and the transient method has been used for several studies (*218–220*). Using a 10 wt% $Cu-Cr/\gamma-Al_2O_3$ ($Cu:Cr = 1:1$) catalyst at 200°C, many step-response experiments have been performed with the aim of understanding the oxidation or reduction of the catalyst as the gas mixture is changed from oxidizing to reducing conditions and vice versa. The results for the reduction of a fully oxidized catalyst by a switch from He to 5% CO in He show two peaks of CO_2, probably indicating that there are both surface and subsurface O associated with the oxidized catalyst (*218*). Additional studies have been performed involving switches from CO/O_2 mixtures in the stoichiometric ratio to mixtures containing excess O_2 or excess CO or to CO/He alone (*219*). Some CO_2 seems to be retained on the catalyst surface.

Finally, for a more detailed understanding, some experiments have been conducted separately using 10% CuO/alumina and 10% Cr_2O_3/alumina (*220*). Figure 38 shows the reduction of the fully oxidized copper catalyst by CO. The first peak comes from an Eley–Rideal (ER) process involving CO_g and weakly bound O on the surface. After some chemisorbed oxygen is removed from the surface, CO can adsorb and react with O by a Langmuir–Hinshelwood (LH) process, producing the second peak of CO_2. As surface O is depleted, O can diffuse from below the surface to prolong the CO_2 peak. The PFR has been modeled by Eqs. (1)–(4). In order to fit the experimentally found second peak of CO_2, it is necessary to consider two

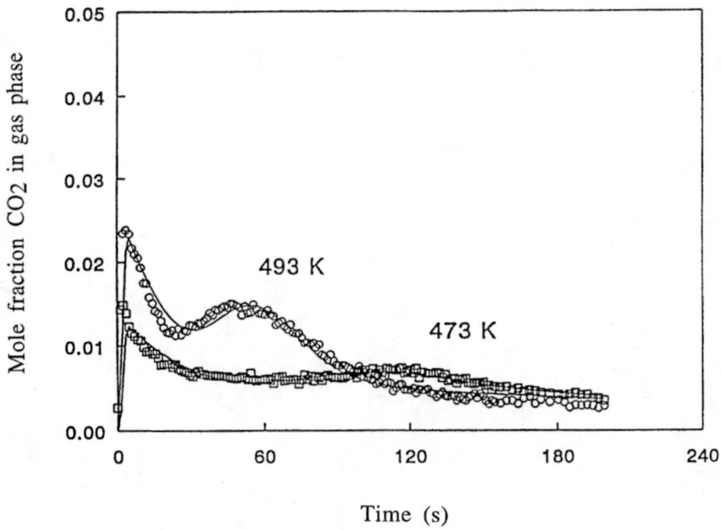

FIG. 38. CO_2 produced by a step change from He to 5% CO/He over a fully oxidized Cu/Al_2O_3 catalyst: symbols, experimental; lines, simulation (after *220*).

kinds of surface O, labeled O_{ER} and O_{LH}, in the surface balance of Eq. (3) and in the elementary steps. The parameters are optimized to give the best fit to the experimental data, and the results are shown as lines in Fig. 38.

CO oxidation over $Pt/CoO_x/SiO_2$, which is supposed to be an attractive catalyst, has been studied in a TAP system (*221*). The qualitative results presented do not seem to provide an improved understanding of this system.

Regarding CO oxidation over oxide catalysts, I consider the work of Kobayashi *et al.* (*222*) on Kadox ZnO (K25–ZnO) compared with Kanto ZnO (Kan–ZnO). Limiting the discussion to K25, the results shown in Fig. 39 were obtained. As shown in the figure, the complex steady-state behavior can be explained by simultaneously occurring ER and LH pathways. The CO_2 responses arising from inlet step changes in the CO pressure are shown and classified in Fig. 39 according to Kobayashi's nomenclature. The steady-state results can be fit by a combination of the usual equations according to the LH and ER mechanisms. To explain and simulate the transient responses, an elementary-step model has been combined with the material balances (Eqs. 2–4). Computed results agree well with the experiments. With respect to the ER process, it is known to occur over oxide catalysts at steady state. However, for the steady oxidation of CO over Pt, it is not usually considered to occur (*77*). Nevertheless, it was shown in the preceding discussion that it may occur in the transient regime even over Pt (*216*).

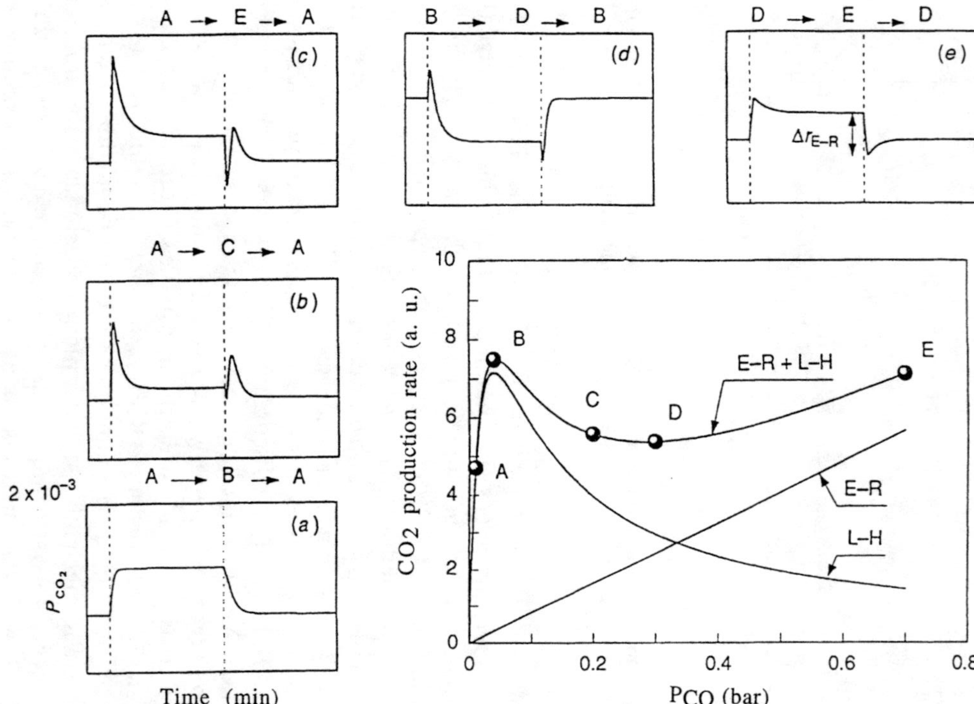

FIG. 39. The large graph is a curve of the steady-state rate of CO_2 production as a function of P_{CO} over K25–ZnO at 150°C and P_{O2} = 0.2 bar. The CO_2 transient response curves (a)–(e) result from stepwise changes of the CO pressure among the points A–E as shown above the response curves. The response modes are classified as (a) monotonic, (b and c) complex, (d) false start, and (e) overshoot (after *222*).

Some other interesting aspects of the reacting system, including the use of an oscillating feed composition, are presented in Kobayashi *et al.* (*223*).

F.   CATALYTIC REMOVAL OF NITROGEN OXIDES

Before considering some studies in the transient regime, it is useful to discuss the practical problems that have led to the great scientific activity in this field (*224*). Modern gasoline-powered automobiles are equipped with three-way catalytic converters to remove CO, hydrocarbons (hc), and oxides of nitrogen (NO, NO_2, and N_2O; collectively called NO_x) from the exhaust gas with good efficiency and durability. This good performance is obtained by precious metals (Pt, Rh, and Pd) supported on a typically alumina washcoat in a monolithic honeycomb-shaped reactor. In order to remove all three types of pollutants, a stoichiometric mixture of fuel and air must be fed to the motor. As already mentioned, the action of the computer-driven control system produces a cycling in the composition of the inlet and thus the exhaust gases that are the feed to the catalytic converter. Thus, much work was done in approximately 1980 by automobile companies to understand the transient behavior of three-way reactors, especially those with an oxygen-adsorbing storage capacity provided by ceria, for example (*225*).

However, diesel engines and stationary power-generation systems typically run with excess air so that the three-way catalysts are not able to reduce NO_x to N_2. In addition, gasoline engines exhibit improved economy (km/liter) when run lean. For large stationary plants selective catalytic reduction by ammonia provides a partial solution, but for all processes the straight decomposition of NO_x would be preferable. In 1986, Iwamoto *et al.* (*226*) discovered the activity of Cu–ZSM-5 for the direct decomposition of NO into O_2 and N_2, leading to many additional studies, but a catalyst with sufficient activity and durability is yet to be found. Of course, it would be better to avoid the production of NO_x by lowering the temperature of reaction, especially in a gas turbine, by catalytic combustion (*227*).

For gasoline engines, systems have been proposed for improving fuel economy by leaner average operation while retaining the required removal of NO_x (*228, 229*). The engine is fed with a lean mixture part of the time, while storing NO_x on a proprietary component added to the precious-metal three-way catalyst. Then the air-to-fuel ratio (AFR) is reduced to approximately stoichiometric, causing the release of NO_x and its decomposition over the catalyst as usual, and the cycling is continued. This procedure is called NO_x storage and reduction (NSR). The results of an engine test are shown in Fig. 40. There is improved fuel efficiency, and the average conversion of NO_x is 85% (*228*). From this discussion, it can be seen that

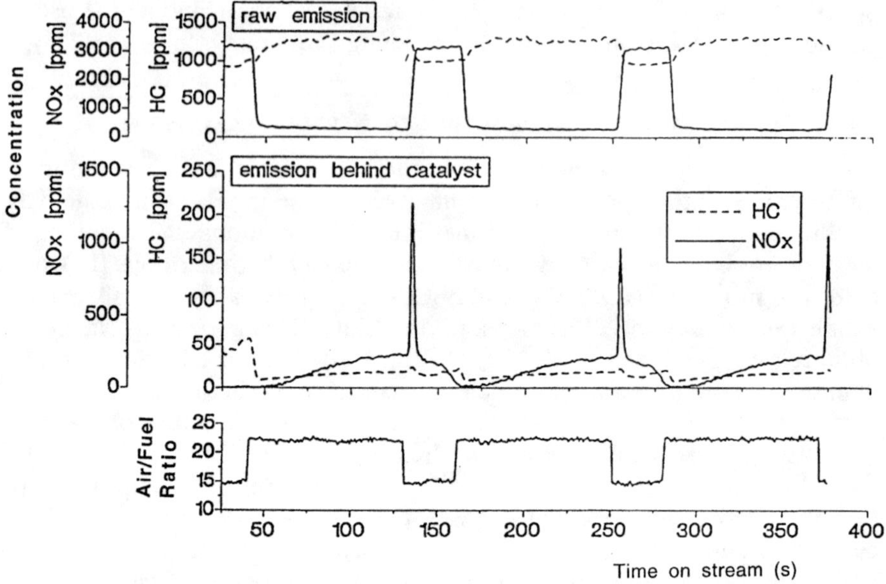

FIG. 40. Nitrogen storage and reduction (NSR) from a lean-burn gasoline engine. Catalyst behavior is in a 30/90-s cycle. Space velocity, 15,000 h^{-1}; inlet gas temperature, 350°C (after *228*).

future catalytic converters may be running in a complicated transient regime, cycling with a period of about 1 s because of the AFR control and with a superposed period of about 1 min arising from NSR However, the catalyst/adsorbent system for NSR does not currently have sufficient durability for application in production cars (*230*). A review with 147 references has appeared (*231*). The technical papers so far mentioned do not treat the underlying kinetics of the process, so I next discuss some (necessarily) transient studies of the problem.

The TPD of preadsorbed $^{15}N^{18}O$, with or without coadsorbed $^{16}O_2$, has been thoroughly studied for Cu^{2+}- and Co^{2+}-exchanged ZSM-5 zeolites (*232*). The graphs of the rate of desorption of the various compounds, obtained from mass spectrometry, are reported quantitatively (mmol/gs), and the contributions of peaks from NO_2 are subtracted from those of NO, for example. This procedure is facilitated by the isotopes chosen. It is unfortunate that many recent results, even those obtained by an expensive TAP system, are reported only in terms of *m/e*, and in arbitrary and/or relative units. The work cited (*232*) shows that the presence of O_2 enhances the amount of NO_x adsorbed and is essential for the decomposition of the NO_x. NO_2 is an important surface intermediate for this process.

The activity of supported noble metals for the reduction of NO_x by hydrocarbons in the presence of oxygen has been studied (*233*). It has been shown by TAP studies that C_3H_6/O_2 is a more effective reductant of NO than H_2 or CO alone (*234, 235*).

The step-response method has been used to study NO reduction over Cu/Al_2O_3 and $Cu–Cr/Al_2O_3$ (*236*). NO is adsorbed on oxidized catalysts as nitrate/nitrite complexes, and the latter decompose only in a slightly reducing atmosphere. These experiments on the transient regime permit the estimation of the surface composition of the catalyst in contact with various gas-phase mixtures. The reaction steps were identified, but no simulation of the observed results was attempted.

The reactivity of absorbed CO and NO on Rh/SiO_2, ceria–Rh/SiO_2, and oxidized Rh/SiO_2 has been studied during step and pulse changes by *in situ* IR spectroscopy and mass spectrometry (*237–240*). The same reaction has been studied over Fe_2O_3/SiO_2 by similar methods (*241*). In the latter work the transient results are simulated by a sequence of steps involving as surface species only CO*, NO*, O*, and*, (* is the empty site) with the experimental packed-bed reactor represented by 20 CSTR in series (i.e., plug flow).

G. Examples Involving Methanol

The synthesis of methanol from a mixture of H_2, CO, and CO_2 (80 : 10 : 10) over the ICI copper/zinc oxide/alumina catalyst has been thoroughly discussed by Waugh (*242*). In particular, it has been shown by using an isotope of C that the carbon in the methanol produced almost all comes from the CO_2. It is possible to make methanol from a feed of CO/H_2 alone but at a rate only 10^{-2} times that obtained from CO_2. The CO in the usual mixture serves to optimize the oxidation state of the catalyst surface. The whole reaction occurs on the copper. Although the copper catalyst was patented in 1928, only recently has its efficient action been understood. Many of the experiments described by Waugh (*242*) are done in the transient regime: TPD, TPRx, step response, transient IR measurements, etc. A recent fundamental study (*243*) concerns the surprising inverted temperature dependence of the decomposition of CO_2 on Cu, studied by "reactive frontal chromatography," which is a step-response done in a PFR. Another study concerns the changes in surface structure of Cu induced by CO adsorption and subsequent TPD (244). This recalls similar studies of CO on Pt (78).

There is interest in the synthesis of methanol from CO/H_2 over Pd, and the reaction has been studied over Pd/SiO_2 and $Li^+/Pd/SiO_2$ (*245*). Only SSITKA experiments have been performed, and it was found that the coverage of surface intermediates is increased for the promoted catalyst. Similar studies have focused on other aspects of the system (*246–248*).

The partial oxidation of methanol to formaldehyde over metal oxides is important industrially, and the results of experiments done in the transient regime are summarized for MoO_3 (*249*). The results in Fig. 41 show that the conversion of methanol and the selectivity toward formaldehyde in a Mars–van Krevelen process are both favored by a relatively oxidized surface. Weber (*250*) has shown by theoretical calculations that the surface methoxy intermediate should lose one of its hydrogens as a hydride ion attracted to Mo^{6+} and not as a hydrogen atom or proton attracted to a doubly bonded or bridging oxygen.

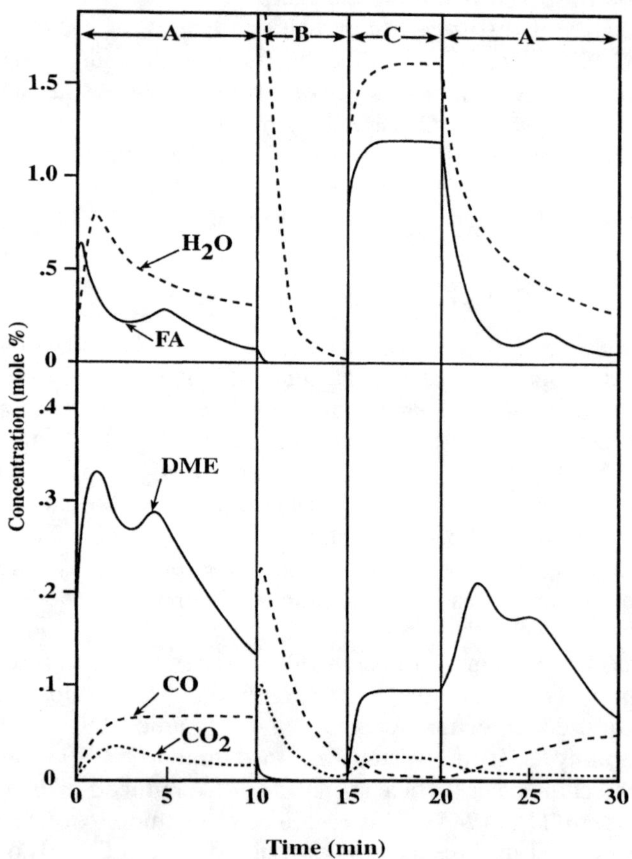

Fig. 41. Methanol partial oxidation over MoO_3 at 300°C. FA, formaldehyde; DME, dimethyl ether. Starting with the approximately stoichiometric oxide, the feed is changed as shown: A, 3.6% methanol/helium; B, 10% oxygen/helium; C, 3.6% methanol/10% oxygen/helium (after *249*).

The advantage of the partial oxidation of methanol of supported metal oxides compared to neat oxides has been shown by transient and steady-state kinetic studies (*251*). Other recent progress in this field has been discussed (*252*). Another transient study involves various copper catalysts (*253*), and particularly noteworthy in this study was the observation of responses to oscillation of the feed composition and the development of self-sustained oscillations.

The previous reaction has been studied over sodium carbonate supported on carbon (*254*). A step in the feed mixture from CH_3OH to CD_3OD has led to useful information. It is suggested that the desorption of H_2 is the rate-controlling step. The reaction has also been studied in a TAP system over a Fe–Cr–Mo Haldor–Topsøe catalyst (*255*). In general, the results of previous studies are confirmed. The removal of H from the adsorbed methoxy is considered to be rate controlling.

Another possibility for the production of formaldehyde is the partial oxidation of methane over V_2O_5 at about 600°C. Koranne *et al.* (*256*), investigated the carbon pathways by replacing $^{12}CH_4$ by $^{13}CH_4$ over 2% V_2O_5/SiO_2 in SSITKA experiments. They propose that methane combines directly with surface oxygen, without appreciable formation of CH_3 radicals or C_2H_6 at approximately 600°C. The carbon in the products changes quickly to ^{13}C, meaning that the surface coverages by reaction products are relatively small. The yield of CH_2O has a maximum at close to zero conversion. Similar experiments have been done (*257*) by replacing ^{16}O with ^{18}O. There is an extended production of oxidized products containing ^{16}O from the original oxide.

H. FINAL EXAMPLES

All but a few of the experiments for which a TAP system is used are interpreted in a qualitative way. However, Creten *et al.* (*258*) analyzed their data on single pulses quantitatively, following the procedures outlined by Gleaves *et al.* (*34, 35*). A pulse of propylene is sent over γ-Bi_2MoO_6 catalyst in the temperature range 369–451°C, and the resulting pulses of propylene, acrolein, and CO_2 are recorded. Then Eqs. (2), (3), and (16) are written for the appropriate sequence of steps, in this case adsorption, surface reaction, and desorption—with all steps reversible. The values of the parameters are optimized, and the study shows a good fit between the simulation and the experiments. An example is shown in Fig. 42. Then, if desired, the steady-state rates can be computed, with no necessity to assume any "rate-determining" step. Step-response experiments are also performed by introducing propylene at 40 pulses per second, and the response of acrolein to this oscillating feed is then measured. It is not clear whether the reactor is

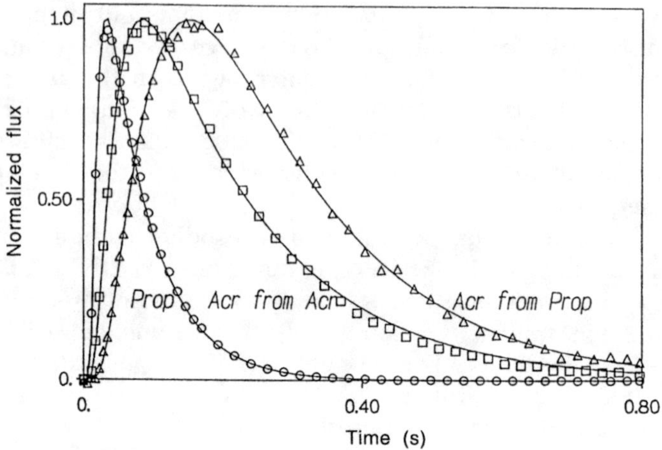

FIG. 42.   Propylene and acrolein responses obtained from a propylene pulse and from an acrolein pulse at 412°C over γ-Bi$_2$MoO$_6$ (after *258*).

in the Knudsen regime or in the plug-flow regime. A more qualitative TAP study appeared earlier (*259*).

Two recent studies of ethylene hydroformylation over Mn–Rh/SiO$_2$ at 240°C and atmospheric pressure have appeared (*260, 261*). The results have been simulated by elementary-step models with Eqs. (2)–(4). Isotopic and IR studies have been performed on the responses to input pulses.

The ammonia synthesis reaction is one of the most studied and best understood reactions in heterogeneous catalysis, but it has been the subject of few papers involving transient methods. SSITKA experiments have been performed at 350–500°C and 204–513 kPa using a commercial Haldor–Topsøe KMIR catalyst, with iron triply promoted by Al$_2$O$_3$, K$_2$O, and CaO (*262*). Similar studies using K-promoted Ru/SiO$_2$ have also been reported (*263*). The promoted Ru catalyst is much more active than Ru alone, and new, very active sites are detected on the promoted catalyst. It seems that the analysis of this type of experiment would benefit from the elementary-step approach, as exemplified by Kao *et al.* (*107*); two kinds of sites can be included in such a model.

IV. Summary

The results of many kinds of experiments in the transient regime commonly appear in the literature, and there is no need to preach to the converted. The following are a few remarks on some of the various ways of doing such experiments.

Modern computer programming has made it possible to use numerical methods to simulate step, pulse, and frequency response experiments in the simplest ideal system, the PFR. Such simulations are also possible in the TAP system with single-pulse or pump–probe experiments at pressures that ensure operation in the Knudsen regime. Obviously, nonideal reactors could also be simulated, as must be done during the design of large-scale reactors, but this would involve the estimation or measurement of quantities such as an axial dispersion coefficient. Thus, in some cases it is no longer advantageous to do experiments in an ideal mixed-flow reactor (CSTR) because they can be simulated by ordinary rather than partial differential equations. Analytical solutions are often elegant and many have been discussed, particularly with respect to low-amplitude frequency response studies. However, numerical methods usually can remove the necessity of linearizations. Temperature-programmed experiments can benefit particularly from numerical methods of analysis (*93*). Note that a CSTR is usually assumed even here.

Experiments at (high) vacuum make it possible, for well-defined surfaces, to obtain rate parameters that may be applicable to real catalysts at higher pressures (*4, 5, 37*). Fast processes can be measured, and this is certainly of fundamental interest in learning about the detailed pathways involved in a global reaction or even in a so-called elementary step. However, for reaction at ordinary pressures, *in situ* transient and other studies are usually sufficient to understand the observed kinetics, which depends essentially on the slow steps in the pathway.

The TAP system provides a way of doing experiments at low pressure with a real granular catalyst. However, for a correct quantitative simulation, the flow should be in the Knudsen regime. Single-pulse responses can lead to quantitative kinetic parameters (*258*). However, the analysis of multiple-pulse methods depends on the time between pulses: If this is long, we return to the single-pulse experiment; if short, we abandon the Knudsen regime and determine what would be obtained by a continuous-step or pulse-response experiment (a relaxed steady state); if in between we need to consider the mathematics needed for an oscillating feed. The most elegant experiment is one for which the pulse is sufficiently small so that its passage does not appreciably affect the surface state of the catalyst.

If we can measure quantitatively the parameters for the sequence of steps for a desired reaction for many different catalyst formulations (e.g., changes of promoter or support), it may be possible to formulate a better catalyst. This approach has been followed in many articles already discussed.

The studies that contribute the most to our understanding of kinetics are usually those that combine many steady-state experiments with *in situ* transient experiments: steps, pulses, (with and without isotopes) temperature programming, gas-phase responses, and infrared and other spectrometric responses of surface species—all with a thorough characterization of the catalyst in as many of its states as possible (*78, 116, 130, 187, 188, 195, 208*).

References

1. Bennett, C. O. *Catal. Rev.-Sci. Eng.* **13,** 121 (1976).
2. Bennett, C. O., *Am. Chem. Soc. Symp. Ser.* **178,** 1 (1982).
3. Che, M., and Bennett, C. O., *Adv. Catal.* **39,** 55 (1989).
4. Ribeiro, F. H., Schach von Wittenau, A. E., Bartholomew, C. H., and Somorjai, G. A., *Catal. Rev.-Sci. Eng.* **39,** 49 (1997).
5. Dumesic, J. A., Rudd, D. F., Aparicio, L. M., Rekoske, J. E., and Treviño, A. A., *in* "The Microkinetics of Heterogeneous Catalysis." American Chemical Society, Washington, DC, 1993.
6. Bennett, C. O., and Che, M., *J. Catal.* **120,** 293 (1989).
7. Fiolikatis, E., Hoffmann, U., and Hofmann, H., *Chem. Eng. Sci.* **34,** 677, 685 (1979).
8. Aris, R., and Amundson, N. R., "Mathematical Methods in Chemical Engineering," Vol. 2. Prentice Hall, Englewood Cliffs, NJ, 1973.
9. Fiolikatis, E., and Hofmann, H., *Catal. Rev.-Sci. Eng.* **24,** 113 (1982).
10. Fiolikatis, E., Hoffmann, U., and Hofmann, H., *Chem. Eng. Sci.* **35,** 1021 (1980).
11. Nibbelke, R. H., Scheerová, J., de Croon, M. H. J. M., and Marin, G. B., *J. Catal.* **156,** 106 (1995).
12. Weekman, V. W., *A. I. Ch. E. J.* **20,** 833 (1974).
13. Froment, G. F., and Bischoff, K. B., "Chemical Reactor Analysis and Design," 2nd ed., p. 85. Wiley, New York, 1990.
14. Berty J. M., paper presented at the 66th annual meeting of the A. I. Ch. E., Philadelphia, 1973.
15. Tajbl, D. C., Simons, J. B., and Carberry, J. J., *Ind. Eng. Chem. Fund.* **5,** 171 (1966).
16. Zielinski, J., *React. Kinet. Catal. Lett.* **17,** 69 (1981).
17. Stockwell, D. M., Chung, J. S., and Bennett, C. O., *J. Catal.* **112,** 135 (1988).
18. Kobayashi, H., and Kobayashi, M., *Catal. Rev.-Sci. Eng.* **10,** 104 (1974).
19. Kobayashi, M., Golman, B., and Kanno, T., *J. Chem. Soc. Faraday Trans.* **91,** 1391 (1995).
20. Eic, M., and Ruthven, D. M., *Zeolites* **8,** 40 (1988).
21. Ruthven, D. M., and Eic, M., *Am. Chem. Soc. Symp. Ser.* **368,** 362 (1988).
22. Kobayashi, M., *Chem. Eng. Sci.* **37,** 393 (1982).
23. Kobayashi, M., *Chem. Eng. Sci.* **37,** 403 (1982).
24. Yang, C. C., Cutlip, M. B., and Bennett, C. O., *in* "Catalysis" (J. W. Hightower, Ed.), Vol. 1, p. 273. Elsevier, New York, 1973.
25. Salmi, T., *Comput. Chem. Eng.* **11,** 83 (1987).
26. Salmi, T., *Chem. Eng. Sci.* **43,** 503 (1988).
27. van der Linde, S. C., Nijhuis, T. A., Dekker, F. H. M., Kaptejn, F., and Moulijn, J. A., *Appl. Catal. A Gen.* **151,** 27 (1997).
28. Arsenova, N., Haag, W. O., and Karge, H. G., *Stud. Surf. Sci. Catal.* **94,** 441 (1995).
29. Tamaru, K., *Adv. Catal.* **15,** 65 (1964).
30. Bennett, C. O., *A. I. Ch. E. J.* **13,** 890 (1967).

31. Masson, A., Bellamy, B., Hadj-Romdhane, Y., Che, M., Roulet, H., and Dufour, G., *Surf. Sci.* **173,** 497 (1986).
32. Boffa, A., Lin, C., Bell, A. T., and Somorjai, G. A., *J. Catal.* **149,** 149 (1994).
33. Bonzel, H. P., and Ku, R., *Surf. Sci.* **33,** 91 (1972).
34. Gleaves, J. T., Ebner, J. R., and Kuechler, T. C., *Catal. Rev.-Sci. Eng.* **30,** 49 (1988).
35. Gleaves, J. T., Jablonskii, G. S., Phanawadee, P., and Schuurman, Y., *Appl. Catal.* **160,** 55 (1997).
36. Kahn, D. R., Petersen, E. E., and Somorjai, G. A., *J. Catal.* **34,** 294 (1974).
37. Rodriguez, J. A., and Goodman, D. W., *Surf. Sci. Rep.* **14,** 1 (1991).
38. Jones, R. H., Olander, D. R., Siekhaus, W. J., and Schwarz, J. A., *J. Vac. Sci. Technol.* **9,** 1429 (1972).
39. Naphtali, L. M., and Polinski, L. M., *J. Phys. Chem.* **67,** 369 (1963).
40. Yasuda, Y., *J. Phys. Chem.* **80,** 1867, 1870 (1976).
41. Yasuda, Y., *J. Phys. Chem.* **86,** 1913 (1982).
42. Hufton, J. R., Brandani, S., and Ruthven, D. M., *Stud. Surf. Sci. Catal.* **84,** 1323 (1994).
43. Kärger, J., and Ruthven, D. M., *Stud. Surf. Sci. Catal.* **105,** 1843 (1997).
44. Lees, L. V. C., *Stud. Surf. Sci. Catal.* **84,** 1133 (1994).
45. Caro, J., Jobic, H., Bülow, M., Kärger, J., and Zibrowius, B., *Adv. Catal.* **39,** 351 (1993).
46. Yasuda, Y., *J. Phys. Chem.* **97,** 3314 (1993).
47. Yasuda, Y., and Nomura, K., *J. Phys. Chem.* **97,** 3319 (1993).
48. Yasuda, Y., *Heterogen. Chem. Rev.* **1,** 103 (1994).
49. Jordi, R. B., and Do, D. D., *J. Chem. Soc. Faraday Trans.* **88,** 2411 (1992).
50. Jordi, R. B., and Do, D. D., *Chem. Eng. Sci.* **48,** 1103 (1993).
51. Li, Y.-E., Willcox, D., and Gonzalez, R. D., *A. I. Ch. E. J.* **35,** 423 (1989).
52. Takagi, N., Yoshinobu, J., and Kawai, M., *Chem. Phys. Lett.* **215,** 120 (1993).
53. Ho, W., *J. Phys. Chem.* **91,** 766 (1987).
54. Sylveston, P., Hudgins, R. R., and Renken, A., *Catal. Today* **25,** 91 (1995).
55. Matros, Y. S. (Ed.), "Studies in Surface Science and Catalysis: Catalytic Processes under Unsteady-State Conditions," Vol. 43. Elsevier, Amsterdam, 1989.
56. Cutlip, M. B., *A. I. Ch. E. J.* **25,** 502 (1979).
57. Lynch, D. T., *Can. J. Chem. Eng.* **61,** 183 (1983).
58. Kaul, D. J., Sant, R., and Wolf, E. E., *Chem. Eng. Sci.* **42,** 1399 (1987).
59. Kaul, D. J., and Wolf, E. E., *J. Catal.* **89,** 348 (1984).
60. Sant, R., and Wolf, E. E., *Chem. Eng. Sci.* **45,** 3137 (1990).
61. Matros, Y. S., and Bunimovich, G. A., *Catal. Rev.-Sci. Eng.* **38, 1,** (1996); *in* "Handbook of Heterogeneous Catalysis" (G. Ertl, G. Knözinger, and J. Weitkamp, Eds.), Vol. 3, p 1464. VCH, Weinheim, 1997.
62. Pinjala, V., Chen, Y. C., and Luss, D., *A. I. Ch. E. J.* **34,** 1663 (1988).
63. Schüth, F., Henry, B. E., and Schmidt, L. D., *Adv. Catal.* **39,** 51 (1993).
64. Hugo, P., *Ber. Bunsenges. Phys. Chem.* **74,** 121 (1970).
65. Wicke, E., Beusch, H., and Fieguth, P., *ACS Symp. Ser.* **109,** 615 (1972).
66. Sheintuch, M., and Schmitz, R. A., *Catal. Rev.-Sci. Eng.* **15,** 107 (1977).
67. Razon, L. F., and Schmitz, R. A., *Chem. Eng. Sci.* **42,** 1005 (1987).
68. Eigenberger, G., *Chem. Eng. Sci.* **33,** 1263 (1978).
69. Ivanov, E. A., Chumakov, G. A., Slin'ko, M. G., Bruns, D. D., and Luss, D., *Chem. Eng. Sci.* **35,** 795 (1990).
70. Pawlicki, P. C., and Schmitz, R. A., *Chem. Eng. Prog.* **83,** 40 (1987).
71. Onken, H. U., and Wolf, E. E., *Chem. Eng. Sci.* **43,** 2251 (1988).
72. Onken, H. U., and Wicke, E. E., *Z. Phys. Chem N. F.* **165,** 23 (1989).
73. Qin, F., and Wolf, E. E., *Catal. Lett.* **39,** 19 (1996).

74. Qin, F., and Wolf, E. E., *Chem. Eng. Sci.* **50,** 117 (1995).
75. Engel, T., and Ertl, G., *Adv. Catal.* **28,** 1 (1979).
76. Ertl, G., Norton, P. R., and Rüstig, J., *Phys. Rev. Lett.* **49,** 177 (1982).
77. Ertl, G., *Adv. Catal.* **37,** 213 (1990).
78. Ertl, G., *in* "Handbook of Heterogeneous Catalysis" (G. Ertl, G. Knözinger, and J. Weitkamp, Eds.), Vol. 3, p. 1032. VCH, Weinheim, 1997.
79. Norton, P. R., Bindner, P. E., Griffiths, K., Jackman, T. E., Davies, J. A., and Rüstig, J., *J. Chem. Phys.* **80,** 3859 (1984).
80. Yeates, R. C., Turner, J. E., Gellman, J. A., and Somorjai, G. A., *Surf. Sci.* **149,** 175 (1985).
81. Imbihl, R., Cox, M. P., and Ertl, G., *J. Chem. Phys.* **84,** 3519 (1986).
82. Cox, M. P., Ertl, G., Imbihl, R., and Rüstig, J., *Surf. Sci.* **134,** L517 (1983).
83. Cox, M. P., Ertl, G., and Imbihl R., *Phys. Rev. Lett.* **54,** 1725 (1985).
84. Imbihl, R., Cox, M. P., Ertl, G. Müller, H., and Brenig, W., *J. Chem. Phys.* **83,** 1578 (1985).
85. Engel, W., Kordesch, M., Rotermund, H. H., Kubala, S., and von Oertzen, A., *Ultramicroscopy* **95,** 6162 (1991).
86. Nettesheim, S., von Oertzen, A., Rotermund, H. H., and Ertl, G., *J. Chem. Phys.* **98,** 9977 (1993).
87. Zhabotinsky, A. M., *Ber. Bunsenges. Phys. Chem.* **84,** 334 (1980).
88. Yokomizo, G. H., Louis, C., and Bell, A. T., *J. Catal.* **120,** 15 (1989).
89. Falconer, J. L., and Schwarz, J. A., *Catal. Rev.-Sci. Eng.* **25,** 141 (1983).
90. Schwarz, J. A., and Falconer, J. L., *Catal. Today* **7,** 1 (1990).
91. Madix, R. J., *Surf. Sci.* **299/300,** 785 (1994).
92. Redhead, P. A., *Vacuum* **12,** 303 (1962).
93. de Jong, A. M., and Niemantsverdriet, J. W., *Surf. Sci.* **233,** 355 (1990).
94. Ibach, H., Erley, W., and Wagner, H., *Surf. Sci.* **92,** 29 (1980).
95. Block, J. H., Cocke, D. L., and Kruse, N., *in* "Handbook of Heterogeneous Catalysis" (G. Ertl, G. Knözinger, and J. Weitkamp, Eds.), Vol. 3, p. 1104. VCH, Weinheim, 1997.
96. Rieck, J. S., and Bell, A. T., *J. Catal.* **85,** 143 (1984).
97. Demmin, R. A., and Gorte, R. J., *J. Catal.* **90,** 32 (1984).
98. Efstathiou, A. M., and Bennett, C. O., *J. Catal.* **124,** 116 (1990).
99. Hathaway, P. E., and Davis, M. E., *Catal. Lett.* **5,** 333 (1990).
100. De Pontes, M., Yokomizo, G. H., and Bell, A. T., *J. Catal.* **104,** 147 (1987).
101. Hoost, T. E., and Goodwin, J. G., Jr., *J. Catal.* **134,** 678 (1992).
102. Shannon, S. L., and Goodwin, J. G., Jr., *Chem. Rev.* **95,** 677 (1995).
103. Le Cardinal, G., Walter, E., Bertrand, P., Zoulalian, A., and Gelus, M., *Chem. Eng. Sci.* **32,** 733 (1977).
104. Shannon, S. L., and Goodwin, J. G., Jr., *Appl. Catal. A* **151,** 3 (1997).
105. Happel, J., "Isotopic Assessment of Heterogeneous Catalysts." Academic Press, Orlando, FL, 1986.
106. Happel, J., Walter, E., and Lecourtier, Y., *J. Catal.* **123,** 12 (1990).
107. Kao, J.-Y., Piet-Lahanier, H., Walter, E., and Happel, J., *J. Catal.* **133,** 383 (1992).
108. Otarod, M., Happel, J., and Walter, E., *Appl. Catal. A* **160,** 3 (1997).
109. Happel, J., Yoshikiyo, M., Yin, F., Cheh, H. Y., Hnatow, M. A., Bajars, L., and Meyers, H. S., *I&EC Prod. Res. Dev.* **25,** 214 (1986).
110. Tamaru, K., *in* "Handbook of Heterogeneous Catalysis" (G. Ertl, G. Knözinger, and J. Weitkamp, Eds.), Vol. 3, p. 1012. VCH, Weinheim, 1997.
111. Yamada, T., Onishi, T., and Tamaru, K., *Surf. Sci.* **133,** 533 (1983).
112. Zhdanov, V. P., *Surf. Sci. Lett.* **157,** L384 (1985).
113. Yates, J. T., and Goodman, D. W., *J. Phys. Chem.* **73,** 6371 (1980).
114. Lombardo, S. J., and Bell, A. T., *Surf. Sci.* **245,** 213 (1991).

115. Nishimura, M., Asakura, K., and Iwasawa, Y., *in* "Proceedings of the 9th International Congress on Catalysis" (M. J. Phillip and M. Ternan, Eds.), Vol. 4, p. 1842. The Chemical Institute of Canada, Ottawa (1988).
116. Coulston, G. W., Bare, S. R., Kung, H., Birkeland, K., Bethke, G. K., Harlow, R., Herron, N., and Lee, P. L., *Science* **275,** 191 (1997).
117. Bianchi, D., Borcar, S., Teulé-Gay, F., and Bennett, C. O., *J. Catal.* **82,** 442 (1983).
118. Jonkers, J., *in* "Handbook of Heterogeneous Catalysis" (G. Ertl, G. Knözinger, and J. Weitkamp, Eds.), Vol. 3, p. 1023. VCH, Weineim, 1997.
119. Anderson, B. G., van Santen, R. A., and van Ijzendoorn, L. J., *Appl. Catal. A* **160,** 125 (1997).
120. Ertl, G., *Appl. Surf. Sci.* **121/122,** 20 (1997).
121. Bowker, M., *Stud. Surf. Sci. Catal.* **101,** 287 (1996).
122. Crew, W. W., and Madix, R. J., *Surf. Sci.* **319,** L34 (1994).
123. Ruen, L., Stensgaard, I., Lægsgaard, E., and Besenbacher, F., *Surf. Sci.* **314,** L873 (1994).
124. Szedlacsek, P., and Guczi, L. (Eds.), "Transient Kinetics," *Appl. Catal. A* **151,** (1997).
125. Chuang, S. S. C., and Fullerton, K. L. (Eds.), Proceedings of the Symposium on Heterogenous Catalysis and Reaction Engineering, *Catal. Today* **35**(4) (1997).
126. Klusàcek, K., and Capek, P. (Eds.), "Dynamics of Catalytic Systems '96," *Catal. Today* **38** (1997).
127. Marin, G. B., and van Santen, R. A. (Eds.), "Kinetic Methods in Heterogenous Catalysis," *Appl. Catal. A* **160** (1997).
128. Froment, G. F., and Waugh, K. C. (Eds.), "Dynamics of Surfaces and Reaction Kinetics in Heterogeneous Catalysis" *Stud. Surf. Sci. Catal.* **109,** (1997).
129. Mills, P. L., and Lerou, J. J., *Rev. Chem. Eng.* **9**(1/2) (1993).
130. Schuurman, Y., and Mirodatos, C., *Appl. Catal. A* **151,** 305 (1997).
131. Efstathiou, A. M., and Verykios, X. E., *Appl. Catal. A* **151,** 109 (1997).
132. Ito, T., Wang, J.-X., Lin, C.-H., and Lunsford, J. H., *J. Am. Chem. Soc.* **107,** 5026 (1985); Lin, C.-H., Wang, J.-X., and Lunsford, J. H., *J. Catal* **111,** 302 (1988).
133. Mims, C. A., Hall, R. B., Rose, K. D., and Myers, G. R., *Catal. Lett.* **2,** 361 (1989).
134. Ekstrom, A., and Lepszewich, J. A., *J. Am. Chem. Soc.* **110,** 5226 (1988).
135. Peil, K. P., Goodwin, J. G., Jr., and Marcelin, G., *J. Am. Chem. Soc.* **112,** 6129 (1990).
136. Peil, K. P., Goodwin, J. G., Jr., and Marcelin, G., *J. Phys. Chem.* **93,** 5977 (1989).
137. Peil, K. P., Goodwin, J. G., Jr., and Marcelin, G., *J. Catal.* **132,** 556 (1991).
138. Lacombe, S., Geantet, C., and Mirodatos, C., *J. Catal.* **151,** 439 (1994).
139. Lunsford, J. H., Yang, X., Haller, K., Laane, J., Mestl, G., and Knözinger, H., *J. Phys. Chem.* **97,** 13810 (1993).
140. Knözinger, H., *Catal. Today* **32,** 71 (1996).
141. Lunsford, J. H., *in* "Handbook of Heterogeneous Catalysis" (G. Ertl, G. Knözinger, and J. Weitkamp, Eds.), Vol. 4, p. 1843. VHC, Weinheim, 1997.
142. Lacombe, S., Geantet, C., and Mirodatos, C., *J. Catal.* **155,** 106 (1995).
143. Statman, D. J., Gleaves, J. T., McNamara, D., Mills, P. L., Fornasari, G., and Ross, J. R. H., *Appl. Catal.* **77,** 45 (1991).
144. Buyevskaya, O. V., Rothaemel, M., Zanthoff, H. W., and Baerns, M., *J. Catal.* **146,** 346 (1994).
145. Peil, K. P., Goodwin, J. G., Jr., and Marcelin, G., *J. Catal.* **131,** 143 (1991).
146. Mallens, E. P. J., Hoebink, G. H. B. J., and Marin, J. B., *J. Catal.* **160,** 222 (1996).
147. Kalenik, Z., and Wolf, E. E., *Catal. Today* **13,** 255 (1992).
148. Kalenik, Z., and Wolf, E. E., *Catal. Lett.* **11,** 309 (1991).
149. Efstathiou, A. M., Papageorgiuo, D., and Verykios, X. E., *J. Catal.* **144,** 352 (1993).

150. Efstathiou, A. M., Lacombe, S., Mirodatos, C., and Verykios, X. E., *J. Catal.* **148,** 639 (1994).
151. Papageorgiuo, D., Efstathiou, A. M., and Verykios, X. E., *Appl. Catal. A* **111,** 41 (1994).
152. E. E. Wolf (Ed.), "Methane Conversion by Oxidative Processes." Van Nostrand Reinhold, New York, 1992.
153. Ekstrom, A., Lepszewicz, J. A., and Campbell, I., *Appl. Catal.* **56,** L29 (1989).
154. Le Van, T., Louis, C., Kermarec, M., Che, M., and Taibouët, J.-M., *Catal. Today* **13,** 321 (1992).
155. Colussi, A. J., and Amorebieta, V. T., *J. Catal.* **169,** 301 (1997).
156. Prettre, M., Eichner, C., and Perrin, M., *Trans. Faraday Soc.* **43,** 335 (1946).
157. Mallens, E. P. J., Hoebink, J. H. B. J., and Marin, G. B., *J. Catal.* **167,** 43 (1997).
158. Hickman, D. A., and Schmidt, L. D., *J. Catal.* **138,** 267 (1992).
159. Hickman, D. A., and Schmidt, L. D., *A. I. Ch. E. J.* **39,** 1164 (1993).
160. Mallens, E. P. J., Hoebink, J. H. B. J., and Marin, G. B., *Catal. Lett.* **33,** 291 (1995).
161. Wang, D., Dewaele, O., De Groote, D. M., and Froment, G. F., *J. Catal.* **159,** 418 (1996).
162. Boucouvalas, Y., Zhang, Z. L., Efstathiou A. M., and Verykios, X. E., *Stud. Surf. Sci. Catal.* **101,** 443 (1996).
163. Buyevskaya, O. V., Wolf, D., and Baerns, M., *Catal. Lett.* **29,** 249 (1994).
164. Walter, K., Buyeskava, O. V., Wolf, D., and Baerns, M., *Catal. Lett.* **29,** 261 (1994).
165. Fischer, F., and Tropsch, H., *Brennstoff Chem.* **39,** 3 (1928).
166. Tenner, S., *Hydrocarbon Process.* **106,** 64 (1985).
167. Kroll, V. C. H., Swaan, H. M., and Mirodatos, C., *J. Catal.* **161,** 409 (1996).
168. Kroll, V. C. H., Swaan, H. M., Lacombe, S., and Mirodatos, C., *J. Catal.* **164,** 387 (1997).
169. Zhang, Z., and Verykios, X. E., *J. Chem. Soc. Chem. Commun.* **71** (1995).
170. Zhang, Z., and Verykios, X. E., *Appl. Catal. A* **138,** 109 (1996).
171. Zhang, Z., Verykios, X. E., MacDonald, S. M., and Affrossman, S., *J. Phys. Chem.* **100,** 744 (1996).
172. Zhang, Z., and Verykios, X. E., *Catal. Lett.* **38,** 175 (1996).
173. Osaki, T., Horiuchi, T., Suzuki, K., and Mori, T., *Catal. Lett.* **44,** 19 (1997).
174. Slagtern, A., Schuurman, Y., Leclercq, C., Verykios, X. E., and Mirodatos, C., *J. Catal.* **172,** 118 (1997).
175. Erdöhelyi, A., Cresényi, J., and Solymosi, F., *J. Catal.* **141,** 612 (1993).
176. Erdöhelyi, A., Cresényi, J., Papp, E., and Solymosi, F., *J. Catal.* **141,** 287 (1993).
177. Tsipouriari, V. A., Efstathiou, A. M., Zhang, Z. L., and Verykios, X. E., *Catal. Today* **21,** 579 (1994).
178. Zhang, Z. L., Tsipouriari, V. A., Efstathiou, A. M., and Verykios, X. E., *J. Catal.* **158,** 51 (1996).
179. Efstathiou, A. M., Kladi, A., Tsipouriari, V. A., and Verykios, X. E., *J. Catal.* **158,** 64 (1996).
180. Tsipouriari, V. A., Efstathiou, A. M., and Verykios, X. E., *J. Catal.* **161,** 31 (1996).
181. Goula, M. A., Lemonidou, A. A., and Efstathiou, A. M., *J. Catal.* **161,** 626 (1996).
182. van Keulen, A. N. J., Seshan, K., Hoebink, J. H. B. J., and Ross, J. R. H., *J. Catal.* **166,** 306 (1997).
183. Bianchi, D., and Gass, J. L., *J. Catal.* **123,** 298 (1990).
184. Bianchi, D., Borcar, S., Teulé-Gay, F., and Bennett, C. O., *J. Catal.* **82,** 442 (1983).
185. Bianchi, D., Tau, L. M., Borcar, S., and Bennett, C. O., *J. Catal.* **84,** 358 (1983).
186. Tau, L. M., Bianchi, D., Borcar, S., and Bennett, C. O., *J. Catal.* **87,** 36 (1984).
187. Winslow, P., and Bell, A. T., *J. Catal.* **86,** 158 (1984).
188. Duncan, T. M., Winslow, P., and Bell, A. T., *J. Catal.* **93,** 1 (1985).
189. Bianchi, D., and Gass, J. L., *J. Catal.* **123,** 310 (1990).

190. Stockwell, D. M., Bianchi, D., and Bennett, C. O., *J. Catal.* **113,** 13 (1988).

191. Winslow, P., and Bell, A. T., *J. Catal.* **91,** 142 (1985).

192. Yamasaki, H., Kobori, Y., Naito, S., Onishi, T., and Tamaru, K., *Chem. Soc. Trans. I* **77,** 2813 (1981).

193. Kobori, Y., Yamasaki, H., Naito, S., Onishi, T., and Tamaru, K., *Chem. Soc. Trans. I* **78,** 1473 (1982).

194. Efstathiou, A. M., and Bennett, C. O., *J. Catal.* **120,** 137 (1989).

195. Efstathiou, A. M., Chafik, T., Bianchi, D., and Bennett, C. O., *J. Catal.* **148,** 224 (1994).

196. Flesner, R. L., and Falconer, J. L., *J. Catal.* **150,** 301 (1994).

197. Balakos, M. W., Chuang, S. S. C., Srinivas, G., and Brundage, M. A., *J. Catal.* **157,** 51 (1995).

198. Stockwell, D. M., and Bennett, C. O., *J. Catal.* **110,** 354 (1988).

199. Krishna, K. R., and Bell, A. T., *J. Catal.* **139,** 104 (1993).

200. Bajusz, I.-G., and Goodwin, J. G., Jr., *J. Catal.* **169,** 157 (1997).

201. Chen, B., and Goodwin, J. G., Jr., *J. Catal.* **158,** 511 (1996).

202. Haddad, J. G., Chen, B., and Goodwin, J. G., Jr., *J. Catal.* **161,** 274 (1996).

203. Vada, S., Chen, B., and Goodwin, J. G., Jr., *J. Catal.* **153,** 224 (1995).

204. Kogelbauer, A., Goodwin, J. G., Jr., and Oukaci, R., *J. Catal.* **160,** 125 (1996).

205. Ali, S., Chen, B., and Goodwin, J. G., Jr., *J. Catal.* **157,** 35 (1995).

206. Marwood, M., Doepper, R., Prairie, M., and Renken, A., *Chem. Eng. Sci.* **49,** 4801 (1994).

207. Marwood, M., Doepper, R., and Renken, A., *Can. J. Chem. Eng.* **74,** 1 (1996).

208. Marwood, M., Doepper, R., and Renken, A., *Appl. Catal. A* **151,** 223 (1997).

209. Herz, R. K., and Marin, S. P., *J. Catal.* **65,** 281 (1980).

210. Lie, A. B. K., Hoebink, J., and Marin, G. B., *Chem. Eng. J.* **53,** 47 (1993).

211. Nievergeld, A. J. L., Hoebink, J. H. B. J., and Marin, G. B., *Stud. Surf. Sci. Catal.* **96,** 909 (1995).

212. Ionnides, T., Efstathiou, A. M., Zhang, Z. L., and Verykios, X. E., *J. Catal.* **156,** 265 (1995).

213. Nijhuis, T. A., Makkee, M., van Langeveld, A. D., and Moulijn, J. A., *Appl. Catal. A.* **164,** 139 (1997).

214. Hoebink, J. H. B. J., Huinink, J. P., and Marin, G. B., *Appl. Catal. A* **160,** 223 (1997).

215. Dwyer, S. M., and Bennett, C. O., *J. Catal.* **75,** 275 (1982).

216. Bennett, C. O., Laporta, L. M., and Cutlip, M. B., *Stud. Surf. Sci. Catal.* **30,** 143 (1987).

217. Dekker, F. H. M., Nazloomian, J. G., Bliek, A., Kapteijn, F., Moulijn, J. A., Coulson, D. R., Mills, P. L., and Lerou, J. J., *Appl. Catal. A* **151,** 247 (1997).

218. Dekker, F. H. M., Dekker, M. C., Kapteijn, F., Moulijn, J. A., and Bliek, A., *in* "Precision Process Technology" (P. C. Weijnen and A. A. H. Drinkenburg, Eds.), p. 473. Kluwer, Dordrecht, 1993.

219. Dekker, F. H. M., Dekker, M. C., Bliek, A., Kapteijn, F., and Moulijn, J. A., *Catal. Today* **20,** 409 (1994).

220. Dekker, F. H. M., Klopper, G., Bliek, A., Kapteijn, F., and Moulijn, J. A., *Chem. Eng. Sci.* **49,** 4375 (1994).

221. Mergler, Y. J., Hoebink, J., and Nieuwenhuys, B. E., *J. Catal.* **167,** 305 (1997).

222. Kobayashi, M., Golman, B., and Kanno, T., *J. Chem. Soc. Faraday Trans.* **91,** 1391 (1995).

223. Kobayashi, M., Golman, B., Kanno, T., and Shintaro, F., *Appl. Catal. A* **151,** 193 (1997).

224. Armor, J. N., *Catal. Today* **26,** 99 (1995).

225. Schlatter, J. C., and Mitchell, P. J., *Ind. Eng. Chem. Prod. Res. Dev.* **19,** 288 (1980).

226. Iwamoto, M., Furukawa, H., Mine, Y., Uemura, F., Mikuriya, S., and Kagawa, S., *J. Chem. Soc., Chem. Commun.,* 1272 (1986).

227. Quick, L. M., and Kamitomai, S., *Catal. Today* **26,** 303 (1995).

228. Bögner, W., Krämer, M., Krutzsch, B., Pischinger, S., Voigtländer, D., Wenninger, G., Wirbeleit, F., Brogan, M. S., Brisley, R. J., and Webster, D. E., *Appl. Catal. B* **7,** 153 (1995).
229. Takahashi, N., Shinjoh, H., Iijima, T., Suzuki, T., Yamazaki, K., Yokota, K., Suzuki, H., Myoshi, N., Matsumoto, S., Tanizawa, T., Tanaka, T., Tateishi, S., and Kasahara, K., *Catal. Today* **27,** 63 (1996).
230. Fekete, N., Kemmler, R., Voigtländer, D., Krutzsch, B., Zimmer, E., Wenninger, G., Strehlau, W., van den Tillaart, J. J. A., Leyrer, J., Lox, E. S., and Müller, W., SAE 970746 (1997).
231. Fritz, A., and Pitchon, V., *Appl. Catal. B* **13,** 1 (1997).
232. Chang, Y., and McCarty, J. G., *J. Catal.* **165,** 1 (1997).
233. Burch, R., and Millington, P. J., *Catal. Today* **26,** 185 (1995).
234. Burch, R., Millington, P. J., and Walker, A. P., *Appl. Catal. B* **4,** 65 (1994).
235. Burch, R., and Watling, T. C., *Catal. Lett.* **37,** 51 (1996).
236. Dekker, F. H. M., Kraneveld, S., Bliek, A., Kapteijn, F., and Moulijn, J. A., *J. Catal.* **170,** 168 (1997).
237. Srinivas, G., Chuang, S. S. C., and Debnath, S., *J. Catal.* **148,** 748 (1994).
238. Krishnamurthy, R., Chuang, S. S. C., and Balakos, M. W., *J. Catal.* **157,** 512 (1995).
239. Krishnamurthy, R., and Chuang, S. S. C., *J. Phys. Chem.* **99,** 16727 (1995).
240. Chuang, S. S. C., and Tan, C.-D., *J. Catal.* **173,** 95 (1997).
241. Randall, H., Döppler, R., and Renken, A., *Can. J. Chem. Eng.* **74,** 1 (1996).
242. Waugh, K. C., *Catal. Today* **15,** 51 (1992).
243. Elliott, A. J., Hadden, R. A., Tabatabaei, J., Waugh, K. C., and Zemicael, F. W., *J. Catal.* **157,** 153 (1995).
244. Hadden, R. A., Sakakini, B., Tabatabaei, J., and Waugh, K. C., *Catal. Lett.* **44,** 145 (1997).
245. Vada, S., and Goodwin, J. G., Jr., *J. Phys. Chem.* **99,** 9479 (1995).
246. Ali, S. H., and Goodwin, J. G., Jr., *J. Catal.* **170,** 265 (1997).
247. Ali, S. H., and Goodwin, J. G., Jr., *J. Catal.* **171,** 333 (1997).
248. Ali, S. H., and Goodwin, J. G., Jr., *J. Catal.* **171,** 339 (1997).
249. Chung, J. S., Miranda, R., and Bennett, C. O., *J. Catal.* **114,** 398 (1988).
250. Weber, R. S., *J. Phys. Chem.* **98,** 2999 (1994).
251. Wachs, I. E., Deo, G., Juskelis, M. V., and Weckhuysen, B. M., *Stud. Surf. Sci. Catal.* **109,** 305 (1997).
252. Holstein, W. L., and Machiels, C. J., *J. Catal.* **162,** 118 (1996).
253. Werner, H., Herein, D., Schulz, G., Wild, U., and Schlögl, R., *Catal. Lett.* **49,** 109 (1997).
254. Zaza, P., Randall, H., Döpper, R., and Renken, A., *Catal. Today* **20,** 325 (1994).
255. Lafyatis, D. S., Creten, G., and Froment, G. F., *Appl. Catal. A* **120,** 85 (1994).
256. Koranne, M. M., Goodwin, J. G., Jr., and Marcelin, G., *J. Phys. Chem.* **97,** 673 (1993).
257. Koranne, M. M., Goodwin, J. G., Jr., and Marcelin, G., *J. Catal.* **148,** 378 (1994).
258. Creten, G., Lafyatis, D. S., and Froment, G. F., *J. Catal.* **154,** 151 (1995).
259. Coulson, D. R., Mills, P. L., Kourtakis, K., Lerou, J. J., and Manzer, L. E., *Stud. Surf. Sci. Catal.* **72,** 305 (1992).
260. Brundage, M. A., Balakos, M. W., and Chuang, S. S. C., *J. Catal.* **173,** 122 (1998).
261. Brundage, M. A., and Chuang, S. S. C., *J. Catal.* **173,** 164 (1998).
262. Nwalor, J. U., Goodwin, J. G., Jr., and Biloen, P., *J. Catal.* **117,** 121 (1989).
263. Nwalor, J. U., and Goodwin, J. G., Jr., *Topics Catal.* **1,** 285 (1994).

Influence of Phosphorus on the Properties of Alumina-Based Hydrotreating Catalysts

RYUICHIRO IWAMOTO

Petroleum Refining Technology Center
Idemitsu Kosan Co. Ltd.
Sodegaura, Chiba, 299-0293, Japan

AND

JEAN GRIMBLOT

Laboratoire de Catalyse Hétérogène et Homogène
URA CNRS 402
Université des Sciences et Technologies de Lille
59655 Villeneuve d'Ascq Cedex, France

On the basis of numerous patents, phosphorus is classified as a second promotor in molybdenum-containing hydrotreating catalysts, which are used for preparing clean-burning fuels. Notwithstanding the substantial work on the influence of phosphorus in catalysts, its role is still a matter of debate, as critically discussed in this review. Preparation of the catalysts, which consists of mixing the selected ingredients (P, Mo, Co, or Ni and alumina) to form solids that meet specific requirements (e.g., specific surface area, dispersion, and loading of the elements), has been extensively examined. The review includes data concerning adsorption (competitive or sequential) of phosphorus oxo-species on alumina. Several states have been identified (e.g., monophosphate, polyphosphate, Al or Ni phosphate, and "Mo–P" heteropoly compounds) on the basis of data characterizing reference compounds and the combined use of techniques such as extended X-ray absorption fine structure, X-ray photoelectron, solid state nuclear magnetic resonance, and vibrational (infrared and Raman) spectroscopies. Texture, thermal stability, and acidity measurements complement the spectroscopic information. Large differences in acidity are reported that might be attributed both to the nature of the catalysts and to the methods of measurement. It is often claimed that the addition of phosphorus does not induce enough acidity to promote hydrogenolysis during cataly-

Abbreviations: AHM, ammonium heptamolybdate; CUS, coordinatively unsaturated sites; DHQ, decahydroquinoline; EXAFS, extended X-ray absorption fine structure; H_0, Hammett acidity function; HDM, hydrodemetalization; HDN, hydrodenitrogenation; HDS, hydrodesulfurization; HREM, high-resolution electron microscopy; HYD, hydrogenation; IEP, isoelectronic point; IR, FTIR, infrared, Fourier transform infrared; NMR, nuclear magnetic resonance; OPA, *ortho*-propylaniline; PD, pore diameter; SSA, specific surface area; TPD, temperature-programmed desorption; TPR, temperature-programmed reduction; UV, ultraviolet; VGO, vacuum gas oil; XPS, X-ray photoelectron spectroscopy; XRD, X-ray diffraction.

417

sis. The dispersion of the elements in the oxide-form catalysts depends not only on the phosphorus content but also on the preparation conditions. Large amounts of phosphorus favor the agglomeration of metal oxides. However, no clear conclusion can be proposed for the influence of phosphorus on the dispersion of the sulfide phases or their sulfur contents. In general, the catalytic influence of phosphorus strongly depends on its content, with an effect that is usually negative at high loadings. Phosphorus shows no effect or only a very small positive effect on hydrodesulfurization of thiophene, and the most significant positive effects have been reported primarily for hydrodenitrogenation reactions. However, the effect of phosphorus depends on the nature of the reactants and intermediates, and different active sites are involved in the various hydrotreating reactions. Structural models of phosphorus-containing hydrotreating catalysts and the main influences of phosphorus on the states of the other elements on the alumina surface are presented.

I. Introduction

The removal of hetero atoms such as sulfur, nitrogen, and metals (mainly V and Ni) from petroleum fractions has become increasingly important because of the need to solve environmental problems associated with combustion of fuels. Catalytic hydrotreating of feedstocks is one of the most effective practical methods for preparing clean-burning fuels. Molybdenum- or tungsten-based catalysts promoted by cobalt or nickel have been widely used in commercial hydrotreating plants. Much attention has been paid to the understanding the structures of catalytically active sites, reaction mechanisms, and effects of promotors in these catalysts. The roles of Co and Ni have been investigated in detail, as have the effects of supports such as Al_2O_3, SiO_2, carbon, TiO_2, ZrO_2, $Al_2O_3-SiO_2$, $Al_2O_3-TiO_2$, and $Al_2O_3-ZrO_2$. The subject has been amply reviewed (*1–3*).

Phosphorus is known as a second promotor (sometimes called an additive or a modifier) to improve molybdenum-containing hydrotreating catalysts. The beneficial effect of phosphorus has been described in numerous patents (*4, 5*), but research investigations began only recently. Some researchers have attempted to understand the role of phosphorus on the basis of structural and spectroscopic characterizations of catalysts, and others have relied on analysis of the kinetics of the catalytic reactions. Notwithstanding the substantial work on the effect of phosphorus, its role is still a matter of debate. Because there is no review of the effects of the role of phosphorus in hydrotreating catalysts, we have attempted here to summarize the subject and present a critical review, focusing on the preparation, physicochemical properties, and performance of phosphorus-containing hydrotreating catalysts. The main goal is to answer questions about the nature of the phosphorus-containing species in the catalysts and how they affect catalytic properties.

Theories and principles of the characterization techniques are not described here. For consistency, all the catalysts described in this review are referred to with the same nomenclature, although a different nomenclature is sometimes used in the cited publications. Each catalyst component (element) separated by the symbol "–" indicates the sequence of its introduction into the catalyst formulation from right to left. Those separated by the symbol "/" between right and left belong to the support material and the elements on the support, respectively. For example, NiMo–P/Al refers to a catalyst prepared such that the phosphorus-containing precursor is loaded on the alumina support first, followed by nickel and molybdenum, which are introduced simultaneously. CoMo/Al—P refers to a catalyst in which cobalt and molybdenum are introduced simultaneously onto an alumina support doped with phosphorus-containing species. Each element may represent its oxide or sulfide forms. In all cases, Al refers to the alumina-based support or to its hydroxide precursor.

II. Properties of Phosphorus-Based Compounds Related to the Co(Ni)–Mo–P–Alumina System

The word "phosphorus" originally came from *phosphoros*, which means "light bearing planet Venus before sunrise" in ancient Greek (*6*). Elemental phosphorus was first obtained from urine by Brand in 1669. It is classified as a metalloid of group V, below nitrogen in the periodic table of the elements, with an atomic number of 15 and an atomic weight of 30.97. During the preparation steps and activation of hydrotreating catalysts and in the course of the different reactions of model compounds or industrial feeds, phosphorus can be incorporated in several types of chemical environments, such as phosphorus oxo-compounds, aluminophosphates, molybdophosphate heteropoly compounds, organic phosphorus-containing compounds, and reduced phosphorus-based species. It is important to keep in mind that handling of such phosphorus-containing compounds sometimes requires special attention because some of them might be toxic, highly corrosive, or dangerous because they spontaneously ignite in air. In this section, properties and structures of phosphorus-containing compounds that are potentially related to hydrotreating catalysts or reactions are examined.

A. ELEMENTAL PHOSPHORUS

Elemental phosphorus can be classified into several allotropic forms, such as white (or yellow), red, violet, and black colored compounds (*6–9*). White phosphorus is a water-insoluble waxy compound consisting of tetrahedral P_4 molecules (P—P bond distance, 221 pm; bond angle, 60°). White phosphorus is found in α and β structures depending on temperature and pressure. When white phosphorus is exposed to sunlight or heated with traces of iodine, it transforms at low pressure into a polymeric amorphous compound called red phosphorus. Violet phosphorus, which has mutually interconnected pentagonal channels, is stable at higher temperature and pressure. Black phosphorus, which consists of parallel double layers, is the most stable form from a thermodynamic point of view. The phase diagram of elemental phosphorus is shown in Fig. 1.

B.   PHOSPHORUS HYDRIDES (PHOSPHANES)

Phosphorus forms many types of hydrides (7–9), called phosphanes (Table I). Phosphine (PH_3) and diphosphine ($H_2P—PH_2$) are the best known. Phosphine is a colorless, toxic gas with a characteristic smell. The higher condensed phosphorus hydrides tend to form chain and ring structures such as cyclopolyphosphine and catenapolyphosphine, respectively. The PH entity has been detected only by spectroscopic analysis, and PH_5, P_3H, and $P_{18}H_2$ have not been found.

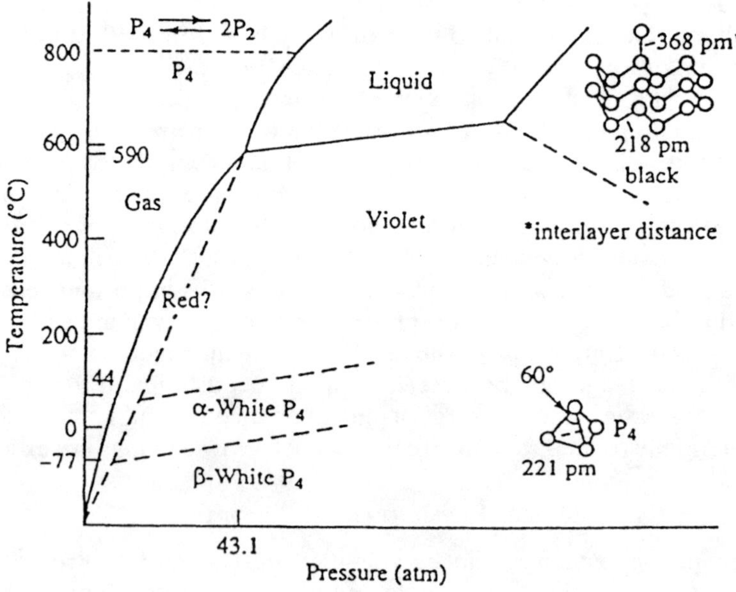

FIG. 1.   Phase diagram of elemental phosphorus [from King (8)].

TABLE I

Known Phosphanes[a]

Phosphane series	Known compounds
P_nH_{n+2}	PH_3, P_2H_4, P_3H_5, P_4H_6, P_5H_7, P_6H_8, P_7H_9, P_8H_{10}, P_9H_{11}
P_nH_n	P_3H_3, P_4H_4, P_5H_5, P_6H_6, P_7H_7, P_8H_8, P_9H_9, $P_{10}H_{10}$
P_nH_{n-2}	P_4H_2, P_5H_3, P_6H_4, P_7H_5, P_8H_6, P_9H_7, $P_{10}H_8$, $P_{11}H_9$, $P_{12}H_{10}$
P_nH_{n-4}	P_5H, P_6H_2, P_7H_3, P_8H_4, P_9H_5, $P_{10}H_6$, $P_{11}H_7$, $P_{12}H_8$, $P_{13}H_9$
P_nH_{n-6}	P_7H, P_8H_2, P_9H_3, $P_{10}H_4$, $P_{11}H_5$, $P_{12}H_6$, $P_{13}H_7$, $P_{14}H_8$, $P_{15}H_9$
P_nH_{n-8}	$P_{10}H_2$, $P_{11}H_3$, $P_{12}H_4$, $P_{13}H_5$, $P_{14}H_6$, $P_{15}H_7$, $P_{16}H_8$, $P_{17}H_9$
P_nH_{n-10}	$P_{12}H_2$, $P_{13}H_3$, $P_{14}H_4$, $P_{15}H_5$, $P_{16}H_6$, $P_{17}H_7$, $P_{18}H_8$, $P_{19}H_9$, $P_{20}H_{10}$
P_nH_{n-12}	$P_{13}H$, $P_{14}H_2$, $P_{15}H_3$, $P_{16}H_4$, $P_{17}H_5$, $P_{18}H_6$, $P_{19}H_7$, $P_{20}H_8$
P_nH_{n-14}	$P_{15}H$, $P_{16}H_2$, $P_{17}H_3$, $P_{18}H_4$, $P_{19}H_5$, $P_{20}H_6$, $P_{21}H_7$
P_nH_{n-16}	$P_{17}H$, $P_{19}H_2$, $P_{20}H_4$, $P_{21}H_5$, $P_{22}H_6$
P_nH_{n-18}	$P_{19}H$, $P_{20}H_2$, $P_{21}H_3$, $P_{22}H_4$

[a] From King (*8*).

C. PHOSPHIDES

The strong chemical reactivity of phosphorus allows the formation of single and mixed phosphides with several metals (*7–10*). Phosphides that are relevant to hydrotreating catalyst formulations (Table II) involve bonding of phosphorus with molybdenum (or tungsten), cobalt, and/or nickel.

TABLE II

Phosphides Relevant to Hydrotreating Catalyst Formulations[a]

Metal	Single phosphides	Mixed phosphides
Mo, W	MoP	Mo_2Ni_3P
	MoP_2, WP_2	$Mo_2Ni_6P_3$
	W_2P	NiMoP
	MoP_4, WP_4	NiWP
	Mo_3P, W_4P	CoMoP
	Mo_8P_5	CoWP
	Mo_4P_3	$NiMoP_2$
Ni, Co	NiP, CoP	$CoWP_2$
	Ni_2P, Co_2P	$NiWP_2$
	Ni_3P	$CoMoP_2$
	NiP_3, CoP_3	$NiMo_2P_3$
	Ni_5P_2, Ni_5P_4	WNi_4P_{16}
	Ni_8P_3, $Ni_{12}P_5$	$NiMoP_8$
	$NiCo_9P_5$	$NiWP_8$
Al	AlP	—

[a] Adapted from King (*8*).

Aluminum phosphide (AlP) also exists, but its presence in hydrotreating catalysts is very unlikely because aluminum oxide is thermodynamically very stable.

D. PHOSPHORUS OXIDES

When phosphorus reacts with molecular oxygen, it forms many types of phosphorus oxides with different oxidation states (7–9). Under low partial pressures of oxygen, phosphorus forms predominantly P_4O_6, a colorless, waxy compound in which the six $P-P$ bonds of P_4 tetrahedra (Fig. 1) are replaced by $P-O-P$ bridges ($P-O$ bond distance; 162 pm). This compound is stable at 25°C in air, but it can be easily transformed into H_3PO_3 in the presence of water,

$$P_4O_6 + 6H_2O \rightarrow 4H_3PO_3. \tag{1}$$

P_4O_{10}, the dimeric form of P_2O_5, is a colorless compound consisting of four phosphorus tetrahedra bridged together through oxygen atoms (Fig. 2). Hydrolysis of P_4O_{10} leads to H_3PO_4:

$$P_4O_{10} \xrightarrow{+2H_2O} H_4P_4O_{12} \xrightarrow{+2H_2O} 2H_4P_2O_7 \xrightarrow{+2H_2O} 4H_3PO_4. \tag{2}$$

In addition to P_4O_6 and P_4O_{10}, which have phosphorus in the formal P^{III} and P^V oxidation states, other phosphorus oxide compounds, such as P_4O_7, P_4O_8, and P_4O_9, also exist. Their structures are shown in Fig. 2.

Furthermore, the reactivity of P_2O_5 allows the preparation of numerous organic phosphorus oxo-complexes with alcoholic solvents (10, 11). For example, with butanol (BuOH), the following reactions are suggested:

$$P_2O_5 + 2\,BuOH + H_2O \rightarrow 2\,PO(OH)_2(OBu) \tag{3}$$

$$P_2O_5 + 4BuOH \rightarrow 2\,PO(OH)(OBu)_2 + H_2O \tag{4}$$

$$P_2O_5 + 6\,BuOH \rightarrow 2\,PO(OBu)_3 + 3\,H_2O. \tag{5}$$

E. PHOSPHORUS OXYSULFIDES AND SULFIDES

Since hydrotreating catalysts are almost always treated under sulfiding conditions (i.e., in the presence of H_2S and organosulfur compounds), phosphorus sulfide or oxysulfide compounds may play an important role both during catalyst activation (reduction–sulfidation) starting from the catalyst in the oxide form and during catalysis. Many kinds of phosphorus sulfide compounds can be formed, e.g., P_4S_3, P_4S_4, P_4S_5, P_4S_7, P_4S_9, and P_4S_{10}, depending on the P/S ratio in the catalyst (7–10, 12). The structures of phosphorus sulfides are basically analogous to those of the phosphorus

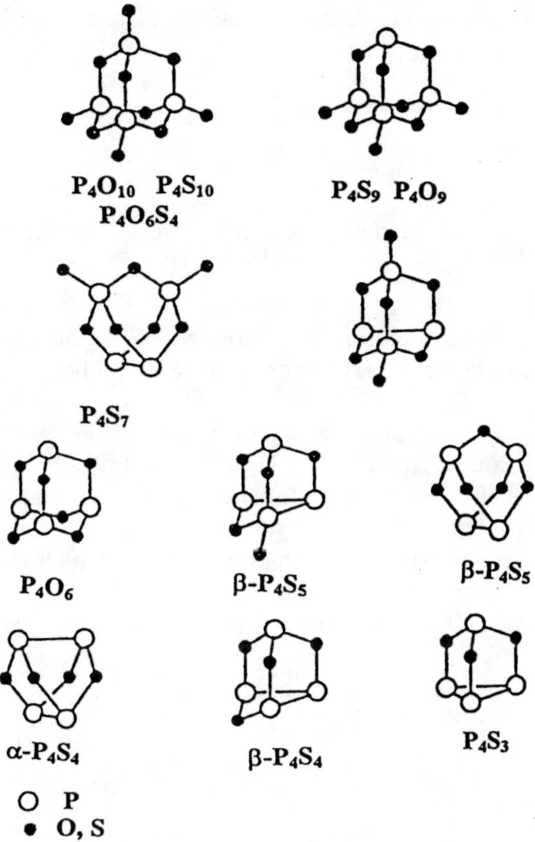

FIG. 2.   Structures of phosphorus oxides, sulfides, and oxysulfides [from King (8)].

oxides shown in Fig. 2. For example, P_4S_{10} has a structure similar to that of P_4O_{10} (P—S bond distance, 209 pm). The structure of P_4S_9 is also similar to that of P_4S_{10}, but one terminal S atom is lacking. These "P—S" compounds are thermally stable heterocycles, but they are chemically reactive; for example, they spontaneously ignite in air or they are hydrolyzed to give H_3PO_4 and H_2S in the presence of water.

F.   Oxy-Acids of Phosphorus and Their Derivatives

The oxy-acids of phosphorus and their derivatives are defined as compounds which contain both P—OH and P=O functions (7–9, 12). Orthophosphoric acid, phosphorus acid, and hypophosphorous acid are well-known as the oxy-acid compounds that contain three, two, and one OH groups associated with the phosphorus atom, respectively. Their schematic structures are as follows:

OH   OH   OH

$O{=}P{-}OH$ $O{=}P{-}H$ $O{=}P{-}H$

OH   OH   H

orthophosphoric acid   phosphorous acid   hypophosphorous acid
(H_3PO_4)   (H_3PO_3)   (H_3PO_2)

Since the $P{-}O{-}H$ group dissociates into $P{-}O^-$ and H^+ ions in the presence of water, these compounds are acidic. The acidity decreases with an increasing number of $P{-}H$ bonds because $P{-}H$ bonds do not dissociate in water.

Orthophosphoric acid is a moderately strong water-soluble triprotic acid (pK_a = 2.1, 7.2, and 12 for H_3PO_4, $H_2PO_4^-$, and HPO_4^{2-}, respectively). The magnitude of its Hammett acidity function (H_0) is less than those of H_2SO_4, HCl, or HNO_3 (Fig. 3) (13). Anhydrous H_3PO_4 is a colorless compound with a molecular structure built of PO_4 layers in which phosphorus has a tetrahedral symmetry.

Stepwise dehydration of phosphoric and phosphorous acids favors forma-

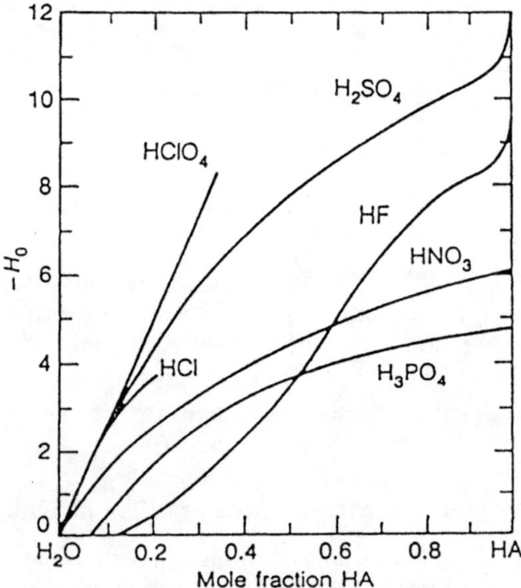

FIG. 3. Hammett acidity function H_0 of H_3PO_4 and those of other mineral acids [from Olah (13); copyright 1973 John Wiley & Sons Inc.].

tion of polymeric $P-O-P$ bonds. Condensation of phosphates then leads to several types of poly oxo-phosphorus compounds such as the straight-chain phosphates, cyclic metaphosphates, and cross-linking ultraphosphates.

Phosphorus thioxy acids and phosphorus thio acids, in which some or all of the oxygen atoms around the phosphorus atom in the oxy-acids are replaced by sulfur atoms, respectively, are also known (8). Physicochemical properties of selected phosphorus compounds described in Section II are shown in Table III.

G. Aluminum Orthophosphates and the $Al_2O_3-P_2O_5-H_2O$ Ternary System

The phosphorus contents of hydrotreating catalysts are typically <10 wt%, with the remaining components including the catalytic components (molybdenum and promotors) and the aluminum oxide support. However, aluminum orthophosphates, which contain ~25.4 wt% phosphorus from the formulation $AlPO_4$, are commonly reported in hydrotreating catalysts. Therefore, a detailed examination of the aluminum orthophosphate struc-

TABLE III

Physical Constants of Phosphorus Compounds[a]

Compound	Density (g/cm³)	Melting point (°C)	Boiling point (°C)	Molecular weight
P_4				
White	1.82	44.1	280	123.89
Red	2.34	590		
Violet	2.36	590		
Black	2.7			
PH_3		−133	−87.7	34
$P_2O_5(P_4O_{10})$	2.39	580		141.94
$P_2O_4(P_4O_8)$	2.54	>100	180 (vac.)	125.95
$P_2O_3(P_4O_6)$	2.16	23.8	175.4	109.95
$P_4O_6S_4$		102	295	348.15
$P_2S_5 (P_4S_{10})$	2.09	290	514	444.54
P_4S_7	2.19	310	523	348.34
P_4S_3	2.03	174	408	222.09
H_3PO_2 (hypophosphorous acid)	1.49	26.5	130 (d)	66
H_3PO_3 (orthophosphorous acid)	1.65	73.6	200 (d)	82
$H_4P_2O_3$ (pyrophosphorous acid)		38	120 (d)	145.98
H_3PO_4 (orthophosphoric acid)	1.7	40	213	98

Note: vac., under vacuum; d, decomposition before boiling.
[a] Adapted from Wells (7).

tures is given here. Crystalline aluminum orthophosphates ($AlPO_4$) have structures analogous to those of all the known SiO_2 modifications, e.g., quartz, crystobalite, and trydimite. The aluminum phosphates of the $Al_2O_3-P_2O_5-H_2O$ system with different $Al-P$ compositions are schematically classified into basic ($Al/P > 1$), neutral ($Al/P = 1$), and acidic phases ($Al/P < 1$). Table IV is a list of crystalline aluminum phosphates (14).

The structures of basic aluminum phosphates, such as trolleite, augelite, senegalite, and wavellite, are shown in Fig. 4. Trolleite and augelite have dimeric and tetrameric units, respectively. Senegalite and wavellite have chain-like structure units. The structures of vashegyite and kingite are not defined, although they might have layered structures. Bolivarite and evansite have been found to be amorphous by X-ray diffraction (XRD).

TABLE IV

Orthophosphates in the Ternary System $Al_2O_3-P_2O_5-H_2O^a$

No.	Composition	Mineral name/ polymorphism	Crystal system
	Basic ($Al/P > 1$)		
1	$Al_4(OH)_3(PO_4)_3$	Trolleite	Monoclinic
2	$Al_2(OH)_3(PO_4)$	Augelite	Monoclinic
3	$Al_2(OH)_3(PO_4) \cdot H_2O$	Senegalite	Orthorhombic
4	$Al_3(OH)_3(PO_4)_2 \cdot 5H_2O$	Wavellite	Orthorhombic
5	$Al_4(OH)_3(PO_4)_3 \cdot 11H_2O$	Vashegyite	Orthorhombic
6	$Al_3(OH)_3(PO_4)_2 \cdot 9H_2O$	Kingite	Triclinic
7	$Al_2(OH)_3(PO_4) \cdot 4.75H_2O$	Bolivarite	(Amorphous to X-ray)
8	$Al_3(OH)_6(PO_4) \cdot 6H_2O$	Evansite	(Amorphous to X-ray, opalescent glass)
	Neutral ($Al/P = 1$)		
9	$AlPO_4 \cdot 2H_2O$	Variscite–lucine	Orthorhombic
		Variscite–messbach	Orthorhombic
		Metavariscite	Monoclinic
	Acid ($Al/P < 1$)		
10	$Al(H_2PO_4)_3$	A	Trigonal
		B	
		C	Rhombohedral
11	$Al(H_2PO_4)(HPO_4) \cdot H_2O$		Monoclinic
12	$H_3O[Al_3(H_2PO_4)_6(HPO_4)_2] \cdot 4H_2O$		Monoclinic
13	$Al_2(HPO_4)_3 \cdot 3.5H_2O$		
14	$Al_2(HPO_4)_3 \cdot 4H_2O$		
15	$Al(H_2PO_4)(HPO_4) \cdot 2.5H_2O$		
16	$Al_2(H_{2-x}PO_4)_3(H_{3x}PO_4] \cdot 6H_2O$	$(0 \leq x \leq 1)$	Trigonal
17	$Al_2(HPO_4)_3 \cdot 6.5H_2O$		
18	$Al_2(HPO_4)_3 \cdot 8H_2O$		

^{a} Adapted from Kniep (14); reprinted with permission, copyright 1986 Wiley–VCH.

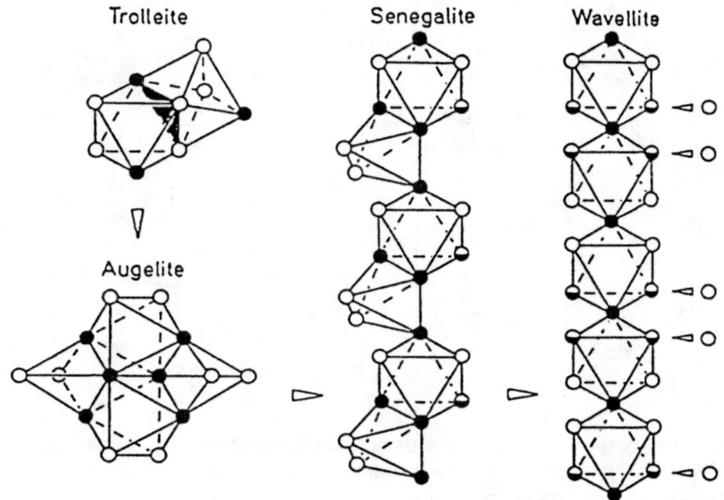

FIG. 4. Structures of basic aluminum orthophosphates. The schemes show only the positions of the oxygen atoms: ●, oxygen atoms of hydroxide ions (O_{OH}); ○, oxygen atoms of phosphate ions (O_p); ◐, oxygen atoms of H_2O molecules (O_{H_2O}) [reprinted with permission from Kniep (*14*); copyright 1986 Wiley–VCH].

Neutral aluminum phosphate compounds such as variscite–lucine and metavariscite have three-dimensional structures built up of $Al(O_p)_4(O_{H_2O})_2$ octahedra and PO_4 tetrahedra (Fig. 5) (where O_p and O_{H_2O} refer to oxygen atoms belonging to phosphate and water, respectively). The structure of variscite–messbach has not been determined.

Five crystal structures have been found for acidic aluminum phosphates. Among them, $Al(H_2PO_4)_3$ — A and $Al(H_2PO_4)_3$ — C are built of $Al(O_p)_6$ octahedra and $PO_2(O_H)_2$ tetrahedra with one-dimensional and three-dimensional linkages, respectively, as shown in Fig. 6 (where O_H refers to oxygen atoms belonging to P–OH groups). $Al_2(H_{2-x}PO_4)_3(H_{3x}PO_4) \cdot 6H_2O$ contains two different units stacked in the direction of the "*c*" crystallographic axis. The first are isolated PO_4 tetrahedra and the second are built of two $Al(O_p)_3(O_{H_2O})_3$ octahedra bound with three PO_4 tetrahedra, as shown in Fig. 7.

$Al(H_2PO_4)(HPO_4) \cdot H_2O$ has a characteristic layered structure which contains isolated $Al(O_p)_5(O_{H_2O})$ octahedra linked by $PO_2(O_H)_2$ and $PO_3(O_H)$ tetrahedra (Fig. 8a). $H_3O[Al_3(H_2PO_4)_6(HPO_4)_2] \cdot 4H_2O$ is a microporous material comparable to some zeolites, with H_3O^+ ionic sites located inside the micropores (Fig. 8b).

Figure 9 presents an estimation of the thermal stability of the crystalline

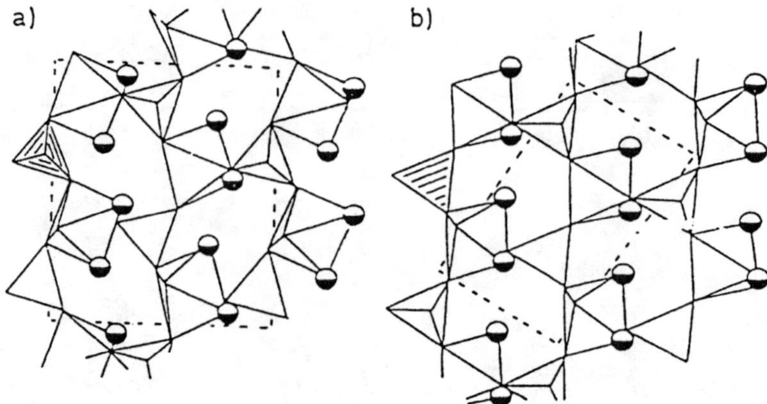

FIG. 5. Parts of the crystal structures of neutral aluminum phosphates. (a) Variscite–lucine (VL); (b) metavariscite (polyhedral representation). The shaded tetrahedra serve to emphasize the difference in the otherwise identical connectivities [reprinted with permission from Kniep (*14*); copyright 1986 Wiley–VCH].

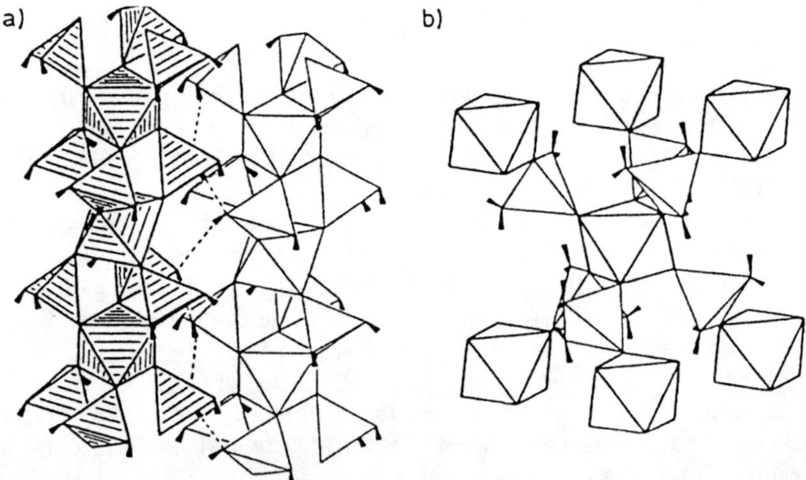

FIG. 6. Parts of the crystal structures of acidic aluminum orthophosphates (polyhedral representation). (a) $Al(H_2PO_4)_3$–C; (b) $Al(H_2PO_4)_3$–A. One structural unit of $Al(H_2PO_4)_3$–C is indicated by shading. The hydrogen atom positions are shown by wedges [reprinted with permission from Kniep (*14*); copyright 1986 Wiley–VCH].

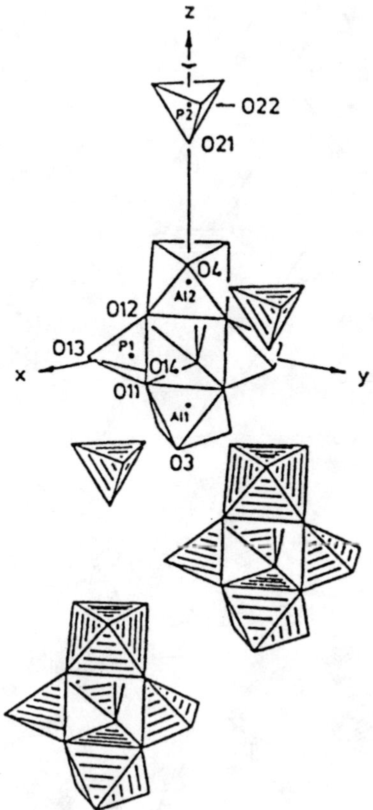

FIG. 7. Parts of the crystal structures of acidic aluminum orthophosphates: $Al_2(H_{2-x}PO_4)_3(H_{3x}PO_4) \cdot 6H_2O$ (polyhedral representation) [reprinted with permission from Kniep (*14*); copyright 1986 Wiley–VCH].

aluminium phosphates listed in Table IV. The aluminum phosphate stability depends predominantly both on the water content and on the nature of the linkages between the polyhedra. Thermal treatments of hydrated aluminum phosphates convert them generally to lower hydrated phases. The highly hydrated phase, such as $Al_2(H_{2-x}PO_4)_3(H_{3x}PO_4) \cdot 6H_2O$, decomposes at 70°C, whereas the least hydrated form of trolleite is stable at temperatures up to 650°C. Note also that the completely dehydrated $AlPO_4$ phase is stable at temperatures up to 1800°C. When this phase is present in the catalysts, it will not decompose under the conventional calcination conditions used to prepare the oxide form starting from the precursors.

In addition, some crystalline compounds called $AlPO_4 - N$ (where $N =$

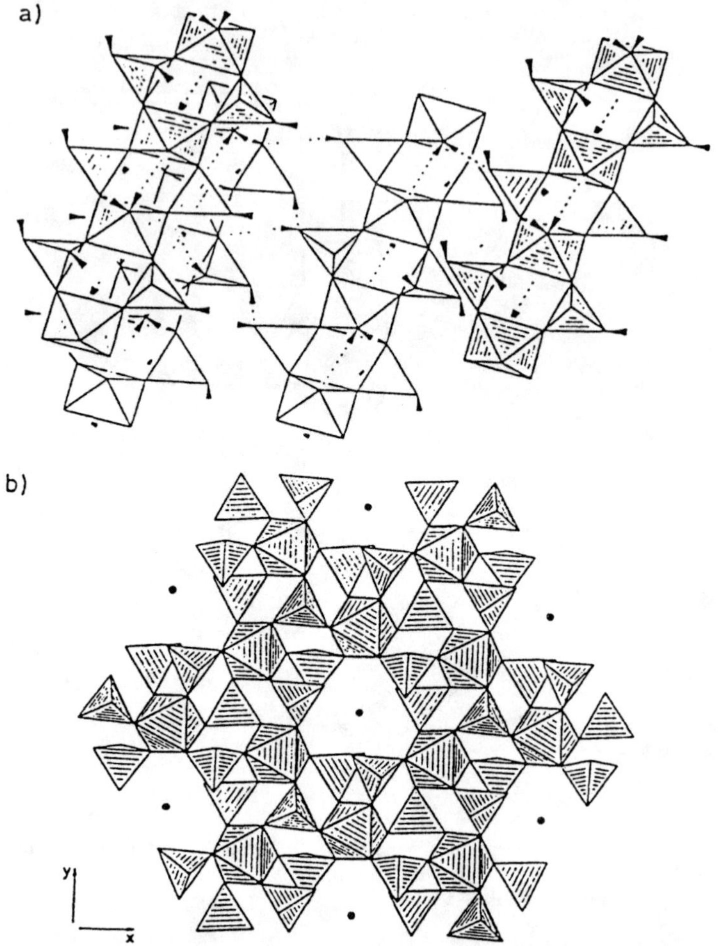

FIG. 8.   Parts of the crystal structures of acidic aluminum orthophosphates (polyhedral representation). (a) $Al(H_2PO_4)(HPO_4) \cdot H_2O$; polyhedron linkages directed upwards in this projection are indicated by shading. The hydrogen atom positions are shown by wedges. Inter- and intramolecular hydrogen bonds are shown by broken lines. (b) $H_3O[Al_3(H_2PO_4)_6(HPO_4)_2] \cdot 4H_2O$; the solid circles represent the H_3O^+ sites [reprinted with permission from Kniep (*14*); copyright 1986 Wiley–VCH].

5, 8, 11, 14, or higher values) which have a zeolite-like structure have already been described (*15, 16*). Another solid, VPI-5, is a crystalline aluminophosphate which has larger pores consisting of 18-membered rings (*16*).

Amorphous aluminum orthophosphate with particular local surface structures could also be present in hydrotreating catalysts. It has been

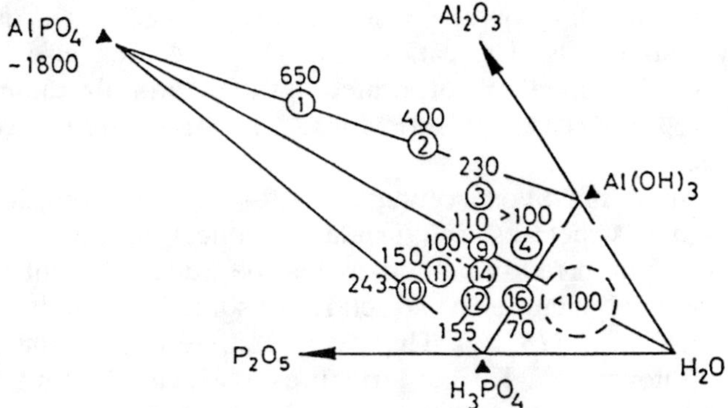

Fig. 9. Thermal stability of some aluminum orthophosphates. The phases are numbered in the different circles according to Table IV, and the other numbers are temperatures (in °C) [reprinted with permission from Kniep (*14*); copyright 1986 Wiley–VCH].

determined (*17*) that amorphous $AlPO_4$ contains four-, five-, and sixfold-coordinated aluminum ions at its surface. The presence of these amorphous phosphates may have an impact on the catalytic properties (e.g., surface acidity).

H. "Mo−P" and "W−P" Heteropoly Compounds

Association of molybdates (tungstates) with phosphate-like structures leads to a class of compounds called heteropoly compounds or heteropoly acids of Mo or W, hereafter schematically written as "Mo−P" or "W−P" heteropoly compounds. The heteropoly anions, which may contain Mo, W, P, and other elements, are paired, both in the solid state and in solution, with cations such as H^+, NH_4^+, and Na^+. Here, only some "Mo−P" heteropoly compounds playing an important role in hydrotreating catalysts during the catalyst preparation (impregnation) or as supported oxidic phases are discussed. The "Mo−P" heteropoly compounds have the following properties (*18, 19*):

1. High solubility in water or polarized solvents such as alcohols, ethers, and ketones. However, larger countercations, such as K^+, Rb^+, Cs^+, and NH_4^+, often decrease their solubility;
2. High stability in acid solutions;
3. Stronger acidity than molybdic acid;
4. Important redox properties.

The phosphomolybdates ("Mo−P" heteropoly compounds) are basically built of PO_4 tetrahedra and MoO_6 octahedra. Several structures can be obtained depending on the P/Mo ratios. Table V is a list of such heteropoly compounds. The cations Co, Ni, or Al can also be used as counter-ions or be directly incorporated in the heteropoly anion structure. $Co_3[PMo_{12}O_{40}]_2 \cdot 34H_2O$ and $Ni_3[PMo_{12}O_{40}]_2 \cdot 34H_2O$ are two compounds of this class.

The corresponding phosphotungstate "P–W" heteropoly compounds have analogous structures and similar chemical properties, but their thermal stability and redox potentials may be quite different from those of the phosphomolybdates. The structures of some heteropoly compounds are shown in Fig. 10 (18, 20). The "$Mo_{12}P$" heteropoly compound (Fig. 10a) is well-known as a Keggin structure in which 12 MoO_6 octahedra surround a central PO_4 tetrahedron. The four oxygen atoms of the central PO_4 are shared with three MoO_6 octahedra, and four of six oxygen atoms in the MoO_6 octahedra are shared with the other MoO_6. Then only one oxygen atom in each MoO_6 unit is individually bound to the molybdenum atom. The structure of the "$Mo_{10}P$" heteropoly anion is unknown, but it is considered to be dimeric. The "$Mo_{11}P$" heteropoly compound (Fig. 10b) is more stable than "$Mo_{12}P$" during reduction treatments, and the "Mo_9P" heteropoly anion (Fig. 10e) is less stable during reduction than is "$Mo_{12}P$." The "$Mo_{18}P_2$" hetero-poly anion (Fig. 10c) has a dimeric structure with two central PO_4 tetrahedra surrounded by 18 MoO_6 octahedra. Two different hydration forms are known for "$Mo_{18}P_2$," e.g., $H_6[P_2Mo_{18}O_{62}] \cdot 33H_2O$ and $H_6[P_2Mo_{18}O_{62}] \cdot 37H_2O$; the latter is unstable at ambient temperature. The precise structures of "Mo_6P" and "$Mo_{17}P_2$" are also unknown. The structure

TABLE V

Chemical Formulations of Some Mo–P
Heteropoly Anions[a]

Type	P:Mo ratio	Typical anions[b]
"$Mo_{12}P$"	1:12	$[PMo_{12}O_{40}]^{3-}$
"$Mo_{11}P$"	1:11	$[PMo_{11}O_{39}]^{7-}$
"$Mo_{10}P$"	1:10	$[PMo_{10}O_{35}]^{5-}$
"$Mo_{18}P_2$"	1:9	$[P_2Mo_{18}O_{62}]^{6-}$
"Mo_9P"	1:9	$[PMo_9O_{31}(HO)_3]^{6-}$
"Mo_6P"	1:6	$[P^{III}Mo_6O_{24}H_6]^{3-}$
"$Mo_{17}P_2$"	2:17	$[P_2Mo_{17}O_{60}]^{8-}$
"Mo_5P_2"	2:5	$[P_2Mo_5O_{23}]^{6-}$

[a] Adapted from Tsigdinos (18). [b] Formal oxidation states of P and Mo are P^V and Mo^{VI}, respectively, unless otherwise stated.

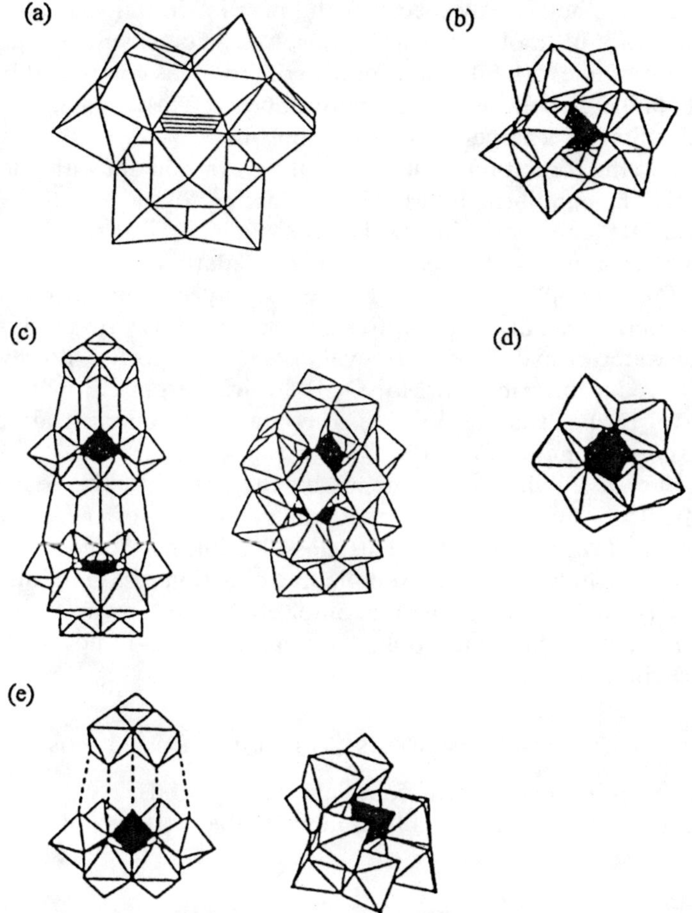

FIG. 10. Structures of some "Mo–P" heteropolyanions. a, "Mo$_{12}$P"; b, "Mo$_{11}$P"; c, "Mo$_{18}$P$_2$"; d, "Mo$_5$P$_2$"; e, "Mo$_9$P." Shadowed and clear polyhedra indicate PO$_4$ tetrahedra and MoO$_6$ octahedra, respectively [adapted from Tsigdinos (*18*) and Pettersson *et al.* (*20*); reprinted with permission from Pettersson *et al.* (*20*); copyright 1986 American Chemical Society].

of "Mo$_5$P$_2$" (Fig. 10d) is interesting because the phosphorus atoms in the "Mo$_5$P$_2$" structure are exposed to the outer surface of the ion, whereas those in "Mo$_{12}$P" and "Mo$_{18}$P$_2$" are surrounded by the molybdenum octahedra.

Alkaline treatments transform these "Mo–P" heteropoly compounds into more stable heteropoly species with lower P/Mo ratios, finally leading to simple "MoO$_4$" molybdate and phosphate. For example, the overall degradation of one "Mo$_{12}$P" heteropoly compound is completed by 20–28 moles of NaOH. In contrast, the transformation can also proceed as a result of the addition of a large amount of phosphoric acid. In this case, it is attributed to the formation of new central PO$_4$ tetrahedra with addition of phosphate. The equilibria of "Mo–P"-containing solutions in the presence of H$^+$ and OH$^-$ are shown in Fig. 11 (21).

Thermal treatments also lead to the degradation of heteropoly compounds. For example, the "Mo$_{12}$P" heteropoly compound decomposes into P$_2$O$_5$ and MoO$_3$ at temperatures exceeding 450°C with gradual elimination of water of hydration. Griboval *et al.* (22) reported, however, that the partly reduced form of [PMo$_{12}$O$_{40}$]$^{3-}$ which permits formation of the Co$_{7/2}$[PMo$_{12}$O$_{40}$] salt is more stable in contact with alumina after drying than the corresponding Co$_{3/2}$[PMo$_{12}$O$_{40}$] compound.

The order of stability also depends on the nature of the central atom (Si > Zr, Ti > Ge > P > As) and on the nature of the surrounding anion groups (W > Mo > V). This classification has some relevance to hydrotreating catalysts; if, for example, introduction of Si into alumina as a support (giving silica–alumina, zeolite, etc.) leads to the formation of "Mo—Si" or "W—Si" heteropoly compounds, they will not be degraded during calcination.

I. OTHER PHOSPHORUS-CONTAINING COMPOUNDS

Other types of phosphorus-containing compounds, e.g., organophosphates, might be present in small amounts on the surfaces of hydrotreating

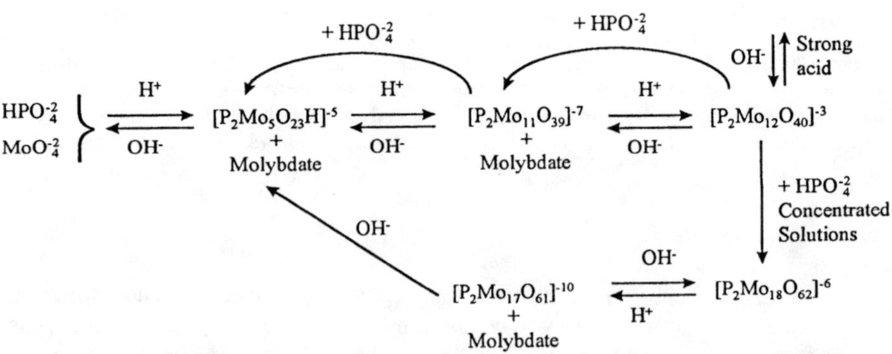

FIG. 11. Equilibria in molybdate–phosphate solutions in solutions of H$^+$ and OH$^-$ [from Souchay (21)].

catalysts under working conditions. Some of these compounds are represented as follows (23):

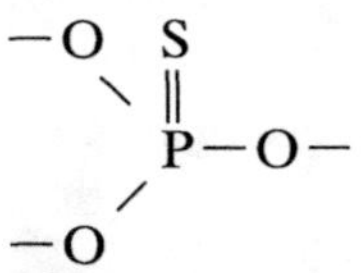

phosphorothioates
(phosphorothionates)

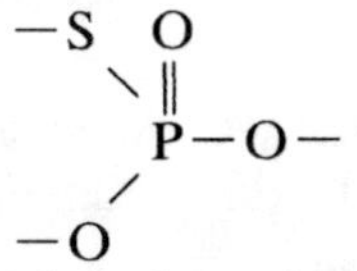

phosphorothiolates

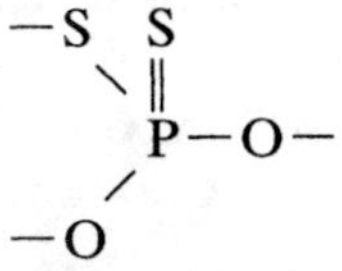

phosphorodithioates
(phosphorothionothiolates)

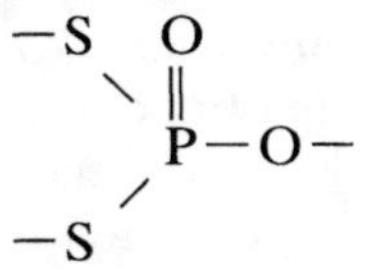

phosphorodithioates

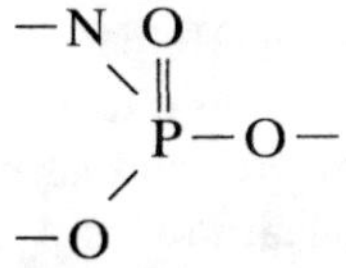

phosphoramides
(phosphoramidates)

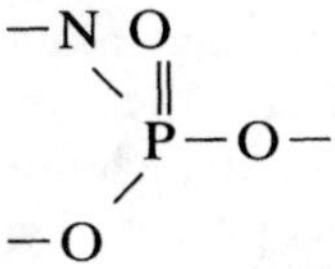

phosphorodiamidates

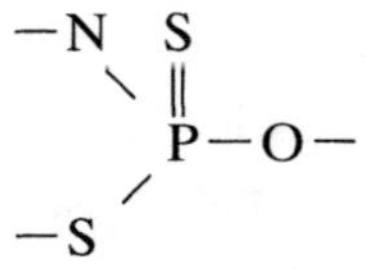

phosphoramidothionates

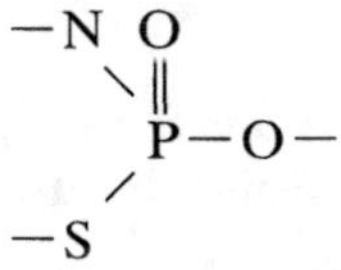

phosphoramidothiolates

It is possible that these compounds participate in hydrotreating reactions as intermediates or as new active sites. For example, Thompson et al. (24) reported that a sulfur radical is predominantly lost from methamidophos (Fig. 12). If a sulfur- (or a nitrogen-) containing compound further reacts

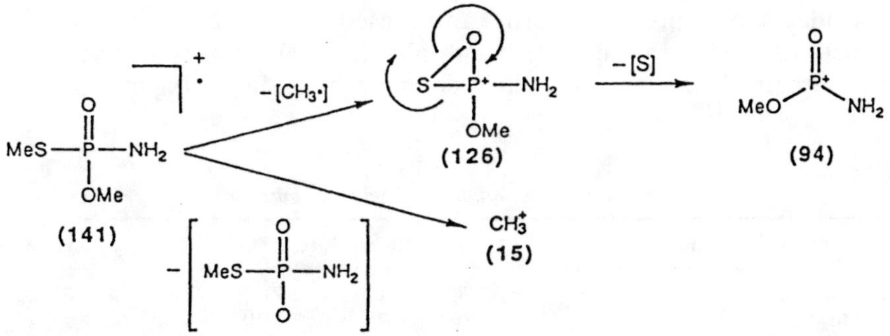

MeS—P(=O)—NH$_2$ / OMe (141) → −[CH$_3$·] → S—P$^+$—NH$_2$ / OMe (126) → −[S] → MeO—P$^+$(=O)—NH$_2$ (94); −[MeS—P(=O)(=O)—NH$_2$] → CH$_3^+$ (15)

FIG. 12. Schematic representation of CH_3^+ and S radical elimination from the "methamidophos" compound. The numbers in parentheses indicate the masses of the different ions, which permitted their identification [reprinted with permission from Thompson et al. (24); copyright 1990 American Chemical Society].

with the methamidophos ion formed after sulfur abstraction, a catalytic desulfurization (or denitrogenation) cycle could result. However, no evidence of such intermediates is known.

J. Vibrational and NMR Data Characterizing Phosphorus-Containing Reference Compounds

To identify the phosphorus-containing compounds described in the previous sections and the related species containing aluminum, molybdenum, cobalt, or nickel which might be present in hydrotreating catalysts, it is convenient to use techniques such as NMR, IR, UV, and Raman spectroscopies and XRD. XRD is useful for characterizing crystalline bulk compounds, and other techniques are appropriate for well-dispersed species and amorphous phases. Typical IR, Raman, and NMR data presented in Tables VI, VII, and VIII, respectively, could be the basis for such identifications.

III. Preparation of Alumina-Based Hydrotreating Catalysts Containing Phosphorus, Molybdenum, and Cobalt or Nickel

The objectives for optimizing the preparation of hydrotreating catalysts are to mix the selected ingredients (phosphorus-, molybdenum-, cobalt-, or nickel-containing compounds and alumina) to form solids that meet specific requirements, such as high metal dispersion, high metal loading, high specific surface area, and low incorporation of some elements (phosphorus, cobalt, or nickel) into the alumina framework. The alumina support can be used as a powder or as a preformed material such as pellets or extrudates, or a precursor form can be used (e.g., aluminium hydroxide, which transforms by calcination into alumina). The aim of this section is to present schematically the most common procedures. Figure 13 shows

TABLE VI

IR Data Characterizing Reference Compounds

Compound	IR bands (cm^{-1})	Reference
AlPO$_4$	3795, 3680, 1130, 735, 715	*25, 26*
"PMo$_{12}$O$_{40}$"	1070, 965, 875, 790, 590, 485	*27, 28*
"PMo$_{11}$O$_{39}$	1110, 1060, 930, 900, 860, 790, 742	*19*
NiMoP heteropoly compound	815, 730	*29*
Aluminum nickel phosphates	1119, 930, 490	*30*
"AlMo$_6$" heteropoly compound	945, 920, 890, 665, 445	*29, 30*

TABLE VII

Raman Data Characterizing Reference Compounds

Band positions (cm^{-1})	Assignment	Reference
990, 820, 670	Bulk MoO_3	31
400–550	Bulk NiO	32
690	Bulk CoO	31
1006, 822, 380	$Al_2(MoO_4)_3$	33
200, 375, 600	$NiAl_2O_4$	32
938, 820, 700	α-$CoMoO_4$	32
952, 946, 938, 870, 820	β-$CoMoO_4$	32
965, 715	$NiMoO_4$	34
1424, 1390, 724, 560, 425	P_2O_5	32
935	P_2Mo_5	31, 25
950	PMo_7 or heptamolybdate	35
963	PMo_9 or PMo_{11}	31, 25
995	PMo_{12}	31
965	$Mo_8O_{26}^{4-}$ anion in aqueous solution	33
218, 359, 895, 938	$Mo_7O_{24}^{6-}$ anion in aqueous solution	33
318, 846	MoO_4^{2-} anion in aqueous solution	33
360, 1058	Polycrystalline boehmite	36

the general preparation procedures used for alumina-based hydrotreating catalysts with phosphorus, molybdenum, and cobalt (or nickel). Each element can be introduced alone or with other elements at any step, and the order of element introduction can be permuted. Between each step, drying or calcination may be carried out. The various preparation methods described in Fig. 13 can be combined to give further variations in the preparation procedures.

TABLE VIII

^{31}*P-NMR Chemical Shift Data Characterizing Phosphorus-Containing Reference Compounds*

Compound	Chemical shift (ppm)	Reference
"$PMo_{12}O_{40}$"	−3.9 to −4.5	19
"PMo_9O_{34}"	−4.1	19
"$P_2Mo_{18}O_{62}$"	−3.1	19
"$P_2Mo_5O_{23}$"	2.2	25
Hydrated $AlPO_4$	−19.2	36
$AlPO_4$	−25 to −30	37
Crystalline $AlPO_4$	−32	38
$AlPO_4$-T	−29.5	39
$AlPO_4$-C	−27.1	39
H_3PO_4	0	—

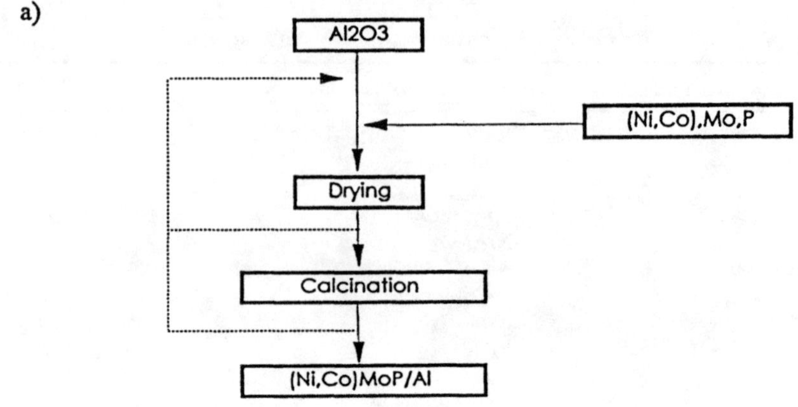

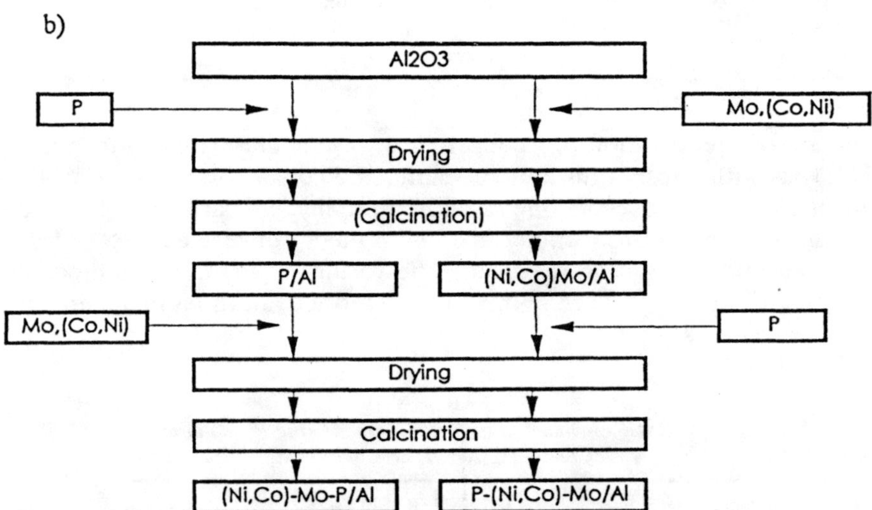

FIG. 13. Procedures for the preparation of alumina-based hydrotreating catalysts containing phosphorus, molybdenum, and cobalt or nickel. a, Impregnation or equilibrium adsorption method (coimpregnation); b, impregnation or equilibrium adsorption method (sequential impregnation); c, precipitation or hydrogel method; d, sol-gel method [adapted from Iwamoto and Grimblot (40)].

In the various preparations, ammonium heptamolybdate, cobalt nitrate, or nickel nitrate are preferred. The preferred phosphorus-containing precursors are phosphorus pentoxide, phosphoric acid, or their anion derivatives such as ammonium dihydrogen phosphate ($NH_4H_2PO_4$) be-c)

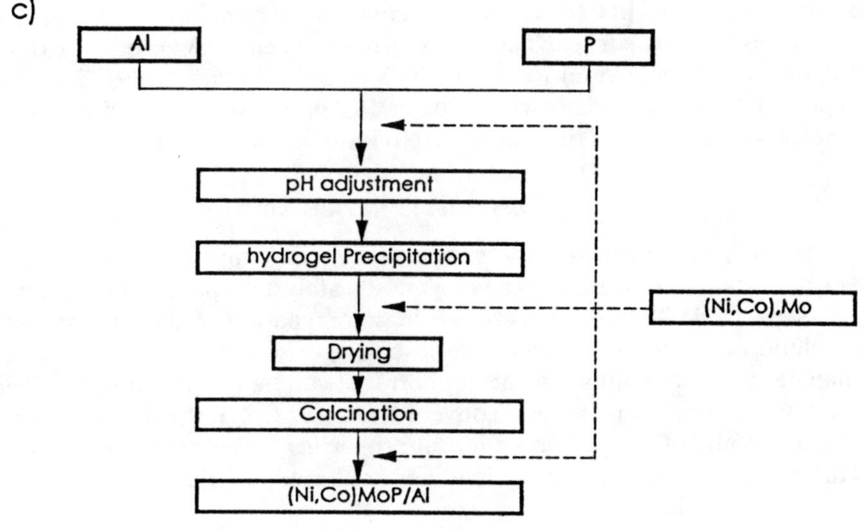

d)

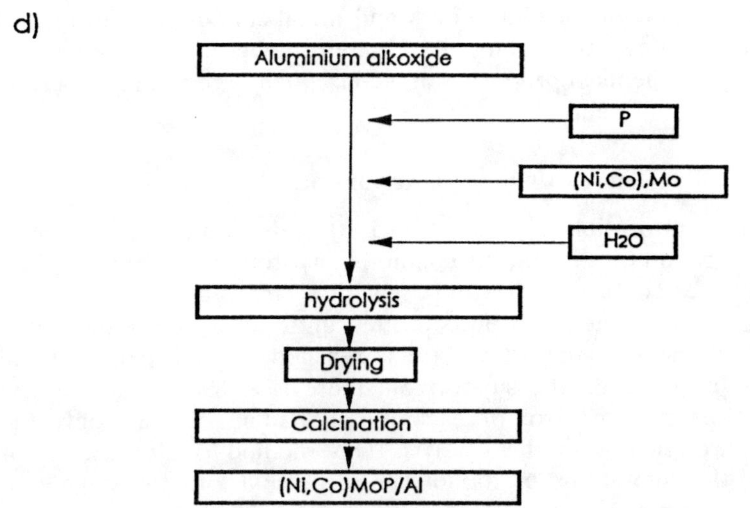

Fig. 13. (*continued*)

cause they are highly water soluble and only phosphorus oxo-species remain after their calcination. $NH_4H_2PO_4$ is a phosphorus source that is reported to react less strongly with the alumina support than H_3PO_4 (*41*).

Some research groups have used heteropoly compounds as "Mo–P" precursors for preparation of hydrotreating catalysts (*22, 25, 42*). As described in Section II.H, the heteropoly compounds have intrinsically unique structures and properties. They have already been recognized to exhibit high activities for several reactions, such as mild oxidation (*19*). They are reported to be potentially valuable for the preparation of hydrotreating catalysts with high activities and selectivities.

A. THE DRY IMPREGNATION METHOD

The incipient wetness impregnation (or "dry" impregnation method) (Figs. 13a and 13b) is one of the most used catalyst preparation techniques (also see Table XIII). It consists of the impregnation of the support with a solution containing the elements to be deposited, the volume of which matches the pore volume of the support that can be filled with water. This method is easy to perform and convenient for control of the metal loadings and also some of the physicochemical properties of the resulting catalysts. However, the maximum metal loadings are limited by the water solubility of the metal salts.

The impregnation method is further classified into coimpregnation and sequential impregnation. When this impregnation method is used, the sequence of introduction of phosphorus and metal-containing species and the pH of the impregnating solutions are important because they affect the resulting physicochemical properties and consequently the catalyst performance (Sections V.D and VI).

B. THE EQUILIBRIUM ADSORPTION METHOD

Equilibrium adsorption (Figs. 13a and 13b) is also a method in general use for preparation of phosphorus-containing hydrotreating catalysts. The method involves contacting the support with a large volume of solution containing the components to be deposited until adsorption equilibrium between ions and the support surface is reached. Well-dispersed metal species are prepared on the support but there is a disadvantage in the lack of independent control of each metal loading because only the adsorbed part remains in the catalyst. This method is also useful for elucidating the limitations of monolayer dispersion and the degree of interaction between phosphorus, the other metals, and the support. As with the incipient wetness method, sequential treatments can be performed.

C. THE PRECIPITATION OR HYDROGEL METHOD

In the precipitation or hydrogel method (Fig. 13c), the metal components and the phosphorus-containing species are directly introduced into a support precursor such as an alumina hydrogel. This method enables one to obtain catalysts with high loadings of metal (molybdenum, cobalt, or nickel) with controlled amounts of phosphorus and also with high specific surface areas, but the metal dispersion is not always optimized. It is also often difficult to control the P—Mo, Co—Mo, or Ni—Mo interactions. Moreover, there may be difficulties in control of the pore structure and mechanical strength of the resulting catalysts.

D.   THE SOL-GEL METHOD

In the sol-gel method (*40*), the phosphorus compound is introduced simultaneously with the other active metal components before or during hydrolysis of an aluminum alkoxide. This method gives catalysts with high molybdenum loadings (>30 wt% Mo, equivalent to $\geq$ 45 wt% MoO_3) with good dispersion and extremely high surface areas (300–600 m^2/g). This method is also useful for preparing samples allowing identification of surface species because the solids have high surface areas and metal loadings. Because the alkoxide precursors are expensive, however, the sol-gel preparation method may be too expensive to be practical in comparison with the other methods described previously.

IV.   Adsorption of Phosphorus-Containing Compounds on Alumina

To understand the complex interaction of phosphorus-containing species with alumina and with the other elements (molybdenum, cobalt, or nickel) introduced in the catalyst formulations, it is important to examine the chemistry involved in the preparation steps. In this section, adsorption of phosphorus-containing species on alumina is discussed.

A.   ADSORPTION OF PHOSPHORUS OXO-SPECIES ON ALUMINA

Phosphorus oxo-species adsorbed on alumina are present in a well-dispersed state up to a surface density of 2.9 $\times$ 10^{-6} P atom/pm^2 (or 2.9 P atom/nm^2) (*31*). IR spectroscopy measurements allowed identification of several types of hydroxyl groups on the γ-alumina surface, such as type Ia (tetrahedral Al—OH, 3780 cm^{-1}), type Ib (octahedral Al—OH, 3795 cm^{-1}), types IIa and IIb (bridged OH between two Al atoms, 3736 cm^{-1}), and type III (bridged OH between three Al atoms, 3697 cm^{-1}) (Fig. 14) (*43*).

At low phosphorus contents, the phosphorus oxo-species interact with both basic and acidic Al—OH groups, whereas the molybdenum oxo-species interact predominantly with basic Al—OH sites. In general, the basic type Ib Al—OH groups should be more reactive with the phosphorus oxo-species than the acidic groups. However, Lewis and Kydd (*44*) found from an IR spectroscopy study that the surface density of basic type Ib Al—OH groups increases with addition of phosphorus up to a value of 1×10^{-6} P atom/pm^2 (which corresponds to 10×10^{13} adsorbed H_3PO_4 molecules/cm^2; Fig. 15a). The authors suggested that Al—OH groups of types II and III react with H_3PO_4 and lead to the formation of a new Al—OH Ib group according to the following reaction:

$$\begin{array}{c} H \\ | \\ Al-O-Al + H_3PO_4 + H_2O \rightarrow Al-OH + Al-\overset{\displaystyle O}{\overset{\|}{O}}-P-(OH)_2 + (H_3O^+). \\ \text{(type II)} \qquad\qquad\qquad \text{(type I)} \end{array}$$

$$(6)$$

In addition, new P—OH groups identified by a peak at 3676 cm^{-1} are generated by loading the alumina support with phosphorus to a surface density of 5×10^{-6} P atom/pm^2 (which corresponds to 50×10^{13} adsorbed H_3PO_4 molecules/cm^2; Fig. 15e). At high phosphorus loadings, the interaction between neighboring P—OH groups induces the formation of polymeric phosphorus oxo compounds, and therefore the intensity of the P—OH band tends to decrease (*44, 45*).

Figure 16 shows schematically the different modes of adsorption occurring during the interaction of phosphoric acid with alumina (*46, 47*). If H_3PO_4 interacts with one surface Al—OH site through a single bond, one surface Al—OH will be replaced by two P—OH groups (Fig. 16a). If two H_3PO_4 molecules interact with two surface Al—OH sites close enough to form P—O—P bridges between the anchored H_3PO_4 molecules, one P—OH group will be generated by one surface Al—OH site (Fig. 16b). However, if H_3PO_4 interacts with the alumina surface through two or three Al—OH groups, it could result in the partial or total loss of hydroxyl groups (Figs. 16c or 16d). IR and NMR data show that H_3PO_4 interacts with several basic Al—OH groups on alumina at low phosphorus contents, whereas single bonding is predominant at high phosphorus contents (*48, 49*).

Different mechanisms of H_3PO_4 adsorption on alumina are proposed in the literature:

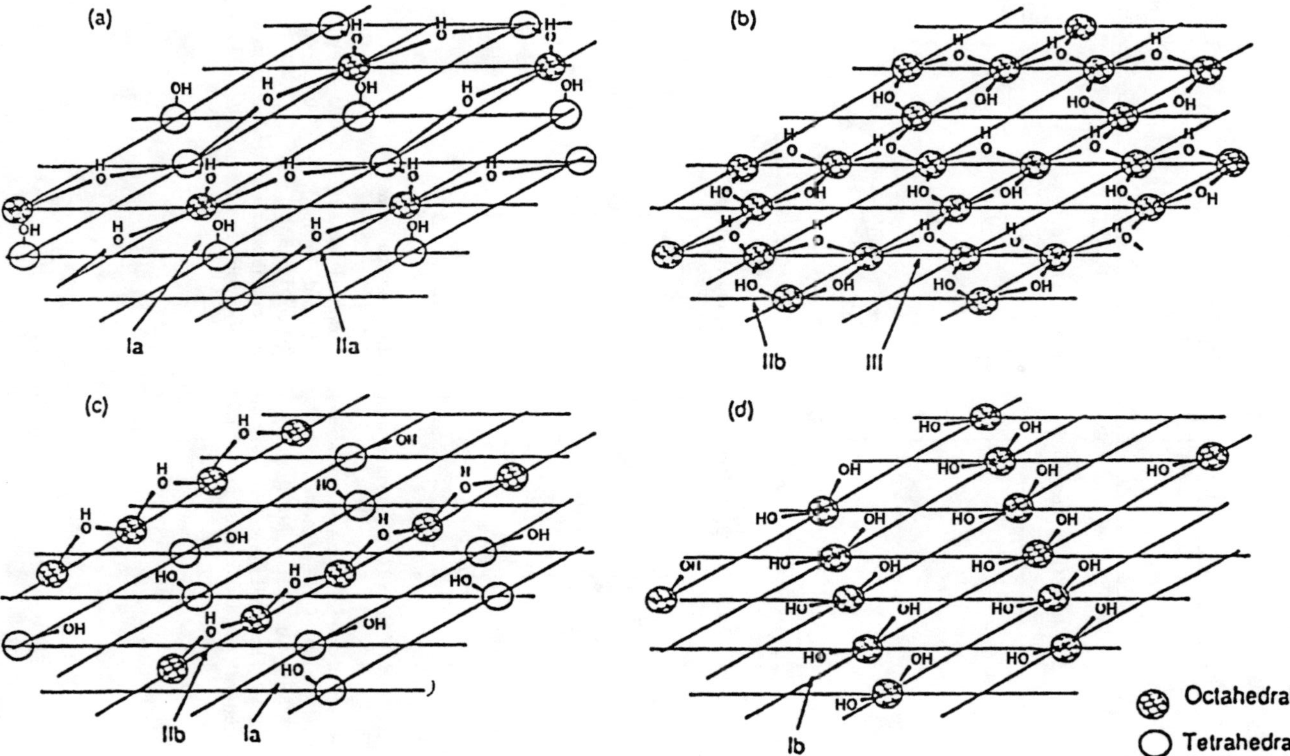

FIG. 14.   Idealized low Miller index surface planes of γ-alumina and nature of the different hydroxyl groups. (a) A layer, parallel to the (111) plane; (b) B layer, parallel to the (111) plane; (c) C layer, parallel to the (110) plane; (d) D layer, parallel to the (110) plane [reprinted with permission from Lewis and Kydd (44); copyright 1991 Academic Press].

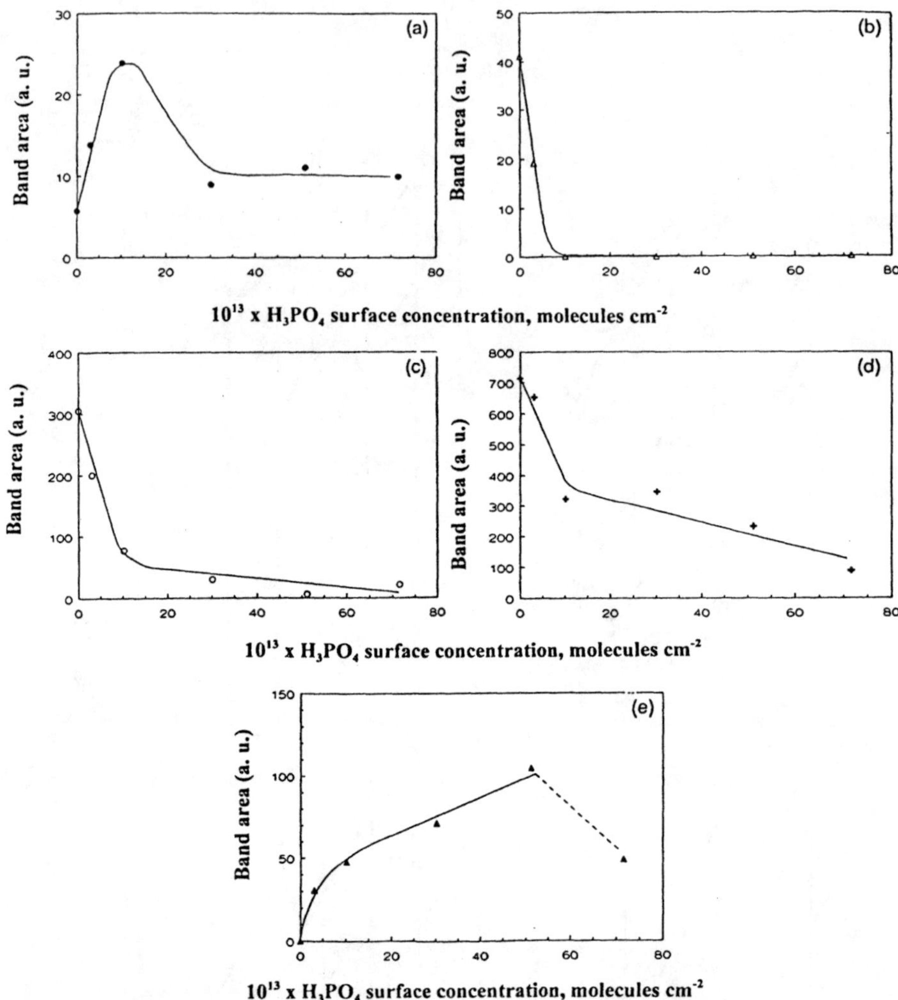

FIG. 15.   Effect of the surface concentration of phosphoric acid on the OH stretching region of the IR spectra of P/Al samples. (a) 3795 cm^{-1} band (Ib sites); (b) 3780 cm^{-1} band (Ia sites); (c) 3736 cm^{-1} band (IIa and IIb sites); (d) 3697 cm^{-1} band (III sites); (e) 3676 cm^{-1} band (P–OH sites). The band areas (arbitrary units) are normalized with respect to sample weight [reprinted with permission from Lewis and Kydd (44); copyright 1991 Academic Press].

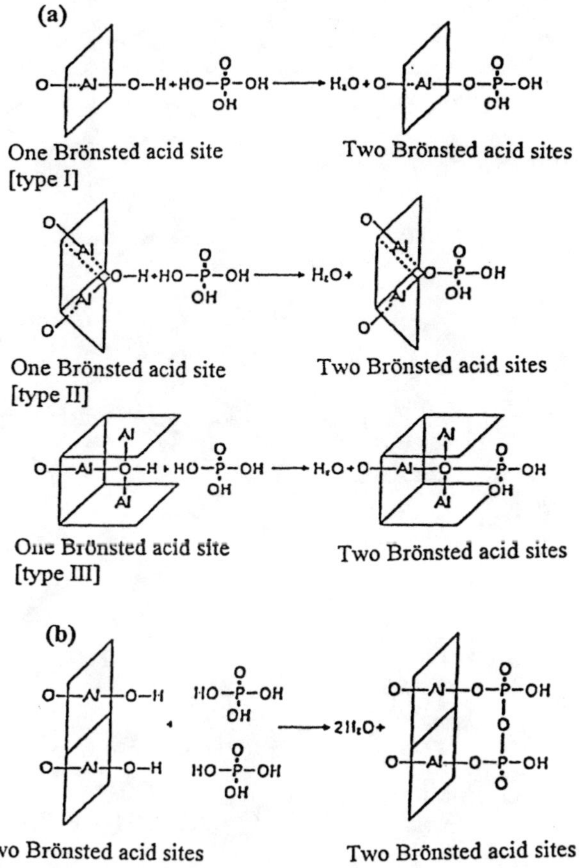

One Brönsted acid site [type I]

Two Brönsted acid sites

One Brönsted acid site [type II]

Two Brönsted acid sites

One Brönsted acid site [type III]

Two Brönsted acid sites

Two Brönsted acid sites

Two Brönsted acid sites

FIG. 16. Schematic representation of the interaction between H_3PO_4 and surface OH groups of alumina. (a) Formation of two Brønsted acid sites (B sites) from the adsorption of one H_3PO_4 on one OH site of alumina; (b) formation of two B sites from two H_3PO_4 and two OH sites (interaction between neighboring P–OH groups); (c) formation of one B site from one H_3PO_4 and two OH sites; (d) formation of one site with no B sites from one H_3PO_4 and three OH sites [adapted from Stanislaus *et al.* (*46*) and Petrakis *et al.* (*47*); reprinted with permission, copyright 1998 Elsevier Science and 1995 the Royal Society of Chemistry].

1. Aqueous solutions of phosphoric acid protonate surface Al—OH groups through hydrogen bonding during impregnation, leading to the formation of Al—O—P bonds by dehydration during the calcination step (*25, 50*):

(c)

(d)

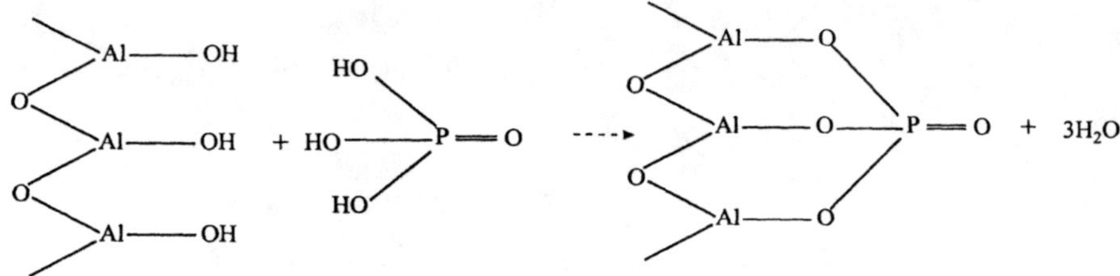

FIG. 16. (*continued*)

$$\begin{array}{c}\diagdown \\ \diagup\end{array}\!\!Al\!-\!OH + H_3PO_4 \text{ (aq.)} \rightleftharpoons \begin{array}{c}\diagdown\quad+\diagup\end{array}\!\!Al\!-\!O\!\!\begin{array}{c}\diagup\\\diagdown\end{array}\!\!\overset{H}{}$$

$$H\!-\!OPO(OH)_2 \text{ (aq.)} \tag{7}$$

$$\text{(calcination)} \quad \underset{\diagup}{\overset{\diagdown}{\longrightarrow}}\quad Al\!-\!O\!-\!PO(OH)_2 + H_2O.$$

2. Phosphorus species adsorb electrostatically on the protonated Al$-$OH surface groups (*51*) through the following equilibria:

$$AlOH_2^+ + H_2PO_4^- \overset{K_1}{\rightleftharpoons} AlOH_2^+ - H_2PO_4^- \tag{8}$$

$$AlOH_2^+ + HPO \overset{K_2}{\rightleftharpoons} AlOH_2^+ - HPO_4^{2-}. \tag{9}$$

3. Phosphorus species adsorb covalently on the protonated Al–OH surface groups (*52, 53*):

$$AlOH_2^+ + H_2PO_4^- \rightarrow AlH_2PO_4 + H_2O \tag{10}$$

$$AlOH_2^+ + HPO_4^{2-} \rightarrow AlHPO_4^- + H_2O. \tag{11}$$

4. "H$_2$PO$_4$" is supported on alumina by a ligand exchange with Al$-$OH groups, especially at low phosphorus contents (*25, 54, 55*):

$$Al-OH + H_2PO_4^- \rightarrow Al-H_2PO_4 + OH^-. \tag{12}$$

Mikami *et al.* (56) concluded from a detailed kinetics investigation that the phosphorus species adsorb electrostatically rather than covalently on the alumina at low phosphorus contents. They calculated the intrinsic value of the adsorption and desorption rate constants at 25°C:

1. For the monovalent phosphate [Eq. (8) with an equilibrium constant K_1]: $k_1 = 4.1 \times 10^5$ mol^{-1} dm^3 s^{-1} and $k_{-1} = 2.3$ s^{-1}.

2. For the divalent phosphate [Eq. (9) with an equilibrium constant K_2]: $k_2 = 1.1 \times 10^7$ mol^{-1} dm^3 s^{-1} and $k_{-2} = 2.7$ s^{-1}.

Because k_2 is two orders of magnitude larger than k_1, while the desorption rate constants (k_{-1} and k_{-2}) are nearly equal, it is suggested that the interaction between the HPO_4^{2-} and $AlOH_2^+$ group is stronger than that between the $H_2PO_4^-$ and $AlOH_2^+$ group. However, by considering the experimental result that the pH of the solution increases during adsorption, we infer that mechanisms 2 and 4 may proceed simultaneously.

Variations of the zeta potential of several supports versus the pH of the impregnating solution and the isoelectronic point (IEP) as a function of phosphorus loading are reported in Figs. 17a and 17b (57). The IEP decreases from ~8.8 to 7.4 up to 2 wt% P_2O_5 and then shows a steady value for loadings exceeding 2 wt% P_2O_5. The total surface density of phosphate ions ($H_2PO_4^-$ and HPO_4^{2-}) adsorbed on alumina decreases with increasing pH of the solution because the alumina surface tends to be negatively charged above its IEP, but the relative amount of HPO_4^{2-} exhibits a maxi-

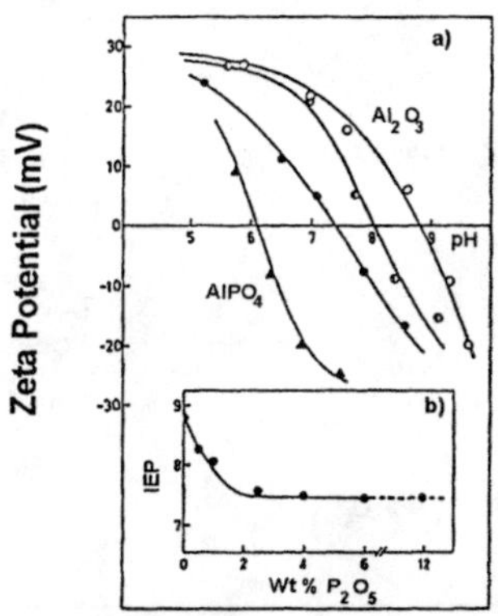

FIG. 17. (a) Variation of the zeta potential at room temperature as a function of suspension pH of the following: O, alumina; ▲, AlPO$_4$; ● or ◑, two different P/Al systems. (b) Variation of the isoelectronic point of P/Al samples as a function of the phosphorus content [reprinted with permission from López-Cordero *et al.* (57); copyright 1989 Elsevier Science].

mum at approximately pH = 8, as shown in Fig. 18 (*56*). Several states of phosphorus oxo-compounds, such as monophosphate, diphosphate, poly-phosphate, and $AlPO_4$, exist on the surface of γ-alumina, depending on the phosphorus content (Fig. 19), as shown by Kraus (*49*) in an extended NMR investigation of the nature of adsorbed phosphorus-containing species.

The amount of polymeric phosphate tends to increase when the phosphorus loading increases. $AlPO_4$ is not detected at low phosphorus loadings, but it can be observed at phosphorus loadings higher than the theoretical phosphorus monolayer coverage. Moreover, Kraus (*49*) did not observe the presence of the crystalline $AlPO_4$ phase upon further addition of phosphorus. On the other hand, the broadening of ^{31}P-NMR lines indicates that several types of $AlPO_4$, with different degrees of hydration and condensation, might also be formed on alumina (*38*). Petrakis *et al.* (*47*) also confirmed that the phosphorus oxo-species prefer to form surface amorphous aluminophosphate rather than the crystalline $AlPO_4$ phase at low phosphorus contents. For example, only 5% of phosphorus can transform into the crystalline $AlPO_4$ phase, even after calcination at 737°C. On the other hand, DeCanio *et al.* (*38*) detected crystalline $AlPO_4$ in dried P/Al and MoP/Al catalysts at phosphorus loadings >9 and >4 wt%, respectively. Crystalline $AlPO_4$ transforms into amorphous $AlPO_4$ after calcination. Cruz Reyes *et al.* (*58*) also observed the presence of hexagonal and orthorhombic crystalline $AlPO_4$ on W−P/Al catalysts by using high-resolution electron microscopy (HREM). Kraus and Prins (*41*) also studied the difference between CoMoP/Al and NiMoP/Al catalysts in regard to the formation of $AlPO_4$.

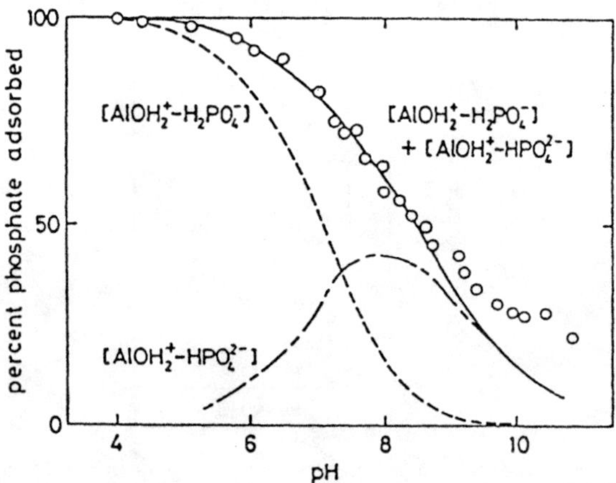

FIG. 18.   Variation of the amount of $H_2PO_4^-$ and HPO_4^{2-} adsorbed on alumina as a function of pH [from Mikami *et al.* (*56*); copyright 1983 American Chemical Society].

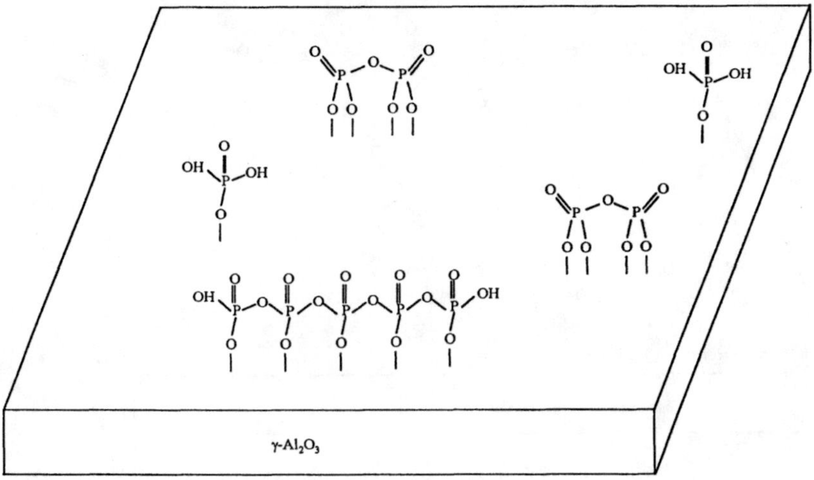

Low Loading

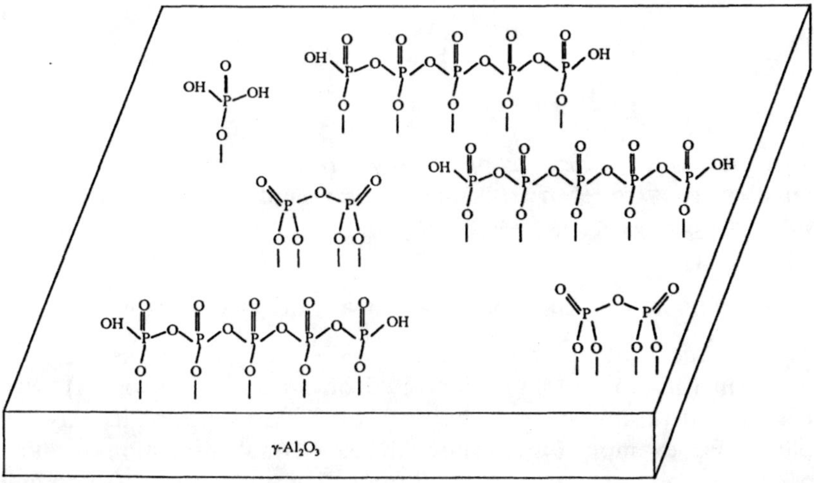

Medium Loading

FIG. 19. Possible phosphate structures on the surface of γ-alumina [reprinted with permission from Kraus (*49*)].

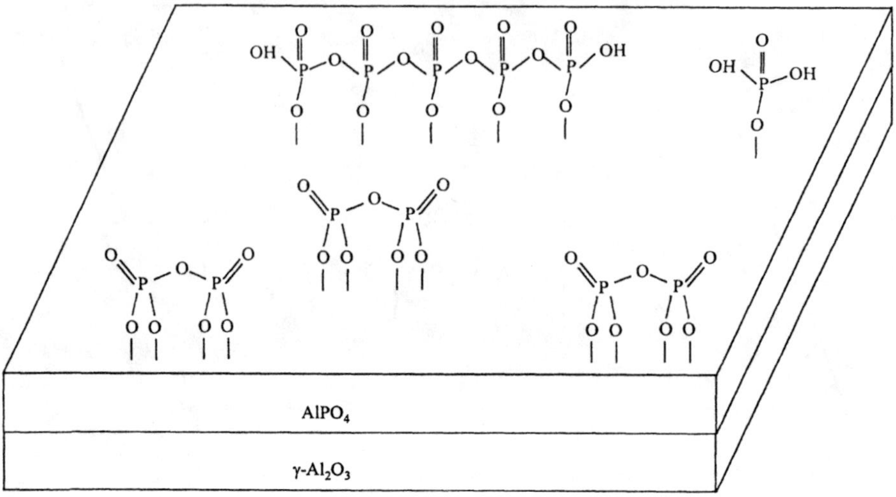

High Loading

Fig. 19. (*continued*)

They reported that AlPO$_4$ is preferably formed in the presence of nickel, whereas some "Co—Mo—P" mixed compounds are mainly formed on CoMoP/Al catalysts.

B. ADSORPTION OF MOLYBDATE AND PHOSPHATE ON ALUMINA

The adsorption of molybdates in the presence of phosphorus oxo-species in solution (or those already deposited on an alumina support) or vice versa (i.e., adsorption of phosphate on a Al—Mo support) has also been explored. For example, Fig. 20 shows the isothermal adsorption on alumina supports of solutions containing different concentrations of ammonium heptamolybdate (AHM). From Fig. 20a, which corresponds to the sequential impregnation steps whereby phosphorus is adsorbed first (before molybdenum), it is clear that molybdenum is adsorbed in lower amounts on the Al—P support than on the phosphorus-free alumina. However, the amount of molybdenum adsorbed on the Al—P support also depends on the pH of the AHM solution (Fig. 20b). Molybdenum oxo-species tend to adsorb in greater amounts on the Al—P support under acidic condition.

IR spectra (*25*) indicate that the heptamolybdate species interact not only with the OH groups of alumina but also with the P—OH groups according to the surface reaction

$$\begin{array}{c} \text{OH} \\ / \\ \text{O-P} \\ \backslash \\ \text{OH} \end{array} + Mo_7O_{24}^{6-} \rightarrow \begin{array}{c} \text{O} \\ / \backslash \\ \text{O-P} \qquad Mo_7O_{23}^{6-} \\ \backslash \quad / \\ \text{O} \end{array} + H_2O \qquad (13)$$

or

$$\begin{array}{c} \text{O-P-OH} \\ \\ + Mo_7O_{24}^{6-} \rightarrow \\ \\ \text{O-P-OH} \end{array} \qquad \begin{array}{c} \text{O-P-O} \\ \backslash \\ \qquad Mo_5O_{15}^{2-} + 2MoO_4^{2-} + H_2O. \\ / \\ \text{O-P-O} \end{array} \qquad (14)$$

The relative amounts of molybdenum and phosphorus adsorbed onto γ-alumina from molybdate and/or phosphate solutions by sequences of pulses are shown in Fig. 21 (*59*). When the solutions contain only molybdenum or phosphorus oxo-species, the amounts of molybdenum or phosphorus retained are high for the first pulses ($\sim$80 and $\sim$100%, respectively) and decrease in further pulses. On the other hand, the evolution during pulses with a "Mo + P" solution, which is equivalent to coimpregnation of molybdenum and phosphorus, is quite different. The molybdenum adsorption decreases significantly as a result of addition of phosphorus due to the strong competition between phosphorus- and molybdenum-containing anions. In addition, the molybdenum adsorption decrease in the presence of phosphorus can also be explained by the formation (Eqs. 15 and 16) of phosphomolybdate species that have lower affinities for alumina:

$$8H^+ + 5MoO_4^{2-} + 2HPO_4^{2-} \rightleftarrows P_2Mo_5O_{23}^{6-} + 5H_2O \qquad (15)$$

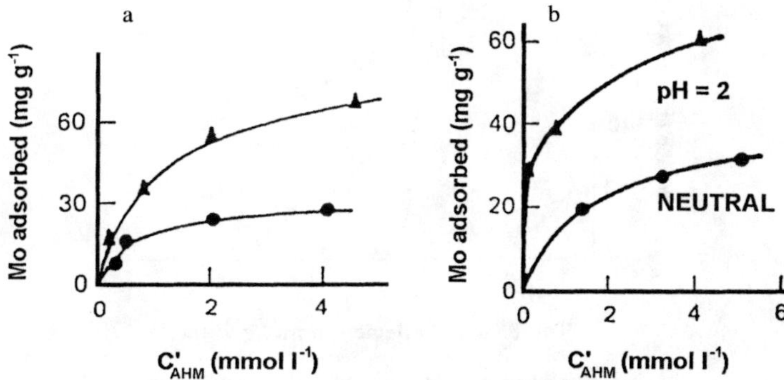

FIG. 20.   Adsorption isotherms of ammonium heptamolybdate (AHM). (a) ▲, on alumina; ●, on alumina after adsorption of $H_2PO_4^-$ (1.8 wt% P); (b) as a function of pH on alumina after adsorption of H_3PO_4; ▲, pH = 2; ●, neutral conditions [reprinted with permission from Van Veen *et al.* (*25*); copyright 1990 American Chemical Society].

and

$$16H^+ + 5Mo_7O_{24}^{6-} + 14H_2PO_4^- \rightleftarrows 7H_2P_2Mo_5O_{23}^{4-} + 15H_2O. \qquad (16)$$

The addition of moderate amounts of phosphorus onto a Mo/Al catalyst increases the formation of molybdate with an octahedral symmetry (polymeric oxo-molybdate) rather than that of tetrahedral MoO_4^{2-}, whereas higher phosphorus loadings lead to the formation of bulk MoO_3 and $Al_2(MoO_4)_3$ (*38, 60*). $Al_2(MoO_4)_3$ might be formed as follows:

$$3MoO_3(\text{bulk}) + Al_2O_3(\text{surface}) \rightarrow Al_2(MoO_4)_3(\text{surface}). \qquad (17)$$

Han *et al.* (*61*) reported that the $Al_2(MoO_4)_3$ phase on alumina is easily hydrated by moisture in air and transforms into amorphous MoO_3, whereas $AlPO_4$ only slightly reacts with water. Further addition of phosphorus decreases the formation of $Al_2(MoO_4)_3$ since competitive adsorption of phosphorus and molybdenum oxo-species occurs on the alumina surface. Phosphorus inhibits the formation of $Al_2(MoO_4)_3$ in the presence of nickel (*62*). The number of deposited polymeric phosphorus–oxo compounds decreases in the presence of molybdenum, probably through the formation of dispersed "Mo−P" heteropoly compounds (*63*).

C. Adsorption of "Mo−P" Heteropoly Compounds on Alumina

As mentioned in Section III, phosphomolybdate heteropoly compounds have recently been used as precursors for preparing hydrotreating catalysts.

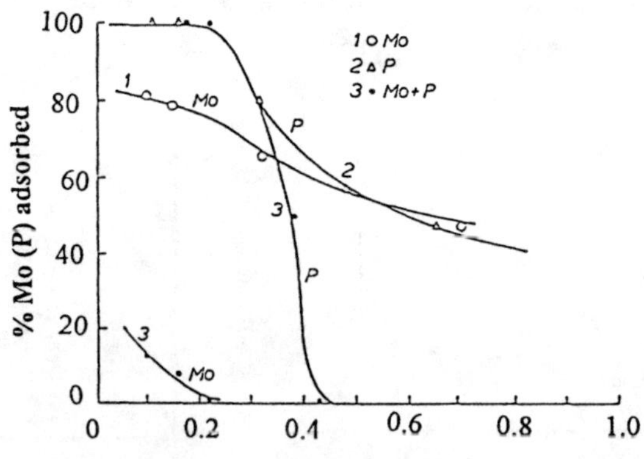

FIG. 21. Amount of molybdenum and phosphorus adsorbed onto γ-alumina from the injection of pulses of molybdate and phosphate solutions. 1, Pulses of molybdenum solution; 2, pulses of phosphorus solution; 3, pulses of "Mo + P" solution [reprinted with permission from Gishti; *et al.* (*59*); copyright 1984 Elsevier Science].

Indeed, the formation of some heteropoly compounds has been effectively detected in impregnation solutions containing phosphorus and molybdenum oxo-species (50, 54). Thus, adsorption of "Mo$-$P" heteropoly compounds on alumina during the impregnation step should also be considered in detail. The chemistry of "Mo$-$P" heteropoly compounds in impregnation solution has been well investigated (20, 64), and characteristic diagrams of ion distributions are reported in Fig. 22.

$P_2Mo_5O_{23}^{6-}$ reacts with basic Al$-$OH groups on alumina but it decomposes gradually into heptamolybdate ($Mo_7O_{24}^{6-}$) and/or poly-oxo-molybdate (25, 50, 54). The pH increase of the adsorption solution also helps the decomposition of heteropoly compounds since it pushes the equilibria to the right:

$$14PMo_{12}O_{40}^{3-} + 114OH^- \rightleftarrows 7P_2Mo_5O_{23}^{6-} + 19Mo_7O_{24}^{6-} + 57H_2O \quad (18)$$

$$7P_2Mo_5O_{23}^{6-} + 16OH^- \rightleftarrows 14HPO_4^{2-} + 5Mo_7O_{24}^{6-} + H_2O. \quad (19)$$

More precisely, the degradation of "Mo$-$P" heteropoly compounds is reported to proceed according to the following sequence (25, 65):

$$\text{"}Mo_{12}P\text{"} \rightarrow \text{"}Mo_{11}P\text{"} \rightarrow \text{"}Mo_5P_2\text{"} \rightarrow \text{"}Mo_7.\text{"} \quad (20)$$

The stability of the "Mo$-$P" heteropoly compounds also depends on the P/Mo ratio in the impregnation solution. For P/Mo ratios lower than 0.4, "Mo_5P_2" decomposes into "Mo_9P" and "$Mo_{11}P$" as follows (66):

$$8H^+ + 15H_2O + 9H_2P_2Mo_5O_{23}^{4-} \rightleftarrows 5PMo_9O_{31}(H_2O)_3^{3-} + 13H_2PO_4^- \quad (21)$$

$$10H_2O + 11H_2P_2Mo_5O_{23}^{4-} \rightleftarrows 8H^+ + 5PMo_{11}O_{39}^{7-} + 17H_2PO_4^-. \quad (22)$$

For P/Mo ratios > 0.4, the amount of phosphorus is high enough to push the equilibria corresponding to Eqs. (21) and (22) to the left and those corresponding to Eqs. (15) or (16) to the right. As a result, all the Mo species in solution exist as stable "Mo_5P_2."

The stability of the "Mo$-$P" compounds also depends on their structures. For example, $PMo_{12}O_{40}^{2-}$ and $P_2Mo_{18}O_{62}^{6-}$ are more stable than $P_2Mo_5O_{23}^{6-}$ on alumina (54). In "Mo_5P_2," the two phosphorus atoms are localized at the exterior of the ion where they are in a position to be in geometrical or electrical contact with the alumina surface, whereas in "$Mo_{12}P$" and "$Mo_{18}P_2$" phosphorus is surrounded by MoO_6 octahedral shells and cannot interact easily with the alumina surface. Indeed, stability of "Mo–P" heteropoly compounds on alumina must depend on both their nature and on the impregnation conditions.

D. Adsorption of Phosphate and Promotor on Alumina

On Ni$-$P/Al catalysts, the presence of phosphorus prevents diffusion of nickel ions into alumina and impedes the formation of $NiAl_2O_4$ after

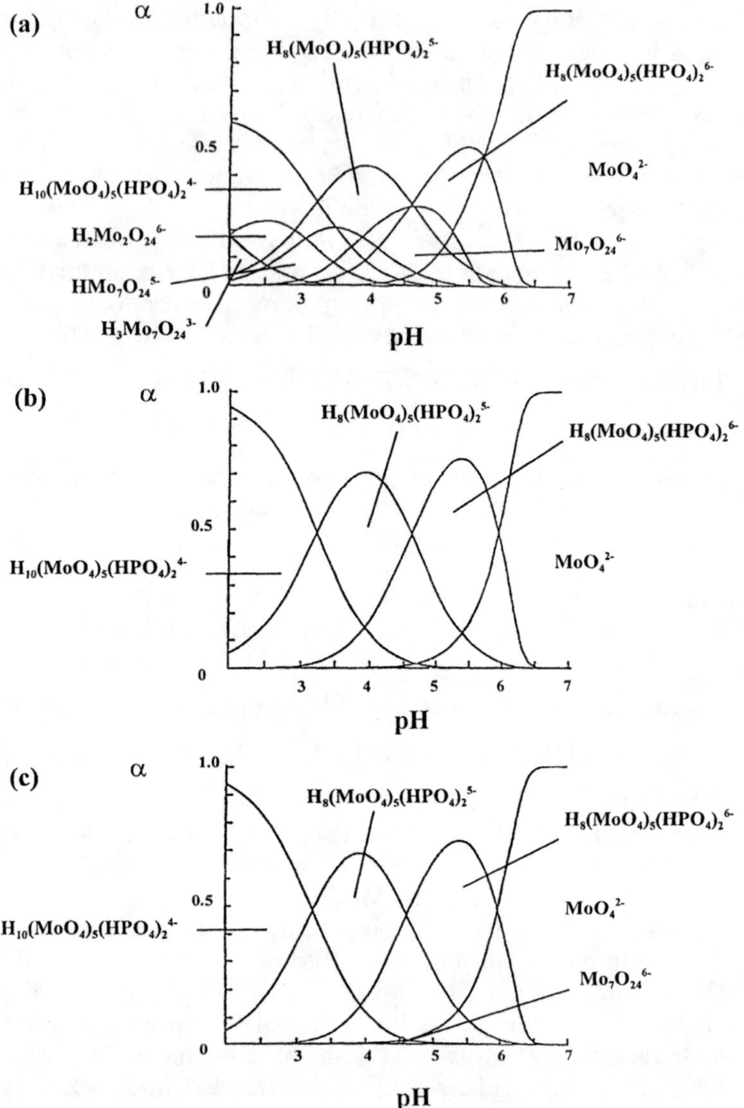

FIG. 22. Distribution diagrams $\alpha = f(\text{pH}, A, B)$ of "Mo–P" oxo-species present in impregnation solution showing; α is defined as the fraction of each molybdenum species. A and B represent initial concentrations of molybdenum and of phosphorus, respectively. Precipitates and species with $\alpha < 0.008$ have been omitted. (a) $A = 40$ mM, $B = 10$ mM; (b) $A = 50$ mM, $B = 20$ mM; (c) $A = 40$ mM, $B = 20$ mM [reprinted with permission from Pettersson (64); copyright 1971 Acta Chemica Scandinavica].

calcination (*38, 67, 68*). The deposited nickel cations may form nickel phosphate or may be present as isolated nickel species on the support surface. López-Agudo *et al.* (*69*), by considering XPS and microanalysis (energy dispersive X-ray analysis) results, reported that nickel and phosphorus species exist separately on NiP/Al catalysts. However, DeCanio *et al.* (*38*) suggested that phosphorus oxo-species preferably form nickel phosphate since their presence drastically decreases the amount of surface Ni^0 obtained after reduction. The metallic nickel particles might be derived from isolated nickel species. However, it is difficult to conclude whether other "Ni—P" or "Co—P" mixed compounds are formed on alumina-based catalysts. Nevertheless, phosphorus is easily incorporated in the $AlPO_4$ phase after calcination at 500°C because thermodynamic data favor its formation (*70*) and because there is a strong interaction between the phosphorus oxo-species and the alumina surface. In contrast, the formation of nickel phosphide Ni_2P is detected on carbon- or silica-based catalysts working at 370°C for quinoline hydrodenitrogenation (HDN) because the nickel species and the support have weak interactions (*71*).

V. Characterization of Phosphorus-Containing Hydrotreating Catalysts

The presence of phosphorus in catalysts significantly affects their physico-chemical properties, such as pore structure, dispersion of active phases, acidity, thermal stability, and reducibility or sulfidability. In this section, relationships between phosphorus content and physicochemical properties of hydrotreating catalysts are presented and discussed.

A. PORE STRUCTURE

The influence of phosphorus on catalyst textural parameters, such as specific surface area (SSA), pore diameter (PD), and pore volume, has been thoroughly investigated (*25, 30, 38, 60, 62, 68, 69, 72, 73*). The SSA decreases with phosphorus loading, irrespective of the preparation procedures. In particular, it was reported that "NiMoP" catalysts obtained by coimpregnation have greater SSA decreases than those prepared by sequential impregnation (*74*).

The effect of phosphorus on PD depends on the catalyst preparation method. The PD of catalysts prepared by impregnation decreases with phosphorus loading, whereas that of catalysts derived from the hydrogel or sol-gel methods tends to increase in some cases. Introduction of phosphorus compounds in the sol-gel procedure may affect the hydrolysis and condensation steps at the origin of the alumina particle formation. Some patents also report that phosphorus increases the PD of catalysts (75, 76).

B.   THERMAL STABILITY

Abbattista *et al.* (26) found that phosphorus addition prevents crystallization of the γ-alumina phase and the transformation from γ- to α-alumina in the system $Al_2O_3 - AlPO_4$ (Fig. 23). More precisely, Morterra *et al.* (77) reported that phosphates do not affect the phase transition from low-temperature spinel alumina (γ-alumina) to high-temperature spinel aluminas (δ and θ phases) but delay the transition of δ and θ to α-alumina (corundum). Stanislaus *et al.* (46) also reported that phosphorus significantly improves the thermal stability of the γ-alumina phase in P/Al catalysts. However, the same authors found that the positive effect of phosphorus seems to be canceled in the presence of molybdenum due to the formation of aluminum molybdate. Thermal treatments of MoP/Al catalysts at temperatures >700°C result in a considerable reduction of SSA and mechanical strength. The presence of phosphorus does not prevent the reaction between the molybdenum oxo-species and alumina since the interaction between molybdates and phosphates is weak. The presence of nickel does not obviously affect the positive effect of phosphorus in terms of thermal stability (46). On the other hand, Hopkins and Meyers (78) reported that the thermal stability of commercial CoMo/Al and NiMo/Al catalysts is improved by the addition of phosphorus.

In summary, it is difficult to conclude whether the thermal stability of

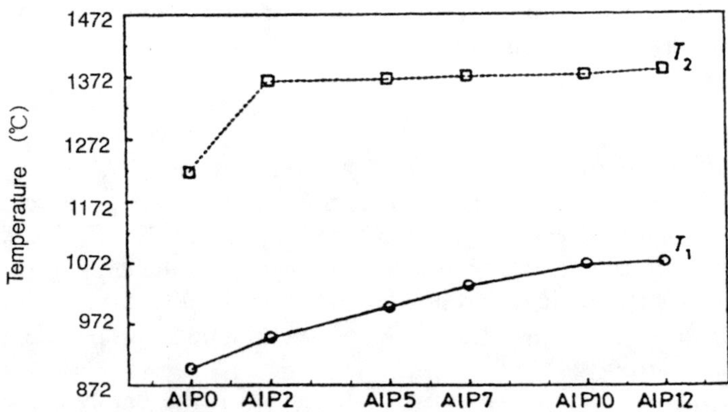

FIG. 23.   Effect of phosphorus content on the transition temperatures of alumina; *x* in AlP*x* refers to the atomic percentage of phosphorus. $\bigcirc$ (T_1), transition from amorphous to γ-alumina; $\square$ (T_2), transition from γ-alumina to α-alumina [reprinted with permission from Abbattista *et al.* (26); copyright 1990 the Royal Society of Chemistry].

alumina is improved by the addition of phosphorus in catalysts containing nickel or cobalt and molybdenum. However, its presence may not affect the thermal stability significantly during conventional hydrotreating performed at temperatures $<450°C$.

C. ACIDITY

1. P/Alumina Catalysts and Aluminophosphate

Petrakis *et al.* (*47*) reported temperature-programmed desorption (TPD) of ammonia and concluded that strong acid sites are created in or on aluminophosphates. Abbattista *et al.* (*26*) observed IR spectra of adsorbed pyridine and inferred that Brønsted acid and strong Lewis acid sites are generated by phosphorus addition to the surface of $Al_2O_3 - AlPO_4$. On the other hand, Morterra *et al.* (*77*) also investigated IR spectra of adsorbed pyridine and found that the presence of phosphorus decreases the number of Lewis acid sites on P/Al samples, but the acid strength was found to increase after calcination at moderate temperatures ($300-1000°C$). The same authors found, by using CO_2 as a probe molecule, that the presence of phosphate does not produce new basic centers on alumina. Instead, the number of basic sites gradually decreases with addition of surface phosphates. Poulet *et al.* (*79*) also supposed that the number of Lewis acid sites decreases with phosphorus addition. Chen *et al.* (*80*) reported, on the basis of NH_3 TPD data, that P/Al samples contain only one kind of site (with medium acid strength), the number of which decreases with increasing deposited phosphorus. Stanislaus *et al.* (*46*) concluded that introduction of phosphorus eliminates strong acid sites and increases the proportion of acid sites of medium strength, with the data determined by NH_3 TPD (Fig. 24). Lewis and Kydd (*44*) indicated that the effect of phosphorus on surface acidity is not very important since cracking of cumene and 1,3-diisopropylbenzene on phosphorus-containing catalysts is very small. Jian *et al.* (*81, 82*) suggested that the addition of phosphorus does not induce enough acidity for C–N bond breaking of alkylamine.

2. Mo + P/Alumina Catalysts

On the basis of an NH_3 TPD investigation, Sajkowski *et al.* (*83*) suggested that phosphorus increases the total acidity but does not increase the acid strength of the Mo/Al system enough to affect the hydrodesulfurization (HDS) or HDN activity. Kim and Woo (*84*) did not observe the formation of Brønsted acid sites on MoP/Al, as indicated by IR spectra of adsorbed pyridine. Iwamoto and Grimblot (*85*), however, found that a certain degree of interaction between phosphorus- and molybdenum-oxo species and alu-

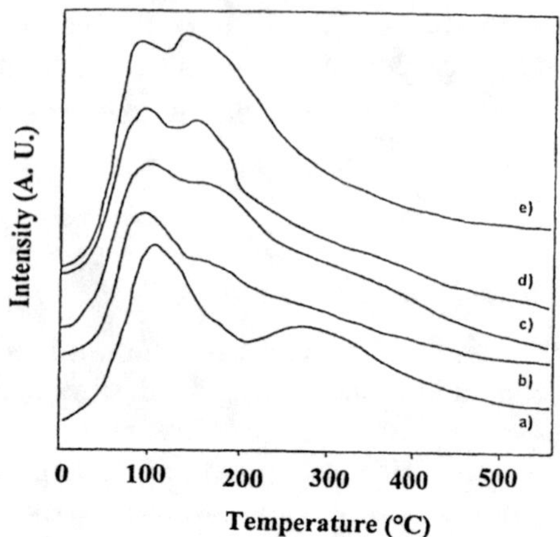

FIG. 24.   TPD of ammmonia from alumina with various phosphate loadings; (a) γ-alumina, (b) 1 wt% P, (c) 2 wt% P, (d) 5 wt% P, and (e) 10 wt% P [reprinted with permission from Stanislaus *et al.* (*46*); copyright 1988 Elsevier Science].

mina significantly affects the acidity of Mo/Al catalysts. For example, the catalysts prepared from P_2O_5 as a phosphorus precursor (by a sol-gel procedure) show lower acidity than those prepared from H_3PO_4 due to less interaction between the phosphorus-oxo species and alumina.

3.   *Mo + Promotor + P/Alumina Catalysts*

Walendziewski (*73*) observed that the total acidity per unit surface area of CoMoP/Al catalysts measured by NH_3 TPD increases with increasing phosphorus loading (Fig. 25). Chadwick *et al.* (*60*) reported that the surface acidity of NiMoP/Al catalysts measured by pyridine adsorption increased slightly as a result of phosphorus addition.

Callant *et al.* (*86*) proposed that two acid sites exist on sulfided Ni–Mo–P/Al catalysts, associated respectively with the alumina support and with the MoS_2 active phase. Iwamoto and Grimblot (*67*) concluded that phosphorus induces formation of at least two acid sites, on alumina and on the molybdenum-containing species. The latter acid sites, due to the OH groups associated with molybdenum oxy-sulfide (in sulfided catalysts), are considered to be stronger than the former, which are attributed to the presence of $AlPO_4$. The same authors found that the acidity measured by cyclopropane cracking increases in the order P/Al = NiP/Al < MoP/Al < NiMoP/Al (Fig. 26). The promotor nickel may enhance the acidity of Mo sites by modifying the electronic charge density of the MoS_2 active phase or by inducing more OH groups to form in the Mo oxy-sulfide species.

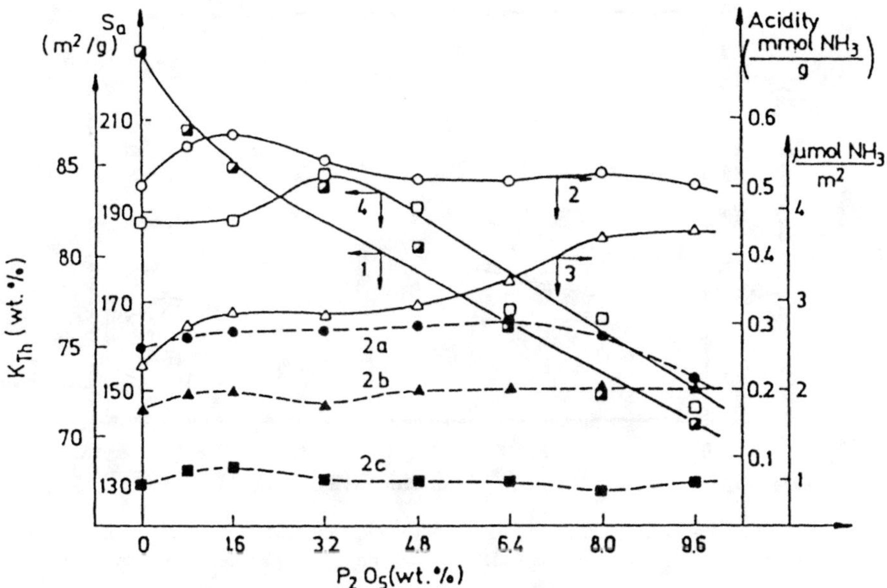

FIG. 25.   Influence of the phosphorus content on the following: 1, SSA (m²/g); 2, total acidity of catalyst (mmol NH$_3$/g) (2a, weak acid; 2b, medium acid; 2c, strong acid sites); 3, total acidity of catalyst (μmol NH$_3$/m²); 4, HDS activity of CoMoP/Al catalysts (conversion of thiophene, K$_{Th}$) [reprinted with permission from Walendziewski (73); copyright 1991 Baltzer Science].

Thus, questions remain; the large differences in acidity reported here might be attributed both to the nature of the catalysts (affected by the method of preparation, the loadings of the different elements, and whether the catalysts are in the oxide or sulfided states) and to the method of acidity measurement (total acidity determination or selective titration of acid sites with a specific strength). Furthermore, the protocols for measurement (e.g., the temperature and duration of outgassing before acidity determination) may also affect the results. Therefore, it can be inferred that some standardization of measurement conditions is required to reach consensus regarding the influence of phosphorus on the acidity of hydrotreating catalysts.

D.   DISPERSION AND DISTRIBUTION OF CATALYST COMPONENTS ON ALUMINA

There is no doubt that both the dispersion and the distribution of the active elements through the catalyst particles influence the performance of hydrotreating catalysts. In this section, the influence of phosphorus on the

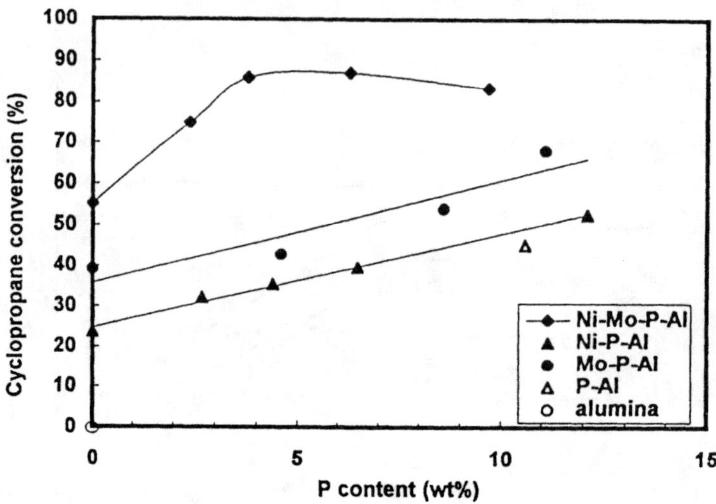

FIG. 26. Comparison of acidity of NiP/Al, MoP/Al, and NiMoP/Al catalysts as measured by cyclopropane (CP) cracking as a function of P content [reprinted with permission from Iwamoto (67)].

dispersion and distribution of the components in hydrotreating catalysts is discussed in light of the parameters used for catalyst preparation.

1. Influence of pH

The pH value of the impregnation solutions for depositing the active elements on alumina is an important variable in catalyst preparation. For example, Jian and Prins (66) reported that the content of bulk molybdenum oxide (detected by XRD) increases in MoP/Al catalysts when the pH of the impregnation solution decreases from 9 to 1 as a result of addition of phosphorus (Fig. 27).

When the pH of the impregnating solution has higher values than the IEP of the support, the amount of PO_4^{3-} and MoO_4^{2-} anions adsorbed decreases because the support surface is negatively charged. Since the driving force for adsorption is mainly imposed by electrostatic interactions, the anions are expected to spread over the entire catalyst surface without formation of MoO_3 aggregates because the interaction between MoO_4^{2-} and the alumina surface is weak. On the other hand, when the solution pH has lower values than the IEP, so that the support is positively charged, the molybdates are present mainly as $Mo_7O_{24}^{6-}$ (according to the equilibrium $7MoO_4^{2-} + 8H^+ \rightleftarrows Mo_7O_{24}^{6-} + 4H_2O$), and phosphates interact strongly with the support (see Section IV.A). A good molybdenum dispersion can

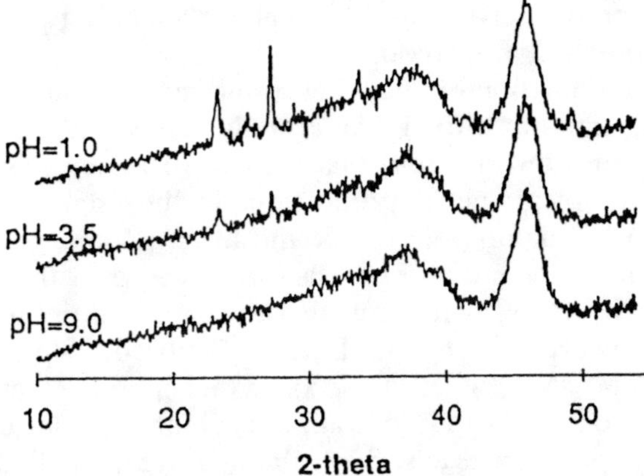

FIG. 27. Effect of the pH of coimpregnation solution on the dispersion of molybdenum species in MoP/Al catalysts. The two sharp diffraction peaks at pH = 1 are due to the presence of MoO_3 [reprinted with permission from Jian and Prins (66); copyright 1995 Comite Van Beheer Van Het Bulletin v.z.w.].

also be obtained in this case unless formation of bulky poly-oxo-molybdate occurs. Indeed, under conditions of low pH, undesirably strong adsorption of anions and precipitation of a hydrated form of MoO_3 (molybdic acid) are often observed, which consequently lead to the characteristic lines of MoO_3 in XRD patterns after calcination. Clearly, the pH of the impregnating solution must be carefully controlled to optimize the molybdenum dispersion.

2. Influence of the Impregnation Method

The impregnation method also affects the dispersion of the deposited elements. In sequential impregnation onto alumina of phosphorus followed by molybdenum (giving Mo–P/Al samples), the SSA of the sample decreases and the Al–OH groups interact first with the phosphates. Therefore, the molybdate ions are not stabilized on the support (see Section IV.B) and, consequently, the molybdenum dispersion tends to decrease as a result of formation of MoO_3. During this sequential impregnation procedure, acidic P—OH groups are also formed on alumina, and they may also enhance the adsorption of Mo species. In such a case, the preferential deposition of molybdenum species on the external surfaces of the catalyst particles has been observed (68). Strong interactions of PO_4^{3-} with alumina may also lead to pore mouth plugging and prevent molybdenum penetration into the support particles. Therefore, it is very difficult to achieve a homogenous distribution of molybdenum throughout a catalyst particle using the sequential impregnation method.

During coimpregnation of phosphorus and molybdenum (MoP/Al samples), the formation of "Mo—P" heteropoly compounds has to be considered. It has been shown (*49, 54*) that the "Mo_5P_2" heteropoly compound is generated in stoichiometric proportions in the impregnating solution. Since the "Mo_5P_2" heteropoly compound has a relatively low affinity for alumina, molybdenum and phosphorus can spread over the entire support surface if the "Mo_5P_2" structure remains unchanged until the impregnation is complete. However, Cheng and Luthra (*54*) used ^{95}Mo-NMR spectroscopy to show that "Mo_5P_2" easily decomposes upon contact with alumina after impregnation. If "Mo_5P_2" decomposes before equilibrium deposition, molybdates and phosphates adsorb on alumina so rapidly that their profiles are not uniform, even under coimpregnation conditions (*68*). Therefore, adjustment of optimum pH and/or the atomic P/Mo ratio of the impregnating solution is needed to prevent the decomposition of "Mo_5P_2" and to give well-dispersed molybdenum species.

3. *Influence of the Phosphorus Content*

The molybdenum dispersion also depends on the phosphorus content of the catalyst. Atanasova *et al.* (*68, 87*) reported that the dispersion of molybdenum and nickel, measured by X-ray photoelectron spectroscopy (XPS), shows a steep increase due to the presence of phosphorus at low loadings. The dispersion of molybdenum in NiMoP/Al catalysts increases further as a result of calcination, whereas that of nickel decreases. In contrast, Sajkowski *et al.* (*83*) reported, on the basis of an extended X-ray absorption fine structure (EXAFS) investigation, that phosphorus does not affect the size of the polymolybdate species. Mangnus *et al.* (*31*) inferred that the stacking of molybdates does not increase as a result of the addition of phosphorus since the height of a temperature-programmed reduction (TPR) peak at ~400°C due to the reduction of deposited multilayered molybdenum oxo-species was found to be independent of the phosphorus content. However, Chadwick *et al.* (*60*) concluded from XPS measurements that the dispersion of molybdenum decreases upon addition of phosphorus.

Morales *et al.* (*88*) reported that the nickel dispersion increases up to a loading of 6 wt% P_2O_5. Mangnus *et al.* (*31*) found that in CoP/Al catalysts the cobalt dispersion increases with phosphorus loading due to the formation of mixed "Co—P" species. However, López-Agudo *et al.* (*69*) reported that phosphorus prevents the homogeneous distribution of the deposited elements throughout NiP/Al catalysts. In NiMo/Al samples, both small and large amounts of phosphorus within the alumina framework lead to aggregation of β-NiMoO$_4$ or bulk MoO$_3$ (*67, 89*). Further addition of phosphorus leads to an increase in the ratio of β-NiMoO$_4$ over α-NiMoO$_4$ (*90*).

Notwithstanding the lack of consistency of the data, there is no doubt that large amounts of phosphorus in the catalyst formulations cause the decrease of the molybdenum and nickel dispersions and favor the formation of bulk compounds such as MoO$_3$, Al$_2$(MoO$_4$)$_3$, and NiO (*60, 67, 80, 84, 91, 92*). McMillan *et al.* (*93*) reported, however, that the formation of Al$_2$(MoO$_4$)$_3$ is suppressed in the presence of nickel.

4. *Influence of Phosphorus on MoS$_2$ and Nickel Sulfide Dispersions*

Hydrotreating catalysts in the oxide form are transformed into the sulfided form either before or during catalysis with model or practical feeds that contain sulfur compounds. The sulfide state is characterized by the presence of MoS$_2$ nanocrystallites (with some oxysulfide species) that are generally described as slabs or patches. Some stacking of the slabs may occur. The promotor nickel or cobalt may form individual sulfides (e.g., Co$_9$S$_8$) or decorate the MoS$_2$ slabs to form the so-called "CoMoS" and "NiMoS" phases (*3*). In this section, the influence of phosphorus on the morphology and dispersion on the sulfide phase is considered.

Eijsbouts *et al.* (*94*) reported XPS data showing that the molybdenum dispersion tends to increase after sulfidation. However, Chadwick *et al.* (*60*) suggested, also on the basis of XPS measurements, that the MoS$_2$ slabs formed after activation of the oxide catalysts tend to grow by stacking of the MoS$_2$ layers. Ramírez *et al.* (*91*) and Hubaut *et al.* (*95*) reported HREM observations indicating that phosphorus induces an increase of the stacking of the layered MoS$_2$ phase. These results were confirmed by Kemp *et al.* (*96*), who reported that phosphorus increases the stacking of MoS$_2$ in both CoMoP/Al and NiMoP/Al catalysts but decreases the lengths of the MoS$_2$ patches. It was suggested that phosphorus hinders the growth of MoS$_2$ crystallites oriented parallel to the alumina surface by suppressing interactions of Al—OH groups and instead causing AlPO$_4$ formation. Van Veen *et al.* (*97*) reported EXAFS data showing that phosphorus increases the formation of the promoted "NiMoS" (type II) phase which has a stacking higher than that obtained in the "NiMoS" (type I) phase. However, Sajkowski *et al.* (*83*) reported conflicting EXAFS data because they claimed that phosphorus does not affect the MoS$_2$ crystallite size.

Eijsbouts *et al.* (*94*) reported XPS data showing that the nickel dispersion tends to decrease as a result of sulfidation. López-Agudo *et al.* (*69*) observed that the dispersion of nickel in NiP/Al samples decreases as a result of sulfidation, although it is not affected by addition of phosphorus to the oxide form of the catalyst.

We conclude that the dispersion of the elements in the oxide-form catalysts depends not only on the phosphorus content but also on the preparation procedures. In any case, large amounts of phosphorus favor the agglomeration of metal oxides. The discrepancies in the literature suggest that both the preparation conditions and the activation conditions (such as the nature of the presulfiding mixture and the sulfidation temperature) may also affect the textural properties and morphology of the sulfided catalysts. The nonuniform distribution of molybdenum in the catalyst particles at low phosphorus loadings may affect the XPS results discussed in terms of phase dispersion.

E.   Effect of Phosphorus on the Adsorption of Probe Molecules

Examination of characteristic IR bands of adsorbed probe molecules such as NO, CO, NH_3, CO_2, and pyridine is useful for providing information about the structure of active sites and acidity and basicity of (sulfided) catalysts. Some IR investigations of phosphorus-containing catalysts with adsorbed probe molecules have also been reported. Because the acidity measurements with pyridine and NH_3 adsorption were considered previously, only the results obtained with NO and CO probe molecules are considered in this section.

Van Veen *et al.* (*97*) found no significant difference between the IR bands of NO adsorbed on (Ni)Mo/Al and (Ni)MoP/Al catalysts. However, Topsøe *et al.* (*98*) observed slight increases in the NO vibrational frequency and intensity increases in the corresponding IR band characterizing MoP/Al and NiMoP/Al catalysts relative to the values for Mo/Al and NiMo/Al, respectively (Fig. 28). These results indicate that the active sites are in a less sulfided state in the catalysts containing phosphorus.

Bouwens *et al.* (*99, 100*) reported that the amount of chemisorbed CO

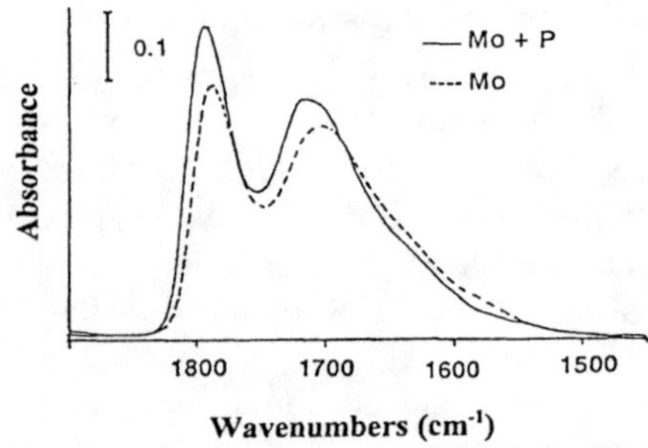

FIG. 28.   IR spectra of NO adsorbed on Mo/Al and MoP/Al catalysts [reprinted with permission from Topsøe *et al.* (*98*); copyright 1989 Elsevier Science].

decreases upon the addition of phosphorus to Co/C$-$P or Co$-$Mo/C$-$P carbon-supported catalysts. They suggested that the structure of the sulfided "Co$-$Mo" phase is modified by interactions between phosphorus and cobalt or by blocking of coordinatively unsaturated sites, although phosphorus does not affect the amount of CO adsorbed on Co/Al. In contrast, Atanasova *et al.* (*101*) found that total adsorption of NO was three times higher on sulfided NiW–P/Al than on the P-free NiW/Al. The phosphorus-containing catalyst was also more active. This comparison shows that the presence of phosphorus leads to an increase in the number of accessible active sites.

F. Effect of Phosphorus on the Activation (Reduction–Sulfidation) of the Catalysts

Since hydrotreating catalysts are usually used in the presence of H_2 and H_2S, it is important to understand the influence of phosphorus on the reduction and sulfidation of the supported metal–oxo-species. It is also important to know whether the phosphates are sensitive to such treatments. In this section, activation of the catalysts is discussed on the basis of XPS, TPR, and temperature-programmed sulfidation results. Note that the bulk of the alumina support is not chemically modified by the reduction–sulfidation treatments. However, some hydrogen-reactive species and surface SH groups have already been detected on it (*31, 70*).

1. *Reduction*

Reduction of bulk $AlPO_4$ and P/Al catalysts starts at temperatures >730°C. However, reduction is not complete even at 1000°C (*31*). The quantitative data indicate that the phosphate is reduced mainly into elemental phosphorus as the P_4 compound, but phosphanes such as PH_3 (Table I) have also been detected as minor products. The volatile P_4 is evolved from the catalysts as expected from its high vapor pressure (Fig. 1). Therefore, it appears that it is necessary to check the phosphorus loadings of catalysts after severe treatments to determine whether some phosphorus has been removed as volatile compounds. Reduction of $AlPO_4$ is accelerated in the presence of reduced cobalt or molybdenum. For P/metal ratios ≤ 1, the $AlPO_4$ phase is completely reduced in H_2 at temperatures >723°C to give metal phosphides (Table II) such as CoP, Co_2P, MoP_x ($x \leq 1$), and Al_2O_3. Arunarkavalli (*102*) showed that phosphorus increases reducibility of nickel, especially at low phosphorus contents, by preventing the formation of nickel aluminate.

Sajkowski *et al.* (*83*) reported XPS measurements showing that the addition of phosphorus to Mo/Al catalysts increases the amount of easily reducible molybdenum species. They presumed that phosphorus weakens the interaction between molybdates and the alumina surface due to the strong interaction between phosphate and alumina. Initial reduction rates (r_0) obtained from kinetics measurements with several MoP/Al catalysts prepared by different procedures were obtained by López-Cordero *et al.* (*72*); r_0 of Mo—P/Al catalysts prepared by sequential impregnation increases rapidly with increasing phosphorus content up to a value of 2.5 wt% P and then decreases only slightly. On the other hand, r_0 of coimpregnated MoP/Al catalysts varies only moderately with phosphorus content. The authors suggested that phosphorus increases the amount of easily reducible octahedral molybdates, such as poly-oxo-molybdate multilayers and bulk MoO_3 clusters, and that this trend is more favored for the Mo—P/Al catalysts (made by sequential impregnation).

On the other hand, Van Veen *et al.* (*25*) reported, on the basis of TPR data, that Mo—P/Al catalysts are rarely reduced at temperatures below 700°C because of the formation of irreducible heteropoly species such as "Mo_5P_2," "Mo_6P," and "Mo_7P." Mangnus *et al.* (*31*) reported TPR results indicating that the reducibility of a fraction of the molybdenum species decreases with increasing phosphorus loading due to the formation of mixed "Mo—O—P" species on MoP/Al catalysts. These authors reported that a new reduction peak appeared at 927°C as a result of the addition of phosphorus, giving CoMoP/Al (Fig. 29). This result indicates that the strong polarization of the Co—O bonds in a "Co—Mo—O—P" mixed phase suppresses the complete reduction of the Co^{2+} ions at low temperature. Atanasova *et al.* (*89*) reported that incorporation of phosphorus in NiMo/Al increases the formation of the β-$NiMoO_4$ phase, which is reduced at high temperatures.

Figure 30 shows thermodynamic data for several cobalt phosphides under various P_4 partial pressures (*31*). Cobalt phosphide CoP_3 reduces in the following three steps:

$$CoP_3 \rightleftarrows CoP + 1/2P_4 \tag{23}$$

$$2CoP \rightleftarrows Co_2P + 1/4P_4 \tag{24}$$

$$Co_2P \rightleftarrows 2Co + 1/4P_4. \tag{25}$$

Under normal TPR measurement conditions, CoP and Co_2P do not decompose, as is apparent from the fact that curves 2 and 3 of Fig. 30 do not cross the dotted band.

2.   *Sulfidation*

It is interesting to know whether phosphorus is sulfided under conventional hydrotreating conditions. Mangnus *et al.* (*70*) showed by TPS that

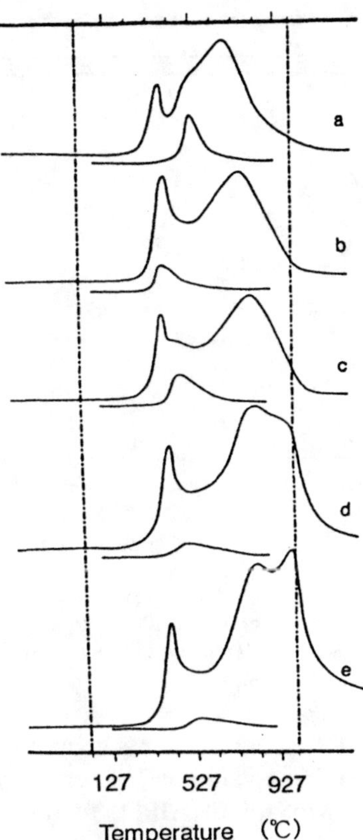

FIG. 29.   TPR patterns of coimpregnated CoMoP/Al catalysts: thermal conductivity detector (top curve) and flame ionization detector signal (bottom curve). a, CoMoP(0)/Al; b, CoMoP(0.4)/Al; c, CoMoP(0.9)/Al; d, CoMoP(1.8)/Al; e, CoMoP(3.8)/Al. P(x) refers to the phosphorus content in wt% [reprinted with permission from Mangnus et al. (31); copyright 1990 Elsevier Science].

P/Al and AlPO$_4$ samples are unreactive with H$_2$S at temperatures up to 723°C. Chadwick et al. (60) found no evidence of formation of sulfided phosphorus from XPS measurements. These results indicate that the phosphorus oxo-species do not transform into phosphorus sulfides or oxysulfides (Fig. 2) during the normal conditions of the hydrotreating reactions.

Van Veen et al. (97) reported EXAFS data showing that phosphorus induces an increase in the formation of the "NiMoS" (type II) phase, which is a more highly sulfided form of the promoted "NiMoS" phase. In contrast, Sajkowski et al. (83) reported that phosphorus does not affect the sulfidation of Mo/Al catalysts. Mangnus et al. (70) found that the S/Mo ratio in the

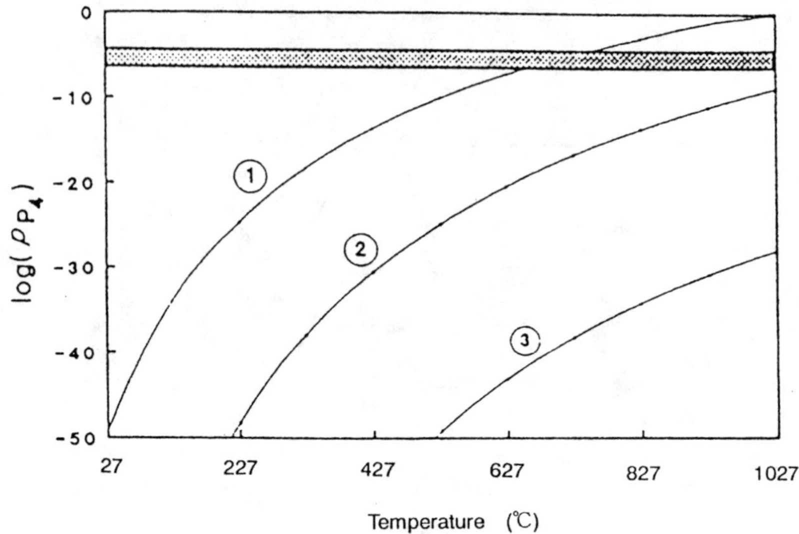

Fig. 30. Thermodynamic equilibrium data of several cobalt phosphides under various P_4 partial pressures. Curves 1, 2, and 3 refer to CoP_2, CoP, and Co_2P, respectively [reprinted with permission from Mangnus *et al.* (*31*); copyright 1990 Elsevier Science].

MoP/Al catalysts after TPS measurements is independent of the phosphorus content, generally having a value of 2. Thermodynamic limitations (*70*) prevent the reaction of MoS_2 with $AlPO_4$ to form MoP_x species (where $x = 1, 2$) at temperatures lower than 1000°C.

However, Poulet *et al.* (*79*) reported that phosphorus induces a decrease of the S/Mo ratio in a MoP/Al catalyst after sulfidation or after a subsequent hydrogen treatment. Topsøe *et al.* (*98*) interpreted the shift of the IR bands of NO adsorbed on (Ni)MoP/Al catalysts toward higher frequency as indicating less sulfidation of the molybdenum (and nickel) species. These authors also confirmed, by using Mössbauer spectroscopy, the presence of a less sulfided state of molybdenum when phosphorus is present.

Iwamoto and Grimblot (*67*) found that the sulfidation of molybdenum depends strongly on the phosphorus and molybdenum contents. Phosphorus does not affect the sulfidation of molybdenum at low phosphorus and molybdenum loadings when these elements are present as individual oxo-species on alumina. On the other hand, phosphorus decreases significantly the sulfidation of molybdenum at higher contents of both elements when they experience interactions with each other (Fig. 31). This result may explain in part the previous discrepancies regarding the different influences of phosphorus on the sulfidation of the deposited species.

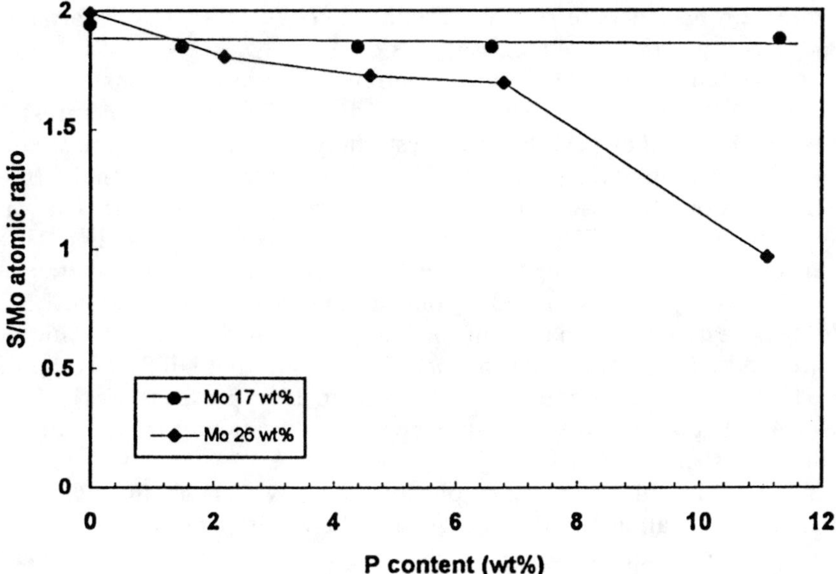

FIG. 31. Variation of the atomic S/Mo ratio measured after thiophene HDS as a function of the phosphorus loading for MoP/Al catalysts [reprinted with permission from Iwamoto (67)].

López-Agudo *et al.* (*69*) reported that the sulfidation of nickel in NiP/Al catalysts, measured by XPS, is not influenced by phosphorus addition. On the other hand, Iwamoto and Grimblot (*67*) found that phosphorus increases sulfidation of nickel in NiP/Al at 400°C because phosphorus prevents the formation of stable nickel aluminate species. A similar explanation was also proposed for nickel reduction (*102*).

At higher temperatures, the influence of phosphorus on the sulfidation of the promotors is completely different. Mangnus *et al.* (*70*) reported that the temperature required to obtain complete sulfidation of cobalt in Co−P/Al increases with increasing phosphorus content. The S/Co atomic ratio after sulfidation at 1000°C decreases from 1.3 to 0.39 as a result of phosphorus addition because of the formation of nonsulfided compounds such as cobalt phosphides, according to the following reactions:

$$CoS_{0.89} + AlPO_4 + 3.39\ H_2 \rightleftharpoons 0.5\ Al_2O_3 + CoP + 2.5\ H_2O + 0.89\ H_2S \quad (26)$$

and/or

$$2\ CoS_{0.89} + AlPO_4 + 4.28\ H_2 \rightleftharpoons 0.5\ Al_2O_3$$
$$+ Co_2P + 2.5\ H_2O + 1.78\ H_2S. \quad (27)$$

Figure 32 represents the relative amount of cobalt present as cobalt phosphides as a function of the phosphorus content.

The formation of Co_2P and Ni_2P requires severe conditions, as stated previously (temperatures between 723 and 1000°C and high pressures). In the case of NiP/Al or NiMoP/Al catalysts, only a treatment with the mixture $PH_3/H_2/H_2S$ is effective to give the Ni_2P phase under conventional sulfidation conditions because deposited phosphorus preferably interacts with the alumina support (71); the gas feed then must contain phosphine. However, carbon and silica supports which interact less with phosphorus can give the Ni_2P phase even under conventional conditions. Andreev *et al.* (103) suggested the formation of the compound $NiPS_3$ after sulfidation of NiMoP/Al. However, Robinson *et al.* (71) showed that $NiPS_3$ decomposes into Ni_2P under hydrotreating conditions, even in the presence of H_2S. Note, however, that $NiPS_3$ is an interesting phase which catalyzes the oxidation of sulfide S^{2-} species (104).

In conclusion, the addition of phosphorus may increase the sulfidation of nickel or cobalt in Ni/Al or Co/Al samples at temperatures up to about 700°C as a consequence of the decreasing amounts of stable aluminates that are formed; at higher temperatures the decrease is caused by the formation of phosphides. However, the Ni_2P and Co_2P phases might be

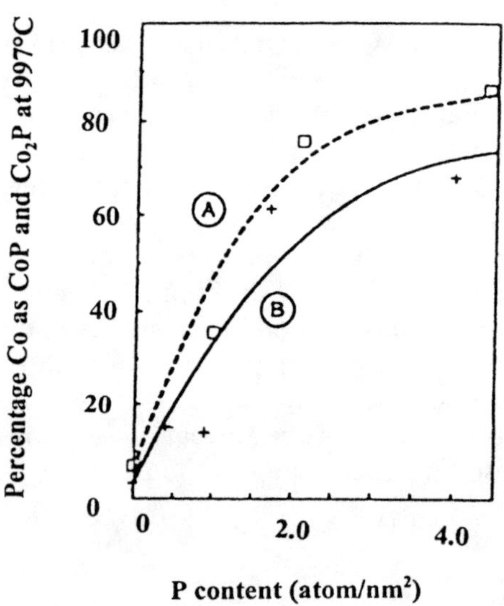

FIG. 32. Variation of the percentage of cobalt reacted with $AlPO_4$ and H_2 to CoP and Co_2P at 1000°C as a function of phosphorus content. a, CoMoP/Al; b, CoP/Al [reprinted with permission from Mangnus *et al.* (70); copyright 1991 Elsevier Science].

present in minor proportions on alumina-based catalysts at conventional hydrotreating temperatures.

Topsøe *et al.* (*98*) found from the IR spectra of adsorbed NO that in sulfided CoMo/Al or NiMo/Al catalysts the phosphorus addition favors a less sulfided environment of the "CoMoS" or "NiMoS" phases. Iwamoto and Grimblot (*67*) determined that phosphorus decreases the sulfidability of molybdenum in NiMo/Al as in the case of Mo/Al, whereas the addition of phosphorus has a less positive effect on nickel sulfidability because the nickel species in NiMo/Al are predominantly associated with molybdenum species rather than with alumina.

G. VIBRATIONAL AND NMR DATA CHARACTERIZING PHOSPHORUS-BASED ALUMINA CATALYSTS

Following Section II.J, which reported some characteristic vibrational and NMR data representing phosphorus-based reference compounds (Tables VI–VIII), it is interesting to propose similar compilations concerning phosphorus-containing alumina catalysts in the dried and calcined forms. These vibrational and NMR data (Tables IX–XII) are more informative in characterizing hydrotreating catalysts than are XRD data because the catalysts are amorphous.

VI. Activities of Phosphorus-Based Catalysts

There are many catalytic reactions involved in hydrotreating processes. Reactions leading to removal of hetero-atoms from the oil fractions, such as HDS, HDN, and hydrodemetalization (HDM), are of primary importance in petroleum refining. In addition, side reactions such as hydrogenation (HYD) and isomerization may considerably improve the quality of the products obtained. Hydrocracking is also of considerable importance for upgrading the heavy fractions. In general, the influence of phosphorus in the catalyst formulations depends on its content. At high loadings, the presence of phosphorus usually leads to negative effects on the hydrotreating reactions. This fact is attributed to the decreases of SSA and of the active phase dispersion; modifications of the pore structure may also be important. At low and medium phosphorus contents, the reported results are often different. The differences may be attributed to the wide variety of physicochemical properties exhibited by the catalysts in the presence of phosphorus, as shown previously. In this section, we examine both the positive and the negative influences of phosphorus in classical hydrotreating catalysts. Table XIII is a summary of the main trends associated with

TABLE IX

IR Data Characterizing Selected Phosphorus-Based Catalysts after Drying and Calcination

IR absorptions, cm^{-1}	Assignment	Reference catalyst	Reference
3785–3800	Al(octa)–OH terminal type Ib	Alumina	105
3760–3780	Al(tetra)–OH type Ia	Alumina	105
3730–3745	Al–OH–Al double-bridge type IIa,b	Alumina	105
3697–3710	Al–OH–Al triple-bridge type III	Alumina	105
3677	P–OH	P/alumina	77
3558	Strongly held water	P/alumina	38
3250	Interaction between neighboring P–OH or between P–OH and Al–OH	P/alumina	44
1700–1300	Physically adsorbed water	Alumina	106
1620	Physically adsorbed water	Alumina	107
1220	Phosphoryl group	P/alumina	77
1180–1130	P–O (wavenumber decreases with decreasing bond order)	P/alumina	77
1127–1152	Formation of polyphosphate	NiMoP/alumina	30
1120	P=O(t)	NiMoP/alumina	29
1119	Monolayer of phosphorus oxo-species	P/alumina	29
1050	Bridge band P–Ob–Mo	NiMoP/alumina	29
1000–1100	P–O–Mo		30
945, 920, 730, 815	P–Ni–Mo heteropoly compounds	NiMoP/alumina	29
943, 912, 893–897	Polymolybdate	Dried NiMoP/alumina	30
936	Asymmetric ($PO_2(H_3)$)	Dried NiMoP/alumina	30
912–918	Weakly bound bridge bands in MoO_4 tetrahedra	Dried NiMoP/alumina	29
900	Asymmetric P–O–P stretching	P/alumina	26
840	Symmetric ($PO_2(H_2)$)	Dried NiMoP/alumina	30
815	Asymmetric P–O–P stretching	P/alumina	26
730–750	P–O–P in polyphosphate	NiMoP/alumina	30
700–550	Al–O stretching	P/alumina	26

phosphorus in catalyst formulations for hydrotreating reactions that have often been conducted with probe molecules as reactants.

A. INFLUENCE OF PHOSPHORUS ON HYDRODESULFURIZATION

At the outset, it is important to know whether a very small quantity of phosphorus leads to detectable changes in the catalytic performance. Ledoux *et al.* (*123*) reported that 120 ppm of phosphorus only slightly affects

TABLE X

Raman Data Characterizing Selected Phosphorus-Based Catalysts after Drying and Calcination

Raman band positions (cm^{-1})	Assignment	Reference catalyst	Reference
200–220	Mo–O–Mo deformation	Mo/alumina, CoMo/alumina,	*33, 108, 109*
310–370	Mo=O bending vibration	or $Mo_6O_{19}^{2-}$, $Mo_7O_{24}^{6-}$,	
500–650	Symmetric Mo–O–Mo stretching	$Mo_8O_{26}^{4-}$	
700–850	Antisymmetric Mo–O–Mo stretching		
900–1006	Terminal Mo=O stretching (depends on Co interaction)		
550	Ni=O stretching vibration	Ni/alumina	*32*
560, 1100	γ-Alumina	Sol-gel alumina	*110*
890, 550	NiMo heteropoly anion	Dried NiMo/alumina	*111*
897–900	Free MoO_4^{2-} ion	Mo/alumina	*33, 112*
918	Monomolybdate	Dried NiMo/alumina	*111*
930	Isolated tetrahedral Mo species	Mo/Alumina	*113*
900	Octamolybdate	Mo/alumina	*111*
950–960	Heptamolybdate	Mo/alumina	*111, 113*
930–940	Hydrated Mo species	Calcined NiMo or CoMo/alumina	*32*
940–965	Mo=O stretching vibration (decrease with degree of polymerization)	Mo/alumina	*33, 114*
914–921	Weakly interacting Mo–O–Mo band	Mo/alumina	*33, 114*
1006	Terminal Mo=O in dehydrated form	Mo/alumina	*32*

the thiophene HDS activity and the physicochemical characteristics of a CoMo/Al-P catalyst. It is concluded, therefore, that the HDS activity of catalysts should not be affected by trace or impurity levels of phosphorus in the catalyst formulations.

Bouwens *et al.* (*100*), Eijsbouts *et al.* (*94*), and Iwamoto and Grimblot (*40*) reported that phosphorus has no detectable promotion effect on thiophene HDS catalyzed by Mo−P/Al and MoP/Al. On the other hand, Fierro *et al.* (*124*) indicated that the thiophene HDS activity of coimpregnated MoP/Al catalysts reaches a maximum for a loading of 4 wt% P_2O_5, whereas the activities of sequentially impregnated Mo−P/Al catalysts are nearly independent of the phosphorus content. Kim and Woo (*84*) reported that

TABLE XI

Al-NMR Chemical Shift Data Characterizing Selected Phosphorus-Based Catalysts after Drying and Calcination

Chemical shift (ppm)[a]	Assignment	Reference catalyst	Measurement conditions					Reference
			Resonance frequency (MHz)	Pulse length (μs)	Recycling time (s)	Spinning frequency (kHz)	Reference sample	
71	Tetrahedral Al	NiMoP/alumina	130.32	5		2.5–4	$AlCl_3$	61
75 to 73	Tetrahedral Al	γ-Alumina						115
72	Tetrahedral Al	P/alumina	78.3	0.2–0.5		2	$KAl(SO_4)_2$	38
67	Tetrahedral Al	Calcined sol-gel alumina	104.26				$Al_2(SO_4)_3$	107
53	Tetrahedral Al	Calcined sol-gel alumina	78.2	3	2	3.5	$Al(H_2O)_6$	116
41, 5	$AlPO_4$	Bulk $AlPO_4$	130.32	5		2.5–4	$AlCl_3$	61
40	Crystalline $AlPO_4$	P/alumina	78.3	0.2–0.5		2	$KAl(SO_4)_2$	38
38	Amorphous $AlPO_4$	P/alumina	78.3	0.2–0.5		2	$KAl(SO_4)_2$	38
30	Penta Al or tetrahedral Al	Al–Si glass						117
29 to 27	Penta $Al(OAl)_5$							117
22	Octahedral Al	NiMoP/alumina	130.32	5		2.5–4	$AlCl_3$	61
20 to 0	Octahedral Al	Sol-gel alumina						116

13	$[Al(OH)_n(H_2O)_{6-n}]_n(MoO_4)$ ($n = 1, 2$)	MoP/alumina	78.42		1	6–7	$KAl(SO_4)_2$	*119*
10	Unreacted Al alkoxide	Sol-gel alumina	78.21	4		3		*120*
10 to 8	Octahedral Al	γ-Alumina						*115*
8	Octahedral Al	NiMoP/alumina	130.32	5	2.5–4		$AlCl_3$	*61*
7	Octahedral Al	Calcined sol-gel alumina	104.26				$Al_2(SO_4)_3$	*107*
5	Octahedral Al	α-Alumina	70.4	1.6		2.6	$Al(H_2O)_6$	*121*
4	Octahedral Al	Sol-gel alumina	78.2	3	2	3.5	$Al(H_2O)_6$	*116*
4	$Al_2(OH)_5$, $Al_3(OH)_8$, etc.	Dried alumina	104.26			3.2	$Al_2(SO_4)_3$	*107*
0	Octahedral Al	P/alumina	78.3	0.2–0.5		2	$KAl(SO_4)_2$	*38*
0	$Al(H_2O)_6$ monomer	Dried alumina	104.26			3.2	$Al_2(SO_4)_3$	*107*
0	$Al(OH_2)_m(OHC_3H_8)_{6-m}$	Sol-gel dried alumina	104.26			3.2	$Al_2(SO_4)_3$	*107*
−1	Amorphous $Al(MoO_4)_3$	NiMoP/alumina	130.32	5		2.5–4	$AlCl_3$	*61*
−8 to 11	Octahedral $Al(OP)_n(OAl)_{6-n}$	$AlPO_4 + Al_2O_3$	104.26	2	2	5–5.5	$Al(H_2O)_6$	*122*
−8 to −13.7	$Al(MoO_4)_3$	MoP/alumina	78.3	0.2–0.5		2	$KAl(SO_4)_2$	*38*
−12		Bulk $Al_2(MoO_4)_3$	130.32	5		2.5–4	$AlCl_3$	*61*
−13 to −22	Octahedral $Al(OP)_6$							*118*
−14	$Al(MoO_4)_3$	Calcined sol-gel alumina	78.2	3	2	3.5	$Al(H_2O)_6$	*116*
−17 to −4.6	Octahedral Al	Sol-gel dried alumina	78.21	4		3		*120*

[a] The chemical shifts reported here are only experimental values; the correction of induced quadrupolar effect has not been taken into account.

TABLE XII

^{31}P-NMR Chemical Shift Data Characterizing Selected Phosphorus-Based Catalysts after Drying and Calcination[a]

Chemical shift (ppm)	Assignment	Reference compound	Measurement conditions				Reference sample	Reference
			Resonance frequency (MHz)	Pulse length (μs)	Recycling time (s)	Spinning frequency (kHz)		
11 to −5	Tetrahedral P(O)$_4$							118
2 to −5	Tetrahedral P(OP)(O)$_3$							34
	Monomeric phosphate or							
−8 to −10	short-chain polyphosphate	Dried P/alumina	121.7	2		6–7	H$_3$PO$_4$	38
−10	Terminal polymeric phosphate	Dried P/alumina	121.7	2		6–7	H$_3$PO$_4$	38
−10 to −20	Polymeric phosphate	Dried P/alumina	121.7	2		6–7	H$_3$PO$_4$	38
−15	Tetrahedral P(OAl)$_2$(OH)$_2$							118
−15 to −25	Tetrahedral P(OP)$_2$(O)$_2$							118
−17 to −20	Internal polymeric phosphate	MoP/alumina	121.7	2		6–7	H$_3$PO$_4$	38
−20	Tetrahedral P(OAl)$_3$(OH)							118
−23 to −34	Tetrahedral P(OAl)$_4$							118
−24	Tetrahedral P−O−Al, P(OAl)$_4$	AlPO$_4$ + Al$_2$O$_3$	161.98	3	3.6	5–5.5	Na$_2$HPO$_4$	122
−25 to −30	AlPO$_4$		200.47	6	2	2.5–4	H$_3$PO$_4$	61
−25 to −26	Crystalline AlPO$_4$	MoP/alumina, P/alumina	121.7	2		6–7	H$_3$PO$_4$	38
−29	AlPO$_4$	P/alumina						25
−30 to −51	Tetrahedral P(OP)$_2$(OAl)$_2$							118
−32	Crystalline AlPO$_4$	Dried P/alumina	121.7	2		6–7	H$_3$PO$_4$	38

[a] Note, for example, that the difference between PO$_4$ and P(OP)O$_3$ species refers to the influence of phosphorus in the second coordination shell of tetrahedral phosphate.

TABLE XIII

Hydrotreating Tests Conducted with Phosphorus-Containing Catalysts

| Reaction | Chemical composition (wt%) | | | | SSA (m²/g)[a] | Method of preparation | Reaction conditions | | | | | | | | Reference |
	Co	Ni	Mo	P			Reactant	T (°C)	P (kg/cm²)	Pretreatment	Additive	Selectivity	P effect	Memo	
HDS			7	4.2	280	SI	Thiophene	400	AP	PS			→		80
			7	1.8	270	SI	Thiophene	400	AP	PS 400°C			→		100
			5.3	0.4–2.6	188	SI	Thiophene	350–400	AP	PS		Butane ↗	→		124
			8	0.75–3	270	SI; Fe(2.8%)–Mo–P	Thiophene	400	AP	PS			→		126
			17	1–11	(503)	SG	Thiophene	300	AP	PS 400°C			→		40
			27	1–11	(503)	SG	Thiophene	300	AP	PS 400°C		Butane ⤵	↘		40
			6.3	0.4–2.7	189	CO	Thiophene	351–400	AP	PS			⌢		124
			10	0.3–3.3	220	SI	Thiophene	300	AP	PS			⌢		84
			0.8–3	1	193	SI	Thiophene	400	AP	PS 500°C			↗		105
		3.2		4.2	280	SI	Thiophene	400	AP	PS			→		94
		7–18		1–16	(503)	SG	Thiophene	400	AP	PS 400°C			↗		67
		1	7	0.6–4.3	280	SI; Ni–Mo–P/Al	Thiophene	400	AP	PS			⌢		94
		3.5	7	0.7–6.2	280	SI	Thiophene	400	AP	PS			↘		94
		2.3	10	0.6–6	266	CO	Thiophene	350	AP	PS			⌢		87
		3	10	0.25–2.5	259	CO	Thiophene	350	AP	PS 300°C			⌢		60
		2.3	10	0.3–7	198	CO	Thiophene	400	AP	PS 500°C			⌢		44, 77
		2.3	10	0.3–7	198	SI; P–Ni–Mo, Ni–Mo–P	Thiophene	400	AP	PS 500°C			⌢	P–Ni–Mo < Ni–Mo–P	44, 77
	2		8	120 (ppm)	240	HG + CO; CoMo/Al–P	Thiophene	198	AP	PS 400°C			→		123
	uk		10	0.3–4	276	SI; Co–MoP/Al, pH = 3	Thiophene	400	AP	nPS		Butane ↘	⌢		73
		6	20	1–11	(503)	SG	Thiophene	300	AP	PS 400°C			⌢		67
	3		8	0.5–5	209	SI; P–Co–Mo, Co–Mo–P	Thiophene	400	AP	PS			↘	P–Co–Mo < Co–Mo–P	125

(continued)

TABLE XIII (*continued*)

Reac-tion	Co	Ni	Mo	P	SSA (m²/g)ᵃ	Method of preparation	Reactant	T (°C)	P (kg/cm²)	Pretreatment	Additive	Selectivity	P effect	Memo	Reference
	(2)	(2)	9.6	2.1	uk	CO	DBT	350–370	35	PS 320°C			↘		97
	3		8	4.5–21	uk	HG + SI	Residual oil	390	78	PS 320°C			⌃		80
		2.3	10	0.3–7	198	SI; Ni–Mo–P	Gas oil	410	69	PS 440°C	Quinoline		⌃		74
		2.3		0.4–2.6	188	SI, CO; NiP < Ni–P	Gas oil + pyridine	325–375	30 bar	PS 350°C	N 0.08%		↗	NiP < Ni–P	69
		3.5		0.8–3.7	uk	CO; pH = 2	(VGO)	360	53	PS 350°C			⌃		88
	12.2		6.5	4–2.5	—	NS	Residual oil	470	100	PS			↗		127
HDN		uk	uk	uk	uk	CO	Indole	350	25–81	PS			↗		98
		3	8	3–4	228	SI; Ni–MoP/Al	OPA	320	30	PS 370°C	DMDS		↗		81
		1–3.5	7	0.6–6.2	280	SI	Quinoline	350–390	30	PS	DMDS		↗		37, 94
		2		1.5–10	uk	CO	HVGO	360	54	PS 350°C			↗		62
		2	10	3.7	uk	CO	HVGO	360	54	PS 350°C			↗		62
			10	3.7	uk	CO	HVGO	360	54	PS 350°C			↗		62
		2.3		0.4–2.6	188	SI	Gas oil + pyridine	325–345	30 bar	PS 350°C	Gas oil (S 1.28%)		↗		69
		2.3		0.4–2.6	188	CO	Gas oil + pyridine	325–345	30 bar	PS 350°C	Gas oil (S 1.28%)		→		69
		2.3	10	0.3–7	198	SI; Ni–Mo–P/Al	Gas oil	410	69	PS 440°C			→		44, 74
		2.3	10	0.3–7	198	SI; Ni–Mo–P/Al	Quinoline	410	69	PS 440°C			⌃		44, 74
		2	9.6	2.1	uk	CO	Quinoline	308	60	PS			↗	Low quinoline	97
		2	9.6	2.1	uk	CO	Quinoline	308	60	PS			↘	High quinoline	97
			9.3	2.8	252	CO	Pyridine	350	60	PS350 + PR350		Unsaturated ↘	↘		79, 95
	3		8	1–4	228	SI; Ni–MoP/Al	Piperidine	320	30	PS 370°C	DMDS		↘	Ni–MoP/Al	81

	3	8	2–4	228	SI; Ni–MoP/Al	DHQ	320	30	PS 370°C	DMDS	↘		81
HYD		27	1–11	(503)	SG	Cyclohexene	350	AP	PS 400°C		↘		85
	3	8	2	228	SI; Ni–MoP/Al	Cyclohexene	220–370	30	PS 370°C	Pentylamine, DMDS	↘		82
	3	8	0.5–5	209	SI; P–Co–Mo, Co–Mo–P	1-Hexene	400	AP	PS	H2S	↘	P–Co–Mo < Co–Mo–P	125
		9.3	2.8	252	CO	Isoprene	50	AP	PS350 + PR547		↘		79, 95
uk		10	0.3–4	276	SI; Co–MoP/Al, pH = 3	Benzene	320–360	30	PS	Thiophene + CS2	↘		73
	2.3	10	0.3–7	198	SI; Ni–Mo–P/Al	Gas oil	410	69	PS 440°C	(Quinoline)	∧		44, 74
		9.3	2.8	252	CO	Toluene	350	60	PS350 + PR350		↗		79, 95
		5.3	0.4–2.6	188	SI	1-Hexene	300	AP	PR 500°C		↗		72
		5.3	0.4–2.6	188	CO	1-Hexene	300	AP	PR 500°C		→		72
HYC	2.3	10	0.3–7	198	CO	DIPB	400	AP	PR 500°C		∧	Ni–Mo–P < P–Ni–Mo < NiMoP	74
	2.3	10	0.3–7	198	SI; P–Ni–Mo	DIPB	400	AP	PR 500°C		∧		74
	2.3	10	0.3	198	SI; Ni–Mo–P	DIPB	400	AP	PR 500°C		↘		74
	2.3	10	0.3–7	198	CO	DIPB	400	AP	PS 500°C		∧	NiMoP < P–Ni–Mo < Ni–Mo–P	74
	2.3	10	0.3–7	198	SI; P–Ni–Mo	DIPB	400	AP	PS 500°C		∧		74
	2.3	10	0.3–7	198	SI; Ni–Mo–P	DIPB	400	AP	PS 500°C		↘		74
	3	8	0.5–5	209	SI; P–Co–Mo, Co–Mo–P	Isooctene	400	AP	PS 500°C		↘	P–Co–Mo < Co–Mo–P	125
	6–18	20–27	1–16	(503)	SG	Cyclopropane	280	AP	PS 400°C	Propane ∧	↗	P < NiP < MoP < NiMoP	67, 85
ISCM		27	1–11	(503)	SG	Cyclohexene	350	AP	PS 400°C		↗		85
		9.3	2.8		CO	Pentadiene	50	AP	PS350 + PR547		↘		79
		2–4.6	0.4–2.5	208	EA	Cyclohexene	400	AP	PR 400°C		↘		59
HDM	12.2	6.5	4–2.5	—	NS	Residual oil	470	100	PS		↗		129

Note. ( ), estimated value; uk, unknown; CO, coimpregnation; SI, sequential impregnation; EA, equilibrium adsorption; HG, hydrogel method; SG, sol-gel method; NS, no supported; AP, atmospheric pressure; PS, presulfiding; PR, prereduction; DBT, dibenzothiophene; DIPB, diisopropylbenzene; DHQ, decahydroquinoline; OPA, ortho-propylaniline; DMDS, dimethyldisulfide; THQ, 5-tetrahydroquinoline.

[a] Specific surface area of support before preparation.

the thiophene HDS activity of Mo—P/Al catalysts containing <5 wt% phosphorus is slightly higher than that of a phosphorus-free catalyst. Lewis and Kydd (*105*) showed that the positive influence of phosphorus on thiophene HDS activity is more significant at higher molybdenum loadings.

Recently, "Mo—P" heteropoly compounds have been used as new precursors of the catalysts. Van Veen *et al.* (*25*) showed that the order of thiophene HDS activity varies with the reducibility of the molybdenum-containing phase according to the following sequence: Monomolybdate < heptamolybdate < octamolybdate < "Mo_9P" < "$Mo_{11}P$." Griboval *et al.* (*22*) reported that chemically reduced "$Mo_{12}P$" allows the preparation of salts such as $Co_{7/2}[PMo_{12}O_{40}]$ which give catalysts with higher thiophene HDS activity than those obtained with the cobalt salt of unreduced "$Mo_{12}P$" because of higher stability of the former after drying on the support surface.

López-Agudo *et al.* (*69*) found that gas oil HDS activity of NiP/Al catalysts prepared by both coimpregnation and sequential impregnation increases with increasing amounts of phosphorus up to a loading of 6 wt% P_2O_5. Morales *et al.* (*88*) reported that activities for HDS of thiophene, vacuum gas oil (VGO), and deasphalted crude oil catalyzed by NiP/Al give a maximum at 7 wt% P_2O_5. A correlation between the HDS performance and the amount of octahedral Ni^{2+} in the oxide form of the catalysts suggests that octahedrally coordinated nickel cations are possible precursors of the active phase. Iwamoto and Grimblot (*67*) indicated that phosphorus addition increases thiophene HDS activity of NiP/Al sol-gel catalysts due to the increasing sulfidability of the nickel species. Andreev *et al.* (*103*) reported that a catalyst prepared with $NiPS_3$ mixed with Al_2O_3 shows high activity for thiophene HDS, whereas selectivity for hydrogenated products is lower than that obtained with NiMo/Al and CoMo/Al. The authors inferred on the basis of this result that the promotion effect of phosphorus in the NiMoP/Al catalyst should be attributed to the formation of the $NiPS_3$ phase during the presulfidation step. However, Robinson *et al.* (*71*) reported that $NiPS_3$ decomposes into Ni_2P under typical hydrotreating conditions even in the presence of H_2S. Therefore, Ni_2P could be the actual active species rather than $NiPS_3$ for HDS if it exists on NiP/Al and NiMoP/Al catalysts.

Muralidhar *et al.* (*125*) reported that the thiophene HDS activities of both Co—Mo—P/Al and P—Co—Mo/Al catalysts do not change as a result of a phosphorus loading of 0.5 wt%, whereas the activity is less at a loading of 5 wt%. Ramselaar *et al.* (*126*) showed that phosphorus has no positive effect on thiophene HDS catalyzed by Fe—Mo—P/Al. Eijsbouts *et al.* (*37, 94*) concluded that thiophene HDS catalyzed by Ni—Mo/Al is not promoted considerably by phosphorus addition, whereas the selectivity for formation of unsaturated hydrocarbons increases slightly. On the other hand, Atanasova *et al.* (*87*) found a maximum activity for thiophene HDS of NiMoP/Al catalysts containing ~2 wt% P_2O_5. Walendzieski (*73*) observed a weak maximum for thiophene HDS catalyzed by a Co—MoP/Al containing 1.5 wt% P, whereas selectivity for hydrogenated product was lower than that measured for the phosphorus-free catalyst. Chadwick *et al.* (*60*) observed that thiophene HDS catalyzed by NiMoP/Al shows a broad maximum for phosphorus loadings of about ~1 wt% P. Lewis *et al.* (*44, 74*) and Jones *et al.* (*127*) observed a large positive effect on HDS of thiophene and quinoline-spiked gas oil at ~1 wt% P, especially with Ni—Mo—P/Al catalysts, whereby phosphorus is impregnated before the other species. Kemp and Adams (*128*) reported that NiMoP/Al—P and CoMoP/Al–P catalysts prepared by the hydrogel method show higher HDS activities for cracked heavy gas oil than a conventional commercial catalyst. Chen *et al.* (*80*) reported that HDS of an atmospheric residue catalyzed by CoMo/Al–P shows a maximum activity at ~5 wt% P. They also correlated reducibility of the molybdenum species with the HDS activity, as in the case of MoP/Al catalysts prepared with "Mo–P" heteropoly compounds.

Kushiyama *et al.* (*129*) reported that phosphorus also promotes the activity of highly divided unsupported molybdenum and cobalt–molybdenum sulfide catalysts for HDS (and HDM) of crude oil, with a maximum at a P/(Co + Mo) atomic ratio of 2. This result warrants attention because it shows that phosphorus may directly modify the active sulfide phase(s) even in the absence of the alumina support.

It is concluded that phosphorus shows no effect or only a very small positive effect on thiophene HDS, but it decreases the selectivity for hydrogenated products with Mo/Al catalysts. However, the presence of phosphorus seems to be beneficial in promoted molybdenum-based catalysts, especially NiMoP/Al.

B. INFLUENCE OF PHOSPHORUS ON HYDRODENITROGENATION

Positive effects of phosphorus have been reported primarily for HDN reactions for converting both model compounds and industrial feeds. Topsøe *et al.* (*98*) indicated that indole HDN is enhanced by phosphorus addition. Similarly, Eijsbouts *et al.* (*37, 94*) reported that quinoline HDN activity increases with the addition of phosphorus, especially to Ni—Mo—P/Al catalysts. They also observed that phosphorus addition increases the selectivity for the production of unsaturated nitrogen-free hydrocarbons (e.g., propylbenzene). On the other hand, Poulet *et al.* (*79, 95*), Jian and Prins (*81*) and Rico Cerda and Prins (*130*) reported that phosphorus has a negative influence on pyridine HDN catalyzed by MoP/Al. The apparently contradictory nature of these results leaves the influence of phosphorus on HDN unclear. Importantly, however, Jian and Prins (*81*) found that the effect of phosphorus depends on the nature of the reactants and intermediates. For example, phosphorus induces a positive effect on the HDN reactions of quinoline, *ortho*-propylaniline (OPA), and indole and a negative effect on pyridine, piperidine, and decahydroquinoline (DHQ) HDN catalyzed by NiMoP/Al.

Eijsbouts *et al.* (*37, 94*) proposed that phosphorus does not modify the active "Ni—Mo—S" phase but rather leads to the formation of new active sites associated with acidic $AlPO_4$ or with other metal phosphates (e.g., nickel phosphate), or it induces modifications of nickel and molybdenum sulfides; however, the effect of phosphorus on the HDN activity does not correspond to changes in the dispersion of these components. In particular, these authors suggested that the improvement of HDN activity is strongly related to the $AlPO_4$ formation since the quinoline HDN performance of a dual catalytic bed of NiMo/Al and P/Al remarkably increases relative to that of NiMo/Al alone. Lewis *et al.* (*44, 74*) and Jones *et al.* (*127*) reported that the addition of 1–3 wt% P to NiMoP/Al catalysts gives the maximum activity for gas oil and quinoline HDN, corresponding to an optimized surface acidity. López-Agudo *et al.* (*69*) proposed that pyridine HDN catalyzed by Ni—P/Al and NiP/Al is promoted by acidic groups of the $AlPO_4$ phase which are effective for C—N bond hydrogenolysis. Ramirez de Agudelo and Morales (*62*) reported that HDN of VGO catalyzed by NiP/Al is also enhanced by improving the activity for the hydrogenolysis. From a kinetics investigation of the HDN of 5-tetrahydroquinoline catalyzed by NiMo(P)/Al, Rico Cerda and Prins (*130*) inferred that the main difference between the phosphorus-containing and the phosphorus-free catalysts is the change of selectivity in the reaction products. The catalyst with phosphorus is more selective for formation of unsaturated products.

However, other results indicate that the P/Al catalyst has very low activity for C—N or C—C bond breaking (*74, 81*). A detailed kinetics study by Jian and Prins (*81*) showed that phosphorus addition decreases the C—N bond cleavage (rate constants k_1' and k_2' in the reaction networks; Figs. 33a and 33b) and the subsequent alkene hydrogenation (rate constant k_3') reactions in piperidine and DHQ HDN (see Table XIV). On the other hand, the presence of phosphorus increases aromatic ring hydrogenation of OPA

FIG. 33.   HDN reaction networks of probe molecules. a, Piperidine; b, decahydroquinoline; c, orthopropylaniline; d, indole [adapted from Jian and Prins (*81*) and Callant *et al.* (*86*); reprinted with permission, copyright 1995 Baltzer Science and 1995 Comite Van Beheer Van Het Bulletin v.z.w.].

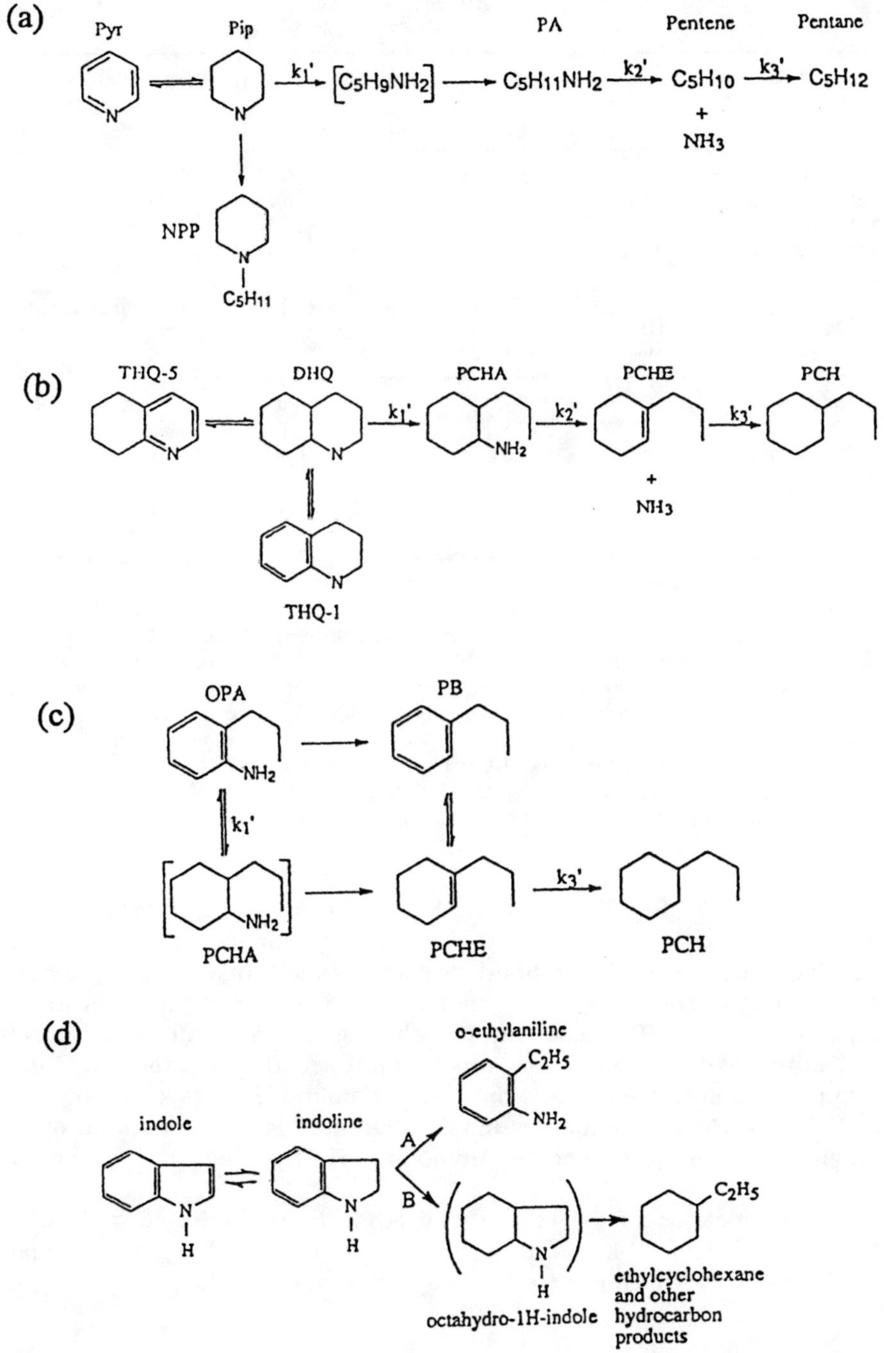

(a)
Pyr
Pip
PA
Pentene
Pentane
NPP
C5H11
k1'
[C5H9NH2]
C5H11NH2
k2'
C5H10
+
NH3
k3'
C5H12
(b)
THQ-5
DHQ
PCHA
PCHE
PCH
k1'
k2'
k3'
NH2
+
NH3
THQ-1
(c)
OPA
PB
k1'
PCHA
NH2
NH2
PCHE
k3'
PCH
(d)
o-ethylaniline
C2H5
NH2
indole
indoline
A
B
N
H
N
H
octahydro-1H-indole
N
H
C2H5
ethylcyclohexane
and other
hydrocarbon
products

TABLE XIV

Kinetics Data for HDN of Probe Molecules[a]

Catalyst	Piperidine			Decahydroquinoline		
	k_1'	k_2'	k_3'	k_1'	k_2'	k_3'
NiMo/Al	0.82	7.4	4.8	0.87	18.7	8.6
NiMoP(1)/Al	0.65	7.5	3.6	0.77	18.0	7.7
NiMoP(2)/Al	0.60	7.3	3.2	0.71	16.8	6.5
NiMoP(4)/Al	0.44	5.4	2.4	0.55	14.4	5.3

Catalyst	HDN (%)	Product composition (%)			Rate constant	
		PCH	PCHE	PB	k_1'	k_3'
P(2)/Al	0	0	0	0	0	0
NiMo/Al	23.5	17.8	4.8	0.9	0.84	14.1
NiMoP(1)/Al	30.2	21.3	7.4	1.5	1.13	11.4
NiMoP(2)/Al	34.6	23.7	9.1	1.8	1.34	10.4
NiMoP(4)/Al	29.9	16.7	11.4	1.8	1.12	6.9

[a] (Top) Effective rate constants for the HDN reaction of piperidine and decahydroquinoline (see reaction network in Figs. 33a and 33b). For piperidine, k_1', k_2', and k_3' correspond to C—N bond cleavage of piperidine to pentylamine, C—N bond cleavage of pentylamine to pentene, and hydrogenation of pentene to pentane, respectively. For decahydroquinoline (DHQ), k_1', k_2', and k_3' correspond to C—N bond cleavage of DHQ to propylcyclohexylamine (PCHA), C—N bond cleavage of PCHA to propylcyclohexene (PCHE), and hydrogenation of PCHE to propylcyclohexane (PCH), respectively. (Bottom) Effect of phosphorus on the HDN reaction of o-propylaniline (see reaction network in Fig. 33c). k_1' and k_3' correspond to hydrogenation of o-propylamine to PCHA and hydrogenation of PCHE to PCH, respectively. All rate constants k have the following units: 10^2 mol of reactant/h.g of catalyst [reprinted with permission from Jian and Prins (81); copyright 1995 Baltzer Science].

(rate constant k_1'; Fig. 33c and Table XIV). This result indicates that phosphorus favors only an increase of aromatic ring hydrogenation but has no positive effect on the C—N bond cleavage. If the former reaction is rate determining, the overall HDN reaction is promoted. For example, phosphorus promotes the HDN reaction only when the intermediate species have an aniline-like structure for which hydrogenation of the aromatic ring is rate controlling. This is the case for aniline HDN through the formation of cyclohexylamine, for indole HDN through the formation of o-ethylaniline (Fig. 33d), and for quinoline HDN through the formation of OPA.

An investigation of the competitive adsorption of selected molecules (131) indicates that at least four distinct sites are involved in the individual HDN reactions, which are (i) the cleavage of aliphatic $C(sp^3)$—N bonds,

(ii) hydrogenation of phenyl groups, (iii) hydrogenation of alkenes, and (iv) hydrogenolysis of $C(sp^2)-N$ bonds of anilines. Phosphorus plays a promoting role of the sites for the last of these reactions, but the rate of the first is negatively affected by phosphorus in HDN of DHQ (*132*).

Jian *et al.* (*133*) investigated the complex influence of H_2S on pyridine and piperidine HDN catalyzed by MoP/Al and Ni—MoP/Al (Table XV). H_2S in the feed decreases the hydrogenation of pyridine into piperidine but increases piperidine HDN, especially on Ni—MoP/Al catalysts. As a consequence, the HDN activity of pyridine increases even if the total conversion is lower. Callant *et al.* (*86*) showed that adsorption of H_2S on a NiMoP/Al catalyst has a positive effect on the C—N bond cleavage but a detrimental effect on indole hydrogenation. An increased H_2S partial pressure favors the transformation of indoline into *o*-ethylaniline by a ring-opening mechanism, whereas a decreased H_2S partial pressure leads to the formation of octahydro-[^{1}H]-indole, which is the totally hydrogenated N-containing compound (Fig. 33d).

The catalyst surface coverage by N-containing compounds also affects the HDN activity. Van Veen *et al.* (*97*) reported that quinoline HDN catalyzed by NiMoP/Al is more strongly influenced by the concentration of N-containing compounds in the feed than is the reaction catalyzed by phosphorus-free NiMo/Al. The authors proposed that quinoline strongly adsorbed on the catalytic sites at high coverage hampers the reaction. At lower concentrations of N-containing compounds, the quinoline HDN activity of NiMoP/Al catalysts is higher than that of NiMo/Al catalysts, and the activities correlate with the catalyst acidities.

All the results suggest that the HDN activity is associated with at least two different active sites, i.e., coordinatively unsaturated hydrogenation

TABLE XV

Pyridine HDN: Influence of H_2S and of Phosphorus[a]

Catalyst	$H_2S/H_2 = 0$				$H_2S/H_2 = 3.0 \times 10^{-3}$ (mol/mol)			
	Conv(%)[b]	pip(%)[c]	C_5(%)[d]	$C_5^0/C_5^=$	Conv(%)[b]	pip(%)[c]	C_5(%)[d]	$C_5^0/C_5^=$
Mo/Al	40	34	3.9	1.7	26	16	7	4.0
MoP(2)/Al	40	34	3.3	1.2	29	19	7	3.9
NiMo/Al	57	49	4.3	4.0	48	18	24	3.8
NiMoP(2)/Al	57	49	4.5	2.9	45	20	20	2.0

[a] Reprinted with permission from Jian *et al.* (*133*); copyright 1995 Comite Van Beheer Van Het Bulletin v.z.w. [b] Conversion of pyridine (%). [c] Piperidine production (%). [d] C_5 production (HDN reaction) (%). [e] Ratio of saturated hydrocarbon to unsaturated hydrocarbon in C_5 products.

sites and acidic sites responsible for $C-N$ bond cleavage. The NiMoP/Al catalyst has fewer surface anionic vacancies than the phosphorus-free NiMo/Al catalysts because it has a worse molybdenum dispersion or sulfidability. Adsorption of H_2S transforms the coordinatively unsaturated sites into acidic sites as follows:

$$\begin{array}{ccc} S \quad \square & & HS \quad SH \\ \diagdown \diagup & & \diagdown \diagup \\ Mo \quad + H_2S \leftrightarrow & & Mo \end{array} \qquad (28)$$

Since both hydrogenation and $C-N$ bond cleavage are necessary for HDN reactions, optimization of the phosphorus loading in the catalyst and of the H_2S partial pressure in the reaction mixture is needed to obtain a maximum HDN activity.

Jian *et al.* (*82*) suggested that the nature of the active sites is not modified by phosphorus addition, but the number of sites or their geometric features are changed. Indeed, phosphorus rarely changes the activation energies or heats of adsorption, but it does increase the adsorption coefficients of N-containing compounds or NH_3 on the catalysts (Fig. 34).

In contrast to the previous interpretation, Topsøe *et al.* (*98*) proposed that the HDS, HDN, and hydrogenation reactions occur on the same catalytic sites and that the effect of phosphorus is probably to decrease the amount of poisoning by adsorbed N-containing species rather than by the number of available hydrogen species on the active sites as shown by equilibrium distributions of the gas phase and adsorbed species. Conse-

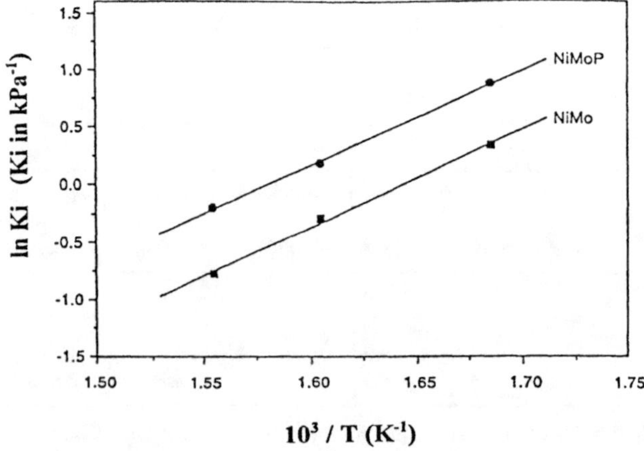

FIG. 34. Temperature variation of the adsorption coefficient of OPA on NiMo(P)/Al catalysts [reprinted with permission from Jian *et al.* (*82*); copyright 1997 Academic Press Inc.].

quently, the overall HDN reaction is effectively activated by phosphorus addition. In short, the effects of phosphorus on HDN reactions are quite complicated and a matter of debate.

C. INFLUENCE OF PHOSPHORUS ON HYDROGENATION

The effect of phosphorus on hydrogenation occurring during HDN has already been considered. Now we examine other results concerning hydrogenation. Fierro *et al.* (*124*) reported that the selectivity for butane formation in thiophene HDS attained a maximum when the MoP/Al catalyst contained 4 wt% P_2O_5. López-Cordero *et al.* (*72*) found that 1-hexene hydrogenation catalyzed by reduced MoP/Al and Mo−P/Al increased with increasing phosphorus content, especially for sequentially impregnated Mo−P/Al catalysts. The addition of 1–3 wt% phosphorus into Ni–Mo/Al gave the maximum aromatic reduction for gas oil (*74, 127*). The optimum surface acidity was obtained for such phosphorus loadings.

On the other hand, Muralidhar *et al.* (*125*) found that the presence of phosphorus increases the rate of the first reaction in aromatic ring hydrogenation but decreases the rates of the successive alkene hydrogenations. These authors also reported that 1-hexene hydrogenation activity of Co−Mo−P/Al catalysts decreases with increasing phosphorus loading. Hubaut *et al.* (*79, 95*) reported that toluene hydrogenation activity increases as a result of phosphorus addition, whereas isoprene hydrogenation does not change as a result of phosphorus addition to MoP/Al. Jian *et al.* (*82, 133*) reported that activity for hydrogenation of cyclohexene decreases with the addition of phosphorus to Ni−MoP/Al catalysts. Iwamoto and Grimblot (*85*) confirmed that the activity for hydrogenation of cyclohexene decreases with the addition of phosphorus to sulfided MoP/Al sol-gel catalysts, whereas the selectivity for propane formation from cyclopropane cracking increases with increasing phosphorus content. Walendziewski (*73*) showed that the activity for benzene hydrogenation decreases as a result of the addition of 3 wt% P_2O_5 to Co−MoP/Al catalysts. Jian *et al.* (*82*) found that N-containing compounds inhibit the alkene hydrogenation relatively more efficiently on NiMoP/Al than on NiMo/Al catalysts because these molecules are more strongly adsorbed on the former catalysts.

These results suggest that the influence on the hydrogenation reactions of the addition of phosphorus to catalysts should depend strongly on the molecules being converted, on the nature of the active sites, and on the reaction conditions (as noted for HDN). However, addition of phosphorus seems predominantly to increase aromatic hydrogenation, as has been shown for the HDN reaction networks. The effect of hydrogen partial pressure is inferred to be relatively important for hydrogenation reactions since the influence of phosphorus is negative when the reactions are performed at only atmospheric pressure.

D. Influence of Phosphorus on Hydrocracking and Isomerization

Lewis *et al.* (*74*) showed that P/Al catalysts have low activities for hydrocracking of diisopropylbenzene, and the activity of NiMoP/Al catalysts increases as a result of addition of 1–3 wt% phosphorus. On the other hand, Muralidhar *et al.* (*125*) reported that isooctene hydrocracking activity of Co—Mo—P/Al catalysts decreases with increasing phosphorus loading. Iwamoto and Grimblot (*67*) reported that cyclopropane hydrocracking increases with increasing phosphorus content, especially with NiMoP/Al catalysts (Fig. 26).

Poulet *et al.* (*79*) reported that the isomerization activity for transforming *trans*- into *cis*-pentadiene on MoP/Al catalysts decreases as a result of phosphorus addition. Gishti *et al.* (*59*) noted that the activity for skeletal isomerization of cyclohexene into methylcyclopentenes on reduced MoP/Al was reduced as a result of phosphorus addition. Iwamoto and Grimblot (*85*) reported, however, that cyclohexene isomerization on sulfided MoP/Al sol-gel catalysts increases with increasing phosphorus content. Since the active sites for hydrocracking and isomerization reactions are predominantly associated with the catalyst acidity, the results should depend strongly on both the nature and surface properties of the catalysts and on the reactants being converted.

E. Influence of Phosphorus on Hydrodemetalization

Results characterizing HDM with phosphorus-containing catalysts are scarce because of the difficulties in conducting experiments with large molecules or heavy feeds and in interpreting the results. Moreover, the influence of catalyst pore size distribution, which may also affect the HDM reaction because of mass transfer effects, makes this kind of investigation difficult. Kushiyama *et al.* (*128*) reported that the addition of oil-soluble phosphorus improves the activity for crude oil HDM catalyzed by unsupported Mo and Co—Mo. They proposed that phosphorus interacts strongly with the vanadium-containing compounds in the feedstock. The positive effect of phosphorus on HDM is also stated in patents (*134*).

F. Influence of Phosphorus on Coke Formation and Catalyst Life

Fitz and Rase (*48*) reported that phosphorus in the catalyst induces a decrease of the total amount of deposited coke for which the H/C ratio

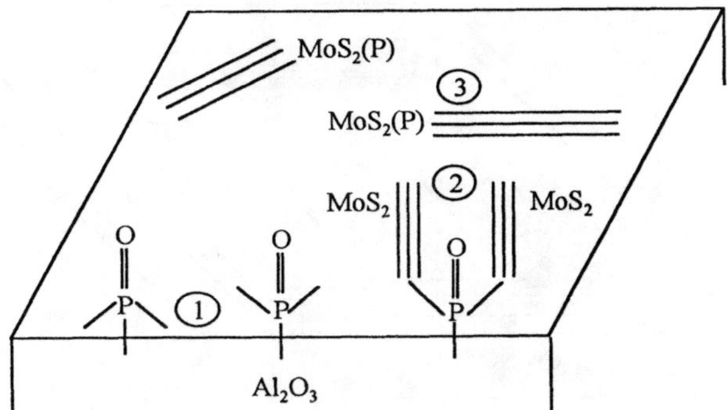

FIG. 35. Schematic representation of the location of phosphorus in the MoP/Al catalyst. (1) Phosphorus oxo-species adsorbed on alumina; (2) phosphorus oxo-species bound to both MoS$_2$ slabs and the alumina support; (3) phosphorus directly incorporated into the MoS$_2$ slab [reprinted with permission from Poulet *et al.* (*79*); copyright 1995 Comite Van Beheer Van Het Bulletin, v.z.w.].

increases during hydrotreating reactions with a thiophene–cyclohexene feed mixture. Spojakina *et al.* (*30*) showed that phosphorus causes a decrease in coke formation on the catalyst during thiophene HDS. Lewis *et al.* (*74*) reported that the catalyst life during HDS of a light gas oil is improved by maintaining the catalyst SSA if the phosphorus loading has been correctly optimized. However, no clear correlation was reported between catalyst SSA and the amount of deposited coke.

VII. Structural Models of Phosphorus-Containing Hydrotreating Catalysts

Structural models of phosphorus-containing hydrotreating catalysts have been proposed from catalyst characterization data and results characterizing the reactions of model compounds. Poulet *et al.* (*79*) reported a structural model of MoP/Al catalysts (Fig. 35) in which the phosphorus-containing species exist in three different states: (i) isolated phosphorus oxo-species interacting with the alumina support, (ii) phosphorus directly incorporated into the MoS$_2$ slab, and (iii) phosphorus oxo-species bound to both MoS$_2$ and the alumina surface.

Mangnus *et al.* (*70*) proposed a structural model for Co−Mo−P/Al catalysts (Fig. 36) whereby phosphorus exists mainly as AlPO$_4$ on the alumina surface. In this model, a fraction of the molybdenum sites at the edges of the MoS$_2$ slabs interact with AlPO$_4$ through complex

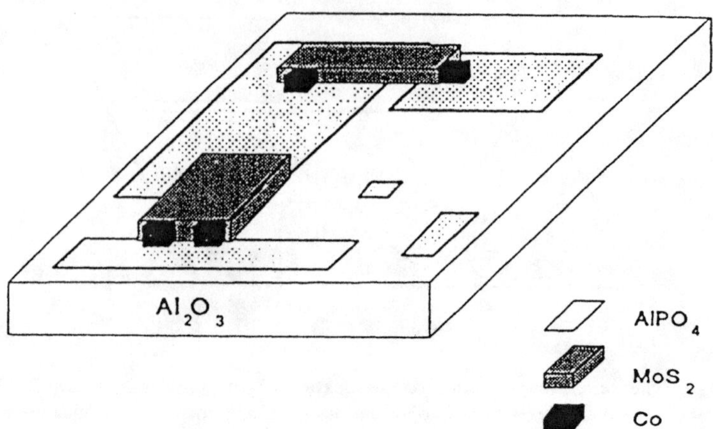

FIG. 36.   Schematic model of sulfided CoMoP/Al catalysts [reprinted with permission from Mangnus *et al.* (*70*); copyright 1991 Elsevier Science].

"Co—Mo—O—P" linkages. On the other hand, it has been reported that there are no obvious structural differences between P-containing and P-free catalysts (*60, 82, 83*). Thus, the influence of the phosphorus would be to affect only the morphology of the MoS_2 slabs and consequently to change the number of active sites and the steric hindrance of adsorption on them. It is clear that more work is needed to clarify the structural issues.

VIII.   Influence of Phosphorus on Other Hydrotreating Catalysts

A.   TUNGSTEN-BASED CATALYSTS

Tungsten-based hydrotreating catalysts have been studied much less than the classical molybdenum-based catalysts. It is expected, however, that phosphorus addition should lead to similar effects in both cases since tungsten is chemically similar to molybdenum. Atanasova *et al.* (*101*) reported that phosphorus increases the thiophene HDS activity, especially that of a sequentially impregnated NiW—P/Al catalyst. Halachev *et al.* (*135*) found that a maximum hydrogenation activity for naphthalene conversion is attained when the catalyst contains 0.6 wt% P_2O_5. Cruz Reyes *et al.* (*58*) reported that phosphorus on a W/Al catalyst notably enhances gas oil HDS and pyridine HDN.

B.   CARBON-SUPPORTED CATALYSTS

Phosphorus in CoMo/C catalysts leads to a significant decrease in thiophene HDS activity (*99, 100*). The weak interaction between the phosphorus oxo-species and the carbon support favors the formation of cobalt phosphate and/or phosphine in the presence of H_2 and leads to poisoning of the active phases. This is not the case with the alumina support because the strong interaction between phosphates and alumina favors the formation of $AlPO_4$, which is not converted into phosphine in the presence of H_2. On the other hand, Robinson *et al.* (*71*) found that Ni—P/C catalysts have high activities for quinoline HDN because of the formation of the active Ni_2P phase.

IX. Impact of Phosphorus Introduction into Industrial Catalyst Formulations

As mentioned in the introduction, the positive influence of phosphorus incorporation into catalysts was first reported on the basis of industrial results. For example, Fig. 37 shows the number of patents filed in Japan concerning phosphorus-containing hydrotreating catalysts during the period 1977–1994. After a period of modest patent activity up to 1982, a marked acceleration followed. This trend indicates that advantages of phosphorus incorporation have been recognized for practical, commercial reasons, notwithstanding the uncertain and controversial scientific basis that is evident from the results presented here. Although the use of phosphorus in commercial catalysts is not well documented in the literature, it is reported in *Oil and Gas Journal* (*136*) that several commercial catalyst suppliers provide phosphorus-containing catalysts (e.g., catalysts TK-551 and TK771 from Haldor Topsøe A/S and IMP-DSD-3 and IMP-DSD-5 from Instituto Mexicano del Petroleo).

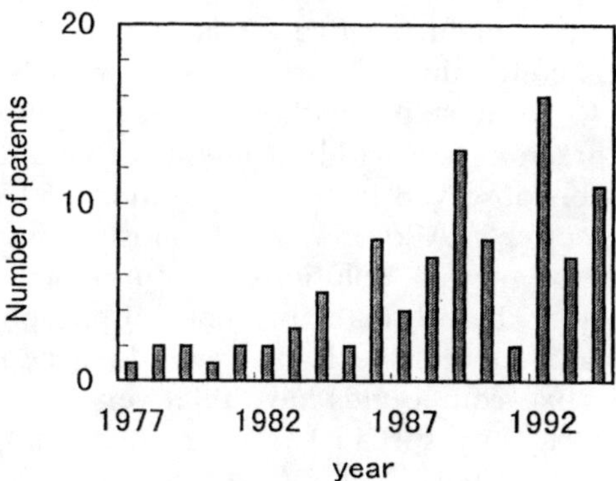

FIG. 37.   Number of patents concerning phosphorus-based catalysts filed in Japan during the period 1977–1994.

The advantages of phosphorus addition to catalyst formulations found in patents can be approximately categorized as follows: (i) optimization of the catalyst pore structure by addition of phosphorus to be applied with certain types of feedstocks such as residual oil, (ii) optimization of the dispersion of Co(Ni) + Mo-containing phases by the presence of phosphorus, (iii) optimization of synergistic effects resulting from complex chemical combinations of phosphorus and other incorporated elements, (iv) optimization of catalyst preparation by use of specific phosphorus precursors, and (v) the use of phosphorus-containing catalysts under specific reaction conditions or processes as well as their use in combination with other hydrotreating catalysts.

Considering the positive effects of phosphorus described here, it is expected that investigations of phosphorus in hydrotreating catalysts will continue and commercial applications will increase.

X. Summary and Conclusions

The catalyst phosphorus content, the nature of the phosphorus precursor used in the preparation, and the method of phosphorus introduction all significantly affect the textural and structural properties of the final catalysts. Therefore, it is important to determine the optimum preparation and activation conditions of hydrotreating catalysts. In particular, pH of the preparative solution and the P/Mo ratio are important factors, but many other factors are also important.

A. INFLUENCE OF PHOSPHORUS ON CATALYST PREPARATION

Phosphorus has two main positive effects on the catalyst preparation. First, the solutions containing the precursors of molybdenum and cobalt (or nickel) tend to be unstable without proper additives. Indeed, light orange or green precipitation within a few minutes after dissolving each metallic salt is often observed in the preparation of Co + Mo or Ni + Mo solutions, respectively. Addition of phosphorus-containing compounds stabilizes such impregnation solutions by formation of "Mo−P" or "Ni−P" oxo-species. Therefore, the presence of phosphorus increases the dispersion of the active metal species supported on the resulting catalysts. Second, "Mo−P" oxo-compounds have relatively weak interactions with the alumina support; they spread over the entire support surface. The coimpregnation method with high P/Mo ratios is especially suitable for preparing high dispersions of the supported components. On the other hand, in sequential impregnations in which phosphorus is introduced first, molybdenum tends to be deposited on the outer surface of the catalyst particles, probably due to the strong interaction between molybdenum and alumina doped with phosphates. Large amounts of phosphorus decrease the dispersion of metal components because the stronger affinity between phosphate and alumina minimizes the interaction between metals and the support. Figure 38 represents schematically the possible interactions relative to the presence of phosphorus oxo-species in the catalyst preparations.

B.   INFLUENCE OF PHOSPHORUS ON TEXTURAL AND STRUCTURAL MODIFICATIONS OF CATALYSTS

One of the most unclear aspects is whether phosphorus can interact with molybdenum species in the final catalyst. In the impregnation steps, phosphorus and molybdenum oxo-species react to form some heteropoly compounds. However, these are usually quite unstable and can decompose during the impregnation step or during the drying and calcination steps. The interaction between the phosphorus and molybdenum species after calcination or sulfidation may depend on the phosphorus content. Phosphorus may not modify the molybdenum-containing species at low phosphorus contents whereby the Al—OH groups reacting with the molybdenum and phosphorus species still remain. In this case, the two elements are supported on alumina independently. However, as a result of increasing the phosphorus loading, its interaction with the molybdenum oxo-species by sharing of oxygen atoms begins to occur. This interaction may be maintained (at least in part) even after sulfidation because phosphorus decreases sulfidation of the molybdenum oxo-species. The oxygen species remaining in the MoS_2

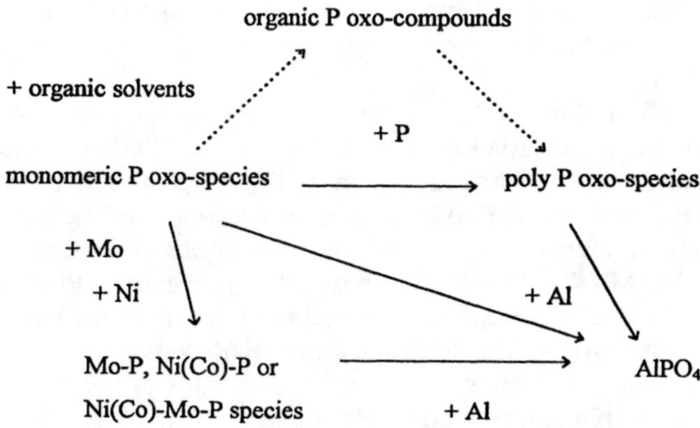

FIG. 38.   Schematic diagram representing the different phosphorus oxo-species present during catalyst preparation.

TABLE XVI

Main Influences of Phosphorus on the States of Nickel, Cobalt, and Molybdenum on the Alumina Surface (Catalyst in the Oxide Form)

Catalyst	−	←——— P ———→	+
Mo/Al		Well-dispersed Mo	Bulk MoO$_3$ Al$_2$(MoO$_4$)$_3$
Ni(Co)/Al	NiAl$_2$O$_4$ or CoAl$_2$O$_4$	Well-dispersed Ni(Co) Ni$_3$(PO$_4$)$_2$ or Co$_3$(PO$_4$)$_2$	Bulk NiO or CoO
Ni(Co)–Mo/Al	Bulk MoO$_3$ β-Ni(Co)MoO$_4$	Well-dispersed Ni(Co) and Mo	Bulk MoO$_3$ β-Ni(Co)MoO$_4$ → α-Ni(Co)MoO$_4$

slabs (as molybdenum oxy-sulfide) contribute to the generation of strong acid sites, such as those found in AlPO$_4$. Addition of a large quantity of phosphorus then favors the segregation of bulk molybdenum oxides such as MoO$_3$ and Al$_2$(MoO$_4$)$_3$. The influence of phosphorus on the states of nickel, cobalt, and molybdenum on the alumina surface in oxide catalysts is shown schematically in Table XVI.

It is remarkable that AlPO$_4$ is not detected at low phosphorus loadings; rather, it is observed only at phosphorus loadings higher than the amount corresponding to the theoretical monolayer coverage of the alumina surface by phosphate ions (*111*). Phosphorus may also interact with nickel and cobalt to create, under certain conditions, new active phases such as Ni$_2$P and Co$_2$P. However, these phases are probably present as minor species on the alumina-based catalysts because phosphorus has a stronger affinity for alumina than for nickel or cobalt species in conventional hydrotreating conditions.

The formation of mixed "Mo−P−S" compounds is thermodynamically restricted at temperatures lower than 1000°C (*70*). This restriction implies that the sulfur atoms in MoS$_2$ are not directly replaced by phosphorus atoms. In the same way, phosphorus does not regularly occupy the edges of the MoS$_2$ platelets through bonds with sulfur atoms as in the case of the promoted "CoMoS" or "NiMoS" phases. The presence of P(white), P(red), and P(black) on catalysts can also be excluded because they have extremely high vapor pressures under hydrotreating conditions.

Phosphorus also modifies the textural properties of the MoS$_2$ slabs (depending on the phosphorus content). It tends to increase the number of stacked MoS$_2$ layers. These modifications might lead to the following positive effects on the catalyst:

Optimization of the steric coordinatively unsaturated sites (CUS) configuration

Increase in the proportion of the "NiMoS" phase (type II) which has a higher HDS activity and less steric hindrance of adsorption than the type I phase

Optimization of the ratio of rim sites to edge sites in the MoS_2 slabs according to the model described by Daage and Chianelli (137).

Considering all the investigations and remarks presented here, we propose (Fig. 39) a model structure of the active phase of MoP/Al and NiMoP/Al catalysts. According to the model, phosphorus creates at least two species on the catalysts. One is isolated phosphorus oxo-species bound to alumina as monomeric phosphate (polymeric phosphorus oxo-species and $AlPO_4$ are not shown in the figure). The other species is phosphorus interacting with both alumina and MoS_2. It prevents the complete sulfidation of molybdenum oxo-species and creates acidic OH groups on the MoS_2 slabs. If cobalt or nickel is present, it can occupy the edges of MoS_2 as in the classical "CoMoS" and "NiMoS" phases.

C. INFLUENCE OF PHOSPHORUS ON HYDROTREATING REACTIONS

Phosphorus has positive effects on isomerization and hydrogenolysis reactions, which require acidic properties. Phosphorus also improves activity for hydrogenation reactions of aromatic rings, especially on NiMoP/Al catalysts. Phosphorus shows minor effects in thiophene HDS, and it may improve the HDS activity of catalysts for heavier sulfur-containing molecules such as those in VGO, in which hydrogenation of aromatic rings may

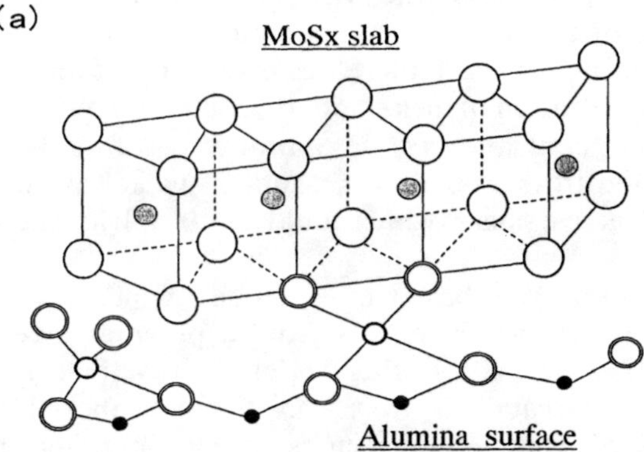

FIG. 39. Model representing the interaction between MoS_2 and alumina through phosphorus oxo-species. (a) Unpromoted MoP/Al; (b) Ni-promoted MoP/Al.

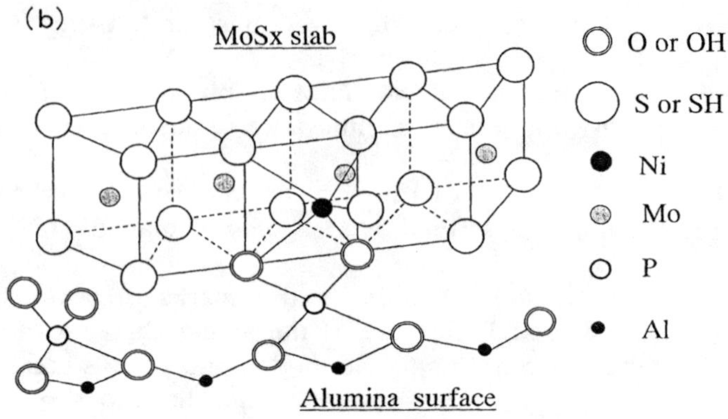

Fig. 39.   (*continued*)

occur before $C-S$ bond scission is effective. Phosphorus also promotes HDN reactions when hydrogenation of aromatics rings is rate determining. In all cases, large amounts of phosphorus are detrimental, no matter what the hydrotreating reaction.

Now, we discuss why phosphorus can enhance hydrogenation of aromatic rings. Daage and Chianelli (*137*) proposed that HDS/HYD selectivity during conversion of dibenzothiophene depends on the MoS_2 slab morphology (number of stacked MoS_2 layers) in unsupported catalysts. The hydrogenation reaction proceeds mainly on rim sites, whereas both rim and edge sites are responsible for the HDS reaction. This means that stacking of the MoS_2 layers enhances HDS rather than hydrogenation. However, this interpretation is not in full agreement with experiment because the addition of phosphorus preferably increases the MoS_2 stacking while also increasing the hydrogenation of aromatic rings. Spojakina *et al.* (*30*) proposed that the presence of phosphorus both increases dispersion of the molybdenum and prevents the penetration of nickel into the alumina support. Thus, phosphorus is inferred to contribute to the optimization of the Ni/Mo ratio in the active phase and to increase the amount of nickel available to form the "$Ni-Mo-S$" active phase, which could also be advantageous for hydrogenation.

Another possibility is based on a model of active sites proposed by Hubaut *et al.* (*138*), which derives from a previous investigation by the same group (*139*) (Fig. 40). In this model, the reactions involved in hydrotreating on Mo/Al catalysts are associated with the CUS configuration. Hydrogenation of aromatic compounds (toluene, pyridine, etc.) and isomerization of dienes require a $^2M-^2M$ or $^2M-^3M$ site, whereas $^3M-^3M$ sites are responsible for complete diene hydrogenation and pyridine HDN (the

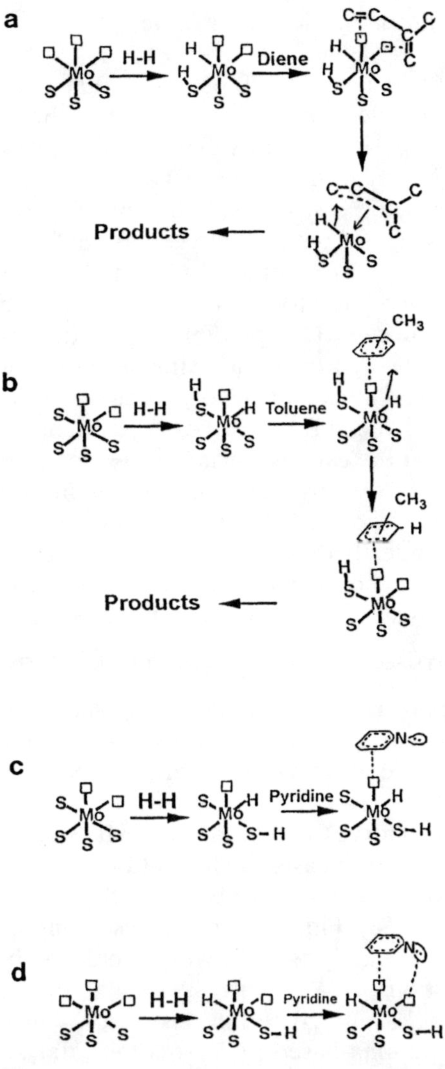

FIG. 40. Probe molecule transformations on coordinatively unsaturated sites (CUS) of MoS_2. (a) Isomerization and hydrogenation of diene; (b) hydrogenation of toluene; (c) hydrogenation of pyridine; (d) HDN of pyridine [adapted from Hubaut *et al.* (*138*); reprinted with permission].

superscript x in ^{x}M refers to the number of CUS around one edge of an Mo atom). As phosphorus decreases the number of exchangable oxygen atoms around a Mo atom (because of the low sulfidability of the MoP/Al catalyst), the proportion of ^{2}M sites may correspondingly increase. Thus, hydrogenation of aromatic rings and some HDN reactions could be promoted.

Furthermore, Moreau *et al.* (*140*) proposed that aromatic hydrogenation is favored on stronger electron-withdrawing sites (those of higher oxidation state) and hydrogenolysis on slightly electron donating sites (those of lower oxidation state). This suggestion is also in good agreement with the influence of phosphorus, which decreases the sulfidability of molybdenum oxo-species leading to the presence of residual oxygen atoms in the MoS_2 slabs.

In addition, there is an important question about whether the effect of phosphorus is mainly related to the increase in acidity or to the generation of new hydrogenation sites. One possible explanation is that Brønsted acid sites may also work as hydrogenation sites. In terms of reaction involving hydrogen species, acid sites and hydrogenation sites could be considered to be the same, but the difference between them is whether hydrogen remains essentially at the catalyst surface or is incorporated in the reactant at the end of reaction. If the hydrogen species on the Brønsted acid sites can be consumed by the reactants and immediately compensated by spillover hydrogen formed under hydrotreating conditions, acid sites can act as hydrogenation sites. Indeed, Kanai *et al.* (*141*) proposed that the Brønsted acid sites develop hydrogenation activity under some conditions.

D.   Other Possible Influences of Phosphorus

Effects of phosphorus have also been proposed from different points of view. First, phosphorus may decrease the polarization of the $Mo-S$ bond and therefore increase its covalent character. Since molybdenum-based catalysts with highly covalent $Mo-S$ bonds are supposed to have high HDS activities, phosphorus can thus improve HDS activity (*84*). Second, the presence of phosphorus increases the formation of octahedral molybdenum, cobalt, and nickel oxo-species which could be the precursors of the catalytically active phase (*38, 88*). Finally, phosphorus strongly promotes hydrogen activation in MoP/Al catalysts (*59*), which could be beneficial for all the hydrotreating reactions.

Finally, much work, summarized in this review, has shown that addition of phosphorus to alumina-based hydrotreating catalysts produces changes in their textural and structural properties or their activities. Some of these changes induce improved performance in hydrotreating reactions, provided that key parameters during preparation are correctly controlled. However, there are still many questions about the exact locations of phosphorus in the catalysts and its roles. Some interesting perspectives and explanations can be found in the literature but, beyond doubt, the understanding is incomplete and there are still opportunities to improve hydrotreating catalysts by phosphorus addition.

REFERENCES

1. Weisser, O., and Landa, S., "Sulphide Catalysts, Their Properties and Applications." Pergamon, Oxford, 1973.
2. Prins, R., de Beer, V. H. J., and Somorjai, G. A., *Catal. Rev. Sci. Eng.* **31,** 1 (1989).
3. Topsøe, H., Clausen, B. S., and Massoth, F. E., "Hydrotreating Catalysis" (and references therein). Springer-Verlag, Berlin, 1996.
4. Colgan, J. P., and Chomitz, N., U.S. Patent No. 327280 (1966).
5. Kearns, B. A., and Larson, O. A., U.S. Patent No. 3446730 (1969).
6. Weast, P. C., and Astle, M. (Eds.), "Handbook of Chemistry and Physics, p. B-29. CRC Press, Boca Raton, Florida, 1981.
7. Wells, A. F., "Structural Inorganic Chemistry," p. 638. Oxford, 1962.
8. King, R. B. (Ed.), "Encyclopedia of Inorganic Chemistry," Vol. 6. Wiley, New York, 1994.
9. Steudel, R., "Chemistry of Non-Metals," p. 307. de Gruyter, Berlin, 1977.
10. Weng, W., and Baptista, J. L., *J. Sol-Gel. Sci. Tech.* **8,** 645 (1997).
11. Livage, J., Barboux, P., Vandenborre, M. T., Schmutz, C., and Taulelle, F., *J. Non-Cryst. Solids* **147/148,** 18 (1992).
12. Rogers, D. E., and Nickless, G., *in* "Inorganic Sulfur Chemistry" (Gnickvess, Ed.), p. 281. Elsevier, Amsterdam, 1968.
13. Olah, G. A., "Friedel–Crafts Chemistry," p. 368. Wiley Interscience, New York, 1973.
14. Kniep, R., *Angew. Chem. Int. Ed. Engl.* **25,** 525 (1986).
15. Fernandez, C., Amoureux, J. P., Chezeau, J. M., Delmotte, L., and Kessler, H., *Microporous Mater.* **6,** 331 (1996).
16. Balakrishnan, I., and Prasad, S., *Appl. Catal.* **62,** L7 (1990).
17. Rebenstorf, B., Lindblad, T., and Andersson, S. L. T., *J. Catal.* **128,** 293 (1991).
18. Tsigdinos, G. A., *Topics Curr. Chem.* **76,** 1 (1978).
19. Okuhara, T., Mizuno, N., and Misono, M., *Adv. Catal.* **41,** 113 (1996).
20. Pettersson, L., Andersson, I., and Öhman, L. O., *Inorg. Chem.* **25,** 4726 (1986).
21. Souchay, P., "Polyanions et Polycations." Gauthier Villars, Paris, 1963.
22. (a) Griboval, A., Blanchard, P., Payen, E., Fournier, M., and Dubois, J. L., *Stud. Surf. Sci. Catal.* **106,** 181 (1997); (b) Griboval, A., doctoral thesis, University of Lille, France, 1998.
23. Chamber, H. W., "Organophosphates" (J. E. Chambers and P. E. Levi, Eds.), p. 3. Academic Press, New York, 1992.
24. Thompson, C. M., Nishioka, T., and Fukuta, T. R., *J. Agric. Food Chem.* **31,** 696 (1983).
25. Van Veen, J. A. R., Hendriks, P. A. J. M., Andréa, R. R., Romers, E. J. G. M., and Wilson, A. E., *J. Phys. Chem.* **94,** 5282 (1990).
26. Abbattista, F., Delmastro, A., Gozzelino, G., Mazza, D., Vallino, M., Busca, G., and Lorenzelli, V., *J. Chem. Soc. Faraday Trans.* **86,** 3653 (1990).
27. Spojakina, A. A., and Damyanova, S., *React. Kinet. Catal. Lett.* **53,** 405 (1994).
28. Shokhireva, T. Kh., Yurieva, T. M., Altynnikov, A. A., Anufrienko, V. F., Plyasova, L. M., Litvak, G. S., Spojakina, A. A., and Kostova, N., *React. Kinet. Catal. Lett.* **47,** 177 (1992).
29. Atanasova, P., and Halachev, T., *Appl. Catal.* **48,** 295 (1989).
30. Spojakina, A. A., Damyanova, S., and Petrov, L., *Appl. Catal.* **56,** 163 (1989).
31. Mangnus, P. J., Van Veen, J. A. R., Eijsbouts, S., de Beer, V. H. J., and Moulijn, J. A., *Appl. Catal.* **61,** 99 (1990).
32. Stencel, J. M., "Raman Spectroscopy for Catalysis, Catalysis Series" (B. Davis, Ed.). Van Nostrand Reinhold, New York, 1989.
33. Jeziorowski, H., and Knözinger, H., *J. Phys. Chem.* **83,** 9 (1979).
34. Gazimzyanov, N. R., Mikhailov, V. I., and Volod'ko, V. V., *Kinet. Catal.* **36,** 694 (1995).

35. Van Veen, J. A. R., Sudmeijer, O., Emeris, C. A., and de Wit, H., *J. Chem. Soc. Dalton Trans.*, 1825 (1986).
36. Bleam, W. F., Pfeffer, P. E., and Frye, J. S., *Phys. Chem. Minerals* **16**, 455 (1989).
37. Eijsbouts, S., Van Gruijthuijsen, L., Volmer, J., de Beer, V. H. J., and Prins, R., *Stud. Surf. Sci. Catal.* **50**, 79 (1989).
38. DeCanio, E. C., Edwards, J. C., Scalzo, T. R., Storm, D. A., and Bruno, J. W., *J. Catal.* **132**, 498 (1991).
39. Fredericci, C., and Domínguez, S. F., *Mater. Res. Bull.* **31**, 235 (1996).
40. Iwamoto, R., and Grimblot, J., *Stud. Surf. Sci. Catal.* **106**, 195 (1997).
41. Kraus, H., and Prins, R., *J. Catal.* **170**, 20 (1997).
42. Okamoto, Y., Gomi, T., Mori, Y., Imanaka, T., and Teranishi, S., *React. Kinet. Catal. Lett.* **22**, 417 (1983).
43. Knözinger, H., and Ratnasamy, P., *Catal. Rev. Sci. Eng.* **17**, 31 (1978).
44. Lewis, J. M., and Kydd, R. A., *J. Catal.* **132**, 465 (1991).
45. Morales, A., Ramírez de Agudelo, M. M., and Hernández, F., *Appl. Catal.* **41**, 261 (1988).
46. Stanislaus, A., Absi-Halabi, M., and Al-Dolama, K., *Appl. Catal.* **39**, 239 (1988).
47. Petrakis, D. E., Hudson, M. J., Pomonis, P. J., Sdoukos, A. T., and Bakas, T. V., *J. Mater. Chem.* **5**, 1975 (1995).
48. Fitz, C. W., Jr., and Rase, H. F., *Ind. Eng. Chem. Res. Dev.* **22**, 40 (1983).
49. Kraus, H., thesis, ETH, Zurich, Switzerland, 1996.
50. Kraus, H., and Prins, R., *J. Catal.* **164**, 251 (1996).
51. Davis, J. A., and Leckie, J. O., *J. Colloid Int. Sci.* **74**, 32 (1980).
52. Breeuwsma, A., and Lyklema, J., *J. Colloid Int. Sci.* **43**, 437 (1973).
53. Huang, C. P., *J. Colloid Int. Sci.* **53**, 178 (1975).
54. Cheng, W. C., and Luthra, N. P., *J. Catal.* **109**, 163 (1988).
55. Van Veen, J. A. R., Hendriks, P. A. J. M., Romers, E. J. G. M., and Andréa, R. R., *J. Phys. Chem.* **94**, 5275 (1990).
56. Mikami, N., Sasaki, M., Hachiya, K., Astumian, R. D., Ikeda, T., and Yasunaga, T., *J. Phys. Chem.* **87**, 1454 (1983).
57. López-Cordero, R., Gil Llambías, F. J., Palacios, J. M., Fierro, J. L. G., and López-Agudo, A., *Appl. Catal.* **56**, 197 (1989).
58. Cruz Reyes, J., Avalos-Borja, M., López-Cordero, R., and López-Agudo, A., *Appl. Catal. A* **120**, 147 (1994).
59. Gishti, K., Iannibello, A., Marengo, S., Morelli, G., and Tittarelli, P., *Appl. Catal.* **12**, 381 (1984).
60. Chadwick, D., Aitchison, D. W., Badilla-Ohlbaum, R., and Josefsson, L., *Stud. Surf. Sci. Catal.* **16**, 323 (1983).
61. Han, O. H., Lin, C. Y., and Haller, G. L., *Catal. Lett.* **14**, 1 (1992).
62. Ramírez de Agudelo, M. M., and Morales, A., *Proc. 9th Int. Cong. Catal.* **1**, 42 (1988).
63. Iwamoto, R., Fernandez, C., Amoureux, J. P., and Grimblot, J., *J. Phys. Chem. B* **102**, 4342 (1998).
64. Petterson, L., *Acta Chem. Scand.* **25**, 1959 (1971).
65. Castillo, M. A., Vázquez, P. G., Blanco, M. N., and Cáceres, C. V., *J. Chem. Soc. Franday Trans.* **92**, 3239 (1996).
66. Jian, M., and Prins, R., *Bull. Soc. Chim. Belges* **104**, 231 (1995).
67. (a) Iwamoto, R., doctoral thesis, University of Lille, France, 1997; (b) Iwamoto, R., and Grimblot, J., submitted for publication.
68. Atanasova, P., Uchytil, J., Kraus, M., and Halachev, T., *Appl. Catal.* **65**, 53 (1990).
69. López-Agudo, A., López-Cordero, R., Palacios, J. M., and Fierro, J. L. G., *Bull. Soc. Chim. Belges* **104**, 237 (1995).

70. Mangnus, R. J., Van Langeveld, A. D., de Beer, V. H. J., and Moulijn, J. A., *Appl. Catal.* **68,** 161 (1991).
71. Robinson, W. R. A. M., Van Gestel, J. N. M., Korányi, T. I., Eijsbouts, S., Van der Kraan, A. M., Van Veen, J. A. R., and de Beer, V. H. J., *J. Catal.* **161,** 539 (1996).
72. López-Cordero, A., Esquivel, N., Lázaro, J., Fierro, J. L. G., and López-Agudo, A., *Appl. Catal.* **48,** 341 (1989).
73. Walendziewski, J., *React. Kinet. Catal. Lett.* **43,** 107 (1991).
74. Lewis, J. M., Kydd, R. A., Boorman, P. M., and Van Rhyn, P. H., *Appl. Catal. A* **84,** 103 (1992).
75. Choca, M. E., U.S. Patent No. 4066572 (1978).
76. Johnson, M. F. L., and Erickson, H., U.S. Patent No. 4202798 (1980).
77. Morterra, C., Magnacca, G., and de Maestri, P. P., *J. Catal.* **152,** 384 (1995).
78. Hopkins, P. D., and Meyers, B. L., *Ind. Eng. Chem. Prod. Res. Dev.* **22,** 421 (1983).
79. Poulet, O., Hubaut, R., Kasztelan, S., and Grimblot, J., *Bull. Soc. Chim. Belges* **100,** 857 (1991).
80. Chen, Y. W., Hsu, W. C., Lin, C. S., Kang, B. C., Wu, S. T., Leu, L. J., and Wu, J. C., *Ind. Eng. Chem. Res.* **29,** 1830 (1990).
81. Jian, M., and Prins, R., *Catal. Lett.* **35,** 193 (1995).
82. Jian, M., Kapteijn, F., and Prins, R., *J. Catal.* **168,** 491 (1997).
83. Sajkowski, D. J., Miller, J. T., and Zajac, G. W., *Appl. Catal.* **62,** 205 (1990).
84. Kim, S. I., and Woo, S. I., *J. Catal.* **133,** 124 (1992).
85. Iwamoto, R., and Grimblot, J., *J. Catal.* **172,** 252 (1997).
86. Callant, M., Holder, K. A., Grange, P., and Delmon, B., *Bull. Soc. Chim. Belges* **104,** 245 (1995).
87. Atanasova, P., Halachev, T., Uchytil, J., and Kraus, M., *Appl. Catal.* **38,** 235 (1988).
88. Morales, A., and Ramírez de Agudelo, M. M., *Appl. Catal.* **23,** 23 (1986).
89. Atanasova, P., López-Cordero, R., Mintchev, L., Halachev, T., and López-Agudo, A., *Appl. Catal. A* **159,** 269 (1997).
90. Rushala, F., Shiriaev, P. A., Kutirev, M. Y., Kushnerev, M. Y., and Shashkin, D. P., *Kinet. Katal.* **22,** 1307 (1981). [In Russian]
91. Ramírez, J., Castaño, V. N., Leclercq, C., and López-Agudo, A., *Appl. Catal. A* **83,** 251 (1992).
92. López-Cordero, R., López-Guerra, S., Fierro, J. L. G., and López-Agudo, A., *J. Catal.* **126,** 8 (1990).
93. McMillan, M., Brinen, J. S., and Haller, G. L., *J. Catal.* **97,** 243 (1986).
94. Eijsbouts, S., Van Gestel, J. N. M., Van Veen, J. A. R., de Beer, V. H. J., and Prins, R., *J. Catal.* **131,** 412 (1991).
95. Hubaut, R., Poulet, O., Kasztelan, S., Payen, E., and Grimblot, J., "Proceedings of the Symposium on Advances in Hydrotreating Catalysts," 208th ACS national meeting, Washington, DC, No. 39, p. 548, 1994; Occelli, M. L., and Chianelli, R. R. (Eds.), "Hydrotreating Technology for Pollution Control," p. 115. Dekker, New York, 1996.
96. Kemp, R. A., Ryan, R. C., and Smegal, J. A., *Proc. 9th Int. Cong. Catal.* **1,** 128 (1988).
97. Van Veen, J. A. R., Colijn, H. A., Hendriks, P. A. J. M., and Van Welsenes, A. J., *Fuel Processing Technol.* **35,** 137 (1993).
98. Topsøe, H., Clausen, B. S., Topsøe, N. Y., and Zeuthen, P., *Stud. Surf. Sci. Catal.* **53,** 77 (1989).
99. Bouwens, S. M. A. M., Van der Kraan, A. M., de Beer, V. H. J., and Prins, R., *J. Catal.* **128,** 559 (1991).
100. Bouwens, S. M. A. M., Vissers, J. P. R., de Beer, V. H. J., and Prins, R., *J. Catal.* **112,** 401 (1988).

101. Atanasova, P., Vladov, C., Halachev, T., Fierro, J. L. G., and López-Agudo, A., *Bull. Soc. Chim. Belges* **104,** 219 (1995).
102. Arunarkavalli, T., *Proc. Indian Acad. Sci.* **108,** 27 (1996).
103. Andreev, A., Vladov, C., Prahov, L., and Atanasova, P., *Appl. Catal. A* **108,** L97 (1994).
104. Manova, E., Severac, C., Andreev, A., and Clement, R., *J. Catal.* **169,** 503 (1997).
105. Lewis, J. M., and Kydd, R. A., *J. Catal.* **136,** 478 (1992).
106. Boorman, P. M., Kydd, R. A., Sorensen, T. S., Chong, K., Lewis, J. M., and Bell, W. S., *Fuel* **71,** 87 (1992).
107. Kurokawa, Y., Kobayashi, Y., and Nakata, S., *Heterogeneous Chem. Rev.* **1,** 309 (1994).
108. Griffith, W. P., and Lesnik, P. J. B., *J. Chem. Soc. A,* 1066 (1969).
109. Mattes, R., Bierbüsse, H., and Fuchs, J., *Z. Anorg. Allg. Chem.* **385,** 230 (1971).
110. Etienne, E., Ponthieu, E., Payen, E., and Grimblot, J., *J. Non-Cryst. Solids* **147/148,** 765 (1992).
111. Van Eck, E. R. H., Kentgens, A. P. M., Kraus, H., and Prins, R., *J. Phys. Chem.* **99,** 16080 (1995).
112. Müller, A., Weinstock, N., Mohan, W., Schläpfer, C. W., and Nakamoto, K., *Appl. Spectrosc.* **27,** 257 (1973).
113. Le Bihan, L., Mauchaussé, C., Duhamel, L., Grimblot, J., and Payen, E., *J. Sol-Gel Sci. Tech.* **2,** 837 (1994).
114. Iannibello, A., Marengo, S., Trifiro, F., and Villa, P. L., *Stud. Surf. Sci. Catal.* **3,** 4 (1978).
115. Tsuchida, T., Ohta, S., and Horigome, K., *J. Mater. Chem.* **4,** 1503 (1994).
116. Cho, I. H., Park, S. B., and Kwak, J. H., *J. Mol. Catal. A* **104,** 285 (1996).
117. Risbud, S. H., Kirkpatrick, R. J., Taglialavore, A. P., and Montez, B., *J. Am. Ceram. Soc.* **70,** C10 (1987).
118. Brow, R. K., Kirkpatrick, R. J., and Turner, G. L., *J. Am. Ceram. Soc.* **73,** 2293 (1990).
119. Edwards, J. C., and DeCanio, E. C., *Catal. Lett.* **19,** 121 (1993).
120. Rezgui, S., and Gates, B. C., *Chem. Mater.* **6,** 2386 (1994).
121. Müller, D., Gessner, W., Behrens, H. J., and Scheler, G., *Chem. Phys. Lett.* **79,** 59 (1981).
122. Bautista, F. M., Campelo, J. M., Garcia, A., Luna, D., Marinas, J. M., and Romero, A., *Appl. Catal. A* **96,** 175 (1993).
123. Ledoux, M. J., Peter, A., Blekkan, E. A., and Luck, F., *Appl. Catal. A* **133,** 321 (1995).
124. Fierro, J. L. G., López-Agudo, A., Esquivel, N., and López-Cordero, R., *Appl. Catal.* **48,** 353 (1989).
125. Muralidhar, G., Massoth, F. E., and Shabtai, J., *J. Catal.* **85,** 44 (1984).
126. Ramselaar, W. L. T. M., Bouwens, S. M. A. M., de Beer, V. H. J., and Van der Kraan, A. M., *Hyperfine Interact.* **46,** 599 (1989).
127. Jones, J. M., Kydd, R. A., Boorman, P. M., and Van Rhyn, P. H., *Fuel* **74**(12), 1875 (1995).
128. Kemp, R. A., and Adams, C. T., *Appl. Catal. A* **134,** 299 (1996).
129. Kushiyama, S., Aizawa, R., Kobayashi, S., Koinuma, Y., Uemasu, I., and Ohuchi, H., *Appl. Catal.* **63,** 279 (1990).
130. Rico Cerda, J. L., and Prins, R., *Bull. Soc. Chim. Belges* **100,** 815 (1991).
131. Jian, M., and Prins, R., *Catal. Lett.* **50,** 9 (1998).
132. Jian, M., and Prins, R., *Ind. Eng. Chem. Res.* **37,** 834 (1998).
133. Jian, M., Rico Cerda, J. L., and Prins, R., *Bull. Soc. Chim. Belges* **104,** 225 (1995).
134. Morales, A., Galiasso, R. E., Ramírez de Agudelo, M. M., Salazar, J. A., and Carrasquel, A. R., U.S. Patent No. 4520128 (1985).
135. Halachev, T., Atanasova, P., López-Agudo, A., Arias, M. G., and Ramírez, J., *Appl. Catal. A* **136,** 161 (1996).
136. Rhodes, A. K., *Oil Gas J.,* 35 (1995, October 2).
137. Daage, M., and Chianelli, R. R., *J. Catal.* **149,** 414 (1994).

138. Hubaut, R., Bonnelle, J. P., and Grimblot, J., *Trends Phys. Chem.* **2,** 277 (1991).

139. Kasztelan, S., Wambeke, A., Jalowiecki, L., Grimblot, J., and Bonnelle, J. P., *J. Catal.* **124,** 12 (1990).

140. Moreau, C., Aubert, C., Durand, R., Zmimita, N., and Geneste, P., *Catal. Today* **4,** 117 (1988).

141. Kanai, J., Martens, J. A., and Jacobs, P. A., *J. Catal.* **133,** 527 (1992).

Skeletal Isomerization of *n*-Butenes Catalyzed by Medium-Pore Zeolites and Aluminophosphates

PAUL MÉRIAUDEAU AND CLAUDE NACCACHE

Institut de Recherches sur la Catalyse—CNRS
2, Avenus A. Einstein 69626 Villeurbanne Cedex, France

The recent literature related to selective skeletal isomerization of *n*-butenes catalyzed by medium-pore zeolites and Me-aluminophosphates is reviewed. In the presence of medium-pore molecular sieve catalysts, *n*-butenes are selectively transformed into isobutylene via a monomolecular mechanism. This is an example of restricted transition state shape selectivity, whereby the space available around the acidic site is restricted, constraining the reaction to proceed mainly through a monomolecular mechanism. Coking of the catalyst that leads to poisoning of the acidic sites located on the external surfaces and to a decrease in the space around the acidic sites located in the micropores renders the catalyst more selective.

I. Introduction

In recent decades the isomerization of olefins has taken an important place in the field of catalysis by solid acids. Since the pioneering work of Haag and Pines (*1*), extensive effort has been dedicated to the isomerization of butenes, and a reasonable understanding of the kinetics and mechanisms of the double-bond shift in *n*-butenes and of the *cis–trans* isomerization in 2-butene has been reached. There is a general agreement in the literature on the carbenium ion mechanism involved in the double-bond migration and in the *cis–trans* conversion. By contrast, the skeletal isomerization of *n*-butenes to give isobutylene, although well studied, was not fully explained, and its mechanism remains controversial. Nevertheless, it is well accepted that the isomerization of *n*-butenes into isobutylene is a highly energetically demanding reaction and as such requires more severe conditions than those generally used in the double-bond migration (*2*).

The reaction of *n*-butenes to give isobutylene is catalyzed by a wide variety of solid acids but requires relatively high temperature. Typical catalysts include alumina, halogenated alumina, amorphous silica–alumina, supported phosphoric acid, and supported tungsten or molybdenum oxide. The most characteristic features of the skeletal isomerization of *n*-butenes

505

have been reviewed by Choudhary (*3*) and by Butler and Nicolaides (*4*). Apparently, the major drawbacks of the previously cited catalysts are the high reaction temperatures needed, which contribute to excessive deactivation, and the difficulty of reaching very high selectivity and yield of isobutylene.

Recently, the use of medium-pore molecular sieves as catalysts for the isomerization of *n*-butenes has allowed the development of new families of catalysts that are highly efficient for this reaction. Several patents claiming the use of zeolitic and nonzeolitic molecular sieves for the skeletal isomerization of $C_4 - C_5$ olefins have been published; the catalysts are ferrierite (*5, 6*) and the nonzeolitic molecular sieves MeAPO and MeAPSO (*7*). Since the appearance of the patents, important scientific contributions have appeared in the literature.

The aim of this review is to describe the most interesting results characterizing the skeletal isomerization of *n*-butenes catalyzed by zeolitic and nonzeolitic molecular sieves and to discuss the state of the art of the isomerization mechanism, the nature and location of the active sites responsible for the selectivity for isobutylene, and the influence of the pore dimensions and pore structures of the molecular sieves.

The importance of isobutylene in the petrochemical industry is well recognized. Isobutylene is used on a large scale for the production of (i) methacrolein by direct oxidation, (ii) polyisobutylene by polymerization, (iii) synthetic rubber (a copolymer of isobutylene and isoprene), and (iv) methyl *tert*-butyl ether (MTBE, a gasoline octane-number enhancer) by reaction with methanol.

The continuous increase in world consumption of MTBE has created a strong incentive to increase the production of isobutylene. Isobutylene can be produced by catalytic dehydrogenation of isobutane. However, the largest production of C_4 olefins comes from the thermal cracking processes for the manufacture of ethylene which generate as by-products C_4 mixtures containing C_4 olefins and C_4 alkanes plus butadiene. Isobutylene is also a product of fluid bed catalytic cracking units.

Several processes are used to upgrade the C_4 fraction. The isobutylene contained in the C_4 cut is removed by reaction with methanol to produce MTBE. The remaining *n*-butenes in the C_4 cut can be alkylated with isobutane catalyzed by liquid HF or H_2SO_4 or isomerized into isobutylene in the presence of acid catalysts.

To better understand why and how microporous molecular sieves (aluminosilicates and aluminophosphates) are highly suitable catalysts for the skeletal isomerization of *n*-butenes, it is important to discuss the mechanisms of the reactions that control the formation of the desired isobutylene and to evaluate the relative importance of secondary reactions that may be responsible for low selectivity. Also, for a clear understanding of the relationships that may exist between the rate and selectivity of the reaction and the chemistry on the one hand and the morphology and internal structure of the molecular sieve catalysts on the other hand, essential properties of the molecular sieves are also reviewed.

II. Mechanisms of Skeletal Isomerization of *n*-Butenes

Generally, the mechanisms of acid-catalyzed reactions involve carbenium ion intermediates, which, in the case of olefin reactants, are formed by the addition of a proton to the double bond:

$$C=C-C-C + H^+ \rightleftarrows C-C^+-C-C. \tag{1}$$

Secondary or tertiary carbenium ions are formed, depending on the branching of the reactant olefin. In the case of 1-butene and 2-butene, the overall reaction scheme for the skeletal isomerization is identical. Protonation of 1-butene or of 2-butene generates *s*-butyl cations.

It has been suggested that the isomerization of *n*-butenes to give isobutylene may proceed either through a monomolecular or through a bimolecular mechanism.

A.   THE MONOMOLECULAR MECHANISM

Brouwer (*8*) studied the isomerization of isotopically labeled butane $^{13}CH_3-CH_2-CH_2-CH_3$. He proposed the conversion of *s*-butyl cation to give a substituted cyclopropyl carbenium ion intermediate:

$$\overset{+}{C}-C-C-C \rightarrow C-C \overset{\overset{\displaystyle C}{\diagup \overset{+}{H} \diagdown}}{-\!\!\!-} C . \tag{2}$$

Such an intermediate explains the ^{13}C scrambling in the butane carbon chain without the formation of isobutane: ring opening at the C_3-C_2 carbon bond results in the formation of *s*-butyl cation, whereby the labeled *C is no longer a terminal carbon:

$$C_1 - C_2 \overset{\overset{\displaystyle C_3}{\diagup \overset{+}{H} \diagdown}}{-\!\!\!-} C_4^* \leftrightharpoons C-{}^+C-C^*-C . \tag{3}$$

Cyclopropyl ring opening at the C_3-C_4 carbon bond results in the formation of a primary carbenium ion:

$$\overset{\displaystyle C_3}{\underset{C_1-C_2 \text{———} C_4}{\overset{H^+}{\diagup\diagdown}}} \to \overset{\displaystyle C_4}{\underset{C_1-C_2-C_3}{|\;\;+}} . \qquad (4)$$

The highly unstable primary carbenium ion rearranges rapidly to give the relatively stable tertiary butyl cation, which gives isobutylene by proton abstraction:

$$\overset{\displaystyle C}{\underset{C-\overset{+}{C}-C}{|\;\;+}} \to \overset{\displaystyle C}{\underset{\underset{+}{C-C-C}}{|}} \rightleftharpoons \overset{\displaystyle C}{\underset{\underset{\displaystyle C}{C=C}}{|}} + H^+ . \qquad (5)$$

The monomolecular mechanism, which involves the formation of a primary carbenium ion, is highly energetically unfavorable. However, such a mechanism would result theoretically in 100% selectivity for the conversion of *n*-butenes into isobutylene.

B.   THE BIMOLECULAR MECHANISM

To overcome the apparently ineffective monomolecular path, it was suggested (*9*) that the skeletal isomerization of linear butenes proceeds through dimerization followed by cracking.

The protonation of butene will form the *s*-butyl cation, and by reaction with another butene molecule a C_8 carbenium ion will be formed:

$$C-\overset{+}{C}-C-C + C=C-C-C \to \underset{}{C-C-\overset{\displaystyle C}{\overset{|}{C}}-C-\overset{+}{C}-C-C} \qquad (6)$$

or

$$C-\overset{+}{C}-C-C + C-C=C-C \to C-C-C-C-\overset{+}{C}-C. \qquad (7)$$

The carbenium ion intermediates rearrange via hydride transfer and alkyl shift steps and through substituted protonated cyclopropane intermediates (*8*). All possible isomers form; and the corresponding carbenium ions crack into smaller fragments, as illustrated for several isomers (*10*):

Trimethylpentene

$$\underset{\overset{\displaystyle |}{\underset{\displaystyle C}{}}}{C}-\overset{\displaystyle C}{\underset{\displaystyle |}{C}}-C-\overset{+}{\underset{\overset{\displaystyle |}{C}}{C}}-C \rightarrow C-\overset{+}{\underset{\overset{\displaystyle |}{C}}{C}}-C + \underset{\overset{\displaystyle |}{C}}{C}=C-C \quad \text{A-type cracking} \quad (8)$$

$$C-\underset{\overset{\displaystyle |}{C}}{\overset{\overset{\displaystyle C}{\displaystyle |}}{C}}-C-\overset{+}{C}-C \rightarrow C-\overset{+}{\underset{\overset{\displaystyle |}{C}}{C}}-C + C-C=C-C \quad \text{B}_1\text{-type cracking} \quad (9)$$

$$C-C-C-\overset{+}{C}-C \rightarrow C-C-C=C + C-\overset{+}{C}-C \quad \text{B}_2\text{-type cracking} \quad (10)$$

Dimethylhexene

$$C-\underset{\overset{\displaystyle |}{C}}{\overset{\overset{\displaystyle C}{\displaystyle |}}{C}}-C-\overset{+}{C}-C-C \rightarrow C-\overset{+}{\underset{\overset{\displaystyle |}{C}}{C}}-C + C=C-C-C \quad \text{B}_1\text{-type cracking} \quad (11)$$

$$C-C-\underset{\overset{\displaystyle |}{C}}{\overset{\overset{\displaystyle C}{\displaystyle |}}{C}}-C-\overset{+}{C}-C \rightarrow C-\overset{+}{\underset{\overset{\displaystyle |}{C}}{C}}-C-C + C=C-C \quad \text{B}_1\text{-type cracking} \quad (12)$$

$$C-\overset{+}{\underset{\overset{\displaystyle |}{C}}{C}}-C-C-C-C \rightarrow C=C-C + C-\overset{+}{\underset{\overset{\displaystyle |}{C}}{C}}-C-C \quad \text{B}_2\text{-type cracking} \quad (13)$$

$$C-C-C-\overset{+}{C}-C-C \rightarrow C-\overset{+}{C}-C + C-C=C-C-C \quad \text{C-type cracking} \quad (14)$$

$$C-C-\overset{+}{C}-C-C-C \rightarrow C-\overset{+}{C}-C + C-C-C=C \quad \text{C-type cracking} \quad (15)$$

Methylheptene

$$C-C-C-\overset{+}{C}-C-C-C \;\rightarrow\; C-\overset{+}{C}-C \;+\; C=C-C-C-C$$
$$\underset{\textstyle C}{|}$$

C-type cracking   (16)

$$C-C-C-C-\overset{+}{C}-C-C \;\rightarrow\; C-C-\overset{+}{C}-C \;+\; C=C-C-C$$
$$\underset{\textstyle C}{|}$$

C-type cracking   (17)

$$C-C-C-C-C-\overset{+}{C}-C \;\rightarrow\; C=C-C \;+\; C-C-C-\overset{+}{C}-C$$
$$\underset{\textstyle C}{|}$$

C-type cracking.   (18)

The cracking of the normal octene is not considered since it occurs via the formation of primary carbeniums ions that are highly unstable so that the reaction is very slow (*11*). Thus, the cracking of all possible isomers generates C_3, C_4, and C_5 olefins (branched and linear), with their respective concentrations depending on the temperature. For example, with linear C_8 feed, isomerization to the 2,4,4-trimethylpentyl isomer with subsequent A-type cracking accounts for 35% of the octane cracking at a temperature of about 500 K (*12*), whereas with 1-octene cracking most of the cracked products are formed via B- and C-type beta scission at 783 K (*13*). It follows from these considerations that the alkylation/cracking (bimolecular pathway) should not give 100% isobutylene.

The existence of a catalytic site different from a proton donor site, e.g., a Lewis acid site, has not been considered. It is known that Lewis acid sites could play a role in such reactions (*14*); however, among the catalysts that are selective for this isomerization, most have little or no Lewis acidity. Moreover, often when the catalysts are nonmicroporous materials, such as halogenated aluminas (*15*) and $SiO_2-Al_2O_3$, the highest isobutylene selectivities are obtained in the presence of a mixture of olefin and water, with the water partial pressure being close to that of the olefin. The existence of Lewis acid sites under these conditions is very unlikely.

III. Trade-off of Selectivity and Activity

From the previous discussion, it is expected that if the two reaction mechanisms operate simultaneously, a change in temperature changes the isobutylene selectivity; the dimerization reaction is favored by low temperatures and consequently the isobutylene selectivity is low at low temperatures, as reported for nonmicroporous materials (*4*). The isobutylene selectivity is also expected to depend on the olefin partial pressure. A high olefin partial pressure favors the bimolecular pathway relative to the monomolecular pathway, as has been observed for non microporous materials (*3, 4*), for which high isobutylene selectivities were obtained at high temperatures (723–793 K) and low butene partial pressures (<101 kPa).

Thermodynamic considerations (*16*) show that the percentage of butene dimerized into octene decreases with increasing temperature and decreasing pressure. The thermodynamic data (*17*) indicate how the equilibrium concentration of isobutylene in a mixture of butenes depends on temperature; for example (Fig. 1), at 600 K isobutylene is almost 45% of the total, whereas at 673 K it is 41%. Recent data (*18*) and experimental measurements (*19*) show higher values of the isobutylene concentration; at 600 K isobutylene is 51.8% of the total butenes; the value is 48% at 673 K. Thus, there is a motivation to work at relatively low temperatures to allow the possibility of achieving high isobutylene yields, although low temperatures favor a mechanism that gives low selectivities and catalyst activities are low at low

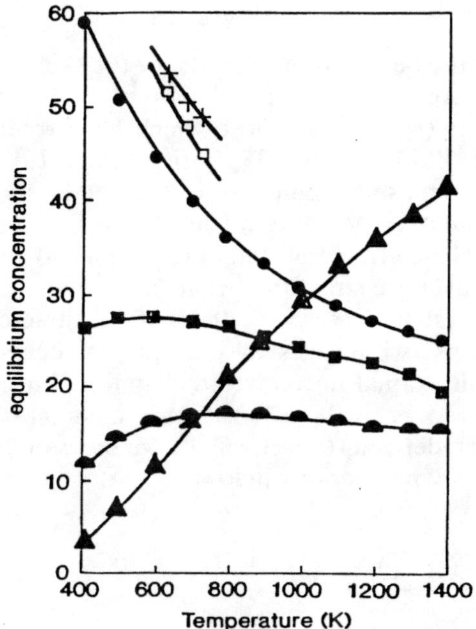

FIG. 1. Distribution of butenes (normal butenes and isobutylene) at the thermodynamic equilibrium as a function of the temperature. ●, isobutylene (*17*); ■, *trans*-2-butene (*17*); ▲, *cis*-2-butene (*17*); ▲, 1-butene (*17*); □, isobutylene (*18*); ×, isobutylene (*19*).

temperatures. Thus, a trade-off is clear, and conditions should be chosen to reach the best compromise of selectivity and activity.

IV. Aluminosilicate and Aluminophosphate Molecular-Sieve Catalysts

Aluminosilicates (zeolites) are widely used as acidic and bifunctional catalysts. The formation of carbocationic intermediates is generally ascribed to the protons present in the open zeolite structure. A newer class molecular-sieve catalyst is the aluminophosphates. These may contain silicon (SAPO) or metal (MeAPO) in their $AiPO_4$ frameworks. These framework substitutions in several cases generate protonic acidity that makes SAPO and MeAPO acid catalysts. There is already an extensive literature of this subject (*20, 21*).

The following sections are a summary of the characteristics of the zeolites and aluminophosphates that have been investigated as catalysts for skeletal isomerization of *n*-butenes.

A. ZEOLITES

It appears that the best zeolite catalysts for the skeletal isomerization of *n*-butenes are silicon-rich (with Si/Al ratios between 10 and 100) with medium pore sizes (<0.6 nm). Those which have received major attention are ZSM-22, ZSM-23, ZSM-35, ferrierite, and (to a smaller extent) ZSM-5. These materials are composed of SiO_4 and AlO_4 tetrahedra, with the negative charge of the AlO_4 tetrahedra being balanced by H^+ ions, (in acidic zeolites). The tetrahedral units are assembled in such geometries that various internal pore structures are formed. In most zeolite structures the internal pore system consists of either two- or three-dimensional interconnected channels (with cages at the channel intersections) or one-dimensional unidirectional unconnected channels. Typically, zeolite pore sizes are categorized as small (0.3–0.45 nm), medium (0.45–0.60 nm), or large (0.7–0.8 nm), defined respectively by windows of 8, 10, and 12 rings. Figure 2 illustrates typical pore windows.

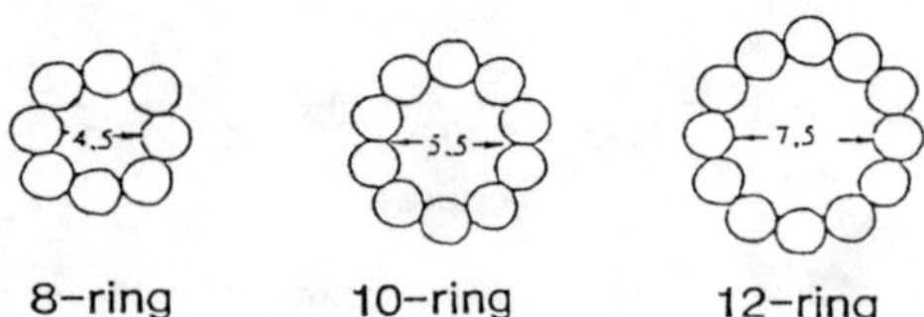

FIG. 2.   Schematic representation of 8, 10, and 12 rings (diameters in Å).

1. *ZSM-5 and ZSM-11 (MFI and MEL Structures)*

The general formula of these zeolites in the hydrogen form is

$$H_xAl_xSi_{96-x}O_{192} \cdot H_2O,$$

where x is in the range 0–7.

ZSM-5 incorporates two systems of intersecting channels lying along the *a* and *b* crystal axes (*22*); the straight channels have elliptical apertures with cross-sections of 0.51 × 0.55 nm, and the sinusoidal channels have nearly circular apertures with cross-sections of 0.53 × 0.56 nm. At the channel intersections there are cages about 1 nm in diameter with 0.6-nm diameter entry ports (Fig. 3).

The ZSM-11 pore structure consists of straight channels in two directions, with intersections and interconnected cavities (cages). The cross section of the 10-ring is 0.54 × 0.53 nm (Fig. 4).

framework viewed along [010]

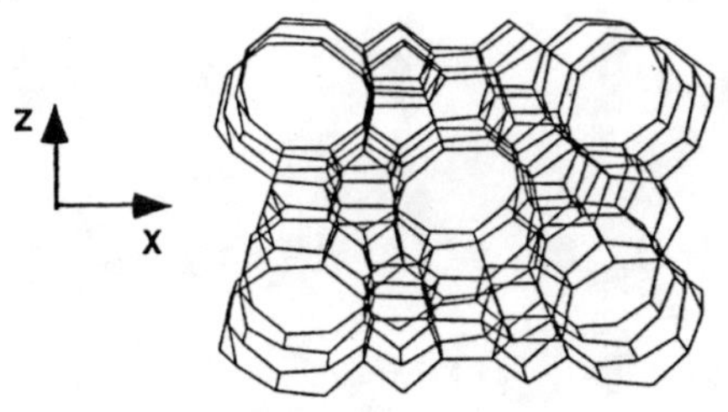

10-ring viewed along [010]
(straight channel)

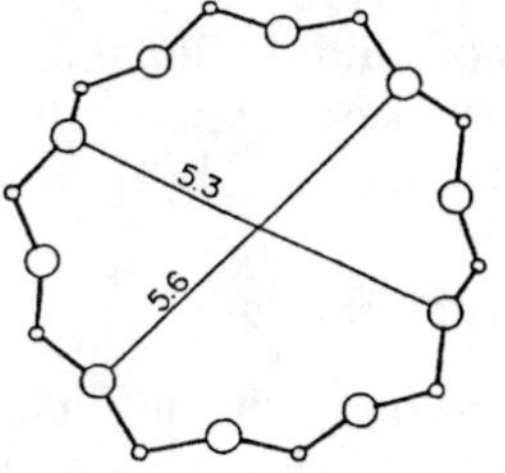

FIG. 3.   View of MFI channel (distances in Å).

framework viewed along [100]

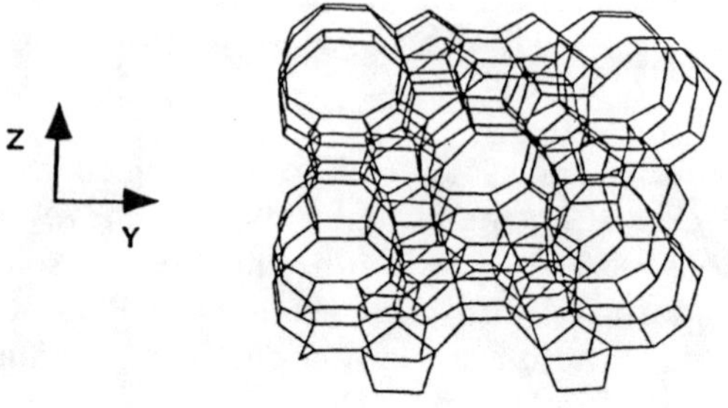

10-ring viewed along [100]

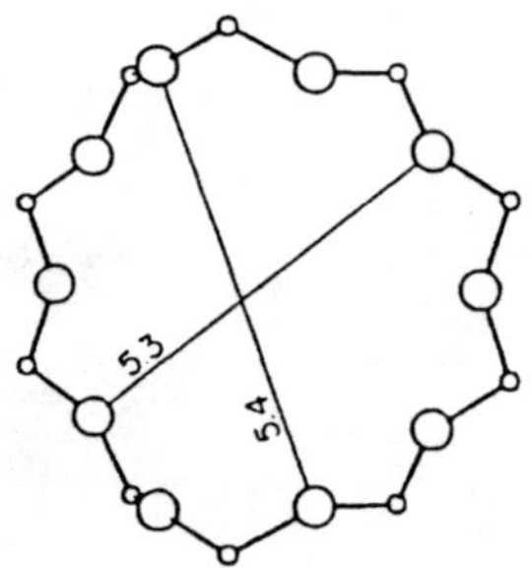

FIG. 4.   View of MEL channel (distances in Å).

2.   ZSM-22 (TON Structure)

The composition of the hydrogen form of ZSM-22 is

$$H_x(Al_xSi_{24-x}O_{48}) \cdot nH_2O; \quad \text{with} \quad x < 2.$$

The structure consists of unidirectional and one-dimensional 10-ring channels. Thus the structure presents no channel intersection and no cage-like void. The cross section of the 10-ring aperture is 0.44 × 0.55 nm (Fig. 5) (23).

3.   ZSM-23 (MTT Structure)

This zeolite has the same composition as ZSM-22. Like ZSM-22, it contains one-dimensional channels. The size of the 10-ring is almost identical to that of ZSM-22, 0.45 × 0.52 nm (Fig. 6) (24).

framework viewed along [001]

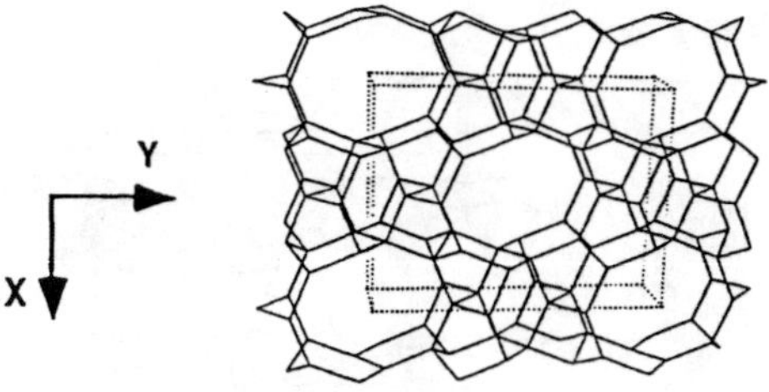

10-ring viewed along [001]

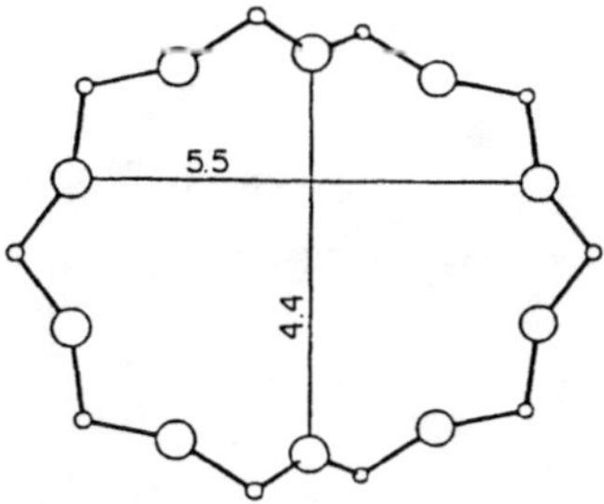

FIG. 5.   View of TON channel (distances in Å).

4.   *Ferrierite–ZSM-35 (FER Structure)*

Ferrierite and ZSM-35 (*25*) have identical orthorhombic structures. The framework contains two perpendicular intersecting channels, a one-dimensional 10-ring channel (0.42 × 0.54 nm) and one-dimensional 8-ring channel (0.35 × 0.48 nm). At the channel intersections there are cage-like voids with a diameter of 0.6–0.7 nm (Fig. 7).

B.   2. SILICO-ALUMINOPHOSPHATES

The silico-aluminophosphate molecular sieves represent a new class of microporous solid acid catalyst. The framework of the aluminophosphate molecular sieves is formed by tetrahedral arrangements of equal numbers of AlO_4 and PO_4 tetrahedra, with Al^{3+} and P^{5+} alternating. The framework of AiPO is neutral. However, the AlO_4 or/and PO_4 may be substituted

framework viewed along [001]

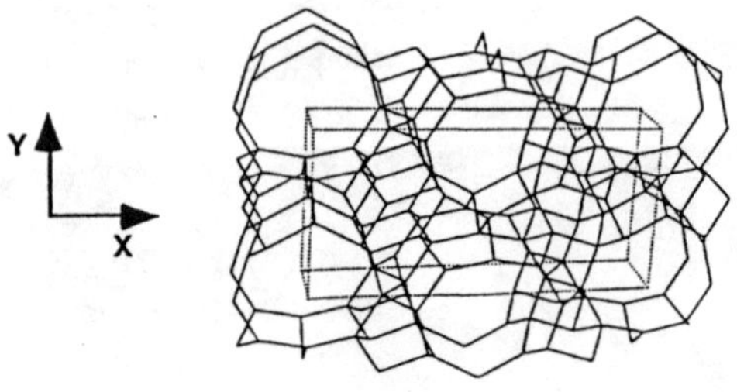

10-ring viewed along [001]

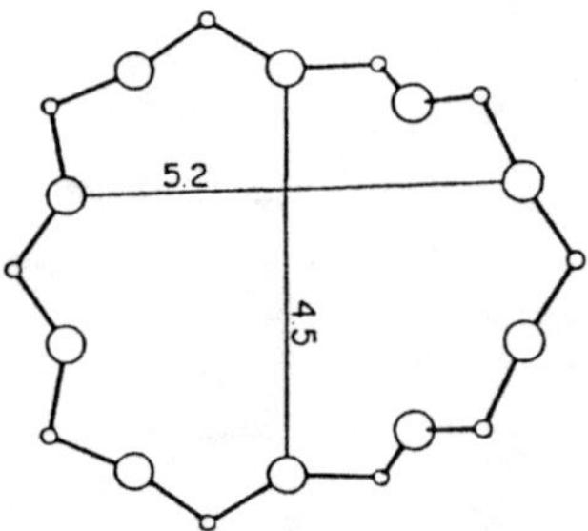

FIG. 6.   View of MTT channel (distances in Å).

partially by SiO_4 tetrahedra. The substitution of PO_4 with SiO_4 introduces a negative charge that must be compensated by a cation, e.g., H^+. Hence, SAPO molecular sieves have protonic acidity and the corresponding catalytic properties.

Metal aluminophosphates (MeAPO) contain framework metal (Me), aluminum, and phosphorus. When the metal is divalent (e.g., Zn^{2+}, Co^{2+}, and Mg^{2+}) and substitutes for aluminum, a negatively charged framework results, with H^+, for example, serving to compensate the charge. Many aluminophosphate molecular sieves have been synthesized. SAPO-11 and MeAPO-11 have interesting catalytic properties. Their structures have one-dimensional 10-ring channels. The 10-ring pore aperture is elliptical with dimensions 0.39×0.63 nm. Table I is a summary of the characteristics of the molecular sieves which have been used for the skeletal isomerization of n-butenes.

framework viewed along [001]

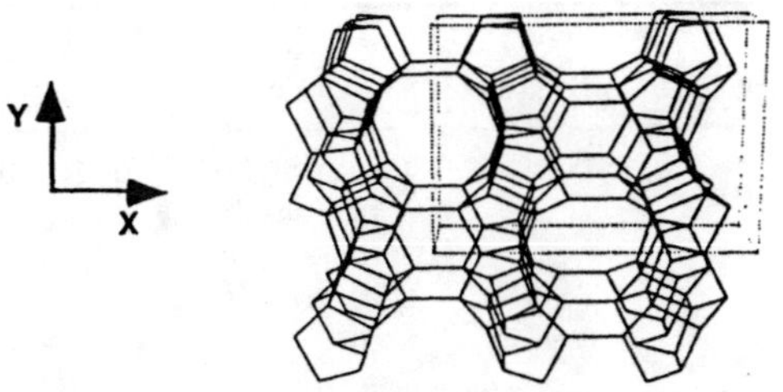

10-ring viewed along [001]

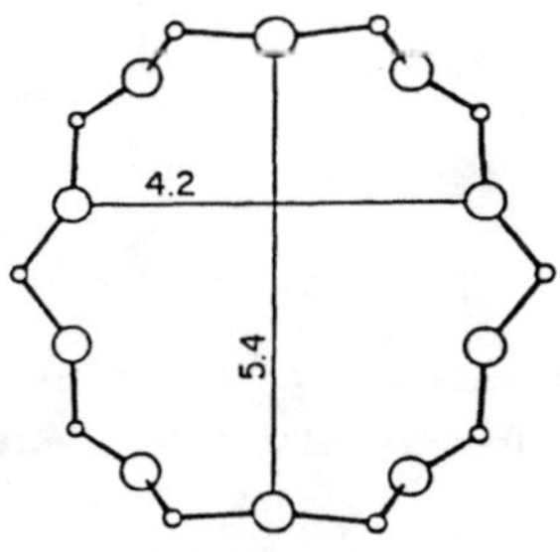

8-ring viewed along [010]

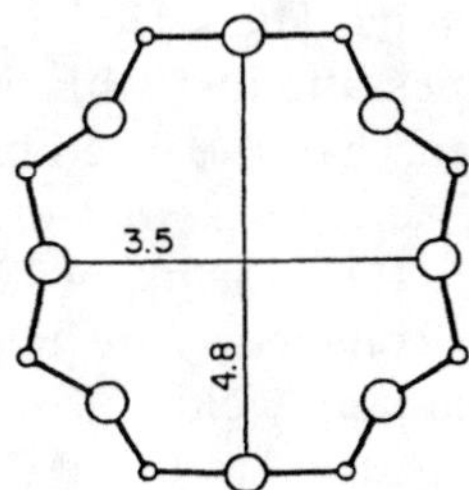

FIG. 7. View of FER channel (distances in Å).

TABLE I

Characteristics of Microporous Molecular Sieves Used as Catalysts for the Conversion of n-Butenes

Type code	Name	Channel system	Ring structure	Pore size (nm)	Isotypic framework
MFI	ZSM-5	Two	10 straight	0.51 × 0.55	
			10 sinusoidal	0.53 × 0.56	—
MEL	ZSM-11	Two	10 straight	0.53 × 0.54	
TON	ZSM-22	One	10	0.44 × 0.55	Theta-1
					KZ-2
					NU-10
MTT	ZSM-23	One	10	0.45 × 0.54	KZ-1
					EU-13
FER	ZSM-35	Two	10	0.42 × 0.54	NU-23
	Ferrierite		8	0.35 × 0.48	FU-9
AEL	AlPO-11	One	10	0.39 × 0.63	SAPO-11
					Mn–APO-11
					Me–APO-11

V. Skeletal Isomerization of *n*-Butenes Catalyzed by Medium-Pore Microporous Molecular Sieves

The skeletal isomerization of *n*-butenes catalyzed by medium-pore molecular sieves has been reported in several patents and articles. It has generally been observed that the initial activity of medium-pore molecular sieves is high but that the conversion decreases during the initial stage of the reaction, while the selectivity for isobutylene increases. Table II is a summary of typical results. The selectivity and yield data given in Table II are the maximum values attained with time on stream. It is important to compare the performances of the listed molecular sieve catalysts at similar *n*-butene conversions since it is well established that the selectivity for isobutylene depends on the conversion. The curves showing yield of isobutylene versus *n*-butene conversion are generally similar to that represented by Fig. 8, with the maximum yield depending on the nature of the catalyst.

The selective catalysts listed in Table II can be classified into two groups, one consisting of ZSM-35 (FER), having bidimensional intersecting pores, and the other consisting of catalysts containing unidimensional pore systems. In addition, there are common features characterizing all these molecular-sieve catalysts:

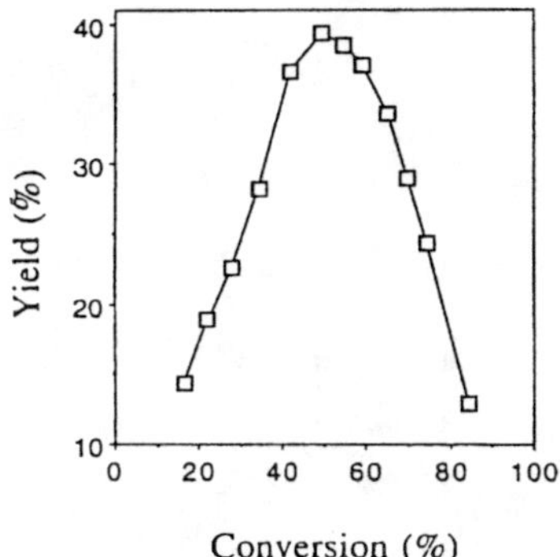

FIG. 8.   Transformation of *n*-butenes catalyzed by ZSM-22: changes in the isobutylene yield as a function of conversion (*35*). Yield is defined as weight of isobutylene formed/weight of *n*-butenes transformed.

1. The *n*-butene conversion decreases with time on stream (TOS) in a flow reactor operated at a constant feed composition, flow rate, and temperature.

2. The isobutylene selectivity at given conversion increases with TOS in such operations. The catalytic behavior is illustrated in Figs. 9 and 10.

Figure 9 indicates that, depending on the nature of the catalyst, the initial selectivity is low, nearly 50% for FER and higher (75%) for ZSM-23. However, after a few hours on stream, both catalysts become highly selective, with the selectivity being as high as 90–95% for FER. Figures 9 and 10 indicate that although the reaction of *n*-butenes follows the same

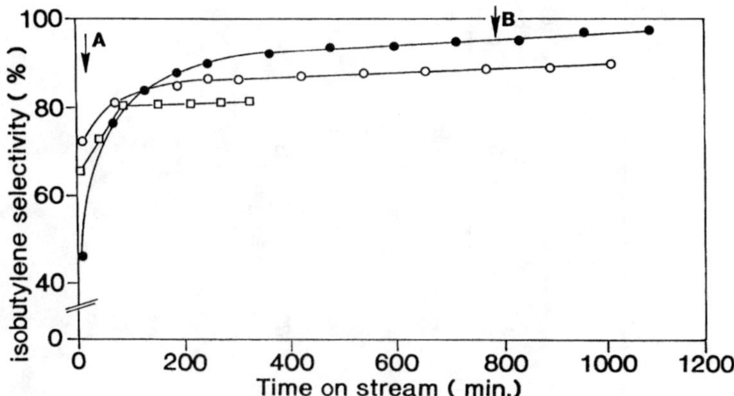

FIG. 9.   Transformation of *n*-butenes: changes in selectivity as a function of time on stream for different catalysts [●, FER (*29*); □, ZSM-22 (*36*); ○, ZSM-23 (*38*)].

TABLE II
n-Butene Isomerization Catalyzed by Medium-Pore Zeolites, SAPOs, and MeAPO

Catalyst	Experimental conditions			n-Butene conversion[b] (%)	Isobutylene			Remark	Reference
	T (K)	P (kPa)[a]	WHSV (h^{-1})		Selectivity (%)	Yield[c] (%)	TOS (h)		
FER	623	141	2	50.9	81	41.2	336		5
FER	703	101	7	48	85	40.8	40		6
FER + binder (Al$_2$O$_3$)	693	50.5/101	8	49.8	78	38.8	44	The binder improves the catalytic performance	26
Deal FER	693	50.5/101	5.3	48.9	81.3	39.8	16	Hydrothermally treated FER + acid treatment + alumina binder	27
ZSM-35	673	101/202	33	45	93	42	96	Treated with oxalic acid	28
FER	693	50.5/101	5.34	34	94	31	20		29
FER	693	50.5/101	5.3	45	93	42	10	Hydrothermally dealuminated and acid treated	30
FER	623	5/101	n.m.[d]	41	82	33.6	2		31
FER	673	20/101	5.6	40	90	36	12		32
MgZSM-35	698	101	n.m.	42	96	40.3	50		33
ZSM-22	823	53/160	21	39.8	94.5	37.6	n.m.		34
ZSM-22	693	20/101	22.9	47	81	38	n.m.		35

ZSM-22	623	5/101	n.m.	43	88	37.8	2	31
ZSM-22	773	50.5/101	150	42.8	81.4	34.8	4.5	36
ZSM-23	773	21/?	n.m.	38	90.6	34.4	?	37
ZSM-23	733	50.5/101	170.9	32.2	86	27.7	10	38
ZSM-23	673	20/101	373	25	65	16.2	0.05	39
SAPO-11	616	101	4.7	47	58	27	n.m.	7
SAPO-11	673	20/101	0.39	58.2	58	33.8	4	40
Mn–APO-11	673	20/101	0.39	42.7	88.6	37.8	4	40
Mn–APO-11	673	10/101	0.6	48	91	42	16	14
Mg–APO-11	673	10/101	0.6	53	77	40.8	16	14
Mg–APO-11	673	20/101	0.39	27.5	90.4	24.9	4	40
Co–APO-11	673	10/101	0.6	54	76	41	16	14
Co–APO-11	673	20/101	0.39	41.2	98	40.4	4	40
Ge–APO-11	673	25/101	2	28	90	26.2	1	41
B · Al–ZSM-11	796	20/101	1.1	44.6	50	22.3	n.m.	42
Al–ZSM-5	833	20/101	n.m.	55.8	36.9	20.6	n.m.	43
Al · B–ZSM-5	796	20/101	n.m.	58.9	49.1	28.9	n.m.	43
Al–ZSM-5	723	18/101	n.m.	40	40	16	n.m.	44
Fe–ZSM-5	773	20/101	12	50	55	27	0.5	45
MCM-22	773	10/101	28	49.8	59.2	29.6	0.5	46

[a] Pressure kPa. A/B means; A, butene partial pressure; B, total pressure. [b] Conversion is calculated by assuming that all n-butenes are reactants even if the true reactant is one particular n-butene. [c] Yield, conversion × selectivity. [d] n.m., not mentioned.

trend when catalyzed by each of the three molecular sieves, the magnitude of the variations in the conversion and selectivity depends to a large extent on the nature of the molecular sieve (whether it is aluminophosphate or aluminosilicate) and on the structural arrangement of the channels (whether they are unidimensional nonintersecting channels or bidimensional intersecting channels).

Many attempts have been made to interpret the origin of the cracked products (propene and pentenes) formed during the reaction of *n*-butenes and to relate the formation of isobutylene, propene, and pentenes to the acidic and structural properties of the molecular sieve catalyst. In the introduction it was noted that the key reaction intermediates are (i) methylcyclopropyl cations (formed in the monomolecular path) and (ii) di- (or tri-) branched methyl C_6 (or C_5) cations (formed in the bimolecular path) and that high selectivity to isobutylene can be achieved only when the monomolecular path predominates.

The overall reaction scheme for the skeletal isomerization of *n*-butenes (including both the mono- and bimolecular processes) is valid for all acidic catalysts. However, the relative amounts of the carbenium ion intermediates involved either in the monomolecular or in the bimolecular reaction paths, as well as their relative rates of conversion, can be dramatically different for the various molecular sieve catalysts and can depend crucially on the sizes of the channels and their structural configurations as well as on the acidity of the catalyst.

On fresh solid catalysts exhibiting selectivities in the range of 50–75%, it is inferred that isobutylene is formed, at least in part, via a bimolecular process (see Section III) because of the formation of by-products such as

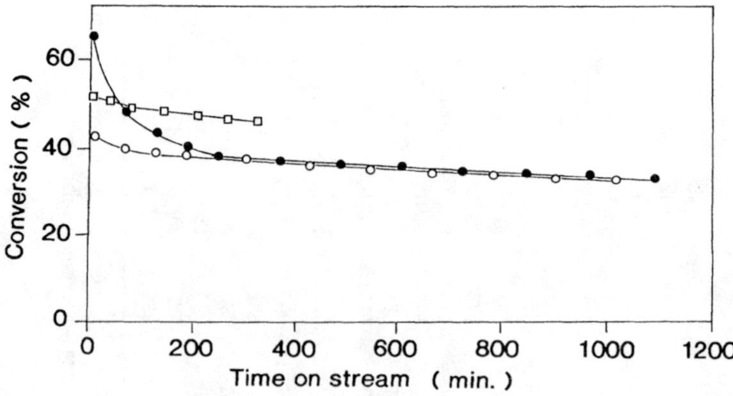

FIG. 10. Transformation of *n*-butenes catalyzed by acidic molecular sieves: changes in isobutylene conversion as a function of TOS [●, FER (*29*); □, ZSM-22 (*36*); ○, ZSM-23 (*38*)].

propene and pentenes. In contrast, for catalysts exhibiting high selectivities (90–95%) (e.g., coked FER or MTT and fresh SAPO-11), the question of the mechanism was not answered. In the first paper published by the Shell group (*47*), it was proposed that isobutylene should be formed from the cracking of 2,2,4-trimethylpentyl cation or 2,4-dimethylhexyl cation, but the monomolecular pathway was not definitely rejected. Nor is it clear why only these two possible intermediates would have been formed, resulting in the observed high selectivity toward isobutylene.

VI. Experimental Evidence for Bimolecular and/or Monomolecular Mechanisms

Additional experiments have enabled researchers to distinguish between the monomolecular and bimolecular mechanisms in the skeletal isomerization of butenes. The experiments were based on two different approaches, one based on the use of ^{13}C-labeled linear butenes and the identification of the ^{13}C label in the isobutylene product and the other based on the identification of the products resulting from the reactions of various possible C_8-olefin intermediates when the bimolecular mechanism prevails.

A. USE OF ^{13}C-LABELED BUTENES

It is clear (see Section III) that the use of ^{13}C-labeled butene containing, for example, one ^{13}C atom will produce isobutylene with only one ^{13}C atom if skeletal isomerization occurs through a monomolecular pathway, whereas in the case of a bimolecular reaction one can predict a random distribution of the labeled carbons in the C_8 olefin isomers (*32*). Consequently, the isobutylene formed by the cracking of the C_8 intermediates will give 25% of the molecules containing no ^{13}C, 50% containing one ^{13}C, and 25% containing two ^{13}C atoms. This is illustrated as follows:

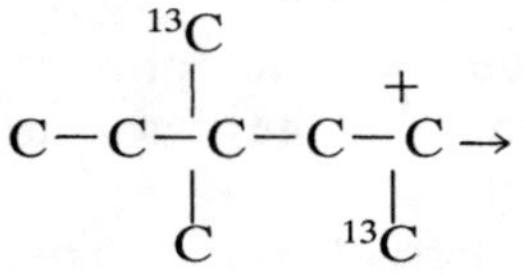

$$
\begin{array}{c}
\overset{\displaystyle ^{13}C}{\underset{\displaystyle C}{\overset{|}{C-C}}} \overset{+}{-C-C} \rightarrow \text{two isobutylene molecules, each with one } ^{13}C \text{ atom}
\end{array}
$$

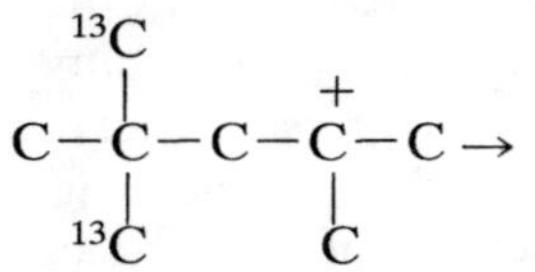

$$
\begin{array}{c}
\overset{\displaystyle ^{13}C}{\underset{\displaystyle ^{13}C}{\overset{|}{C-C}}} \overset{+}{-C-C} -C \rightarrow \text{one isobutylene molecule with no } ^{13}C \text{ + one}
\end{array}
$$
isobutylene molecule with two ^{13}C atoms.

The reaction of $^{13}C-C=C-C$ was performed with a ferrierite catalyst in a batch reactor at 673 K and $P_{C_4H_8} = 10$ kPa, with the catalyst being either in its fresh state (Fig. 9, point A) or in its selective state (Fig. 9, point B); the conversion was kept low ($\sim$20%) to avoid possible secondary reactions.

The results indicate that for FER in its selective state, more than 90% of isobutylene molecules contain only one ^{13}C atom, showing that isobutylene is formed via a monomolecular reaction. In contrast, and as expected, it was observed that for the fresh FER, a nonselective catalyst, the ^{13}C distribution in isobutylene was as follows: 25% without a ^{13}C atom, 50% with one ^{13}C atom, and 25% with two ^{13}C atoms. This confirms that isobutylene is formed mainly via a bimolecular mechanism. The conclusion is that the bimolecular reaction is not selective for isobutylene, and that undesirable cracking reactions occur simultaneously with the formation of isobutylene. When this bimolecular path is impeded, the monomolecular path becomes the only possible path for the skeletal isomerization, and high selectivity to isobutylene results. Similar results were reported by de Jong *et al.* (48).

B. USE OF DIFFERENT C_8 ISOMERS

An idea that was tested experimentally is that if the bimolecular mechanism prevails (*19*), the products formed from *n*-butene reactants, on the one hand, and from any of the possible isomers formed by dimerization of *n*-butenes (such as 3,4-dimethylhex-1-ene), on the other hand, should be similar. Thus, the transformations of 2,2,4-trimethylpent-2-ene, 3,4-dimethylhex-2-ene, and methylheptenes were investigated with microporous catalysts such as MnAPO-11 and SAPO-11. The results are summarized in Fig. 11, in which the ratio (propene + pentene)/*n*-butenes is plotted for different reactants. The data show that with a selective isomerization catalyst, this ratio is quite low ($<$0.1) for *n*-butene reactants; in contrast, it is quite high (approximately 0.8) when 3,4-dimethylhex-2-ene or methylheptenes are the reactants, indicating that these compounds are not intermediates in the selective isomerization of *n*-butenes. Consequently, the isobutylene formed on selective catalysts results from a monomolecular process. These results are considered to be good indirect evidence that the bimolecular reaction is not selective for isobutylene formation.

Additional indirect experimental evidence was reported by the same authors (*19, 49*). They observed that when *n*-butenes reacted at different pressures in the presence of a catalyst exhibiting moderate isobutylene selectivity (50–70%), the highest isobutylene selectivity was observed at the lowest butene pressure; this result was interpreted by considering that a decrease in butene pressure causes a greater decrease in the bimolecular reaction than in the monomolecular reaction. Similar results and a similar

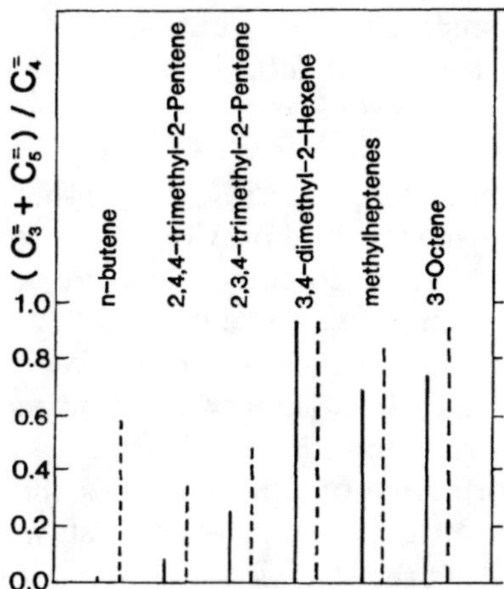

FIG. 11. Transformation of different possible C_8 olefin isomers (*19*). Change in $C_3^= + C_5^=/C_4^=$ as a function of the nature of the reactant and of the catalyst; —, Mn–APO-11; ---, B-ZSM-5. Note: $C_4^=$ means the summation of all linear butenes.

interpretation were reported for different catalysts, such as MFI (*45*), TON (*35*), and MCM-22 (*46*). The authors also reported that moderately selective catalysts gave isobutylene selectivities that increased with temperature. Again, this result was considered as indirect evidence that isobutylene was formed via both a monomolecular and a bimolecular path, the latter being disfavored by the increase in temperature (*45, 46, 49*). For example, it was shown that with ZSM-5 at 673 K and $P_{C_4H_8}$ = 20 kPa, the isobutylene selectivity was 22% at a conversion of 55%, compared with the 42% selectivity observed at 773 K and the same butene pressure at 50% conversion (*45*).

VII. Nature and Location of the Active Sites

The acidity of H-form molecular sieves is generally ascribed to pure Brønsted acid sites. Lewis acidity may also exist when extraframework aluminum species are present. The number of Brønsted acid sites is directly associated with the amount of framework Al (or other trivalent metal ions), and in general these outnumber the Lewis acid sites. In hydrocarbon reactions proceeding via carbenium ion intermediates, the active sites in aluminophosphates (SAPO) and zeolites are the Brønsted acid sites. Several factors affect the strengths of these sites. As a general rule, zeolites exhibit higher acid strengths than SAPO (or MeAPO), and the acid strengths of zeolites decrease in the sequence of substituted trivalent metal: Al > Ga > Fe > B.

The Brønsted acid sites were found to be important for the catalytic activity for skeletal isomerization of n-butenes. Boron aluminosilicate zeolites are more active for the skeletal isomerization when water is cofed with n-butenes. It was suggested that water generates additional protonic sites, increasing the Brønsted acidity (*43*).

The mechanism of skeletal isomerization of n-butenes may be rationalized in terms of the steps presented previously; the key reaction intermediate is the s-butyl cation. The predominent structure of the adsorbed intermediate was recently considered to be an alkoxy (*50*), which either adds to one butene molecule and cracks into C_3, C_4, or C_5 fragments (the bimolecular mechanism) or rearranges into isobutylene (the monomolecular mechanism) via a primary carbenium ion.

To overcome the formation of a primary carbenium ion intermediate, it has been proposed for an aged ferrierite catalyst, which is highly selective for isobutylene, that a carbenium ion trapped within the carbonaceous residues formed in or on the zeolite could be the active site. The authors suggested that the structure of such an active site could be

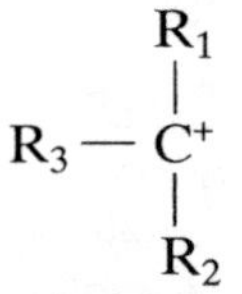

$$
\begin{array}{c}
R_1 \\
| \\
R_3 - C^+ \\
| \\
R_2
\end{array}
\tag{51}
$$

or

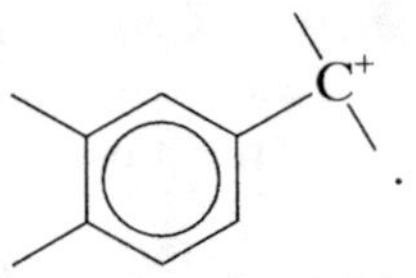

$$
\tag{52}
$$

The monomolecular mechanism would transform n-butenes into isobutylene via the formation of a secondary carbenium ion as follows:

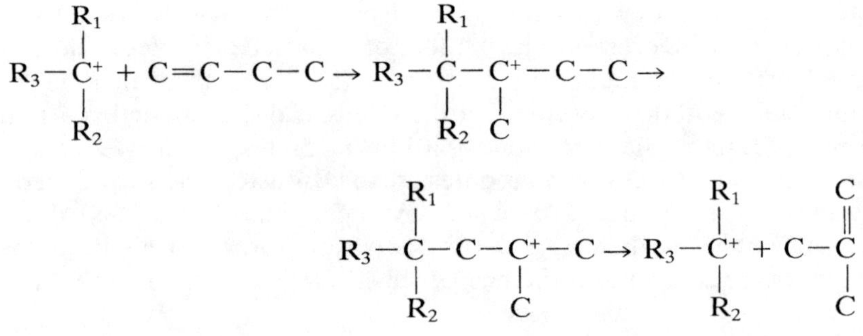

$$
\begin{array}{c}
R_1 \\
| \\
R_3 - C^+ \\
| \\
R_2
\end{array}
+ C{=}C{-}C{-}C \rightarrow
\begin{array}{c}
R_1 \\
| \\
R_3 - C - C^+ - C - C \\
| \quad | \\
R_2 \quad C
\end{array}
\rightarrow
$$

$$
\begin{array}{c}
R_1 \\
| \\
R_3 - C - C - C^+ - C \\
| \quad\quad | \\
R_2 \quad\quad C
\end{array}
\rightarrow
\begin{array}{c}
R_1 \\
| \\
R_3 - C^+ \\
| \\
R_2
\end{array}
+
\begin{array}{c}
C \\
\| \\
C - C \\
| \\
C
\end{array}
\tag{19}
$$

athway will be more favorable kinetically than that sug-
ue monomolecular process, whereby a primary carbenium
To further test the idea that carbonaceous residues are the
ctive sites for the skeletal isomerization of *n*-butenes, the
ted results showing that the rate of isobutylene formation
errierite passed through a maximum as the conversion contin-
ased (Fig. 12) (*51*).

he validity of the hypothesis that the active site is a carbenium
ue was carried out in the following manner (*53*): If the transformation of *n*-butenes into isobutylene via a monomolecular reaction is initiated
by a simple proton transfer to the reactant, this transformation will involve
an unstable primary carbenium ion, as shown previously. In contrast, for
a larger olefin like pentene, the monomolecular reaction will occur via a
secondary carbenium ion rather than a primary carbenium ion and take
place more rapidly as follows:

$$C-C-C-C-C \underset{}{\overset{H^+}{\rightleftharpoons}} C-\overset{+}{C}-C-C-C \rightleftharpoons$$

$$\overset{C}{\underset{C-C \,\overline{\quad H^+ \quad}\, C-C}{\diagup \diagdown}} \rightarrow \overset{C}{\underset{C-C-C-C}{|}}\overset{}{{}^+} \quad . \quad (20)$$

Thus, it is expected that the rate of *n*-butene transformation into isobutylene
will be much less than the rate of *n*-pentene transformation into isopentenes.

In the case of the reaction proceeding via a carbenium ion residue (*51*)
(the active site being a tertiary carbenium ion), the transformation of
n-butenes into isobutylene should involve a secondary carbenium ion,

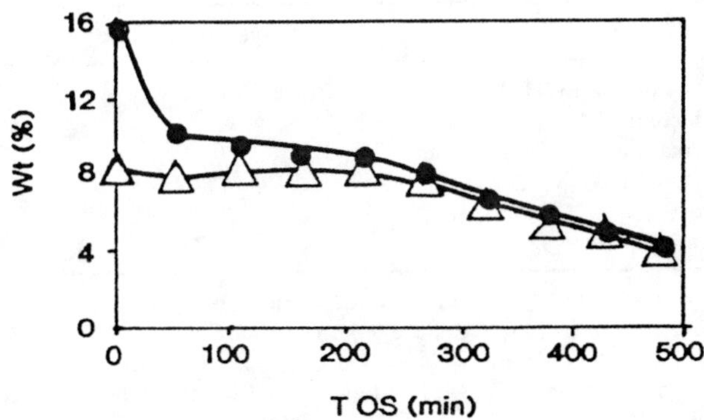

Fɪɢ. 12.   Change in butene converted into isobutylene and butene conversion as a function
of TOS (*51*): △, isobutylene; ●, conversion, *T* = 623 K.

the same kind of intermediate formed in the skeletal isomeriz
pentenes. Thus, with this hypothesis, it is expected that the skeleta
ization rates of n-butenes and n-pentenes would be similar. Howev
results (Table III) indicate that in the presence of the aged selective fe.
ite the rate of n-butene skeletal isomerization is much less than that o.
pentenes. On the basis of these results it was concluded that the active siı
is a proton and not the carbenium ion residue. Results obtained for fresh
catalysts also indicate that the rate of skeletal isomerization of n-butenes
is much less than that of n-pentenes, with this result being interpreted
as previously discussed. In addition, a comparison between the rate of
isobutylene formation and the rate of isopentene formation for fresh and
aged catalysts indicated a decrease by a factor of nearly 73 for isobutylene
and by a factor of 64 for isopentenes, these values being in satisfactory
agreement with the decrease in the number of acid sites between fresh and
aged solids (see next section). Thus, the role of the carbonaceous residues
as catalytic sites is discounted.

The role of the carbonaceous residues in the skeletal isomerization of
n-butenes was also questioned by Houzvicka and Ponec (49). These authors
observed that in a continuous flow reactor and at high n-butene conversions,
the relative concentration of isobutylene increased with TOS, with the
results being similar to those reported by Guisnet $et\ al.$ (51). However, the
ratio isobutylene/all butenes decreased with TOS. It was concluded that
at the initial stage of the reaction, a fraction of the isobutylene product
reacted subsequently with n-butenes to form byproducts including carbona-

TABLE III

Catalytic Properties of Fresh and Aged FER for 1-butene and (cis + trans)-2-pentene Conversions at 673 K[a]

Property	Aged catalyst[b]	Fresh catalyst
n-Butene conversion, %[c] (WHSV, h^{-1})[d]	2 (45)	4 (3,400)
Isobutylene selectivity (%)[e]	92	45
Isobutylene formation rate[f]	0.015	1.1
n-Pentene conversion %[c] (WHSV, h^{-1})[d]	3 (1700)	8 (48,000)
Isopentene selectivity (%)[e]	98	92
Isopentene formation rate[f]	0.7	48

Note: $P_{olefin} = 26$ kPa, in N_2; pressure, atmospheric.
[a] From Mériaudeau $et\ al.$ (53). [b] Solid aged for 16 h, $T = 673$ K, WHSV = 6 h^{-1}. [c] For butenes (or pentenes) the conversion is defined as $[(C_{in} - C_{out})/C_{in}] \times 100$, where C_{in} is the number of moles of 1-butene a $C_{out} = $ (1-butene + cis-2-butene + $trans$-2-butene). The same assumptions are made for linear pentenes. [d] WHSV adjusted to give conversions $\leq 8\%$. [e] Carbon basis; all linear butenes are considered to be reactants. The same assumption is made for pentene. [f] Rate in mol · h^{-1} (g of catalyst)$^{-1}$.

ceous deposits. These subsequent reactions lowered the measured rate of isobutylene production. With increasing TOS, the carbonaceous byproducts increasingly poison the active sites responsible for the consecutive reactions, and the loss of isobutylene due to the consecutive reactions becomes less. Consequently, the rate of isobutylene formation apparently increases. Pulse reactor experiments confirmed the previous conclusions. In these experiments, a small amount of *n*-butene was fed in each pulse, and the amount of carbon deposited after each pulse far exceeded the total number of Brønsted acid sites. Hence, the technique allowed the authors to follow accurately the change in the apparent rate of isobutylene formation with the number of pulses and to relate this change to secondary reactions. Since no significant difference in the reactivity of isobutylene and *n*-butenes to give carbonaceous deposits was observed, the ratio isobutylene/total butenes should not change with the pulse number, as illustrated in Fig. 13. The inhibition of secondary reactions is probably due to the deposition of carbonaceous residues on the proton-donor sites involved in the formation of these residues. It was concluded that the active sites for selective isomerization of *n*-butenes to give isobutylene are not the carbonaceous deposits proposed by Guisnet *et al.* (*51*) but rather very probably the naked Brønsted acid sites.

A.   Monomolecular Mechanism: General Considerations

It is clear from the previously discussed results that on a selective catalyst the skeletal isomerization of *n*-butenes proceeds through the monomolecu-

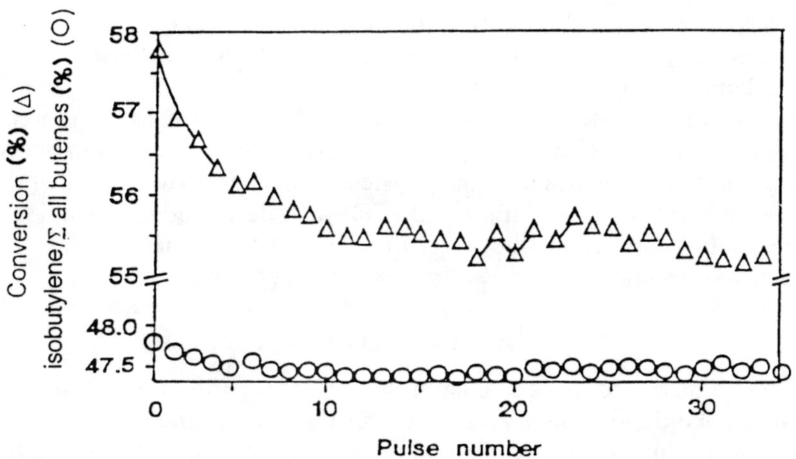

FIG. 13.   Butene reaction catalyzed by ferrierite in a pulse reactor at 623 K. Conversion (△) and isobutylene/Σ all butenes (○) versus the number of pulses (*49*).

lar mechanism, whereas on nonselective catalysts both mechanisms, mono- and bimolecular, operate, with their respective rates determining the isobutylene selectivity; the larger the contribution of the monomolecular pathway, the higher the isobutylene selectivity.

The remaining important question is the nature of the intermediates by which the monomolecular mechanism operates. It has been shown (see Section II.A) that the formation of a primary carbenium ion is energetically unfavorable [the difference in stabilization energy between a tertiary and a secondary carbenium ion is 40 kJ/mol and that between a secondary and a primary carbenium ion is 104.5 kJ/mol (*11*)]. Thus, in order to bypass the difficult step of primary carbenium ion formation, it has been suggested that the selective skeletal *n*-butene isomerization could proceed via an alkoxy intermediate (*50*). However, there is no any experimental evidence of its participation in isobutylene formation. Another possible intermediate, an allyl, was also proposed (*54, 55*), but since on the catalysts that are selective for this reaction allyl species were not detected (*56*), their participation is regarded as unlikely. It would be of interest to consider activation energies for isobutylene formation as a basis for determining more precisely the nature of the intermediates, but it is probable that intrinsic activation energies will be difficult to determine because of diffusion limitations in the microporous catalysts.

VIII. The Coke Deposits and Their Effects

It has been shown (Fig. 9) that for most of the molecular sieve catalysts the selectivity for isobutylene improves with TOS. Furthermore, it has been demonstrated with ferrierite that the change with TOS from the nonselective form (the noncoked state) to the selective form (the coked state) is accompanied by a change in the reaction mechanism from bimolecular to monomolecular. Here, the physicochemical properties of the molecular sieves and their variation with TOS are described and analyzed in terms of the correlations between the state of the molecular sieves and their activities and selectivities for *n*-butene isomerization.

A. Coke Deposits

When a ferrierite is used to catalyze the conversion of *n*-butenes, there is rapid coke deposition, as illustrated in Fig. 14. Several authors (*51, 57*) have shown that the coke deposition is rapid during the first hours of reaction in a flow system and much slower thereafter. The nature of the coke changes with TOS; during the initial period, the coke is more aromatic

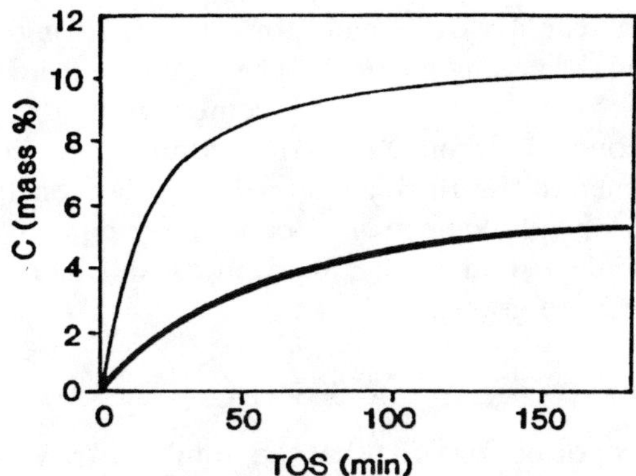

FIG. 14.   Coke deposit as a function of TOS on FER catalyst: thin line, static conditions, T = 693 K (*57*); thick line, dynamic conditions, T = 623 K (*51*).

than during the latter period, during which the H/C ratio decreases (Table IV). Similar results, including kinetics of the coke deposition and changes in the nature of the coke as a function of TOS, were reported by de Jong *et al.* (*48*).

The coke deposits affect the physical properties of the catalysts, as described in the following section.

B.   A PORE BLOCKING

Xu *et al.* (*57*) compared the pore size distributions of fresh and aged HFER on alumina catalysts, showing that the mesoporosity does not change much with TOS. In contrast, the microporosity decreased substantially

TABLE IV

Change in the Coke Deposition and Coke Composition on a FER Catalyst as a Function of TOS

TOS (h)	Total carbon content (mmol/g)	H/C ratio
2	2.46	0.93
11	2.70	0.84
18	3.84	0.47

Experimental conditions: T = 693 K, $P_{C_4H_6}$ = 50 kPa, WHSV = 8 h^{-1}, from Xu *et al.* (*57*).

during operation (Table V); the microporosity was strongly reduced by the coke deposited in the zeolite channels. The results indicating the formation of coke in the micropores are in good agreement with those reported by Seo *et al.* (*44*), who concluded from Xe NMR measurements that carbonaceous deposits are formed in the 10-ring channels. Similar conclusions were also derived from X-ray photoelectron spectroscopy data (*47*) showing that carbon was first deposited in the micropores and later on the external surfaces of the zeolite grains.

C. Poisoning of the Acidic Sites by Carbonaceous Deposits

Xu *et al.* (*57*) reported for FER that the number of acidic sites decreased with TOS (Table VI). The decrease in the number of detected acidic sites could be due to poisoning and/or to pore blocking. Table VI indicates that the larger the probe molecule used to detect acidic sites, the fewer the sites detected; evidently, the accessibility of acidic sites in ferrierite supported on alumina depends on the size of the adsorbate.

Evidence of poisoning of acidic sites was also obtained by IR spectroscopy. The number of protons on the ferrierite surface was estimated by the intensity of the OH band at 3600 cm^{-1}. After a few hours on stream, the absorbance of this OH vibration was reduced by a factor of 2.4; in contrast, the acidity of the solid probed by ammonia temperature-programmed desorption (TPD) decreased by a larger factor (*53, 58*). These results were considered as evidence of the poisoning of a large number of acidic sites and a partial blocking of pores, rendering some acidic sites inaccessible to the reactant.

The poisoning of the strongest acid sites was also shown indirectly by ammonia adsorption and TPD or butene adsorption and TPD. The coked sample adsorbed a smaller amount of each base than did the fresh catalyst,

TABLE V

Pore Size Distributions of Fresh and Aged FER Catalyst

Porosity[a]	Pore volume (μL/g)	
	Fresh catalyst	Aged catalyst (TOS: 18 h)
Meso-macropores >10 A	218	189
micropores <10 A	96	0.8

Experimental conditions as in Table IV [a] As measured by nitrogen adsorption and desorption, from Xu *et al.* (*57*).

TABLE VI

Change in the Number of Acidic Sites on a FER Catalyst (as Probed by Ammonia and by 1-Butene) as a Function of TOS[a]

Catalyst treatment	Number of acid sites as probed by	
	Ammonia[b]	1-Butene[c]
None (fresh catalyst)	100	74.1
2 h on stream	62.1	8.8
11 h on stream	47.7	4.7
18 h on stream	35.1	4.2

[a] From Xu *et al.* (*57*). Experimental conditions as in Table IV. [b] In arbitrary units. [c] Relative to fresh catalyst, as probed by ammonia.

and the maximum of the TPD peak was shifted to lower temperature for the aged catalyst (*57*). This result was interpreted as evidence of poisoning of the strongest acidic sites. Combining this result with the previously mentioned changes in the catalyst microporosity, and loss of catalytic activity and an increase of selectivity with TOS, it was concluded that the improvement of the selectivity for isobutylene is due to the following:

1. A shape-selective effect due to coke deposits (*29, 35, 44, 57*)

2. Poisoning of the strongest acid sites, which could have been responsible for unselective isomerization via the bimolecular mechanism (*19, 29, 35, 57*)

3. A decrease in acid site density (*57*), which leads to a decrease in the undesired consecutive reactions (*46*)

Similar behavior was observed for ZSM-23 (*38*), MeAPO-11 (*14*), and ZSM-22 (*35*). For these unidimensional pore systems (MTT, AEL, and TON), the coking rate was less than that for FER catalysts; furthermore, the magnitude of the differences in activity and selectivity between fresh and coked samples was less than that for catalysts with bidimensional pore systems. For example, when ZSM-23 was used at 693 K and a WHSV of 171 h^{-1}, the micropore volume decreased from 58.4 to 11 μl/g after 20 h on stream. Also, the number of acidic sites estimated by butene TPD decreased from 0.45 to 0.06 mol per unit cell, whereas the butene conversion decreased only from 41 to 31%; the isobutylene selectivity increased from 72 to 92% (*38*). Evidently, the catalyst pore geometry significantly affects the coke deposition and thus the selectivity. The relevant literature is discussed in the following section.

IX. Factors Affecting Isobutylene Selectivity

A. INFLUENCE OF THE ACID SITE DENSITY

To investigate the influence of the acid site density, a HFER catalyst was prepared with increasing amounts of Li^+ exchange ions, thus decreasing the number of protons, finally to a low level. These Li^+-exchanged FER samples were used for n-butene isomerization. The results are summarized in Table VII (*58*). The results indicate that at a constant conversion ($\sim$30%), the isobutylene selectivity is not much improved by a decrease of the acid site density; the small increase of isobutylene selectivity resulting from a large decrease in acid site density was attributed to the preferential poisoning of the external acid sites.

B. INFLUENCE OF SHAPE SELECTIVITY

In the same investigation (*58*), other cations (i.e., Na^+, K^+, and Cs^+) were exchanged into the hydrogen form of the ferrierite. The results are illustrated in Table VIII for the catalyst exchanged with various amounts of Cs^+. The selectivity for isobutylene formation increases with increases in the degree of ion exchange, with the sample having a small number of residual protons being highly selective. In the reported experiments, at very short TOSs, these selectivities were considered to have been obtained for noncoked catalysts. The results obtained for catalysts partially exchanged by Na^+ or K^+ (*58*) are between those obtained for the sample partially exchanged by Li^+ and Cs^+.

The remarkable effect of the size of the exchange cation was interpreted in terms of shape selectivity; the larger the exchange cation, the smaller

TABLE VII

Catalytic Properties of Ferrierite Modified by Ion Exchange with Li^{+a}

Ion exchange[b] (%)	Normalized[c] remaining acidity	Butene conversion (%)	Isobutylene selectivity (%)
0	100	40	49
23	75	36.7	51.7
82	15	40.8	53
94	5	43	56

[a] Reaction conditions: $T = 673$ K; $P_{C_4H_8} = 26$ kPa; P_{total}, 1 bar (remainder N_2). Time on stream, 2 min (for these low TOS, results were considered to represent noncoked FER); WHSV adjusted to give nearly identical conversions. [b] From elemental analysis. [c] As measured by IR spectroscopy (absorbance of the OH vibration at 3600 cm^{-1}); in percentage of exchangeable protons.

TABLE VIII

Catalytic Properties of Ferrierite Modified by Ion Exchange with Cs^{+a}

Catalyst	Ion exchange (%)	Normalized remaining acidity	Butene conversion (%)	Isobutylene selectivity (%)
HFER	0	100	40	49
	45	55	41	68
	55	43	36	78
	80	18	37	84.5

[a] Experimental conditions as in Table VII (*58*). WHSV adjusted to give nearly identical conversions.

the free space around the remaining protons and the higher the isobutylene selectivity. In other words, when the space around the active site decreases because of the increasing size of the exchangable cation, the rate of *n*-butene dimerization/cracking (bimolecular process) decreases sharply, whereas the rate of isomerization of *n*-butenes via the monomolecular pathway is little affected. Thus, there is a simultaneous decrease in the total butene conversion and an increase in the isobutylene selectivity.

These results are important because they clearly demonstrate the role of shape selectivity in the selective skeletal isomerization of *n*-butene. Similar results and interpretation were obtained independently by Kwak *et al.* (*33*), who modified the ferrierite catalyst by replacing part of the protons with Mg^{2+}. They observed a small decrease in microporosity which was related to an increase in the isobutylene selectivity. Again, it was demonstrated that the micropore size is a key factor governing the selectivity for isobutylene.

The previous results underline the importance of shape selectivity effects even for the transformation of small olefins such as butene. The results are in agreement with the early, related work by Haag *et al.* (*59*), who investigated cracking of olefins and paraffins catalyzed by the zeolite HZSM-5 and distinguished between restricted transition state shape selectivity and mass transport shape selectivity. It is clear that the effects discussed here are best described in terms of restricted transition state shape selectivity.

It has been shown by using a methodology combining molecular dynamics and an energy minimization technique (*60*) that in the FER pores (cavities and channels), the formation of C_8 olefin intermediates is inhibited. These theoretical results agree well with the experimental indications of restricted transition state shape selectivity. Indeed, for materials such as zeolites and MeAPOs, most of the sites are expected to be located inside the micropores. Molecular sieves with large mesopores and/or large external surface areas with the acid sites located on these external surfaces will not provide the benefits of restricted transition state shape selectivity; results discussed in the following section demonstrate this point.

C.   Influence of the External Sites

Treatment with oxalic acid has been described as a method for selective removal of the external acid sites of medium-pore zeolites (*61*). FER and ZSM-23 zeolites were treated with a 1-*M* solution of oxalic acid at 353 K overnight (*39, 62*). The characterization of the acid sites showed that the treated materials had a low number of external acid sites compared with the untreated materials and, when used in *n*-butene isomerization, they exhibited an improved isobutylene selectivity. It was also observed that acid-treated FER does not have a high selectivity for isobutylene formation. It was inferred (*62*) that the cavities in ferrierite at the intersections of 8- and 10-ring channels are large enough to accommodate butene dimer intermediates, thus favoring the unselective bimolecular path. In contrast, when the external acid sites are removed from a zeolite with a unidimensional pore system (e.g., ZSM-23), the initial isobutylene selectivity is higher (nearly 80%) than that of the untreated sample.

Again, this difference in isobutylene selectivity between ZSM-23 and FER, when both were treated with oxalic acid and had few external acid sites, was attributed to the presence of cavities in FER. These cavities, which are not accessible to oxalic acid, are too large to restrict the formation of C_8 olefinic intermediates.

D.   The Role of Pore Size

A family of catalysts with undimensional pores was compared for *n*-butene conversion. Data were obtained at short TOS so that coking was negligible and the catalysts could be compared on the basis of their intrinsic pore geometries (*63*). The results, summarized in Fig. 15, show that the isobutylene selectivity increases with decreasing pore dimensions as follows:

Catalyst	SAPO-11	>	ZSM-23	>	SAPO-41	>	SAPO-31
Pore cross section (Å)	3.9 × 6.3		4.5 × 5.2		4.3 × 7		5.4 × 5.4

The authors concluded that the major factor governing the isobutylene selectivity is the free space around the active site, as was proposed for FER (*62*).

This concept accounts for the performance of other MeAPO microporous catalysts as well. The evidence is as follows: There are numerous

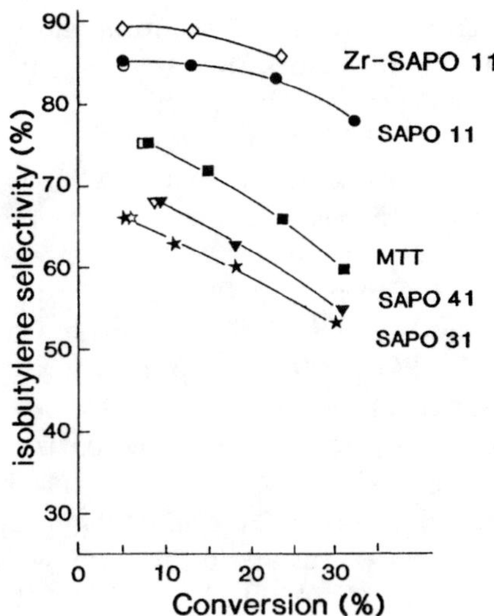

FIG. 15. Transformation of *n*-butenes into isobutylene in the presence of various catalysts (*39, 61, 62*); $T = 673$ K, $P_{C_4H_8} = 26$ kPa in N_2, $P_{total} = 1$ bar. WHSV was varied to give different conversions. Time on stream: 2 min. After one run (2 min), the samples were flushed with N_2 (5 min) and then oxygen (30 min) and N_2 (10 min) before the next run. Results are considered to represent noncoked catalysts. ◇, ZrSAPO-11; ●, SAPO-11; ■, MTT; ▼, SAPO-41; *, SAPO-31.

reports indicating that MeAPO-11 catalysts (e.g., MgAPO-11, MnAPO-11, and CoAPO-11) are highly selective for *n*-butene skeletal isomerization. For example, Gielgens *et al.* (*14*) observed that the isobutylene selectivity of MnAPO-11 was 91% after a few minutes on stream and that the rate of activity loss with TOS was very low. The high isobutylene selectivity was attributed to the low strength of the acidic sites and to the restricted pore dimensions. Zubowa *et al.* (*64*) compared MnAPO-31 and SAPO-31 and reported that after 6 h on stream MnAPO-11 was more selective than SAPO-31 (but no explanation was given for the higher selectivity of the Mn-containing catalysts). The selective skeletal isomerization of *n*-butene catalyzed by SAPO-11 and ZrSAPO-11 was compared (*65*). Zirconium was not incorporated in the framework of the SAPO, but it was, in part, deposited within the micropores; the initial selectivity of ZrSAPO-11 for isobutylene was higher than that of SAPO-11 (Fig. 15). It was concluded that the Zr deposited in the channels decreased the porosity of the SAPO-11, increasing isobutylene selectivity.

Furthermore, GeAPO-11 and SAPO-11 were similarly compared as catalysts (*42*). Only part of the Ge^{4+} in the GeAPO-11 had substituted P^{5+}, and the acidic sites created by Ge substitution were identical in strength to those of SAPO-11. It was also shown that part of the Ge was located in the micropores of the catalyst. The selectivity of the GeAPO-11 catalyst was greater than that of SAPO-11 catalyst (90% selectivity at 30% conversion with GeAPO-11 vs. 83% selectivity with SAPO-11 at the same conversion), and the difference was also attributed to the reduced microporosity of GeAPO-11 due to the presence of extraframework Ge. A comparison was also made between different MeAPO-11 catalysts (40). The isobutylene yields after 1 h on stream are in the following order: CoAPO-11 > ZnAPO-11 > MnAPO-11 > SAPO-11. It is tempting to assume that in all these MeAPO-11 catalysts, part of the Me, was extraframework; this assumption is not unlikely since the Me loadings (atomic basis) are higher than that of the corresponding SAPO-11. Moreover, it has been mentioned for CoAPO-11 that no IR vibration due to acidic OH groups is observed, again suggesting that only a few Co ions are in the framework. According to this suggestion, these materials (CoAPO-11, MnAPO-11, etc.) will obey the same rule as that obeyed by the others; transition-state shape selectivity governs the isobutylene selectivity.

E.   THE ROLES OF HYDROTHERMAL TREATMENT AND EXTRAFRAMEWORK SPECIES

Hydrothermally dealuminated FER and sample that were subsequently acid treated exhibited better selectivities for isobutylene formation than an untreated FER catalyst (27). Furthermore, hydrothermally dealuminated FER exhibited a lower activity than untreated FER but higher selectivity for isobutylene (30, 62, 66). A subsequent acid treatment (with 5% HCl solution) further decreased the conversion and increased the isobutylene selectivity. The hydrothermal treatment created mesoporosity by Al extraction. The Al extraframework species were located in the mesopores and/or in the micropores. The HCl treatment removed part of the extraframework Al, leaving part in the micropores. The elimination of extraframework Al from the mesopores was evidently beneficial for isobutylene selectivity. Evidently, the active sites associated with extraframework Al located in large voids are nonselective; in contrast, extraframework Al located in the micropores (and not removed by acid treatment) does not contribute to catalytic activity. The steamed and acid-washed ferrierite exhibits excellent isobutylene selectivity and catalytic stability (30).

In an investigation of FER synthesized without a template (66), it was shown that an acid treatment causes the removal of non-shape-selective acidic sites located in pores larger than the zeolitic pores or on the external surfaces of the ferrierite grains (66). Consequently, isobutylene selectivity increased. Thus, the hydrothermal treatment plus the subsequent acid treatment evidently lower the number of acid sites, decrease interactions between butene intermediates located at adjacent acid sites, and improve the catalytic selectivity and stability (*66*).

For such treated samples it is not easy to discriminate between two possible effects of dealumination, namely, the removal of some acid sites and the decrease in microporosity due to the deposition of aluminum-containing debris in the pores. Thus, hydrothermally dealuminated FER, hydrothermally dealuminated acid-washed FER, acid-washed FER, and CsFER were compared under the same experimental conditions (*62*). The results indicate the following order of isobutylene selectivities: untreated FER < acid-treated FER < hydrothermally treated FER < hydrothermally acid-treated FER < CsFER (*61*). These results, obtained with noncoked catalysts, reinforce the interpretation in terms of shape selectivity. The hydrothermally acid-treated sample has acid sites located only in the micropores, and the aluminium debris in the micropores creates an additional constraint playing a role identical to that of Cs^+ in FER.

F.   INFLUENCE OF ACID STRENGTH

The question of the influence of the Brønsted acid strength on the isobutylene selectivity has been reinvestigated with two microporous catalysts, AlZSM-23 and FeZSM-23 (*39*). The trivalent Al and Fe were shown to be in the frameworks, with the possible existence of a small amount of extraframework Fe. The acid strength, as probed by ammonia TPD, was much less for the Fe sample than for the Al sample. The comparison of the two catalysts under the same conditions indicated only a very small difference in their isobutylene selectivities. It was concluded that within the range of acid strengths studied, the change in Brønsted acid strength is not an important factor governing selectivity.

X.   Deactivation of the Catalysts with Time on Stream

The deactivation of the molecular sieve catalysts during the skeletal isomerization of *n*-butenes has not received much attention with regard to how it affects the reaction mechanism. In general, medium-pore molecular sieves deactivate slowly (Fig. 16) and much less than catalysts with open surfaces. In this section, the magnitude of deactivation is represented by the ratio yield of isobutylene/WHSV of *n*-butenes. According to this criterion, the stability with respect to TOS of the most efficient molecular sieves is as follows:

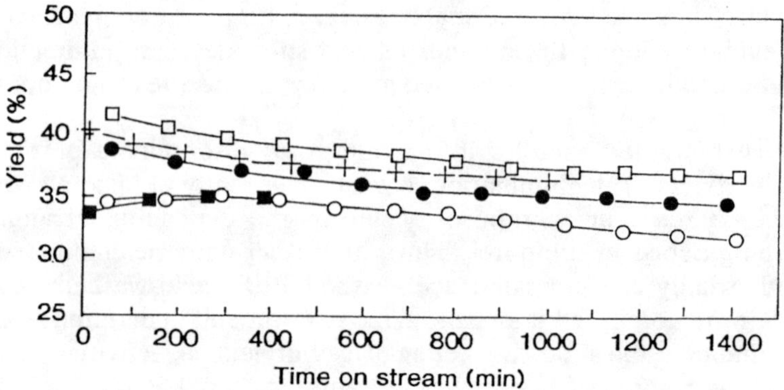

FIG. 16. Change in isobutylene yield for various catalysts as a function of TOS. □, CoAPO-11, T = 673 K (40); +, ZSM-23, T = 733 K (38); ●, MnAPO-11, T = 673 K (40); ○, SAPO-11, T = 673 K (40); ■, ZSM-22, T = 773 K (36). Isobutylene yield: n-butenes isobutylene/n-butenes converted.

$$ZSM\text{-}23 > ZSM\text{-}22 \sim FER > MeAPO.$$

Among the main factors which could affect the catalyst lifetime are the following:

1. The surface acidic sites: These sites, which are not selective for isobutylene formation, will also be active sites for coke formation. It has been reported (28) that the removal of these sites by oxalic acid treatment improves the catalyst lifetime.

2. The catalyst crystal sizes: It has been observed that the best ZSM-22 catalysts are those consisting of small crystals (largest dimension <0.5 μm). For these unidimensional pore systems, the smaller the crystal size, the lower will be the aging rate by pore mouth plugging.

3. Other parameters such as acid site density could also have an influence on catalyst aging, but there are no reports on this point.

XI. Conclusions

The literature related to selective skeletal isomerization of n-butenes catalyzed by medium-pore zeolites and Me–aluminophosphates leads to the following conclusions:

1. Ten-ring zeolites and Me–aluminophosphates are good catalysts for selective isomerization of n-butenes.

2. The FER zeolite (ZSM-35) with a bidimensional pore system (8 and

10 rings) having small cavities at channel intersections is also a good catalyst for this reaction, but only after a few hours on stream or after postsynthesis modifications.

3. The selective reaction occurs through a monomolecular mechanism, and the bimolecular (alkylation/cracking) mechanism is unselective. Both reactions are catalyzed by Brønsted acids.

4. The main factor governing the isobutylene selectivity is restricted transition-state shape selectivity. The space available around the acid site governs the isobutylene selectivity by allowing the reaction to proceed mainly through the monomolecular and not the bimolecular mechanism.

5. Acidic sites on the external surfaces of the catalysts are not shape selective and have to be removed or poisoned.

6. The acid strength does not have an important influence on isobutylene selectivity, and the acid site density seems to play only a minor role.

7. Coke is deposited during the catalytic reaction, and it has two effects:
   a. Poisoning of the acidic sites located on the external surfaces of the catalyst and in the mesopores and poisoning of a fraction of those sites located in the micropores.
   b. A decrease of the micropore volume, rendering the catalysts increasingly shape selective.

8. These microporous catalysts have good lifetimes and are regenerable in air.

ACKNOWLEDGMENT

We are very grateful to Dr. Werner Haag, who stimulated us to prepare this review.

REFERENCES

1. Haag, W. O., and Pines, H., *J. Am. Chem. Soc.* **82,** 2488 (1960).
2. Pines, H., *in* "Chemistry of Catalytic Hydrocarbon Conversions." Academic Press, New York, 1962.
3. Choudhary, V. R., *Chem. Ind. Dev.* **8,** 32 (1974).
4. Butler, A. C., and Nicolaides, C. P., *Catal. Today* **18,** 443 (1993).
5. Grandvallet, P., de Jong, K. P., Mooiweer, H. H., Kortbeek, A. G., and Kraushaar-Czarnetzi, B., European Patent No. 0501577A1 (1992).
6. Powers, D. H., Murray, B. D., Winquist, B. H., Callender, E. M., and Varner, J. H., European Patent No. 0523 838 A2 (1993).
7. Gajda, G. J., and Barger, P. T., U.S. Patent No. 5,191,146 (1993).
8. Brouwer, D. M., *Rec. Trav. Chim. Pays-Bas* **87,** 1435 (1968).
9. Karabatsos, G. J., Vane, F. M., and Meyerson, S., *J. Am. Chem. Soc.* **83,** 4297 (1961).
10. Weitkamp, J., Jacobs, P. A., and Martens, J. A., *Appl. Catal.* **8,** 123 (1983).
11. Germain, J. E., *in* "Catalytic Conversion of Hydrocarbons." Academic Press, New York, 1969.
12. Martens, J. A., Jacobs, P. A., and Weitkamp, J., *Appl. Catal.* **20,** 283 (1986).

13. Buchanan, J. S., Santiesteban, J. G., and Haag, W. O., *J. Catal.* **158,** 279 (1996).
14. Gielgens, L. H., Veenstra, I. H., Ponec, V., Haanepen, M. J., and van Hooff, J. H., *Catal. Lett.* **32,** 195 (1995).
15. Adams, J. R., Gelbein, A. P., Hansen, R., Peress, J., and Sherwin, M. B., European Patent No. 071198 (1982).
16. Szabo, J., Perrotey, J., Szabo, G., Duchet, J. C., and Cornet, D., *J. Mol. Catal.* **67,** 79 (1991).
17. Kilpatrick, J., Prosen, E., Pitzer, K., and Rossini, F., *J. Res. Natl. Bur. Standards* **36,** 559 (1946).
18. Stull, D. R., Westrum, E. F., and Sinke, G. C., *in* "The Chemical Thermodynamics of Organic Compounds." Wiley, New York, 1969.
19. Houzvicka, J., Diefenbach, O., and Ponec, V., *J. Catal.* **164,** 288 (1996).
20. Flanigen, E. M., Lok, B. M., Patton, R. L., and Wilson, S. T., *Stud. Surf. Sci. Catal.* **28,** 103 (1986).
21. Flanigen, E. M., Patton, R. L., and Wilson, S. T., *Stud. Surf. Sci. Catal.* **37,** 13 (1988).
22. Kokotailo, G. T., Chu, P., Lawton, S. L., and Meier, W. M., *Nature* **275,** 119 (1978).
23. Barri, S. A., Smith, G. W., White, D., and Young, D., *Nature* **312,** 533 (1984).
24. Rohrman, A. C., La Pierre, R. B., Schlenker, J. L., Wood, J. D., Valyocsik, E. W., Rubin, M. K., Higgins, J. B., and Rohrbaugh, W. J., *Zeolites* **5,** 352 (1985).
25. Plank, C. J., Rosinski, E. J., and Rubin, M. K., U.S. Patent No. 4,016,245 (1977).
26. O'Young, C. L., Pellet, R. J., Hadowanetz, A. E., Hazen, J., and Browne, J. E., U.S. Patent No. 5,510,560 (1996).
27. Pellet, R. J., O'Young, C. L., Hazen, J., Hadowanetz, A. E., and Browne, J. E., U.S. Patent No. 5,523,510 (1996).
28. Apelian, R., Fung, S. L., Huss, A., and Rahmim, I., W.O. Patent No. 93/23353.
29. Xu, W. Q., Yin, Y. G., Suib, S. L., Edwards, J. C., and O'Young, C. L., *J. Phys. Chem.* **99,** 9443 (1995).
30. Pellet, R. J., Casey, D. J., Huang, H. M., Kessler, R. V., Kuhlman, E. J., O'Young, C. L., Sawicki, R. A., and Ugolini, J. R., *J. Catal.* **157,** 423 (1995).
31. Houzvicka, J., Hansildaar, S., and Ponec, V., *J. Catal.* **167,** 273 (1997).
32. Mériaudeau, P., Bacaud, R., Ngoc Hung, L., and Vu Tuan, A., *J. Mol. Catal.* **110,** L117 (1996).
33. Kwak, B. S., Jeong, H. S., Lee, J. M., and Ahn, B. J., *Stud. Surf. Sci. Catal.* **105,** 1423 (1997).
34. Rahmim, I., Huss, A., Lissy, N., Klocke, D., and Johnson, I. D., U.S. Patent No. 5,157,194 (1992).
35. Simon, M. W., Suib, S. L., and O'Young, C. L., *J. Catal.* **147,** 484 (1994).
36. Byggningsbacka, R., Lindfors, L. E., and Kumar, N., *Ind. Eng. Chem. Res.* **36,** 2990 (1997).
37. Haag, W. O., Harandi, M. N., and Owen, H., U.S. Patent No. 5,132,467 (1992).
38. Xu, W. Q., Yin, Y. G., Suib, S. L., and O'Young, C. L., *J. Catal.* **150,** 34 (1994).
39. Mériaudeau, P., Vu Tuan, A., Le Hung, N., and Naccache, C., *J. Chem. Soc. Faraday Trans.* **94,** 467 (1997).
40. Yang, S. M., Guo, D. H., Lin, J. S., and Wang, G. T., *Stud. Surf. Sci. Catal.* **84,** 1677 (1994).
41. Mériaudeau, P., Vu Tuan, A., Le Hung, N., and Szabo, G., *Zeolites* **19,** 449 (1997).
42. Simon, M. W., Xu, W. Q., Suib, S. L., and O'Young, C. L., *Microporous Mater.* **2,** 477 (1994).
43. Bianchi, D., Simon, M. W., Nam, S. S., Xu, W. Q., Suib, S. L., and O'Young, C. L., *J. Catal.* **145,** 551 (1994).
44. Seo, G., Jeong, H. S., Jang, D. L., Cho, D. L., and Hong, S. B., *Catal. Lett.* **41,** 189 (1996).
45. Mériaudeau, P., Vu Tuan, A., Le Hung, N., and Naccache, C., *Stud. Surf. Sci. Catal.* **105,** 1373 (1997).
46. Asensi, M. A., Corma, A., and Martinez, A., *J. Catal.* **158,** 561 (1996).
47. Mooiweer, H. H., De Jong, K. P., Kraushaar-Czarnetzki, B., Stork, W. H., and Krutzen, B. C., *Stud. Surf. Sci. Catal.* **84,** 2327 (1994).

48. De Jong, K. P., Mooiweer, H. H., Buglass, J. G., and Maarsen, P. K., *Stud. Surf. Sci. Catal.* **111,** 127 (1997).
49. Houzvicka, J., and Ponec, V., *Ind. Eng. Chem. Res.* **36,** 1424 (1997).
50. Kazansky, V. B., *Acc. Chem. Res.* **24,** 379 (1991).
51. Guisnet, M., Andy, P., Gnep, G. S., Travers, C., and Benazzi, E., *J. Chem. Soc. Chem. Commun.,* 1685 (1995).
52. Guisnet, M., Andy, P., Gnep, G. S., Travers, C., and Benazzi, E., *Stud. Surf. Sci. Catal.* **105,** 1365 (1997).
53. Mériaudeau, P., Vu Tuan, A., Le Ngoc, H., and Szabo, G., *J. Catal.* **169,** 397 (1997).
54. Cheng, Z. X., and Ponec, V., *Catal. Lett.* **25,** 337 (1994).
55. Cheng, Z. X., and Ponec, V., *J. Catal.* **148,** 607 (1994).
56. Meijer, S., Ponec, V., Finocchio, E., and Busca, G., *J. Chem. Soc. Faraday Trans.* **91,** 1861 (1995).
57. Xu, W. Q., Yin, Y. G., Suib, S. L., and O'Young, C. L., *J. Phys. Chem.* **99,** 758 (1995).
58. Hung Le, N., PhD thesis, Lyon University, 1997; Mériaudeau, P., Naccache, C., Le Hung, N., Vu Tuan, A., and Szabo, G., *J. Mol. Catal.* **123,** L1 (1997).
59. Haag, W. O, Lago, R. M., and Weisz, P. B., *Faraday Disc. Chem. Soc.* **72,** 317 (1982).
60. Freeman, C. M., Catlow, C. R., Thomas, J. M., and Brode, S., *Chem. Phys. Lett.* **186,** 137 (1991).
61. Apelian, M. R., Fung, A. S., Kennedy, G. J., and Thomas, F. D., preprint of the 14th North American Meeting of Catalysis Society, June 11–16, 1995, Snowbird, UT, p. 116.
62. Mériaudeau, P., Vu Tuan, A., Le Hung, N., Naccache, C., and Szabo, G., *J. Catal.* **171,** 329 (1997).
63. Mériaudeau, P., Vu Tuan, A., Le Hung, N., and Szabo, G., *Catal. Lett.* **47,** 71 (1997).
64. Zubowa, H. L., Richter, M., Roost, U., Parlitz, B., and Fricke, R., *Catal. Lett.* **19,** 67 (1993).
65. Mériaudeau, P., Vu Tuan, A., Le Hung, N., Lefebvre, F., and Nguyen, H. P., *J. Chem. Soc. Faraday Trans.* **93,** 4201 (1997).
66. Xu, W. Q., Yin, Y. G., Suib, S. L., Edwards, J. C., and O'Young, C. L., *J. Catal.* **163,** 232 (1996).

F

Face-centered cubic (FCC) metal, 37
Fermi contact interaction, 20–21
Fermi energy
  orbitals near, 7, 101
  and work function, 9–13
Fine structure, 124, 160–162
Fischer–Tropsch reactions, and methanation using transient method, 385–394
Fluorescence, 125–127
Franck–Condon analysis, 124
  and CO photoreduction of catalyst, 164–170
  of photoluminescence spectrum, 162–164
Frequency response methods, in the transient method, 344–349
Frontier orbitals, 15–19, 101
  chemisorption on metals and, 3
  history of expression, 16–17
  LDOS-based model, 18
  in ^{195}Pt NMR, 98–101

G

Galvanostatic oxidation, 73

H

H_2, adsorption of ZrO_2 on, 152–153
Hartree approximation, 9–11
Heine–Friedel invariance rule, 16
Heterodiffusion, on surfaces of metals, 80
Heterogeneous catalysis, transient regime for, 329–410
Heteropoly compounds, 431–434
  adsorption on alumina, 452–453
^{1}H NMR, 28–59
  bulk palladium hydride line shifts in, 37–42
  on bimetallics, 56–59
  on copper, 54–56
  on palladium, 42–46
  on platinum, 46–50
  on rhodium, 50–51
  on ruthenium, 51–54
  spillover of line intensities in, 28–37
H_2O, adsorption in automotive catalytic converters, 272–273
Hydrocracking, phosphorus and, 488

Hydrodenitrogenation, phosphorus and, 481–487
Hydrodesulfurization, phosphorus and, 472–481
Hydroformylation, of ethylene, 408
Hydrogel method, for preparation of alumina-based hydrotreating catalysts, 440–441
Hydrogen. *See also* ^{1}H NMR
  adsorption in automotive catalytic converters, 264–265
Hydrogenation, phosphorus and, 487–488
Hydrodemetalization, phosphorus and, 488–489
Hydrothermal treatment, isobutylene selectivity and, 538–539
Hydrotreating reactions, phosphorus and, 496–497

I

Impregnation method, dispersion of phosphorus-containing catalyst components on alumina, 461–462
Industrial catalyst formulations, 491–492
Internal conversion, 130–131
Intersystem crossing, in photoluminescence, 128–129
Isobutylene selectivity, 534–539
  acid site density and, 534
  acid strength and, 539
  external sites and, 536
  hydrothermal treatment and extraframework species and, 538–539
  pore size and, 536–538
  shape selectivity and, 534–536
Isomerization
  phosphorus, influence on, 488
  skeletal, of *n*-butenes, 505–541
Isotopic tracers
  adsorption-assisted desorption, 365–367
  steady-state tracing, 359–365
  in the transient method, 359–367

J

Jellium calculations, 12–13

Printed in the United States
120201LV00002B/381/A